AF352897

Sharks and Skates: Ecology, Distribution and Conservation

Sharks and Skates: Ecology, Distribution and Conservation

Editor

Martina Francesca Marongiu

MDPI • Basel • Beijing • Wuhan • Barcelona • Belgrade • Manchester • Tokyo • Cluj • Tianjin

Editor
Martina Francesca Marongiu
University of Cagliari
Cagliari, Italy

Editorial Office
MDPI
St. Alban-Anlage 66
4052 Basel, Switzerland

This is a reprint of articles from the Special Issue published online in the open access journal *Animals* (ISSN 2076-2615) (available at: https://www.mdpi.com/journal/animals/special_issues/Sharks_and_Skates_Ecology_Distribution_and_Conservation).

For citation purposes, cite each article independently as indicated on the article page online and as indicated below:

LastName, A.A.; LastName, B.B.; LastName, C.C. Article Title. *Journal Name* **Year**, *Volume Number*, Page Range.

ISBN 978-3-0365-8262-7 (Hbk)
ISBN 978-3-0365-8263-4 (PDF)

Contents

About the Editor

Martina Francesca Marongiu

Martina Francesca Marongiu graduated with a degree in applied bio-ecology from the University of Cagliari in 2009, an M.Sc. degree in marine biology in 2011, and a Ph.D. degree in the biology and biochemistry of humans and the environment in 2015, with a thesis about Chondrichthyes' reproduction in Sardinian waters (central–western Mediterranean). She participated in a seasonal experimental trawl survey project known as "Mediterranean International Trawl Survey" (MEDITS), as well as commercial landings through the Data Collection Framework (European Union Regulation 199/2008). Her research interests include cartilaginous fish, conservation biology and ecology, reproductive biology, macro- and micro-anatomy, and histology. Martina Francesca has contributed to 24 research papers, mostly focused on Chondrichthyan fishes and principally regarding their life history traits (e.g., reproduction, age, and growth), diet, feeding behaviors, distribution, and morphological descriptions of skate eggcases. She also coauthored several chapters of the book *Microscopic Anatomy of Animals*, regarding the macro- and micro-analysis of the blackmouth catshark *Galeus melastomus*. She contributed to the "Atlas of the maturity stages of Mediterranean fishery resources" for the General Fisheries Commission for the Mediterranean (GFCM). She also participated in several national and international congresses and in addition to the "Workshop on Sexual Maturity Staging of Elasmobranchs" (WKSEL3).

Editorial

Editorial: Sharks and Skates—Ecology, Distribution and Conservation

Martina Francesca Marongiu

Dipartimento di Scienze della Vita e dell'Ambiente, Università degli Studi di Cagliari, Via Tommaso Fiorelli 1, 09126 Cagliari, Italy; mfmarongiu@unica.it

The class Chondrichthyes (sharks, rays and chimeras) is one of the three lineages of fishes and the most evolutionary distinct radiation of vertebrates [1]. It has survived at least five mass extinctions in its 420-million-year history [2,3] and has radiated throughout the major marine (and some freshwater) habitats, dominating upper trophic levels and imposing predation risk in many food webs [4,5].

Chondrichthyes are among the most threatened vertebrates on the planet due to several reasons, such as overfishing, habitat degradation and slow life histories [6–8]. According to the International Union for Conservation of Nature (IUCN) Red List criteria, one-third of the world's chondrichthyan fishes (sharks, rays and chimaeras) are now threatened with extinction [9].

The current observed number of threatened species is more than twice (391 of 1199) [9] that of the first global assessment in 2014, which reported that 181 of 1041 species were threatened [10]. If we assume that Data Deficient (DD) species are threatened in proportion to other species, then over one-third (37.5%) of chondrichthyans are threatened, with a lower estimate of 32.6% (assuming DD species are all Least Concern or Near Threatened) and an upper estimate of 45.5% [9]. Among the three chondrichthyan fish groups, the most threatened are: rays (41% of the 611 assessed species), sharks (35.9% of the 536 assessed species) and chimaeras (9.3% of the 52 assessed species) [9].

The depletion of chondrichthyan populations could lead to worrying ecosystem-level consequences [5,11,12] because many of these fishes are apex or mesopredators that range widely and may affect ecosystem processes through predation and associated risk effects, competition, nutrient transport and bioturbation [13–16].

In these regards, collecting information about these species is essential to formulate conservation strategies to preserve this precious marine resource.

In this Special Issue, we aimed to expand the knowledge about these important predators, collecting data about the following topics: abundance and distribution, genetic information, life histories (e.g., reproduction, age and growth), demographic analysis, stock assessment and behavior.

Acknowledgments: I would like to thank all of the authors who contributed their papers to this Special Issue and the reviewers for their helpful recommendations. I am also grateful to all members of the *Animals* Editorial Office, for giving me this opportunity and for their continuous support in managing and organizing this Special Issue.

Conflicts of Interest: The author has no conflict of interest to declare.

Citation: Marongiu, M.F. Editorial: Sharks and Skates—Ecology, Distribution and Conservation. *Animals* **2023**, *13*, 2222. https://doi.org/10.3390/ani13132222

Received: 28 June 2023
Accepted: 4 July 2023
Published: 6 July 2023

References

1. Stein, R.W.; Mull, C.G.; Kuhn, T.S.; Aschliman, N.C.; Davidson, L.N.K.; Joy, J.B.; Smith, G.J.; Dulvy, N.K.; Mooers, A.Ø. Global priorities for conserving the evolutionary history of sharks, rays and chimaeras. *Nat. Ecol. Evol.* **2018**, *2*, 288–298. [CrossRef] [PubMed]
2. Cappetta, H. Extinctions and faunic renewals on post-Jurassic Selachians. *Mem. Soc. Geol. Fr.* **1987**, *150*, 113–131.
3. Sibert, E.C.; Rubin, L.D. An early Miocene extinction in pelagic sharks. *Science* **2021**, *372*, 1105–1107. [CrossRef] [PubMed]
4. Heithaus, M.R.; Wirsing, A.J.; Dill, L.M. The ecological importance of intact top-predator populations: A synthesis of 15 years of research in a seagrass ecosystem. *Mar. Freshw. Res.* **2012**, *63*, 1039–1050. [CrossRef]
5. Ferretti, F.; Worm, B.; Britten, G.L.; Heithaus, M.R.; Lotze, H.K. Patterns and ecosystem consequences of shark declines in the ocean. *Ecol. Lett.* **2010**, *13*, 1055–1071. [CrossRef] [PubMed]
6. Worm, B.; Davis, B.; Kettemer, L.; Ward-Paige, C.A.; Chapman, D.; Heithaus, M.R.; Kessel, S.T.; Gruber, S.H. Global catches, exploitation rates, and rebuilding options for sharks. *Mar. Policy* **2013**, *40*, 194–204. [CrossRef]
7. Cortés, E. Life history patterns and correlations in sharks. *Rev. Fish. Sci.* **2000**, *8*, 299–344. [CrossRef]
8. Talwar, B.S.; Anderson, B.; Avalos-Castillo, C.G.; del Pilar Blanco-Parra, M.; Briones, A.; Cardeñosa, D.; Carlson, J.K.; Charvet, P.; Cotton, C.F.; Crysler, Z.; et al. Extinction risk, reconstructed catches and management of chondrichthyan fishes in the Western Central Atlantic Ocean. *Fish Fish.* **2022**, *23*, 1150–1179. [CrossRef]
9. Dulvy, N.K.; Pacoureau, N.; Rigby, C.L.; Pollom, R.A.; Jabado, R.W.; Ebert, D.A.; Finucci, B.; Pollock, C.M.; Cheok, J.; Derrick, D.H.; et al. Overfishing drives over one-third of all sharks and rays toward a global extinction crisis. *Curr. Biol.* **2021**, *31*, 4773–4787. [CrossRef] [PubMed]
10. Dulvy, N.K.; Fowler, S.L.; Musick, J.A.; Cavanagh, R.D.; Kyne, P.M.; Harrison, L.R.; Carlson, J.K.; Davidson, L.N.K.; Fordham, S.V.; Francis, M.P.; et al. Extinction risk and conservation of the world's sharks and rays. *eLife* **2014**, *3*, e00590. [CrossRef] [PubMed]
11. Burkholder, D.A.; Heithaus, M.R.; Fourqurean, J.W.; Wirsing, A.; Dill, L.M. Patterns of top-down control in a seagrass ecosystem: Could a roving apex predator induce a behaviour-mediated trophic cascade? *J. Anim. Ecol.* **2013**, *82*, 1192–1202. [CrossRef] [PubMed]
12. Estes, J.A.; Heithaus, M.; McCauley, D.J.; Rasher, D.B.; Worm, B. Megafaunal impacts on structure and function of ocean ecosystems. *Annu. Rev. Environ. Resour.* **2016**, *41*, 83–116. [CrossRef]
13. Flowers, K.I.; Heithaus, M.R.; Papastamatiou, Y.P. Buried in the sand: Uncovering the ecological roles and importance of rays. *Fish Fish.* **2021**, *22*, 105–127. [CrossRef]
14. Heithaus, M.R.; Frid, A.; Wirsing, A.J.; Worm, B. Predicting ecological consequences of marine top predator declines. *Trends Ecol. Evol.* **2008**, *23*, 202–209. [CrossRef] [PubMed]
15. Heithaus, M.R.; Frid, A.; Vaudo, J.J.; Worm, B.; Wirsing, A.J. Unraveling the ecological importance of elasmobranchs. In *Sharks and Their Relatives II: Biodiversity, Adaptive Physiology, and Conservation*; Carrier, J.C., Musick, J.A., Heithaus, M.R., Eds.; CRC Press: Boca Raton, FL, USA, 2010; pp. 627–654. [CrossRef]
16. Heupel, M.R.; Knip, D.M.; Simpfendorfer, C.A.; Dulvy, N.K. Sizing up the ecological role of sharks as predators. *Mar. Ecol. Prog. Ser.* **2014**, *495*, 291–298. [CrossRef]

 *animals*

Article

Vulnerability Assessment of Pelagic Sharks in the Western North Pacific by Using an Integrated Ecological Risk Assessment

Kwang-Ming Liu [1,2,3,*], Lung-Hsin Huang [1], Kuan-Yu Su [1] and Shoou-Jeng Joung [2,4]

[1] Institute of Marine Affairs and Resource Management, National Taiwan Ocean University, Keelung 20224, Taiwan; resumption2@gmail.com (L.-H.H.); supipi76@gmail.com (K.-Y.S.)
[2] George Chen Shark Research Center, National Taiwan Ocean University, Keelung 20224, Taiwan; f0010@mail.ntou.edu.tw
[3] Center of Excellence for the Oceans, National Taiwan Ocean University, Keelung 20224, Taiwan
[4] Department of Environmental Biology and Fisheries Science, National Taiwan Ocean University, Keelung 20224, Taiwan
* Correspondence: kmliu@mail.ntou.edu.tw

Simple Summary: A new integrated ecological risk assessment (ERA) including the IUCN Red List category, the body weight variation trend of 1989–2011 with large sample size ($n > 678{,}000$), and the inflection point of population growth curve coupled with the ERA was developed to assess the impact of longline fishery on the pelagic sharks in the western North Pacific. The intrinsic rate of population growth was used to estimate the productivity, and the susceptibility was estimated by the multiplication of the catchability, selectivity, and post-capture mortality. Five groups were identified based on the cluster analysis coupling with non-parametric multi-dimensional scaling. Rigorous management measures are recommended for the scalloped hammerhead, silky, and spinner shark at highest risk, setting total allowable catch quota is recommended for the bigeye thresher, and sandbar shark, and a consistent monitoring scheme is suggested for the smooth hammerhead, shortfin mako, pelagic thresher, oceanic whitetip, and dusky shark.

Abstract: The vulnerability of 11 pelagic shark species caught by the Taiwanese coastal and offshore longline fisheries in the western North Pacific were assessed by an ecological risk assessment (ERA) and 10 of the 11 species was assessed by using an integrated ERA developed in this study. The intrinsic rate of population growth was used to estimate the productivity of sharks, and the susceptibility of sharks was estimated by the multiplication of the catchability, selectivity, and post-capture mortality. Three indices namely, the IUCN Red List category, the body weight variation trend, and the inflection point of population growth curve coupled with ERA were used to conduct an integrated ERA. The results indicated that the scalloped hammerhead is at the highest risk (group 1), followed by the silky shark, and the spinner shark at high risk (group 2). The bigeye thresher, and sandbar shark fall in group 3, the smooth hammerhead falls in group 4, and the shortfin mako, pelagic thresher, oceanic whitetip, and dusky shark fall in group 5. Rigorous management measures for the species in groups 1 and 2, setting total allowable catch quota for group 3, and consistent monitoring schemes for groups 4 and 5 are recommended.

Keywords: demographic analysis; productivity; susceptibility; intrinsic rate of population growth; fishing impact

Citation: Liu, K.-M.; Huang, L.-H.; Su, K.-Y.; Joung, S.-J. Vulnerability Assessment of Pelagic Sharks in the Western North Pacific by Using an Integrated Ecological Risk Assessment. *Animals* **2021**, *11*, 2161. https://doi.org/10.3390/ani11082161

Academic Editor: Martina Francesca Marongiu

Received: 26 June 2021
Accepted: 19 July 2021
Published: 21 July 2021

Publisher's Note: MDPI stays neutral with regard to jurisdictional claims in published maps and institutional affiliations.

1. Introduction

Most pelagic sharks are apex predators in the ocean, they can maintain the balance of marine ecosystem and thus play an important role in the ecosystem [1–3]. Several studies have indicated the trend of pelagic fish species are shrinking [4–8] and the latest study indicated that global abundance of 74% of 31 pelagic shark and ray species have declined by 71% since 1970 [9]. Increasing fishing efforts and expanding fishing grounds, not only led to the reduction of target species, habitat destruction, and decreased the size at catch, but also resulted in dramatic changes in marine ecosystems directly or indirectly [10–12]. In addition, the marine ecosystem structure may be altered when constantly removing the by-catch species such as sharks, rays, and marine mammals [13]. In view of this, how to effectively manage marine resources has become an important issue for tuna regional fisheries management organizations (tRFMOs).

In the past, most fish stock assessment studies focused on single species assessment. However, for stocks with poor data or limited biological information, risk assessment methods were usually used by applying a qualitative or semi-quantitative approach. In recent years, the ecological risk assessment (ERA) method such as productivity-susceptibility analysis (PSA) has been proposed for fisheries management, marine animals can be analyzed based on existing biological and fishery information [14]. It is a useful methodology for assisting fisheries management from an ecosystem perspective [15]. The method is calculated by combining the productivity of animals and the susceptibility to fisheries to assess the stock's relative vulnerability [16]. Based on the results from the ERA, the relative risk of animals, and the prioritization of management and conservation can be identified.

The ERA method has been proposed to use by various tRFMOs to evaluate the vulnerability of animals, including the species caught by tuna fisheries in the Western and Central Pacific Fishery Commission (WCPFC) waters [17], sharks caught by various fisheries in the Indian Ocean Tuna Commission (IOTC) waters [18,19], and sharks in the International Commission for the Conservation of Atlantic Tunas (ICCAT) waters [20,21] as well as national organization such as the Australian Fisheries Management Authority (AFMA) [22]. Furthermore, the ERA methods have been used to assess the vulnerability of target and by-catch species stocks to fishing gear, including the Atlantic tuna fisheries by the United States and European Union [15], Alaska demersal fishery [16], and Australian trawl fishery [23]. In addition, Stelzenmüller et al. [24] proposed the spatial risk assessment framework for the UK continental shelf fish and shellfish. Chin et al. [25] analyzed the vulnerability of sharks and rays in Australia's Great Barrier Reef by taking account the climate change effect. Gallagher et al. [26] reviewed the ERA and its application to elasmobranch conservation and management.

Although ERA has been used by WCPFC on assessing the risk of species caught in the western and central Pacific Ocean (WCPO) tuna fisheries, such an approach has never been applied to the multi-species management in the western North Pacific Ocean despite a recent study by Lin et al. [27]. The authors used a semi-quantitative PSA on 52 species including three shark species—the silky, *Carcharhinus falciformis*, blue shark, *Prionace glauca*, and shortfin mako shark, *Isurus oxyrichus* in five fisheries in eastern Taiwanese waters. However, other pelagic shark species commonly caught by the Taiwanese coastal and offshore longline fishing vessels were not included in their study. Annual yield of pelagic sharks (excluding blue shark), caught in the western North Pacific, landed at Nanfangao fishing port, northeastern Taiwan increased from 1724 tons in 1989 to the peak of 2876 tons in 1996 and decreased thereafter to around 1500 tons after 2005 based on the sales records. Species composition in terms of weight indicated that the blue shark was the dominant species, followed by the shortfin mako and the scalloped hammerhead. Although stock assessments of some of these species such as pelagic thresher [28,29], bigeye thresher [30,31], shortfin mako [32–34], and smooth hammerhead [35] in the region have been conducted using demographic or per recruit models. The relative risk on the pelagic sharks from fisheries in the region has not been evaluated, thus, this study attempts to assess the vulnerabilities of 11 pelagic shark species commonly caught by the Taiwanese coastal and offshore longline fisheries in the western North Pacific.

In addition to conventional PSA, an integrated ERA combining the ERA with species' endangered status and the inflection point of population growth curve has been developed to better describe the ecological risk of sharks and rays [20]. However, the size variation of these elasmobranchs has not been taken into account due to unavailability of the data. In this study, as the fishing ground and fishing technique of the Taiwanese longline fishing vessels in the western North Pacific did not change, the long-term body weight variation trend (1989–2011) of pelagic sharks can better represent the fishing impact on species population. Hence, an integrated ERA combining the ERA, IUCN Red List index, annual body weight variation trend, and the inflection point of population growth curve was developed to assess the vulnerability of pelagic sharks in the western North Pacific. It is hoped that the results derived from the present study can provide useful information for prioritizing management measures to achieve the goal of sustainable use of shark resources in this region.

2. Materials and Methods

2.1. Study Area and Species

This study covered the waters in the western North Pacific ranging from 20° N to 30° N, and from 120° E to 140° E where was the conventional fishing ground of the Taiwanese coastal and offshore longline fishing vessels (<100 gross tonnage) (Figure 1). Most of these longline fishing vessels mainly target dolphinfish *Coryphaena hippurus*, tunas, and billfishes (sharks are the bycatch) from April to October and switch to targeting sharks (seasonal shark targeting) by changing gear configuration from November to next March. Eleven pelagic shark species, commonly caught by these fishing vessels, namely, the bigeye thresher *Alopias superciliosus*, pelagic thresher, *A. pelagicus*, silky shark, spinner shark, *C. brevippina*, dusky shark, *C. obscurus*, oceanic whitetip, *C. longimanus*, sandbar shark, *C. plumbeus*, shortfin mako shark, blue shark, scalloped hammerhead, *Sphyrna lewini*, and smooth hammerhead, *S. zygaena* were analyzed in this study.

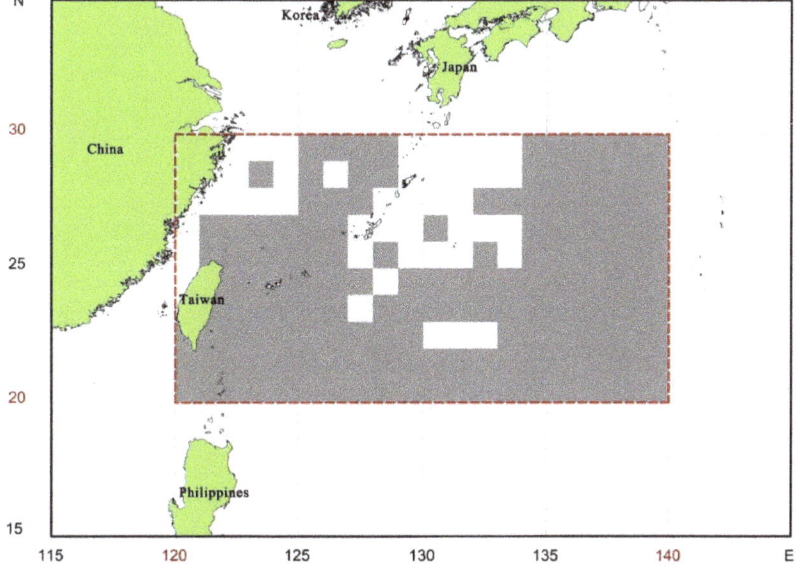

Figure 1. The study area (from 20° N to 30° N, 120° E to 140° E) in the present study. Gray areas indicate fishing grounds of the Taiwanese coastal and offshore longline fishing vessels.

2.2. Source of Data

Species-specific landing data, including the whole (body) weight of each individual fish (except very few small-size individuals that were weighed in group and only average weights were available) at the Nanfangao fishing port, northeastern Taiwan from 1989–2011, were collected from the sales records. These data were used to estimate the species-specific catch in number, the species composition, and the mean and median weight. The individual body weight data were converted to length via an existing length–weight relationship in this region. The age of each individual was then estimated by substituting the converted length into the growth equation of each species (Table 1).

Table 1. Age and growth parameters of 11 shark species in the western North Pacific Ocean.

Species	L_{max} (cm TL)	L_∞ (cm)	k (Year^{-1})	t_0	t_{max} (Year)	References
Pelagic thresher	365.18	382.94	0.09	−7.67	27.57	[36]
Bigeye thresher	422.00	422.00	0.09	−4.21	28.35	[37]
Spinner	274.00	288.20	0.15	−1.99	17.85	[38]
Silky	256.00	332.00	0.08	−2.76	32.99	[39]
Oceanic whitetip	268.00	323.80	0.11	−0.37	12.30	[40]
Dusky	364.00	415.70	0.06	−3.42	50.08	[41]
Sandbar	210.00	210.00	0.17	−2.30	15.32	[42]
Shortfin mako	375.00	413.80	0.05	—	40.04	[43,44]
Scalloped hammerhead	324.00	319.72	0.25	−0.41	11.62	[45]
Smooth hammerhead	324.00	375.20	0.11	−1.31	25.73	[46]
Blue	323.00	322.70	0.16	−1.33	17.24	[47]

L_{max}: maximum observed length, L_∞: asymptotic length, k: growth coefficient, t_0: age at length 0, t_{max}: longevity. Pelagic thresher, *Alopias pelagicus*; Bigeye thresher, *A. superciliosus*; Spinner, *Carcharhinus brevipinna*; Silky, *C. falciformis*; Oceanic whitetip, *C. longimanus*; Dusky shark, *C. obscurus*; Sandbar, *C. plumbeus*; Shortfin mako, *Isurus oxyrinchus*; Blue, *Prionace glauca*; Scalloped hammerhead, *Sphyrna lewini*; Smooth hammerhead shark, *S. zygaena*.

2.3. Demographic Analysis

Demographic parameters of each species were estimated using Krebs [48] equations assuming sex ratio was 1:1 for the embryos and the population was in equilibrium condition ($\sum \frac{1}{2} m_x l_x e^{-rx} = 1$) based on existed life history parameters (Tables 1 and 2). The demographic parameters were calculated as follows:

$$R_0 = \sum_{x=0}^{t_{max}} \frac{1}{2} m_x l_x, \ G = \frac{\sum_{x=0}^{t_{max}} \frac{1}{2} x l_x m_x}{R_0}, \ r = \frac{\ln R_0}{G}, \ \lambda = e^r \tag{1}$$

where R_0: the net reproductive value per generation, x: age, t_{max}: the maximum age, m_x: the fecundity of age x, l_x: survival rate until age x, G: generation time, r: intrinsic rate of population growth, λ: finite rate of population increase, t_{x2}: theoretical population doubling or halving time.

Table 2. Reproductive parameters of 11 pelagic shark species in the western North Pacific Ocean.

Species	R	L_b (cm)	L_m (cm)	t_m (Year)	f	G_p (Month)	R_c (Year)	References
Pelagic thresher	av	174.0	287.0	8.6	2	12	1	[36]
Bigeye thresher	av	148.7	336.6	12.85	2	12	1	[49]
Spinner	v	67.5	222.5	7.8	8.5	11	2	[38]
Silky	v	69.5	215.0	9.7	9	12	2	[39]
Oceanic whitetip	v	64.0	194.7	8.23	10	12	2	[40]
Dusky	v	101.0	281.0	16.4	11	13	2	[41]
Sandbar	v	62.5	172.5	7.85	7.5	11	2	[50]
Shortfin mako	av	74.0	278.0	20	11.1	24	3	[43,44]
Scalloped hammerhead	v	48.5	230.0	4.7	25.8	10	2	[51]
Smooth hammerhead	v	55.0	259.4	11	30	10	2	[52]
Blue	v	45.0	189.0	4.2	29	10	2	[53]

R: reproductive strategy (av: aplacental viviparity, v: viviparity), L_b: size at birth, L_m: size at maturity, t_m: age at maturity, f: littler size, G_p: gestation period, R_c: reproductive cycle, sex ratio of embryos (F/(F + M)) was set as 0.5.

2.4. Ecological Risk Assessment (ERA)

ERA considers productivity and susceptibility of sharks. Productivity is the ability of withstanding the exploitation by replacing individuals through the reproduction, survival, and growth of individuals. Usually, the intrinsic rate of population growth (r) is expressed as the productivity. Susceptibility is the impact of a fishery on a species population and is estimated by the multiplication of the probabilities of the following three parameters: (1) catchability: the species composition (percentage of catch in weight) of 11 shark species based on the landing data from 1989–2011; (2) selectivity, the ratio of age range of catch (the maximum age at catch–minimum age at catch) and the longevity [22]; and (3) post-capture mortality, including the mortality of retention and those after being discarded or live released. Based on the fisherman interview, despite the releasing of oceanic whitetip and silky shark which were non-retention species in the WCPO after 2013 and 2014, only very small portion of other shark catch was discarded or released. Thus, the post-capture mortality was referred to Cortés et al. [54] that estimated from the observed data of US far sea fishery.

Susceptibility was estimated from the multiplication of the aforementioned three parameters. When the productivity and susceptibility were estimated, a modified Euclidean distance (D) [20] was used to estimate the vulnerability as:

$$D = \sqrt{(p-0.5)^2 + (s-0)^2} \tag{2}$$

where p is productivity, s is susceptibility. Index of body weight variation trend (S_w).

IUCN Red list index (C)

The endangered status of the 11 pelagic shark species was evaluated based on the latest IUCN Red List assessment of pelagic sharks [55]. To quantify the risk of Red List species, we followed the definition proposed by Simpfendorfer et al. [20], C ranges from 0 and 1 based on the endanger condition. The C = 1 for critical endangered (CR) species, C = 0.8 for endangered (EN) species, C = 0.6 for vulnerable (VU) species, C = 0.4 for near threatened (NT) species, and C = 0.2 for species of least concern (LC) (Table 3).

Table 3. IUCN red list global status of 11 shark species in the western North Pacific Ocean. The IUCN Red List index (C) indicates the relative risk of species. LC = 0.2, NT = 0.4, VU = 0.6, EN = 0.8 and CR = 1.0. (IUCN [55]).

Species	IUCN Status	IUCN Status	IUCN Red List Index (C)
Pelagic thresher	Endangered	EN	0.8
Bigeye thresher	Vulnerable	VU	0.6
Spinner	Near Threatened	NT	0.4
Silky	Vulnerable	VU	0.6
Oceanic whitetip	Critically Endangered	CR	1.0
Dusky	Endangered	EN	0.8
Sandbar	Vulnerable	VU	0.6
Shortfin mako	Endangered	EN	0.8
Scalloped hammerhead	Critically Endangered	CR	1.0
Smooth hammerhead	Vulnerable	VU	0.6
Blue	Near Threatened	NT	0.4

Body weight variation trend (S_w)

Annual median weight of each shark species was calculated from the landing data of Nanfangao fish market from 1989 to 2011. The slope of the simple linear regression between annual median weight and year was used as the index of body weight variation trend (S_w). The negative value of slope indicated a decrease of size at catch. The value of S_{wa} ranged from 0–1 by taking an absolute value of S_w.

Inflection point of population growth curve (I)

The inflection point of population growth curve (I) is the ratio of the biomass at maximum sustainable yield (B_{MSY}) and the initial biomass (B_0). The larger the value is, the higher the exploitation is and the less biomass can be utilized. The I value can be estimated as: I = 0.633–0.187(ln(rG)) [56], where G is the generation time in years.

2.5. Integrated Ecological Risk Assessment

An integrated ERA was developed based on the combined information of the ERA, C, I, and S_w. As most blue sharks were processed on the sea, the individual weight data only available for very small portion of the blue shark landings. Although the sales data of frozen meat, internal organs, and fins were available after 2001, the catch in number data were still lacking which hindered the estimation of individual weight of blue sharks. Due to lack of S_w information, this species was not included in the integrated ERA assessment (Figure 2). Multivariate analyses including principal component analysis (PCA), and cluster analysis (CA) and non-parametric multi-dimensional scaling (NMDS) have been used in grouping of sharks, skates, and rays based on their life history parameters and habitat information [57,58]. The CA coupled with NMDS was used to group the 10 species based on their similarity of the four indices (ERA, C, S_{wa}, and I) to conduct the integrated ERA using Primer V. 6 [59].

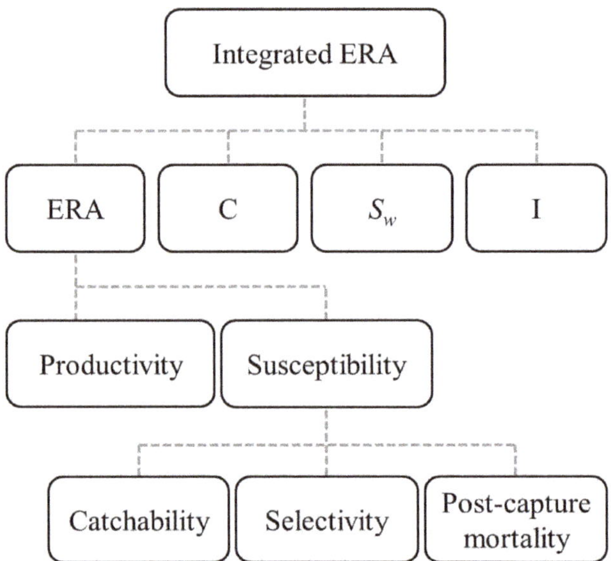

Figure 2. The flow chart of integrated ERA methods used in this study. Four indices were used to conduct the integrated assessment on 10 shark species: ecological risk assessment (ERA), IUCN Red List (C), body weight variation trend (S_W), and reflection point of population growth curve (I).

3. Results

3.1. Demographic Parameters

Estimated demographic parameters of 11 species are in Table 4. The blue shark has the highest intrinsic rate of population growth (r = 0.351 year^{-1}), followed by the scalloped hammerhead (r = 0.243 year^{-1}), while the bigeye thresher has the lowest value (r = 0.008 year^{-1}), and other species range from 0.016–0.154 year^{-1}. Without fishing mortality, λ ranged from 1.008 for the bigeye thresher to 1.420 for the blue shark with only two species (the blue shark and scalloped hammerhead) were greater than 1.20. The values of λ for the silky, smooth hammerhead, dusky, and oceanic whitetip shark ranged from 1.088

to 1.166, while λ was smaller than 1.07 for the pelagic thresher, bigeye thresher, spinner, sandbar, and shortfin mako sharks (Table 4).

Table 4. Estimated demographic parameters for 11 pelagic shark species in the western North Pacific Ocean without fishing mortality.

Species	r	λ	t_{x2}	G
Pelagic thresher	0.050	1.051	13.9	14.3
Bigeye thresher	0.008	1.008	82.6	18.0
Spinner	0.061	1.063	11.3	11.1
Silky	0.115	1.122	6.0	16.4
Oceanic whitetip	0.087	1.091	7.9	10.7
Dusky	0.085	1.088	8.2	26.5
Sandbar	0.016	1.016	43.5	10.5
Shortfin mako	0.030	1.031	23.0	26.7
Scalloped hammerhead	0.243	1.275	2.9	7.1
Smooth hammerhead	0.154	1.166	4.5	15.6
Blue	0.351	1.420	2.0	8.5

r: finite rate of population increase; λ: population increase rate; t_{x2}: population doubling time; G: generation time in years.

3.2. Ecological Risk Assessment (ERA)

The values of susceptibility ranged from 0.0073 (oceanic whitetip) to 0.2387 (blue shark) and can be categorized as three groups: (1) susceptibility = 0.2387 (blue shark), (2) susceptibility = 0.1069–0.1754 (shortfin mako, scalloped hammerhead), (3) susceptibility = 0.0073–0.0645 (pelagic thresher, bigeye thresher, spinner, oceanic whitetip, dusky, sandbar, silky, and smooth hammerhead shark) (Table 5).

Table 5. Parameters of susceptibility in the ecological risk assessment of the 11 shark species in the Northwest Pacific pelagic species. The susceptibility indicators (S) reflect the risk of fisheries development.

Species	Catchability	Selectivity	Post-Capture Mortality	Susceptibility
Pelagic thresher	0.0647	1.0000	0.7800	0.0505
Bigeye thresher	0.0827	1.0000	0.7800	0.0645
Spinner	0.0396	0.9970	0.7700	0.0304
Silky	0.0432	0.9960	0.7230	0.0311
Oceanic whitetip	0.0108	0.9350	0.7200	0.0073
Dusky	0.0576	1.0000	0.9200	0.0529
Sandbar	0.0396	0.9480	0.8600	0.0323
Shortfin mako	0.1906	1.0000	0.9200	0.1754
Scalloped hammerhead	0.1331	0.9680	0.8300	0.1069
Smooth hammerhead	0.0360	0.9490	0.8500	0.0290
Blue	0.3022	1.0000	0.7900	0.2387

The risk based on Euclidean distance (D) ranged from the highest (D = 0.5017) for the shortfin mako to the lowest (D = 0.2784) for the scalloped hammerhead (Table 6). The higher risk of the shortfin mako, bigeye thresher, sandbar, pelagic thresher, and spinner sharks is mainly due to their low productivity despite various susceptibility. On the other hand, the silky, and oceanic whitetip sharks have lower risk because of their higher productivity and lower susceptibility (Table 6; Figure 3).

Table 6. Ecological risk assessments of 11 pelagic sharks in the western North Pacific Ocean using the Euclidean distance (D).

Species	Productivity	Susceptibility	ERA (D)	Risk Rank
Pelagic thresher	0.0500	0.0505	0.4528	4
Bigeye thresher	0.0080	0.0645	0.4962	2
Spinner	0.0610	0.0304	0.4400	5
Silky	0.1150	0.0311	0.3863	8
Oceanic whitetip	0.0870	0.0073	0.4131	7
Dusky	0.0850	0.0529	0.4184	6
Sandbar	0.0160	0.0323	0.4851	3
Shortfin mako	0.0300	0.1754	0.5017	1
Scalloped hammerhead	0.2430	0.1069	0.2784	11
Smooth hammerhead	0.1540	0.0290	0.3472	9
Blue	0.3510	0.2387	0.2814	10

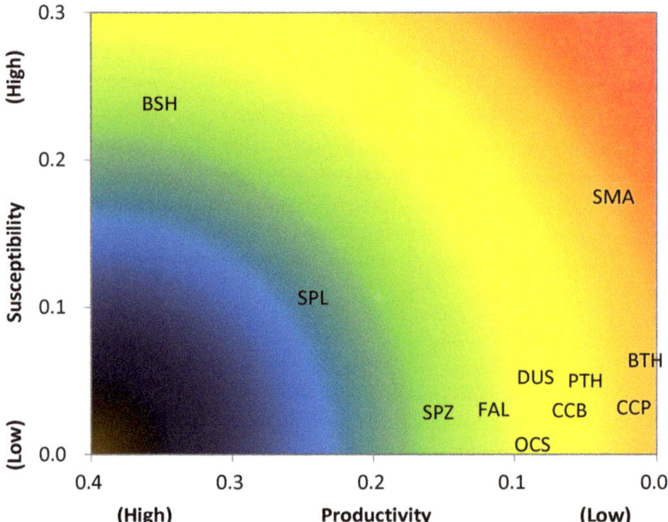

Figure 3. Productivity and susceptibility plot for 11 pelagic shark species in the western North Pacific Ocean. Productivity is expressed as r (intrinsic rate of population growth) and susceptibility as the catchability, selectivity, and post-capture mortality. PTH: pelagic thresher, BTH: bigeye thresher, CCB: spinner, FAL: silky, OCS: oceanic whitetip, DUS: dusky, CCP: sandbar, SMA: shortfin mako, BSH: blue, SPL: scalloped hammerhead, SPZ: smooth hammerhead.

3.3. IUCN Red List Index (C)

Based on the latest IUCN red list assessment of pelagic sharks [55], the blue, and spinner shark fall in the near threshed (NT, C = 0.4), bigeye thresher, sandbar, silky sharks, and smooth hammerhead fall in vulnerable (VN, C = 0.6), pelagic thresher, shortfin mako, and dusky shark fall in endanger (EN, C = 0.8), while the scalloped hammerhead and oceanic whitetip shark fall in critical endangered (CR, C = 1.0), which have the highest extinction risk (Table 7).

Table 7. Integrated ecological risk assessment of 11 pelagic sharks in the western North Pacific Ocean. ERA: ecological risk assessment; C: IUCN Red List index; *Sw*: index of body weight variation trend, inside parenthesis is the standard error (SE); I: position of the inflection point of the population growth curve. The integrated ERA did not include the blue shark as its *Sw* was not available.

Species	ERA	C	S_W	I
Pelagic thresher	0.4528	0.8	−0.0828 (0.1084)	0.6955
Bigeye thresher	0.4962	0.6	−0.4360 (0.0788)	0.9870
Spinner	0.4400	0.4	−0.9410 (0.4563)	0.7050
Silky	0.3863	0.6	−1.2762 (0.2110)	0.5139
Oceanic whitetip	0.4131	1.0	−0.1319 (0.0781)	0.6454
Dusky	0.4184	0.8	−0.0163 (0.3635)	0.4820
Sandbar	0.4851	0.6	−0.3079 (0.1383)	0.9669
Shortfin mako	0.5017	0.8	−0.2818 (0.1477)	0.6740
Scalloped hammerhead	0.2784	1.0	−1.8279 (0.1894)	0.5323
Smooth hammerhead	0.3472	0.6	−0.2404 (0.0860)	0.4697
Blue	0.2814	0.4	—	0.4286

3.4. Index of Body Weight Variation Trend (S_w)

The variations of annual median weight for each species estimated from over 678,000 individual weight records indicated that the scalloped hammerhead shark had the largest decline ($S_w = -1.8279$) (Table 7), dropping from the peak of 63.4 kg in 1989 to less than 21.2 kg after 2010 (Supplementary Figure S1). The silky shark had the second largest decline in median weight ($S_w = -1.2762$) (Table 7), dropping from 40–50 kg before 1997 to less than 30 kg after 2010 (Supplementary Figure S1). The spinner shark had large variations in annual median weight, while the oceanic whitetip and pelagic thresher shark did not have remark changes in their median weights during the period of 1989–2011 (Supplementary Figure S1).

3.5. Inflection Point of the Population Growth Curve (I)

The estimated I for 11 shark species ranged from 0.4286 for the blue shark to 0.9870 for the bigeye thresher shark (Table 7).

3.6. Integrated Ecological Risk Assessment

Five groups were categorized based on the ERA, C, S_{wa}, and I using the cluster analysis: group (1) the highest risk including the scalloped hammerhead shark, group (2) high risk including the silky, and spinner shark, group (3) median risk including the bigeye thresher, and sandbar shark, group (4) less risk including the smooth hammerhead, and group (5) least risk including the oceanic whitetip, pelagic thresher, shortfin mako, and dusky shark (Figure 4). Similar results were obtained from the NMDS indicating five groups of ecological risk (Figure 5) even the weighting of S_{wa} was set as 0.5.

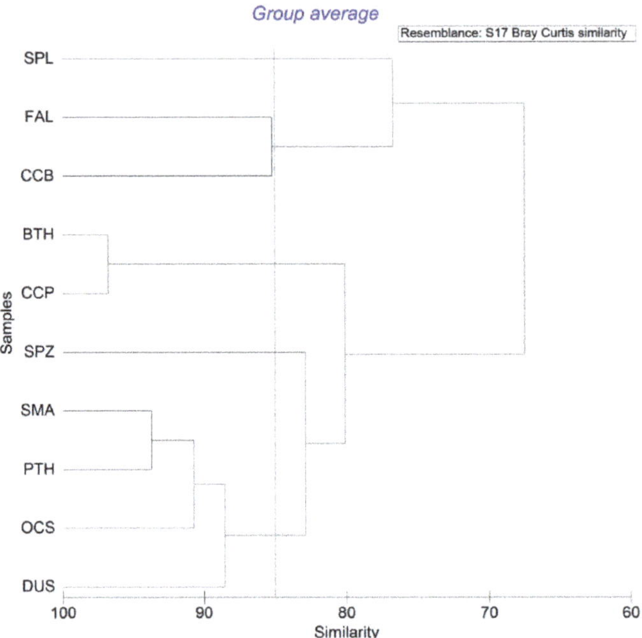

Figure 4. Results of the cluster analysis of the integrated ecological risk assessment for the 10 pelagic shark species in the western North Pacific Ocean. The scalloped hammerhead (SPL) has the highest risk (group 1), followed by silky (FAL) and spinner (CCB) (group 2). The bigeye thresher (BTH) and sandbar shark (CCP) fall in the group 3. Group 4 includes smooth hammerhead (SPZ), and group 5 includes shortfin mako (SMA), pelagic thresher (PTH), oceanic whitetip (OCS), and dusky shark (DUS).

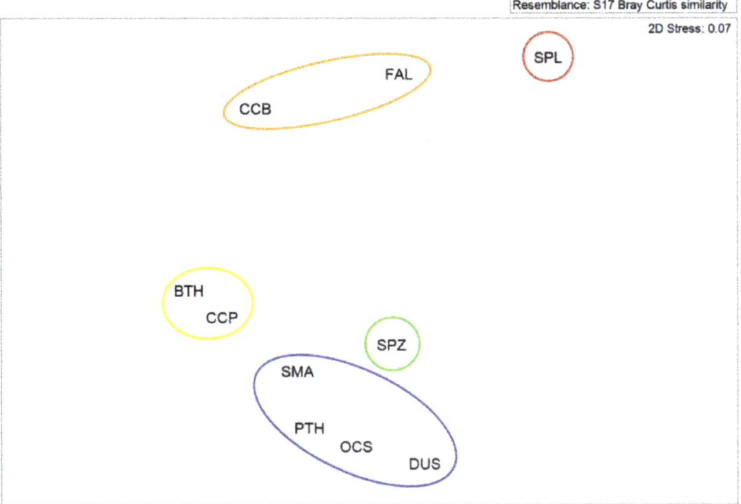

Figure 5. Results of the non-parametric multi-dimensional scaling (NMDS) analysis of the integrated ecological risk assessment for the 10 pelagic shark species in the western North Pacific Ocean. PTH: pelagic thresher, BTH: bigeye thresher, CCB: spinner, FAL: silky, OCS: oceanic whitetip, DUS: dusky, CCP: sandbar, SMA: shortfin mako, SPL: scalloped hammerhead, SPZ: smooth hammerhead.

4. Discussion

This study provides the first assessment of ERA for the 11 shark species (integrated ERA for 10 shark species) in the western North Pacific. The results derived from the present study can be used as reference for setting the priority of conservation and management measures of these species in this region.

4.1. Assumptions in This Study

The historical species composition of shark catch, sex ratio, length–weight relationship, age-structure, and proportion of maturity were assumed to be the same as those used in this study. The results derived from this study may be affected if these assumptions are violated. Two stocks—Northern and southern populations of blue sharks in the Pacific Ocean were identified based on tag–recapture data [60]. A single stock of shortfin mako shark was assumed in the North Pacific Ocean based on evidence from genetics, and tagging studies but catch and biological data indicated regional sub-stocks may exist [61]. Although two management stocks (east and west) of the silky shark in the Pacific Ocean were suggested based on genetic analysis [62], one Pacific-wide stock was assumed in the stock assessments of the silky shark [63] and the oceanic whitetip shark [64] without solid evidence. Apart from the blue and shortfin mako shark, the assumption of unit stock for the remaining 9 species in the present study should be further validated using tag-recapture study or molecular techniques.

4.2. Analysis of Landing Data

As the body weight variation trend was not available, the blue shark was excluded in the integrated ERA analysis. However, we believed that the blue shark was still at the least risk among all shark species if it was included in the integrated ERA analysis (Table 7).

Besides the Taiwanese fishing vessels, some Japanese longline fishing vessels also operate in the study area. As the fisheries information including species-specific catch in number and weight of these vessels were not available, their impact on sharks could not be assessed. The results of the present study can be improved if the aforementioned data can be incorporated in future analysis.

4.3. Analysis of Life History Parameters

The uncertainty of life history parameter estimations was often due to small sample size as the collection of shark sample was difficult. The life history parameters used in this study were adopted from the literature with considerable sample size collected in the western North Pacific. Therefore, we believe these values can represent the life history characteristics of these 11 species. Age and growth parameters of 11 species used in this study was based on vertebral band counting but only bigeye and pelagic threshers were verified with length-frequency analyses. Campana [65] pointed out that bias of age estimation may occur using hard part as ageing character. In addition, the von Bertalanffy growth function may not necessarily provide the best fit for all shark species [66]. Therefore, a multi-model approach with larger sample size and various ageing techniques should be considered in future age and growth study for sharks [67]. In addition, there was uncertainty on vertebral band pair deposition period for the scalloped hammerhead and shortfin mako shark. We adapted the biannual deposition for scalloped hammerhead from Chen et al. [51] and this was supported by Anislado-Tolentino and Robinson-Mendoza [68] in Mexican waters. However, other authors reported annual deposition in other regions of the Pacific Ocean [69–71]. As for the shortfin mako shark, annual deposition used in this study was adopted from Hsu [72] and Semba et al. [73] in Northwest Pacific. Ribot-Carballal et al. [74] also concluded annual deposition of growth band pair but Wells et al. [75] concluded a biannual cycle of vertebral band-pair deposition up to 5 years old, whereas Kinney et al. [76] concluded an annual cycle for those older than 5 years old in the eastern North Pacific Ocean. If aforementioned uncertainties in life history parameters were taken account in the integrated ERA as different scenarios, the scalloped hammerhead was still in the highest ecological risk group and the shortfin mako shark fell in the least risk group suggesting that our results were robust.

The intrinsic rate of population growth (r), estimated from demographic analysis, was used as productivity in this study. However, uncertainties of life history parameters and natural mortality that may lead to a biased estimation of r [77] by using this approach. Therefore, stochastic approach should be considered to obtain a more robust estimate of intrinsic population growth rate in the future.

4.4. Ecological Risk Assessment

ERA indicated that the shortfin mako, bigeye thresher, sandbar, and pelagic thresher shark were at the highest ecological risk. Similar findings have been reported by Simpfendorfer et al. [20] and Cortés et al. [54]. These authors found that the silky, shortfin mako, and bigeye thresher shark had the highest risk although the dusky and sandbar sharks were not included in their analysis. The high risk of these four species is very likely due to their low productivity. Particularly, the late age at maturity and three-year reproductive cycle yielded the second lowest productivity for the shortfin mako shark. This low productivity coupled with high susceptivity make it rank the highest risk species. Comparing with other teleost species, Lin et al. [27] concluded that sharks (silky, blue, and shortfin mako shark) have the highest vulnerability among 52 fish species caught in eastern Taiwan waters based on a semi-quantitative ERA.

The present study used ERA to assess the risk of exploitation for the 11 shark species in the western North Pacific but some these sharks are highly migratory species such as blue shark, shortfin mako, and oceanic whitetip that widely distribute in the ocean and they may migrate beyond the study area. However, due to lacking of specific fishing location by set, the fishing ground was based on the records by the sampling vessels and fisherman interview. Therefore, the availability—overlapping between the geographic distribution of sharks and the longline fishing ground assumed to be the same for all species was not used in this study. The information of vertical movement range for the 11 shark species in the western North Pacific is still little known despite Musyl et al.'s [78] description of the bigeye thresher, blue, oceanic whitetip, silky, and shortfin mako in the eastern/central Pacific by using tagging experiment. Due to the uncertainty, the availability and encounterability was replaced by the catchability estimated based on the mean percentage of shark catch in weight from 1989 to 2011. We believe this long-term historical landing data of large sample size (n > 678,000) can better described the species-specific vulnerability to longline fishery. Future study should focus on the tagging research to understand the vertical and horizontal distribution and movement of the sharks to improve the parameter estimation of availability and encounterability.

The selectivity was estimated based on the ratio of age range of catch and the longevity. The age range was estimated by converting individual shark landing data (body weight) to age which were then used to estimate the minimum and maximum ages of each species. We believed that our estimates based on long-term data of large sample size were representative. As individual body weight data were not available for most blue sharks, the selectivity of blue shark being set as 1 was assumed due to its wide range of size at catch. We believe this was a reasonable assumption.

The post capture mortality used in this study was adopted from Cortés et al. [54] based on the US observer's data on Atlantic sharks except the oceanic whitetip and silky shark. However, these values may not be representative for the sharks in the western North Pacific. The post release mortality of several pelagic shark species using satellite popup tagging techniques has been documented in recent years. Musyl et al. [78] reported the post release mortality and vertical movement of five pelagic shark species in the central North Pacific. Musyl and Gilman [79] and Schaefer et al. [80,81] demonstrated the post release mortality of blue shark and silky shark in the Pacific Ocean. Campana et al. [82] and Santos et al. [83] also documented the information of shortfin mako in the Atlantic Ocean. However, these updated estimations of post release mortality were much smaller than those reported by Simpfendorfer [20] and Cortés et al. [54] based on the onboard observer's records. The possible reason was due to different definitions among studies. The post capture mortality defined in this study include retention and mortality of live release (post release mortality). It is likely that the tagging experiments were based on live (healthy or minor injury) individuals but the observer's estimation was from all individuals landed on deck. To improve our results, further study on the post release mortality for other species should be conducted in the future.

4.5. IUCN Red List Index

The IUCN red list assessment of pelagic sharks was based on life history parameters, abundance trend, fishery data, and expert opinions. Dulvy et al. [84] suggested that the assessment result derived by the IUCN Red List is a good index that can represent the stock status of sharks when the full stock assessment is lacking. In the present study, the endanger index was estimated based on the latest global pelagic shark assessment results [55] and was believed to be representative. However, this information was in global scale but was not specifically for the western North Pacific. If regional (Indo-west Pacific) assessment results are available in the future, that information should be used in an updated analysis.

4.6. Integrated ERA

Simpfendorfer et al. [20] suggested that the scalloped hammerhead falls in the group of low risk in the Atlantic Ocean. The authors reported that selectivity being 0.11, susceptibility being 0.06, IUCN index of 0.4 (near threatened) [55] with ERA of 0.42 for scalloped hammerhead. In the present study, we estimated the selectivity to be 0.968, susceptibility of 0.1069, ERA of 0.2784 with IUCN index of 1.0 [55] combined with the largest decline of median weight among 10 species concluding this species has the highest risk. The recent promoting of IUCN red list from EN to CR for scalloped hammerhead in global scale [85] is another reason that resulted in different results in these two studies. We believed our integrated ERA assessment of scalloped hammerhead can better represent the stock status of this species in the western North Pacific.

The decline of median weight from 43 kg in 1989 to 33 kg in 2010 for the silky shark suggesting over-exploitation of this species is likely the reason that this species had the second highest risk among the 10 shark species. Cortés et al. [54] conducted the ERA on the pelagic sharks and reported that the silky shark had the highest risk which was comparable with the result derived from this study. The management measure of ban retention for this species has been taken in WCPFC and ICCAT.

Simpfendorfer et al. [20] suggested that the shortfin mako and bigeye thresher had the highest risk among the pelagic sharks in the Atlantic Ocean. However, we concluded the shortfin mako and bigeye thresher in the least risk group in the western North Pacific. The possible reason is the body weight of these two species in the western North Pacific did not have significant decline ($S_w = -0.436$ and -0.282) compared with the scalloped hammerhead and silky shark and the body weight variation was not considered by Simpfendorfer et al. [20]. In addition, different stock status of the shortfin mako shark in the two oceans was likely another reason. The latest North Pacific shortfin mako shark stock assessment indicated that there was no overfished and overfishing was not occurring [59] suggesting the assessment of shortfin mako derived from integrated ERA in this study was reasonable. However, the recent stock assessment indicated the North Atlantic shortfin mako was overfished and overfishing was continuing, similar situation may occur in South Atlantic stock [86]. In summary, the inclusion of the index of body weight variation trend in the integrated ERA used in this study can provide better assessment on the risk of over-exploitation for sharks in this region.

4.7. Uncertainty

Accurate stock assessment is difficult as it is hard to collect accurate biological and fisheries information. Therefore, management regulations based on various management schemes may be set by fishery managers [87]. If the uncertainty was resulted from artificial factors, this can be reduced by collecting more accurate data [88]. Therefore, to reduce the uncertainty, the biological and fisheries data should be updated to improve the accuracy of ERA and integrated ERA.

4.8. Current Management Measures

Of the 11 species analyzed in this study, the oceanic whitetip shark and silky shark have been banned for commercial retention in the WCPO by WCPFC since 2013 and 2014, respectively. No specific management measures have been taken for other shark species in this region besides the catch reporting scheme. Based on the vital parameter analysis of 38 species of sharks, Liu et al. [57] suggested that protection of adults or TAC management measure should be taken for shark species of slow growing and small litter size such as bigeye thresher, pelagic thresher, silky, and spinner shark. However, for the late-maturing species such as dusky, sandbar, shortfin mako, oceanic whitetip, bigeye thresher, and pelagic thresher shark, a reduce of catch or TAC management measure has been suggested [57]. Tsai et al. [30,31] demonstrated that the bigeye thresher stock in the Northwest Pacific was declining in population size under current fishery condition and overfishing was likely occurring. Similar conclusion was also made for the shortfin mako shark [32–34,43] using various approaches. The pelagic thresher and female smooth hammerhead shark also have been reported as overfishing at current fishing effort [28,29,35]. Based on aforementioned single species assessment results, it is suggested, for precautionary purpose, the management plan for each species should be developed.

In addition to the conventional PSA, quantitative ERA tools such as sustainability assessment for fishing effects (SAFE) [89–93] and Ecological Assessment of Sustainable Impacts of Fisheries (EASI-Fish) [94,95] have been used to derive a proxy for fishing mortality (F) based on the productivity and susceptibility of fish. The F values estimated from SAFE were comparable to those derived from data-rich quantitative stock assessments in most cases although overestimation of F may occur [78]. Therefore, the SAFE should be considered in future ecological risk assessment to provide more solid recommendations for management measures.

5. Conclusions

The integrated ERA method developed in this study can prioritize the risk of pelagic sharks in the western North Pacific. However, this approach cannot provide concrete management information such as total allowable catch (TAC), biological reference points (BRPs), and optimum fishing effort until further quantitative ERA is conducted. Even though the integrated ERA cannot replace the conventional stock assessment method, it can provide useful information for precautionary management measures. In addition to the ban retention on the silky and oceanic whitetip shark, for the species in high risk group (groups 1 and 2), stock assessment as well as rigorous management measures such as catch quota, and size limit are recommended. Setting total allowable catch quota is recommended for the species in group 3 and a consistent monitoring scheme is suggested for the species in groups 4 and 5. The integrated ERA should be updated regularly according to the availability of new information of the productivity and susceptibility of sharks.

Supplementary Materials: The following are available online at https://www.mdpi.com/article/10.3390/ani11082161/s1, Figure S1: Annual median weight (kg) of 10 pelagic shark species landed at Nanfangao fishing port from 1989 to 2011.

Author Contributions: K.-M.L. conceived and supervised the study and had substantial inputs into the analysis and all drafts. L.-H.H. conducted the study and created a first draft of the paper. K.-Y.S. had substantial inputs into the data analysis and S.-J.J. co-supervised the study and drafts of the paper. All authors have read and agreed to the published version of the manuscript.

Funding: This study was financially supported by the Ministry of Science and Technology, Taiwan under Contracts No. MOST 105-2313-B-019-005-MY3.

Conflicts of Interest: The authors declare no conflict of interest.

References

1. Cortés, E. Standardized diet compositions and trophic levels of sharks. *ICES J. Mar. Sci.* **1999**, *56*, 707–717. [CrossRef]
2. Stevens, J.D.; Bonfil, R.; Dulvy, N.; Walker, P.A. The effects of fishing on sharks, rays, and chimaeras (chondrichthyans), and the implications for marine ecosystems. *ICES J. Mar. Sci.* **2000**, *57*, 476–494. [CrossRef]
3. Schindler, D.E.; Essington, T.E.; Kitchell, J.F.; Boggs, C.; Hilborn, R. Sharks and tunas: Fisheries impacts on predators with contrasting life histories. *Ecol. Appl.* **2002**, *12*, 735–748. [CrossRef]
4. Baum, J.; Myers, R.A.; Kehler, D.G.; Worm, B.; Harley, S.J.; Doherty, P.A.; Baum, J.; Myers, R.A.; Kehler, D.G.; Worm, B.; et al. Collapse and Conservation of Shark Populations in the Northwest Atlantic. *Science* **2003**, *299*, 389–392. [CrossRef] [PubMed]
5. Myers, R.A.; Worm, B. Rapid worldwide depletion of predatory fish communities. *Nat. Cell Biol.* **2003**, *423*, 280–283. [CrossRef]
6. Baum, J.; Myers, R.A. Shifting baselines and the decline of pelagic sharks in the Gulf of Mexico. *Ecol. Lett.* **2004**, *7*, 135–145. [CrossRef]
7. Burgess, G.H.; Beerkircher, L.R.; Cailliet, G.M.; Carlson, J.K.; Cortés, E.; Goldman, K.J.; Grubbs, R.D.; Musick, J.A.; Musyl, M.K.; Simpfendorfer, C.A. Is the collapse of shark populations in the Northwest Atlantic Ocean and Gulf of Mexico real? *Fisheries* **2005**, *30*, 19–26. [CrossRef]
8. Dulvy, N.; Baumb, J.K.; Clarkec, S.; Compagno, L.J.V.; Cortés, E.; Domingo, A.; Fordhamg, S.; Fowler, S.; Francisi, M.P.; Gibsonh, C.; et al. You can swim but you can't hide: The global status and conservation of oceanic pelagic sharks and rays. *Aquat. Conserv. Mar. Freshw. Ecosyst.* **2008**, *18*, 459–482. [CrossRef]
9. Pacoureau, N.; Rigby, C.L.; Kyne, P.M.; Sherley, R.B.; Winker, H.; Carlson, J.K.; Fordham, S.V.; Barreto, R.; Fernando, D.; Francis, M.P.; et al. Half a century of global decline in oceanic sharks and rays. *Nat. Cell Biol.* **2021**, *589*, 567–571. [CrossRef]
10. Haedrich, R.; Barnes, S. Changes over time of the size structure in an exploited shelf fish community. *Fish. Res.* **1997**, *31*, 229–239. [CrossRef]
11. Walker, T.I. Can shark resources be harvested sustainably? A question revisited with a review of shark fisheries. *Mar. Freshw. Res.* **1998**, *49*, 553–572. [CrossRef]
12. Yemane, D.; Field, J.G.; Leslie, R.W. Exploring the effects of fishing on fish assemblages using Abundance Biomass Comparison (ABC) curves. *ICES J. Mar. Sci.* **2005**, *62*, 374–379. [CrossRef]
13. Myers, R.A.; Baum, J.K.; Shepherd, T.D.; Powers, S.P.; Peterson, C.H. Cascading Effects of the loss of apex predatory sharks from a coastal ocean. *Science* **2007**, *315*, 1846–1850. [CrossRef] [PubMed]
14. Scandol, J.; Ives, A.R.; Lockett, M.M. *Development of National Guidelines to Improve the Application of Risk-Based Methods in the Scope, Implementation and Interpretation of Stock Assessments for Data-Poor Species, Cronulla Fisheries Research Centre of Excellence*; Industry and Investment NSW: Canberra, Australia, 2009.
15. Arrizabalaga, H.; De Bruyn, P.; Diaz, G.A.; Murua, H.; Chavance, P.; De Molina, A.D.; Gaertner, D.; Ariz, J.; Ruiz, J.; Kell, L.T. Productivity and susceptibility analysis for species caught in Atlantic tuna fisheries. *Aquat. Living Resour.* **2011**, *24*, 1–12. [CrossRef]
16. Ormseth, O.A.; Spencer, P.D. An assessment of vulnerability in Alaska groundfish. *Fish. Res.* **2011**, *112*, 127–133. [CrossRef]
17. Kirby, D.S. Ecological risk Assessment for species caught in WCPO tuna fisheries: Inherent risk as determined by productivi-ty-susceptibility analysis. *Meet Sci. Commun. West Cent. Pac. Fish. Commun.* **2006**, *25*, 7–18.
18. Murua, H.; Coelho, R.; Santos, M.N.; Arrizabalaga, H.; Yokawa, K.; Romanov, E.; Zhu, J.F.; Kim, Z.G.; Bach, P.; Chavance, P.; et al. Preliminary ecological risk assessment (ERA) for shark species caught in fisheries managed by the Indian Ocean Tuna Commission (IOTC). *IOTC* **2012**, *13*, 16.
19. Murua, H.; Santiago, J.; Coelho, R.; Zudaire, I.; Neves, C.; Rosa, D.; Zudaire, I.; Semba, Y.; Geng., Z.; Bach, P.; et al. Updated Ecological Risk Assessment (ERA) for shark species caught in fisheries managed by the Indian Ocean Tuna Commission (IOTC). *IOTC* **2018**, *14*, 28.
20. Simpfendorfer, C.A.; Cortés, E.; Heupel, M.; Brooks, E.; Babcock, E.; Baum, J.; McAuley, R.; Dudley, S.; Stevens, J.D.; Fordham, S.; et al. An integrated approach to determining the risk of over-exploitation for data-poor pelagic Atlantic sharks. *Lenfest Ocean Program* **2008**, *140*, 1–22.
21. Cortés, E.; Domingo, A.; Miller, P.; Forselledo, R.; Mas, F.; Arocha, F.; Campana, S.; Coelho, R.; Da Silva, C.; Hazin, F.H.V.; et al. Expanded ecological risk assessment of pelagic sharks caught in Atlantic pelagic longline fisheries. *ICCAT* **2012**, *167*, 56.
22. Webb, H.; Hobday, A.; Dowdney, J.; Bulman, C.; Sporcic, M.; Smith, T.; Furlani, D.; Fuller, M.; Williams, A.; Stobustzki, I. Ecological risk assessment for effects of fishing. Report for eastern tuna & billfish fishery: Longline sub-fishery. *Aust. Gov. Aust. Fish. Manag. Auth.* **2007**, *4*, 230.
23. Griffiths, S.P.; Brewer, D.T.; Heales, D.S.; Milton, D.A.; Stobutzki, I.C. Validating ecological risk assessments for fisheries: Assessing the impacts of turtle excluder devices on elasmobranch bycatch populations in an Australian trawl fishery. *Mar. Freshw. Res.* **2006**, *57*, 395–401. [CrossRef]
24. Stelzenmüller, V.; Ellis, J.R.; Rogers, S.I. Towards a spatially explicit risk assessment for marine management: Assessing the vulnerability of fish to aggregate extraction. *Biol. Conserv.* **2010**, *143*, 230–238. [CrossRef]
25. Chin, A.; Kyne, P.; Walker, T.I.; McAuley, R.B. An integrated risk assessment for climate change: Analysing the vulnerability of sharks and rays on Australia's Great Barrier Reef. *Glob. Chang. Biol.* **2010**, *16*, 1936–1953. [CrossRef]

26. Gallagher, A.J.; Kyne, P.; Hammerschlag, N. Ecological risk assessment and its application to elasmobranch conservation and management. *J. Fish. Biol.* **2012**, *80*, 1727–1748. [CrossRef]
27. Lin, C.-Y.; Wang, S.-P.; Chiang, W.-C.; Griffiths, S.; Yeh, H.-M. Ecological risk assessment of species impacted by fisheries in waters off eastern Taiwan. *Fish. Manag. Ecol.* **2020**, *27*, 345–356. [CrossRef]
28. Liu, K.-M.; Chang, Y.-T.; Ni, I.-H.; Jin, C.-B. Spawning per recruit analysis of the pelagic thresher shark, *Alopias pelagicus*, in the eastern Taiwan waters. *Fish. Res.* **2006**, *82*, 56–64. [CrossRef]
29. Tsai, W.-P.; Liu, K.-M.; Joung, S.-J. Demographic analysis of the pelagic thresher shark, *Alopias pelagicus*, in the northwestern Pacific using a stochastic stage-based model. *Mar. Freshw. Res.* **2010**, *61*, 1056–1066. [CrossRef]
30. Tsai, W.-P.; Chang, Y.-J.; Liu, K.-M. Development and testing of a Bayesian population model for the bigeye thresher shark, *Alopias superciliosus*, in an area subset of the western North Pacific. *Fish. Manag. Ecol.* **2019**, *26*, 269–294. [CrossRef]
31. Tsai, W.-P.; Liu, K.-M.; Chang, Y.-J. Evaluation of biological reference points for conservation and management of the bigeye thresher shark, *Alopias superciliosus*, in the Northwest Pacific. *Sustainable* **2020**, *12*, 8646. [CrossRef]
32. Tsai, W.-P.; Sun, C.-L.; Wang, S.-P.; Liu, K.-M. Evaluating the impacts of uncertainty on the estimation of biological reference points for the shortfin mako shark, Isurus oxyrinchus, in the north-western Pacific Ocean. *Mar. Freshw. Res.* **2011**, *62*, 1383–1394. [CrossRef]
33. Tsai, W.-P.; Sun, C.-L.; Punt, A.E.; Liu, K.-M.; Tsai, W.-P.; Sun, C.-L.; Punt, A.E.; Liu, K.-M. Demographic analysis of the shortfin mako shark, Isurus oxyrinchus, in the Northwest Pacific using a two-sex stage-based matrix model. *ICES J. Mar. Sci.* **2014**, *71*, 1604–1618. [CrossRef]
34. Tsai, W.-P.; Liu, K.-M.; Punt, A.E.; Sun, C.-L. Assessing the potential biases of ignoring sexual dimorphism and mating mechanism in using a single-sex demographic model: The shortfin mako shark as a case study. *ICES J. Mar. Sci.* **2015**, *72*, 793–803. [CrossRef]
35. Tsai, W.-P.; Wu, J.-R.; Yan, M.-Z.; Liu, K.-M. Assessment of biological reference points for management of the smooth hammerhead shark, *Sphyrna zygaena*, in the Northwest Pacific Ocean. *J. Fish. Soc. Taiwan* **2018**, *45*, 29–41. [CrossRef]
36. Liu, K.-M.; Chen, C.-T.; Liao, T.-H.; Joung, S.-J. Age, growth, and reproduction of the pelagic thresher shark, *Alopias pelagicus* in the northwestern Pacific. *Copeia* **1999**, *1999*, 68–74. [CrossRef]
37. Liu, K.-M.; Chiang, P.-J.; Chen, C.-T. Age and growth estimates of the bigeye thresher, *Alopias superciliosus*, in northwestern Taiwan waters. *Fish. Bull.* **1998**, *96*, 262–271.
38. Joung, S.-J.; Liao, Y.-Y.; Liu, K.-M.; Chen, C.-T.; Leu, L.-C. Age, growth, and reproduction of the spinner shark, *Carcharhinus brevipinna*, in the northeastern waters of Taiwan. *Zool. Stud.* **2005**, *44*, 102–110.
39. Joung, S.-J.; Chen, C.-T.; Lee, H.-H.; Liu, K.-M. Age, growth, and reproduction of silky sharks, *Carcharhinus falciformis*, in northeastern Taiwan waters. *Fish. Res.* **2008**, *90*, 78–85. [CrossRef]
40. Joung, S.-J.; Chen, N.-F.; Hsu, H.-H.; Liu, K.-M. Estimates of life history parameters of the oceanic whitetip shark, *Carcharhinus longimanus*, in the Western North Pacific Ocean. *Mar. Biol. Res.* **2016**, *12*, 758–768. [CrossRef]
41. Joung, S.-J.; Chen, J.-H.; Chin, C.-P.; Liu, K.-M. Age and growth of the dusky shark, *Carcharhinus obscurus*, in the western North Pacific Ocean. *Terr. Atmos. Ocean Sci.* **2015**, *26*, 153–160. [CrossRef]
42. Joung, S.-J.; Liao, Y.-Y.; Chen, C.-T. Age and growth of sandbar shark, *Carcharhinus plumbeus*, in northeastern Taiwan waters. *Fish. Res.* **2004**, *70*, 83–96. [CrossRef]
43. Chang, J.-H.; Liu, K.-M. Stock assessment of the shortfin mako shark (*Isurus oxyrinchus*) in the Northwest Pacific Ocean using per recruit and virtual population analyses. *Fish. Res.* **2009**, *98*, 92–101. [CrossRef]
44. Joung, S.-J.; Hsu, H.-H. Reproduction and embryonic development of the shortfin mako, *Isurus oxyrinchus* Rafinesque, 1810, in the northwestern Pacific. *Zool. Stud.* **2005**, *44*, 487–496.
45. Chen, C.-T.; Leu, T.-C.; Joung, S.-J.; Lo, N.C.-H. Age and growth of the scalloped hammerhead, *Sphyrna lewini*, in northeastern Taiwan waters. *Pac. Sci.* **1990**, *44*, 156–170.
46. Chou, Y.-T. Studies on Age and Growth of Smooth Hammerhead, *Sphyrna zygaena* in Northeastern Taiwan Waters. Master's Thesis, Department of Environment Biology and Fisheries Science, National Taiwan Ocean University, Keelung, Taiwan, 2004; p. 66.
47. Huang, C.-C. Age and Growth of the Blue shark, *Prionace glauca*, in the Northwestern Pacific Ocean. Master's Thesis, Department of Environment Biology and Fisheries Science, National Taiwan Ocean University, Keelung, Taiwan, 2006; p. 76.
48. Krebs, C.J. *Ecology: The Experimental Analysis of Distribution and Abundance*, 3rd ed.; Harper and Row: New York, NY, USA, 1985; p. 800.
49. Chen, C.-T.; Liu, K.-M.; Chang, Y.-C. Reproductive biology of the bigeye thresher shark, *Alopias superciliosus* (Lowe, 1839) (Chondrichthyes: Alopiidae), in the northwestern Pacific. *Ichthyol. Res.* **1997**, *44*, 227–235. [CrossRef]
50. Joung, S.-J.; Chen, C.-T. Reproduction in the sandbar shark, *Carcharhinus plumbeus*, in the waters off northeastern Taiwan. *Copeia* **1995**, *1995*, 659–665. [CrossRef]
51. Chen, C.-T.; Leu, T.-C.; Joung, S.-J. Notes on reproduction in the scalloped hammerhead, *Sphyrna lewini*, in northeastern Taiwan waters. *Fish. Bull.* **1988**, *86*, 389–393.
52. Liu, S.C. Reproductive Biology of Smooth Hammerhead, *Sphyrna zygaena* in Northeastern Taiwan Waters. Master's Thesis, Department of Environment Biology and Fisheries Science, National Taiwan Ocean University, Keelung, Taiwan, 2002; p. 84.
53. Wu, T.-Y. Reproductive Biology of Blue Shark, *Prionace glauca* in the Northwestern Pacific Ocean. Master's Thesis, Department of Environment Biology and Fisheries Science, National Taiwan Ocean University, Keelung, Taiwan, 2003; p. 109.

54. Cortés, E.; Arocha, F.; Beerkircher, L.; Carvalho, F.; Domingo, A.; Heupel, M.; Holtzhausen, H.; Santos, M.N.; Ribera, M.; Simpfendorfer, C. Ecological risk assessment of pelagic sharks caught in Atlantic pelagic longline fisheries. *Aquat. Living Resour.* **2009**, *23*, 25–34. [CrossRef]
55. IUCN. The IUCN Red List of Threatened Species. Available online: https://www.iucnredlist.org (accessed on 18 January 2021).
56. Fowler, C.W. Population dynamics as related to rate of increase per generation. *Evol. Ecol.* **1988**, *2*, 197–204. [CrossRef]
57. Liu, K.-M.; Chin, C.-P.; Chen, C.-H.; Chang, J.-H. Estimating finite rate of population increase for sharks based on vital parameters. *PLoS ONE* **2015**, *10*, e0143008. [CrossRef]
58. Liu, K.-M.; Huang, Y.-W.; Hsu, H.-H. Management Implications for skates and rays based on analysis of life history parameters. *Front. Mar. Sci.* **2021**, *8*, 664611. [CrossRef]
59. Clarke, K.R.; Gorley, R.N. PRIMER v6: User Manual/Tutorial (Plymouth Routines in Multivariate Ecological Research). 2006, p. 182. Available online: https://www.scienceopen.com/document?vid=2cd68314-640b-4288-8316-532e8932d7a1 (accessed on 21 July 2021).
60. Sippel, T.; Wraith, J.; Kohin, S.; Taylor, V.; Holdsworth, J.; Taguchi, M.; Matsunaga, H.; Yokawa, K. A summary of blue shark (*Prionace glauca*) and shortfin mako shark (*Isurus oxyrinchus*) tagging data available from the North and Southwest Pacific Ocean. In *Working Document Submitted to the ISC Shark Working Group Workshop*; WCPFC: California, CA, USA, 2011; p. 8.
61. ISC. *Stock Assessment of Shortfin Mako Shark in the North Pacific Ocean through ISC*; ISC/18/ANNEX/15; ISC: San Diego, CA, USA, 2018; p. 180.
62. Galván-Tirado, C.; Díaz-Jaimes, P.; León, F.J.G.-D.; Galván-Magaña, F.; Uribe-Alcocer, M.; Galván-Tirado, C.; Díaz-Jaimes, P.; León, F.J.G.-D.; Galván-Magaña, F.; Uribe-Alcocer, M. Historical demography and genetic differentiation inferred from the mitochondrial DNA of the silky shark (*Carcharhinus falciformis*) in the Pacific Ocean. *Fish. Res.* **2013**, *147*, 36–46. [CrossRef]
63. Clarke, S.; Langley, A.; Lennert-Cody, C.; Aires-da-Silva, A.; Maunder, M. Pacific-wide silky shark (*Carcharhinus falciformis*) stock status assessment. *WCPFC Sci. Comm. 14th Regul. Sess.* **2018**, *WCPFC-SC14-2018/SA-WP-08*, 137.
64. Tremblay-Boyer, L.; Carvalho, F.; Neubauer, P.; Pilling, G. Stock assessment for oceanic whitetip shark in the Western and Central Pacific Ocean. *WCPFC Sci. Comm. 15th Regul. Sess. Pohnpei Fed. States Micronesia* **2019**, *WCPFC-SC15-2019/SA-WP-06*, 99.
65. Campana, S.E. Accuracy, precision and quality control in age determination, including a review of the use and abuse of age validation methods. *J. Fish. Biol.* **2001**, *59*, 197–242. [CrossRef]
66. Goldman, K.; Cailliet, G. Age determination and validation in Chondrichthyan fishes. In *Biology of Sharks and Their Relatives*; Carrier, J., Musick, J.A., Heithaus, M.R., Eds.; CRC Press LLC.: Boca Raton, FL, USA, 2004; Volume 14, pp. 399–447. [CrossRef]
67. Liu, K.-M.; Wu, C.-B.; Joung, S.-J.; Tsai, W.-P.; Su, K.-Y. Multi-model approach on growth estimation and association with life history trait for elasmobranchs. *Front. Mar. Sci.* **2021**, *8*, 591692. [CrossRef]
68. Anislado-Tolentino, V. Age and growth for the scalloped hammerhead shark, *Sphyrna lewini* (griffith and smith, 1834) along the central pacific coast of Mexico. *Cienc. Mar.* **2001**, *27*, 501–520. [CrossRef]
69. Klimley, A.P.; Klimley, A.P. The determinants of sexual segregation in the scalloped hammerhead shark, *Sphyrna lewini*. *Environ. Biol. Fishes* **1987**, *18*, 27–40. [CrossRef]
70. White, W.T.; Bartron, C.; Potter, I.C. Catch composition and reproductive biology of *Sphyrna lewini* (Griffith & Smith) (Carcharhiniformes, Sphyrnidae) in Indonesian waters. *J. Fish. Biol.* **2008**, *72*, 1675–1689. [CrossRef]
71. Harry, A.V.; Macbeth, W.G.; Gutteridge, A.N.; Simpfendorfer, C.A. The life histories of endangered hammerhead sharks (Carcharhiniformes, Sphyrnidae) from the east coast of Australia. *J. Fish. Biol.* **2011**, *78*, 2026–2051. [CrossRef] [PubMed]
72. Hsu, H.-H. Age, Growth, and Reproduction of Shortfin Mako, *Isurus oxyrinchus*, in the Northwestern Pacific. Master's Thesis, Department of Environmental Biology and Fisheries Science, National Taiwan Ocean University, Keelung, Taiwan, 2003; p. 107.
73. Semba, Y.; Nakano, H.; Aoki, I. Age and growth analysis of the shortfin mako, *Isurus oxyrinchus*, in the western and central North Pacific Ocean. *Environ. Biol. Fishes* **2009**, *84*, 377–391. [CrossRef]
74. Ribot-Carballal, M.; Galván-Magaña, F.; Quiñónez-Velázquez, C. Age and growth of the shortfin mako shark, *Isurus oxyrinchus*, from the western coast of Baja California Sur, Mexico. *Fish. Res.* **2005**, *76*, 14–21. [CrossRef]
75. Wells, R.J.D.; Smith, S.E.; Kohin, S.; Freund, E.; Spear, N.; Ramon, D.A. Age validation of juvenile shortfin mako (*Isurus oxyrinchus*) tagged and marked with oxytetracycline off southern California. *Fish. Bull.* **2013**, *111*, 147–160. [CrossRef]
76. Kinney, M.J.; Wells, R.J.D.; Kohin, S. Oxytetracycline age validation of an adult shortfin mako shark *Isurus oxyrinchus* after 6 years at liberty. *J. Fish. Biol.* **2016**, *89*, 1828–1833. [CrossRef]
77. Tsai, W.-P.; Chin, C.-P.; Liu, K.-M. Estimate the intrinsic population growth rate of the blue shark in the North Pacific using demographic analysis. *J. Fish. Soc. Taiwan* **2013**, *40*, 231–238.
78. Musyl, M.K.; Brill, R.W.; Curran, D.S.; Fragoso, N.M.; McNaughton, L.M.; Nielsen, A.; Kikkawa, B.S.; Moyes, C.D. Post-release survival, vertical and horizontal movements, and thermal habitats of five species of pelagic sharks in the central Pacific Ocean. *Fish. Bull.* **2011**, *109*, 341–368.
79. Musyl, M.K.; Gilman, E.L. Post-release fishing mortality of blue (*Prionace glauca*) and silky shark (*Carcharhinus falciformes*) from a Palauan-based commercial longline fishery. *Rev. Fish. Biol. Fish.* **2018**, *28*, 567–586. [CrossRef]
80. Schaefer, K.M.; Fuller, D.W.; Aires-Da-Silva, A.; Carvajal, J.M.; Martínez-Ortiz, J.; Hutchinson, M.R. Post-release survival of silky sharks (*Carcharhinus falciformis*) following capture by longline fishing vessels in the equatorial eastern Pacific Ocean. *Bull. Mar. Sci.* **2019**, *95*, 355–369. [CrossRef]

81. Schaefer, K.; Fuller, D.; Castillo-Geniz, J.L.; Godinez-Padilla, C.J.; Dreyfus, M.; Aires-Da-Silva, A.; Schaefer, K.; Fuller, D.; Castillo-Geniz, J.L.; Godinez-Padilla, C.J.; et al. Post-release survival of silky sharks (*Carcharhinus falciformis*) following capture by Mexican flag longline fishing vessels in the northeastern Pacific Ocean. *Fish. Res.* **2021**, *234*, 105779. [CrossRef]
82. Campana, S.E.; Joyce, W.; Fowler, M.; Showell, M. Discards, hooking, and post-release mortality of porbeagle (*Lamna nasus*), shortfin mako (*Isurus oxyrinchus*), and blue shark (*Prionace glauca*) in the Canadian pelagic longline fishery. *ICES J. Mar. Sci.* **2015**, *73*, 520–528. [CrossRef]
83. Santos, C.C.; Domingo, A.; Carlson, J.; Natanson, L.; Travassos, P.; Macias, D.; Cortés, E.; Miller, P.; Hazin, F.; Mas, F.; et al. Updates on the habitat use and migrations of shortfin mako in the Atlantic using satellite telemetry. *Collect. Vol. Sci. Pap. ICCAT* **2020**, *76*, 235–246.
84. Dulvy, N.K.; Jennings, S.; Goodwin, N.B.; Grant, A.; Reynolds, J.D. Comparison of threat and exploitation status in North-East Atlantic marine populations. *J. Appl. Ecol.* **2005**, *42*, 883–891. [CrossRef]
85. Rigby, C.L.; Dulvy, N.K.; Barreto, R.; Carlson, J.; Fernando, D.; Fordham, S.; Francis, M.P.; Herman, K.; Jabado, R.W.; Liu, K.-M.; et al. *Sphyrna lewini. IUCN Red List Threat Species* 2019, eT39385A2918526. Available online: https://www.iucnredlist.org/ja/species/39385/2918526 (accessed on 21 July 2021).
86. Anon. Report of the 2019 shortfin mako shark stock assessment update meeting. In Proceedings of the ICCAT Shortfin mako shark stock assessment intersessional meeting, Madrid, Spain, 20–24 May 2019; p. 41.
87. Linder, E.; Patil, G.; Vaughan, D.S. Application of event tree risk analysis to fisheries management. *Ecol. Model.* **1987**, *36*, 15–28. [CrossRef]
88. Fogarty, M.; Mayo, R.; O'Brien, L.; Serchuk, F.; Rosenberg, A. Assessing uncertainty and risk in exploited marine populations. *Reliab. Eng. Syst. Saf.* **1996**, *54*, 183–195. [CrossRef]
89. Zhou, S.; Griffiths, S. Sustainability Assessment for Fishing Effects (SAFE): A new quantitative ecological risk assessment method and its application to elasmobranch bycatch in an Australian trawl fishery. *Fish. Res.* **2008**, *91*, 56–68. [CrossRef]
90. Zhou, S.; Griffiths, S.; Miller, M. Sustainability assessment for fishing effects (SAFE) on highly diverse and data-limited fish bycatch in a tropical prawn trawl fishery. *Mar. Freshw. Res.* **2009**, *60*, 563–570. [CrossRef]
91. Zhou, S.; Smith, A.D.; Fuller, M. Quantitative ecological risk assessment for fishing effects on diverse data-poor non-target species in a multi-sector and multi-gear fishery. *Fish. Res.* **2011**, *112*, 168–178. [CrossRef]
92. Zhou, S.; Hobday, A.J.; Dichmont, C.M.; Smith, A.D. Ecological risk assessments for the effects of fishing: A comparison and validation of PSA and SAFE. *Fish. Res.* **2016**, *183*, 518–529. [CrossRef]
93. Zhou, S.; Daley, R.M.; Fuller, M.; Bulman, C.; Hobday, A.J. A data-limited method for assessing cumulative fishing risk on bycatch. *ICES J. Mar. Sci.* **2019**, *76*, 837–847. [CrossRef]
94. Griffiths, S.P.; Kesner-Reyes, K.; Garilao, C.; Duffy, L.M.; Roman, M.H. Development of a flexible ecological risk assessment (ERA) approach for quantifying the cumulative impacts of fisheries on bycatch species in the eastern Pacific Ocean. In Proceedings of the Inter-American Tropical Tuna Commission, Scientific Advisory Committee, 9th Meeting, IATTC, La Jolla, CA, USA, 14–18 May 2018; p. 61.
95. Griffiths, S.; Kesner-Reyes, K.; Garilao, C.; Duffy, L.; Román, M. Ecological assessment of the sustainable impacts of fisheries (EASI-Fish): A flexible vulnerability assessment approach to quantify the cumulative impacts of fishing in data-limited settings. *Mar. Ecol. Prog. Ser.* **2019**, *625*, 89–113. [CrossRef]

 *animals* 

Article

Batoid Abundances, Spatial Distribution, and Life History Traits in the Strait of Sicily (Central Mediterranean Sea): Bridging a Knowledge Gap through Three Decades of Survey

Michele Luca Geraci [1,2], Sergio Ragonese [2,*], Danilo Scannella [2], Fabio Falsone [2], Vita Gancitano [2], Jurgen Mifsud [3], Miriam Gambin [3], Alicia Said [3] and Sergio Vitale [2]

[1] Geological and Environmental Sciences (BiGeA)–Marine Biology and Fisheries Laboratory, Department of Biological, University of Bologna, Viale Adriatico 1/n, 61032 Fano, PU, Italy; micheleluca.geraci2@unibo.it

[2] Institute for Marine Biological Resources and Biotechnology (IRBIM), National Research Council–CNR, Via Luigi Vaccara, 61, 91026 Mazara del Vallo, TP, Italy; danilo.scannella@irbim.cnr.it (D.S.); fabio.falsone@irbim.cnr.it (F.F.); vita.gancitano@cnr.it (V.G.); sergio.vitale@cnr.it (S.V.)

[3] Department of Fisheries and Aquaculture, Ministry for Agriculture, Fisheries and Animal Rights (MAFA), Ghammieri Government Farm, Triq l-Ingiered, Malta; jurgen.a.mifsud@gov.mt (J.M.); miriam.gambin@gov.mt (M.G.); alicia.bugeja-said@gov.mt (A.S.)

* Correspondence: sergio.ragonese@cnr.it

Citation: Geraci, M.L.; Ragonese, S.; Scannella, D.; Falsone, F.; Gancitano, V.; Mifsud, J.; Gambin, M.; Said, A.; Vitale, S. Batoid Abundances, Spatial Distribution, and Life History Traits in the Strait of Sicily (Central Mediterranean Sea): Bridging a Knowledge Gap through Three Decades of Survey. *Animals* **2021**, *11*, 2189. https://doi.org/10.3390/ani11082189

Academic Editor: Christopher Hoagstrom

Received: 26 May 2021
Accepted: 14 July 2021
Published: 23 July 2021

Publisher's Note: MDPI stays neutral with regard to jurisdictional claims in published maps and institutional affiliations.

Simple Summary: Batoid species are cartilaginous fish commonly known as rays, but they also include stingrays, electric rays, guitarfish, skates, and sawfish. These species are very sensitive to fishing, mainly because of their slow growth rate and late maturity; therefore, they need to be adequately managed. Regrettably, information on life history traits (e.g., length at first maturity, sex ratio, and growth) and abundance are still scarce, particularly in the Mediterranean Sea. In this regard, the present study focuses on the Strait of Sicily (Central Mediterranean) and aims to improve knowledge gained through scientific survey data. In particular, abundance data, spatial distribution, and some life history traits are herein presented. In the investigated area, the biomass trends of the batoids indicated a slight recovery even if few species showed a depletion. Considering the importance of this taxon for maintaining the marine ecosystem equilibrium, management measures are desirable.

Abstract: Batoid species play a key role in marine ecosystems but unfortunately they have globally declined over the last decades. Given the paucity of information, abundance data and the main life history traits for batoids, obtained through about three decades of bottom trawl surveys, are presented and discussed. The surveys were carried out in two areas of the Central Mediterranean (South of Sicily and Malta Island), in a timeframe ranging from 1990 to 2018. Excluding some batoids, the abundance trends were stable or increasing. Only *R. clavata*, *R. miraletus*, and *D. oxyrinchus* showed occurrence and abundance indexes notable enough to carry out more detailed analysis. In particular, spatial distribution analysis of these species highlighted the presence of two main hotspots in Sicilian waters whereas they seem more widespread in Malta. The lengths at first maturity (L_{50}) were 695 and 860, 635 and 574, and 364 and 349 mm total length (TL), respectively, for females and males of *D. oxyrinchus*, *R. clavata*, and *R. miraletus*. The asymptotic lengths ($L\infty$) and the curvature coefficients (K) were 1365 and 1240 (K = 0.11 and 0.26), 1260 and 1100 (K = 0.16 and 0.26), and 840 and 800 mm TL (K = 0.36 and 0.41), respectively, for females and males of *D. oxyrinchus*, *R. clavata*, and *R. miraletus*. The lack of detailed quantitative historical information on batoids of Sicily and Malta does not allow to analytically judge the current status of the stocks, although the higher abundance of some species within Malta raises some concern for the Sicilian counterpart. In conclusion, suitable actions to protect batoids in the investigated area are recommended.

Keywords: Mediterranean Sea; bottom trawl survey; spatial distribution; length at first maturity; Linfinity; sex ratio; length–weight relationship

1. Introduction

Batoidea is an infraclass of cartilaginous fish commonly known as batoids or rays, but it also includes stingrays, electric rays, guitarfish, skates, and sawfish. Batoid fishes are moderately to greatly flattened and are distinguished from other elasmobranchs (non-batoid sharks) by their ventral gill slits, their lack of an anal fin, and having the pectoral fins connected to the sides of the head and trunk to form a disk [1,2]. In the Mediterranean Sea (herein Mediterranean), at least 38 species of batoid fishes have been reported [3]. However, long-term sources of information to assess the batoids' exploitation status are very rare in this region [4,5]. Up until the 1980s, these animals were considered as nuisance species having low economic value to Mediterranean fishers, and consequently most of them were discarded. Therefore, these species were neither recorded in the official landing statistics nor in the first experimental surveys (see, in [5,6]). Thereafter, facing the decline in the more productive target bony and shellfish species and the increasing demand of markets, fishers began to retain and land large batoids, returning to the sea only the small or damaged specimens [7].

The lack of significant and representative historical data (reflecting also the misclassification of these species) makes it almost impossible to gauge the current standing stock condition and to detect depletion (if any) [5,6,8]. Moreover, there is evidence that Mediterranean batoid stocks (as well as their oceanic counterparts) are extremely vulnerable to fishing intensity, even when the resulting fishing mortality remains at a low level and they are exposed to relatively short periods of exploitation [9,10]. As a result, nowadays there is general agreement on the generalized rarefaction and bad exploitation status of almost all Mediterranean batoid stocks (see, in [11,12]), even though few quantitative assessments have been conducted. Furthermore, considering that they play an important ecological role as they are meso-predator [13], and, being very sensitive to any change in the ecosystem, they are often used as biological indicators [14–16]. In the last decades, the IUCN (International Union for Conservation of Nature) has focused on these vulnerable taxa and provided a relative estimation of the likelihood of extinction, which is summarized in the "European Red list of Fishes" (see, in [11,17]). Regarding the GFCM (General Fisheries Commission for the Mediterranean) geographical sub-areas (GSA16 and GSA15), named South of Sicily and Malta Island, information has been made available through a standardized scientific international program launched in 1994, the MEDITS (International bottom trawl survey in the Mediterranean [18,19]). In addition, the MEDITS surveys were integrated between 1990 and 2006 by the homologous Italian surveys program called GRUND (GRUppo Nazionale Demersali).

Only scattered data on distribution and abundance, and limited information on growth and unit stock identification, have been published for the Strait of Sicily sensu Jereb et [20]. In particular, contributions on batoids can be found for the Sicilian waters [21–27], the Malta Island [28], and for Tunisian waters [29–36].

In this context, the aim of this paper is to provide a summary of the current knowledge on batoid fishes occurring in GSA16 and GSA15. This will be done through the use of the frequency of occurrences, abundance indices, and some relevant biological information provided through the MEDITS and GRUND surveys, performed in a wide time interval following, as closely as possible, the same methodologies. The present work provides a basic tool for implementing a specific management plan for these endangered species.

2. Materials and Methods

2.1. Study Area and Sampling Methodology

The analyzed data refer to a wide area located between the Southern coasts of Sicily and the Northern coasts of Africa. According to the GFCM classification [37], the study areas are (i) South of Sicily, GSA16, and (ii) Malta Island, GSA15 (Figure 1).

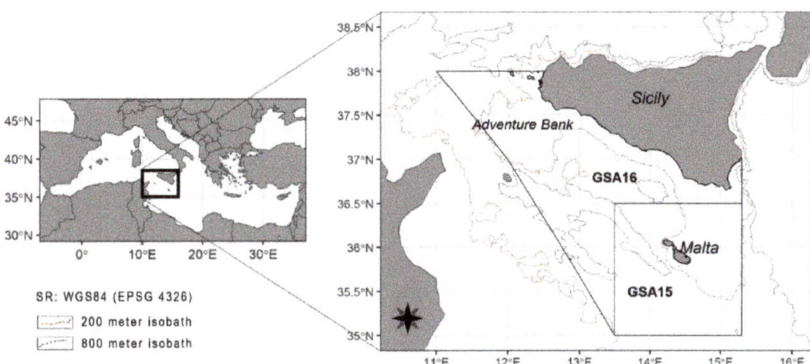

Figure 1. The study area with the two considered Geographical Sub-Areas: South of Sicily (GSA16) and Malta Island (GSA15). Red and blue dashed lines denote the 200 and 800 m isobaths, respectively.

A database collecting about three decades of experimental (scientific) bottom trawl surveys in GSA16 and GSA15 was used. For the surveys, the analyzed data refer to the batoids directly sampled mainly in spring and summer during the MEDITS survey from 1994 to 2018 in GSA16 (herein MEDITS16) and from 2005 to 2018 in GSA15 (herein MEDITS15), and in autumn during the GRUND survey from 1990 to 2006 in both GSAs (herein GRUND15 and GRUND16, survey carried out respectively in the GSA15 and GSA16). The changes to survey protocols/specifications over the years are documented in the manual [18] and descriptive papers [19,38–40]. In particular, the foreseen average sampling rate was one station per 60 square nautical miles in both areas. As closely as possible, the same stations were visited each year. Within MEDITS, sampling at sea has always been conducted with the bottom trawl net GOC73 [19,40], whereas in GRUND a typical commercial trawl net, locally called "Mazarese" or "tartana di banco", was used [38,39]. Specifically, the two gears mainly differ in the vertical mouth opening (2.4–2.9 in MEDITS vs. 0.6–1.3 m in GRUND), but both gears mount a 20 mm side diamond stretched mesh in the cod end. In both surveys, the haul durations were 30′ and 60′ at stations between 10 and 200 m (shelf) and 201–800 m (slope), respectively. MEDITS16 and MEDITS15 have been carried out consistently, with MEDITS15 starting in 2005, whereas GRUND was not carried out in 1993 and 1999 because of administrative constraints (Tables S1 and S2). The stations were distributed through a stratified sampling scheme with random drawings inside each stratum. The stratification criterion adopted was the depth, with the following bathymetric limits being 10 to 50 m (a stratum), 51 to 100 m (b stratum), 101 to 200 m (c stratum), 201 to 500 m (d stratum), and 501 to 800 m (e stratum). However, in order to simplify the presentation of abundance indices, the previous microstrata were also pooled into two macrostrata: shelf (10–200 m) and slope (201–800 m).

2.2. Abundance Data and Biological Parameter Analyses

The biological samples, from sampling at the sea, were identified, frozen, and brought to the laboratories for successive biometric (Total Length, TL, in mm), gravimetric (Total body weight, W, in g), and, after dissection, reproductive analyses by distinguishing macroscopically the sex (F, females; M, males) and maturity stages. The data collected and elaborated through the surveys were presented as follows (the symbols refer to the work in [41], if not otherwise specified):

(a) frequency of occurrence as percentage of positive hauls (f%) and survey (s%);

(b) two abundance indices: in weight, Biomass Index (BI; kg/km^2), and number, Density Index (DI; N/km^2), expressed as grand mean (the mean of the mean) with their relative standard error; f% and both abundance indices were estimated for the continental shelf (10–200 m), slope (200–800 m), and overall (10–800 m) depth stratum;

(c)   depth presence among the identified five microstrata;

(d)   correlation among survey abundance indexes and years were assessed by GSA, species, and macrostratum, by computing the nonparametric Spearman linear rank coefficient;

(e)   overall standing stock expressed in weight (tons) and number (thousand);

(f)   biological information in terms of overall sex ratio (sex ratio = F/M), median length, and length–weight relationship (LWR, power function; k will indicate the steepness; b will indicate the positive ($b > 3$) or negative ($b < 3$) allometric coefficient; isometric when $b = 3$) only from MEDITS16 because of the longer time series.

Sex ratio deviations from the expected 1:1 were checked by applying the nonparametric Chi-square test [42]; $p < 0.05$ was considered statistically significant.

In addition, for the most representative and abundant species, identified according to the few, although arbitrary, thresholds (f% $\geq$ 10 and 50 tons in standing stock at least in two surveys), the following biological parameters were also estimated for females and males from MEDITS16:

(i)   length at first sexual maturity (L_{50}, also known as at onset of sexual maturity), i.e., the length at which 50% of specimens resulted mature (stage 3a onward according to MEDITS gonadic scale [18]), or in case of lack of fit of the logistic using the median length of the 3a stage as a proxy of L_{50} [43];

(ii)   Length–Frequency Distribution (LFD) "stability" in the different years (performed via the Kolmogorov–Smirnov test);

(iii)   L∞ (the infinite/asymptotic length at which the average growth rate of the oldest cohort becomes close to zero) and K (Brody or curvature coefficient), both related to the von Bertalanffy growth function or;

(iv)   Z/K ratio by sex [44] at 95% confidence interval, where Z denotes the instantaneous rate of total mortality (as fishing, F, plus natural mortalities, M). The previous parameters were estimated using the ELEFAN (Electronic Length Frequencies Analysis) and Powell–Wetherall procedures in the TropFishR package [45]. All the size measurements were expressed as the Total Length (TL) in mm.

Furthermore, the life history traits estimated in the present study were compared to those of the other Mediterranean studies (Table S3).

Finally, yearly DI plots of the surveys were represented. Differences between the means of the DI, BI, and frequency of occurrence (%) in the two considered GSA for these species were tested using the nonparametric Wilcoxon's signed rank test (or Mann–Whitney U test) for surveys datasets; $p < 0.05$ was considered statistically significant. All the data were checked for normality using the empirical distribution of the data (the histogram), Shapiro–Wilk test, and qqplot (quantile-quantile plot).

2.3. Spatial Analysis

Geo-statistical analysis [46,47] was applied to the DI in order to obtain the horizontal distribution map of the three species identified from the above prefixed thresholds, i.e., *D. oxyrinchus*, *R. clavata*, and *R miraletus*. Overall, the datasets used consisted of 2750 and 1737 hauls from MEDITS and GRUND, respectively. Considering that geo-statistic techniques use spatial interpolation procedures to obtain spatially continuous variables from isolated station measurements [48,49], to avoid any sampling area "weighing" the estimates more than another location with few hauls, a data manipulation series was carried out. First, a 4 × 4 nautical mile grid covering the whole study area (GSA15 and 16) was constructed (1689 cells), and then a WGS84 UTM 33N (World Geographic System 1984, Universal Transverse Mercator 33 Nord) projected coordinate system was set. Second, for each survey, sampled stations within each cell were managed in order to determine a new point with geographical coordinates that minimized the Euclidean distance of all sampled stations within the same cell, where the median DI value was used. At the end of this step, the new points were 421 and 435 from MEDITS and GRUND, respectively. However, the reduction in the number of stations requires a minimum of 30–50 pairs of points to ensure statistical consistency and representativeness of the sampling space [50]. Therefore, these data were log-transformed (DI + 1) to improve the normality and were used to fit asymptotic models and to compute standard parameters of experimental semivariograms (i.e., range, nugget, and sill). To estimate these parameters, among Gaussian, Exponential, and Spherical models the one reducing the residual sum of squares was selected. Then, following the estimation of variogram parameters, ordinary kriging was applied to map the spatial distribution of the DI for each species and survey using the geoR library [51]. Spatial interpolation was performed using a 0.5×0.5 km^2 grid. In addition, maps of the positive hauls for the three above-mentioned species were provided (Figures S1–S3). All analyses were carried out in R studio, version 3.6.2 [52].

3. Results

3.1. General

A total of 4575 scientific hauls performed over 25 and 17 years, respectively, for MEDITS and GRUND were analyzed, and the results are summarized in Tables 1–4. The numbers of batoid species reported during MEDITS were 22 in the South of Sicily (GSA16) and 16 in the Malta Island (GSA15), whereas during GRUND they were 20 (in the South of Sicily (GSA16) and 17 in the Malta Island (GSA15), overall belonging to three orders and four families. From Tables 1–4, it is possible to notice that only three species, namely, *R. clavata*, *R. miraletus*, and *D. oxyrinchus*, showed values above the prefixed thresholds, (f% > 10; SSw > 50 t at least in two surveys). Excluding these species, in MEDITS16 only *Leucoraja melitensis* (Clark, 1926), *Raja asterias* Delaroche, 1809, *Raja montagui* Fowler, 1910, and *Torpedo marmorata* Risso, 1810 showed overall mean DI higher than 2 N/km^2 (a very low density), whereas in MEDITS15 *Dasyatis pastinaca* (Linnaeus, 1758), *Myliobatis aquila* (Linnaeus, 1758), *Leucoraja circularis* (Couch, 1838), *L. melitensis*, *R. montagui*, *Raja radula* Delaroche, 1809, and *T. marmorata* showed a higher DI. More worrying, only *L. melitensis*, for GRUND16, and *D. pastinaca* and *L. melitensis*, for GRUND15, exceeded a DI of 2 N/km^2. It is worth remarking that nine species with a standing stock less than 10 tons might be considered close to local extinction. Synthetic results concerning abundance and life history traits (when available) of the taxa sampled during surveys are herein provided.

3.1.1. Great Torpedo Ray—*Tetronarce nobiliana* (Bonaparte, 1835)

The species is diffused over the investigated area but in scant numbers. Even if it was sampled across all strata, it seemed to inhabit the slope as the preferential macrostratum. In spring–summer, *T. nobiliana* was found with higher frequency of occurrence (2.1%), abundance indices (slope: BI = 0.7 ± 0.4 kg/km^2, DI = 0.3 ± 0.1 N/km^2), and standing stock (12.5 ± 6.7 tons, 9 ± 2 thousand) in South of Sicily (Tables 1 and 2). In fact, in autumn it was sampled more frequently in South of Sicily (1.2%) with the highest values on the slope (Tables 3 and 4); however, no significant correlations over the years were found (Tables 1–4). The TL ranged from 115 to 970 mm for females, while for males it ranged from 120 to 610 mm. The LWR parameters were 3.3×10^{-5} (k) and 2.89 (b), thus showing negative allometry, for sex combined. The median length was between 395 and 330 mm (TL) for females and males. The sex ratio was 0.63:1 (X^2 = 4.84, p = 0.02781) (Table S3).

Table 1. Synoptic table of batoid abundance sampled during the MEDITS survey (spring–summer) in the GSA16 (South of Sicily), from 1994 to 2018. Percent frequency of occurrence for number of surveys (s%) and positive hauls (f%). Standing Stock (SS) expressed in weight (SSw, tons) and number (SSn, thousand). Stratum distribution and mean biomass index kg/km^2 (BI) expressed by depth of the macrostratum (10–200 m, shelf; 201–800 m, slope; 10–800 m, overall), mean density index N/km^2 (DI) by depth of the macrostratum, and Spearman coefficients for BI and DI by depth of the macrostratum.

Species	%N. Survey	% Positive Hauls	SSw Tons	SSn Number (Thousand)	10–50	50–100	100–200	200–500	500–800	BI Shelf	BI Slope	BI Overall	Spearman BI Shelf	Spearman BI Slope	DI Shelf	DI Slope	DI Overall	Spearman DI Shelf	Spearman DI Slope
Tetronarce nobiliana	76	2.1	12.5 ± 6.7	9 ± 2		●	●	●	●	<0.1	0.7 ± 0.4	0.4 ± 0.2	0.05	0.20	0.2 ± 0.1	0.3 ± 0.1	0.3 ± 0.1	−0.03	−0.10
Torpedo marmorata	100	8.7	20.8 ± 2.5	69 ± 7	●	●	●	●	●	1.2 ± 0.1	0.2 ± 0.0	0.7 ± 0.1	−0.22	−0.17	3.6 ± 0.4	1.0 ± 0.2	2.2 ± 0.2	−0.29	0.08
Torpedo torpedo	40	0.8	5.8 ± 3.2	9 ± 3	●		●	●		0.1 ± 0.0	0.3 ± 0.2	0.2 ± 0.1	0.49	−0.09	0.6 ± 0.2	<0.1	0.3 ± 0.1	0.49	−0.09
Dipturus batis	4	0.1	2.0 ± 2.0	0.2 ± 0.2					●	//	0.1 ± 0.1	<0.1	//	NC	//	<0.1	<0.1	//	NC
Dipturus oxyrinchus	100	10.6	92.3 ± 9.4	73 ± 7		●	●	●		<0.1	5.5 ± 0.6	2.9 ± 0.3	NC	0.54 *	<0.1	4.3 ± 0.4	2.3 ± 0.2	NC	0.36
Leucoraja circularis	40	0.6	2.6 ± 1.0	2 ± 1		●	●	●		<0.1	0.2 ± 0.1	0.1 ± 0.0	NC	0.37	<0.1	0.1 ± 0.0	0.1 ± 0.0	NC	0.38
Leucoraja fullonica	4	0.1	0.5 ± 0.5	0.1 ± 0.1			●			//	<0.1	<0.1	//	NC	//	<0.1	<0.1	//	NC
Leucoraja melitensis	96	6.9	26.7 ± 3.0	132 ± 15	●	●	●	●	●	0.4 ± 0.1	1.2 ± 0.2	0.8 ± 0.1	0.54 *	0.04	1.6 ± 0.5	6.5 ± 0.8	4.2 ± 0.5	0.54 *	0.03
Leucoraja naevus	4	0.2	0.1 ± 0.1	0.4 ± 0.4			●			//	<0.1	<0.1	//	NC	//	<0.1	<0.1	//	NC
Raja asterias	96	4.3	32.7 ± 4.7	71 ± 12	●	●	●	●	●	2.1 ± 0.3	0.1 ± 0.1	1.0 ± 0.1	−0.37	0.24	4.4 ± 0.6	0.4 ± 0.3	2.3 ± 0.4	−0.29	0.24
Raja brachyura	28	0.9	9.1 ± 5.6	10 ± 6	●	●	●			0.3 ± 0.2	//	0.1 ± 0.1	0.41	//	0.3 ± 0.2	//	0.1 ± 0.1	0.42	//
Raja clavata	100	19.8	603.6 ± 66.4	609 ± 70	●	●	●	●	●	33.3 ± 3.8	7.2 ± 1.0	19.2 ± 2.1	0.86 *	0.59 *	31.8 ± 3.6	8.8 ± 1.4	19.4 ± 2.2	0.85 *	0.66 *
Raja miraletus	100	18.9	371.9 ± 34.2	2282 ± 224	●	●	●	●	●	25.4 ± 2.4	0.2 ± 0.1	11.8 ± 1.1	−0.00	0.46	156.2 ± 15.5	1.9 ± 0.4	35.2 ± 3.7	−0.21	0.42
Raja montagui	96	6.0	19.8 ± 2.6	73 ± 13	●	●	●	●		1.3 ± 0.2	0.1 ± 0.0	0.6 ± 0.1	−0.09	0.07	4.7 ± 0.9	0.3 ± 0.1	2.3 ± 0.4	−0.16	0.08
Raja polystigma	28	0.5	1.2 ± 0.5	4 ± 1	●	●		●		0.1 ± 0.0	<0.1	<0.1	0.45	0.49	0.2 ± 0.1	<0.1	0.1 ± 0.0	0.38	0.49
Raja radula	12	0.4	1.2 ± 0.9	1 ± 1	●					0.1 ± 0.1	//	<0.1	0.07	//	0.1 ± 0.1	//	<0.1	0.06	//
Rostroraja alba	64	1.6	57.1 ± 15.3	11 ± 3		●		●	●	3.0 ± 1.1	0.8 ± 0.3	1.8 ± 0.5	0.73 *	0.13	0.6 ± 0.2	0.1 ± 0.0	0.3 ± 0.1	0.69 *	0.06
Dasyatis pastinaca	64	1.2	51.7 ± 13.0	11 ± 2		●	●	●		3.6 ± 0.8	<0.1	1.6 ± 0.4	0.73 *	NC	0.8 ± 0.2	<0.1	0.4 ± 0.1	0.67 *	NC
Pteroplatytrygon violacea	4	0.1	1.9 ± 1.9	0.2 ± 0.2			●			0.1 ± 0.1	//	<0.1	NC	//	<0.1	//	<0.1	NC	//
Aetomylaeus bovinus	12	0.4	41.7 ± 23.0	3 ± 2	●					2.2 ± 1.6	//	1.0 ± 0.7	0.01	//	0.2 ± 0.1	//	0.1 ± 0.1	−0.02	//
Myliobatis aquila	44	0.7	31.0 ± 9.7	8 ± 3	●	●				1.0 ± 0.3	//	0.5 ± 0.1	0.71 *	//	0.6 ± 0.2	//	0.3 ± 0.1	0.73 *	//

Significant correlations are noted by an asterisk (Spearman coefficients > ±0.50); NC: Not Considered because of the low sample size; //: Not Available. In bold, the species is above the fixed thresholds (f% ≥ 10; SSw ≥ 50 tons at least in two surveys). Black dots indicate the depth range within which the species was sampled.

Table 2. Synoptic table of batoid abundance sampled during the MEDITS survey (spring–summer) in the GSA15 (Malta Island), from 2005 to 2018. Percent frequency of occurrence for number of surveys (s%) and positive hauls (f%). Standing Stock (SS) expressed in weight (SSw, tons) and number (SSn, thousand). Stratum distribution and mean biomass index kg/km^2 (BI) expressed by depth of the macrostratum (10–200 m, shelf; 201–800 m, slope; 10–800 m, overall), mean density index N/km^2 (DI) by depth of the macrostratum, and Spearman coefficients for BI and DI by depth of the macrostratum.

Species	Frequency of Occurrence		SSw (t) and SSn (Mean ± SE)		Stratum Distribution (Depth Interval in m)				BI (kg/km^2) Mean ± SE			Spearman Coeff. BI		DI (N/km^2) Mean ± SE			Spearman Coeff. DI	
	% N. Survey	% Positive Hauls	Tons	Number (Thousand)	50–100	100–200	200–500	500–800	Shelf	Slope	Overall	Shelf	Slope	Shelf	Slope	Overall	Shelf	Slope
Tetronarce nobiliana	29	1.4	1.5 ± 1.1	3 ± 1		●	●		<0.1	0.2 ± 0.2	0.1 ± 0.1	−0.38	−0.48	0.1 ± 0.1	0.4 ± 0.2	0.3 ± 0.1	−0.38	−0.49
Torpedo marmorata	100	11.1	9.2 ± 2.2	35 ± 7	●	●	●		1.8 ± 0.5	0.1 ± 0.0	0.9 ± 0.2	−0.33	0.11	6.5 ± 1.3	0.7 ± 0.3	3.3 ± 0.6	−0.16	0.05
Dipturus oxyrincus	100	30.94	262.4 ± 26.5	202 ± 36.8		●	●	●	5.1 ± 1.1	40.5 ± 5.4	24.8 ± 2.5	0.15	−0.15	2.5 ± 0.5	32.3 ± 4.8	19.0 ± 2.9	0.53 *	−0.34
Leucoraja circularis	79	5.7	247.5 ± 233.8	20 ± 7		●	●	●	0.3 ± 0.3	41.8 ± 39.8	23.4 ± 22.1	−0.49	−0.27	0.2 ± 0.2	3.3 ± 1.2	1.9 ± 0.7	−0.49	−0.09
Leucoraja fullonica	43	1.8	2.1 ± 1.5	3 ± 1			●	●	//	0.4 ± 0.2	0.2 ± 0.1	//	−0.19	//	0.5 ± 0.2	0.3 ± 0.1	//	−0.12
Leucoraja melitensis	93	10.5	20.9 ± 3.8	105 ± 19	●	●	●	●	0.3 ± 0.1	3.3 ± 0.6	2.0 ± 0.4	0.05	0.34	1.9 ± 1.0	16.3 ± 2.9	9.9 ± 1.8	0.04	0.24
Raja asterias	7	0.2	<0.1	1 ± 1			●		//	<0.1	<0.1	//	NC	//	<0.1	<0.1	//	NC
Raja brachyura	14	0.3	2.2 ± 1.8	1 ± 1	●	●			0.4 ± 0.4	<0.1	0.2 ± 0.2	NC	NC	0.2 ± 0.2	<0.1	0.1 ± 0.1	NC	NC
Raja clavata	100	52.8	706.8 ± 61.6	660 ± 49	●	●	●	●	90.7 ± 15.0	47.7 ± 5.2	66.8 ± 5.8	0.80 *	−0.44	59.0 ± 7.6	65.1 ± 9.1	62.4 ± 4.7	0.74 *	−0.41
Raja miraletus	100	24.3	70.3 ± 8.0	277 ± 27	●	●	●	●	14.1 ± 1.7	0.7 ± 0.3	6.6 ± 0.8	0.26	−0.29	53.8 ± 5.4	4.0 ± 2.0	26.2 ± 2.6	−0.13	−0.23
Raja montagui	93	9.4	28.6 ± 8.4	103 ± 38	●	●	●	●	1.5 ± 0.6	3.6 ± 1.1	2.7 ± 0.8	−0.02	0.83 *	3.3 ± 1.4	14.9 ± 5.8	9.7 ± 3.6	0.03	0.83 *
Raja polystigma	29	1.3	4.3 ± 2.6	5 ± 3	●		●		0.1 ± 0.1	0.6 ± 0.4	0.4 ± 0.2	NC	−0.41	<0.1	0.8 ± 0.5	0.5 ± 0.3	NC	−0.44
Raja radula	50	1.7	22.4 ± 13.6	24 ± 13	●				4.8 ± 2.9	//	2.1 ± 1.3	−0.42	//	5.2 ± 2.7	//	2.3 ± 1.2	−0.42	//
Rostroraja alba	7	0.3	0.4 ± 0.4	4 ± 4			●		//	<0.1	<0.1	//	NC	<0.1	0.7 ± 0.7	0.4 ± 0.4	//	NC
Dasyatis pastinaca	71	4.3	60.2 ± 17.6	24 ± 6	●	●		●	12.8 ± 3.7	<0.1	5.7 ± 1.7	−0.97 *	NC	5.1 ± 1.4	<0.1	2.3 ± 0.6	−0.78 *	NC
Myliobatis aquila	64	4.6	56.5 ± 16.8	23 ± 7	●	●			12.0 ± 3.6	//	5.3 ± 1.6	−0.29	//	4.8 ± 1.5	//	2.1 ± 0.7	−0.40	//

Significant correlations are noted by an asterisk (Spearman coefficients > ±0.50); NC: Not Considered because of the low sample size; //: Not Available. In bold, the species is above the fixed thresholds (f% ≥ 10; SSw ≥ 50 tons at least in two surveys). Black dots indicate the depth range within which the species was sampled.

Table 3. Synoptic table of batoid abundance sampled during the GRUND survey (autumn) in the GSA16 (South of Sicily), from 1990 to 2006. Percent frequency of occurrence for number of surveys (s%) and positive hauls (f%). Standing Stock (SS) expressed in weight (SSw, tons) and number (SSn, thousand). Stratum distribution and mean biomass index kg/km^2 (BI) expressed by depth of the macrostratum (10–200 m, shelf; 201–800 m, slope; 10–800 m, overall), mean density index N/km^2 (DI) by depth of the macrostratum, and Spearman coefficients for BI and DI by depth of the macrostratum.

Species	Frequency of Occurrence		SSw (t) and SSn (Mean ± SE)		Stratum Distribution (Depth Interval in m)					BI (kg/km^2) Mean ± SE			Spearman Coeff. BI		DI (N/km^2) Mean ± SE			Spearman Coeff. DI	
	% N. Survey	% Positive Hauls	Tons	Number (Thousand)	10–50	50–100	100–200	200–500	500–800	Shelf	Slope	Overall	Shelf	Slope	Shelf	Slope	Overall	Shelf	Slope
Tetronarce nobiliana	40	1.2	18.7 ± 15.7	4 ± 2	●	●	●	●	●	<0.1	1.1 ± 0.9	0.6 ± 0.5	NC	−0.23	0.1 ± 0.1	0.2 ± 0.1	0.1 ± 0.1	NC	−0.26
Torpedo marmorata	80	8.2	9.2 ± 2.5	29 ± 5	●	●	●	●	●	0.6 ± 0.2	0.1 ± 0.0	0.3 ± 0.1	−0.39	0.44	1.6 ± 0.3	0.3 ± 0.1	0.9 ± 0.2	−0.26	0.49
Torpedo torpedo	40	0.7	1.1 ± 0.8	3 ± 1	●			●	●	<0.1	<0.1	<0.1	0.27	NC	0.2 ± 0.1	<0.1	0.1 ± 0.0	0.19	NC
Dipturus batis	13	0.2	1.4 ± 0.9	1 ± 1				●	●	//	0.1 ± 0.1	<0.1	//	NC	//	<0.1	<0.1	//	NC
Dipturus oxyrinchus	73	6.3	47.7 ± 11.0	47 ± 19			●	●	●	<0.1	2.8 ± 0.7	1.5 ± 0.4	−0.20	0.05	<0.1	2.8 ± 1.1	1.5 ± 0.6	−0.20	0.00
Leucoraja circularis	27	0.7	3.6 ± 1.9	2 ± 1					●	//	0.2 ± 0.1	0.1 ± 0.1	//	0.35	//	0.1 ± 0.1	0.1 ± 0.0	//	0.34
Leucoraja melitensis	73	4.9	14.2 ± 4.0	77 ± 25		●	●	●	●	0.2 ± 0.1	0.7 ± 0.2	0.5 ± 0.1	0.24	0.11	0.8 ± 0.3	3.9 ± 1.4	2.5 ± 0.8	0.31	0.07
Leucoraja naevus	7	0.1	<0.1	0.1 ± 0.1			●			//	<0.1	<0.1	//	NC	//	<0.1	<0.1	//	NC
Raja asterias	73	2.5	4.9 ± 1.4	10 ± 2	●	●		●		0.3 ± 0.1	<0.1	0.2 ± 0.0	0.31	0.31	0.7 ± 0.2	<0.1	0.3 ± 0.1	0.35	0.31
Raja brachyura	7	0.1	0.2 ± 0.2	0.2 ± 0.2					●	//	<0.1	<0.1	//	0.36	//	<0.1	<0.1	//	0.36
Raja clavata	100	18.6	304.4 ± 31.8	341 ± 51	●	●	●	●	●	16.5 ± 2.3	3.8 ± 0.7	9.7 ± 1.0	0.81 *	−0.15	26.2 ± 5.8	4.7 ± 1.1	14.6 ± 3.0	0.86 *	−0.16
Raja miraletus	100	23.2	196.2 ± 18.7	1211 ± 125	●	●	●	●		13.3 ± 1.3	0.3 ± 0.1	6.3 ± 0.6	0.67 *	−0.20	81.7 ± 8.7	1.6 ± 0.8	38.6 ± 4.0	0.58 *	−0.33
Raja montagui	73	5.2	8.2 ± 1.8	30 ± 6		●	●	●		0.5 ± 0.1	0.1 ± 0.0	0.3 ± 0.1	−0.07	−0.12	1.8 ± 0.4	0.2 ± 0.1	1.0 ± 0.2	0.09	0.16
Raja polystigma	7	0.1	<0.1	0.1 ± 0.1		●				<0.1	//	<0.1	NC	//	<0.1	//	<0.1	NC	//
Raja radula	13	0.2	0.1 ± 0.1	1 ± 0.4	●	●				<0.1	//	<0.1	−0.08	//	<0.1	//	<0.1	−0.08	//
Rostroraja alba	73	2.5	26.1 ± 9.5	9 ± 2		●	●	●	●	1.6 ± 0.6	0.2 ± 0.2	0.8 ± 0.3	−0.11	−0.41	0.6 ± 0.1	0.1 ± 0.0	0.3 ± 0.1	0.04	−0.39
Dasyatis pastinaca	80	2.3	36.9 ± 9.2	9 ± 2	●	●	●			2.5 ± 0.6	//	1.2 ± 0.3	0.36	//	0.7 ± 0.2	//	0.3 ± 0.1	0.36	//
Pteroplatytrygon violacea	7	0.1	1.5 ± 1.5	0.3 ± 0.3			●			0.1 ± 0.1	//	<0.1	NC	//	<0.1	//	<0.1	NC	//
Aetomylaeus bovinus	13	0.1	0.4 ± 0.3	0.5 ± 0.3				●		<0.1	<0.1	<0.1	NC	NC	<0.1	<0.1	<0.1	NC	NC
Myliobatis aquila	27	0.4	1.5 ± 0.9	1 ± 1	●	●				0.1 ± 0.1	//	<0.1	0.21	//	0.1 ± 0.0	//	<0.1	0.31	//

Significant correlations are noted by an asterisk (Spearman coefficients > ±0.50); NC: Not Considered because of the low sample size; //: Not Available. In bold, the species is above the fixed threshold (f% ≥ 10; SSw ≥ 50 tons at least in two surveys). Black dots indicate the depth range within which the species was sampled.

Table 4. Synoptic table of batoid abundance sampled during the GRUND survey (autumn) in the GSA15 (Malta Island), from 1990 to 2006. Percent frequency of occurrence for number of surveys (s%) and positive hauls (f%). Standing Stock (SS) expressed in weight (SSw, tons) and number (SSn, thousand). Stratum distribution and mean biomass index kg/km^2 (BI) expressed by depth of the macrostratum (10–200 m, shelf; 201–800 m, slope; 10–800 m, overall), mean density index N/km^2 (DI) by depth of the macrostratum, and Spearman coefficients for BI and DI by depth of the macrostratum.

Species	Frequency of Occurrence		SSw (t) and SSn (Mean ± SE)		Stratum Distribution (Depth Interval in m)					BI (kg/km^2) Mean ± SE			Spearman Coeff. BI		DI (N/km^2) Mean ± SE			Spearman Coeff. DI	
	% N. Survey	% Positive hauls	Tons	Number (thousand)	10–50	51–100	101–200	201–500	501–800	Shelf	Slope	Overall	Shelf	Slope	Shelf	Slope	Overall	Shelf	Slope
Tetronarce nobiliana	20	0.8	1.4 ± 0.8	2 ± 1			●	●	●	<0.1	0.1 ± 0.0	<0.1	NC	NC	<0.1	0.1 ± 0.1	0.1 ± 0.0	NC	NC
Torpedo marmorata	67	10.8	9.0 ± 4.0	24 ± 7	●	●	●	●		0.8 ± 0.4	<0.1	0.3 ± 0.1	0.1	0.71 *	1.9 ± 0.6	0.1 ± 0.1	0.8 ± 0.2	0.01	0.74 *
Torpedo torpedo	7	0.4	1.2 ± 1.2	1 ± 1		●	●			0.1 ± 0.1	//	<0.1	NC	//	0.1 ± 0.1	//	<0.1	NC	//
Dipturus batis	7	0.6	13.8 ± 13.8	4 ± 4				●		//	0.7 ± 0.7	0.4 ± 0.4	//	NC	//	0.2 ± 0.2	0.1 ± 0.1	//	NC
Dipturus oxyrichus	73	19.4	283.2 ± 71.5	199 ± 43			●	●	●	1.6 ± 0.8	13.2 ± 3.3	9.0 ± 2.3	0.75 *	0.38 *	0.5 ± 0.2	9.7 ± 2.1	6.4 ± 1.4	0.75 *	0.22
Leucoraja circularis	40	3.1	9.7 ± 5.5	7 ± 3				●	●	//	0.5 ± 0.3	0.3 ± 0.2	//	0.41	//	0.4 ± 0.1	0.2 ± 0.1	//	0.25
Leucoraja melitensis	60	4.7	9.7 ± 4.7	51 ± 22			●	●	●	<0.1	0.5 ± 0.2	0.3 ± 0.1	NC	0.29	0.1 ± 0.1	2.5 ± 1.1	1.6 ± 0.7	NC	0.21
Raja asterias	13	0.4	0.7 ± 0.7	1 ± 1				●	●	//	<0.1	<0.1	//	NC	//	0.1 ± 0.1	<0.1	//	NC
Raja brachyura	7	0.2	0.8 ± 0.8	1 ± 1					●	//	<0.1	<0.1	//	NC	//	<0.1	<0.1	//	NC
Raja clavata	100	49	521.3 ± 67.3	627 ± 93		●	●	●	●	13.0 ± 3.1	18.6 ± 3.2	16.6 ± 2.1	0.78 *	0.12	8.1 ± 2.2	26.6 ± 4.6	20.0 ± 2.9	0.87 *	0.09
Raja miraletus	93	27.6	135.8 ± 24.1	741 ± 156		●	●	●	●	11.6 ± 2.1	0.2 ± 0.2	4.3 ± 0.8	0.16	0.23	63.4 ± 13.9	1.3 ± 1.0	23.6 ± 5.0	0.18	0.34
Raja montagui	47	3.2	4.0 ± 1.6	13 ± 5		●		●		0.2 ± 0.1	0.1 ± 0.1	0.1 ± 0.1	0.50 *	−0.21	0.3 ± 0.3	0.5 ± 0.2	0.4 ± 0.2	0.50 *	−0.21
Raja radula	16	2.4	20.5 ± 11.7	55 ± 36	●	●	●			1.8 ± 1.0	//	0.7 ± 0.4	0.75 *	//	4.8 ± 3.2	//	1.7 ± 1.1	0.75 *	//
Raja undulata	7	0.1	0.2 ± 0.2	1 ± 1		●				<0.1	//	<0.1	NC	//	0.1 ± 0.1	//	<0.1	NC	//
Rostroraja alba	33	1.8	7.1 ± 5.4	6 ± 3				●	●	//	0.4 ± 0.3	0.2 ± 0.2	//	0.18	//	0.3 ± 0.1	0.2 ± 0.1	//	0.22
Dasyatis pastinaca	47	10.1	108.4 ± 59.5	43 ± 27	●	●	●			9.6 ± 5.3	//	3.5 ± 1.9	0.56 *	//	3.8 ± 2.4	//	1.4 ± 1.0	0.55 *	//
Myliobatis aquila	33	1.4	43.1 ± 23.6	12 ± 6	●	●				3.8 ± 2.1	//	1.4 ± 0.8	0.64 *	//	1.1 ± 0.6	//	0.4 ± 0.2	0.67 *	//

Significant correlations are noted by an asterisk (Spearman coefficients > ±0.50); NC: Not Considered because of the low sample size; //: Not Available. In bold, the species is above the fixed threshold (f% ≥ 10; SSw ≥ 50 tons at least in two surveys). Black dots indicate the depth range within which the species was sampled.

3.1.2. Marbled Electric Ray—*Torpedo marmorata* Risso, 1810

The surveys highlighted its wide spatial, as well as depth, distribution. In fact, this species was recorded in all strata in the spring–summer and autumnal surveys in the South of Sicily, although it seemed to be more abundant on the continental shelf. In spring–summer, the species showed the highest frequency of occurrence (11%) and abundance indices (shelf: BI = 1.8 ± 0.5 kg/km^2, DI = 6.5 ± 1.3 N/km^2) in Malta Island, although the standing stock was higher in South of Sicily (20.8 ± 2.5 tons, 69 ± 7 thousand) (Tables 1 and 2). Similarly, in autumn, it was caught more frequently (10.8%) with slightly higher abundance indices (shelf: BI = 0.8 ± 0.4 kg/km^2, DI = 1.9 ± 0.6 N/km^2) in Malta Island, although the standing stock was higher in South of Sicily (9.2 ± 2.5 tons, 29 ± 5 thousand). The abundance indices were positively correlated over the years (on the slope) only in Malta Island (Tables 3 and 4). The TL ranged from 105 to 495 mm for females and from 100 to 370 mm for males. The median lengths were 430 and 445 mm TL for females and males, respectively. The LWR parameters were 3.3×10^{-5}, 1×10^{-4}, and 4.5×10^{-5} (k) and 2.96, 2.67, and 2.85 (b), respectively, for females, males, and sex combined. The sex ratio was 0.85:1 ($X^2 = 0.64$, $p = 0.4237$) (Table S3).

3.1.3. Common Torpedo—*Torpedo torpedo* (Linnaeus, 1758)

T. torpedo has been recorded in a wide bathymetric range from the surface (10 m) up to 800 m. In general, it was more abundant on the slope. In spring-summer, it was caught only in South of Sicily with a frequency of occurrence of 0.8%. Note that, in this area, the highest DI values were observed on the shelf (0.6 ± 0.2 N/km^2), whilst the highest BI value was recorded on the slope (0.3 ± 0.2 kg/km^2), suggesting a positive relationship between size/age and depth (i.e., juveniles live in shallower water than adults) (Table 1). In autumn, the species was rarely caught, even though a slightly higher abundance was observed in South of Sicily. No significant correlation between abundance and time was found (Tables 3 and 4). The TL ranged from 105 to 275 mm for females and from 135 to 355 mm for males. The median lengths were 170 and 190 mm TL for females and males, respectively. The LWR parameters were 8.8×10^{-6} (k) and 3.13 (b) for sex combined. The sex ratio was 1.86:1 ($X^2 = 9$, $p = 0.0027$) (Table S3).

3.1.4. Gray Skate—*Dipturus batis* (Linnaeus, 1758)

The surveys indicated few records on the slope, in the e stratum (501–800 m), in Adventure Bank. In spring–summer, it was caught only in GSA16 once in 2011, thus with very low values in terms of frequency of occurrence, BI, and DI (Table 1). In autumn, the highest frequency (0.6%), BI (0.7 ± 0.7 kg/km^2), and DI (0.2 ± 0.2 N/km^2) were recorded in Malta Island (Tables 3 and 4). Very few specimens and biological data were collected. The maximum length was 1260 mm TL (female) (Table S3).

3.1.5. Norwegian Skate—*Dipturus nidarosiensis* (Storm, 1881)

Only a single juvenile male specimen was caught in 2017 in Pantelleria bank during spring–summer survey in South of Sicily (see [53] for details).

3.1.6. Longnosed Skate—*Dipturus oxyrinchus* (Linnaeus, 1758)

D. oxyrinchus appeared mainly distributed on the slope in surveys. In spring-summer, the highest frequency of occurrence (30%), BI (40.53 ± 5.4 kg/km^2), and DI (32.3 ± 4.8 N/km^2) were recorded in Malta Island (Tables 1 and 2). Similarly, in autumn it was caught more frequently in Malta Island (Tables 3 and 4). In spring–summer, a slight, although significant, positive correlation was recorded in South of Sicily between the BI over the years on the slope (Table 1). Conversely, a strong, positive correlation was found on the shelf only in autumnal survey in Malta Island (Table 4). As matter of fact, the yearly DI plot showed that a higher index was always recorded from the spring–summer in Malta Island, with a clear peak in 2008. The Mann–Whitney U test showed a significant difference between the DI ($W = 196$, $p < 0.001$), BI ($W = 196$, $p < 0.001$), and frequency of occurrence ($W = 194$,

$p < 0.001$) between areas in autumn. Similarly, from the autumnal survey of Malta Island, except for 1990, higher values of DI were recognized with two peaks in 1994 and 2005 (Figure 2).

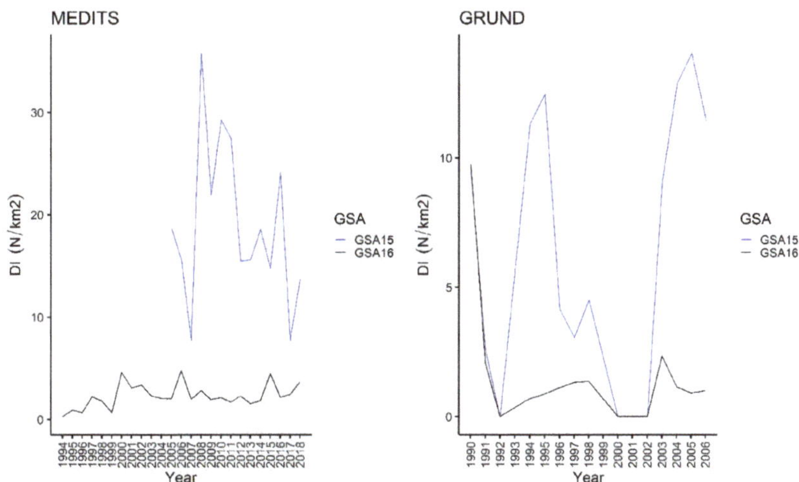

Figure 2. Density Index (DI, N/km^2) of *Dipturus oxyrinchus* by overall deep interval (10–800 m) in the South of Sicily (GSA16) and Malta Island (GSA15) between 1994 and 2018 during MEDITS–spring summer–(**left**) and between 1990 and 2006 during GRUND–autumn–(**right**).

The Mann–Whitney U test detected a significant difference between the DI (W = 167, $p = 0.02376$), BI (W = 179, $p = 0.001319$), and frequency of occurrence (W = 166, $p = 0.02644$) from the autumnal surveys of both areas. The DI map showed the presence of a little spot located at West of Marettimo Island, and other two bigger spots located in SE Linosa waters and on the NW border of the Malta Island (Figure 3 and Figure S1)with higher occurrences of this species. In these waters the species appeared more widespread, in the NW Malta Island spot and on the slope in the South of Sicily S-SW spots (Figure 3).

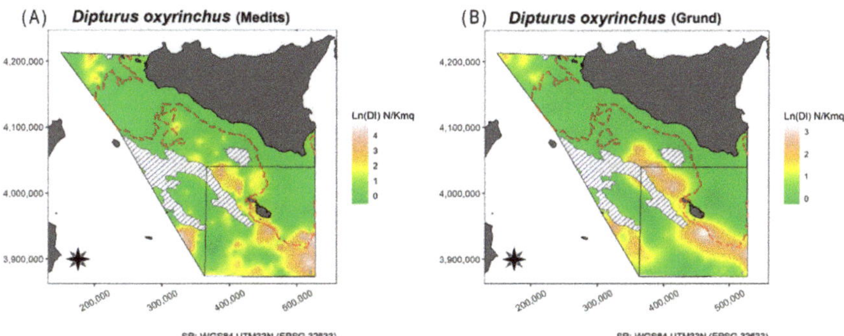

Figure 3. Spatial distribution, in terms of Density Index (DI, N/Km^2), of *Dipturus oxyrinchus* in the South of Sicily (GSA 16) and Malta island (GSA15) during (**A**) MEDITS–spring summer–and (**B**) GRUND–autumn–surveys. The dashed areas denote depths below 800 m not explored in present surveys.

The TL ranged from 230 to 1240 mm and 250 to 1100 mm for females and males, respectively. The median length was slightly higher in males (620 mm TL) compared to females (600 mm TL). The LFD showed significant differences between 2010, 2011, and 2014. The LWR parameters were 5.4×10^{-7}, 1×10^{-6}, and 7.2×10^{-7} (k) and 3.29, 3.20, and 3.25 (b), respectively, for females, males, and sex combined. The growth parameters, as estimated through ELEFAN, indicated that females attained a larger $L\infty$ (1365 mm TL) than males (1240 mm TL), while the K results were 0.11 and 0.26 for females and males, respectively. The Z/K values were 1.66 (Confidence Interval: 1.64–1.68) and 1.21 (Confidence Interval: 1.19–1.23) for females and males, respectively. The Z values obtained were 0.18 and 0.31 for females and males, respectively. The estimated L_{50} values (median stage approach) were 695 and 860 mm TL for females and males, respectively. The sex ratio was 1.56:1 ($X^2 = 4.84$, $p = 0.02781$) (Table S3).

3.1.7. Sandy Ray—*Leucoraja circularis* (Couch, 1838)

The bathymetric distribution of *L. circularis* was mainly focused on the slope at depths ranging from 100 to 800 m. In spring–summer, the highest values in terms of frequency of occurrence (6%), abundance indices (in slope BI = 41.8 ± 39.8 kg/km^2, DI = 3.3 ± 1.2 N/km^2), standing stock (247.5 ± 233.8 tons, 20 ± 7 thousand) were recorded from the Malta Island (Tables 1 and 2). Similarly, in autumn, the highest values were recorded in Malta Island (Tables 3 and 4). Very few specimens and biological data were collected. The TL ranged from 310 to 880 mm for females, and from 180 to 750 mm for males (Table S3).

3.1.8. Shagreen Ray—*Leucoraja fullonica* (Linnaeus, 1758)

The present study confirms the rarity of this ray in South of Sicily. In fact, it was never caught in autumn and was very rarely caught in spring-summer. From this study, this species was more frequent (2%) and abundant (standing stock: 2.1 ± 1.5 tons, 3 ± 1 thousand, slope: BI = 0.4 ± 0.2 kg/km^2, DI = 0.5 ± 0.2 N/km^2) in Malta Island (Tables 1 and 2). Very few specimens and no biological data were collected.

3.1.9. Maltese Ray—*Leucoraja melitensis* (Clark, 1926)

The Maltese ray was caught in a wide range of depths, from the surface to 800 m, but it was more common on the slope. During spring–summer, the highest frequency of occurrence (11%) and abundance indices (slope: BI = 3.3 ± 0.6 kg/km^2, DI = 16.3 ± 2.9 N/km^2) were observed in Malta Island; however, the highest standing stock values were observed in South of Sicily (26.7 ± 3.0 tons, 132 ± 15 thousand) (Tables 1 and 2). Conversely, in autumn, the species was more frequent (4.9%) and abundant in the South of Sicily (slope: BI = 0.7 ± 0.2 kg/km^2; DI = 3.9 ± 1.4 N/km^2) (Tables 3 and 4). A slight, positive correlation was observed from the spring-summer survey in South of Sicily over the years in the shelf (Table 1). The TL ranged from 140 to 450 mm for females and from 100 to 520 mm for males, while the median lengths were 340 and 350 mm TL for females and males, respectively. The LWR parameters were 5.4×10^{-7}, 6.2×10^{-3}, and 2.4×10^{-6} (k) and 3.37, 2.95, and 3.12 (b), respectively, for females, males, and sex combined. The sex ratio was 1.13:1 ($X^2 = 0.36$, $p = 0.5485$) (Table S3).

3.1.10. Cuckoo Ray—*Leucoraja naevus* (Müller and Henle, 1841)

The rarity of the species was supported in this study, given that it was only recorded in South of Sicily, with few scattered samples on the slope (Tables 1 and 3). No biological data were collected.

3.1.11. Starry Ray—*Raja asterias* Delaroche, 1809

The surveys indicated a wide distribution along the sampling area with a low number of specimens. It also showed a wide depth distribution, even though the species was mainly concentrated on the continental shelf. The starry ray was recorded more frequently in South of Sicily, during spring–summer (4.3%) (Tables 1 and 2), and autumn (2.5%) (Tables 3 and 4). The TL ranged from 205 to 590 mm and 220 to 550 mm, while the median length was 320 and 390 mm for females and males, respectively. The LWR parameters were 2.4×10^{-6}, 4.9×10^{-6}, and 3.2×10^{-6} (k) and 3.14, 3.02, and 3.09 (b), respectively, for females, males, and sex combined. The sex ratio was 1.13:1 ($X^2 = 0.36$, $p = 0.5485$) (Table S3).

3.1.12. Blonde Ray—*Raja brachyura* Lafont, 1873

In spring–summer, the blonde ray was caught with high frequency of occurrence (0.9%), abundance indices (shelf: BI = 1.2 ± 0.2 kg/km^2; DI = 0.3 ± 0.2 N/km^2), and standing stock (9.1 ± 5.6 tons, 10 ± 6 thousand) in South of Sicily (Tables 1 and 2). Conversely, in autumn, even if was sampled only in 2006 in both areas, it was caught slightly more frequently in Malta Islands (0.2%) (Tables 3 and 4). The TL ranged from 310 to 410 mm for females and from 300 to 320 mm for males. The median lengths were 360 and 310 mm TL for females and males, respectively. The sex ratio was 0.85:1 ($X^2 = 0.64$, $p = 0.4237$) (Table S3).

3.1.13. Thornback Ray—*Raja clavata* Linnaeus, 1758

The species has been recorded across all the considered strata, albeit it seemed more abundant on the shelf (Tables 1–3). In spring–summer, the highest frequency of occurrence (53%), abundance (slope: BI = 90.7 ± 15.0 kg/km^2, DI = 65.1 ± 9.1 N/km^2), and standing stock (706.8 ± 61.6 tons, 660 ± 49 thousand) were recorded in Malta Island (Tables 1 and 2). Similarly, in autumn, the species seemed more common and abundant in Malta Island (Tables 3 and 4). The abundance indices of the thornback ray was positively correlated over the years on the slope during spring-summer in South of Sicily and on the shelf from the other surveys (Tables 1–4). The yearly DI plot showed that the DI obtained was overall higher from Malta Island during spring-summer; this did not apply to 2007 (Figure 4).

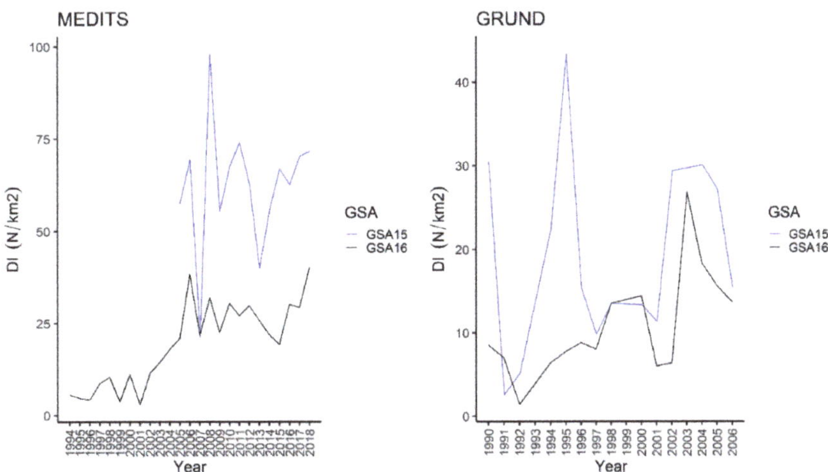

Figure 4. Density Index (DI, N/km^2) of *Raja clavata* by overall depth in the South of Sicily (GSA16) and Malta Island (GSA15) between 1994 and 2018 during MEDITS–spring summer–(**left**) and between 1990 and 2006 during GRUND–autumn–(**right**).

In addition, the Mann–Whitney U test highlighted significant differences for DI (W = 183, $p < 0.001$), BI (W = 194, $p < 0.001$), and frequency of occurrence (W = 196, $p < 0.001$) between areas during spring–summer. The highest values of DI were recorded during autumn in Malta Island, except for 1991, 1998, and 2000; a clear peak was observed in 1995 (Figure 4). In addition, the Mann–Whitney U test showed significant differences between the DI (W = 114, $p = 0.01449$), BI (W = 175, $p = 0.008642$), and frequency of occurrence (W = 144, $p < 0.001$) between areas during autumn. The spatial distribution of the species in South of Sicily was mainly concentrated in two areas, namely Adventure Bank and in the southernmost sector of the GSA16 (NE Lampedusa Island), whereas in Malta Island it seemed to be more widespread, particularly in the SE part (Figure 5).

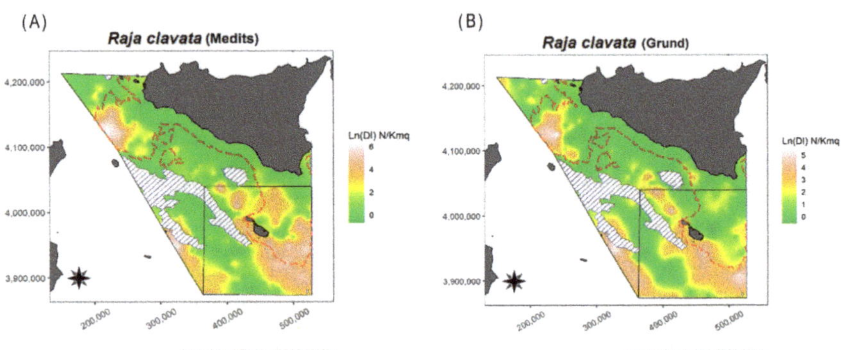

Figure 5. Spatial distribution, in terms of Density Index (DI, N/Km^2), of *Raja clavata* in South of Sicily (GSA 16) and Malta island (GSA15) during (**A**) MEDITS–spring summer–and (**B**) GRUND–autumn–surveys. The dashed areas denote depths below 800 m not explored in present surveys.

The TL ranged from 95 to 1100 and from 110 to 810 mm TL for females and males, respectively. The LFD was quite comparable between sexes, although the median length was slightly lower for females (480 mm) than males (500 mm). The Kolmogorov–Smirnov test showed no significant statistical differences in the LFDs over the years for females, whilst a significant difference was observed for males only in 1995. The LWR parameters were 6.0×10^{-7}, 7.6×10^{-7}, and 6.8×10^{-7} (k) and 3.36, 3.33, and 3.35 (b), respectively, for females, males, and sex combined. The ELEFAN growth estimates confirmed a larger size (L∞: 1260 vs. 1100 mm) and lower growth rate (K: 0.16 vs. 0.26) in females than males. The Z/K ratios were 1.32 (and 0.95 (Confidence Interval: 0.94–0.96) for females and males, respectively. The estimated values of L$_{50}$, through the logistic approach, were 635 mm for females and 574 mm for males (Figure 6). The sex ratio was 1:1 (Table S3).

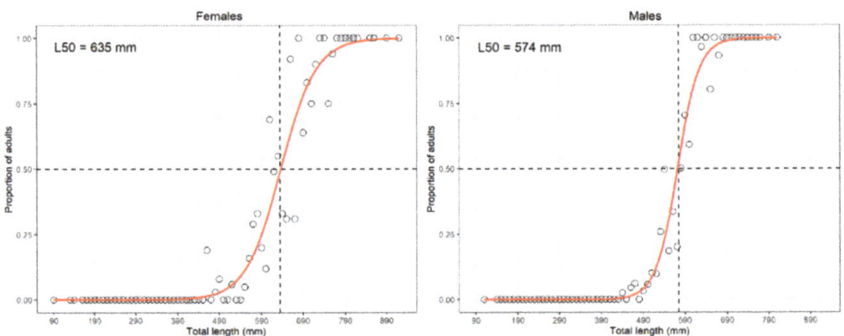

Figure 6. Logistic curves describing the proportion of *Raja clavata* matures by length (Total Length, mm) for females and males; L$_{50}$ denotes the estimated size at first maturity from the MEDITS–spring summer–survey carried out in the South of Sicily (GSA16).

3.1.14. Brown Ray—*Raja miraletus* Linnaeus, 1758

The species was recorded from 10 to 800 m; however, it was more abundant in the b stratum, 50–100 m, namely, the inner shelf. The brown ray was more frequent in Malta Island (spring–summer = 24%; autumn = 27.6%), although it was more abundantly caught in South of Sicily during spring-summer and autumn (e.g., South of Sicily: spring–summer 371.9 ± 34.2 tons, 2282 ± 224 thousand; autumn 1211 ± 125 thousand) (Tables 1–4). The abundance indices showed a significant, positive correlation only during autumn in South of Sicily on the shelf (Table 3). The yearly DI plot showed that the DI was always higher in MED16 (Figure 7).

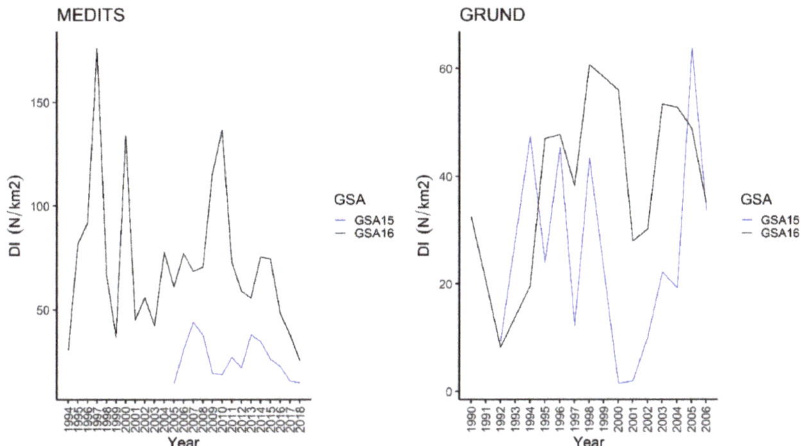

Figure 7. Density Index (DI, N/km^2) of *Raja miraletus* by overall depth in the South of Sicily (GSA16) and Malta island (GSA15) between 1994 and 2018 during MEDITS–spring summer–(**left**) and between 1990 and 2006 during GRUND–autumn–(**right**).

The Mann–Whitney U test highlighted significant differences in DI ($W = 9$, $p < 0.001$), BI ($W = 31$, $p = 0.001433$), and frequency of occurrence ($W = 142.5$, $p = 0.04232$) between areas in spring summer. As for autumnal survey the DI was generally higher in South of Sicily, except for 1992, 1994, and 2005 (Figure 7). Significant differences were observed for DI ($W = 58$, $p = 0.0411$) but not for BI ($W = 66$, $p = 0.09322$) and frequency of occurrence ($W = 136.5$, $p = 0.176$) between areas in autumn. The map of the spatial distribution showed that the brown ray in South of Sicily was mainly distributed on Adventure Bank, whereas in Malta Island it was mainly concentrated in the NE part. In both areas, the species was more present mainly on the shelf (Figure 8).

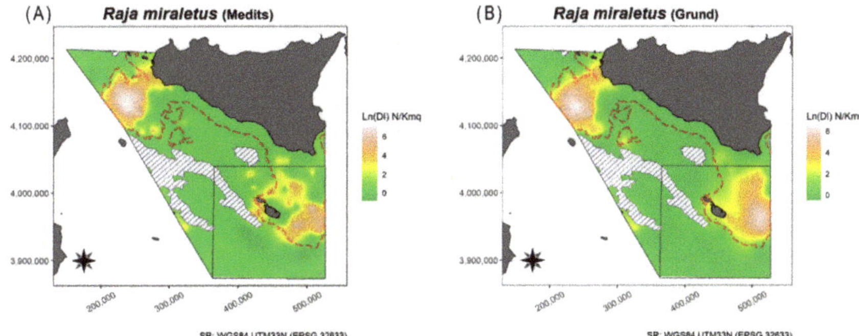

Figure 8. Spatial distribution, in terms of Density Index (DI, N/Km^2), of *Raja miraletus* in South of Sicily (GSA 16) and Malta island (GSA15) during (**A**) MEDITS–spring summer–and (**B**) GRUND–autumn–surveys. The dashed areas denote depths below 800 m not explored in present surveys.

The TL ranged from 135 to 470 and from 135 to 460 mm for females and males, respectively. The LFD stability showed no significant statistical difference between years for both sexes, while the median length was 320 and 330 mm for females and males, respectively. The LWR parameters were 2.7×10^{-7}, 4.2×10^{-7}, and 3.4×10^{-7} (steepness) and 3.36, 3.33, and 3.35, respectively, for females, males, and sex combined. ELEFAN resulted in L∞ of 840 and 800 mm TL, while K was 0.36 and 0.41 for females and males, respectively. Differently from other rays, the Z/K value was quite higher than expected for both sexes, that is, 2.96 and 3.13 for females and males, respectively (the corresponding Z would be around 1.2, a value which looks more realistic). The estimated L_{50} values were 364 and 349 mm TL (logistic approach; Figure 9) for females and males, respectively. The sex ratio was 0.96:1 ($X^2 = 0.04$, $p = 0.8415$) (Table S3).

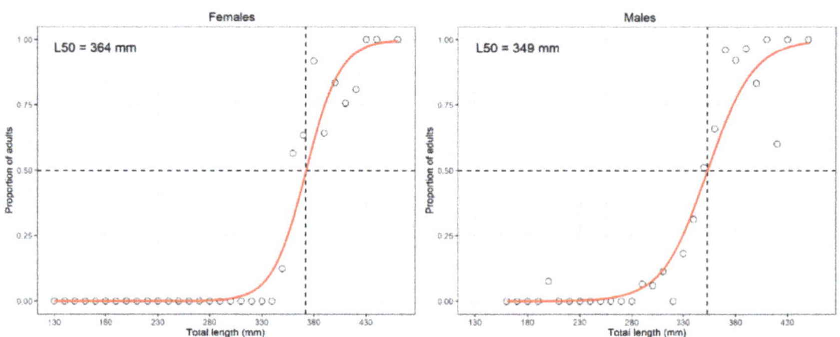

Figure 9. Logistic curves describing the proportion of *Raja miraletus* matures by length (Total Length, mm) for females and males; L_{50} denotes the estimated size at first maturity from the MEDITS–spring summer–survey carried out in the South of Sicily (GSA16).

3.1.15. Spotted Ray—*Raja montagui* Fowler, 1910

Survey data showed a wide spatial distribution and a preference for coastal and shelf areas (mainly inner shelf), although the species was found up to 500 m deep. In spring-summer, the highest frequency of occurrence (9%), abundance indices (shelf: BI = 15.0 ± 0.6 kg/km^2, DI = 3.3 ± 1.4 N/km^2), and standing stock (28.6 ± 8.4 tons; 103 ± 38 thousand) were observed from Malta Island (Tables 1 and 2). In Malta Island, the BI on the shelf was higher than that on the slope, while the DI was higher on the slope (Table 2). This could be attributed to the bathymetric segregation of the species across ontogeny, i.e., juveniles are more abundant on the slope. In autumn, the spotted ray was more frequently caught in South of Sicily (5.2%), with the highest abundance indices on the shelf (BI = 0.5 ± 0.1 kg/km^2; DI = 1.8 ± 0.4 N/km^2) (Tables 3 and 4). In Malta Island, the abundance indices were positively correlated over the years on the slope in spring–summer (Table 2), whereas in autumn the abundance indices were only positively correlated to the years on the shelf, i.e., 0.50 (Table 4). The TL ranged from 205 to 520 mm for females and from 200 to 495 mm for males, with the median lengths being 325 and 345 mm for females and males, respectively. The LWR parameters were 2.8×10^{-7}, 2.0×10^{-7}, and 2.8×10^{-7} (k) and 3.51, 3.55, and 3.50 (b, i.e., a strong, positive allometry) for females, males, and sex combined, respectively. The sex ratio was 1.22:1 ($X^2 = 1$, df = 1, $p = 0.3173$) (Table S3).

3.1.16. Speckled Ray—*Raja polystigma* Regan, 1923

The speckled ray seemed to inhabit the shelf. In spring–summer, the species was most frequently encountered (1%) and most abundant (BI 0.6 ± 0.4 kg/km^2; DI 0.8 ± 0.5 N/km^2) on the slope in the Malta Island (Tables 1 and 2). In autumn, it was caught only in South of Sicily in 1995; therefore, a very low abundance indices were recorded (Table 3). The TL ranged from 225 to 405 mm for females and from 300 to 475 mm for males. Considering the combined sexes, the b parameter of the LWR showed a strong, positive allometric growth (3.45), while the steepness (k) was 4.4×10^{-7}. The sex ratio was 0.64:1 (X^2 = 4.84, df = 1, p = 0.02781) (Table S3).

3.1.17. Rough Ray—*Raja radula* Delaroche, 1809

The surveys suggested that the rough ray showed preference towards the shelf (mainly the outer shelf, 100–200 m deep). In spring–summer, the species was frequently encountered in Malta Island (1.7%), particularly on the shelf (BI = 4.8 $\pm$ 2.9 kg/km^2; DI = 5.2 $\pm$ 2.7 (N/km^2) (Tables 1 and 2). Similarly, in autumn, the highest frequency of occurrence (2.4%) and standing stock (20.5 $\pm$ 11.7 tons; 55 $\pm$ 36 thousand) were observed from the Malta Island; here the species was found only on the shelf (BI = 1.8 $\pm$ 1.0 kg/km^2; DI = 4.8 $\pm$ 3.2 N/km^2) (Tables 3 and 4). The abundance of the rough rays was only positively correlated to the shelf during autumn in Malta Island (Table 4). The median lengths were 430 and 445 mm TL, respectively, for females and males. The sex ratio was 0.51:1 (X^2 = 10.24, p = 0.001374) (Table S3).

3.1.18. Undulate Ray—*Raja undulata* Lacepède, 1802

In surveys, it was encountered only in 2004 during autumn survey in Malta Island within the b stratum (51–100 m). Therefore, the resulting abundance indices were very low, i.e., <0.1 (Table 4). No biological data were collected.

3.1.19. White Skate—*Rostroraja alba* (Lacepède, 1803)

The white skate was most frequently encountered on the shelf, but it could reach depths of 800 m. In spring–summer, the highest frequency of occurrence (1.6%), abundance indices (shelf: BI = 3.0 $\pm$ 1.1 kg/km^2, DI = 0.6 $\pm$ 0.2 N/km^2), and standing stock (57.1 $\pm$ 15.3 tons, 11 $\pm$ 3 thousand) of the white stake were observed from South of Sicily (Tables 1 and 2). Similarly, in autumn, the species was caught more frequently from South of Sicily (2.5%). The abundance indices of the white skate were positively correlated across years on the shelf only during spring–summer in South of Sicily (Table 1). The TL ranged from 325 to 1450 mm for females and from 315 to 1320 mm for males. The median lengths were 890 and 740 mm (TL), respectively, for females and males. The steepness values of the LWR were 1.9×10^{-6}, 1.7×10^{-6}, and 1.7×10^{-6} for females, males, and sex combined, and the b parameter had positive allometry for females (3.18), males (3.09), and sex combined (3.14). The sex ratio was 0.75:1 (X^2 = 1.96, p = 0.1615) (Table S3).

3.1.20. Common Stingray—*Dasyatis pastinaca* (Linnaeus, 1758)

The common stingray was more frequent and abundant in Malta Island in both seasons, in comparison to South of Sicily (Tables 1–4). In spring–summer, a positive correlation for the abundance was observed in South of Sicily on the shelf, while a negative correlation was observed from Malta Island (Tables 1 and 2). The TL ranged from 315 to 970 mm for females and from 510 to 710 mm for males, respectively. The median lengths were 830, 680, and 710 mm TL for females, males, and for sex combined, respectively. The steepness of the LWR was 3.0×10^{-4}, whereas the b parameter was 2.48 (a quite extreme negative allometry) for the sexes combined. The sex ratio was 1.94:1 (X^2 = 10.24, p = 0.001374) (Table S3).

3.1.21. Blue Stingray—*Pteroplatytrygon violacea* (Bonaparte, 1832)

The surveys show that the species in South of Sicily was only encountered once in spring-summer (2012) and once in autumn (1996); in both cases, very low values of abundance indices were observed (Tables 1 and 3). The rarity could be related to the prevalent pelagic behavior of this species; for the surveys, it was classified as incidental. No biological data were collected.

3.1.22. Bull Ray—*Aetomylaeus bovinus* (Geoffroy Saint-Hilaire, 1817)

Few scattered records were recorded during spring–summer in South of Sicily, in one spot next to Pozzallo (southeast Sicilian coast) and along the northeast of Linosa Island. *A. bovinus* has only been recorded in South of Sicily at depths ranging from 10 to 50 m during spring-summer, whereas it was caught between 200 and 500 m in autumn (Tables 1 and 3); this could be related to the seasonal migration of the species. The frequency of occurrence was 0.4% and 0.1, respectively, during spring-summer and autumn from the South of Sicily. On the shelf, in South of Sicily, DI and BI were 0.2 ± 0.1 N/km^2 and 2.2 ± 1.6 kg/km^2, respectively, in spring–summer, while in autumn both abundance indices resulted lower than 0.1. In South of Sicily, the standing stock in weight were 41.7 ± 23.0 and 0.4 ± 0.3 tons, while standing stock in number were 3 ± 2 and 0.5 ± 0.3 thousand, respectively, in spring–summer and autumn. This suggests that it reached large sizes and also indicates a seasonality trend in the catches. No significant correlation for the abundance indices over the years was observed (Tables 1 and 3). The median lengths were 2650 and 820 mm TL for females and males; the marked differences between sexes may be due to the few sampled specimens. The sex ratio was 0.96:1 ($X^2 = 0.04$, $p = 0.8415$) (Table S3).

3.1.23. Common Eagle Ray—*Myliobatis aquila* (Linnaeus, 1758)

The common eagle ray's preferential stratum was the shelf. It was more frequent and abundant in Malta Island than South of Sicily in both surveys (Tables 1–4). In spring–summer, a significant, positive correlation was observed between the abundance indices and years from South of Sicily (Table 1). In autumn, a significant, positive correlation between abundance indices and years was only observed from Malta Island (Table 4). The TL ranged from 425 to 1110 mm in females and from 435 to 775 mm in males. The median lengths were 530, 580, and 542 mm TL for females, males, and sex combined, respectively. The steepness was 4×10^{-9}, and the b parameter of the LWR was 4.01 for the sexes combined (a quite extreme positive allometry, likely reflecting the low representativeness of the present samples). The sex ratio was 0.72:1 ($X^2 = 2.56$, $p = 0.1096$) (Table S3).

4. Discussion

4.1. General Consideration

In the Mediterranean, according to the most recent assessment of the IUCN, 50% of the considered rays (16 of 32 species) are facing a high risk of extinction [11]. In addition, the lack of most basic life-history data and taxonomic uncertainties make it harder to assess their status (see, e.g., in [54,55]). Bradai et al. [29] underlined that some neritic species have almost locally disappeared (e.g., *R. alba*) or should be considered as highly depleted (*R. polystigma*), whereas few species are quite stable (e.g., *R. clavata* and *R. miraletus*), although considered by the authors in a depleted status (i.e., quite low standing stock in weight). In particular, *R. clavata*, *R. radula*, and *R. miraletus* are the species most commonly caught and landed by trawlers [36,56].

In the Strait of Sicily, Giudicelli [57] in the middle of the 70th explored 10 fishing grounds between 70 and 400 m depths and reported a maximum of 23.3 kg/h of "Rays" (this corresponds to around 233 kg/km^2). Recently, in the GSA16, Geraci et al. [26] found quite stable trends by analyzing the pooled BI and DI of the recorded batoids from 1994 to 2013, and the BI/DI ratio remained heterogeneous.

In the present study, among all the species analysed only three taxa (*R. clavata*, *R. miraletus*, and *D. oxyrinchus*) showed relatively high abundance indices so much that they mightbe considered profitable for fisheries. For some coastal species, e.g., *T. marmorata*, recreational fisheries (i.e., spear fishing) could be an important cause of mortality (Ragonese pers. obs.). Besides, differences in the DI of *D. oxyrinchus*, *R. clavata*, and *R. miraletus* between the studied areas were recognized attributable to the presence of a large protected area, namely the Maltese Economic Exclusive Zone, in Malta Island (GSA15) with limited access to fishing activities (i.e., a limited number of trawlers). Differences may also be due to the great abundance of Fish Aggregating Devices (FADs), which make large portions of the seabed non-trawable as several thousands of heavy stones used as anchors are left at the bottom at the end of the fishing season. This hypothesis could be corroborated by the fact that the largest catches of *R. clavata*, the most landed species, are mainly produced by boats with bottom trawls and set nets, while minimum quantities are caught with long-lines [25]. Considering that from 2009 onwards only MEDITS (spring–summer) was maintained, the re-implementation of a second MEDITS Autumn survey (as the Italian GRUND) could greatly improve the monitoring of these important animals. Overall, the spatial distribution of these taxa highlighted a widespread distribution in Malta Island (GSA15) and a quite unbalanced distribution in the South of Sicily (GSA16), with two main hotspots, i.e., Adventure Bank and the waters close to Linosa Island (Figures 3, 5 and 8). Generally, our results, especially for Adventure Bank and the waters of Linosa Island, are comparable to the spatial analysis reported by previous studies in the area (see, e.g., in [23,26,58,59]). Herein, present findings are summarily compared among species with data reported in the literature, in general for the Mediterranean Sea and in particular for the Strait of Sicily.

4.1.1. Great Torpedo Ray—*Tetronarce nobiliana* (Bonaparte, 1835)

In the North Mediterranean, the highest frequency (9.5%) and abundances were recorded in GSA01 (Northern Alboran Sea; DI = 1 ± 0.8 N/km^2, BI = 3.5 ± 3.7 kg/km^2) [27]. In the Strait of Sicily, Lanfranco [60] stated that this species is usually rare or occasionally frequent off the Southern coast of Sicily, and the same information has been reported by other authors [61,62]. Arena and Li Greci [63] reported isolated records off west of the Egadi Islands. In Tunisia, this species was reported in the north [64,65]. The maximum TL was slightly inferior to the other Mediterranean areas. The slope of the LWR was lower when compared with Turkish specimens [66,67] whereas the sex ratio was higher when compared to Basusta et al. [66] and lower than those reported by Kaya and Basusta [67] (Table S3).

4.1.2. Marbled Electric Ray—*Torpedo marmorata* Risso, 1810

The distribution of this species in the Mediterranean does not seem to be homogeneous. In the Adriatic Sea the species seems to be more common than the other two indigenous Torpedo species (*T. torpedo* and *T. nobiliana*) [68], and it is also relatively abundant in the northern Mediterranean [27], compared to other batoid species [69]. In the Strait of Sicily, it has been historically described as common or very common [60–62]. This species is caught mainly off the north coast of Tunisia and more rarely in the Gulf of Tunis and the Gulf of Gabes [70]. The maximum TL was bigger when compared to specimens analyzed by Duman and Başusta [71] in Turkey, whereas it was lower when compared to the other areas. Similar slope values of the LWR to Turkish [71–73] and Italian [74,75] specimens were found, whereas a quite low value was recorded in Lebanon [76], which may be due to the small sample size. The sex ratio was quite similar to that found by Consalvo et al. [75] and Yeldan and Gundogdu [72], whereas it was inferior when compared to Duman and Basusta [71], Abdel-Aziz [77], and Bellodi et al. [74] (Table S3).

4.1.3. Common Torpedo—*Torpedo torpedo* (Linnaeus, 1758)

This was an infrequent species, recorded with the highest frequency (3.7%) in GSA19 (Western Ionian Sea), while the highest abundances were detected in GSA01 (DI = 3.0 ± 3.5, BI = 1.9 ± 2.1) [27]. The maximum TL recorded in the present study was lower when compared to Tunisian [78] and Italian (regions of central western Italy) specimens, whereas it was bigger than Egyptian specimens. The slope of the LWR was higher and lower when compared, respectively, to male and female specimens from Latium, Italy [75]. The sex ratio was higher than that for Tunisian [78] and Italian specimens [74,75] (Table S3).

4.1.4. Gray Skate—*Dipturus batis* (Linnaeus, 1758)

Off the Strait of Sicily, taxonomic uncertainties make the available information on abundance not fully reliable. Indeed, the low consistency, as well as the doubts concerning historical records, poses questions over the possible misidentification of this species with regard to its congeneric species, *D. nidarosiensis*. The maximum TL recorded during surveys was lower than those reported by Serena et al. [79,80] (Table S3).

4.1.5. Norwegian Skate—*Dipturus nidarosiensis* (Storm, 1881)

Very little information about the Mediterranean abundance of the species is available. In GSA11 (Sardinia), 14 specimens were caught between 2005 and 2008 [81]. More recently, *D. nidarosiensis* was found also in the Ionian Sea and the Adriatic Sea [82–84], as well as in the Alboran Sea where eight specimens were caught between 2013 and 2016 [85]. Recently, Carugati et al. [86] provided new evidence of the occurrence of *D. nidarosiensis* in the Central-Western Mediterranean Sea and stated the lack of Atlantic-Mediterranean genetic divergence.

4.1.6. Longnosed Skate—*Dipturus oxyrinchus* (Linnaeus, 1758)

This species was recorded in all areas covered by the MEDITS survey, with exception of GSA06 (Northern Spain) and 17 (Northern Adriatic Sea). The highest values of frequency, DI, and BI were recorded in GSA08 (Corsica), respectively, 82.7%, $26 \pm 1\,\mathrm{N/km^2}$, and $24.4 \pm 5.7\,\mathrm{kg/km^2}$ [27]. Off the South Coasts of Sicily, this species is reported as ubiquitarian [22]. In the present study, *D. oxyrinchus* was more abundant close to Linosa Island, NW and SE Malta Island, in deeper water (Figure 3). Concerning the biological aspects of *D. oxyrinchus*, the maximum size appears to be smaller than that reported by Serena et al. [79]; however, this is similar to other Mediterranean areas [31,87]. The positive allometry indicated by the b parameter of the LWR, estimated in the present study, was also comparable to other areas [73,88,89]. On the contrary, the L_{50} was higher and lower when compared, respectively, to male and female specimens from the Gulf of Gabes [31], North Aegean Sea [88] and Syrian waters [89], while it was lower when compared to those from Sardinian waters [87]. Strangely, the L_{50} estimates, resulted lower for females than males; this probably maybe due to low representativity of the data about males.The sex ratio was quite unbalanced towards females, as reported in Syrian waters [89], whereas, in other areas, it was quite balanced (e.g., Tunisia [31]; North Aegean Sea [88]). Estimates of L∞ in the the South of Sicily were quite similar to the other Mediterranean areas, with the exception of the Aegean Sea [88] (Table S3). On the other hand, the calculated K was higher. From these estimates, it is possible to observe that stocks might be affected by fishing activities in a similar manner to natural, undisturbed conditions. The shape of the LFD indicates a Z/K value less than or around 1, i.e., K-oriented life history traits [44,90].

4.1.7. Sandy Ray—*Leucoraja circularis* (Couch, 1838)

In the North Mediterranean, higher frequencies were recorded in GSA11 (8.6) and GSA28 (Marmara Sea; 8.3%), while the highest abundance was recorded in GSA05 (Balearic Islands; $DI = 2 \pm 3\,\mathrm{N/km^2}$, $BI = 2.1 \pm 2.6\,\mathrm{kg/km^2}$) ([27]). Off the South coasts of Sicily, the species was more prevalent in the southeastern area, but it was also caught in the western area [22]. The maximum TL recorded during surveys (F: 880, M: 750 mm TL) was lower than that reported by Serena [79] and Ebert & Stehmann [91] but resulted quite similar to the other Mediterranean areas [92,93] (Table S3).

4.1.8. Shagreen Ray—*Leucoraja fullonica* (Linnaeus, 1758)

Follesa et al. [27], based on the MEDITS data from 2012 to 2015, found that the species was only present in GSA25 (Cyprus).

4.1.9. Maltese Ray—*Leucoraja melitensis* (Clark, 1926)

This species seems to be very rare. Indeed, Follesa et al. [27] recorded its presence only in GSA16 and GSA22 (Aegean Sea). The maximum TL of the present study was slightly higher compared to the Mediterranean studies (420 mm TL [22]; 500 mm TL [79,94]. In the present study, the parameters of the LWR were for the first time estimated in the Mediterranean. The slope (C: 3.12) was quite similar to that estimated globally by Froese et al. [95], i.e., (C: 3.22).

4.1.10. Cuckoo Ray—*Leucoraja naevus* (Müller and Henle, 1841)

In the North Mediterranean, this species was found with the highest frequency and BI (25.7 ± 11.7 kg/km^2) in GSA05, while the highest DI (26.0 ± 29.0 N/km^2) was recorded in GSA01 [27].

4.1.11. Starry Ray—*Raja asterias* Delaroche, 1809

This is a frequently found species, recorded in almost all the areas covered by MEDITS, with the exception of GSA20 (Eastern Ionian Sea), GSA23 (Crete), and GSA25 [27]. Abella et al. [24] analyzed MEDITS data from 1994 to 2013 and found the highest DI (6.4 N/km^2) and BI (5.3 kg/km^2) in GSA11, followed by GSA16 (DI = 3 N/km^2). The highest average individual weight (g) was reported, in decreasing order, for GSA 11 and GSA10 (Southern and Central Tyrrhenian Sea) as well as GSA19 and GSA09 (Ligurian Sea and Northern Tyrrhenian Sea). Ferrà et al. [54], analyzed the data obtained from a modified beam trawl in GSA17 (the so-called "rapido" survey, i.e., Solemon) held between 2005 and 2014, and they found an increasing trend in the occurrence of this species over the years. The maximum TL was lower when compared to other Mediterranean specimens [79,96]. The slopes of the LWR were similar to those reported in the Mediterranean [97–99] except for the GSA09 (Ligurian & North Tyrrhenian Sea, [100]). The sex ratio was quite balanced between sexes as reported in other areas [101,102] (Table S3).

4.1.12. Blonde Ray—*Raja brachyura* Lafont, 1873

In the North Mediterranean, this species was the most frequent (18.3%) and abundant (DI = 41.0 ± 38.0 N/km^2, BI = 26.0 ± 28.6 kg/km^2) in GSA11 [27]. In the Strait of Sicily, it appears to be relatively rare [22,102–104], even though it is not known whether this species was always this rare. Little is known about the abundance of this species along the northwest African coastlines; however, Serena et al. [69] reported its presence from Algeria to Egypt. The maximum TL was lower compared to Mediterranean specimens. Similarly to the present study, Catalano et al. [104] found the sex ratio slightly unbalanced towards males, whereas Porcu et al. [105] reported a sex ratio skewed towards females (Table S3).

4.1.13. Thornback Ray—*Raja clavata* Linnaeus, 1758

This species was recorded across all the North Mediterranean, with exception of GSA19 [27]. On the basis of data collected in Italian seas during 1994–2013 MEDITS, this species was more abundant in Sardinia waters (BI = 27.2 kg/km^2, DI = 40.5 N/km^2) and in South of Sicily (BI = 17 kg/km^2, DI = 17 N/km^2). The highest mean weight values were reported for Northern Adriatic Sea followed by South of Sicily [25]. Surveys conducted on the Italian and African sides of the Strait of Sicily between 1985 and 2002 revealed negative abundance trends in the past, although Garofalo et al. [58] reported signs of recovery. The abundance on the Italian side of the Strait of Sicily between 1997 and 2004 was approximately five times lower than the abundance on the African side; this was based on higher historical fishing effort on Sicilian than on N African fishing grounds [106]. In the present study, the spatial distribution of *R. clavata* confirms it is a more eurybathic species. Indeed, in the GSA16 it was recorded in Adventure Bank as well as off Linosa Island, whereas in the GSA15 it seems more widespread, particularly concentrated in the SE part of Malta Island (Figure 5). The higher abundances on the edge of GSA16 seem to confirm the analysis performed by Bottari et al. [106] suggesting low connectivity among the stocks occurring there. The TL range of *R. clavata* falls within those reported in the Mediterranean (see, e.g., in [21,32]). The b parameter of the LWR does not differ significantly from those reported by other Mediterranean studies, except for the Balearic Island specimens [107]. Generally, the L_{50} was lower compared to other studies [32,108–111], except for the Adriatic Sea population which was very similar to the South of Sicily. The regular distribution of the proportions of specimens classified as mature around the logistics indicates a good resolution capability of the used maturity scale and the suitability of the 3a stage as discriminant between immature/mature specimens. In the GSA16, the sex ratio was quite balanced, as was previously reported in the same area [21,66] as well as in Turkey. Conversely, in the Adriatic Sea the sex ratio was slightly unbalanced [112]. The estimates of L∞ confirmed that female specimens are larger than males. The estimates of the L∞ in the South of Sicily were higher when compared to the Adriatic Sea [112], whereas they were similar to the Tunisian area [32] as well as to a previous study carried out in the South of Sicily through vertebrae reading [21]. In the present study, the estimates of the K parameter were higher for both sexes (especially females) than Cannizzaro et al. [21] (Table S3). The Z/K ratios were estimated at 1.32 and 0.95 for females and males, respectively. The values, closer to the expected (undisturbed) value of 1, seem to suggest a high resistance of *R. clavata* (especially in males) to fishing compared to the other batoids; being smaller than females, it is more likely that males are returned to the sea after the sorting operation onboard of fishing vessels and, thus, may survive. Last, the Z/K values are in agreement with the LFD shape, whereas the K estimates are too low; the corresponding Z values (i.e., 0.21 and 0.25) are in agreement with the Z estimates in Abella et al. [25] for South of Sicily (0.25) and Northern Adriatic Sea (0.29). Considering the homogenity of the Z estimation, it is reasonable that the F estimation is also similar. Consequently, given that Abella et al. [25] estimated $F_{current}$ = 0.09 and $F_{0.1}$ = 0.11, the state of *R. clavata* stock is not so bad.

4.1.14. Brown Ray—*Raja miraletus* Linnaeus, 1758

This is a common species, recorded across all the Mediterranean except for Northern Alborean Sea (GSA01) and Gulf of Lion (GSA07) [27]. In the Strait of Sicily, the biomass has fluctuated over the years; however, it remains relatively stable. An overall increase in biomass of 55.6% was reported by Garofalo et al. [58] between 1986 and 2002. However, overall levels of biomass in the northern part of the Strait of Sicily were lower than those off Libya, owing to intensive fishing activities carried out since the 1970s. The overall increase in abundance of brown rays in this area is, therefore, in accordance with the decreasing trend in fishing efforts observed at the beginning of the 21st century due to the displacement of many large trawlers [58]. In the present study, the spatial distribution of *R. miraletus* showed that it is a shallower species, in fact in GSA16 it is mainly present in Adventure Bank and in the NE part of the GSA15 (Figure 8). The size range of *R. miraletus* sampled in the GSA16 was smaller than in other Mediterranean areas. The slope of the LWR was similar to the estimates from the Gulf of Gabes [33] and the Adriatic Sea [113], but they were different from those from the Aegean Sea [73], which reported high values for this parameter. The value was also different from that for Lebanese samples, in which an anomalous, negative allometric growth was estimated [76]. In the present study, the L_{50} estimates of the male specimens were similar to those in other Mediterranean areas, whilst the L_{50} from female specimens was smaller [108,114]. Moreover, for *R. miraletus*, the logistic model fitted well the data (proportion of mature specimens) therefore the L_{50} was quite reliable. The sex ratio, as already noted for *R. clavata*, was well balanced. In the GSA16 the estimates of L∞ were higher when compared to the Tunisian waters [30] and were lower than those in Egyptian specimens [114]). The K parameter was higher, suggesting a faster growth rate compared to other areas [30,114] (Table S3). The obtained Z/K value was quite high for both sexes (~3); this could be due to fishing pressure, which could affect the stock. The early sexual maturity of the males, when compared to females, might explain both the lower L∞ and the higher K observed. In fact, the shape of the Z/K plot for males suggested a discontinuity in the growth pattern with a reduction in the growth rate after sexual maturity.

4.1.15. Spotted Ray—*Raja montagui* Fowler, 1910

In the North Mediterranean, this species appears most frequently (25.0%) in Crete (GSA23), while it seems more abundant (D = 17.0 N/km^2, B = 11.1 kg/km^2) in the Agean sea (GSA22) [27]. In the Strait of Sicily, it appears to be ubiquitarian, with concentrations in the south and east of Malta, and between Egadi and Pantelleria [22]. The maximum TL was lower than that reported in the Mediterranean [80] (Table S3). The slope of the LWR was never estimated in the Mediterranean, whereas globally a positive allometry was recorded (C: 3.23 [95]).

4.1.16. Speckled Ray—*Raja polystigma* Regan, 1923

In the North Mediterranean, this species was most frequently caught in Cyprus (GSA25), while it was most abundant in Balearic Island (GSA05) (DI = 13.0 ± 8.0 N/km^2, BI = 37.9 ± 35.6 kg/km^2) [27]. The maximum TL was lower compared to other Mediterranean studies [79,115,116]. The parameters of the LWR were for the first time presented in the Mediterranean, but the slope of the present study when compared to the global estimate (C: 3.27, [95]) was higher. The sex ratio (0.64:1) was quite unbalanced towards males, whereas in Sardinia it was quite balanced [115,116] (Table S3).

4.1.17. Rough Ray—*Raja radula* Delaroche, 1809

In the North Mediterranean, the species was rarely caught and was in low abundance [27]. Off the Strait of Sicily, limited information is available on the presence and abundance of this species [22,28]. The sex ratio in the present study was very unbalanced, which may be due to the small sample size, whereas in other areas it seems more balanced [34,117,118] (Table S3).

4.1.18. Undulate Ray—*Raja undulata* Lacepède, 1802

This is a very rare species; its occurrence was recorded in the North Mediterranean from 2012 to 2015 only in Agean Sea (GSA22) and Crete (GSA23) [27]. In the Strait of Sicily, the first specimen of the species was reported by Bini [68].

4.1.19. White Skate—*Rostroraja alba* (Lacepède, 1803)

In the Mediterranean, this species is rare; it was recorded most frequently in Corsica (7.1%) and South of Sicily (4.4%), and more abundantly in South od Sicily (DI = 0.7 ± 0.8 N/km^2) and Balearic Islands (BI = 5.5 ± 11 kg/km^2) [27]. R. *alba* was not reported in the list compiled by Bombace and Sarà [119] and Arena and Li Greci [63] for the Strait of Sicily. The maximum TL in the present study was inferior compared to that sampled in Tunisia [35]). The *b* parameter of the LWR was inferior when compared to other estimates [35,120]). The sex ratio was very similar to that found by Kadri et al. [35] (i.e., Tunisia: 0.79:1 vs. present study: 0.89:1) (Table S3).

4.1.20. Common Stingray—*Dasyatis pastinaca* (Linnaeus, 1758)

This species has a low frequency of occurrence in all areas covered by the MEDITS survey [69]. In Turkey, the reported frequency of occurrence was 56.0%. The overall mean DI and BI were 55.3 (N/km^2) and 107.5 (kg/km^2), respectively [121]. In the Strait of Sicily, historical information on the presence and abundance of this species was reported by Schembri et al. [28]. Frequency in the northern Mediterranean ranged from a minimum of 0.3% (Spain) to 51.8% (Cyprus), while the highest DI (44.0 ± 20.0 N/km^2) and BI (23.3 ± 9.6 kg/km^2) were recorded in Sardinia [27]. The maximum TL was very similar to that found by Yeldan & Gundogdu [72], higher compared to Girgin & Basusta [122], and lower compared to that reported by Serena [123], Yeldan et al. [124] and Yigin & Ismen [125]. The slope of the LWR showed a negative allometry, as reported by Filiz & Bilge [73], whereas other studies in the Aegean Sea reported a positive allometry [66,72,121,122,125,126]. The sex ratio was higher compared to the specimens from the Aegean Sea [66,72,122,124–126] and Tunisia [127] (Table S3).

4.1.21. Blue Stingray—*Pteroplatytrygon violacea* (Bonaparte, 1832)

Very low frequencies have been recorded in the Mediterranean, with the highest values (2.7%) being observed from Northern Alboran Sea [27].

4.1.22. Bull Ray—*Aetomylaeus bovinus* (Geoffroy Saint-Hilaire, 1817)

From the MEDITS (2012–2015), the species was recorded only in South of Sicily and in Western Ionian Sea (GSA19), with frequency of 1.0% and 4.6%, respectively. The highest DI was observed in GSA19 (1 ± 0.7 N/km^2), while the highest BI was observed in GSA16 (2.3 ± 4.6 kg/km^2) [27]. The sex ratio (0.96:1) was very similar to that estimated in Turkey (1:1) by Başusta & Aslan [128].

4.1.23. Common Eagle Ray—*Myliobatis aquila* (Linnaeus, 1758)

In the Mediterranean Sea, the highest frequency (7.1%) was recorded in Crete, while the highest DI (14 ± 14.4 N/km^2) and BI (42.5 ± 49.6 kg/km^2) were observed in Northern Adriatic Sea [27]. The maximum TL was lower than that reported in the Mediterranean (C: 2600 mm; [123]). The *b* parameter of the LWR for the sexes combined was higher compared to the estimate provided by Filiz & Bilge [73].

5. Conclusions

Despite the general consensus on the bad exploitation status of almost all Mediterranean batoid stocks, the abundance indices of most of them from the Strait of Sicily showed stable or even increasing trends depicting signs of slight recovery. However, few species seem to be in an ongoing depletion (e.g., *D. pastinaca, R. radula, L. circularis,* and *R. asterias*) although with different orders of magnitude. This pattern may be owed to the combination of, on one hand, reduced fishing capacity and efforts of the South of Sicily fleet, and, on the other, to the presence of the Maltese Economic Exclusive Zone in Malta Island as well as to the great abundance of abandoned Fishing Aggregating Devices which act as dissuaders to trawling and providing further shelters.

A true rebuilding (i.e., local abundances approaching the level corresponding to maximum sustainable yield) could, however, require more time, as demonstrated by the Z/K parameter estimates. The picture provided by the present study indicates the magnitude of the combined effect of fisheries and natural mortality on this vulnerable taxonomic group. However, as reported by Schwamborn [129], some biases may occur in the estimation of Z/K ratio due to (i) the perfect linearization of the data, (ii) the extremely narrow confidence intervals, and (iii) to a possible artifact, i.e., the spurious autocorrelation between cutofflengths and mean lengths. Furthermore, lower values of L_{50} in the GSA16 were observed when compared to other Mediterranean studies for all species, except for male of *D. oxyrinchus* which are likely biased because of the low number of mature specimens. This trend could be an adaptive response of this taxon to the high fishing effort in the area [130]. In conclusion, the lack of detailed quantitative previous information on the batoid of South of Sicily and Malta Island does not allow to judge analytically the current status of the stocks, although the higher abundance of some species within Malta raises some concern for the Sicilian counterpart.

In spite of this relevant historical limitation, at a precautionary level, it seems wise to adopt measures finalised at the protection of these species such as (i) maintenance or a reduction of fishing effort, (ii) improvement of the selectivity of trawl gears (implementation of an ad-hoc modified Turtle Excluder Device) [131,132], (iii) to improve education on responsible fishing by maximizing the number of cartilaginous fish returned to the sea alive (see, e.g., in [133]), (iv) increase collaboration among enterprises and, in general, among stakeholders to define innovative technical solutions (see, e.g., in [53,134–136]) and, (v) exploring the feasibility of creating ground aquaculture facilities to raise the spawners, and then taking the ovigerous capsules and implanting them on artificial substrates.

Supplementary Materials: The following are available online at https://www.mdpi.com/article/10.3390/ani11082189/s1, Figure S1: Overall positive hauls of *Dipturus oxyrinchus* in the GSA16 and GSA15. Blue and Red crosses indicate, respectively, GRUND (GRU) and MEDITS (MED). The size of the cross is proportional to the number of positive hauls. The dashed areas represent depths below 800 m not explored in the present surveys., Figure S2: Overall positive hauls of *Raja clavata* in the GSA16 and GSA15. Blue and Red crosses indicate, respectively, GRUND (GRU) and MEDITS (MED). The size of the cross is proportional to the number of positive hauls. The dashed areas represent depths below 800 m not explored in the present surveys, Figure S3: Overall positive hauls of *Raja miraletus* in the GSA16 and GSA15. Blue and Red crosses indicate, respectively, GRUND (GRU) and MEDITS (MED). The size of the cross is proportional to the number of positive hauls. The dashed areas represent depths below 800 m not explored in the present surveys. Table S1:Calendar of the MEDITS (MED) survey carried out in the GSA16 and GSA15. Blue color depicted the surveys timeframe, Table S2. Calendar of the GRUND (GRU) survey carried out in the GSA16 and GSA15. Blue color depicted the surveys timeframe, **Table S3**: Synopsis of the batoids life history traits estimated from the South of Sicily compared to the other Mediterranean areas. Table S4. Taxonomic and systematic features of the 38 species of batoids documented in the Mediterranean Sea following Dulvy et al. [11], integrated by Last et al. [137], and Fricke et al. [138]. In the Family column, the most accepted common/vernacular name is also presented, Video S1: not applicable. References [21,22,30,31,33,34,66,67,71–81,83–85,87,88,91–94,96–102,104,105,107–118,120–128,139–144] are cited in the supplementary materials.

Author Contributions: Conceptualization: S.V., S.R.; Data Curation: V.G., M.G., J.M., A.S.; Formal Analysis: M.L.G., D.S.; Investigation: M.L.G., D.S., F.F.; Visualization: M.L.G., S.R.; Writing—Original Draft: M.L.G.; Writing—Review & Editing: M.L.G., S.R., D.S., F.F., V.G., J.M., M.G., A.S., S.V. All authors have read and agreed to the published version of the manuscript.

Funding: This research was funded by the European Data Collection Framework (DCF), module trawl surveys MEDITS, funded by the European Union and the Italian Ministry for Agricultural, Food, and Forestry Policies.

Institutional Review Board Statement: Ethical review and approval were waived for this study because animals were sampled during trawl surveys and were analyzed when they were already dead.

Data Availability Statement: The data presented in this study are available on request from the corresponding author.

Acknowledgments: We are grateful to all the technical staff of SS CNR-IRBIM of Mazara del Vallo and of DFA-Malta who have collected data and processed samples from the trawl surveys.

Conflicts of Interest: The authors declare no conflict of interest.

References

1. McEachran, J.D.; de Carvalho, M.R.; Carpenter, K.E. (Eds.) Batoid fishes. In *The Living Marine Resources of the Western Central Atlantic*; Introduction, Molluscs, Crustaceans, Hagfishes, Sharks, Batoid Fishes, and Chimaeras; FAO Species Identification Guide for Fishery Purposes; American Society of Ichthyologist and Herpetologists Special Publication No. 5; FAO: Rome, Italy, 2002; Volume 1, pp. 507–589.
2. Iglesias, S.P. Handbook of the Marine Fishes of Europe and Adjacent Waters (A Natural Classification Based on Collection Specimens, with DNA Barcodes and Standardized Photographs). Volume I (Chondrichthyans and Cyclostomata). Provisional Version 08. 2014, p. 105. Available online: http://www.mnhn.fr/iccanam (accessed on 7 March 2019).
3. Serena, F.; Abella, A.J.; Bargnesi, F.; Barone, M.; Colloca, F.; Ferretti, F.; Fiorentino, F.; Jenrette, J.; Moro, S. Species diversity, taxonomy and distribution of Chondrichthyes in the Mediterranean and Black Sea. *Eur. Zool. J.* **2020**, *87*, 497–536. [CrossRef]
4. Guijarro, B.; Quetglas, A.; Moranta, J.; Ordines, F.; Valls, M.; González, N.; Massutí, E. Inter- and intra-annual trends and status indicators of nektobenthic elasmobranchs off the Balearic Islands (northwestern Mediterranean). *Sci. Mar.* **2012**, *76*, 87–96. [CrossRef]
5. Ligas, A.; Osio, G.C.; Sartor, P.; Sbrana, M.; De Ranieri, S. Long-term trajectory of some elasmobranch species off the Tuscany coasts (NW Mediterranean) from 50 years of catch data. *Sci. Mar.* **2013**, *77*, 119–127. [CrossRef]
6. Ferretti, F.; Myers, R.A.; Serena, F.; Lotze, H.K. Loss of Large Predatory Sharks from the Mediterranean Sea. *Conserv. Biol.* **2008**, *22*, 952–964. [CrossRef]
7. Walls, R.H.L.; Dulvy, N.K. Predicting the conservation status of Europe's Data Deficient sharks and rays. *bioRxiv* **2019**. [CrossRef]
8. Cardinale, M.; Bartolino, V.; Llope, M.; Maiorano, L.; Sköld, M.; Hagberg, J. Historical spatial baselines in conservation and management of marine resources. *Fish Fish.* **2010**, *12*, 289–298. [CrossRef]

9. Abella, A. General review on the available methods for stock assessment of Elasmobranchs, especially in data shortage situations. In *Sub-Committee on Stock Assessment (SCSA) Report of the Workshop on Stock Assessment of Selected Species of Elasmobranchs in the GFCM Area DG-MARE*; GFCM-FAO: Brussels, Belgium, 2011; pp. 12–16.

10. Booth, H.; Squires, D.; Milner-Gulland, E.J. The neglected complexities of shark fisheries, and priorities for holistic risk-based management. *Ocean Coast. Manag.* **2019**, *182*, 104994. [CrossRef]

11. Dulvy, N.K.; Allen, D.J.; Ralph, G.M.; Walls, R.H. *The Conservation Status of Sharks, Rays and Chimaeras in the Mediterranean Sea [Brochure]*; IUCN Centre for Mediterranean Cooperation: Malaga, Spain, 2016; p. 14.

12. Milazzo, M.; Cattano, C.; Al Mabruk, S.A.A.; Giovos, I. Mediterranean sharks and rays need action. *Science* **2021**, *371*, 355–356. [CrossRef]

13. Flowers, K.I.; Heithaus, M.R.; Papastamatiou, Y.P. Buried in the sand: Uncovering the ecological roles and importance of rays. *Fish Fish.* **2021**, *22*, 105–127. [CrossRef]

14. Pratt, H.L., Jr.; Gruber, S.H.; Taniuchi, T. Elasmobranchs as living resources: Advances in the biology, ecology, systematics and the status of the fisheries. In *NOAA Technical Report NMFS*; National Marine Fisheries Service: Silver Spring, MD, USA, 1990; 518p.

15. Dayton, P.K.; Thrush, S.F.; Agardy, M.T.; Hofman, R.J. Environmental effects of marine fishing. *Aquat Conserv.* **1995**, *5*, 205–232. [CrossRef]

16. Stevens, J. The effects of fishing on sharks, rays, and chimaeras (chondrichthyans), and the implications for marine ecosystems. *ICES J. Mar. Sci.* **2000**, *57*, 476–494. [CrossRef]

17. Nieto, A.; Ralph, G.M.; Comeros-Raynal, M.T.; Heessen, H.J.; Rijnsdorp, A.D. *European Red List of Marine Fishes*; Publications Office of the European Union: Luxembourg, 2015; p. 61.

18. MEDITS Working Group. *International Bottom Trawl Survey in the Mediterranean*; Instruction manual; Version 9. [MEDITS–handbook. Version n. 9.]; MEDITS Working Group, 2017; p. 106.

19. Spedicato, M.T.; Massutí, E.; Mérigot, B.; Tserpes, G.; Jadaud, A.; Relini, G. The MEDITS trawl survey specifications in an ecosystem approach to fishery management. *Sci. Mar.* **2019**, *83*, 9–20. [CrossRef]

20. Jereb, P.; Cannizzaro, L.; Norrito, G.; Ragonese, S. Strait of Sicily': Sensu stricto vs sensu lato. Setting a baseline definition for an important Mediterranean fisheries management area. *Biol. Mar. Mediterr.* **2016**, *69*, 49–56. (In Italian)

21. Cannizzaro, L.; Garofalo, G.; Levi, D.; Rizzo, P.; Gancitano, S. *Raja clavata* (Linneo, 1758) nel Canale di Sicilia: Crescita, distribuzione e abbondanza. *Biol. Mar. Mediterr.* **1995**, *2*, 257–262. (In Italian)

22. Ragonese, S.; Cigala Fulgosi, F.; Bianchini, M.L.; Norrito, G.; Sinacori, G. Annotated check list of the skates (Chondrichthyes, Rajidae) in the Strait of Sicily (Central Mediterranean). *Biol. Mar. Mediterr.* **2003**, *10*, 874–881. (In Italian)

23. Lauria, V.; Gristina, M.; Attrill, M.J.; Fiorentino, F.; Garofalo, G. Predictive habitat suitability models to aid conservation of elasmobranch diversity in the central Mediterranean Sea. *Sci. Rep.* **2015**, *5*, 13245. [CrossRef]

24. Abella, A.; Mancusi, C.; Serena, F. Raja asterias. In *Sintesi delle Conoscenze di Biologia, Ecologia e pesca delle Specie Ittiche dei mari Italiani*; Sartor, P., Mannini, A., Carlucci, R., Massaro, E., Queirolo, S., Sabatini, A., Scarcella, G., Simoni, R., Eds.; Società Italiana di Biologia Marina (S.I.B.M.): Genoa, Italy, 2017; pp. 144–149.

25. Abella, A.; Mancusi, C.; Serena, F. Raja clavata. In *Sintesi delle Conoscenze di Biologia, Ecologia e pesca delle Specie Ittiche dei mari Italiani*; Sartor, P., Mannini, A., Carlucci, R., Massaro, E., Queirolo, S., Sabatini, A., Scarcella, G., Simoni, R., Eds.; Società Italiana di Biologia Marina (S.I.B.M.): Genoa, Italy, 2017; pp. 150–156.

26. Geraci, M.L.; Ragonese, S.; Norrito, G.; Scannella, D.; Falsone, F.; Vitale, S. A Tale on the Demersal and Bottom Dwelling Chondrichthyes in the South of SicilySouth Sicily through 20 Years of Scientific Survey. In *Chondrichthyes—Multidisciplinary Approach*; Rodrigues-Filho, L.F., de Luna Sales, J.B., Eds.; IntechOpen: London, UK, 2017; pp. 13–37. [CrossRef]

27. Follesa, M.C.; Marongiu, M.F.; Zupa, W.; Bellodi, A.; Cau, A.; Cannas, R.; Colloca, F.; Djurovic, M.; Isajlovic, I.; Jadaud, A.; et al. Spatial variability of Chondrichthyes in the northern Mediterranean. *Sci. Mar.* **2019**, *83*, 81–100. [CrossRef]

28. Schembri, T.; Schembri, P.J.; Fergusson, I.K. Revision of the records of shark and ray species from the Maltese Islands (Chordata: Chondrichthyes). *Cent. Medit. Nat.* **2003**, *4*, 71–104.

29. Bradai, M.N.; Saidi, B.; Enajjar, S. *Elasmobranchs of the Mediterranean and Black Sea: Status, Ecology and Biology*; Bibliographic analysis; Studies and Reviews-General Fisheries Commission for the Mediterranean; No. 91; FAO: Rome, Italy, 2012.

30. Kadri, H.; Marouani, S.; Sadi, B.; Bradai, M.N.; Ghorbel, M.; Bouan, A.; Morize, E. Age, growth and reproduction of *Raja miraletus* (Linnaeus, 1758) (Chondrichthyes: Rajidae) of the Gulf of Gabès (Tunisia, Central Mediterranean Sea). *Mar. Biol. Res.* **2012**, *8*, 388–396. [CrossRef]

31. Kadri, H.; Marouani, S.; Bradai, M.N.; Bouaïn, A.; Morize, E. Age, growth, longevity, mortality and reproductive biology of *Dipturus oxyrinchus*, (Chondrichthyes: Rajidae) off the Gulf of Gabès (Southern Tunisia, central Mediterranean). *J. Mar. Biol. Assoc. UK* **2014**, *95*, 569–577. [CrossRef]

32. Kadri, H.; Marouani, S.; Saïdi, B.; Bradai, M.N.; Bouaïn, A.; Morize, E. Age, growth, sexual maturity and reproduction of the thornback ray, *Raja clavata* (L.), of the Gulf of Gabès (south-central Mediterranean Sea). *Mar. Biol. Res.* **2014**, *10*, 416–425. [CrossRef]

33. Kadri, H.; Marouani, S.; Bradai, M.N.; Bouaïn, A.; Morize, E. Distribution and Morphometric Characters of the Mediterranean Brown Ray, *Raja miraletus* (Chondrichthyes: Rajidae) in the Gulf of Gabes (Tunisia, Central Mediterranean). *Am. J. Agric. For.* **2014**, *2*, 45–50. [CrossRef]

34. Kadri, H.; Marouani, S.; Bradai, M.N.; Bouaïn, A.; Morize, E. Age, growth and length-weight relationship of the rough skate, *Raja radula* (Linnaeus, 1758) (Chondrichthyans: Rajidae), from the Gulf of Gabes (Tunisia, Central Mediterranean). *J. Coast. Life Med.* **2014**, *2*, 344–349.

35. Kadri, H.; Marouani, S.; Bradai, M.N.; Bouaïn, A.; Morize, E. Age, growth and length-weight relationship of the white skate, *Rostroraja alba* (Linnaeus, 1758) (Chondrichthyans: Rajidae), from the Gulf of Gabes (Tunisia, Central Mediterranean). *J. Coast. Life Med.* **2014**, *14*, 193–204. [CrossRef]

36. Bradai, M.N.; Saidi, B.; Enajjar, S. Overview on Mediterranean shark's fisheries: Impact on the biodiversity. In *Marine Ecology–Biotic and Abiotic Interactions*; Turkoglu, M., Onal, U., Ismen, A., Eds.; IntechOpen: London, UK, 2018; pp. 211–230. [CrossRef]

37. GFCM. Resolution 31/2007/2: On the Establishment of GeographicalSub-Areas in the GFCM Area. 2007. Available online: http://www.gfcm.org/figis/pdf/gfcm/topic/16162/en?title=GFCM%20-%20Geographical%20Sub-Areas (accessed on 15 December 2013).

38. Levi, D.; Ragonese, S.; Andreoli, M.G.; Norrito, G.; Rizzo, P.; Giusto, G.B.; Gancitano, S.; Sinacori, G.; Bono, G.; Garofalo, G.; et al. Sintesi delle ricerche sulle risorse demersali dello Stretto di Sicilia (Mediterraneo centrale) negli anni 1985–1997 svolte nell'ambito della legge 41/82. *Biol. Mar. Mediterr.* **1998**, *5*, 130–139. (In Italian)

39. Relini, G. Demersal trawl surveys in Italian Seas: A short review. *Actes Colloq. Ifremer* **2000**, *26*, 76–93.

40. Papaconstantinou, C.; Relini, G.; Souplet, A. The general specifications of the Medits surveys. *Sci. Mar.* **2002**, *66*, 9–17. [CrossRef]

41. Ragonese, S.; Vitale, S. Desirability of a standard notation for fisheries assessment. *Agric. Sci.* **2013**, *04*, 399–432. [CrossRef]

42. Zar, J.H. *Biostatistical Analysis*, 4th ed.; Prentice Hall: Upper Saddle River, NJ, USA, 1999; 662p.

43. Lembo, G. *SAMED Stock Assessment in the Mediterranean*; EC project n° 99/047; COISPA: Bari, Italy, 2002.

44. Wetherall, J.A.; Polovina, J.J.; Ralston, S. Estimating growth and mortality in steady-state fish stocks from length-frequency data. In *ICLARM Conference Proceedings*; WorldFish: Penang, Malaysia, 1987; Volume 13, pp. 53–74.

45. Mildenberger, T.K.; Taylor, M.H.; Wolff, M. TropFishR: An R package for fisheries analysis with length-frequency data. *Methods Ecol. Evol.* **2017**, *8*, 1520–1527. [CrossRef]

46. Petitgas, P. Geostatistics and their applications to fisheries survey data. In *Computers in Fisheries Research*; Megrey, B.A., Moksness, E., Eds.; Chapman &Hall: London, UK, 1996; pp. 113–142.

47. Goovaerts, P. *Geostatistics for Natural Resources Evaluation*; Oxford University Press: New York, NY, USA, 1997.

48. Matheron, G. *La theorie des Variables Regionalisees et ses Applications. Cahiers du Centre de Morphologie Mathématique de Fontainebleau*; École Nationale Supérieure des Mines: Paris, France, 1970; Volume 5, pp. 1–212. (In French)

49. Isaaks, E.H.; Srivastava, R.M. *Applied Geostatistics*; Oxford University Press: New York, NY, USA, 1989.

50. Rossi, R.E.; Mulla, D.J.; Journel, A.G.; Franz, E.H. Geostatistical tools for modeling and interpreting ecological spatial dependence. *Ecol. Monogr.* **1992**, *62*, 277–314. [CrossRef]

51. Pebesma, E.J. Multivariable geostatistics in S: The gstat package. *Comput. Geosci.* **2004**, *30*, 683–691. [CrossRef]

52. R Core Team. *R: A Language and Environment for Statistical Computing*; Version 3.6.2; R Foundation for Statistical Computing: Vienna, Austria, 2019; Available online: https://www.R-project.org/ (accessed on 15 March 2020).

53. Geraci, M.L.; Di Lorenzo, M.; Falsone, F.; Scannella, D.; Di Maio, F.; Colloca, F.; Vitale, S.; Serena, F. The occurrence of Norwegian skate, *Dipturus nidarosiensis* (Elasmobranchii: Rajiformes: Rajidae), in the Strait of Sicily, central Mediterranean. *Acta Ichthyol. Piscat.* **2019**, *49*, 203–208. [CrossRef]

54. Ferrá, C.; Fabi, G.; Polidori, P.; Tassetti, A.N.; Leoni, S.; Pellini, G.; Scarcella, G. *Raja asterias* population assessment in FAO GFCM GSA17 area. *Mediterr. Mar. Sci.* **2016**, *17*, 651–660. [CrossRef]

55. Cashion, M.S.; Bailly, N.; Pauly, D. Official catch data underrepresent shark and ray taxa caught in Mediterranean and Black Sea fisheries. *Mar. Policy.* **2019**, *105*, 1–9. [CrossRef]

56. Abella, A.J.; Serena, F. Comparison of Elasmobranch Catches from Research Trawl Surveys and Commercial Landings at Port of Viareggio, Italy, in the Last Decade. *J. Northwest Atl. Fish. Sci.* **2005**, *35*, 345–356. [CrossRef]

57. Giudicelli, M. *Malta: Simulated Commercial Trawling and Scouting Operations in the Central Mediterranean (January 1976–June 1977)*; FI: MAT/75/001/1; FAO: Rome, Italy, 1978; 93p.

58. Garofalo, G.; Gristina, M.; Fiorentino, F.; Cigala Fulgosi, F.; Norrito, G.; Sinacori, G. Distributional pattern of rays (Pisces, Rajidae) in the Strait of Sicily in relation to fishing pressure. *Hydrobiologia* **2003**, *503*, 245–250. [CrossRef]

59. Ragonese, S.; Vitale, S.; Dimech, M.; Mazzola, S. Abundances of demersal sharks and chimaera from 1994-2009 scientific surveys in the central Mediterranean Sea. *PLoS ONE* **2013**, *8*, e74865. [CrossRef]

60. Lanfranco, G. *The Fish around Malta (Central Mediterranean)*; Progress Press: Valletta, Malta, 1993.

61. Farrugia Randon, S.; Sammut, R. *Fishes of Maltese Waters*; The Authors: Malta, 1999.

62. Sammut, R. *Mediterranean Sea Fishes Central Region*; The Authors: Malta, 2001.

63. Arena, P.; Li Greci, F. Indagine sulle condizioni faunistiche e sui rendimenti di pesca dei fondali batiali della Sicilia occidentale e della bordura settentrionale dei banchi della soglia Siculo-Tunisina. *Quad. Lab. Tecnol. Pesca* **1973**, *1*, 157–201. (In Italian)

64. Maurin, C. Etude des fonds chalutables de la Mediterranee occidentale (Ecologie et Peche). *Rev. des Trav. de l'Institut des Pêches Marit.* **1968**, *26*, 163–218. (In French)

65. Quignard, J.P.; Capapé, C. Liste commentée des Sélaciens de Tunisie 2. *Bull. Inst. Océanogr. Pêche Salammbô* **1971**, *2*, 131–142. (In French)

66. Başusta, A.; Başusta, N.; Sulikowski, J.A.; Driggers, W.B.; Demirhan, S.A.; Çiçek, E. Length-weight relationships for nine species of batoids from the Iskenderun Bay, Turkey. *J. Appl. Ichthyol.* **2012**, *28*, 850–851. [CrossRef]

67. Kaya, G.; Başusta, N. A study on age and growth of juvenile and semi adult *Torpedo nobiliana* Bonaparte, 1835 inhabiting Iskenderun Bay, northeastern Mediterranean Sea. *Acta Biol. Turc.* **2016**, *29*, 143–149.

68. Bini, G. *Atlante dei Pesci delle Coste Italiane. Milano: Mondo Sommerso*; Vol. 1 Leptocardi, Ciclostomi, Selaci, Mondo Sommerso: Roma, Italy, 1967; 206p. (In Italian)

69. Baino, R.; Serena, F.; Ragonese, S.; Rey, J.; Rinelli, P. Catch composition and abundance of elasmobranchs based on the MEDITS program. *Rapp. Comm. Int. Mer. Médit.* **2001**, *36*, 234.

70. Bradai, M.N. Diversité du Peuplement Ichtyque et Contribution à la Connaisance des Sparidés du Golfe de Gabès. Ph.D. Thesis, University of Sfax, Sfax, Tunisia, 2000. (In French).

71. Duman, Ö.V.; Başusta, N. Age and growth characteristics of marbled electric ray *Torpedo marmorata* (Risso, 1810) inhabiting Iskenderun Bay, North-eastern Mediterranean Sea. *Turk. J. Fish Aquat. Sci.* **2013**, *13*, 541–549. [CrossRef]

72. Yeldan, H.; Gundogdu, S. Morphometric relationships and growth of common stingray, *Dasyatis pastinaca* (Linnaeus, 1758) and marbled stingray, *Dasyatis marmorata* (Steindachner, 1892) in the northeastern Levantine Basin. *J. Black Sea/Mediterr. Environ.* **2018**, *24*, 10–27.

73. Filiz, H.; Bilge, G. Length-weight relationships of 24 fish species from the North Aegean Sea, Turkey. *J. Appl. Ichthyol.* **2004**, *20*, 431–432. [CrossRef]

74. Bellodi, A.; Mulas, A.; Carbonara, P.; Cau, A.; Cuccu, D.; Marongiu, M.F.; Mura, V.; Pesci, P.; Zupa, W.; Porcu, C.; et al. New insights into life–history traits of Mediterranean Electric rays (Torpediniformes: Torpedinidae) as a contribution to their conservation. *Zoology* **2021**, *146*, 125922. [CrossRef]

75. Consalvo, I.; Scacco, U.; Romanelli, M.; Vacchi, M. Comparative study on the reproductive biology of *Torpedo torpedo* (Linnaeus, 1758) and *T. marmorata* (Risso, 1810) in the central Mediterranean Sea. *Sci. Mar.* **2007**, *71*, 213–222. [CrossRef]

76. Lteif, M.; Mouawad, R.; Jemaa, S.; Khalaf, G.; Lenfant, P.; Verdoit-Jarraya, M. The length-weight relationships of three sharks and five batoids in the Lebanese marine waters, eastern Mediterranean. *Egypt. J. Aquat. Res.* **2016**, *42*, 475–477. [CrossRef]

77. Abdel-Aziz, S. Observations on the biology of the common torpedo (*Torpedo torpedo*, Linnaeus, 1758) and marbled electric ray (*Torpedo marmorata*, Risso, 1810) from Egyptian Mediterranean waters. *Mar. Freshw. Res.* **1994**, *45*, 693–704. [CrossRef]

78. El Kamel-Moutalibi, O.; Mnasri, N.; Boumaïza, M.; Ben Amor, M.M.; Reynaud, C.; Capapé, C. Maturity, reproductive cycle and fecundity of common torpedo, *Torpedo torpedo* (Chondrichthyes, Torpedinidae) from the Lagoon of Bizerte (Northeastern Tunisia, central Mediterranean). *J. Ichthyol.* **2013**, *53*, 758–774. [CrossRef]

79. Serena, F.; Mancusi, C.; Barone, M. (Eds.) Field guide to the identification of the skates (Rajidae) of the Mediterranean Sea. Guidelines for data collection and analysis. *Biol. Mar. Mediterr.* **2010**, *17*, 1–204.

80. Serena, F.; Mancusi, C.; Barone, M. MEDiterranean Large Elasmobranchs Monitoring. In *Protocollo di Acquisizione Dati*; SharkLife Program: Roma, Italy, 2014; 130p. (In Italian)

81. Cannas, R.; Follesa, M.C.; Cabiddu, S.; Porcu, C.; Salvadori, S.; Iglésias, S.P.; Deiana, A.M.; Cau, A. Molecular and morphological evidence of the occurrence of the Norwegian skate *Dipturus nidarosiensis* (Storm, 1881) in the Mediterranean Sea. *Mar. Biol. Res.* **2010**, *6*, 341–350. [CrossRef]

82. Cariani, A.; Messinetti, S.; Ferrari, A.; Arculeo, M.; Bonello, J.J.; Bonnici, L.; Cannas, R.; Carbonara, P.; Cau, A.; Charilaou, C.; et al. Improving the Conservation of Mediterranean Chondrichthyans: The ELASMOMED DNA Barcode Reference Library. *PLoS ONE* **2017**, *12*, e0170244. [CrossRef]

83. Fang, K.; Chen, L.; Zhou, J.; Yang, Z.P.; Dong, X.F.; Zhang, H.B. On the presence of *Dipturus nidarosiensis* (Storm, 1881) in the Central Mediterranean area. *PeerJ* **2019**, *7*, e7009. [CrossRef]

84. Isajlović, I.; Dragičević, B.; Manfredi, C.; Vrgoč, N.; Piccinetti, C.; Dulčić, J. Additional records of Norwegian skate *Dipturus nidarosiensis* (Storm, 1881) (Pisces: Rajidae) in the Adriatic Sea. *Acta Adriat.* **2020**, *61*, 217–222. [CrossRef]

85. Ramírez-Amaro, S.; Ordines, F.; Puerto, M.Á.; García, C.; Ramon, C.; Terrasa, B.; Massutí, E. New morphological and molecular evidence confirm the presence of the Norwegian skate *Dipturus nidarosiensis* (Storm, 1881) in the Mediterranean Sea and extend its distribution to the western basin. *Mediterr. Mar. Sci.* **2017**, *18*, 251–259. [CrossRef]

86. Carugati, L.; Melis, R.; Cariani, A.; Cau, A.; Crobe, V.; Ferrari, A.; Follesa, M.C.; Geraci, M.L.; Iglésias, S.P.; Pesci, P.; et al. Combined COI barcode-based methods to avoid mislabelling of threatened species of deep-sea skates. *Anim. Conserv.* **2021**. [CrossRef]

87. Bellodi, A.; Porcu, C.; Cannas, R.; Cau, A.; Marongiu, M.F.; Mulas, A.; Vittori, S.; Follesa, M.C. Life-history traits of the long-nosed skate *Dipturus oxyrinchus*. *J. Fish Biol.* **2016**, *90*, 867–888. [CrossRef]

88. Yığın, C.Ç.; Ismen, A. Age, growth, reproduction and feed of longnosed skate, *Dipturus oxyrinchus* (Linnaeus, 1758) in Saros Bay, the north Aegean Sea. *J. Appl. Ichthyol.* **2010**, *26*, 913–919. [CrossRef]

89. Alkusairy, H.; Saad, A. Some morphological and biological aspects of longnosed skate, *Dipturus oxyrinchus* (Elasmobranchii: Rajiformes: Rajidae), in Syrian marine waters (eastern Mediterranean). *Acta Ichthyol. Piscat.* **2017**, *47*, 371–383. [CrossRef]

90. Powell, D.G. Estimation of mortality and growth parameters from the length frequency of a catch [model]. *Rapp. Proc. Verb. Reun.* **1979**, *175*, 167–169.

91. Ebert, D.A.; Stehmann, M.F. *Sharks, Batoids and Chimaeras of the North Atlantic*; FAO: Roma, Italy, 2013.

92. Deval, M.C.; Güven, O.; Saygu, İ.; Kabapçioğlu, T. Length-weight relationships of 10 fish species found off Antalya Bay, eastern Mediterranean. *J. Appl. Ichthyol.* **2013**, *30*, 567–568. [CrossRef]

93. Mnasri, N.; Boumaïza, M.; Capapé, C. Morphological data, biological observations and occurrence of a rare skate, *Leucoraja circularis* (Chondrichthyes: Rajidae), off the northern coast of Tunisia (central Mediterranean). *Pan-Am. J. Aquat. Sci.* **2009**, *4*, 70–78.

94. Stehmann, M.; Bürkel, D.L. Rajidae. In *Fishes of the Northeastern Atlantic and Mediterranean*; Whitehead, P.J.P., Bauchot, M.L., Hureau, J.C., Nielsen, J., Tortonese, E., Eds.; UNESCO: Paris, France, 1984; pp. 163–196.

95. Froese, R.; Thorson, J.; Reyes, R.B., Jr. A Bayesian approach for estimating length-weight relationships in fishes. *J. Appl. Ichthyol.* **2014**, *30*, 78–85. [CrossRef]

96. Bono, L.; De Ranieri, S.; Fabiani, O.; Mancusi, C.; Serena, F. Studio sull'accrescimento di *Raja asterias* (Delaroche, 1809) (Chondrichthyes, Raidae) attraverso l'analisi delle vertebre. *Biol. Mar. Mediterr.* **2005**, *12*, 470–474. (In Italian)

97. Serena, F.; Abella, A. Sintesi delle conoscenze sulle risorse da pesca dei fondi del Mediterraneo centrale (Italia e Corsica). *Biol Mar. Medit.* **1999**, *6*, 82–86.

98. Cau, A.; Follesa, M.C.; Pesci, P.; Cuccu, D.; Sabatini, A. Campionamento Biologico Delle Catture. Sezioni C ed E. Dipartimento di Scienze della Vita e dell'Ambiente. In *Programma Nazionale Italiano per la Raccolta Dati Alieutici 2012*; Università di Cagliari: Cagliari, Italy, 2013. (In Italian)

99. Carbonara, P.; Spedicato, M.T.; Casciaro, L.; Bitetto, I.; Facchini, M.T.; Zupa, W.; Gaudio, P.; Surico, A.; Gallo, P.; Fazi, S.; et al. Campionamento Biologico delle catture. Sezioni C ed E. Rapporto Finale. GSA19 Mar Ionio. In *Programma Nazionale Italiano per la Raccolta Dati Alieutici 2012*; COISPA Tecnologia Ricerca: Italy, 2012. (In Italian)

100. De Ranieri, S. Campionamento Biologico delle catture. Sezioni C ed E. Rapporto Finale. In *Programma Nazionale Italiano per la Raccolta Dati Alieutici 2011*; Consorzio per il Centro Interuniversitario di Biologia Marina ed Ecologia Applicata "G. Bacci": Italy, 2012. (In Italian)

101. Minervini, R.; Giannotta, M.; Bianchini, M.L. Observations on the fishery of Rajiformes in Central Tyrrhenian Sea. *Oebalia* **1985**, *11*, 583–591.

102. Abella, A.J.; Auteri, R.; Baino, R.; Lazzaretti, A.; Righini, P.; Serena, F.; Silvestri, R.; Volani, A.; Zucchi, A. Reclutamento di forme giovanili nella fascia costiera toscana (Juveniles recruitment along the Tuscan coastal area). *Biol. Mar. Mediterr.* **1997**, *4*, 172–181. (In Italian)

103. Follesa, M.C.; Addis, P.; Murenu, M.; Saba, R.; Sabatini, A. Annotated check list of the skates (Chondrichthyes, Rajidae) in the Sardinian seas. *Biol. Mar. Mediterr.* **2003**, *10*, 828–833.

104. Catalano, B.; Dalù, M.; Scacco, U.; Vacchi, M. New biological data on *Raja brachyura* (Chondrichthyes, Rajidae) from around Asinara Island (NW Sardinia, Western Mediterranean). *Ital. J. Zool.* **2007**, *74*, 55–61. [CrossRef]

105. Porcu, C.; Bellodi, A.; Cannas, R.; Marongiu, M.F.; Mulas, A.; Follesa, M.C. Life-history traits of a commercial ray, *Raja brachyura* from the central western Mediterranean Sea. *Med. Mar. Sci.* **2014**, *16*, 90–102. [CrossRef]

106. Bottari, T.; Rinelli, P.; Bianchini, M.L.; Ragonese, S. Stock identification of *Raja clavata* L. (Chondrichthyes, Rajidae) in two contiguous areas of the Mediterranean. *Hydrobiologia* **2013**, *703*, 215–224. [CrossRef]

107. Ramírez-Amaro, S.; Ordines, F.; Terrasa, B.; Esteban, A.; García, C.; Guijarro, B.; Massutí, E. Demersal chondrichthyans in the western Mediterranean: Assemblages and biological parameters of their main species. *Mar. Freshw. Res.* **2016**, *67*, 636. [CrossRef]

108. Barone, M. Life History of *Raja clavata* (Chondrichthyes, Rajiformes). Ph.D. Thesis, Faculty of Mathemathics, Physics and Natural Science of the University of Pisa, Pisa, Italy, 2009.

109. Saglam, H.; Ak, O. Reproductive biology of *Raja clavata* (Elasmobranchii: Rajidae) from Southern Black Sea coast around Turkey. *Helgol. Mar. Res.* **2011**, *66*, 117–126. [CrossRef]

110. Bilgin, S.; Onay, H. Spawning Period and Size at Maturity of the Thornback ray, *Raja clavata* (Linnaeus, 1758), (Elasmobranchii: Rajidae) in the Black Sea. *Acta Aquat. Turc.* **2020**, *16*, 525–534. [CrossRef]

111. Krstulović Šifner, S.; Vrgoč, N.; Dadić, V.; Isajlović, I.; Peharda, M.; Piccinetti, C. Long-term changes in distribution and demographic composition of thornback ray, *Raja clavata*, in the northern and central Adriatic Sea. *J. Appl. Ichthyol.* **2009**, *25*, 40–46. [CrossRef]

112. Carbonara, P.; Bellodi, A.; Palmisano, M.; Mulas, A.; Porcu, C.; Zupa, W.; Donnaloia, M.; Carlucci, R.; Sion, L.; Follesa, M.C. Growth and Age Validation of the Thornback Ray (*Raja clavata* Linnaeus, 1758) in the South Adriatic Sea (Central Mediterranean). *Front. Mar. Sci.* **2020**, *7*, 586094. [CrossRef]

113. Ungaro, N. Biological parameters of the brown ray, *Raja miraletus*, in the Southern Adriatic basin. *Cybium* **2004**, *28*, 174–176.

114. Abdel-Aziz, S.H. The use of vertebral rings of the brown ray *Raja miraletus* (Linnaeus, 1758) off Egyptian Mediterranean coast for estimation of age and growth. *Cybium* **1992**, *16*, 121–132.

115. Bellodi, A.; Cau, A.; Marongiu, M.F.; Mulas, A.; Porcu, C.; Vittori, S.; Follesa, M.C. Life history parameters of the small Mediterranean-endemic skate, *Raja polystigma* Regan 1923, from Sardinian seas. In Proceedings of the 5th International Otolith Symposium ICES-CIEM, Mallorca, Spain, 20–24 October 2014; pp. 230–231.

116. Sin, T.M.; Ang, H.P.; Buurman, J.; Lee, A.C.; Leong, Y.L.; Ooi, S.K.; Steinberg, P.; Teo, S.L. Uncommon biological patterns of a little known endemic Mediterranean skate, *Raja polystigma* (Risso, 1810). *Reg. Stud. Mar. Sci.* **2020**, *34*, 101065. [CrossRef]

117. Yığın, C.Ç.; Işmen, A. Age, growth and reproduction of the rough ray, Raja radula (Delaroche, 1809) in Saros Bay (North Aegean Sea). *J. Black Sea/Mediterr. Environ.* **2014**, *20*, 213–227.

118. Tiralongo, F.; Messina, G.; Gatti, R.C.; Tibullo, D.; Lombardo, B.M. Some biological aspects of juveniles of the rough ray, *Raja radula* Delaroche, 1809 in Eastern Sicily (central Mediterranean Sea). *J. Sea Res.* **2018**, *142*, 174–179. [CrossRef]

119. Bombace, G.; Sara, R. La pesca a strascico sui fondali da-500 a-700 metri nel settore a sud-est di Pantelleria. *Mem MIN Mar Merc.* **1972**, *33*, 63–77. (In Italian)

120. Yiğin, C.Ç.; Işmen, A. Age, growth, reproduction and feed of bottlenosed skate, *Rostroraja alba* (Lacepède, 1803) in Saros Bay, the north Aegean Sea. In Proceedings of the ICES Annual Science Conference, Gdańsk, Poland, 20–24 September 2010; pp. 20–24.

121. Özbek, E.Ö.; Çardak, M.; Kebapçioğlu, T. Spatio-temporal patterns of abundance, biomass and length-weight relationships of *Dasyatis* species (Pisces: Dasyatidae) in the Gulf of Antalya, Turkey (Levantine Sea). *J. Black Sea/Mediterr. Environ.* **2015**, *21*, 169–190.

122. Girgin, H.; Başusta, N. Testing staining techniques to determine age and growth of *Dasyatis pastinaca* (Linnaeus, 1758) captured in Iskenderun Bay, northeastern Mediterranean. *J. Appl. Ichthyol.* **2016**, *32*, 595–601. [CrossRef]

123. Serena, F. Field identification guide to the sharks and rays of the Mediterranean and Black Sea. In *FAO Species Identification Guide for Fishery Purposes*; FAO: Rome, Italy, 2005; 97p.

124. Yeldan, H.; Avsar, D.; Manaşırlı, M. Age, growth and feeding of the common stingray (*Dasyatis pastinaca*, L., 1758) in the Cilician coastal basin, northeastern Mediterranean Sea. *J. Appl. Ichthyol.* **2009**, *25*, 98–102. [CrossRef]

125. Yigin, C.C.; Ismen, A. Age, growth and reproduction of the common stingray, *Dasyatis pastinaca* from the North Aegean Sea. *Mar. Biol. Res.* **2012**, *8*, 644–653. [CrossRef]

126. Ismen, A. Age, growth, reproduction and food of common stingray (*Dasyatis pastinaca* L., 1758) in İskenderun Bay, the eastern Mediterranean. *Fish Res.* **2003**, *60*, 169–176. [CrossRef]

127. Saadaoui, A.; Saidi, B.; Enajjar, S.; Bradai, M.N. Reproductive biology of the common stingray *Dasyatis pastinaca* (Linnaeus, 1758) off the Gulf of Gabès (Central Mediterranean Sea). *Cah. Biol. Mar.* **2015**, *56*, 389–396.

128. Başusta, N.; Aslan, E. Age and growth of bull ray Aetomylaeus bovinus (Chondrichthyes: Myliobatidae) from the northeastern Mediterranean coast of Turkey. *Cah. Biol. Mar.* **2018**, *59*, 107–114. [CrossRef]

129. Schwamborn, R. How reliable are the Powell–Wetherall plot method and the maximum-length approach? Implications for length-based studies of growth and mortality. *Rev. Fish Biol Fish.* **2018**, *28*, 587–605. [CrossRef]

130. Russo, T.; Carpentieri, P.; D'Andrea, L.; De Angelis, P.; Fiorentino, F.; Franceschini, S.; Garofalo, G.; Labanchi, L.; Parisi, A.; Scardi, M.; et al. Trends in Effort and Yield of Trawl Fisheries: A Case Study From the Mediterranean Sea. *Front. Mar. Sci.* **2019**, *6*, 153. [CrossRef]

131. Kynoch, R.J.; Fryer, R.J.; Neat, F.C. A simple technical measure to reduce bycatch and discard of skates and sharks in mixed-species bottom-trawl fisheries. *ICES J. Mar. Sci.* **2015**, *72*, 1861–1868. [CrossRef]

132. Chosid, D.M.; Pol, M.; Szymanski, M.; Mirarchi, F.; Mirarchi, A. Development and observations of a spiny dogfish *Squalus acanthias* reduction device in a raised footrope silver hake *Merluccius bilinearis* trawl. *Fish Res.* **2012**, *114*, 66–75. [CrossRef]

133. Scannella, D.; Geraci, M.L.; Falsone, F.; Colloca, F.; Zava, B.; Serena, F.; Vitale, S. A new record of a great white shark, *Carcharodon carcharias* (Chondrichthyes: Lamnidae) in the Strait of Sicily, Central Mediterranean Sea. *Acta Adriat.* **2020**, *61*, 181–186. [CrossRef]

134. ICES. ICES Workshop on Innovative Fishing Gear (WKING). *ICES Sci. Rep.* **2020**, *2*, 96. [CrossRef]

135. Sardo, G.; Okpala, C.O.; Geraci, M.L.; Fiorentino, F.; Vitale, S. The effects of different artificial light wavelengths on some behavioural features of juvenile pelagic atlantic horse mackerel, *Trachurus trachurus* (actinopterygii: Perciformes: Carangidae). *Acta Ichthyol. Piscat.* **2020**, *50*, 85–92. [CrossRef]

136. Sardo, G.; Geraci, M.L.; Scannella, D.; Falsone, F.; Vitale, S. New records of two uncommon species, *Calappa tuerkayana* Pastore, 1995 (Decapoda, Calappidae) and Parasquilla ferrussaci (Roux, 1828) (Stomatopoda, Parasquillidae), from the Strait of Sicily (central Mediterranean Sea). *Arx. Miscel·Lània Zoològica* **2020**, *18*, 113–121. [CrossRef]

137. Last, P.R.; White, W.; De Carvalho, M.; Séret, B.; Stehmann, M.; Naylor, G.J.P. *Rays of the World*; CSIRO Publishing: Clayton, Australia, 2016.

138. Fricke, R.; Eschmeyer, W.N.; Van der Laan, R. Eschmeyer's Catalog of Fishes: Genera, Species, References [Internet]. Electronic Version. 2020. Available online: http://researcharchive.calacademy.org/research/ichthyology/catalog/fishcatmain.asp (accessed on 10 September 2020).

139. Capapé, C.; Guélorget, O.; Vergne, Y.; Quignard, J.P.; Amor, M.M.B.; Bradaï, M.N. Biological observations on the black torpedo, *Torpedo nobiliana* Bonaparte 1835 (Chondrichthyes: Torpedinidae), from two Mediterranean areas. *Ann. Ser. Hist. Nat.* **2006**, *16*, 19.

140. Barone, M.; De Ranieri, S.; Fabiani, O.; Pirone, A.; Serena, F. Gametogenesis and maturity stages scale of *Raja asterias* Delaroche, 1809 (Chondrichthyes, Raijdae) from the South Ligurian Sea. *Hydrobiologia* **2007**, *580*, 245–254. [CrossRef]

141. Bakiu, R.; Kolitari, J.; Lleshaj, A. Biological Characteristics and Length-Weight Relationships of landed Thornback Ray (Raja clavata, Linnaeus 1758) in the Fishing Port of Durres, Albania. *Albanian J. Agric. Sci.* **2021**, *20*, 14–23.

142. Capapé, C.; Quignard, J.P. Contribution ò lo biologie des Rajidae. *Raja polystigma. Cah. Biol. Mar.* **1978**, *19*, 233–244.

143. Deval, M.C.; Saygu, İ.; Güven, O.; Özgen, G. Elasmobranch species caught by demersal trawl fisheries in Gulf of Antalya, eastern Mediterranean. In Proceedings of the GFCM First Transversal Expert Meeting on the status of Elasmobranches in the Mediterranean and the Black Sea, Sfax, Tunisia, 20–22 September 2010.

144. Hemida, F.; Seridji, R.; Ennajar, S.; Bradai, M.N.; Collier, E.; Guelorget, O.; Capape, C. New observations on the reproductive biology of the pelagic stingray, *Dasyatis violacea* Bonaparte, 1832 (Chondrichthyes: Dasyatidae) from the Mediterranean Sea. *Acta Adriat.* **2003**, *44*, 193–204.

Article

Passive Prey Discrimination in Surface Predatory Behaviour of Bait-Attracted White Sharks from Gansbaai, South Africa

Primo Micarelli [1,*], **Federico Chieppa** [1], **Antonio Pacifico** [1,2], **Enrico Rabboni** [1] **and Francesca Romana Reinero** [1]

[1] Sharks Studies Center—Scientific Institute, 58024 Massa Marittima, Italy; chieppafederico@gmail.com (F.C.); antonio.pacifico86@gmail.com (A.P.); ENRICO.RABBONI@gmail.com (E.R.); ricerca@centrostudisquali.org (F.R.R.)

[2] Department of Political Science and CEFOP-LUISS, LUISS Guido Carli University, 00197 Rome, Italy

[*] Correspondence: direzione@centrostudisquali.org; Tel.: +39-3896732796 or +39-0566919529

Simple Summary: White sharks, in surface passive prey predatory behaviour, are initially attracted by the olfactory trace determined by the bait and then implement their predatory choices to energetical richer prey, especially thanks to their visual ability, which plays an important role in adults and immatures with dietary shifts in their feeding patterns. Gansbaai represents a hunting training area for white sharks who are changing their diet.

Abstract: Between the years 2008 and 2013, six annual research expeditions were carried out at Dyer Island (Gansbaai, South Africa) to study the surface behaviour of white sharks in the presence of two passive prey: tuna bait and a seal-shaped decoy. Sightings were performed from a commercial cage-diving boat over 247 h; 250 different white sharks, with a mean total length (TL) of 308 cm, were observed. Of these, 166 performed at least one or more interactions, for a total of 240 interactions with bait and the seal-shaped decoy. In Gansbaai, there is a population of transient white sharks consisting mainly of immature specimens throughout the year. Both mature and immature sharks preferred to prey on the seal-shaped decoy, probably due to the dietary shift that occurs in white sharks whose TL varies between 200 cm and 340 cm. As it is widely confirmed that white sharks change their diet from a predominantly piscivorous juvenile diet to a mature marine mammalian diet, it is possible that Gansbaai may be a hunting training area and that sharks show a discriminate food choice, a strategy that was adopted by the majority of specimens thanks to their ability to visualize energetically richer prey, after having been attracted by the odorous source represented by the tuna bait.

Keywords: white shark; behaviour; Gansbaai; prey choice; shark vision; shark olfaction

Citation: Micarelli, P.; Chieppa, F.; Pacifico, A.; Rabboni, E.; Reinero, F.R. Passive Prey Discrimination in Surface Predatory Behaviour of Bait-Attracted White Sharks from Gansbaai, South Africa. *Animals* **2021**, *11*, 2583. https://doi.org/10.3390/ani11092583

Academic Editors: Martina Francesca Marongiu and Tyrone Lucon-Xiccato

Received: 21 July 2021
Accepted: 1 September 2021
Published: 3 September 2021

1. Introduction

The white shark (WS) *Carcharodon carcharias* [1] is an important top predator and the largest fish predator in existence, reaching about 6 m in length, combining many particular features including large size, regional endothermy (restricted to swimming muscles, viscera, and brain), and coarsely serrated dentition [2]. Predatory behaviour was recently described by Martin et al. [3], who provided ethograms with frequency and event sequence analyses of behavioural units for the Seal Island white shark population in South Africa, and by Hammerschlag et al. [4], who reported the effects of environmental factors on the frequency and success rate of predatory attacks. The white shark is an interesting species in the study of shark behaviour thanks to the relative ease with which it can be observed from the surface, especially near pinniped colonies on rocky islands where sharks congregate [5]. White sharks have the largest olfactory bulb among sharks, and thanks to their perceptions, they are able to trace wounded prey, whale carcasses, seal colonies or sea lions even at great distances [3]. During hunting, sharks refer mainly to odorous stimuli from prey [5], but many sharks are also thought to rely on their visual system for prey detection, predator avoidance, navigation, and communication [6]. For many sharks, vision plays a vital role in their ecology, particularly in detecting and identifying prey [7–9]. It is widely accepted that white sharks undergo an ontogenetic shift in prey preference [2,10–13]; both stomach-content and stable-isotope analyses indicate that an ontogenetic shift is expressed by a change in the trophic level, passing from a predominantly piscivorous diet when immature to a marine mammalian diet when adult [10,12–15]. The estimated total length (TL) at which they undergo this dietary shift varies between 200 cm and 340 cm [2,10–13,16,17]. White sharks' teeth also reflect their ontogenetic diet change: as adults, the teeth of their upper jaw have a single large cusp, which is triangular and has serrated edges, while those of the lower jaw are tighter, smaller, narrower, and slightly sharper. The lower teeth penetrate and hold the prey, the upper ones cut the flesh: this enables both the predation of large prey such as pinnipeds, and the detachment of large pieces of meat from carcasses [18]. Martin et al. [3] observed that Sea Island white sharks appear to select the age class of Cape fur seal *Arctocephalus pusillus pusillus* [19], a group with a defined size and a clear direction of movement, as well as a choice of hunting during times and in locations that maximize their probability for predatory success. Regarding the ontogenetic shift, French et al. [20] suggested that gender and individual specialization are also key drivers in white sharks' ecological variation, and that they remain important throughout ontogeny; in fact, individuals may learn a variety of different movement tactics for encountering and catching prey, and they may develop a preference for a particular tactic based on their experiences [21,22]. The individual surface behaviour of white sharks in the presence of a bait, in the same manner as their predatory [3,23,24] and social [25] behaviour, is not a simple stimulus response reflex, but rather a complex tactical situation with plastic responses [24]. Only a few studies have described the mechanisms that underlie patterns of prey selection in white sharks [9,26–28]. The present study carried out along the South African coasts in Gansbaai was aimed at providing a contribution to confirm whether, by simulating the natural condition of scavenging, white sharks, in the presence of artificial passive prey, such as a seal-shaped decoy and tuna bait, implemented a real food choice based on vision. In the attempt to gather new information on the topic, this study on the Gansbaai white shark transient population aimed to research the following: (i) the main surface predatory behaviours of white sharks when using both baits and seal-shaped decoys; (ii) the existence of possible interlinkages between predatory behaviour and other observed endogenous factors, such as maturity; and (iii) white sharks' tendency to exhibit a real food choice, based on vision, rather than indiscriminate attacks on the two target passive preys, which are the bait and seal-shaped decoy.

2. Materials and Methods

Observations and data collection were recorded in the Dyer Island Natural Reserve, located 7.5 km south-east of Gansbaai, South Africa (34°41′ S; 19°24′ E). The reserve includes Dyer Island and Geyser Rock (Figure 1): the first is a low-profile island ca. 1.5 km long and 0.5 km wide, and it is characterised by the presence of different seabird colonies; the second is ca. 0.5 km long and 180 m wide, and it hosts a colony of Cape fur seals, *Arctocephalus pusillus pusillus* [19].

In Gansbaai (South Africa), at Dyer Island Nature Reserve, a large white shark population is present and can be observed thanks to the support of local ecotourism operators authorised to reach the field observation sites, with a prevalence of immature individuals [29]. The greatest proportion of sub-adults and potentially mature sharks also occurs in Seal Island, False Bay, South Africa [30,31]. The 2008–2013 study periods occurred in the autumn season between March and May and required a total of 247 h. Seal-shaped decoys and floating baits of tuna pieces, of similar size, were the tested passive target preys; in line with the research protocol adopted by Sperone et al. [24], the two baits were chosen first on the basis of the odourless seal shape, and second, the production of odour. The decoy's size was as close as possible to that of a juvenile Cape fur seal (*A. pusillus pusillus*): 70 cm long and 32 cm wide, with a diameter of about 60 cm (Figure 2a,b).

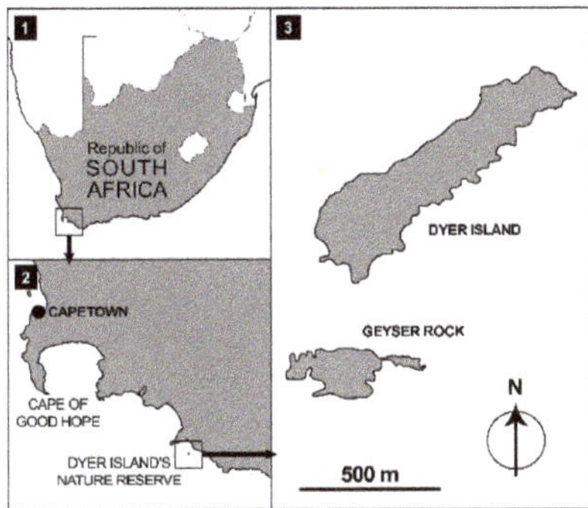

Figure 1. Dyer Island and Geyser Rock Nature Reserve.

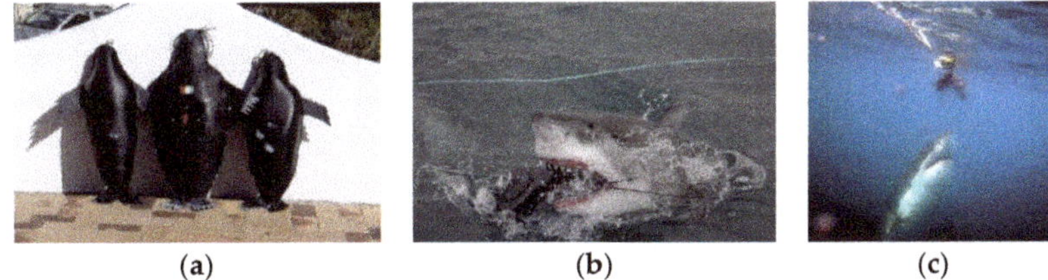

(a)   (b)   (c)

Figure 2. (a) Decoys, (b) interaction shark versus seal-shaped decoy, and (c) tuna bait.

The ethological observations on the sharks were made from the 13 m "Barracuda" boat owned by "Shark diving unlimited", which was anchored at 100–150 m off Dyer Island. The boat was equipped with an upper deck from which it was possible to observe and photograph the sighted sharks. The observations were also carried out in an anti-shark cage, which was fixed, for the duration of the observations, to the side of the boat. Throughout this study, sharks were identified by the same research team (Sharks Studies Center— Centro Studi Squali—Istituto Scientifico, Massa Marittima, Italy), and the identification was based on the recognition of different anatomical features such as the dorsal fin, the caudal fin, and the presence of scars and ectoparasites and their arrangement [29,32,33]. It was also possible that one specimen was observed at various times throughout the day and that all exhibited behaviours were recorded. All sharks' total lengths were estimated from the boat, referring to structures of a size known as the length of the cage. Inside the cage, operators were equipped with a mask, boots, a semi-dry suit and weights, remaining on the surface and maintaining a vertical position. When white sharks approached, the operators descended in a free dive to the bottom of the cage, in order not to disturb the sharks with air bubbles and to be able to observe, up close, the behaviour of the animals, the sex of the specimen and other body features. The cage was, therefore, an excellent and useful tool for identifying the sex of each individual with precision and for supporting the identification made on board. The sex of each shark was determined by the surface and by cage-diving observations, and also with underwater video recordings of the pelvic fin area: the males were recorded if claspers were seen and the females if the lack of claspers was verified and their pelvic fin area was filmed [29]. All other specimens were categorised as being of unknown sex. In our study, we estimated the white shark's size at sexual maturity according to Hewitt et al. [31] and Micarelli et al. [29]: a mature male if the TL was ≥350 cm and a mature female if the TL was ≥450 cm. To attract the sharks, olfactory stimulants (chum) were used, following the methods described in Laroche et al. [34], Ferreira and Ferreira [35], and Sperone et al. [25]. The chumming was composed of sea water, cod liver oil (*Gadus* sp.), tuna blood and small pieces of fish [25,36]. Shark predatory behaviour was induced using two types of surface passive prey placed into the water: the bait, consisting of tuna pieces tied to a floating buoy positioned at the stern of the boat, and a seal-shaped decoy, positioned at the bow at a distance from the tuna bait of at least 10 m. A constant distance between the two-surface passive prey was maintained to isolate the seal-shaped decoy from the odorous and bloody trail coming from the tuna bait. In order to analyse the surface behaviour of white sharks in the presence of passive preys, tuna baits and an odourless seal-shaped decoy, we performed a Chi-squared independence test to check whether the types of prey (bait and seal-shaped decoy) were associated, and whether the choice of prey was not causal. To investigate the presence of independency between two causal variables, we used Pearson's Chi-squared test, where H0 tests the null hypothesis of independency between the variables, and H1 the alternative hypothesis of dependency. To strengthen the results regarding the presence of causality between behaviour and prey, we also used Cochran's Q test. This is a non-parametric statistical test used to verify whether k treatments (or number of studies) have identical effects. Generally, the test statistic refers to two-way randomized block designs, where the response variable takes only two possible outcomes coded as 0 or 1 denoting failure or success, respectively. It is often used to assess whether different observers of the same phenomenon have consistent results (interobserver variability). Cochran's Q test is as follows: the null hypothesis (H_0), where there is no difference in the effectiveness of treatments (the choice is causal) and the alternative hypothesis (H_1), where there is a difference in the effectiveness of treatments (the choice is not causal).

3. Results

3.1. Descriptive Analysis

Overall, 250 white sharks were sighted, with 166 having at least one or more interactions for a total of 240 interactions: 41 adults, 183 immatures and 16 unsexed. The variables denoting maturity and prey were transformed into dichotomous variables in order to be quantitatively evaluated (Table 1). In the statistical analysis, the re-interaction (more precisely, we observed 64 reinteractions) was counted as additional information investigating the frequency among white sharks' individual characteristics. During the discriminant analysis, when studying the independency and heterogeneity among units, potential re-interactions were counted as unique observations. The same occurred in the empirical analysis accounting for non-linear methods, to investigate (potential) interactions among the observed variables. Multiple interactions were dealt with by shrinking the dataset with respect to every re-sighting. More precisely, the sample size 'n'—used in the analysis—refers to the 'individuals' (counted only one time for any observation) and was weighted by the factors (e.g., behaviour and maturity). In this way, any (potential) bias—affecting the robustness of estimates—was sufficiently avoided in the statistical analyses. For the shrinking, a discriminant analysis involved in constructing hypotheses for a significance test on an individual was used before the estimation procedure.

Table 1. Description of variables. Own computations.

Variable	Label
Maturity	(=1) adult, (=0) immature
Prey	(=1) bait, (=0) seal-shaped decoy

3.2. Statistical Analysis

In the analysis of the white sharks (individual) and the two types of prey (bait and seal-shaped decoy) (Figure 3), the preference of white sharks for the seal-shaped decoy rather than the bait was proven through appropriate test statistics. More precisely, we ran a Chi-squared test (Table 2) and Cochran's test (Table 3) rejecting any causality concerning the shark's choice under the null hypothesis. In this way, the descriptive statistics were confirmed, and it was proven that the preference for the seal-shaped decoy was not causal.

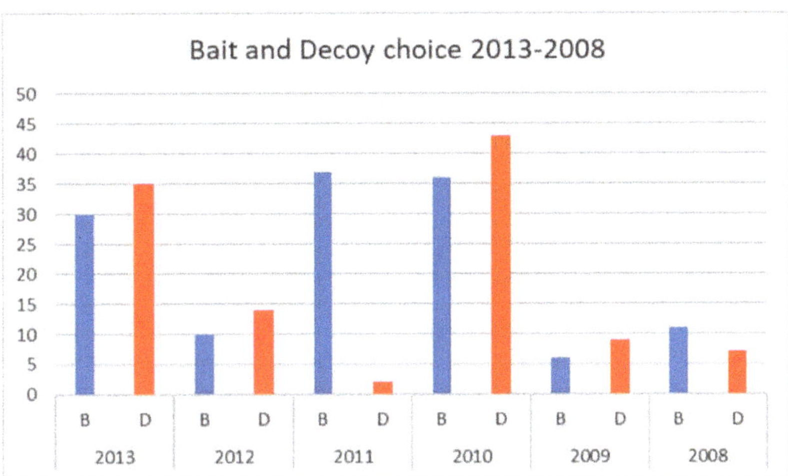

Figure 3. Bait and seal-shaped decoy shark choice, between 2013 and 2008.

Table 2. Pearson's Chi-squared test of independence. The first column denotes the sub-categories, and the second column displays the corresponding statistical results in terms of *p*-values. The significant codes are as follows: * significance at 10%, ** significance at 5%, and *** significance at 1%.

Chi-Squared Test of Independence on Prey	
Full Sample	0.00 ***
Adult	0.00 ***
Immature	0.00 ***

Table 3. Cochran's Q test (asymptotic *p*-value).

Test Statistic	175.98
Chi-squared distribution (critical value)	3.840
Df	1
p-value (one-tailed)	0.027
Significance level	0.050

All the test statistics are significant (at 1%) and consistent (*p*-value < 0.001).

3.3. Frequency Distribution

In order to investigate potential relationships between white sharks and types of prey, the previous preliminary analysis was expanded to account for related frequency distributions. More precisely, the relationship was maturity–prey, highlighting possible interactions between each shark's life stage and hunting patterns. The relationship maturity–prey highlighted that adult and even immature sharks preferred the seal-shaped decoy, confirming previous results. We performed a Chi-squared test with each frequency distribution accounted for. The main results, summarised in Table 2, highlight that the null hypothesis of independency can be rejected according to the whole sample. More precisely, the relationship between the types of prey (bait and seal-shaped decoy) is associated with a p-value close to zero (=0.00). This finding shows that their choice is not causal. In Cochran's test, the treatments denote how the choice of different prey affects the sharks' behaviour, as shown in Table 3. Let k −1 = 1 be the degrees of freedom, with k denoting the types of prey; as the computed p-value is lower than the significance level ($\alpha = 5\%$), one should reject the null and then accept the alternative hypothesis.

4. Discussion

Carcharodon carcharias is a "top predator" that has a wide distribution in temperate and tropical areas [5,18] with a higher concentration in eight spots around the world [18], including South Africa. Studies aimed at investigating the various methods of white sharks' surface artificial prey approach or their food choices are scarce: between 1989 and 1992, Anderson et al. [32] examined the predatory behaviour of the Californian Farallon Islands sharks towards decoys with different shapes, with an approximate size of prey represented by pinnipeds and also including the reproduction of a sea lion shape, for a total of 159 h of observation. In that study, it was observed that vision played a major role in the approach and predation activities: the sharks were attracted, not by smell, electric fields or vibrations coming from the baits, but only by the presence of odourless prey on the surface. Strong [9] carried out an experiment comparing two floating shapes, one with the shape of a seal and the other with a square shape: during the first experiences with these preys, the sighted sharks showed a significant preference for the seal shape. He also stated that in his study there was little doubt that the sharks initially located the bait via olfaction, and vision was clearly used to orient their actual approaches, but its relative importance to most shark species remains, however, poorly understood. The present study carried out along the South African coasts in Gansbaai was aimed at providing a contribution to confirm whether, by simulating the natural condition of scavenging, white sharks, in the presence of artificial passive prey, such as a seal-shaped decoy and tuna bait, implemented a real food choice. Micarelli et al. [29] stated that, in Gansbaai, the white shark population was always made up of a prevalence of immature individuals, and that the mean TL recorded was 308 cm. It is important to remember that the estimated length at which white sharks undergo dietary shifts varies between 200 cm and 340 cm in TL [2,10–13,16,17]; between 2008 and 2013, only six specimens showed a total length <200 cm, of 250 specimens. Tricas and McCosker [14] showed a clear prevalence of an ichthyophagous diet in immature specimens, while adult sharks preferred marine mammals at Dangerous Reef, South Australia. Moreover, in our study, considering the tests performed to distinguish the adult from the immature individuals, it emerged that the adult sharks seemed to prefer the seal-shaped decoy, and the frequency of attacks on tuna bait by the adult sharks was not significantly greater than that of the immature ones. However, since the Gansbaai population of white sharks showed a mean TL of 308 cm, it is possible that the majority of Gansbaai white sharks' transient population had already undergone the dietary shift, and this would explain why immature specimens also showed a similar interest towards the seal-shaped decoy (Figure 4).

Regarding our question (which are the white sharks' main surface predatory behaviours when using both baits and seal-shaped decoys?), it is possible to conclude from our studies that, at least as far as the behaviour of adult specimens is concerned, the same predatory trend observed near the Californian coasts emerged, but also included most immatures with dietary shift. White sharks must be selective when there is an abundance of food; therefore, according to the Optimal Foraging Theory [37], they prefer the most caloric sources over low-energy ones [38]. In response to the question "Do white sharks tend to exhibit a real food choice or indiscriminate attacks on the two passive prey (bait and seal shaped decoy)?", it is possible to conclude that there was a real food choice rather than an indiscriminate attack, and that this strategy was adopted by the majority of specimens, helped by the ability to visualize the energetically richer preys, also with respect to the odorous source represented by the tuna bait. In this regard, it was possible to observe, in this study, that, in the contemporary presence of surface prey, characterized by different stimulating conditions, the white sharks preferred the odourless seal-shaped decoy rather than the tuna bait, which produces a strong odour stimulus. In only two of the six observation years, 2011 and 2008, we recorded the prevalence of the bait choice instead of the seal-shaped decoy. We suppose that this different behaviour, as proposed by Sperone et al. [39], could be linked to environmental factors, such as cloud cover and water visibility. This allows us to confirm that vision, in the Gansbaai transient population of white sharks, as suggested by Anderson [32] and Strong [9], plays an important role during the investigation and choice of prey, helping adults and immatures, which are in progress with dietary shift, to optimize the result of their predatory activity.

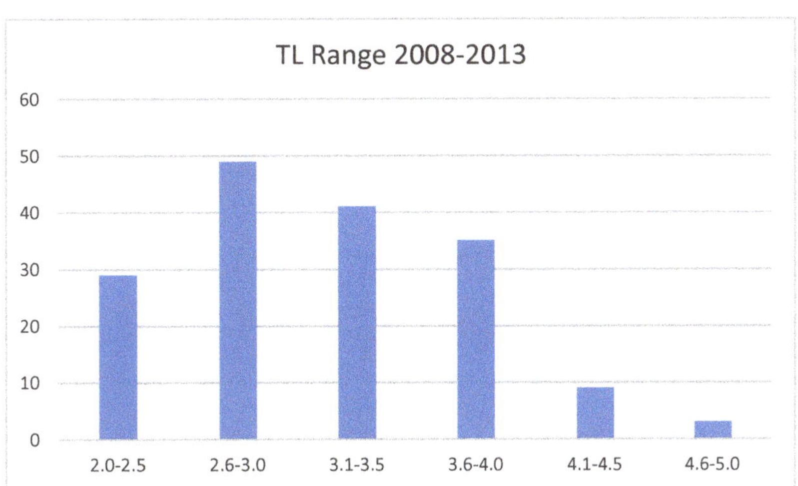

Figure 4. Total length range (meters) of white sharks, between 2008 and 2013.

5. Conclusions

It is possible to confirm, based on the data collected in the expeditions between 2008 and 2013, that the Gansbaai white shark transient population shows food preferences that are similar to those already observed in adult specimens in similar studies carried out along the Australian and Californian coasts [14,32]. The presence of a white shark population showing a mean TL of 308 cm, and only six specimens out of 250 showing a TL <200 cm, implies that for the majority of the Gansbaai white shark population, the change in diet has already occurred or is in progress and, therefore, the interest in energetically richer prey can be important for the overall population. Gansbaai, therefore, could represent a hunting training area for white sharks who are changing or have already changed their diet. It can be assumed that white sharks are initially attracted by the olfactory trace [5] determined by the bait and then implement their predatory choices to energetical richer prey, especially thanks to their visual ability, which plays an important role in adults and immatures with dietary shifts in their feeding patterns. The behaviour shown in the presence of the two preys examined in this work is, therefore, linked to the different dietary needs of white sharks in different stages of development. These needs are linked both to the modification of the dental system and to the different energy necessary in the immature and adult stages. As demonstrated by Laroche et al. [34] and Sperone et al. [24], the results presented in this article are based on the assumption that white sharks do not respond to the presence of the boat, but focus their attention mainly on the floating object, because the presence of a boat has very little influence on the white sharks' behaviour.

Author Contributions: Conceptualization, P.M.; methodology, P.M.; software, A.P.; validation, P.M., F.R.R., F.C. and A.P.; formal analysis, A.P., P.M. and F.R.R.; investigation, P.M., F.R.R., E.R. and F.C.; writing—original draft preparation, P.M.; writing—review and editing, P.M., F.R.R., F.C. and A.P.; funding acquisition, P.M. All authors have read and agreed to the published version of the manuscript.

Funding: This research was funded by the Sharks Studies Center—Scientific Institute, University of Calabria and University of Siena.

Institutional Review Board Statement: The study and experimentation protocols were reviewed and approved in accordance 166 with the Directive 2010/63/EU.

Data Availability Statement: https://www.researchgate.net/project/Great-White-Shark-Carcharodon-carcharias-Behaviour-Ecology-and-cotoxicology/update/61312b952897145fbd6df39c.

Acknowledgments: This research was made possible by the financial support of the Sharks Studies Center—Scientific Institute, University of Calabria and University of Siena. We are grateful to the CSS team members that carried out 11 expeditions for their indirect financial support of this research, and thanks are also due to Shark Diving Unlimited for the logistical assistance and all other field assistance with data collection. All research protocols were approved by the University of Calabria and Siena.

Conflicts of Interest: The authors declare no conflict of interest.

References

1. Linnaeus, C. *Systema Naturae per Regna Tria Naturae, Secundum Classes, Ordines, Genera, Species, Cum Characteribus, Differentiis, Synonymis, Locis, Tomus 1*, 10th ed.; Holimae: Laurentii, Salvii, Stockholm, Sweden, 1758; p. 823.
2. Compagno, L.J.V. *Sharks of the World: An Annotated and Illustrated Catalogue of Shark Species Known to Date, Volume 2: Bullhead, Mackerel, and Carpet Sharks (Heterodontiformes, Lamniformes and Orectolobiformes)*; FAO: Rome, Italy, 2001; p. 269.
3. Martin, A.R.; Hammerschlag, N.; Collier, R.S.; Fallows, C. Predatory behaviour of white sharks (Carcharodon carcharias) at Seal Island, South Africa. *J. Mar. Biol. Assoc. UK* **2005**, *85*, 1121–1135. [CrossRef]
4. Hammerschlag, N.; Martin, R.A.; Fallows, C. Effects of environmental conditions on predator-prey interactions between white sharks (Carcharodon carcharias) and Cape fur seals (Arctocephalus pusillus pusillus) at Seal island, South Africa. *Environ. Biol. Fishes* **2006**, *76*, 341–350. [CrossRef]
5. Compagno, L.J.V.; Dando, M.; Fowler, S. *A Field Guide to the Sharks of the World*; Harper Collins: London, UK, 2005; p. 368.
6. Ryan, L.A. A Shark Eye View: Motion Vision and Behaviour in Apex Predators. Ph.D. Thesis, School of Animal Biology and Ocean Institute, Western Australia University, Perth, Australia, 2016.
7. Hobson, E.S. Feeding behavior in three species of sharks. *Pac. Sci.* **1963**, *17*, 171–194.
8. Gilbert, P.W. *Studies on the Anatomy, Physiology, and Behavior of Sharks*; Office of Naval Research: Washington, VA, USA, 1970.
9. Strong, W.R. Shape discrimination and visual predatory tactics in white sharks. In *Great White Sharks: The Biology of Carcharodon Carcharias*; Klimley, P., Ainley, D.G., Eds.; Academic Press: New York, NY, USA, 1996; pp. 229–240.
10. Cliff, G.; Dudley, S.F.J.; Davis, B. Sharks caught in the protective gill nets off Natal, South Africa. 2. The great white shark Carcharodon carcharias (Linnaeus). *S. Afr. J. Mar. Sci.* **1989**, *8*, 131–144. [CrossRef]
11. Bruce, B.D. Preliminary observations on the biology of the white shark, Carcharodon carcharias, in south Australian waters. *Mar. Freshw. Res.* **1992**, *43*, 1–11. [CrossRef]
12. Estrada, J.A.; Rice, A.N.; Natanson, L.J.; Skomal, G.B. Use of isotopic analysis of vertebrae in reconstructing ontogenetic feeding ecology in white sharks. *Ecology* **2006**, *87*, 829–834. [CrossRef]
13. Hussey, N.E.; McCann, H.M.; Cliff, G.; Dudley, S.F.J.; Wintner, S.P.; Fisk, A.T. Size-based analysis of diet and trophic position of the white shark (Carcharodon carcharias) in South African waters. In *Global Perspectives on the Biology and Life History of the White Shark*; CRC Press: Florida, FL, USA, 2012; pp. 27–45.
14. Tricas, T.C.; McCosker, J.E. Predatory behaviour of the white shark (Carcharodon carcharias), with notes on its biology. *Proc. Calif. Acad. Sci.* **1984**, 221–238.
15. Klimley, A.P. The areal distribution and autoecology of the white shark, Carcharodon carcharias, off the west coast of North America. *South. Calif. Acad. Sci. Mem.* **1985**, *9*, 15–40.
16. Malcolm, H.; Bruce, B.D.; Stevens, J.D. *A Review of the Biology and Status of White Sharks in Australian Waters*; CSIRO Marine Research: Hobart, Tasmania, Australia, 2001.
17. Bruce, B.D.; Stevens, J.D.; Malcolm, H. Movements and swimming behaviour of white sharks (Carcharodon carcharias) in Australian waters. *Mar. Biol.* **2006**, *150*, 161–172. [CrossRef]
18. Martin, A.R. *Field Guide to the Great White Shark*; RCSR: Vancouver, BC, Canada, 2003.
19. Schreber, J.C.D. Die Saugethiere in abbildungen nach der nature mit beschreibungen. *Erlangen Wolfgang Walther* **1775**, *2*, 223–230.
20. French, G.C.A.; Rizzuto, S.; Sturup, M.; Inger, R.; Barker, S.; van Wyk, J.H.; Towner, A.V.; Hughes, W.O.H. Sex, size and isotopes: Cryptic trophic ecology of an apex predator, the white shark Carcharodon carcharias. *Mar. Biol.* **2018**, *165*, 1–11. [CrossRef] [PubMed]
21. Towner, A.V.; Leos-Barajas, V.; Langrock, R.; Schick, R.S.; Smale, M.J.; Kaschke, T.; Jewell, O.J.D.; Papastamatiou, Y.P. Sex-specific and individual preferences for hunting strategies in white sharks. *Funct. Ecol.* **2016**, *30*, 1397–1407. [CrossRef]
22. Micarelli, P.; Spinetti, S.; Andreotti, S.; Muià, C.; Vitaro, M.; Leone, A.; Camigliano, G.; Antonucci, G.; Sperone, E.; Tripepi, S. Is it good-looking or does it smell good? Preliminary observations about great white shark discriminatory patterns. In Proceedings of the 13th European Elasmobranch Association, Palma de Mallorca, Palma, Spain, 19–22 November 2009; p. 81.

23. Micarelli, P.; Spinetti, S.; Tripepi, S.; Sperone, E. Observations about surface behavior of the great white shark Carcharodon carcharias (L.) in presence of passive preys at Dyer Island (South Africa). *Biol. Mar. Mediterr.* **2006**, *13*, 278–279.

24. Sperone, E.; Micarelli, P.; Andreotti, S.; Brandmayr, P.; Bernabò, I.; Brunelli, E.; Tripepi, S. Surface behaviour of bait-attracted white sharks at Dyer Island (South Africa). *Mar. Biol. Res.* **2012**, *8*, 982–991. [CrossRef]

25. Sperone, E.; Micarelli, P.; Andreotti, S.; Spinetti, S.; Andreani, A.; Serena, F.; Brunelli, E.; Tripepi, S. Social interactions among bait-attracted white sharks at Dyer Island (South Africa). *Mar. Biol. Res.* **2010**, *6*, 408–414. [CrossRef]

26. Tricas, T.C. Feeding ethology of the white shark Carcharodon carcharias. *Mem. South. Calif. Acad. Sci.* **1985**, *9*, 81–91.

27. Myrberg, A.A.; Nelson, D.R. The behavior of sharks: What have we learned. In *Discovering Sharks*; Gruber, S.H., Ed.; American Littoral Society: Highlands, NJ, USA, 1991; pp. 92–100.

28. Sperone, E.; Circosta, A.L.; Giglio, G.; Tripepi, S.; Micarelli, P. Preliminary observations on Great White Shark (Carcharodon carcharias) surface predatory activity. *Biol. Mar. Mediterr.* **2015**, *22*, 184.

29. Micarelli, P.; Bonsignori, D.; Compagno, L.J.V.; Pacifico, A.; Romano, C.; Reinero, F.R. Analysis of sightings of white sharks in Gansbaai (South Africa). *Eur. Zool. J.* **2021**, *88*, 363–374. [CrossRef]

30. Kock, A.A.; Johnson, R.L. White shark abundance: Not a causative factor in numbers of shark bite incidences. In *Finding a Balance: White Shark Conservation and Recreational Safety in Inshore Waters of Cape Town, South Africa: Proceedings of a Specialist Workshop*; Nel, D.C., Peschak, T.P., Eds.; WWF: Washington, DC, USA, 2006; pp. 1–19.

31. Hewitt, A.M.; Kock, A.A.; Booth, A.J.; Griffiths, C.L. Trends in sightings and population structure of white sharks, Carcharodon carcharias, at Seal Island, False Bay, South Africa, and the emigration of subadult female sharks approaching maturity. *Environ. Biol. Fishes* **2017**, *101*, 39–54. [CrossRef]

32. Anderson, S.D.; Henderson, R.P.; Pyle, P. White shark reaction to unbaited decoys. In *Great White Sharks: The Biology of Carcharodon Carcharias*; Klimley, A.P., Ainley, D.G., Eds.; Academic Press: San Diego, SD, USA, 1996; pp. 223–228.

33. Micarelli, P.; Sperone, E.; Giglio, G.; Pecchia, J.; Romano, C.; Scuderi, A.; Vespaziani, L.; Mele, F. Dorsal fin photoidentification: Tool for long term studies of White shark (Carcharodon carcharias) behavior. *Biol. Mar. Mediterr.* **2015**, *22*, 144–145.

34. Laroche, R.; Kock, A.A.; Dill, L.M.; Oosthuizen, W. Effects of provisioning ecotourism activity on the behaviour of white sharks, Carcharodon carcharias. *Mar. Ecol. Prog. Ser.* **2007**, *338*, 199–209. [CrossRef]

35. Ferreira, C.A.; Ferreira, T.P. Population dynamics of white sharks in South Africa. In *Great White Sharks: The Biology of Carcharodon Carcharias*; Klimley, A.P., Ainley, D.G., Eds.; Academic Press: San Diego, SD, USA, 1996; pp. 381–391.

36. Strong, W.R.; Murphy, R.C.; Bruce, B.D.; Nelson, D.R. Movements and associated observations of bait-attracted white sharks, Carcharodon carcharias: A preliminary report. *Aust. J. Mar. Freshw. Res.* **1992**, *43*, 13–20. [CrossRef]

37. Gerking, S.D. *Feeding Ecology of Fish*; Elsevier: Amsterdam, The Netherlands, 2014; p. 416.

38. Helfman, G.; Collette, B.B.; Facey, D.E.; Bowen, B.W. *The Diversity of Fishes: Biology, Evolution, and Ecology*; John Wiley and Sons: Hoboken, NJ, USA, 2009; p. 736.

39. Sperone, E.; Micarelli, P.; Romano, C.; Giglio, G.; Giovannelli, P.; Reinero, F.R.; Rijillo, G.; Leporati, L.; Fabietti, Y.; Chiriaco, S. Biological and ecological factors affecting individual surface behaviours of bait attracted White Sharks. In Proceedings of the 3rd Sharks International Conference, João Pessoa, Brazil, 3–8 June 2018.

A Taxonomic Survey of Female Oviducal Glands in Chondrichthyes: A Comparative Overview of Microanatomy in the Two Reproductive Modes

Martina Francesca Marongiu [1,2,*,†], Cristina Porcu [1,2,*,†], Noemi Pascale [1], Andrea Bellodi [1,2], Alessandro Cau [1,2], Antonello Mulas [1,2], Paola Pesci [1,2], Riccardo Porceddu [1] and Maria Cristina Follesa [1,2]

[1] Dipartimento di Scienze della Vita e dell'Ambiente, Università degli Studi di Cagliari, Via Tommaso Fiorelli 1, 09126 Cagliari, Italy; pascalenoemi3@gmail.com (N.P.); abellodi@unica.it (A.B.); alessandrocau@unica.it (A.C.); amulas@unica.it (A.M.); ppesci@unica.it (P.P.); riccardo.porceddu@unica.it (R.P.); follesac@unica.it (M.C.F.)

[2] Consorzio Nazionale Interuniversitario per le Scienze Mare (CoNISMa), Piazzale Flaminio 9, 00196 Roma, Italy

* Correspondence: mfmarongiu@unica.it (M.F.M.); cporcu@unica.it (C.P.)

† The authors contributed equally to this work.

Citation: Marongiu, M.F.; Porcu, C.; Pascale, N.; Bellodi, A.; Cau, A.; Mulas, A.; Pesci, P.; Porceddu, R.; Follesa, M.C. A Taxonomic Survey of Female Oviducal Glands in Chondrichthyes: A Comparative Overview of Microanatomy in the Two Reproductive Modes. *Animals* 2021, 11, 2653. https://doi.org/10.3390/ani11092653

Academic Editor: Michele Deflorio

Received: 2 August 2021
Accepted: 7 September 2021
Published: 9 September 2021

Publisher's Note: MDPI stays neutral with regard to jurisdictional claims in published maps and institutional affiliations.

Simple Summary: The oviducal gland (OG) is a specialized region of the reproductive female system in cartilaginous fish located in the anterior oviduct. Its biological importance is closely related to the reproductive modalities of these species, and its basic function is the production of the egg jellies, the tertiary envelope formation (egg case in oviparous and candle case in viviparous) and sperm storage. Since knowledge on the overall process of Chondrichthyes reproduction is still scarce, in this study we conducted morphological and morphometrical analysis on the OGs belonging to several cartilaginous fish displaying two different reproductive modalities (oviparity and viviparity). Moreover, we paid particular attention to the fate of spermatozoa in the female reproductive tract, which would be useful to better understand the ecology and population dynamics of these species.

Abstract: Oviducal glands (OGs) are distinct expanded regions of the anterior portion of the oviduct, commonly found in chondrichthyans, which play a key role in the production of the egg in-vestments and in the female sperm storage (FSS). The FSS phenomenon has implications for understanding the reproductive ecology and management of exploited populations, but little information is available on its taxonomic extent. For the first time, mature OGs from three lecithotrophic oviparous and four yolk-sac viviparous species, all considered at risk from the fishing impacts in the central western Mediterranean Sea, were examined using light microscopy. The OG microanatomy, whose morphology is generally conserved in all species, shows differences within the two reproductive modalities. Oviparous species show a more developed baffle zone in respect to viviparous ones because of the production of different egg envelopes produced. Among oviparous species, *Raja polystigma* and *Chimaera monstrosa* show presence of sperm, but not sperm storage as observed, instead, in *Galeus melastomus* and in all the viviparous sharks, which preserve sperm inside of specialized structures in the terminal zone.

Keywords: oviducal gland; histology; sperm storage; oviparity; viviparity

1. Introduction

Chondrichthyes, as well as being the only anamniote vertebrate class that exclusively employs internal fertilization during reproduction [1], have several features in common (low growth rate, delayed maturity and long gestation), all resulting in low reproductive potential that makes them highly vulnerable to overfishing (e.g., [2,3]). Chondrichthyan fish exhibit two main modes of reproduction, spanning both oviparity and viviparity. Oviparity (i.e., eggs are enclosed within an egg case and deposited in the sea) is restricted to the orders Carcharhiniformes, Heterodontiformes, Orectolobiformes, Rajiformes and Chimaeriformes, while viviparity is prevailing in all other chondrichthyan orders [4]. Viviparous species are further subcategorized as lecithotrophic and matrotrophic [5,6]. Depending on the pattern of embryonic nutrition, in the lecithotrophic mode, embryos do not receive any maternal nourishment, which is granted, instead, by a yolk-sac (yolk reserves); in the matrotrophic mode, embryos receive nourishment through ingestion of lipids or mucous (histotrophy) produced by the uterine walls of the mother [5,6]. Oophagy and intrauterine cannibalism (those that eat eggs or other embryos) are also enclosed in the matrotrophic reproductive mode [5,6].

The morphology of the female reproductive system is generally conserved, and the genital duct displays a variety of specializations linked to the different reproduction modes [7]. All Chondrichthyes show a glandular enlargement of the oviduct known as oviducal gland (OG) interposed between the oviduct and the uterus, except for the Narcinidae family in which the OG is lacking [8] and some species of Myliobatiformes which have a vestigial OG [9,10]. The OG is responsible mainly for producing secondary and tertiary egg coats [11], revealing four distinct secretory zones (e.g., [11–16]): club and papillary zones that produce mucus surrounding the fertilized ovum in the early stages of embryogenesis, the baffle zone that forms the tertiary envelope of the egg (egg case in oviparous and candles in viviparous species) and the terminal zone where the eggs' ornamentations are produced and the sperm can be stored (e.g., [16–21]). A small number of chondrichthyan species, such as the yellow stingray *Urobatis jamaicensis* (Cuvier 1816), do not produce an egg envelope, and the baffle zone is modified accordingly [17]. The size and structural complexity of the OG, as well as the reproductive relevance of each glandular region, are correlated with the mode of reproduction (e.g., [5,22–24]). The shape of the OG is also highly variable at the gross level, but as a general rule, oviparous species have the largest OGs among chondrichthyans, since they produce not only the jellies that envelop the egg, but also an external capsule and all its ornamentations (e.g., [15,25]). Conversely, viviparous sharks show a uniform breakdown of OG volume due to a small dimension of the baffle zone (in Triakidae members the OG is virtually indistinguishable) (e.g., [7]).

Female sperm storage is the extended maintenance of viable sperm prior to its use in fertilization and likely occurs in most internally fertilizing animals [26]. Female sperm storage has been described in nematodes, annelids, squids and arthropods and in all major groups of vertebrates, including amphibians, cartilaginous fishes (sharks and rays), bony fish, squamates, reptiles, birds and mammals [26]. Female sperm storage makes delayed fertilization possible and facilitates a temporal separation between copulation and fertilization, even if not all stored sperm is used for this purpose. It can be viewed as an evolutionarily conserved strategy to ensure reproductive efficiency and increase the probability of insemination of nomadic species or in those with low population density [27], impacting many aspects of their biology, including life histories, mating systems, cryptic female choice and sperm competition, and ultimately may lead to sexual conflict [26]. Sperm storage has been found in the terminal zone of multiple chondrichthyans, mainly sharks and holocephalans (e.g., [7,16,18–20,28]).

Of the 88 chondrichthyan species recorded in the Mediterranean Sea [29], more than half (at least 53%; [30]) are threatened because of overfishing and are classified by the International Union for Conservation of Nature classification as vulnerable, endangered or critically endangered [30]. In this sense, understanding the overall process of reproduction would be useful for assessing the population status of these species [31] by investigating male–female interactions, physiology, biochemistry and anatomy [7,18], thus providing scientifically reliable advice for understanding their ecology and population dynamics [7,18]. Although chondrichthyans tend to be apex predators and fulfill many key roles in their ecosystems [32], there has been little global research on basic biology and population composition for many species [33].

For these reasons, the goal of this study is to provide a broader taxonomic survey of female oviducal glands across different reproductive modes present in some demersal chondrichthyans inhabiting the Mediterranean Sea. More specifically, we investigated (i) the microanatomy and the carbohydrate distribution pattern in the different zones of the OGs through morphological, histological and histochemical observations for three oviparous species and four yolk-sac viviparous ones and (ii) sperm storage and sperm distribution in the two reproductive modes.

2. Materials and Methods

2.1. Sampling

Female specimens of seven chondrichthyan species were collected between 2012 and 2020 around Sardinian waters (central western Mediterranean) during the Mediterranean International Trawl Survey (MEDITS; [34]) and commercial hauls along with data collected monthly from commercial landings through the Data Collection Framework (European Union Regulation 199/2008). Collection and handling of animals took into account the ethical and welfare considerations approved by the ethics committee of the University of Cagliari (Sardinia, Italy).

For each individual, the total length (TL, in centimeters) and the oviducal gland width (OGW, in millimeters considering the maximum breadth) were recorded. Only for the holocephalan *Chimaera monstrosa*, the anal length (AL, in centimeters) was taken. Maturity stages were established following the scale for oviparous and viviparous elasmobranchs [35]. According to these scales, oviparous females were classified into six stages on the basis of ovary structure and OG and uterus dimension and texture as follows: stage 1, immature (F1); stage 2, developing (F2); stage 3a, spawning capable (F3A); stage 3b, actively spawning (F3B); stage 4a, regressing (F4A); and stage 4b, regenerating (F4B). Instead, viviparous females were classified into seven stages: stage 1, immature (F1); stage 2, developing (F2); stage 3a, capable of reproducing (F3A); stage 3b, early pregnancy (F3B); stage 3c, mid-pregnancy (F3C); stage 3d, late pregnancy (F3D); stage 4a, regressing (F4A); and stage 4b, regenerating (F4B).

2.2. Histological Procedures

For the histological analysis and, in particular for the analysis regarding the sperm storage, were used only mature OGs fully developed from stage 3a to 4b. A subsample of five OGs for each maturity stage (when available) was processed for histological analysis. Whole OGs were first fixed in 5% buffered formaldehyde (0.1 mol L^{-1}, pH 7.4) for a maximum period of 48 h. They were then dissected by cutting through the center from the oviduct to the uterus, which provided a piece of tissue representing the sagittal plane of the gland. The tissues were embedded in a synthetic resin (GMA, Technovit 7100, Bio-Optica, Milan, Italy) following routine protocols and sectioned at 3.5 µm with a rotating microtome (ARM3750, Histo-Line Laboratories, Pantigliate, Italy). Slides were stained with hematoxylin and eosin (H&E) for standard histology and with periodic acid–Schiff (PAS) and Alcian blue (AB) in combination to assess the production of neutral and sulfated acid mucins [36]. Subsequently, sections were dehydrated in graded ethanol (96–100%), cleared in Histolemon (Carlo Erba Reagents, Cornaredo, Italy) and mounted in resin (Eukitt, Bio-Optica). Selected sections were observed and photographed using a Nexcope NE600 optical microscope equipped with a digital camera (MD6iS) at different magnifications (40×, 100× and 400×) and edited with Adobe Photoshop CS6 /(http://www.adobe.com/products/photoshop.html (22 July 2021)).

2.3. Macroscopic and Microscopic Measurements and Statistical Analysis

Analysis of variance (ANOVA, multiple range test) was used to test for statistical differences in OG width among all the maturity stages [37] using the software Stat-graphics Centurion XVI to test differences between the several reproductive cycle phases.

To evaluate the surface (expressed in mm^2 and in %) occupied by each OG zone (club, papillary, baffle and terminal), ImageJ software [38] was used. Each zone was measured considering the entire area covered by tubules and lamellae, excluding the connective tissue that borders the OG (Figure S1).

TpsDig software [39] was used to collect the detailed measurements of the OG and check differences between the four zones. The dimension of gland tubules was evaluated measuring their maximum length (Figure S1). Lamellae from each zone were counted and measured starting from the base towards the lumen (Figure S1).

3. Results

In Table 1, the analyzed species, grouped based upon their mode of reproduction (oviparous and yolk-sac viviparous), are presented.

Table 1. List of the selected chondrichthyan species. Reproductive mode, depth, size ranges and IUCN status in the Mediterranean are presented.

Reproduction Mode	Order	Family	Species	Depth Range (m)	Size Range (cm)	IUCN Status
Oviparity	Carcharhiniformes	Pentanchidae	*Galeus melastomus* (Rafinesque, 1810)	100–730	6.9–89.7 [†]	LC
	Chimaeriformes	Chimaeridae	*Chimaera monstrosa* (Linnaeus, 1758)	320–1132	3.6–27.8 *	NT
	Rajiformes	Rajidae	*Raja polystigma* (Regan, 1923)	33–600	15.6–59.3 [†]	LC
Viviparity (yolk-sac)	Hexanchiformes	Hexanchidae	*Heptranchias perlo* (Bonnaterre, 1788)	336–600	54.5–113.3 [†]	DD
	Squaliformes	Centrophoridae	*Centrophorus uyato* (Rafinesque, 1810)	445–600	38.7–106.5 [†]	CR
		Dalatidae	*Dalatias licha* (Bonnaterre, 1788)	300–800	26.2–108.6 [†]	VU
		Oxynotidae	*Oxynotus centrina* (Linnaeus, 1758)	175–600	25.6–78.2 [†]	CR

CR, critically endangered; DD, data deficient; LC, least concern; NT, near threatened; VU, vulnerable. * AL, anal length. [†] TL, total length.

3.1. Macroscopic Development of the OG

Comparing the morphology of the oviducal glands of the oviparous species, they showed different shapes. *C. monstrosa* and *G. melastomus* glands had an oval shape (Figure S2), while *R. polystigma* glands had a heart shape (Figure S2).

Considering the maturation cycle of each species, the OGs become larger when females attained maturity, reaching the maximum values at stages 3A and 3B with a slight decline in the postspawning phase in all examined species (Figure 1). Statistically significant differences during the evolution of the OGs of all species were found (*C. monstrosa*, ANOVA, F-ratio = 167.40, statistically significant differences between all stages except stages 3A–3B, *p*-value = 0; *G. melastomus*, ANOVA, F-ratio = 97.40, statistically significant differences between all stages except stages 3A–4A, 3A–4B, 4A–4B, *p*-value = 0; *R. polystigma*, ANOVA, F-ratio = 379.48, statistically significant differences between all stages except stages 3B–4A, *p*-value = 0).

The oviducal glands of all viviparous species had a barrel shape with considerable lateral extension visible (Figure S3). In *O. centrina*, a dark stripe in the external part of the gland was visible (Figure S3D). The OG of *C. uyato*, *D. licha* and *O. centrina* reached the maximum development in capable of reproducing females (stage 3A) with a subsequent decline in the maternal stages (Figure 1, ANOVA, *C. uyato*, F-ratio = 1340.41, statistically significant differences between all stages, *p*-value = 0; *D. licha*, F-ratio = 60.26, statistically significant differences between all stages, *p*-value = 0; *O. centrina*, F-ratio = 13.68, statistically significant differences between all stages except stages 3B–3C, 3C–3D, 4A–4B, *p*-value = 0.0001). Only few specimens of *H. perlo* were caught during the sampling period, and for this reason, it was not possible to perform a statistical analysis.

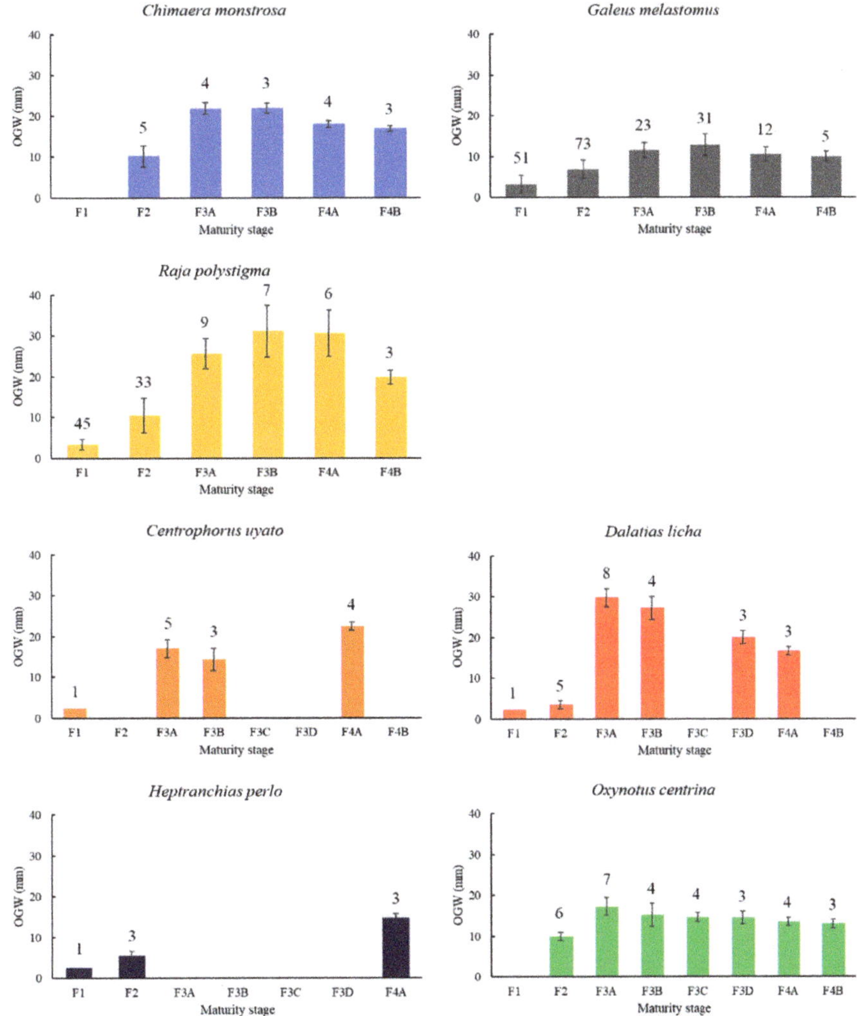

Figure 1. Changes of the oviducal gland width (OGW) through maturity stages in oviparous (*C. monstrosa*, *G. melastomus*, *R. polystigma*) and viviparous (*C. uyato*, *D. licha*, *H. perlo*, *O. centrina*) species. Above the bars, the number of the OGs analyzed is reported.

3.2. Microscopic Development of the OG

3.2.1. Oviparous Species

Chimaera monstrosa

On the basis of AB-PAS staining, each different zone displayed unique staining affinities (Figure 2A).

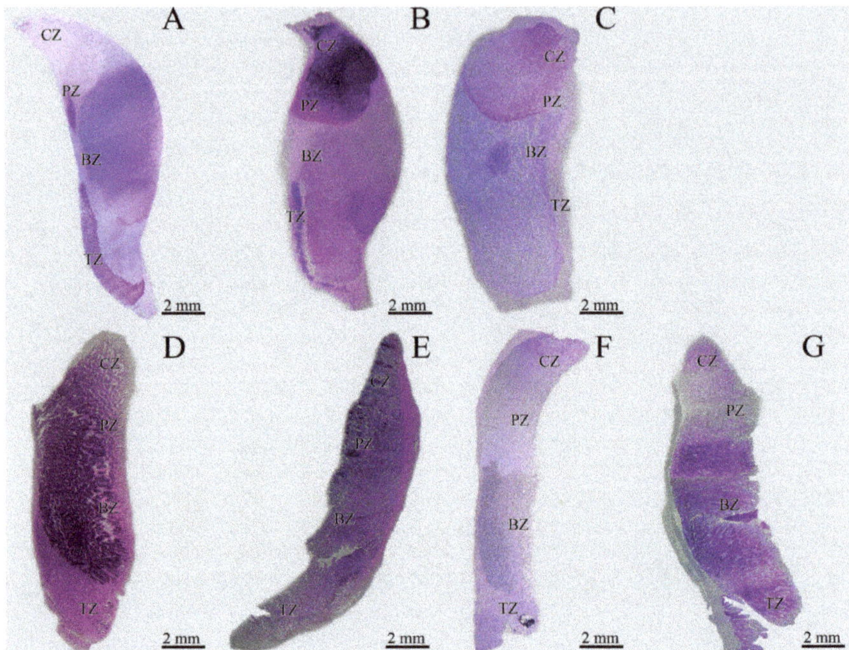

Figure 2. Sagittal section of the oviducal gland (OG) belonging to oviparous and viviparous species. (**A**) *C. monstrosa*; (**B**) *G. melastomus*; (**C**) *R. polystigma*; (**D**) *C. uyato*; (**E**) *D. licha*; (**F**) *H. perlo*; (**G**) *O. centrina*. BZ, baffle zone; CZ, club zone; PZ, papillary zone; TZ, terminal zone.

The club zone (Table 2) showed the club-shaped lamellae (Table 3), characterized by a simple, ciliated and columnar epithelium (Figure 3A). Tubules were very elongated near the lumen and were composed of ciliated columnar cells with basal nuclei and secretory cells containing a mixture of AB+ and PAS+ secretory material (Figure 3B, Table 4). The papillary zone, which occupied a similar surface to the club zone (Table 2), showed sever-al digitiform lamellae (Table 3; Figure 3C). As with the club zone, the tubular glands were elongated near the lamellae (Figure 3C), while in the deep recesses, they were more rounded and showed less affinity for dyes (Figure 3D). Secretory cells of the caudal-most papillary tubules (Figure 3E), adjacent to the baffle zone, produced a less AB+/PAS+ secretory material than the papillary zone. The baffle zone, the most conspicuous part of the OG (Table 2), consisted of 34 apical expanded lamellae (Table 3) that alternated with transversal grooves (Figure 3F). The glandular duct ended in a pair of spinnerets, then passing to the transverse grooves (Figure 3F). Tubular glands, PAS/AB-negative (Figure 3G), had cells with round cytoplasmatic granules, basal nucleus and evident nucleolus. The terminal zone was the smallest part of the gland (Table 2), and it had no lamellae but had crypts where the gland ended (Figure 3H). In this zone, the tubular glands (Table 3) mainly had PAS-positive mucous cells (Figure 3I). Accumulation of secretions in the lumen of secretory glands and between lamellae was observed in all maturity stages (from mature to regenerating females), being more evident in the baffle and terminal zones.

Table 2. Oviducal gland area expressed in mm^2 and % of the total gland for the four zones in the oviparous (*C. monstrosa*, *G. melastomus*, *R. polystigma*) and viviparous (*C. uyato*, *D. licha*, *H. perlo*, *O. centrina*) species examined.

Species	Club		Papillary		Baffle		Terminal	
	mm^2	%	mm^2	%	mm^2	%	mm^2	%
C. monstrosa	12.8–13.9	13.2–13.6	13.0–14.2	13.3–13.8	63.0–67.2	63.1–66.4	8.5–9.3	8.2–9.0
G. melastomus	10.1–11.2	18.0–19.5	4.9–6.3	9.9–10.8	34.7–35.9	60.1–65.0	4.1–4.9	7.8–8.5
R. polystigma	7.2–8.4	15.0–16.2	7.3–8.2	15.0–16.2	28.6–34.3	61.1–66.7	1.9–2.3	4.2–4.8
C. uyato	18.2–21.9	20.6–220	23.5	23.2	38.2	37.7	18.1	17.9
D. licha	3.8–5.1	10.6–11.9	9.6–10.7	24.5–26.0	21.5–22.9	55.1–56.9	1.9–2.9	5.8–6.9
H. perlo	14.9	27.8	12.4	23.1	20.7	38.5	5.6	10.6
O. centrina	9.9–11.0	13.2–15.1	8.6–10.4	12.2–14.0	35.4–36.9	48.5–50.8	14.4–15.7	20.4–21.9

Table 3. Number, minimum, maximum, mean length (μm) and standard deviation (S.D.) of lamellae belonging to different OG zones in oviparous (*C. monstrosa*, *G. melastomus*, *R. polystigma*) and viviparous (*C. uyato*, *D. licha*, *H. perlo*, *O. centrina*) species.

Species	Club			Papillary			Baffle		
	N	Mean ± S.D.	Min–Max	N	Mean ± S.D.	Min–Max	N	Mean ± S.D.	Min–Max
C. monstrosa	37	94.3 ± 45.5	56.3–235.4	33	119.6 ± 19.9	89.5–169.5	34	407.7 ± 88.0	264.5–537.8
G. melastomus	15	177.7 ± 34.4	128.9–212.7	8	108.8 ± 22.3	63.3–141.2	28	506.6 ± 29.5	476.3–550.9
R. polistygma	14	296.9 ± 165.3	101.4–786.9	12	252.8 ± 59.7	158.5–373.1	22	584.3 ± 105.5	399.4–792.4
C. uyato	31	857.9 ± 69.0	725.7–982.2	14	828.8 ± 207.6	503.7–1407.1	43	304.6 ± 151.7	136.8–572.4
D. licha	17	1555.5 ± 233.2	1046.6–1701.0	26	1804.4 ± 38.2	1756.2–1892.3	56	1013.0 ± 452.4	191.4–1962.8
H. perlo	19	547.7 ± 174.5	193.2–803.0	21	442.6 ± 88.9	307.8–575.6	22	982.8 ± 327.9	534.7–1459.2
O. centrina	16	532.4 ± 92.5	368.5–617.4	16	695.8 ± 180.4	519.0–1167.4	44	888.0 ± 200.7	600.0–1239.6

Table 4. Minimum, maximum, mean length (μm) and standard deviation (S.D.) of secretory tubules belonging to different OG zones in oviparous (*C. monstrosa*, *G. melastomus*, *R. polystigma*) and viviparous (*C. uyato*, *D. licha*, *H. perlo*, *O. centrina*) species.

Species	Club		Papillary		Baffle		Terminal	
	Min–Max	Mean ± S.D.	Min–Max	Mean ± S.D.	Min–Max	Mean ± S.D.	Min–Max	Mean ± S.D.
C. monstrosa	71.0–354.5	193.6 ± 67.8	82.3–566.6	228.5 ± 86.3	103.8–443.6	191.5 ± 60.3	89.0–614.2	267.9 ± 115.5
G. melastomus	65.1–332.4	132.0 ± 82.7	126.1–353.9	175.4 ± 46.4	174.0–429.1	262.4 ± 89.0	39.3–140.8	73.9 ± 22.1
R. polistygma	64.9–408.2	179.1 ± 75.1	64.4–483.0	169.9 ± 85.1	101.8–768.4	223.8 ± 106.9	52.0–428.6	119.0 ± 84.4
C. uyato	136.9–509.8	279.8 ± 70.8	140.7–1022.5	353.0 ± 122.9	190.8–1027.3	513.5 ± 179.8	84.2–1141.9	174.5 ± 166.2
D. licha	58.1–568.8	194.7 ± 108.7	89.4–683.8	269.2 ± 137.3	59.7–539.7	198.7 ± 84.7	63.0–434.1	176.7 ± 88.8
H. perlo	43.1–496.5	161.7	58.0–335.5	117.5 ± 41.7	60.2–343.1	128.9 ± 60.4	30.9–198.1	71.59 ± 36.0
O. centrina	173.3–1133.6	432.9 ± 195.9	222.3–971.0	444.0 ± 177.9	143.6–1230.9	494.6 ± 264.4	281.9–869.6	461.8 ± 130.0

Galeus melastomus

The blackmouth catshark OG displayed a similar structure compared to *C. monstrosa* (Figure 2B). The club zone presented very developed lamellae in width rather than in length (Figure 4A, Table 3). The granules of the secretory cells were PAS+ and AB+ (neutral and sulfated acid mucins) (Figure 4B, Table 4). The papillary zone, smaller than the club zone (Table 2) and characterized by few lamellae (Table 3, Figure 4C), showed an affinity only for PAS (Figure 4D). Tubules, surrounded by a basement membrane, are simple, and secretory materials produced inside are PAS+ (Figure 4D), identifying the production of neutral mucopolysaccharides. Differently from *C. monstrosa*, *G. melastomus* did not show the caudal-most papillary zone. The baffle zone is very large (Table 2) causing the gland to be much larger, with an increased luminal profile (Figure 4E, Table 3). The secretory ducts open into the lumen through the spinneret region, composed of two baffle plates. The baffle plates are surrounded by another pair of large folds, the plateau projections, lining the transverse groove (Figure 4E). Secretory tubules did not react to any of the special staining techniques (PAS-, AB-) (Figure 4F, Table 4), while the cells near the lamellae had a weak affinity for PAS (PAS+). The structural organization of the terminal zone (Table 2) displayed no lamellae and a regular surface epithelium with elongated tubular glands opening into the lumen (Figure 4G). The elongated secretory tubules (Table 4) consisted mainly of mucous secretions with large vesicle cells intensely blue (AB+) (Figure 4H).

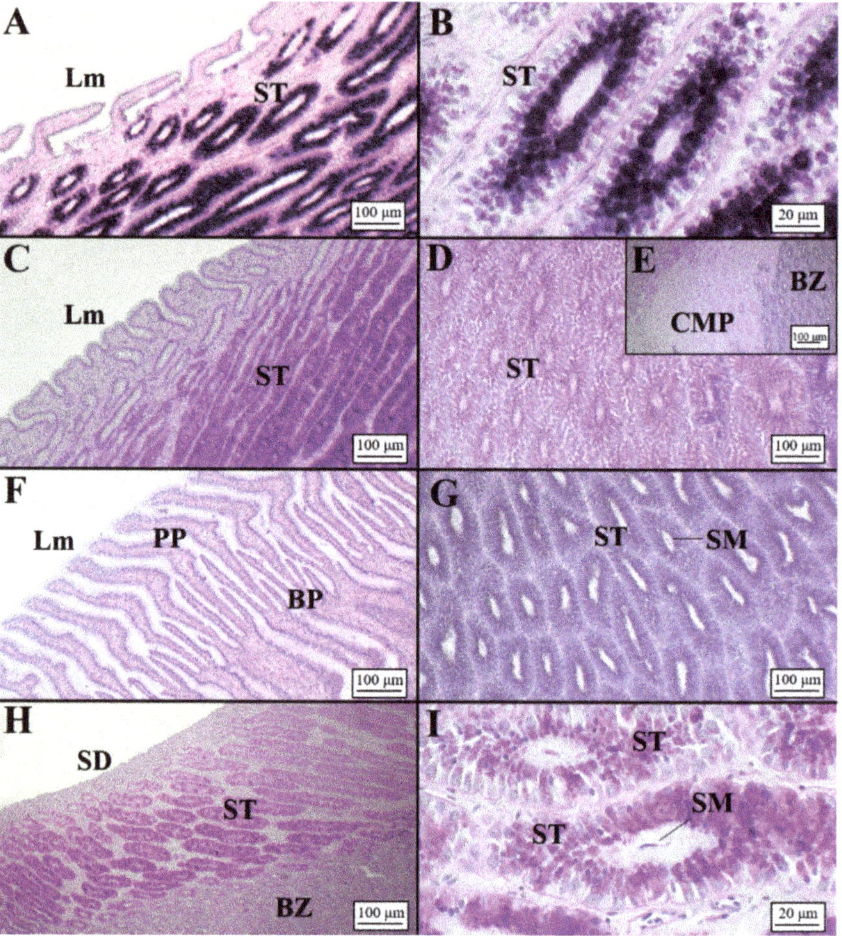

Figure 3. Microarchitecture of a *C. monstrosa* mature female (AB/PAS staining method applied to all slides). (**A**) Club-shaped lamellae and secretory tubules of the club zone. (**B**) High magnification of the club tubules having secretory granules stained AB+/PAS+ (deep purple color). (**C**) Digitiform lamellae of the papillary zone and secretory tubules near them showing a PAS+ affinity. (**D**) Secretory tubules of the papillary zone in the deep recesses: tubules are rounded showing a light affinity for PAS. (**E**) Transitional zone between papillary and baffle zone known as caudal-most papillary in which it is possible to observe a slight affinity only for PAS. (**F**) Baffle lamellae constituted by the typical plateau projections and baffle plates. (**G**) High magnification of the baffle zone tubules with secretory material distinguishable inside them and no affinity for dyes (AB-/PAS-). (**H**) Terminal zone showing secretory ducts opened towards the lumen. Tubules are elongated, showing an affinity for PAS (PAS+). (**I**) High magnification of the terminal zone secretory tubules in which the secretory material is visible. BP, baffle plates; BZ, baffle zone; CMP, caudal-most papillary; Lm, lamellae; PP, plateau projections; SM, secretory material; SD, secretory ducts; ST, secretory tubules.

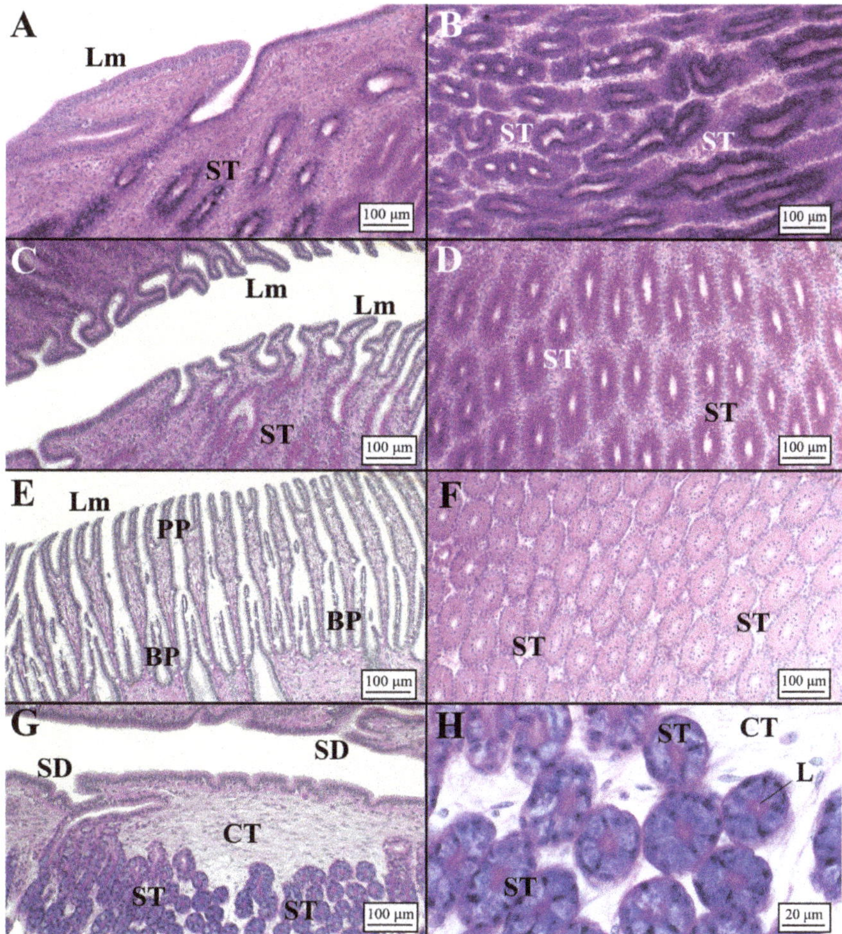

Figure 4. Microarchitecture of a *G. melastomus* mature female (AB/PAS staining method applied to all slides). (**A**) Club-shaped lamellae and secretory tubules of the club zone. (**B**) High magnification of the club tubules having secretory granules stained in AB+/PAS+ (purple color). (**C**) Digitiform lamellae of the papillary zone and secretory tubules near them showing a PAS+ affinity. (**D**) Secretory tubules of the papillary zone showing affinity for PAS. (**E**) Baffle lamellae constituted by the typical plateau projections and baffle plates. (**F**) High magnification of the baffle zone tubules showing no affinity for dyes (AB-/PAS-). (**G**) Terminal zone showing secretory ducts opened towards the lumen. Tubules are elongated, showing an affinity for acid mucins (AB+). (**H**) High magnification of the terminal zone secretory tubules showing an intense affinity for Alcian blue (AB+). BP, baffle plates; CT, connective tissue; L, lumen; Lm, lamellae; PP, plateau projections; SM, secretory material; SD, secretory ducts; ST, secretory tubules.

Raja polystigma

In females spawning capable of and actively spawning, the OG was fully developed with the four zones clearly recognizable (Figure 2C), enclosed by a serosa of dense connective tissue. The luminal surface of the lamellae of all zones was composed of a simple ciliated columnar epithelium supported by loose connective tissue (lamina propria). The club zone (Table 2) had the characteristic club-shaped lamellae (Figure 5A, Table 3), and the glandular epithelium tubules (Figure 5B, Table 4) were composed of ciliated and secretory cells. CZ reacted strongly to PAS/AB in the area near the lamellae (deep purple color) and slightly less in the deep recesses (purple color). The papillary zone, recognizable by the digit-shaped lamellae (Figure 5C, Table 3), had simple or occasionally ramified tubules and secretory tubules producing mucins PAS+/AB+ (Figure 5D, Table 4). Adjacent to the baffle zone, a row of secretory tubules (caudal-most papillary zone) (Figure 5E) reacted slightly less than the papillary zone. The BZ, the most conspicuous and extensive segment (Table 2), did not react to the PAS/AB stains. It contained the characteristic lamellae (Figure 5F, Table 3) called "baffle plates" in the spinneret region and "plateau projections" extending toward the lumen; each lamella was surrounded by two short epithelial folds (spinnerets) that had two ciliated baffle plates. This region was composed of secretory and ciliated cells, and the presence of blood vessels, between tubules (Figure 5G, Table 4), was consistent, especially in the distal part near the smooth muscle tissue. The brown material was detected only between the secretory tubules of the baffle zone (Figure 5G). Finally, the terminal zone (Table 2) did not contain real lamellae (Figure 5H,I), and it consisted of serous and mucous tubules. The luminal tubular glands were a mixture of serous tubules (PAS-/AB-) with small granules and mucous tubules (AB+) with large vesicle cells (Figure 5H,I). Some gland tubules had a mixture of the two types of mucins, showing an apical region stained in slight blue (AB+), while the remaining were serous and similar to those of the baffle zone (Figure 5H,I).

Presence of Sperm

Sperm was detected in all examined species (Figure 6). In particular, unaggregated sperm was principally found within the lamellae (Figure 6A,D,G) and tubules (Figure 6B,E,H) of the baffle zone. Aggregated sperm bundles were found within an invagination between the terminal zone and the distal oviduct only in *C. monstrosa* (Figure 6C).

In *G. melastomus*, instead, bundles of sperm involved in acid mucopolysaccharides (AB+) were observed in the secretory tubules of the terminal zone (Figure 6F).

3.2.2. Viviparous Species
Centrophorus uyato

In this species, it is possible to discern the club, papillary and baffle zones mainly on the basis of the luminal profile (Figure 2D). The club zone (Table 2) presented the typical club-shaped lamellae (Figure 7A, Table 3), and the glandular tubules, rounded in shape both near the lamellae and in the deep recesses of the gland (Table 4), strongly reacted to the AB/PAS staining, showing an intense purple coloration (Figure 7B). The papillary zone (AB+/PAS+), similar in extension to CZ (Table 2), possessed tubules (Table 4) showing different shapes: near the lamellae they were rounded (Table 4; Figure 7C), while towards the connective tissue they assumed a more elongated shape (Figure 7D). The baffle zone occupied the most part of the OG (Table 2) and showed very elongated lamellae (Figure 7E, Table 3) and tubules (Figure 7F, Table 4) reacting to the AB/PAS stain (AB+/PAS+). The terminal zone (Table 2) did not show very developed lamellae (Figure 7G, Table 3). It displayed several elongated tubules (Table 4) that showed affinity for PAS (PAS+, Figure 7H) and shorter ones (near the distal oviduct) with an affinity AB+/PAS+ (Figure 7G).

Dalatias licha

The OG of *D. licha* (Figure 2E) showed long lamellae in club, papillary and baffle zones (Figure 8A,C,E, Table 3); instead, the terminal zone possessed short lamellae (Figure 8G, Table 3). Club and papillary zones showed rounded and elongated tubules (Table 4), respectively, with a simple ciliated cuboidal epithelium (Figure 8B,D). More specifically, club zone tubular glands reacted slightly to PAS (PAS+) Figure 8D), while the papillary zone showed a strong affinity for neutral mucins (PAS+) (Figure 8B). In mature females, a great amount of secretory material was found in both the club and papillary zones (Figure 8B,D). The baffle and terminal zones (Table 2) showed secretory tubules with a simple ciliated columnar epithelium (Figure 8F,H). In particular, the baffle zone also showed secretory material within the tubules and the lamellae with PAS affinity (PAS+) (Figure 8E,F). The terminal zone showed few tubules (Table 4) in respect to the other zones (Figure 8H).

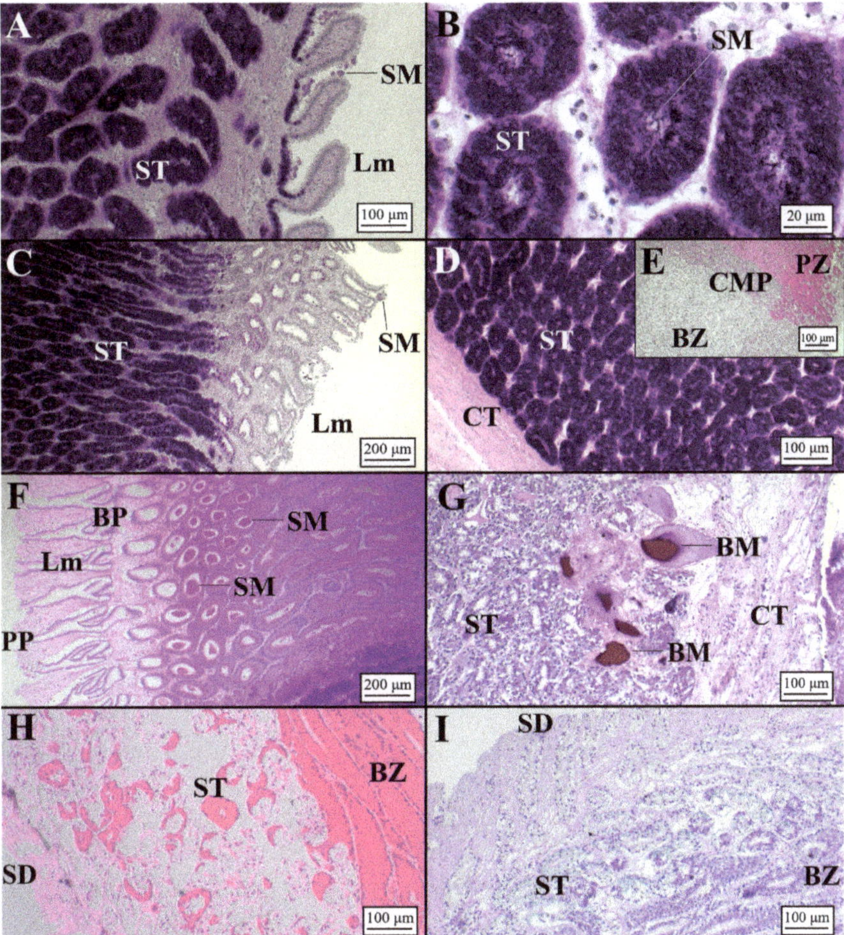

Figure 5. Microarchitecture of a *R. polystigma* mature female (AB/PAS staining method applied to all slides, except to (**E,H**) stained in (**H,E**)). (**A**) Club-shaped lamellae with secretory material found within them and secretory tubules of the club zone. (**B**) High magnification of the club tubules stained AB+/PAS+ (deep purple color) having secretory material inside them. (**C**) Digitiform lamellae of the papillary zone with secretory material within them and secretory tubules. (**D**) Secretory tubules of the papillary zone showing affinity for AB+/PAS+ (deep purple color). (**E**) Transitional zone between papillary and baffle zone (caudal-most papillary) in which it is possible to observe a slight affinity for dyes. (**F**) Baffle lamellae constituted by the typical plateau projections and baffle plates and tubules of the baffle zone containing a great amount of secretory material. (**G**) Deep recesses of the baffle zone near the connective tissue in which tubules do not show affinity for dyes (AB-/PAS-). Brown material was detected. (**H**) Terminal zone showing secretory ducts opened towards the lumen. Tubules are rounded, showing a mixture of mucins. (**I**) Terminal zone secretory tubules showing a slight affinity for Alcian blue (AB+) and also for PAS (PAS+). BM, brown material; BP, baffle plates; BZ, baffle zone; CMP, caudal-most papillary; CT, connective tissue; Lm, lamellae; PP, plateau projections; SM, secretory material; SD, secretory ducts; ST, secretory tubules.

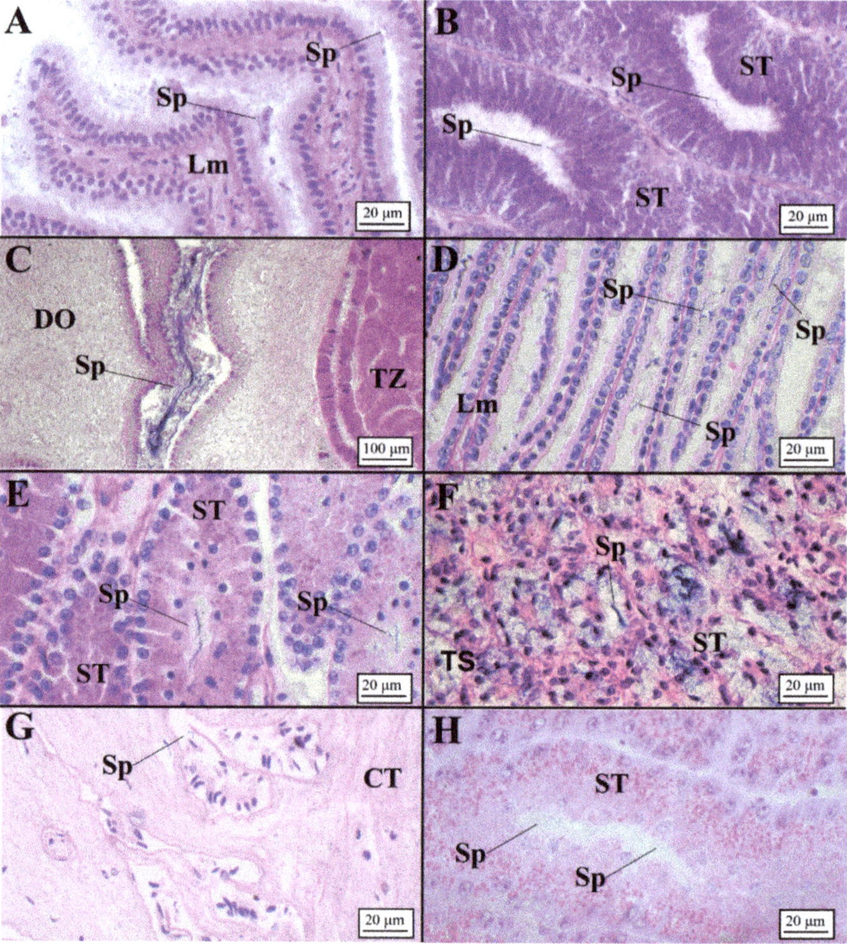

Figure 6. Presence of sperm in oviparous species (AB/PAS staining method applied to all slides). (**A,B**) Unaggregated sperm found between lamellae (**A**) and tubules (**B**) of the baffle zone in *C. monstrosa*. (**C**) Aggregated bundles of sperm found between the terminal zone and the distal oviduct in *C. monstrosa*. (**D,E**) Unaggregated sperm found between lamellae (**D**) and tubules (**E**) of the baffle zone in *G. melastomus*. (**F**) Bundles of sperm involved in AB+ matrix found in terminal zone tubules in *G. melastomus*. (**G,H**) Unaggregated sperm found between lamellae (**G**) and tubules (**H**) of the baffle zone in *R. polystigma*. CT, connective tissue; DO, distal oviduct; Lm, lamellae; Sp, sperm; ST, secretory tubules; TZ, terminal zone.

Heptranchias perlo

The only two OGs analyzed (belonging to regressing females) showed the typical zonation displayed in the other viviparous glands (Figure 2F) with club and papillary occupying 27.8% and 23.1% of the entire surface, respectively (Table 2). These zones did not show specifical affinity for dyes, but inside the secretory tubules (especially those near the connective tissue), secretory granules stained in magenta (PAS+) were detected (Figure 9B,D). Neither the baffle nor the terminal zone showed a reaction to the staining method (Figure 9E–H). The baffle zone (occupying 38.5% of the gland, Table 2) showed long plateau projection and baffle plates (Table 3; Figure 9E). Secretory granules stained in magenta (PAS+) were found within the tubules (Figure 9F) and the lamellae. The terminal zone was the smallest one (10.6% of the OG surface); it showed scantier and small tubules and few lamellae (Tables 3 and 4; Figure 9G–H) in respect to the other zones.

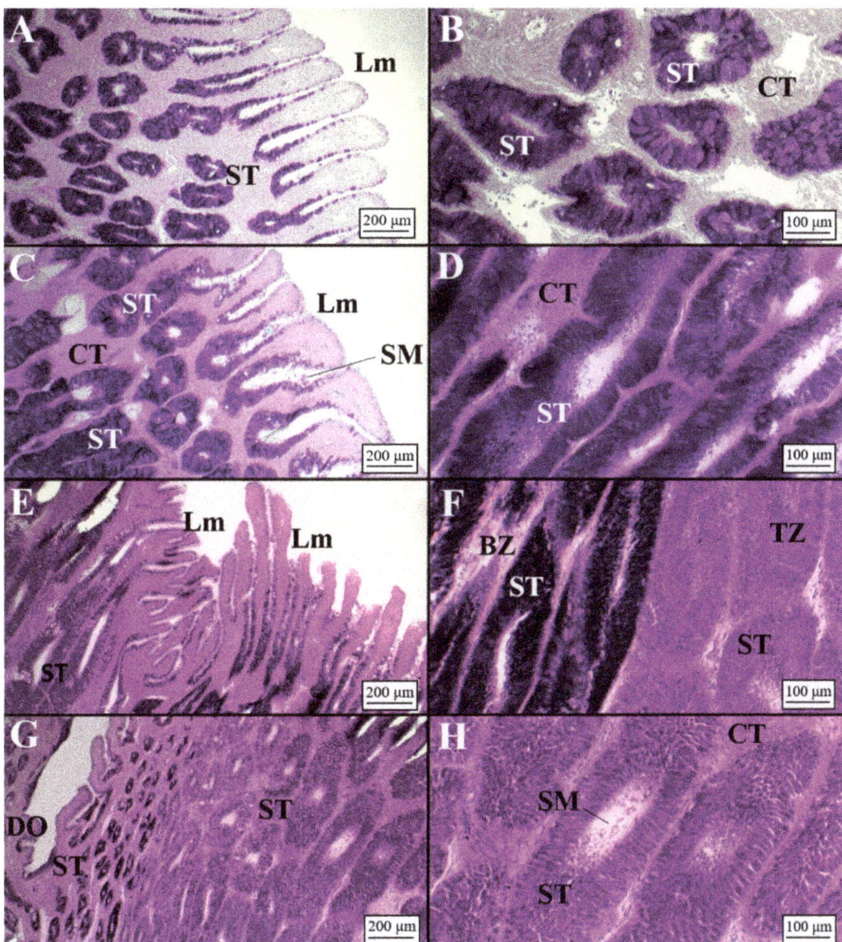

Figure 7. Microarchitecture of a *C. uyato* mature female (AB/PAS staining method applied to all slides). (**A**) Club-shaped lamellae and secretory tubules of the club zone showing AB+/PAS+ affinity. It is possible to observe the affinity for dyes also in the epithelium bordering the lamellae. (**B**) High magnification of the club tubules having different shapes (rounded and elliptical) stained AB+/PAS+ (purple color). (**C**) Digitiform lamellae of the papillary zone and secretory tubules near them showing AB+/PAS+ affinity. The epithelium bordering the lamellae reacted to AB/PAS staining and secretory material detected within them. Tubules near the lamellae have a rounded shape. (**D**) Elongated secretory tubules of the papillary zone showing affinity for AB+/PAS+. (**E**) Baffle lamellae that are thin and long showing affinity for dyes in the bordering epithelium. (**F**) High magnification of the baffle zone tubules bordering the terminal zone. Tubules of the baffle zone strongly reacted to AB/PAS. (**G**) Terminal zone showing short lamellae. The majority of tubules are elongated, showing an affinity for neutral mucins (PAS+), and only a few of them, near the distal oviduct, are smaller and displayed an affinity for AB+/PAS+. (**H**) High magnification of the terminal zone secretory tubules showing an intense affinity for PAS (PAS+) and secretory material within them. BZ, baffle zone; CT, connective tissue; DO, distal oviduct; Lm, lamellae; SM, secretory material; SD, secretory ducts; ST, secretory tubules; TZ, terminal zone.

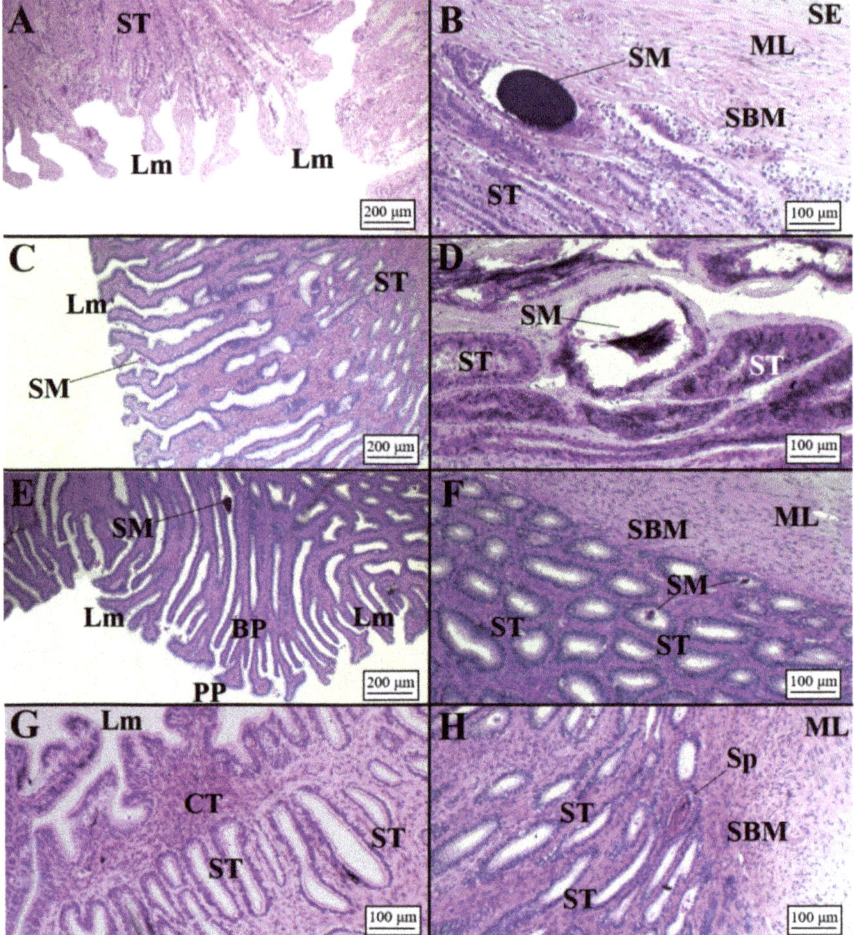

Figure 8. Microarchitecture of a *D. licha* mature female (AB/PAS staining method applied to all slides). (**A**) Long lamellae of the club zone showing a slight affinity for PAS+. (**B**) Club zone secretory tubules, rich in secretory material, located in the deep recesses. (**C**) Papillary lamellae and secretory tubules near them. Secretory material was detected between the lamellae. The epithelium bordering the lamellae reacted to AB/PAS staining, and secretory material was detected within them. Tubules near the lamellae have a rounded shape. (**D**) Secretory tubules in the deep recesses of the OG showing affinity for PAS. A great amount of secretory material was found inside them. (**E**) Thin and long baffle lamellae composed of baffle plates and plateau projections. (**F**) High magnification of the baffle zone tubules bordering the connective tissue. Secretory material was detected. (**G**) Terminal zone showing short lamellae and elongated secretory tubules composed of secretory ciliated cells near them. (**H**) Deep recesses of the terminal zone in which sperm, involved in a PAS+ matrix, were detected. BP, baffle plates; CT, connective tissue; Lm, lamellae; ML, muscular layer; PP, plateau projections; SBM, submucosa; SD, secretory ducts; SE, serosa; SM, secretory material; Sp, sperm; ST, secretory tubules.

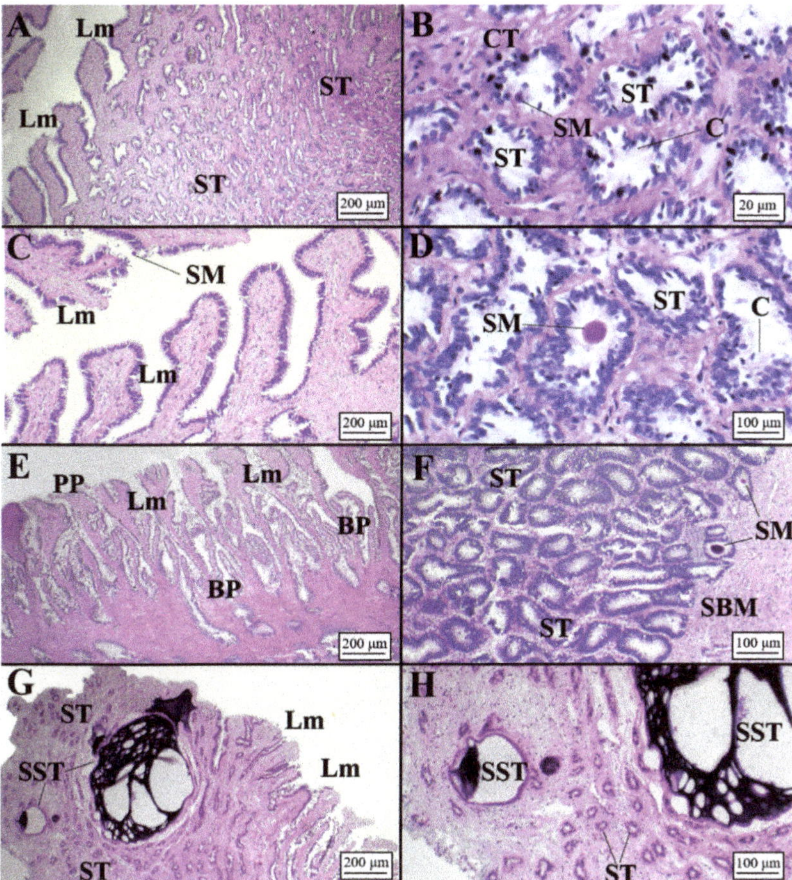

Figure 9. Microarchitecture of an *H. perlo* regressing female (AB/PAS staining method applied to all slides). (**A**) Club zone lamellae and rounded secretory tubules without any affinity for dyes. (**B**) High magnification of club zone glandular tubules with secretory material weakly PAS+. (**C**) Lamellae of the papillary zone. (**D**) Detail of glandular tubules with secretory material PAS+. (**E**) Plateau projections and baffle plates of the baffle zone. (**F**) Tubular glands of the baffle zone. The secretory tubules found in the deep recesses showed affinity for PAS. (**G**) Lamellae and tubular glands of the terminal zone. (**H**) Terminal zone in which sperm storage tubules are visible. BP, baffle plates; C, cilia; CT, connective tissue; LM, lamellae; PP, plateau projections; SBM, submucosa; SM, secretory material; SST, sperm storage tubules; ST, secretory tubules.

Oxynotus centrina

The microscopic distinction of the zones of *O. centrina* mature OG appeared less marked in respect to oviparous species (Figure 2). In females capable of reproducing, tubular glands of the club and papillary zones showed affinity for PAS+, as the epithelium of their lamellae (Figure 10A–D). The baffle zone consisted of many elongated tubules rich in secretory material (PAS+ secretions), while the epithelium of lamellae was positive to AB (Figure 10E,F). The terminal zone consisted of elongate tubules PAS+ that ran adjacent to the baffle zone and terminated in recesses beyond the peripheral tubules of the baffle zone (Figure 10G,H). Accumulation of secretory material, PAS+, in the epithelium of secretory tubules and lamellae was observed, while brown material was detected between the secretory tubules (Figure 11E).

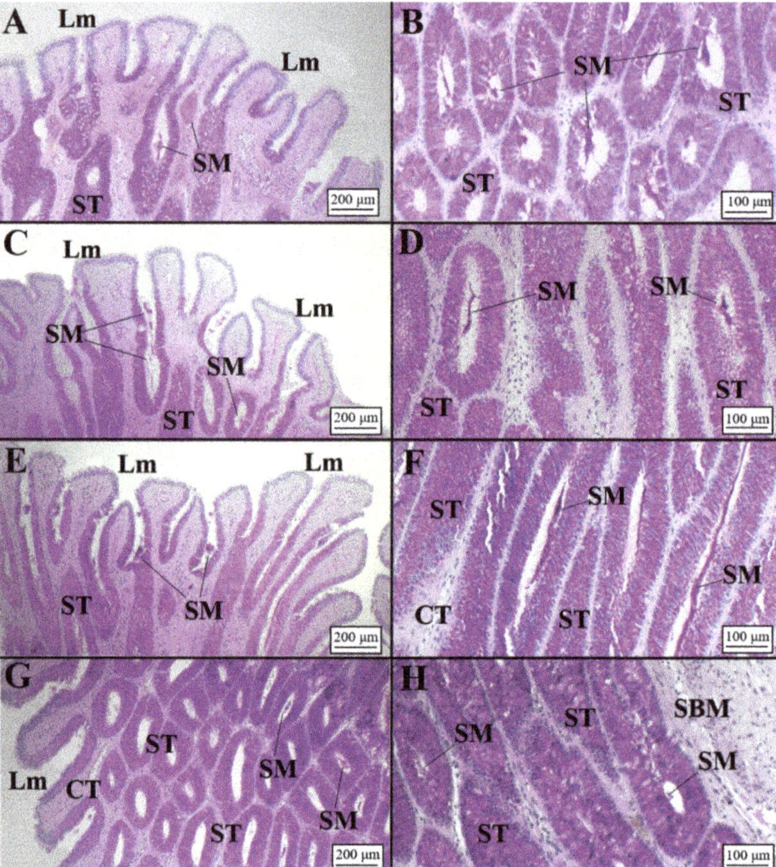

Figure 10. Microarchitecture of an *O. centrina* mature female (AB/PAS staining method applied to all slides). (**A**) Lamellae of the club zone. Secretory material (PAS+) is visible inside and between lamellae. (**B**) Secretory tubules of the club zone full of secretory material PAS+. (**C**) Lamellae of the papillary zone. Secretory material (PAS+) is visible between lamellae and inside tubules. (**D**) Secretory tubules of the baffle zone full of secretory material (PAS+). (**E**) Lamellae of the baffle zone with secretory material (PAS+) between them. (**F**) High magnification of elongated secretory tubules in the deep recesses of the baffle zone. (**G**) Terminal zone with lamellae visible. Secretory tubules showed PAS+ secretory material inside. (**H**) High magnification of secretory tubules of the terminal zone. CT, connective tissue; Lm, lamellae; SBM, submucosa; SM, secretory material; ST, secretory tubules.

Presence of Sperm

All examined species showed the presence of sperm storage tubules (SSTs), with the exception of *C. uyato*, in which spermatozoa were observed only in secretory tubules (Figure 11A). SSTs were clearly distinguishable from secretory ones in dimension and structure being composed of a simple cuboidal epithelium with ciliated and secretory cells, containing sperm without any specific pattern in their disposition, involved in a PAS+ matrix in its lumen (Figure 11). These structures were usually located in the terminal zone towards the lumen and/or near the distal oviduct (Figure 11B,D–F) in all species, except for *O. centrina* and *D. licha*, which also showed SSTs in the deep recesses of the terminal zone. Moreover, in *D. licha*, sperm were also observed inside the secretory tubules near the serosa of the OG (Figure 11C). Most females assigned to the capable of reproducing, pregnant, regressing and regenerating stages presented sperm within the SSTs (Figure 11A–E). Only in one *O. centrina* female in mid-pregnancy (stage 3C) were SSTs found empty of sperm (Figure 11F).

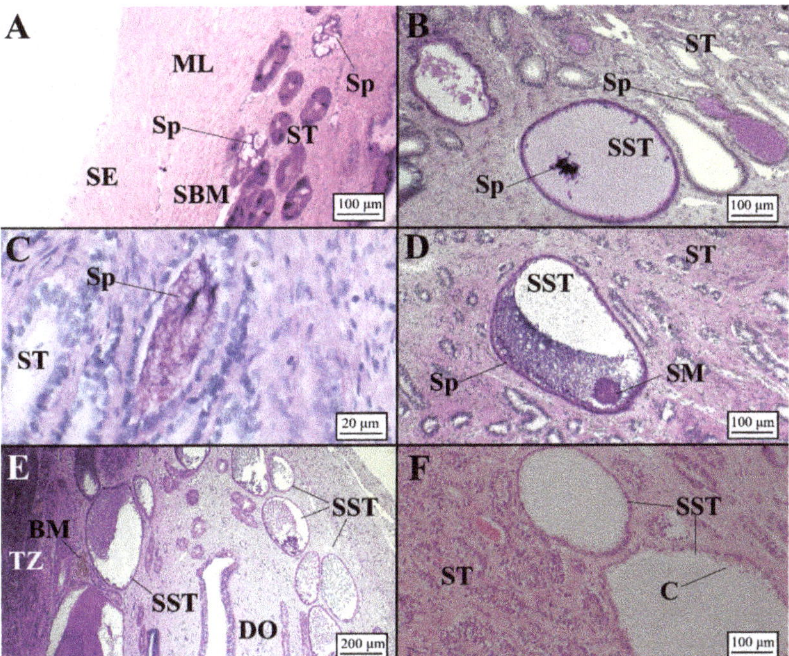

Figure 11. Presence of sperm in viviparous species (AB/PAS staining method applied to all slides, except to F stained in H&E). (**A**) Sperm found inside the secretory tubules in the terminal zone near the serosa, which enclosed the gland in *C. uyato*. (**B**) Detail of a sperm storage tubule containing spermatozoa, involved in PAS+ matrix, located in the deep recesses of the terminal zone and (**C**) high magnification of a secretory tubule with sperm in the terminal zone, near the distal oviduct, in *D. licha*. (**D**) High magnification of a sperm storage tubule with sperm and secretory material visible located in the terminal zone of *H. perlo*. (**E**) Several sperm storage tubules full of sperm located in the terminal zone and near the distal oviduct and (**F**) empty sperm storage tubules, bordered by a simple cuboidal epithelium with cilia inside, found in mid-pregnancy female of *O. centrina*. C, cilia; DO, distal oviduct; ML, muscular layer; SBM, submucosa; SE, serosa; SM, secretory material; Sp, sperm; SST, sperm storage tubule; ST, secretory tubule; TZ, terminal zone.

4. Discussion

To the best of our knowledge, in this work, for the first time, information on the OG morphology and microarchitecture of the holocephalan *C. monstrosa* and sharks belonging to the orders Hexanchiformes (*H. perlo*) and Squaliformes (*C. uyato*, *D. licha* and *O. centrina*) distributed in the Mediterranean [40,41] has been reported. The structure and organization of this organ are retained in all examined species, independently of their reproductive modality. However, the reproductive relevance of each glandular region, the secretion types and the conservation of sperm varied among them.

In the three oviparous species (*C. monstrosa*, *G. melastomus* and *R. polystigma*), the OGs were greatly extended since they produce not only the jellies that envelop the egg, but also an external capsule and all its ornamentations, reaching the maximum dimensions during the mature phase (e.g., [15,22,25,42]). The considerable enlargement of the OG in these species was due to the most extensive baffle zone (highly specialized in producing the egg case), which accounted for up to 62–66% of the total gland volume according to other oviparous species analyzed (e.g., [7,14,15,24]), differently from the three other zones which accounted for the remaining gland volume with the terminal zone occupying the smallest part. The morphometric analysis conducted in this study seemed to confirm this pattern. In fact, in *G. melastomus* and *R. polystigma*, lamellae and glandular tubules of the baffle zone were the most developed. On the contrary, in *C. monstrosa*, the most developed tubules were those of the terminal zone. This fact is probably due to the peculiar structure of its egg case characterized by a tear-drop shape with an elongated peduncle ending in a filament, produced by the terminal zone, that the species inserts into the substrate [42]. In addition, the microscopic structure of the *C. monstrosa* OG differed from other holocephalans such as the Australian ghostshark *Callorhinchus milii* [43]. The mucous secretions produced from the latter displayed an affinity for Alcian blue (acid mucopolysaccharides), while *C. monstrosa* produced neutral mucopolysaccharides (strong affinity for PAS). This different pattern could be explained by the different substrata in which the two species lay the egg cases. In fact, *C. milii* inhabits environments at depths <200 m [44], while *C. monstrosa* is distributed exclusively in bathyal ecosystems [40,41,45]. Regarding the other zones of the OG, club and papillary displayed a similar secretion pattern and occupied an analogous surface in the three species, except the "caudal-most papillary" (an intersection area between papillary and baffle zone) being found present in *C. monstrosa* and *R. polystigma* as in other oviparous species [14,15,23,45,46] and not detected in *G. melastomus*. The production of mucus that surrounds the fertilized egg and covers the interior walls of the different egg cases could be the cause of the structural differences [17,43].

The presence of spermatozoa was detected in all oviparous species. More specifically, in *R. polystigma* and *C. monstrosa*, disaggregated individual spermatozoa were found exclusively in the lamellae and tubules of the baffle zone. This fact could be the result of a recent mating episode or due to a "short-term storage" as reported for other oviparous species, mainly skates (e.g., [14,15,17]). In *R. polystigma*, as reported by Porcu et al. [47], mating occurs in coastal waters (<100 m) where populations are sexually segregated, behavior that does not necessarily require sperm storage. The observation of disaggregated sperm in *C. monstrosa*, instead, seemed to be unusual considering its deep bathymetric distribution. However, the presence of a great amount of sperm involved in an acid mucous matrix, detected in the distal oviduct, could be a clear sign of keeping spermatozoa, as also detected in other holocephalans [43,48,49]. Further investigations should be performed to verify this statement. The blackmouth catshark *G. melastomus* seemed to be the only oviparous species analyzed that retains sperm for a long time (long-term storage) showing sperm bundles involved in an AB+ matrix found in secretory glands of the terminal zone at different maturity stages (spawning capable, actively spawning and regressing). Long-term storage could be advantageous for deep species as *G. melastomus*, in which evidence of segregation by sex with females more frequent at greater depths [45] was found. In this way, the species would ensure fertilization occurs asynchronously with mating events [18,27]. In this sense, sperm storage is believed to be a particularly important strategy for oviparous species, such as catsharks, where the timing between mating and ovulation is often decoupled [50].

Differently from the oviparous species, viviparous species (*C. uyato*, *D. licha*, *H. perlo* and *O. centrina*) showed a smaller and less developed OG with respect to the body dimensions reached by these sharks. The volume breakdown of the baffle zone observed in the viviparous sharks (37–57% of the total gland) could be justified by the production of a thin, not durable, candle case in respect to a sturdy egg case produced by the oviparous species [1]. Some viviparous sharks such as *Centroscymnus coelolepis* [19] showed in the OG a distinct dark band, externally visible, that corresponds to the baffle zone and is probably associated with the production of the candle case. We found this brown coloration in adult OG females of *O. centrina*, and we can affirm surely that this species produces a candle case.

Club and papillary zone displayed a similar pattern already found in other viviparous species (e.g., [16,18–20]). Furthermore, our results highlight that the terminal zone was more developed in viviparous species (max 21.9% of total gland dimension) than in oviparous ones (max 9%). A possible explanation of this fact is the presence of sperm storage tubules (SSTs), specialized structures where spermatozoa are stored for a long time, which probably require a more extended surface. All the analyzed species presented SSTs in the terminal zone and in the posterior oviduct, suggesting long-term storage [28]. Only *C. uyato* did not seem to have these structures, even if bundles of sperm were found within terminal zone tubules in the deep recesses as also found in *D. licha* and other viviparous species [16,18]. The location of the SSTs, found in all mature stages, might give an indication of the elapsed time for mating. The finding of SSTs near the surface of the OG could indicate a recent formation, while those observed in the deep recesses (with more secretory vesicles) near the connective tissue could represent a longer retention of sperm as observed in *C. coelolepis* [19].

Long-term storage, in association with the presence of SSTs, here reported for the first time for the deep-sea sharks analyzed, represents a phenomenon clearly linked to the viviparity, both aplacental and placental (e.g., [18–20,51]). It is an advantageous mechanism that could bring several benefits: (i) ensuring the fertilization in systems where males and females are, or can be, largely solitary or separated [18,27,52] with females that may migrate away from males to more favorable habitats to release the litter [53–55]; (ii) guaranteeing the reproductive success and ensuring fertilization for copulation, fertilization and parturition [18,19,56], especially in species displaying low population densities such as the species analyzed in this work [42,43]; (iii) allowing the female to maximize the genetic quality of offspring whilst also ensuring a maximum number of offspring [18,27]; and (iv) avoiding energetically expensive and potentially damaging multiple mating events and possible injury caused by aggressive male mating behavior [16,21,57–59]. Sperm storage could represent an adaptive response to shark mating behavior that may also have benefits in the relatively low productivity environment of the deep sea.

5. Conclusions

Information reported here represents a step to expand the knowledge concerning the reproductive biology of chondrichthyan species showing different reproductive modalities. In fact, understanding the entire reproductive processes of Chondrichthyes and, in particular, the sperm storage mechanism which influences the reproductive potential of a species will constitute an input to understand the behavior, ecology and population dynamics of these species. It would be useful to conduct analysis on the spatial distribution, especially for species showing long-term storage, to test if females migrate away from males to find optimal environments to release their offspring and to identify possible spawning grounds.

Supplementary Materials: The following are available online at https://www.mdpi.com/article/10.3390/ani11092653/s1. Figure S1: Example of the measures (surface of zones, tubules and lamellae length) taken from each oviducal gland. CZ, club zone; BZ, baffle zone; Ltu, length of gland tubules; Llm, length of the lamellae; PZ, papillary zone; TZ, terminal zone. Figure S2: Body cavity of the three oviparous species (A) C. monstrosa, (B) G. melastomus and (C) R. polystigma in which the oviducal gland (OG) is visible. Figure S3: Body cavity of the four viviparous species (A) C. uyato, (B) D. licha, (C) H. perlo and (D) O. centrina in which the oviducal gland (OG) is visible.

Author Contributions: Conceptualization: M.F.M., C.P., M.C.F.; methodology: M.F.M., C.P., N.P.; software: M.F.M., C.P., R.P.; formal analysis: M.F.M., C.P., N.P.; writing—original draft preparation: M.F.M., C.P., M.C.F.; writing—review and editing: M.F.M., C.P., A.B., A.C., A.M., P.P.; visualization: A.B., A.C., A.M., P.P., R.P.; supervision, M.C.F.; project administration, M.C.F.; funding acquisition, M.C.F. All authors have read and agreed to the published version of the manuscript.

Funding: Funding for this research was provided by the European Union and the Italian Ministry for Agriculture and Forestry.

Institutional Review Board Statement: Ethical review and approval was not required for the animal study because the vertebrate animals we worked with for this study were all dead before research began. We can assure that all the animals analyzed in this study were already dead when they came on board.

Data Availability Statement: The data presented in this study are available on request from the corresponding authors.

Acknowledgments: We are grateful to the captain Marco Giordano and crew of the *Gisella* and many other colleagues onboard for their assistance with sample collection during the cruises.

Conflicts of Interest: The authors declare that they have no conflict of interest.

References

1. Conrath, C.L.; Musick, J.A. Reproductive Biology of Elasmobranchs. In *Biology of Sharks and their Relatives*, 2nd ed.; Carrier, J.C., Musick, J.A., Heithaus, M.R., Eds.; CRC Press: Boca Raton, FL, USA, 2012; pp. 291–312.
2. Grogan, D.E.; Lund, R.; Greenfest-Allen, E. The Origin and Relationships of Early Chondrichthyans. In *Biology of Sharks and Their Relatives*, 2nd ed.; Carrier, J.C., Musick, J.A., Heithaus, M.R., Eds.; CRC Press: Boca Raton, FL, USA, 2012; pp. 1–29. [CrossRef]
3. Awruch, C.A. Reproductive Strategies. In *Physiology of Elasmobranch Fishes: Structure and Interaction with Environment*; Shadwick, R.E., Farrell, A.P., Brauner, C.J., Eds.; Academic Press: Cambridge, MA, USA, 2015; Volume 34A, Chapter 7; pp. 255–310.
4. Nakaya, K.; White, W.T.; Ho, H.-C. Discovery of a new mode of oviparous reproduction in sharks and its evolutionary implications. *Sci. Rep.* **2020**, *10*, 12280. [CrossRef]
5. Hamlett, W.C.; Koob, T.J. Female Reproductive System. In *Sharks, Skates and Rays the Biology of Elasmobranch Fishes*; Hamlett, W.C., Ed.; John Hopkins University Press: Baltimore, MD, USA, 1999; pp. 398–443.
6. Carrier, J.C.; Pratt, H.L.; Castro, J.I. Elasmobranch Reproduction. In *Biology of Sharks and their Relatives*, 1st ed.; Carrier, J.C., Musick, J.A., Heithaus, M.R., Eds.; CRC Press: Boca Raton, FL, USA, 2012; pp. 269–286.
7. Jordan, R.P.; Graham, C.T.; Minto, C.; Henderson, A.C. Assessment of sperm storage across different reproductive modes in the elasmobranch fishes. *Environ. Biol. Fish.* **2021**, *104*, 27–39. [CrossRef]
8. Prasad, R.R. The structure, phylogenetic significance, and function of the nidamental glands of some elasmobranchs of the Madras coast. *Proc. Natl. Inst. Sci. India Part. B Biol. Sci.* **1944**, *11*, 282–302.
9. Colonello, J.H.; Christiansen, H.E.; Cousseau, M.B.; Macchi, G.J. Uterine dynamics of the southern eagle ray *Myliobatis goodei* (Chondrichthyes: Myliobatidae) from the southwest Atlantic Ocean. *Ita. J. Zool.* **2013**, *80*, 187–194. [CrossRef]
10. Burgos-Vázquez, M.I.; Chávez-García, V.E.; Cruz-Escalona, V.H.; Navia, A.F.; Mejía-Falla, P.A. Reproductive strategy of the Pacific cownose ray *Rhinoptera steindachneri* in the southern Gulf of California. *Mar. Freshwater Res.* **2018**, *70*, 93–106. [CrossRef]
11. Hamlett, W.C.; Hysell, M.K.; Jezior, M.; Rozycki, T.; Brunette, N.; Tumilty, K. Fundamental zonation in elasmobranch oviducal glands. In Proceedings of the 5th Indo-Pacific Fish Conference, Nouméa, Society of French Ichthyology, France, Paris, 3–8 November 1997; pp. 271–280.
12. Metten, H. Studies on the reproduction of the dogfish. *Philos. Trans. R. Soc. London Ser. B* **1939**, *230*, 217–238.
13. Knight, D.P.; Feng, D.; Stewart, M.; King, E. Changes in macromolecular organization in collagen assemblies during secretion in the nidamental gland and formation of the egg capsule wall in the dogfish *Scyliorhinus caniculia*. *Trans. R. Soc. Lond. Ser. B* **1993**, *341*, 419–436.
14. Serra-Pereira, B.; Afonso, F.; Farias, I.; Joyce, P.; Ellis, M.; Figueiredo, I.; Gordo, L.S. The development of the oviducal gland in the Rajid thornback ray, *Raja clavata*. *Helgol. Mar. Res.* **2011**, *65*, 399. [CrossRef]
15. Marongiu, M.F.; Porcu, C.; Bellodi, A.; Cuccu, D.; Mulas, A.; Follesa, M.C. Oviducal gland microstructure of two Rajidae species, *Raja miraletus* and *Dipturus oxyrinchus*. *J. Morphol.* **2015**, *76*, 1392–1403. [CrossRef]
16. Marongiu, M.F.; Porcu, C.; Bellodi, A.; Cannas, R.; Carbonara, P.; Cau, A.; Coluccia, E.; Moccia, D.; Mulas, A.; Pesci, P.; et al. Abundance, distribution and reproduction of the Data-Deficient species (*Squalus blainville*) around Sardinia Island (central western Mediterranean Sea) as a contribution to its conservation. *Mar. Freshwater Res.* **2020**, 19372. [CrossRef]
17. Hamlett, W.C.; Knight, D.P.; Koob, T.J.; Jezior, M.; Luong, T.; Rozycki, T.; Brunette, N.; Hysell, M.K. Survey of oviducal gland structure and function in elasmobranchs. *J. Exp. Zool.* **1998**, *282*, 399–420. [CrossRef]
18. Storrie, M.T.; Walker, T.I.; Laurenson, L.J.; Hamlett, W.C. Microscopic organization of the sperm storage tubules in the oviducal gland of the female gummy shark (*Mustelus antarcticus*), with observations on sperm distribution and storage. *J. Morphol.* **2008**, *269*, 1308–1324. [CrossRef] [PubMed]
19. Moura, T.; Serra-Pereira, B.; Gordo, L.S.; Figueiredo, I. Sperm storage in males and females of the deepwater Portuguese dogfish with notes on oviducal gland microscopic organization. *J. Zool.* **2011**, *283*, 210–219. [CrossRef]
20. Porcu, C.; Marongiu, M.; Follesa, M.; Bellodi, A.; Mulas, A.; Pesci, P.; Cau, A. Reproductive aspects of the velvet belly lantern shark *Etmopterus spinax* (Condrichthyes: Etmopteridae), from the central western Mediterranean Sea. Notes on gametogenesis and oviducal gland microstructure. *Mediterr. Mar. Sci.* **2014**, *15*, 313–326. [CrossRef]
21. Dutilloy, A.; Dunn, M.R. Observations of sperm storage in some deep-sea elasmobranchs. *Deep-Sea Res. PT I* **2020**, *166*, 103405. [CrossRef]

22. Hamlett, W.C. Reproductive Biology and Phylogeny of Chondrichthyes. In *Sharks, Batoids, and Chimaeras*, 1st ed.; Hamlett, W.C., Ed.; CRC Press: Boca Raton, FL, USA, 2005; Volume 3, p. 576. [CrossRef]

23. Galíndez, E.J.; Díaz-Andrade, M.C.; Avaca, M.S.; Estecondo, S. Morphological study of the oviductal gland in the smallnose fanskate *Sympterygia bonapartii* (Müller and Henle, 1841) (Chondrichthyes, Rajidae). *Braz. J. Biol.* **2010**, *70*, 325–333. [CrossRef]

24. Moya, A.C.; Acuña, F.; Díaz Andrade, M.C.; Barbeito, C.G.; Galíndez, E.J. Glycan expression as a tool for a deeper understanding of a reproductive gland in a skate of economic importance. *J. Fish. Biol.* **2020**, *98*, 537–547. [CrossRef]

25. Porcu, C.; Marongiu, M.F.; Bellodi, A.; Cannas, R.; Cau, A.; Melis, R.; Mulas, A.; Soldovilla, G.; Vacca, L.; Follesa, M.C. Morphological descriptions of the eggcases of skates (Rajidae) from the central–western Mediterranean, with notes on their distribution. *Helgol. Mar. Res.* **2017**, *71*, 10. [CrossRef]

26. Orr, T.J.; Brennan, P.L.R. Sperm Storage and Delayed Fertilization. In *Encyclopedia of Reproduction*, 2nd ed.; Skinner, M.K., Ed.; Academic Press: Cambridge, MA, USA, 2018; Volume 6, pp. 350–355.

27. Holt, W.V.; Lloyd, R.E. Sperm storage in the vertebrate female reproductive tract: How does it work so well? *Theriogenology* **2010**, *73*, 713–722. [CrossRef] [PubMed]

28. Pratt, H.L. The storage of spermatozoa in the oviducal glands of western North Atlantic sharks. *Environ. Biol. Fish.* **1993**, *38*, 139–149. [CrossRef]

29. Otero, M.D.; Serena, F.; Gerovasileiou, V.; Barone, M.; Bo, M.; Arcos, J.M.; Vulcano, A.; Xavier, J. *Identification Guide of Vulnerable Species Incidentally Caught in Mediterranean Fisheries*; IUCN: Malaga, Spain, 2019; 203p.

30. Dulvy, N.K.; Allen, D.J.; Ralph, G.M.; Walls, R.H.L. *The Conservation Status of Sharks, Rays and Chimaeras in the Mediterranean Sea*; IUCN: Malaga, Spain, 2016.

31. Cortés, E. Incorporating uncertainty into demographic modeling: Application to shark populations and their conservation. *Conserv. Biol.* **2002**, *16*, 1048–1062. [CrossRef]

32. Ferretti, F.; Worm, B.; Britten, G.L.; Heithaus, M.R.; Lotze, H.K. Patterns and ecosystem consequences of shark declines in the ocean. *Ecol. Lett.* **2010**, *13*, 1055–1071. [CrossRef] [PubMed]

33. Frisk, M.G.; Miller, T.J.; Fogarty, M.J. Estimation and analysis of biological parameters in elasmobranch fishes: A comparative life history study. *Can. J. Fish. Aquat. Sci.* **2001**, *58*, 969–981. [CrossRef]

34. Spedicato, M.T.; Massutí, E.; Mérigot, B.; Tserpes, G.; Jadaud, A.; Relini, G. The MEDITS trawl survey specifications in an ecosystem approach to fishery management. *Sci. Mar.* **2019**, *83*, 9–20. [CrossRef]

35. Follesa, M.C.; Agus, B.; Bellodi, A.; Cannas, R.; Capezzuto, F.; Casciaro, L.; Cau, A.; Cuccu, D.; Donnaloia, M.; Fernandez-Arcaya, U. The MEDITS maturity scales as a useful tool for investigating the reproductive traits of key species in the Mediterranean Sea. *Sci. Mar.* **2019**, *83S1*, 235–256. [CrossRef]

36. Cerri, P.S.; Sasso-Cerri, E. Staining methods applied to glycol methacrylate embedded tissue sections. *Micron* **2003**, *34*, 365–372. [CrossRef]

37. Zar, J.H. *Biostatistical Analysis*, 4th ed.; Prentice Hall: Upper Saddle River, NJ, USA, 1999; 663p.

38. Schneider, C.A.; Rasband, W.S.; Eliceiri, K.W. NIH Image to ImageJ: 25 years of image analysis. *Nat. Methods* **2012**, *9*, 671–675. [CrossRef] [PubMed]

39. Rohlf, F.J. TpsDig, Digitize Landmarks and Outlines, Version 2.16. In *Department of Ecology and Evolution*; State University of New York at Stony Brook: New York, NY, USA, 2005.

40. Marongiu, M.F.; Porcu, C.; Bellodi, A.; Cannas, R.; Cau, A.; Cuccu, D.; Mulas, A.; Follesa, M.C. Temporal dynamics of demersal chondrichthyan species in the central western Mediterranean Sea: The case study in Sardinia Island. *Fish. Res.* **2017**, *193*, 81–94. [CrossRef]

41. Follesa, M.C.; Marongiu, M.F.; Zupa, W.; Bellodi, A.; Cau, A.; Cannas, R.; Colloca, F.; Djurovic, M.; Isajlovic, I.; Jadaud, A.; et al. Spatial variability of Chondrichthyes in the northern Mediterranean. *Sci. Mar.* **2019**, *83*, 81–100. [CrossRef]

42. Mancusi, C.; Massi, D.; Baino, R.; Cariani, A.; Crobe, V.; Ebert, D.A.; Ferrari, A.; Gordon, C.A.; Hoff, G.R.; Iglesias, S.P.; et al. An identification key for Chondrichthyes egg cases of the Mediterranean and Black Sea. *Eur. Zool.* **2021**, *88*, 436–448. [CrossRef]

43. Smith, R.M.; Walker, T.I.; Hamlett, W.C. Microscopic organisation of the oviducal gland of the holocephalan elephant fish, *Callorhynchus milii*. *Mar. Freshw. Res.* **2004**, *55*, 155–164. [CrossRef]

44. Cox, G.; Francis, M. *Sharks and Rays of New Zealand*; Canterbury University Press: Christchurch, New Zealand, 1997; 68p.

45. Porcu, C.; Marongiu, M.F.; Olita, A.; Bellodi, A.; Cannas, R.; Carbonara, P.; Cau, A.; Mulas, A.; Pesci, P.; Follesa, M.C. The demersal bathyal fish assemblage of the Central-Western Mediterranean: Depth distribution, sexual maturation and reproduction. *Deep-Sea Res. PT I* **2020**, *166*, 103394. [CrossRef]

46. Galíndez, E.J.; Estecondo, S. Histological remarks of the oviduct and the oviducal gland of *Sympterygia acuta* Garman, 1877. *Braz. J. Biol.* **2008**, *68*, 359–365. [CrossRef]

47. Porcu, C.; Bellodi, A.; Cau, A.; Cannas, R.; Marongiu, M.F.; Mulas, A.; Follesa, M.C. Uncommon biological patterns of a little known endemic Mediterranean skate, *Raja polystigma* (Risso, 1810). *Reg. Stud. Mar. Sci.* **2020**, *34*, 101065. [CrossRef]

48. Dean, B. *Fishes, Living and Fossil. An Outline of Their Forms and Probable Relationships*; Macmillan and Co.: New York, NY, USA, 1895.

49. Dean, B. *Chimaeroid Fishes and Their Development*; Carnegie Institution of Washington: Washington, DC, USA, 1906.

50. Awruch, C.A. Reproductive endocrinology in chondrichthyans: The present and the future. *Gen. Comp. Endocrinol.* **2013**, *192*, 60–70. [CrossRef]

51. Hamlett, W.C.; Fishelson, L.; Baranes, A.; Hysell, C.K.; Sever, D.M. Ultrastructural analysis of sperm storage and morphology of the oviducal gland in the Oman shark, *Iago omanensis* (Triakidae). *Mar. Freshwater Res.* **2002**, *53*, 601–613. [CrossRef]
52. Bernal, M.A.; Sinai, N.L.; Rocha, C.; Gaither, M.R.; Dunker, F.; Rocha, L.A. Long-term sperm storage in the brownbanded bamboo shark *Chiloscyllium punctatum*. *J. Fish. Biol.* **2015**, *86*, 1171–1176. [CrossRef]
53. Kimley, A.P. The determinants of sexual segregation in the scalloped hammerhead shark, *Sphyrna lewini*. *Environ. Biol. Fish.* **1987**, *18*, 27–40. [CrossRef]
54. Economakis, A.E.; Lobel, P.S. Aggregation behavior of the grey reef shark, *Carcharhinus amblyrhynchos*, at Johnston Atoll, Central Pacific Ocean. *Environ. Biol. Fish.* **1998**, *51*, 129–139. [CrossRef]
55. Finucci, B.; Dunn, M.R.; Jones, E.G. Aggregations and associations in deep-sea chondrichthyans. *ICES J. Mar. Sci.* **2018**, *75*, 1613–1626. [CrossRef]
56. Figueiredo, I.; Moura, T.; Neves, A.; Gordo, L.S. Reproductive strategy of leafscale gulper shark *Centrophorus squamosus* and the Portuguese dogfish *Centroscymnus coelolepis* on the Portuguese continental slope. *J. Fish. Biol.* **2008**, *73*, 206–225. [CrossRef]
57. Pratt, H.; Carrier, J.C. A Review of Elasmobranch Reproductive Behavior with a Case Study on the Nurse Shark, *Ginglymostoma cirratum*. *Environ. Biol. Fish.* **2001**, *60*, 157–188. [CrossRef]
58. Sims, D.W. Differences in Habitat Selection and Reproductive Strategies of Male and Female Sharks. In *Sexual Segregation in Vertebrates: Ecology of the Two Sexes*; Ruckstuhl, K.E., Neuhaus, P., Eds.; Cambridge University Press: Cambridge, UK, 2005; pp. 127–147.
59. Finucci, B.; Duffy, C.A.J.; Francis, M.P.; Gibson, C.; Kyne, P.M. The extinction risk of New Zealand chondrichthyans. *Aquat. Conserv.* **2019**, *29*, 783–797. [CrossRef]

Article

Spatial–Temporal Distribution of Megamouth Shark, *Megachasma pelagios*, Inferred from over 250 Individuals Recorded in the Three Oceans

Chi-Ju Yu [1,2], Shoou-Jeng Joung [1,3,*], Hua-Hsun Hsu [4], Chia-Yen Lin [1], Tzu-Chi Hsieh [1], Kwang-Ming Liu [3,5,6] and Atsuko Yamaguchi [2]

[1]   Department of Environmental Biology and Fisheries Science, National Taiwan Ocean University, Keelung 20224, Taiwan; wing13260@gmail.com (C.-J.Y.); skinomotohiko@gmail.com (C.-Y.L.); andy760327@gmail.com (T.-C.H.)
[2]   Graduate School of Fisheries Science and Environmental Studies, Nagasaki University, Nagasaki 852-8521, Japan; y-atsuko@nagasaki-u.ac.jp
[3]   George Chen Shark Research Center, National Taiwan Ocean University, Keelung 20224, Taiwan; kmliu@mail.ntou.edu.tw
[4]   Marine Studies Section, Center for Environment and Water, Research Institute, King Fahd University of Petroleum and Minerals, Dhahran 31261, Saudi Arabia; hsuhuahsun@yahoo.com.tw
[5]   Institute of Marine Affairs and Resource Management, National Taiwan Ocean University, Keelung 20224, Taiwan
[6]   Center of Excellence for the Oceans, National Taiwan Ocean University, Keelung 20224, Taiwan
*   Correspondence: f0010@mail.ntou.edu.tw

Citation: Yu, C.-J.; Joung, S.-J.; Hsu, H.-H.; Lin, C.-Y.; Hsieh, T.-C.; Liu, K.-M.; Yamaguchi, A. Spatial–Temporal Distribution of Megamouth Shark, *Megachasma pelagios*, Inferred from over 250 Individuals Recorded in the Three Oceans. *Animals* **2021**, *11*, 2947. https://doi.org/10.3390/ani11102947

Academic Editor: Martina Francesca Marongiu

Received: 6 September 2021
Accepted: 4 October 2021
Published: 12 October 2021

Publisher's Note: MDPI stays neutral with regard to jurisdictional claims in published maps and institutional affiliations.

Simple Summary: In this study, we integrate *Megachasma pelagios* records from the three oceans, refine previous results, add more individual data, solve the problem of uncertain body size estimations, and provide additional information on the horizontal and vertical distributions. A checklist of over 250 *M. pelagios* is integrated in this study based on numerous public sources, published papers, personal communication, and unreleased information, especially the catch records from Taiwanese waters. The conversion equations among different length measurements are provided. In addition, the spatial–temporal movement of *M. pelagios* is inferred from the integrated data, and the results may provide important information on the vertical and geographic migration behavior of the mysterious species.

Abstract: The megamouth shark (*Megachasma pelagios*) is one of the rarest shark species in the three oceans, and its biological and fishery information is still very limited. A total of 261 landing/stranding records were examined, including 132 females, 87 males, and 42 sex unknown individuals, to provide the most detailed information on global megamouth shark records, and the spatial–temporal distribution of *M. pelagios* was inferenced from these records. The vertical distribution of *M. pelagios* ranged 0–1203 m in depth, and immature individuals were mostly found in the waters shallower than 200 m. Mature individuals are not only able to dive deeper, but also move to higher latitude waters. The majority of *M. pelagios* are found in the western North Pacific Ocean (>5° N). The Indian and Atlantic Oceans are the potential nursery areas for this species, immature individuals are mainly found in Indonesia and Philippine waters. Large individuals tend to move towards higher latitude waters (>15° N) for foraging and growth from April to August. Sexual segregation of *M. pelagios* is found, females tend to move to higher latitude waters (>30° N) in the western North Pacific Ocean, but males may move across the North Pacific Ocean.

Keywords: horizontal movement; vertical movement; elasmobranchs; sex segregation; western North Pacific; eastern Taiwan waters

---

1. Introduction

The megamouth shark, *Megachasma pelagios* [1], is one of the most spectacular discoveries of a new shark species in the late 20th century [2]. The first record of this remarkable species was an individual caught accidentally from an entanglement in parachutes that had been deployed from a research vessel operated by the Naval Undersea Center, Kaneohe, Hawaii, on 15 November 1976 [1]. The specimen was retained and sent to the Bernice P. Bishop Museum, Honolulu, Hawaii, for further study. The specimen measured 446 cm (14.6 ft) total length (TL) and weighed 750 kg (1653 lbs.). The species was considered rare for decades since fewer than 50 individuals were recorded until 2010 [3]. However, it has become clear that this species is more common than previously thought in recent years. In fact, to date, new records extending its geographic range continue to surface [4–8].

Megachasma pelagios reaches a maximum length of at least 820 cm TL, with females maturing at nearly 600 cm TL and males at 425 to 450 cm TL [5,9]. Its size at birth based on the smallest known individuals has been estimated at approximately 170 cm TL [10–12]. Although its fecundity or reproductive mode is not clear, scientists suggest that it may be viviparous with oophagy due to its similarity to other lamniform sharks [9,13]. It appears to feed by swimming through dense aggregations of krill and shrimp [14,15]. This is primarily an oceanic species usually found offshore in very deep water from 0 to 1500 m deep, but may also occasionally occur over continental shelf waters at 5–40 m depth [9]. Due to the scarcity of biological and catch data, it has been categorized as of least concern on the red list by the IUCN [16].

On the other hand, a resource management strategy has been followed and *M. pelagios* retention has been prohibited in United States Pacific fisheries since 2004, but the rule was refined for scientific or educational use in 2015, indicating the lack of information regarding this species. Afterward, the Taiwan Fisheries Agency announced a ban fishing management measure on *M. pelagios* on 10 November 2020 for conservation purposes (the catching is forbidden; fisherman have to release the shark whether it is dead or alive); however, further effectiveness and study remain to be elucidated. Studies to date have included its morphology, movement, molecular biology, physiology, and chemical analysis [17–25].

Although *M. pelagios* was mainly recorded in the western North Pacific Ocean in previous studies, a recent study using molecular technology indicated low genetic diversity of this species in the Pacific Ocean, suggesting that it may have the ability to move across ocean basins [23]. Mature individuals are able to move to higher latitudes and may be segregated by sex, and gravid females may pup in warmer waters [3,6]. However, these studies were based on a small sample size, very few individuals between 250 and 400 cm TL were included, individuals were uncommon at lower latitudes from April to October in the Northern Hemisphere, and there was insufficient information on latitudinal distribution to reach any conclusion.

In addition, recent *M. pelagios* landing records were also found in different public sources or published papers, but these data are very scattered and incomplete and need to be integrated [7,8]. There were many unpublished data or unreleased information that needed to be included; for example, numerous individuals were recorded by the mandatory catch and report system of the Taiwan Fisheries Agency, and some individuals were recorded by the Japanese Society for Elasmobranch Studies in Japanese. Therefore, the present study aims to (1) integrate records of *M. pelagios* from the three oceans, (2) refine previous results and solve problems such as small sample sizes or uncertain body size estimations, and (3) provide additional information on the horizontal and vertical distributions of *M. pelagios*. It is hoped that the complete global landing data and spatial–temporal distribution of *M. pelagios* derived from these records can provide useful information for better understanding the ecology of this mysterious species.

2. Materials and Methods

2.1. Data Collection and Integration

The data used in this study were collected from published scientific articles, gray literature, online information, news, social network service (SNS) resources, private contact with research institutes, and interviews with fishermen and researchers. In addition, the following public websites were reviewed for *M. pelagios* records: Florida Museum [26], Sharkman's World [27], Summary of Megamouth Sharks [28], Japanese Society for Elasmobranch Studies [29], and catch and report data from the Taiwan Fisheries Agency, Council of Agriculture [30]. To confirm the accuracy of these data, we cross-validated and checked each record from the above sources, including date, time, method (fishing gear, sighting, or stranded), location, operation depth, record country, length, sex, and maturity, if available. The data published in journals would be the most convincing, others (e.g., unpublished online resources and personal communication) were cross-checked from different sources. This information was used for further estimation of the spatial–temporal distribution of *M. pelagios.*

2.2. Bycatch in Taiwan Fisheries

There has been a mandatory catch and report system for *M. pelagios* in Taiwan since 2013, and fishermen have to report their catch information to scientific institutions and the Fishery Agency when they catch megamouth sharks. Almost all *M. pelagios* records from Taiwan were from the bycatch of large-mesh (mesh size = 90 cm) drift net vessels, which operated in the eastern waters of Taiwan, targeting ocean sunfish (*Mola mola*), sharptail mola (*Masturus lanceolatus*), and Indo-Pacific sailfish (*Istiophorus platypterus*), especially during April–August (Figure 1). According to interviews with fishermen, the large-mesh drift net fishery operates primarily in the evening (18:00–24:00) with net deployment ranging from 10 to 140 m in depth and ~2000 m wide, soaking for 2–3 h. Only a few landed specimens were accidentally caught by trawl nets and longline vessels.

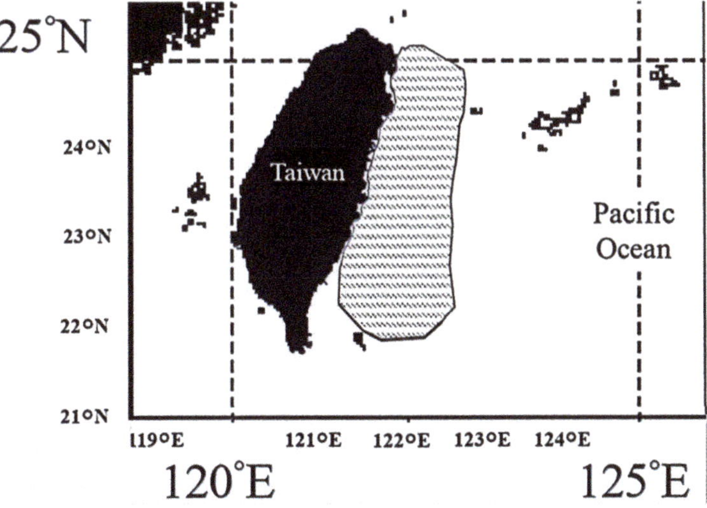

Figure 1. Fishing ground of *M. pelagios*, in the eastern Taiwan waters (shaded area is the operation area of large-mesh drift net fishery).

2.3. Meristic Measurement

Measurements were taken on total length (TL in cm), precaudal length (PCL), fork length (FL), body weight (BW in kg), mouth width (MW), 1st dorsal fin anterior margin, (D1A), 1st dorsal fin height (D1H), 1st dorsal fin base (D1B), pectoral fin anterior margin

(P1A), and caudal fin dorsal margin (CDM) of those sharks landed at Taiwanese fish markets following the protocol described by Ebert et al. [9]. These data were used to develop conversion equations between different measurements and length–weight relationships.

2.4. Maturity Stage and Distribution

To understand the monthly horizontal distribution of *M. pelagios* in different life stages, the maturity stage was identified by macroscopic examination of reproductive organs if possible. Three maturity stages of *M. pelagios* were categorized as: immature, maturing, and mature. The maturity stage of the individuals not in Taiwanese waters was based on the descriptions of other resources. The maturity stage of *M. pelagios* in Taiwanese waters was determined by the following criteria: Stage I (immature): immature males and females have undeveloped gonads, testis, and ovaries were small or nondistinguishable, vas deferens and oviducts were small in diameter, clasper were uncalcified, and uteri were threadlike. Stage II (maturing): developing (transitional) reproductive organs were observed in males by clasper development (could be slightly rotated) and the presence or absence of semen, and inflating ovaries or uteri were observed in females. Stage III (mature): developed claspers (could be rotated), inflated testis, and semen were found in males; mating scars, inflated ovaries, and large uteri were found in females. The vertical distribution of *M. pelagios* at different sizes and time was plotted, and depth data included reliable catch depths or operation depths from fishermen and observers of NOAA (National Ocean and Atmospheric Administration) Fisheries.

2.5. Data Analysis

A linear regression analysis was used to describe relationships for TL-FL, TL-PCL, TL-MW, TL-D1A, TL-D1H, TL-D1B, TL-P1A, and TL-CDM. An allometric equation (BW = aTL^b) was used to describe the relationship between BW and TL, where a and b are parameters. The maximum likelihood ratio test was used to examine the difference in the BW–TL relationships among sexes.

3. Results

3.1. Global Distribution

A total of 261 *M. pelagios* individuals recorded from 15 November 1976 to 7 August 2020 were analyzed in this study (Table S1). The majority of sharks were recorded from the western Pacific (n = 214), followed by the eastern Pacific (n = 35), with only six specimens being recorded each from the Atlantic and Indian oceans (Figure 2). Females represented slightly more than half (51%) of all sharks recorded with a breakdown by sex, if known, revealing a total of 132 females (226–710 cm TL) and 87 males (176.7–690.2 cm TL), with the sex unknown for 42 individuals (180–530 cm TL) (Figure 3a). The TL for all females was mostly between 400 and 500 cm, followed by 501 and 600 cm and 301 and 400 cm (Figure 3a). The TL for all males was mostly between 401 and 500 cm, followed closely by 301 and 400 cm (Figure 3a). Females represented the majority of records for the western Pacific (female:male = 125:65) compared with males, but more males (female:male = 4:14) were reported in the eastern Pacific (Figure 3).

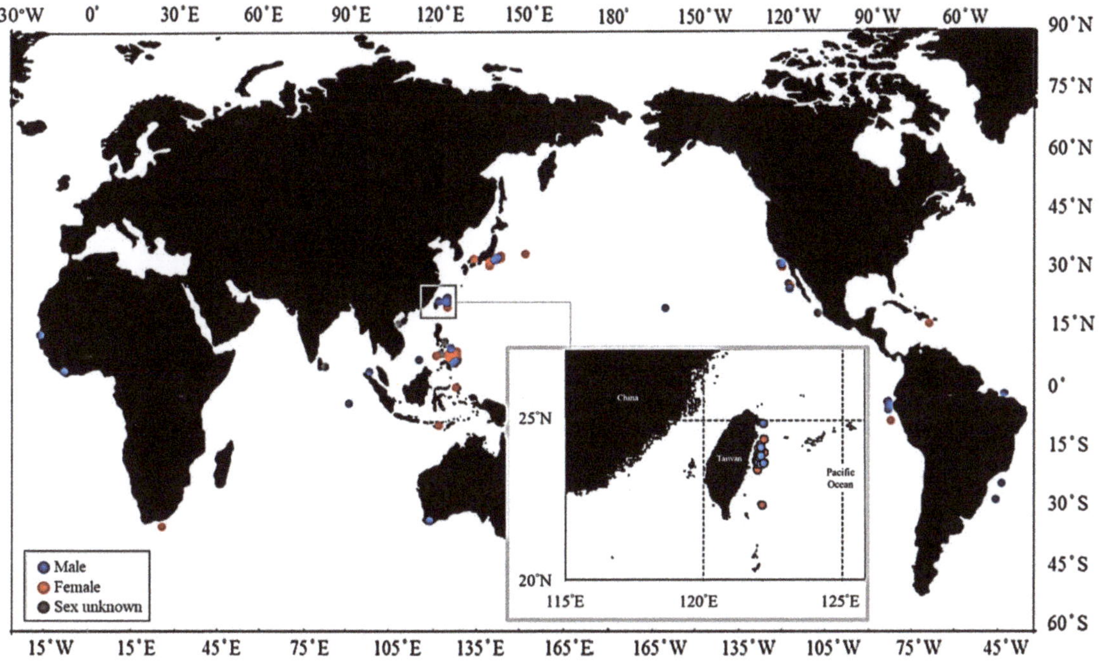

Figure 2. Records of *M. pelagios* in the world and in the eastern Taiwan waters.

3.2. Size and Sex Distribution in the Three Oceans

The length frequency of *M. pelagios* was estimated by three oceans, different groups of size and sex found tended to occur in different waters. The small individuals (<200 cm TL) were found more often but middle–large size individuals (250–400 cm TL) were few in the Indian and Atlantic Oceans (Figure 3b,c). More large males ($\geq$500 cm TL) appeared in the eastern Pacific Ocean, but more large females ($\geq$500 cm TL) in the western Pacific Ocean (Figure 3d,e).

Furthermore, because there was no record from the Southwestern Pacific, the records from the western North Pacific Ocean were divided by latitude, (i) $\leq$15° N, (ii) 15–30° N, and (iii) >30° N, and the ratios of females from zones (i) to (iii) were 43% ($n = 28$), 58% ($n = 158$), and 75% ($n = 28$), respectively (Figure 4). In zone (i), the length of *M. pelagios* ranged from ~300 to 549 cm TL for males ($n = 5$), 226 to 550 cm TL for females ($n = 12$), and 213 to 480 cm TL for sex unknown ($n =11$); the mean TL was 418 $\pm$ 114 cm TL (Figure 4). In zone (ii), the length ranged from 250 to 570 cm TL for males ($n = 57$), 247 to 710 cm TL for females ($n = 92$), and 470 to 700 cm TL for unknown sex ($n = 9$); the mean TL was 446 $\pm$ 80 cm TL (Figure 4). In zone (iii), the length ranged from 400 to 425 cm TL for males ($n = 3$), 346.6 to 577 cm TL for females ($n = 21$), and 247 to 710 cm TL for sex unknown ($n = 4$); the mean TL was 496 $\pm$ 74 cm TL (Figure 4).

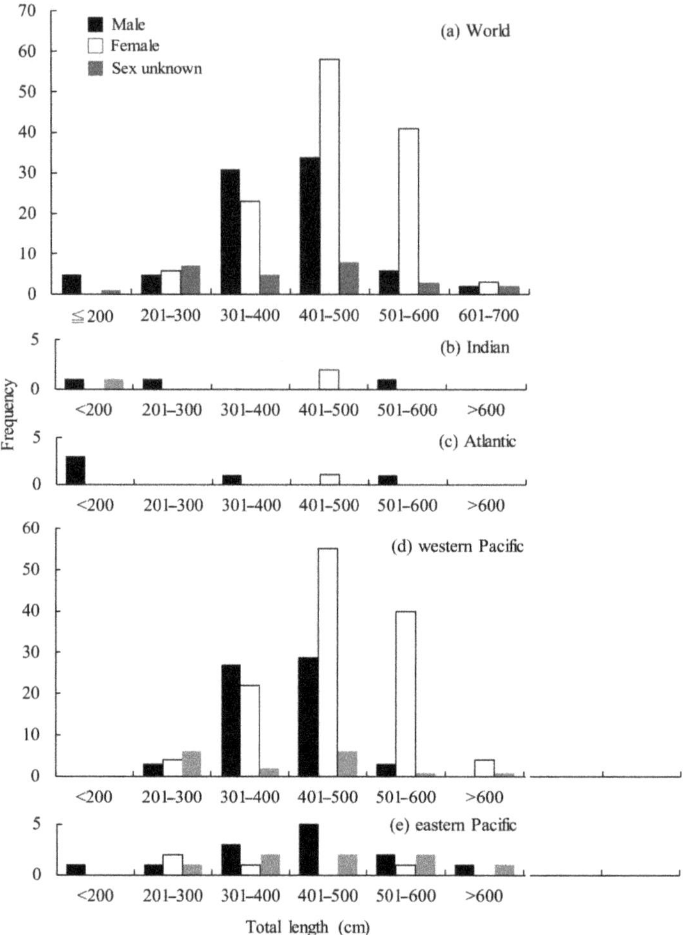

Figure 3. (a) Total length frequency of *M. pelagios* in the (a) three oceans, (b) Indian Ocean, (c) Atlantic Ocean, (d) Western Pacific Ocean, and (e) Eastern Pacific Ocean.

3.3. BW–TL Relation and Conversion Equations

The maximum likelihood test indicated that there was a significant difference in the BW–TL between sexes (Chi-square = 7.92, critical value = 5.99, $p < 0.05$), and the sex-specific BW–TL relationships were estimated as follows (Figure 5):

$$BW = 0.014TL^{1.74} \ (\text{females}, \ r^2 = 0.94, \ n = 93, \ p < 0.001) \tag{1}$$

$$BW = 0.057TL^{1.49} \ (\text{males}, \ r^2 = 0.95, \ n = 58, \ p < 0.001) \tag{2}$$

The linear relationships among measurements were expressed as follows:

$$TL = 1.131PCL + 86.731 \ (r^2 = 0.865, \ n = 126, \ p < 0.05) \tag{3}$$

$$TL = 1.257FL + 1.407 \ (r^2 = 0.975, \ n = 10, \ p < 0.05) \tag{4}$$

$$TL = 4.236MW - 22.461 \ (r^2 = 0.946, \ n = 3, \ p = 0.149) \tag{5}$$

$$TL = 3.520D1A + 330.06 \ (r^2 = 0.068, \ n = 61, \ p < 0.05) \tag{6}$$

$$TL = 4.948D1B + 280.04 \ (r^2 = 0.206, n = 54, p < 0.05) \tag{7}$$

$$TL = 5.959P1A - 35.27 \ (r^2 = 0.876, n = 9, p < 0.05) \tag{8}$$

$$TL = 2.988CDM + 123.62 \ (r^2 = 0.861, n = 6, p < 0.05) \tag{9}$$

where BW is the body weight, TL is the total length, PCL is the precaudal length, FL is the fork length, MW is the mouth width, D1A is the 1st dorsal fin anterior margin, D1H is the 1st dorsal fin height, D1B is the 1st dorsal fin base, P1A is the pectoral fin anterior margin, and CDM is the caudal fin dorsal margin. For consistency, the lengths from other studies that were not reported in TL were converted to TL using above equations (Table S1).

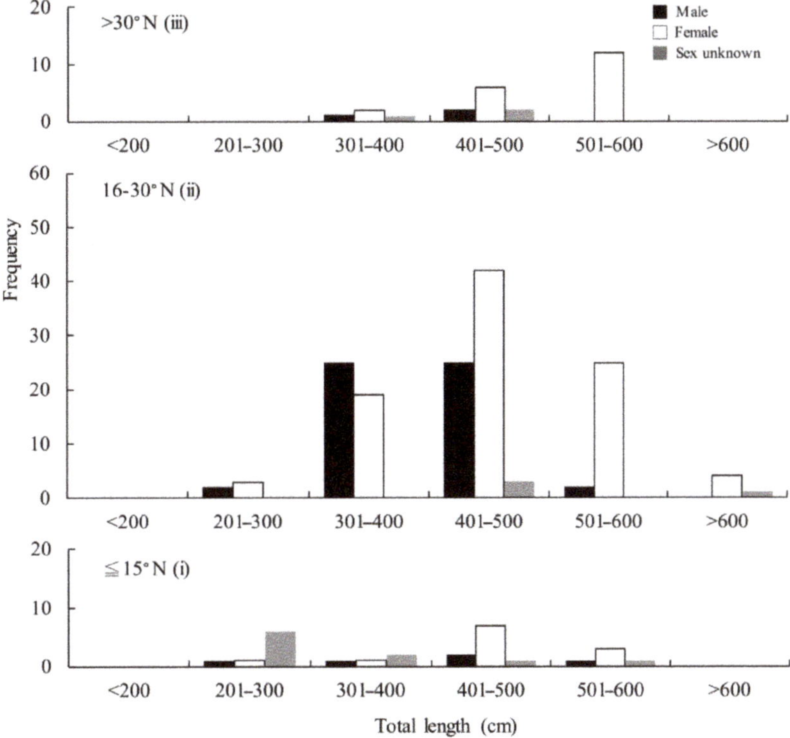

Figure 4. The length frequency of *M. pelagios* in the western Pacific Ocean.

3.4. Spatial–Temporal Distribution

3.4.1. Horizontal Distribution

According to the landing records of *M. pelagios* around the world, no immature individual landed at latitudes higher than 30°, while mature individuals were widely distributed (35° N–7° S). To further investigate the spatial–temporal distribution of *M. pelagios*, we eliminated data from the Indian and Atlantic Oceans and uncertain data from the Pacific Ocean. Figure 6 shows the monthly latitudinal occurrence of sex-specific *M. pelagios* at different maturity stages in the Pacific Ocean. Females appeared sporadically in the western Pacific Ocean from January to March and appeared in the higher latitude area (mainly in zone ii) from April to August (Figure 6a). In September, only one female was found in the high latitude area, and the distribution separated in the eastern and western Pacific Ocean after October (Figure 6a). On the other hand, male *M. pelagios* were found mainly in lower latitude waters in both the eastern and western Pacific Oceans from January to March. Males were mostly found in the middle latitude area from April to August (Figure 6b). There was no record for males in September, but mature males occurred in the eastern Pacific Ocean, and immature males occurred in the western Pacific Ocean in October and November (Figure 6b).

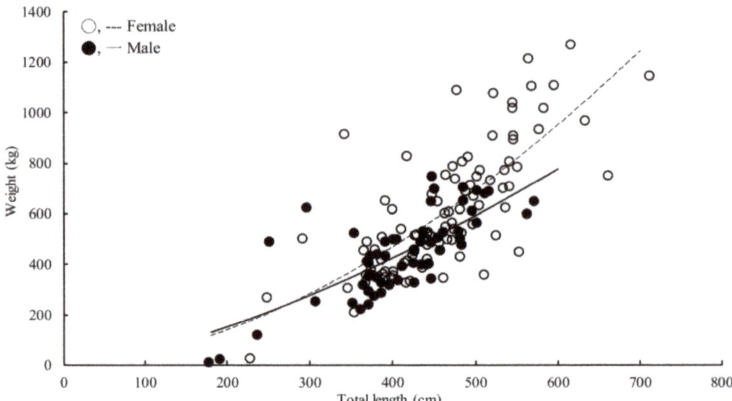

Figure 5. Total length–weight relationship of *M. pelagios*.

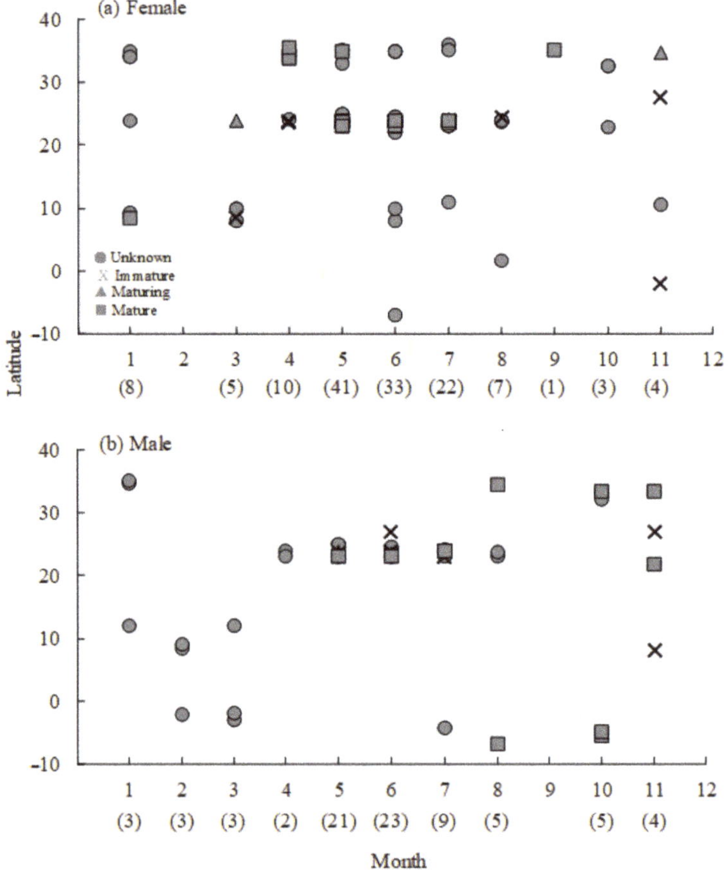

Figure 6. The monthly latitudinal occurrence in the Pacific Ocean of *M. pelagios* for (**a**) females and (**b**) males, number in the parentheses brackets was individual number.

3.4.2. Vertical Distribution

Figure 7a shows the 64 *M. pelagios* caught (*n* = 60) or sighted (*n* = 4) from different depths in size. Individuals < 300 cm TL were only found in the shallower water column (no deeper than 200 m), and large individuals were found at all water depths. In addition, the temporal vertical movements of 23 *M. pelagios* indicated that sharks tended to occur in deep water at dawn (00:00–06:00 a.m.) and then appeared in shallow water at dusk (18:00–00:00) (Figure 7b). However, one female was recorded around noon, which was a sighting event in which the individual was attacked by a whale.

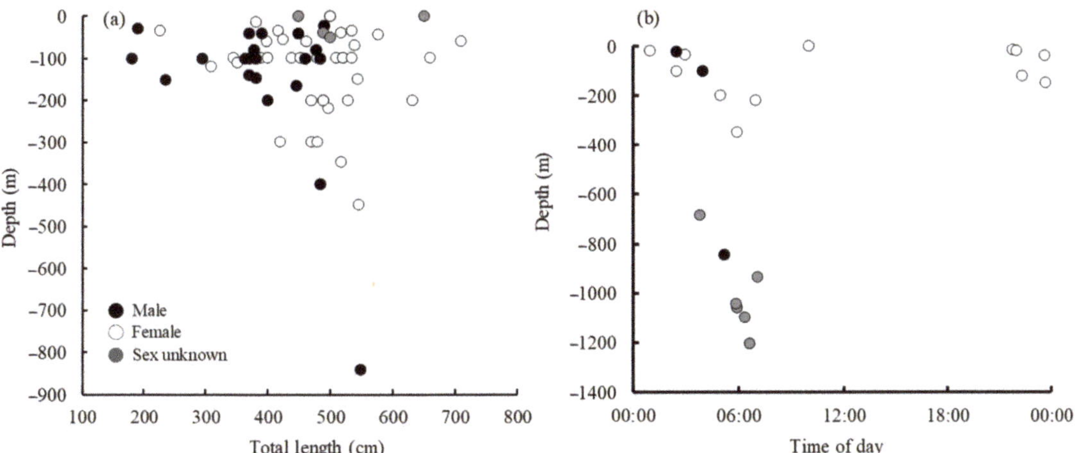

Figure 7. Vertical movement of *M. pelagios*. (**a**) The *M. pelagios* caught or sighted from different depths in size (*n* = 64). (**b**) Temporal vertical movement of *M. pelagios* (*n* = 23).

4. Discussion

This study provided the first complete and detailed landing records of *M. pelagios* and its spatial–temporal distribution information. The results derived from this study can be used as a reference for future studies on the ecology, conservation, and management of this species.

In this study, we found that the body size of female *M. pelagios* was larger than that of males. One possible reason for this is that females need more space in the coelom to carry large and well-developed pups. Another reproductive strategy was also considered: females will be more reproductively fit through their growth, larger females may delivery more pups [31,32]. Although no pregnant *M. pelagios* has ever been found, the observation of its gonad structure showed that it was very similar to *C. maximus* [33]. Additionally, the smallest free-swimming *M. pelagios* was 176.7 cm TL. One convincing inference was that *M. pelagios* is an aplacental viviparous species, delivering a few well-developed pups, which is similar to *C. maximus* [11,34]. The ovary and uterus of female *M. pelagios* may become heavy when they reach the mature stage, leading to a length–weight relationship difference between sexes (Figure 4).

Parameter b of the length–weight equation of *M. pelagios* is far smaller than the value (2.5–3.0) commonly known from sharks [35]. According to the data for allometric equations having been weighed by scientists and fishermen associations, one possible reason for this is that *M. pelagios* is an engulfment filter feeder, the mechanism of energy use such as the metabolic rate and growth may differ from other shark species, leading to different allometric equation results compared with other species [14,36]. The linear regression analysis showed that TL had a good correlation with PCL, FL, P1A, and CDM, but the small sample size for FL, P1A, and CDM remains to be further enforced. On the other hand, the formulas of TL-MW, TL-D1A, and TL-D1B were for reference only due to either a limited sample size or low correlation. However, considering the rarity of *M. pelagios*, this information would still be useful. Furthermore, we attempted to estimate the "~9 m TL male" from Martínez-Ortiz et al. [5] by using the TL-CDM formula; pursuant to "1700 mm measured at the dorsal margin of the caudal fin", this male had a 631.58 cm TL rather than ~9 m TL. This result indicated that the regression formulas from the present study provided an opportunity to validate uncertain data regarding *M. pelagios*. Further regression data should be collected from more individuals to be more convincing.

In this study, no evidence indicated that the population of *M. pelagios* is female biased due to 16% (*n* = 42) sexually unknown records. Some studies have indicated that the sexual segregation of elasmobranchs usually leads to sexual bias when under local investigation [37,38]. In this study, we integrated diverse record resources, including academic journals, conference reports, and public online resources, and the female ratio was 51%, close to half of the records. Additionally, *M. pelagios* is considered a panmictic population with no genetic structure, showing the ability to move across oceans [23]. The length frequency showed that small (≤200 cm TL) free-swimming *M. pelagios* (*n* = 4) were found only in the Indian and Atlantic Oceans, indicating a potential nursery area in these waters (Figure 3b,c). On the other hand, the males (*n* = 14) were more than three times as abundant as females (*n* = 4) in the eastern Pacific Ocean, but females were notably more abundant in the western North Pacific. Both the mean body size and the ratio of large females increased when the latitude was higher. Recent studies have shown that large shark species usually have sexual segregation behavior because (1) females escape forced mating by mature males, (2) to avoid consuming the same prey resources, and (3) gravid females move to habitats that offer stable resources, through which they can gain more energy for offspring from predation [39,40]. A significant sexual bias of reef manta rays (*Mobula alfredi*) was found in southern Mozambique, and female *M. alfredi* uses this habitat as the breeding and birthing grounds [41]. During the mating season, male shortfin mako sharks (*Isurus oxyrinchus*) harass females that lead to fitness consequences, which reflect avoidance behavior [40]. White sharks (*Carcharodon carcharias*) near the Neptune Islands experience segregation due to different physiological strategies, females are absent in the breeding season and only return in the feeding period to increase the growth rate of pups [37]. Large shark species usually have large spatial-scale segregation behavior, and our results showed that female *M. pelagios* mainly inhabit the western Pacific Ocean, while males prefer to inhabit the eastern Pacific Ocean. However, the mechanisms resulting in sex segregation need further investigation in the future.

Immature individuals were found only between 30° N and 30° S, but mature *M. pelagios* could not only move toward higher latitude waters, but also have the ability to dive deeper, where the water temperature is lower (Figures 6 and 7). Previous studies have suggested that many Lamnidae sharks, such as *I. oxyrinchus*, big-eye thresher (*Alopias superciliosus*), and pelagic thresher (*A. pelagicus*), have some capability of endothermic regulation; they could conserve heat and arrange their body temperature well to protect against low temperatures (high latitude or winter) environments [42,43]. The most famous endothermic species are the porbeagle (*Lamna nasus*) and salmon shark (*L. ditropis*), which are usually distributed at latitudes higher than 40°; however, limited by thermal inertia, young individuals can only remain in moderate temperature areas until they mature [44,45]. The distribution pattern of *M. pelagios* based on latitude or water depth in this study showed that mature individuals have the ability to protect themselves against low-temperature water compared with immature individuals, although there is a lack of evidence to verify whether *M. pelagios* is an endothermic species.

Records from the western North Pacific Ocean were mostly bycatch from fisheries (85%). According to fishermen, the occurrence of *M. pelagios* seems to be seasonal, some large-mesh drift net fishing vessels in Taiwan operate year-round but do not catch any *M. pelagios* from September to March. On the other hand, *M. pelagios* was mainly recorded between October and March in the eastern Pacific Ocean (74%). These results may indicate that *M. pelagios* move between eastern and western Pacific waters. Large sharks are able to move across the Ocean, a tracking study provided evidence of the trans-Pacific migration of the whale shark (*Rhincodon typus*). One female *R. typus* individual was tracked from the eastern (Panama) to the western (Mariana Trench) Pacific Ocean [46]. In addition, the monthly latitudinal occurrence of *M. pelagios* was slightly different between females and males. In the spring, both sexes from lower latitudes (zone i) moved toward middle latitude water (zone ii), but females also went further to the high latitude area (zone iii) in the western North Pacific Ocean. In the summer, *M. pelagios* was dispersed but mainly found in zone ii around Taiwanese waters in the western North Pacific Ocean. Afterward, *M. pelagios* was absent in early autumn (September); by October, females appeared in zone iii and the eastern Pacific, and males mainly appeared in the eastern Pacific. The records from winter were few, but included *M. pelagios* from both the eastern and western North Pacific Oceans and different latitudes.

Geographically, the movement of *M. pelagios* seems to be related to the current flow in the North Pacific Ocean. Many studies have suggested that the migration or movement of large marine animals relates to the current, and they benefit from the current, such as for moving, spawning, or foraging [47,48]. There are different names of current in the North Pacific Gyre (NPG) according to their position and characteristics, including the Kuroshio Current (KC), which is warm, less productive flows pass through the western Pacific Ocean, and extends from the Philippines to Taiwan and Japan year-round. The intensity of the KC increases from May to August; during this period, many migrating fish species are transported by KC, passing through eastern Taiwanese waters toward northern Japan. The KC merges with the Oyashio Current (OC) at approximately 35° N, forming a good feeding and fishing ground and, finally, turning toward the east across the North Pacific Ocean [49]. There was evidence shown that large sharks such as *R. typus* prefer to give birth or aggregate in warm waters [50,51]. *Megachasma pelagios* migrate northward with the KC in the late spring to summer offshore of Taiwan. Tang et al. [52] suggested that the subsurface Kuroshio water on the shelf along the east coast of Taiwan indicated upwelling and nutrient transport, which could explain why the *M. pelagios* we observed from Taiwan were mostly full with prey in their stomach. Therefore, the middle western Pacific Ocean may be the feeding grounds of *M. pelagios* in the spring and summer; moving northward to 35° N, some females remain and other individuals change direction toward the eastern Pacific Ocean in early autumn and arrive after October. When arriving in the eastern Pacific Ocean, *M. pelagios* move southward with the California Current (CC) until they meet the Peru Current (PC) and then turn to cross the Pacific again toward the western side, which returns to Indonesian or Philippine waters. This inference was based on the lack of genetic structure and panmictic populations in *M. pelagios*, but further study in addition to the genetic study of Liu et al. [23], such as tagging or analyzing more specimens from different oceans, should be conducted to verify this hypothesis.

Vertical migration behavior has been verified for *M. pelagios* since 1997. One 490 cm TL male was attached with acoustic transmitters and tracked for 50.5 h in the eastern Pacific (southern California), and the results indicated that *M. pelagios* has a very specific vertical movement during dawn and dusk [19]. Sharks make vertical movements for different purposes, e.g., *R. typus* spends time daily at the surface to gain energy for thermoregulation [53]; the basking shark (*Cetorhinus maximus*) spends half a day in deep water (800–1000 m) and reduces depths gradually, indicating foraging behavior [54]. However, many studies have shown that even the same species may have different horizontal or vertical movement patterns [48,53,55]. As only one *M. pelagios* individual was successfully tracked in the past, we integrated historical time–depth records of *M. pelagios*. In the present study, the daily vertical movement of *M. pelagios* was found in multiple individuals. The shallow–deep water movement was extremely significant from dusk to dawn, but one record was found at approximately 10 am (Figure 7b). Amorim et al. [56] noted that one *M. pelagios* was sighted with three sperm whales (*Physeter macrocephalus*), and there was some scarring on the fin and gills of the shark, indicating that it may have been attacked or traced by the sperm whales; therefore, it came to the surface. In addition to the sighting of this individual, other studies have shown a similar temporal vertical movement pattern of *M. pelagios* with acoustic techniques [19]. However, most data were operating depth recorded by the Taiwan large-mesh drift net fishery and NOAA Fisheries; the actual catching depth remains to be further elucidated. According to previous studies, as a filter-feeding shark species, *M. pelagios* seems to prefer euphausiid shrimp. Sawamoto and Matsumoto [15] observed the stomach composition of one female *M. pelagios*, which was caught by a seine net near Japan, and euphausiids (*Euphausia pacifica*) were the main prey of *M. pelagios* [20]. Nakagawa et al. [57] found that *E. pacifica* migrate to the surface at night (20:30), move down to approximately 100 m at midnight (00:30), and move toward deeper water (150–300 m) after dawn (06:00). The vertical movement patterns of *M. pelagios* and *E. pacifica* seem very similar, indicating that the vertical movement of *M. pelagios* may be related to its foraging behavior.

The notable number of *M. pelagios* landing records from Taiwanese waters compared with those elsewhere may be attributed to the cooperation between fishermen, research institutes, and the Fishery Agency (Figure 2). Due to the catch and report system, we were able to measure *M. pelagios* at the market. The large-mesh drift net fishery usually operates on the east coast of Taiwan from April to August and targets *M. mola* and *M. lanceolatus* at night. The fishermen change different fishing gear in other months in order to catch other species, e.g., striped bonito (*Sarda orientalis*). Previous studies have suggested that oceanic sunfish movement vertically depends on the temperature and depth of the mixed layer. Moreover, oceanic sunfish also move to shallower water during the night and back to deeper water at dawn, which is similar to *M. pelagios* [58,59]. Therefore, *M. pelagios* may be accidentally caught by the drift net due to sharing the same vertical movement as molas. Additionally, the catch and report system plays an important role; nearly 40% of the records from the Philippines and Indonesia are either stranded or sexually unknown because of the scattered islands, which prevents the information from being transmitted effectively. To better understand the information of *M. pelagios*, the reporting system or open platform should be designed and propagated, especially for waters with potential nursery grounds.

The data collected on the spatial–temporal movement of *M. pelagios* provide important insights into their vertical and geographic migration behaviors. This study was the first to include different body part measurements of multiple *M. pelagios* individuals using the same standards. Additionally, we integrated the results from previous studies, refined the data records presented in supplementary materials (Table S1), and established conversion equations for future research. Furthermore, we updated the catch records from Taiwanese waters, including the 250–400 cm TL individuals, integrated the missing records from April to October, and included the vertical movement data in this study.

5. Conclusions

One hypothesis was proposed in this study: *M. pelagios* give birth in the eastern Indian Ocean near the Philippines and Indonesia; during growth, they move northward to the western Pacific Ocean, joining the NPG in the spring and arriving in Taiwanese waters, foraging on planktonic prey from late spring to summer (Figure 8). Some mature females, which can withstand cool temperatures, keep following the KC and arrive in Japanese waters in the spring. In late summer, females remain in the water around Japan, and males remain across the Pacific Ocean toward the eastern side by following the North Pacific Current (NPC), indicating sexual segregation. *M. pelagios* arrive in Californian waters in late summer or autumn by following the CC and PC south to Mexico, Ecuador, and Peru. Afterward, some *M. pelagios* may follow the North Equatorial Current (NEC) across the Pacific again toward the western side, thereby returning to Indonesian or Philippine waters. However, (1) how does the Atlantic Ocean serve as a potential nursery area for *M. pelagios*? (2) Where do males and females mate? (3) Where do the gravid females go? These questions remain poorly known and need further study. The catch and retention of *M. pelagios* have been banned in Taiwan, fishermen have to release the shark no matter if it is alive or dead [30]. Therefore, data collection and biological study, such as reproduction and age growth, will be difficult in the future. Future studies, such as satellite tracking or international data exchange, would help confirm our hypothesis.

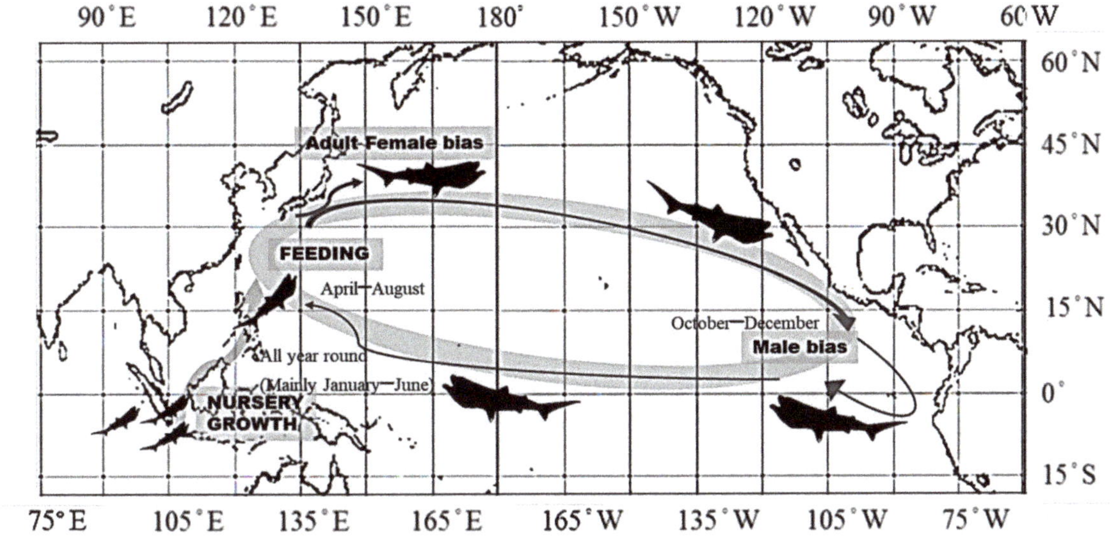

Figure 8. Schematic *M. pelagios* spatial–temporal distribution model in the Pacific Ocean proposed by this study.

Supplementary Materials: The following are available online at https://www.mdpi.com/article/10.3390/ani11102947/s1, Table S1: complete megamouth shark records from November 1976 to August 2020 [4,60–79].

Author Contributions: Conceptualization and methodology, C.-J.Y. and S.-J.J.; investigation, C.-J.Y., S.-J.J., H.-H.H., C.-Y.L. and T.-C.H.; resources, C.-J.Y., S.-J.J. and K.-M.L.; data curation, C.-J.Y.; writing—original draft preparation, C.-J.Y.; writing—review and editing, S.-J.J., K.-M.L., H.-H.H. and A.Y. All authors have read and agreed to the published version of the manuscript.

Funding: This study was supported in part by the Fisheries Agency, Council of Agriculture, Taiwan, R. O. C. (grant numbers FA104-11.1.4-F1-2), Ministry of Science and Technology, Taiwan, R. O. C. (grant numbers MOST 104-2313-B-019-002, MOST107-2611-M-019-013), and Ocean Conservation Administration, Ocean Affairs Council, Taiwan, R. O. C. (grant numbers 108-C-43, 109-C-24).

Acknowledgments: We would like to express our sincere thanks to the fishermen from Yilan, Hualien, and Taitung, especially captain K. H. Wu, C. P. Wu, C. B. Wu, T. H. Lin, K. P. Li, I H. Huang, and W. H. Hung, etc., for their help with catch and report information of the megamouth shark from eastern Taiwan waters. We thank P. L. Lin, Y. W. Chang, and C. H. Wang, observers from the Taiwan Fisheries Sustainable Development Association and Taiwan Ocean Conservation and Fisheries Sustainability Foundation, and all the crew from the Nanfangao and Chenggong fish market, for making contact with the research institution and collecting the samples. We would also like to express our gratitude to the crews of CANG HAI ENTERPRISE CO. for dissection and sample collecting. Thank you to NOAA Fisheries, Paul Clerkin, Sharkman's World, and the Japanese Society for Elasmobranch Studies for parts of data integration; also, special thanks to David A. Ebert for the English review with helpful comments. Finally, we would like to thank for the assistance of Y. H. Lin, Y. H. Li, and members from the National Taiwan Ocean University and George Chen Shark Research Center, who assisted with the catch and report information, sample collecting, and experiment.

Conflicts of Interest: The authors declare no conflict of interest.

References

1. Taylor, L.R.; Compagno, L.J.V.; Struhsaker, P.J. Megamouth—A new species, genus, and family of lamnoid shark (*Megachasma pelagios*, family Megachasmidae) from the Hawaiian Islands. *Proc. Calif. Acad. Sci.* **1983**, *43*, 87–110.
2. Compagno, L.J.V. *Sharks of the World: An Annotated and Illustrated Catalogue of Shark Species Known to Date: Bullhead, Mackerel and Carpet Sharks (Heterodontiformes, Lamniformes and Orectolobiformes)*; FAO Species Identification Guides for Fishery Purposes; Food and Agriculture Organization of the United Nations: Rome, Italy, 2001; Volume 2, p. 269.
3. Nakaya, K. Biology of the megamouth shark, *Megachasma pelagios* (Lamniformes: Megachasmidae). In Proceedings of the International Symposium, into the Unknown, Researching Mysterious Deep-Sea Animals, Okinawa, Japan, 23–24 February 2007; pp. 69–83.
4. Rodriguez-Ferrer, G.; Wetherbee, B.M.; Schärer, M.; Lilyestrom, C.; Zegarra, J.P.; Shivji, M. First record of the megamouth shark, *Megachasma pelagios*, (family Megachasmidae) in the tropical western North Atlantic Ocean. *Mar. Biodivers. Rec.* **2017**, *10*, 20. [CrossRef]
5. Martínez-Ortiz, J.; Mendoza-Intriago, D.; Tigrero-Gonzalez, W.; Flores-Rivera, G.; López-Parraga, R. New records of megamouth shark, *Megachasma pelagios* off Ecuador, Eastern Pacific Ocean. *Cienc. Pesq.* **2017**, *25*, 27–30.
6. Watanabe, Y.Y.; Papastamatiou, Y.P. Distribution, body size and biology of the megamouth shark *Megachasma pelagios*. *J. Fish Biol.* **2019**, *95*, 992–998. [CrossRef]
7. Kelez, S.; Napuri, R.M.; Pfennig, A.M.; Martinez, O.C.; Carrasco, A.T. First reports of Megamouth Shark, *Megachasma pelagios* Taylor, Compagno & Struhsaker, 1983 (Lamniformes, Megachasmidae), in Peru. *Check List* **2020**, *16*, 1361–1367. [CrossRef]
8. Acuña-Perales, N.; Córdova-Zavaleta, F.; Alfaro-Shigueto, J.; Mangel, J.C. First records of the megamouth shark *Megachasma pelagios* (Taylor, Compagno & Struhsaker, 1983) as bycatch in Peruvian small-scale net fisheries. *Mar. Biodivers. Rec.* **2021**, *14*, 1. [CrossRef]
9. Ebert, D.A.; Dando, M.; Fowler, S. *Sharks of the World: A Complete Guide*; Princeton University Press: Princeton, NJ, USA, 2021; p. 607.
10. Séret, B. Première capture d'un requin grande gueule (Chondrichthyes, Megachasmidae) dans l'Atlantique, au large du Sénégal. *Cybium* **1995**, *19*, 425–427.
11. White, W.T.; Fahmi, M.A.; Sumadhiharga, K. A juvenile megamouth shark *Megachasma pelagios* (Lamniformes: Megachasmidae) from Northern Sumatra, Indonesia. *Raffles Bull. Zool.* **2004**, *52*, 603–607.
12. Fernando, D.; Perera, N.; Ebert, D.A. First record of the megamouth shark, *Megachasma pelagios*, (Chondrichthyes: Lamniformes: Megachasmidae) from Sri Lanka, Northern Indian Ocean. *Mar. Biodivers. Rec.* **2015**, *8*, e75. [CrossRef]
13. Castro, J.I.; Clark, E.; Yano, K.; Nakaya, K. The Gross anatomy of the female reproductive tract and associated organs of the Fukuoka megamouth shark (*Megachasma pelagios*). In *Biology of the Magamouth Shark*; Yano, K., Morrissey, J.F., Yabumoto, Y., Nakaya, K., Eds.; Tokai University Press: Tokyo, Japan, 1997; pp. 115–119.
14. Nakaya, K.; Matsumoto, R.; Suda, K. Feeding strategy of the megamouth shark *Megachasma pelagios* (Lamniformes: Megachasmidae). *J. Fish Biol.* **2008**, *73*, 17–34. [CrossRef]
15. Sawamoto, S.; Matsumoto, R. Stomach contents of a megamouth shark *Megachasma pelagios* from the Kuroshio Extension: Evidence for feeding on a euphausiid swarm. *Plankton Benthos Res.* **2012**, *7*, 203–206. [CrossRef]
16. Kyne, P.M.; Liu, K.M.; Simpfendorfer, C. Megachasma Pelagios. In *The IUCN Red List of Threatened Species*; IUCN Global Species Programme Red List Unit: Cambridge, UK, 2018. [CrossRef]
17. Ishida, H.; Miyamoto, H.; Kajino, T.; Nakayasu, H.; Nukaya, H.; Tsuji, K. Study on the Bile Salt from Megamouth Shark. I. The Structures of a New Bile Alcohol, 7-Deoxyscymnol, and Its New Sodium Sulfates. *Chem. Pharm. Bull.* **1996**, *44*, 1289–1292. [CrossRef]
18. Martin, A.P.; Naylor, G.J. Independent origins of filter-feeding in megamouth and basking sharks (order Lamniformes) inferred from phylogenetic analysis of cytochrome b gene sequences. In *Biology of the Magamouth Shark*; Yano, K., Morrissey, J.F., Yabumoto, Y., Nakaya, K., Eds.; Tokai University Press: Tokyo, Japan, 1997; pp. 39–50.
19. Nelson, D.R.; McKibben, J.N.; Strong, W.R.; Lowe, C.G.; Sisneros, J.; Schroeder, D.M.; Lavenberg, R.J. An acoustic tracking of a megamouth shark, *Megachasma pelagios*: A crepuscular vertical migrator. *Environ. Boil. Fishes* **1997**, *49*, 389–399. [CrossRef]
20. Yano, K.; Toda, M.; Uchida, S.; Yasuzumi, F. Gross anatomy of the viscera and stomach contents of a megamouth shark, Megachasma pelagios, from Hakata Bay, Japan, with a comparison of the intestinal structure of other planktivorous elasmobranchs. In *Biology of the Magamouth Shark*; Yano, K., Morrissey, J.F., Yabumoto, Y., Nakaya, K., Eds.; Tokai University Press: Tokyo, Japan, 1997; pp. 105–113.
21. Chang, C.-H.; Shao, K.-T.; Lin, Y.-S.; Chiang, W.-C.; Jang-Liaw, N.-H. Complete mitochondrial genome of the megamouth shark *Megachasma pelagios* (Chondrichthyes, Megachasmidae). *Mitochondrial DNA* **2013**, *25*, 185–187. [CrossRef]
22. Tomita, T.; Tanaka, S.; Sato, K.; Nakaya, K. Pectoral Fin of the Megamouth Shark: Skeletal and Muscular Systems, Skin Histology, and Functional Morphology. *PLoS ONE* **2014**, *9*, e86205. [CrossRef] [PubMed]
23. Liu, S.Y.V.; Joung, S.J.; Yu, C.-J.; Hsu, H.-H.; Tsai, W.-P.; Liu, K.M. Genetic diversity and connectivity of the megamouth shark (*Megachasma pelagios*). *PeerJ* **2018**, *6*, e4432. [CrossRef]
24. Duchatelet, L.; Moris, V.C.; Tomita, T.; Mahillon, J.; Sato, K.; Behets, C.; Mallefet, J. The megamouth shark, *Megachasma pelagios*, is not a luminous species. *PLoS ONE* **2020**, *15*, e0242196. [CrossRef]

25. Ju, Y.-R.; Chen, C.-F.; Chen, C.-W.; Wang, M.-H.; Joung, S.-J.; Yu, C.-J.; Liu, K.-M.; Tsai, W.-P.; Liu, S.Y.V.; Dong, C.-D. Profile and consumption risk assessment of trace elements in megamouth sharks (*Megachasma pelagios*) captured from the Pacific Ocean to the east of Taiwan. *Environ. Pollut.* **2020**, *269*, 116161. [CrossRef]

26. Florida Museum. Available online: https://www.floridamuseum.ufl.edu/discover-fish/sharks/megamouths/ (accessed on 1 September 2021).

27. Sharkmans-World. Available online: https://sharkmans-world.blogspot.com/ (accessed on 1 September 2021).

28. Summary of Megamouth Sharks. Available online: http://elasmollet.org/Mp/Mplist.html (accessed on 1 September 2021).

29. Japanese Society for Elasmobranch Studies. Available online: http://www.jses.info/index.html (accessed on 1 September 2021).

30. Taiwan Fisheries Agency; Council of Agriculture. Available online: https://www.fa.gov.tw/cht/index.aspx (accessed on 1 September 2021).

31. Baremore, I.E.; Hale, L.F. Reproduction of the Sandbar Shark in the Western North Atlantic Ocean and Gulf of Mexico. *Mar. Coast. Fish.* **2012**, *4*, 560–572. [CrossRef]

32. Goodwin, N.B.; Dulvy, N.K.; Reynolds, J.D. Life-history correlates of the evolution of live bearing in fishes. *Philos. Trans. R. Soc. Ser. B Biol. Sci.* **2002**, *357*, 259–267. [CrossRef]

33. Matthews, L.H. Reproduction in the basking shark, *Cetorhinus maximus* (Gunner). *Philos. Trans. R. Soc. Ser. B Biol. Sci.* **1950**, *234*, 247–316.

34. Tanaka, S.; Yano, K. Histological observations on the reproductive organs of a female megamouth shark, *Megachasma pelagios*, from Hakata Bay, Japan. In *Biology of the Magamouth Shark*; Yano, K., Morrissey, J.F., Yabumoto, Y., Nakaya, K., Eds.; Tokai University Press: Tokyo, Japan, 1997; pp. 121–129.

35. Wigley, S.E.; McBride, H.M.; McHugh, N.J. *Length-Weight Relationships for 74 Fish Species Collected during NEFSC Research Vessel Bottom Trawl Surveys*; NOAA Technical Memorandum NMFS-NE 171; Northeast Fisheries Science Center: Woods Hole, MA, USA, 2003; pp. 1992–1999.

36. Hsu, H.H.; Joung, S.J.; Liu, K.M. Fisheries, management and conservation of the whale shark *Rhincodon typus* in Taiwan. *J. Fish Biol.* **2012**, *80*, 1595–1607. [CrossRef] [PubMed]

37. Robbins, R.L. Environmental variables affecting the sexual segregation of great white sharks *Carcharodon carcharias* at the Neptune Islands South Australia. *J. Fish Biol.* **2007**, *70*, 1350–1364. [CrossRef]

38. Borrell, A.; Aguilar, A.; Gazo, M.; Kumarran, R.P.; Cardona, L. Stable isotope profiles in whale shark (*Rhincodon typus*) suggest segregation and dissimilarities in the diet depending on sex and size. *Environ. Boil. Fishes* **2011**, *92*, 559–567. [CrossRef]

39. Klimley, A.P. The determinants of sexual segregation in the scalloped hammerhead shark, *Sphyrna lewini*. *Environ. Boil. Fishes* **1987**, *18*, 27–40. [CrossRef]

40. Mucientes, G.R.; Queiroz, N.; Sousa, L.L.; Tarroso, P.; Sims, D. Sexual segregation of pelagic sharks and the potential threat from fisheries. *Biol. Lett.* **2009**, *5*, 156–159. [CrossRef] [PubMed]

41. Marshall, A.D.; Bennett, M.B. Reproductive ecology of the reef manta ray *Manta alfredi* in southern Mozambique. *J. Fish Biol.* **2010**, *77*, 169–190. [CrossRef]

42. Carey, F.G.; Teal, J.M. Mako and porbeagle: Warm-bodied sharks. *Comp. Biochem. Physiol.* **1969**, *28*, 199–204. [CrossRef]

43. Carey, F.G.; Casey, J.G.; Pratt, H.L.; Urquhart, D.; McCosker, J.E. Temperature, heat production, and heat exchange in lamnid sharks. *Mem. South. Calif. Acad. Sci.* **1985**, *9*, 92–108.

44. Goldman, K.J.; Anderson, S.D.; Latour, R.J.; Musick, J.A. Homeothermy in adult salmon sharks, *Lamna ditropis*. *Environ. Boil. Fishes* **2004**, *71*, 403–411. [CrossRef]

45. Carlisle, A.B.; Goldman, K.J.; Litvin, S.Y.; Madigan, D.J.; Bigman, J.S.; Swithenbank, A.M.; Thomas, C.; Kline, T.C.; Block, B.A. Stable isotope analysis of vertebrae reveals ontogenetic changes in habitat in an endothermic pelagic shark. *Proc. R. Soc. Ser. B Biol. Sci.* **2015**, *282*, 20141446. [CrossRef] [PubMed]

46. Guzman, H.M.; Gomez, C.G.; Hearn, A.; Eckert, S.A. Longest recorded trans-Pacific migration of a whale shark (*Rhincodon typus*). *Mar. Biodivers. Rec.* **2018**, *11*, 8. [CrossRef]

47. Bayliff, W.H.; Ishizuki, Y.; Deriso, R.B. Growth, movement, and attrition of northern bluefin tuna, *Thunnus thynnus*, in the Pacific Ocean, as determined by tagging. *Inter Am. Trop. Tuna Comm. Bull.* **1991**, *20*, 1–94.

48. Dewar, H.; Thys, T.; Teo, S.L.H.; Farwell, C.; O'Sullivan, J.; Tobayama, T.; Soichi, M.; Nakatsubo, T.; Kondo, Y.; Okada, Y.; et al. Satellite tracking the world's largest jelly predator, the ocean sunfish, Mola mola, in the Western Pacific. *J. Exp. Mar. Biol. Ecol.* **2010**, *393*, 32–42. [CrossRef]

49. Teague, W.J.; Shiller, A.M.; Hallock, Z.R. Hydrographic section aeross the Kuroshio near 35° N, 143° E. *J. Geophys. Res. Ocean.* **1994**, *99*, 7639–7650. [CrossRef]

50. Chen, C.-T.; Liu, K.-M.; Chang, Y.-C. Reproductive biology of the bigeye thresher shark, *Alopias superciliosus* (Lowe, 1839) (Chondrichthyes: Alopiidae), in the northwestern Pacific. *Ichthyol. Res.* **1997**, *44*, 227–235. [CrossRef]

51. Wilson, S.G.; Taylor, J.G.; Pearce, A.F. The Seasonal Aggregation of Whale Sharks at Ningaloo Reef, Western Australia: Currents, Migrations and the El Niño/Southern Oscillation. *Environ. Boil. Fishes* **2001**, *61*, 1–11. [CrossRef]

52. Tang, T.; Tai, J.-H.; Yang, Y.-J. The flow pattern north of Taiwan and the migration of the Kuroshio. *Cont. Shelf Res.* **2000**, *20*, 349–371. [CrossRef]

53. Thums, M.; Meekan, M.; Stevens, J.; Wilson, S.; Polovina, J. Evidence for behavioural thermoregulation by the world's largest fish. *J. R. Soc. Interface* **2013**, *10*, 20120477. [CrossRef]

54. Gore, M.A.; Rowat, D.; Hall, J.; Gell, F.R.; Ormond, R.F. Transatlantic migration and deep mid-ocean diving by basking shark. *Biol. Lett.* **2008**, *4*, 395–398. [CrossRef] [PubMed]
55. Araujo, G.; Labaja, J.; Snow, S.; Huveneers, C.; Ponzo, A. Changes in diving behaviour and habitat use of provisioned whale sharks: Implications for management. *Sci. Rep.* **2020**, *10*, 16951. [CrossRef] [PubMed]
56. Amorim, A.F.; Arfelli, C.A.; Castro, J.I. Description of a Juvenile Megamouth Shark, *Megachasma pelagios*, Caught off Brazil. *Environ. Boil. Fishes* **2000**, *59*, 117–123. [CrossRef]
57. Nakagawa, Y.; Endo, Y.; Sugisaki, H. Feeding rhythm and vertical migration of the euphausiid *Euphausia pacifica* in coastal waters of north-eastern Japan during fall. *J. Plankton Res.* **2003**, *25*, 633–644. [CrossRef]
58. Potter, I.F.; Howell, W.H. Vertical movement and behavior of the ocean sunfish, Mola mola, in the northwest Atlantic. *J. Exp. Mar. Biol. Ecol.* **2010**, *396*, 138–146. [CrossRef]
59. Chang, C.-T.; Lin, S.-J.; Chiang, W.-C.; Musyl, M.K.; Lam, C.-H.; Hsu, H.-H.; Chang, Y.-C.; Ho, Y.-S.; Tseng, C.-T. Horizontal and vertical movement patterns of sunfish off eastern Taiwan. *Deep Sea Res. Part II Top. Stud. Oceanogr.* **2019**, *175*, 104683. [CrossRef]
60. Lavenberg, R.J.; Seigel, J.A. The Pacific's megamystery—Megamouth. *Terra* **1985**, *23*, 29–31.
61. Berra, T.M.; Hutchins, J.B. A specimen of megamouth shark, *Megachasma pelagios* (Megachasmidae) from Western Australia. *Rec. West. Aust. Mus.* **1990**, *14*, 651–656.
62. Nakaya, K. Discovery of a megamouth shark from Japan. *Jpn. J. Ichthyol.* **1989**, *36*, 144–146. [CrossRef]
63. Miya, M.; Hirosawa, M.; Mochizuki, K. Occurrence of a megachasmid shark in Suruga Bay: Photographic evidence. *J. Nat. Hist. Mus. Inst.* **1992**, *2*, 41–44.
64. Takada, K.; Hiruda, H.; Wakisaka, S.; Mori, T.; Nakaya, K. Capture of the first female megamouth shark, *Megachasma pelagios*, from Hakata Bay, Fukuoka, Japan. In *Biology of the Magamouth Shark*; Yano, K., Morrissey, J.F., Yabumoto, Y., Nakaya, K., Eds.; Tokai University Press: Tokyo, Japan, 1997; pp. 3–9.
65. Yano, K.; Yabumoto, Y.; Tanaka, S.; Tsukada, O.; Furuta, M. Capture of a mature female megamouth shark, *Megachasma pelagios*, from Mie, Japan. In Proceedings of the 5th Indo-Pacific Conference, Nouméa, New Caledonia, 3–8 November 1997; pp. 335–349.
66. Smale, M.J.; Compagno, L.J.V.; Human, B.A. First megamouth shark from the western Indian Ocean and South Africa: News & views. *S. Afr. J. Sci.* **2002**, *98*, 349–350.
67. Lee, P.F.; Shao, K.T. Two new records of Lamniform shark from the waters adjacent to Taiwan. *J. Fish. Soc. Taiwan* **2009**, *36*, 303–311.
68. Tanaka, T.; Noguchi, F.; Tanaka, S. Dentition of a male megamouth shark, *Megachasma pelagios* from Suruga Bay, Japan, with a comparison of the fossil shark teeth from Chile. *Rep. Jpn. Soc. Elasmobranch Stud.* **2004**, *40*, 31–37.
69. Iida, M. Catch of megamouth shark by set net in Sagami Bay. *Rep. Jpn. Soc. Elasmobranch Stud.* **2004**, *40*, 38–40.
70. Castillo-Géniz, J.L.; Ocampo-Torres, A.I.; Shimada, K.; Rigsby, C.K.; Nicholas, A.C. Juvenile megamouth shark, *Megachasma pelagios*, caught off the Pacific coast of Mexico, and its significance to chondrichthyan diversity in Mexico. *Cienc. Mar.* **2012**, *38*, 467–474. [CrossRef]
71. De Moura, J.F.; Merico, A.; Montone, R.C.; Silva, J.; Seixas, T.G.; de Oliveira Godoy, J.M.; Pierre, T.D.; Hauser-Davis, R.A.; Di Beneditto, A.P.M.; Reis, E.C.; et al. Assessment of trace elements, POPs, 210Po and stable isotopes (15N and 13C) in a rare filter-feeding shark: The megamouth. *Mar. Pollut. Bull.* **2015**, *95*, 402–406. [CrossRef]
72. Senou, H.; Taru, H.; Tanaka, S. Record of a megamouth shark, *Megachasma pelagios* (Elasmobranchii: Megachasmidae), from Sagami Bay. *Rep. Jpn. Soc. Elasmobranch Stud.* **2012**, *48*, 21–27.
73. Asahi Shimbun. Available online: https://www.asahi.com/ (accessed on 1 September 2021).
74. Misawa, R.; Wada, J.; Kitadani, Y.; Nishida, K.; Kai, Y.; Mizumachi, K.; Endo, H. A checklist of sharks based on voucher specimens and photographs from Kochi Prefecture (southern Shikoku Island, Japan). *Rep. Jpn. Soc. Elasmobranch Stud.* **2019**, *55*, 31–54.
75. Elusive Megamouth Shark Snared in Mexico. Available online: http://weareseaborn.blogspot.com/2011/09/elusive-megamouth-shark-snared-in.html (accessed on 1 September 2021).
76. Tanaka, S.; Horie, T.; Yuki, Y. Occurrence and catch records of megamouth shark, *Megachasma pelagios* in Japan from 2013 to 2014. *Rep. Jpn. Soc. Elasmobranch Stud.* **2014**, *50*, 35–39.
77. Senou, H. The largest megamouth shark in Japan collected from Sagami Bay. *Rep. Jpn. Soc. Elasmobranch Stud.* **2013**, *49*, 18–20.
78. Fujii, M. Bycatch of megamouth shark by a set net in the east coast of Izu. *Rep. Jpn. Soc. Elasmobranch Stud.* **2015**, *51*, 21–23.
79. Chukyo TV. Available online: https://www.ctv.co.jp/indexmenu.html (accessed on 1 September 2021).

Article

Growth Modeling of the Giant Electric Ray *Narcine entemedor* in the Southern Gulf of California: Analyzing the Uncertainty of Three Data Sets

Pablo Mora-Zamacona [1], Felipe N. Melo-Barrera [1], Víctor H. Cruz-Escalona [1,*], Andrés F. Navia [2], Enrique Morales-Bojórquez [3], Xchel A. Pérez-Palafox [1] and Paola A. Mejía-Falla [2]

[1] Centro Interdisciplinario de Ciencias Marinas, Instituto Politécnico Nacional, Baja California Sur, La Paz 23096, Mexico; pablomoraz90@gmail.com (P.M.-Z.); fmelo@ipn.mx (F.N.M.-B.); xapp39@gmail.com (X.A.P.-P.)

[2] Fundación Colombiana para la Investigación Y Conservación de Tiburones Y Rayas, SQUALUS, Calle 10A No 72-35, Apto. 310 E, Cali 760001, Colombia; anavia@squalus.org (A.F.N.); pmejia@squalus.org (P.A.M.-F.)

[3] Centro de Investigaciones Biológicas del Noroeste, Av. Instituto Politécnico Nacional 195, Col. Playa Palo de Santa Rita Sur, La Paz 23096, Mexico; emorales@cibnor.mx

* Correspondence: vescalon@ipn.mx; Tel.: +52-1-612-12-253-44

Simple Summary: The study of age and growth patterns in skates and rays can be conducted by analyzing mineral deposition patterns inside the vertebrae as biological features may influence age estimation. For the giant electric ray (*Narcine entemedor*), age was estimated by analyzing the vertebrae and an annual deposition pattern was found. After considering additional biological features such as birth date and date of capture, a more precise description of growth pattern was made. We concluded that this species is a moderate body size elasmobranch with moderate longevity and fast growth. Our results provide useful information for the future management of this exploited species.

Abstract: The age and growth rate of the giant electric ray, *Narcine entemedor*, was estimated using growth bands deposited in the vertebral centra of 245 specimens. Differences in size and age distribution were found between the sexes, a pattern that suggests the annual deposition of band pairs, possibly occurring in April. Multimodel inference and back-calculation were performed to three age data sets of females considering their reproductive cycle and time of capture, among which the von Bertalanffy growth function was found to be the most appropriate (L_∞ = 81.87 cm TL, k = 0.17 year^{-1}). Our research supports the idea that age can be determined via biological features such as birth date and growth band periodicity. We concluded that *N. entemedor* is of a moderate body size, moderate longevity and is a fast-growing elasmobranch species.

Keywords: elasmobranchs; individual growth modelling; artisanal fisheries; multimodel inference; back-calculation

Citation: Mora-Zamacona, P.; Melo-Barrera, F.N.; Cruz-Escalona, V.H.; Navia, A.F.; Morales-Bojórquez, E.; Pérez-Palafox, X.A.; Mejía-Falla, P.A. Growth Modeling of the Giant Electric Ray *Narcine entemedor* in the Southern Gulf of California: Analyzing the Uncertainty of Three Data Sets. *Animals* **2022**, *12*, 19. https://doi.org/10.3390/ani12010019

Academic Editor: Martina Francesca Marongiu

Received: 5 November 2021
Accepted: 10 December 2021
Published: 23 December 2021

Publisher's Note: MDPI stays neutral with regard to jurisdictional claims in published maps and institutional affiliations.

1. Introduction

Traditionally, it has been recognized that elasmobranchs, due to their biological characteristics (late maturation, low fecundity, slow growth), are especially vulnerable to overfishing [1–4]. Nevertheless, evidence has proven that life-history traits of this group may vary between K and r strategies, allowing some species to respond differently to fishing pressure [5–7]. In this sense, it is important to understand the life-history parameters of the species, particularly those that are related to the degree of vulnerability and risk to fishing pressure, providing basic information for demographic models [8–10]. Among the most important parameters for this purpose are the ones related to the age and growth of the species [11].

The number of age and growth studies in elasmobranchs has increased significantly in recent years [12]; however, the techniques and structures used for these purposes have remained constant, based mainly on the identification and count of opaque and translucent banding patterns present in hard anatomical structures such as vertebral centra, denticles and dorsal spines [13,14].

Many studies on the age and growth of elasmobranchs have been encouraged by the increasing exploitation of this group, which has been documented by numerous researchers, reporting the important effects of fishing mortality on batoid populations [15–18]. In Mexican elasmobranch fisheries, the main species of batoids captured are benthic, such as *Pseudobatos productus*, *Zapteryx exasperate* and *Hypanus dipterurus* [19–22], although some pelagic species such as *Pteroplatytrygon violacea* and *Myliobatis californica* have been found frequently in the catch [22,23]. In all these study cases, *Narcine entemedor* (Jordan and Starks, 1895) have been reported with low abundances; nevertheless, Villavicencio-Garayzar [24] and Márquez-Farías [25] reported that they are commonly captured in artisanal fisheries of Mexico, especially during spring and summer. In Bahía de La Paz, an artisanal fishery on batoids captures around 14 species, among which *N. entemedor* ranks third in captures [26].

The giant electric ray *N. entemedor* is an endemic batoid of the eastern tropical Pacific Ocean distributed from southern Baja California and the Gulf of California to Peru. While the diet and feeding ecology [27] and reproductive biology [24,28] of the giant electric ray have been investigated, there is little information on the age and growth of this species. Moreover, information on the species is limited to the Pacific coast of Mexico, while life-history traits could vary in Central and South America. The objective of this study was to estimate age and growth parameters for *N. entemedor* in the southern Gulf of California using a multimodel inference approach, and to consider the influence of biological features in the estimates. We conclude that using data such as birth date may result in a more precise description of individual growth.

2. Materials and Methods

2.1. Collection of Samples

The specimens were captured in collaboration with a fisherman who has a commercial fishing permit under Mexican fishing regulations and laws (CONAPESCA-103053993316-1); therefore, the data in this study are fishery-dependent. Monthly samplings were made from October 2013 through December 2015 in the south of Bahía de La Paz, located in the southern portion of the Gulf of California (24°25′ N, 110°18′ W). The organisms were captured using monofilament gill nets (200–300 m long, 1.5 m high, 20–25 cm stretch mesh) set in the afternoon at depths between 10 and 30 m over sandy bottoms and recovered the next morning. Individuals were measured for total length (TL in cm), and the sex was determined by the presence of copulatory organs in males (claspers; [28]). Vertebrae were collected from the abdominal region of each specimen and placed in a freezer.

2.2. Vertebral Preparation

Vertebrae were thawed, excess tissue manually removed and individual centra were separated using a scalpel. Cleaned vertebrae centra were dried at room temperature. A qualitative analysis to test the effectiveness of diverse treatments to enhance the visibility of growth bands was evaluated. Three cutting thicknesses (0.3, 0.4 and 0.5 mm) and two dyes, Alizarin Red S (0.01 g/250 mL water) and Bismarck Brown Y (0.01 g/250 mL alcohol 95%), were tested to enhance the visibility of the centra growth bands. Staining was performed at different times of exposure, from 1 min to saturation. However, none of the dyes tested presented an improvement in the clarity of growth bands; thus, vertebrae were instead prepared as follows. Centra were fixed on wooden structures using a cyanoacrylate-based adhesive. Sagittal sections of 0.4 mm thick were made using double saws fitted with diamond-impregnated blades, ensuring that the focus of the centrum was included. Subsequently, sections were cleaned with scalpel and water and then dried at room temperature.

2.3. Reading of Growth Bands

Thin sections were observed under a stereo microscope (Olympus SZX9) and digitized using a video camera (Sony CCD-IRIS-RGB). The sections were illuminated using reflected light on a dark background and submerged in a thin layer of water to improve the observation. The birth band was defined as the angle change on the centrum side [29]. Additionally, a centrum of a 12.4 cm TL near-term embryo [28] was polished and compared with a thin section of an adult. The size of the embryo's centra and the birth band coincided; furthermore, no pre-birth bands were observed. The radius of each vertebra was measured on the corpus calcareum long a straight line through the focus of each vertebra with Sig-maScan Pro 5.0.0 Software (Systat Software, Palo Alto, CA, US). The vertebral radius (VR) was plotted against TL and tested for a linear relationship to determine if these vertebrae provided a suitable structure for age determination and for back-calculated estimation of length at previous ages. The influence of sex in the VR-TL relationship was evaluated with a test of slopes and elevations [30].

A training exercise counting the bands of a subsample (n = 50) was performed by two readers to refine the identification and growth band counts criteria. The following criteria were established: (1) identification of the presence of pairs of growth bands (one translucent and one opaque; [31]), (2) identification of a birth band through a change in the angle of the corpus calcareum in the place closest to the focus of the vertebra [29] and (3) counting of translucent bands (Figure 1). The two readers (reader 1 was most experienced) then did a simultaneous and independent band count without knowing the sex or size of the specimens. Readings of bands were performed on the corpus calcareum due to the poor visibility in the intermedialia zone. This procedure was repeated twice. Any vertebra yielding an age estimate that differed between counts was re-examined by both readers jointly; if no consensus was reached, the sample was discarded.

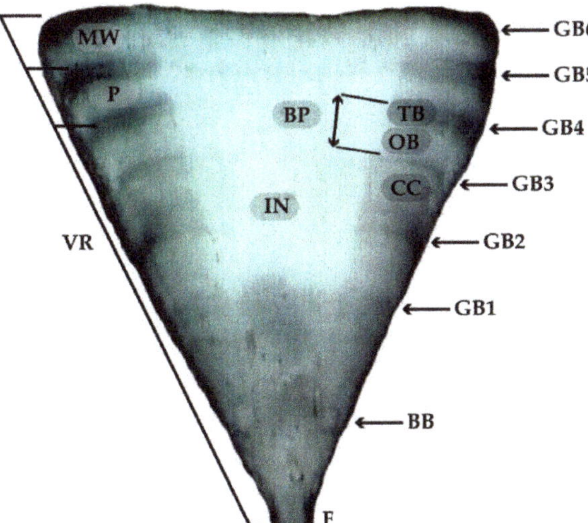

Figure 1. Sagittal section of a centrum of a six-year-old *Narcine entemedor* specimen, showing MW, marginal width; P, previous band pair width; VR, the vertebral radius; F, focus; CC, corpus calcareum; IN, intermedialia zone; BP, band pair; OB, opaque band; TB, translucent band; BB, birth band; GB, growth bands.

2.4. Precision and Accuracy

According to Campana [31], aging errors can be expressed as follows: (a) discrepancies in the reproducibility of repeated measurements on a given structure (precision), and (b) differences between the closeness of the age estimate to the true value (accuracy). Thus, count reproducibility, as indicated by reader variability, was determined by calculating the percent of agreement (PA) by ± 1 mark [32], average percent error (APE; [33]) and coefficient of variation, which is an alternative precision analysis that uses the standard deviation rather than the absolute deviation (CV; [34]). Each method was applied to the total of the sample. Additionally, to assess whether there are systematic differences between the readings made by the readers, age-bias plots of band counts [35] and Bowker's test of symmetry were performed [36].

2.5. Periodicity of Band Formation

The periodicity of band pair formation was evaluated using two methods. The "centrum edge analysis" considers whether the last deposited band was translucent or opaque and relates it to the month of capture [37]. The "marginal increment analysis" was undertaken by measuring the distance from the last band to the edge of the centrum (marginal width) as a proportion of the distance between the last and the penultimate band pair (previous band pair width; [38]).

$$MIR = \frac{MW}{P} = \frac{VR - R_n}{R_n - R_{n-1}} \tag{1}$$

where MIR = the marginal increment ratio, MW = marginal width, P = previous band pair width, VR = the vertebral radius, R_n = the distance from the focus to the last complete growth mark band and R_{n-1} = the distance from the centrum origin to the penultimate complete growth mark(Figure 1). Those distances were measured using SigmaScan Pro 5.0.0 Software (SPSS Inc., Chicago, IL, USA). Individuals that presented only one translucent band (referred to as the birth band) were not considered for MIR analysis. Mean MIRs were plotted against months to examine trends in band formation. A Kruskal–Wallis test was used to examine for differences among months, followed by a nonparametric multiple comparison test to find the months among these differences were presented [30]. In addition, a Kolmogorov–Smirnov distribution test was used to examine for differences in age structures between sexes [39].

2.6. Age Adjustment

For the individual age estimation, three data sets were analyzed, one using the growth band counts information (unadjusted) and two more performing an adjustment to age, considering reproductive cycle (Adjustment 1) and time of capture (Adjustment 2). For the unadjusted analysis, it was considered that the birth band formed shortly after parturition, irrespective of reproductive seasonality. Adjustment 1 was adjusted to the period between birth and first band formation. In the study area, *N. entemedor* has two birth peaks per year, a major one during August and a minor one during January [28]; the month of band formation is April (see Section 3). We considered the birth month to be August; thus, the time between birth and formation of the first band was assumed to be seven months (0.58 years). Adjustment 2, the second age adjustment integrated the capture date; therefore, age was adjusted with the time between the month of band formation and capture month.

2.7. Back-Calculation

Due to the small sample size of juvenile giant electric rays, back-calculated estimates of length at previous ages were calculated for each of the three data sets. Back-calculated lengths were calculated using the proportion-based back-calculation equation proposed by Francis [40], modified from Hile [41]. The equation used to back-calculate the lengths at presumed ages was:

$$L_i = -\left(\frac{a}{b}\right) + \left(L_c + \frac{a}{b}\right)\left(\frac{VR_i}{VR_c}\right) \tag{2}$$

where L_i is the TL at time i, a and b are parameters obtained from the linear relation between total length and vertebral radius, L_c is the TL at capture; VR_c is the VR at capture and VR_i is the VR at age i.

2.8. Growth Estimation

A multimodel inference approach was used to determine the most appropriate candidate growth model [42]. The candidate set of models consisted of the traditional 3-parameter von Bertalanffy growth model (VBG-3; [43]); a 2-parameter modified form of the VBG forced through the length-at-birth (L_0) (VBG–2; [44]), where L_0 was estimated using the largest near-term embryo (i.e., 14.5 cm TL) reported by Burgos-Vázquez et al. [28]; the 3-parameter Gompertz growth model (GG-3) and the logistic model with three parameters (LG-3; [45]).

The four growth models were fitted to a combination of back-calculated lengths and sample data for the three data sets previously described (unadjusted and adjusted ages), and the resulting parameters were estimated and compared. The parameters in the candidate growth models were estimated when the negative log-likelihood was maximized with a nonlinear fit using the generalized reduced gradient method, assuming a multiplicative error in the residuals [46,47]. The objective function is expressed as follows:

$$- logL(\theta_i | data) = \sum_n \left[-\frac{1}{2} ln(2\pi) \right] - \left[-\frac{1}{2}\left(\sigma^2\right) - \frac{(lnTL_O - lnTL_E)^2}{2\sigma^2} \right] \tag{3}$$

where n is number of data, i indicates the number of parameters for each candidate growth model selected (VBG-3, VBG-2, GG-3, and LG-3), TL_O is the total length observed, and TL_E is the total length estimated. For σ, the following analytical solution [48] is proposed:

$$\sigma = \sqrt{\frac{1}{n} \sum_{t=1}^{n} (lnTL_O - lnTL_E)^2} \tag{4}$$

2.9. Confidence Intervals

To estimate the confidence intervals (CI) of the θ_i parameters in the candidate growth models, two approaches were used: (1) the likelihood profile method [48] for the parameter t_0 because there is no correlation between parameters and (2) the likelihood contour method when there is a correlation between parameters [49–51], which is observed in the parameters L_∞ and k. For the likelihood profile method, a chi-square distribution with 1 degree of freedom (df) was used, and therefore, values equal to or less than 3.84 were accepted within the CI. For the likelihood contour method, a chi-square distribution with 2 df was used, and values equal to or less than 5.99 were accepted within the CI [30]. The CIs were estimated based on Haddon [49].

2.10. Model Selection

Model performance was evaluated using Akaike's Information Criterion (AIC), where the best model was the one with the lowest AIC_c value. For model comparisons, the delta AIC (ΔAIC) and Akaike weights (w_i) were calculated [42]. The ΔAIC is a measure of each model relative to the best model and is calculated as $\Delta AIC = AIC_i - minAIC$, where AIC_i is the AIC value for model i and $minAIC$ is the AIC value of the best model. Models with ΔAIC of 0–2 had substantial support, while models with ΔAIC of 4–7 had considerably less support, and models with $\Delta AIC > 10$ had essentially no support. Akaike weights (w_i) represent the probability of choosing the correct model from the set of R-candidate models and was calculated as:

$$w_i = \frac{e^{-0.5\Delta AIC}}{\sum_{i=1}^{R} e^{-0.5\Delta AIC}} \tag{5}$$

3. Results

3.1. Collection of Samples

A total of 305 specimens (260 females and 45 males) were initially used for the aging study. Of the processed vertebrae, 245 (80%) were readable from 209 females ranging in size from 49 to 84 cm TL and 36 males ranging from 41.5 to 58.8 cm TL, with females being significantly larger than males (D = 0.84, p < 0.001) and having a well-represented sample for each month of the year (Table 1). Growth bands were poorly visible in the intermedialia zone (Figure 2); therefore, reading was performed on the corpus calcareum. Significant linear relationships between VR and TL (p = 0.001) were found for both sexes (females: TL = 0.08VR−0.65, r^2 = 0.81; males: TL = 0.07VR−0.2, r^2 = 0.67), verifying that these vertebrae were suitable structures for age determination. The mean radius of the observed birth band was 0.79 ± 0.08 mm (mean and S.E.). Similarly, the mean VR of the near-term embryo was 0.69 ± 0.02 mm TL, indicating that the birth band was identified correctly.

Table 1. Monthly sample size of *Narcine entemedor* specimens analyzed in the present study.

Month	January	February	March	April	May	June	July	August	September	October	November	December
Sample size	18	33	7	3	29	10	42	37	20	14	10	22

Figure 2. Sagittal sections of centra where growth bands were not evident and could not be read.

3.2. Precision and Accuracy

Age estimates agreed closely between readers. Age band counts resulted in an APE between readers of 3.3% and CV of 4.7%, with a PA of 69%, PA ± by one band of 95% and PA ± by two bands of 100%. Both age-bias plots (Figure 3) and Bowker's test of symmetry (X^2 = 14.5, df = 9, p = 0.89) indicated no systematic differences between readers. These precision and accuracy values indicate a high level of reproducibility.

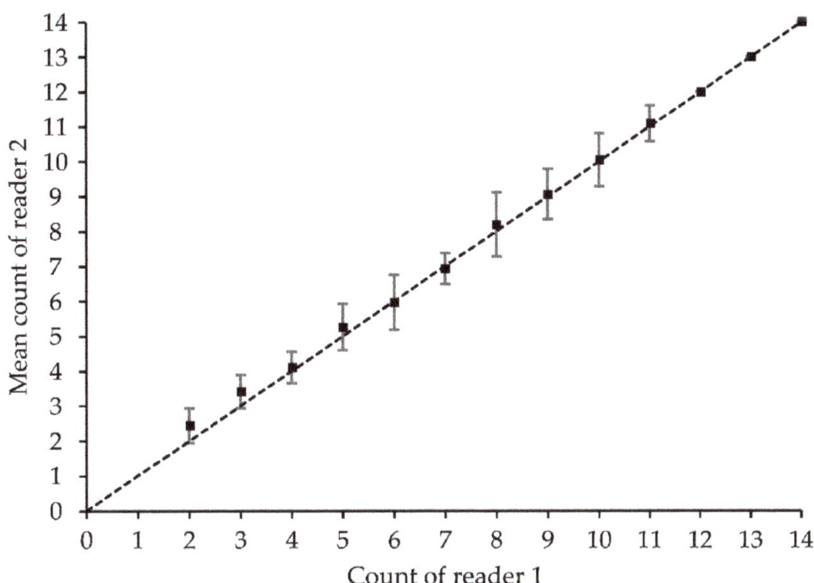

Figure 3. Age-bias plot of reader band-pair counts. Dots with standard deviation bars are the mean counts of reader 2 relative to reader 1. The diagonal line indicates a one-to-one relationship.

3.3. Periodicity of Band Formation

The categorization of growth bands at the edges of the vertebrae as opaque or translucent was possible. There were no differences in monthly marginal increments (Kruskal–Wallis: $H_{11, 246} = 16.7$, $p = 0.11$), although a pattern was observed in both marginal increment analysis and edge analysis (Figure 4). The mean MIR and the translucent edge percentage were highest during April and lowest during June, suggesting that a single band pair is formed annually on the vertebral centra of *N. entemedor* during April. Assuming the formation of a pair of bands each year, 14 age groups were identified for *N. entemedor* (Tables 2 and 3), as well as the formation of a birth band right after birth. The fourth and fifth age groups were predominant, whereas the 14–15 year groups were poorly represented (Figure 5a,c,e). The one-year age group was represented only by males (n = 2), and in the 2-year age group, males were more frequent than females. For the 3–6 year age groups, females were increasingly more frequently than males, which were absent in the remaining age groups. The age structure was different between sexes (D = 0.66, $p < 0.001$).

3.4. Growth Estimation and Model Selection

Due to a low sample size of males, growth models were only adjusted to the observed and back calculated age-length data of females. Growth models fitted to the data are shown in Table 4, with CIs for growth parameters L_∞ and k for each one of the data sets analyzed. Their likelihood contours are shown in Figure 5b,d,f. Based on *AIC* values, the VBG-3 presented the best fit to the data to describe the growth of *N. entemedor* females for every data set analyzed (unadjusted and adjusted ages; Table 4 and Figure 5a,c,e). Furthermore, the VBG-3 was the only fitted model with empirical support ($\Delta_i = 0$) for every data set analyzed, while the rest of the models had no support ($\Delta_i > 10$). In each dataset, VBG-3 obtained the maximum $-\log$-likelihood values. The VBG-3 unadjusted-age data set showed the maximum $-\log$-likelihood value among the three, while the rest of the parameters varied slightly between data sets, L_∞ ranged from 81.5 to 82.1 cm TL, and the estimates were close to the observed maximum length (TL = 84.0 cm); k remained constant at 0.17 cm year^{-1}.

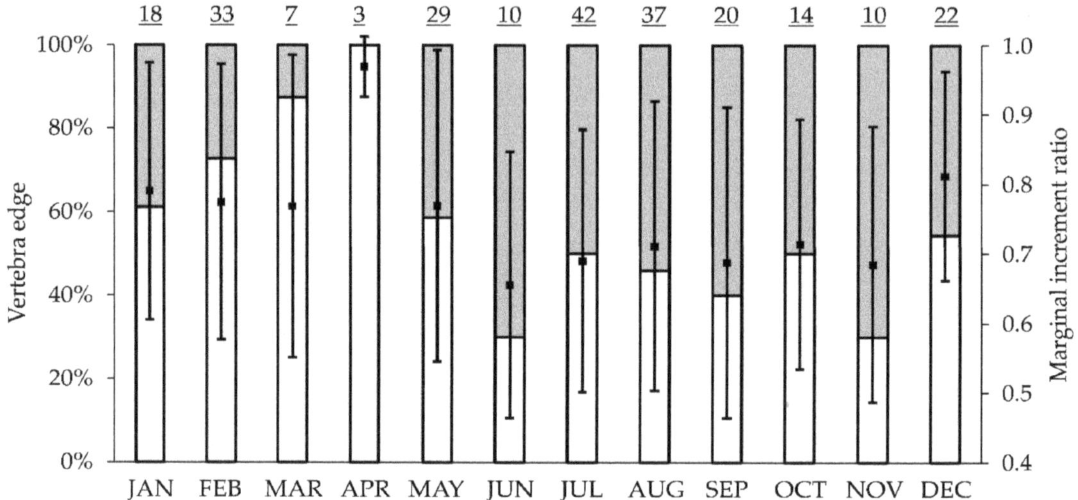

Figure 4. Monthly variations in vertebra edge and marginal increment ratios (MIR). Monthly frequency of translucent (□) and opaque bands (■) determined from thin sections and monthly variation of mean MIR. Sample size is indicated on top of the graph; (■) mean monthly MIR; bars show the standard error for monthly values.

Table 2. Age–length distribution for *Narcine entemedor* obtained from growth band counts of vertebrae sections. Age estimation follows the age adjustment 2 (adjusted to reproductive cycle and date of capture).

Total Length (cm)	Age Group (Down-Rounded)														Total
	1	2	3	4	5	6	7	8	9	10	11	12	13	14	
42		1													1
44		2													2
46															0
48	2	1	1												4
50		1	3		1										5
52		5		5											10
54		1	4	5		1									11
56		1	1	10	3										15
58			4	6	3	3	1								17
60			3	7	9	3									22
62			1	3	12	6									22
64			1	3	7	2	2	1							16
66				3	5	7	3	1							19
68				1		3	5	1	3	1					14
70						3	5	2	5	6	1				22
72							4	1	6	8	2	1			22
74								3	4	8	5	3			23
76								2	2	1	2	2			9
78										2	4	2			8
80															0
82													1	1	2
84														1	1
Total	2	12	18	40	38	26	24	13	21	26	14	8	1	2	245

Table 3. Age–length distribution for *Narcine entemedor* obtained from growth band counts of vertebrae sections and back-calculated. Age estimation follows the age adjustment 2 (adjusted to reproductive cycle and date of capture).

Total Length (cm)	\n Age Group (Down-Rounded)															Total
	0	1	2	3	4	5	6	7	8	9	10	11	12	13	14	
16										1						1
18	3															3
20	10															10
22	59															59
24	78															78
26	67	4														71
28	14	19														33
30	9	37	2													48
32	5	48	3													56
34		63	10													73
36		31	25	2												58
38		27	34	5												66
40		6	48	7	1											62
42		7	49	16	2											74
44		2	25	31	5											63
46		1	23	41	8	2										75
48		2	16	44	19	4										85
50			3	38	32	7										80
52			8	22	45	11	4									90
54			1	13	33	24	8									79
56			1	9	40	31	8	3								92
58				6	21	32	25	7	1							92
60				3	13	34	19	12	2							83
62				1	4	23	26	19	6	1						80
64				1	3	14	22	21	18	1						80
66						3	16	23	15	12	1					70
68					1		7	12	21	15	5					61
70							3	11	14	17	14	3				62
72								5	5	16	12	5	1			44
74									3	8	12	7	4			34
76									2	2	4	4	3			15
78											2	5	3	1		11
80																0
82													1	1	1	3
84														1	1	2
Total	245	247	248	239	227	185	138	113	87	73	50	24	12	3	2	1893

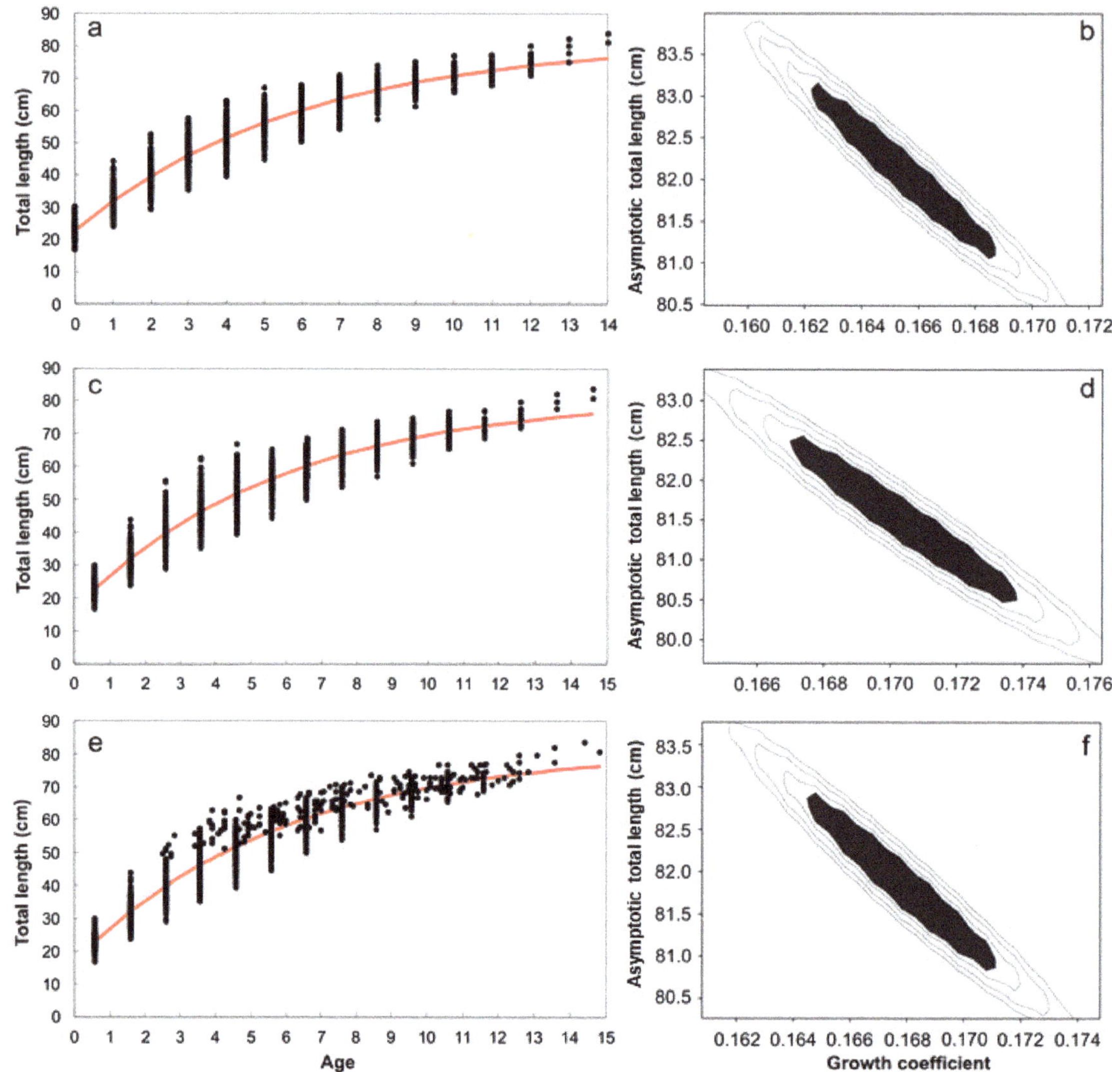

Figure 5. Growth curves and likelihood contours for L_∞ (asymptotic total length) and k (growth coefficient), estimated for VBG-3 for three different data sets: unadjusted (**a,b**), adjusted to reproductive cycle (**c,d**) and adjusted to reproductive cycle and date of capture (**e,f**). For growth curves, black dots represent the observed and back-calculated data, and the red lines represent estimated growth. For likelihood contours, shaded areas denote the confidence region for L_∞ and k, assuming a chi-square distribution with n = 2 df.

Table 4. Growth model parameter estimates of female *N. entemedor* using three different data sets (unadjusted, adjusted to reproductive cycle (RC) and adjusted to the reproductive cycle and date of capture (DC); see text for details). L_∞ is the theoretical asymptotic size, k is the growth coefficient, L_0 is the birth size estimated at age 0, t_0 is the hypothetical age at length zero, $-$LL is negative log-likelihood, *AIC* is Akaike's Information Criterion, Δ_i is the Akaike difference, and W_i is the *AIC* weight estimates. CIs estimated from log-likelihood contours are shown in parentheses.

Dataset	Model	L_∞	k	t_0	L_0	$-$LL	*AIC*	Δ_i	W_i
Unadjusted	VBG-3	82.09 (80.59–83.69)	0.165 (0.170–0.173)	−1.988 (−2.008––1.968)	23.01	3507.23	−7008.45	0.00	1.00
	VBG-2	69.35 (67.25–71.65)	0.314 (0.294–0.334)	-	14.50	2158.51	−4313.03	2695.43	0.00
	GG-3	73.49 (72.89–74.14)	0.301 (0.292–0.310)	0.46 (0.437–0.477)	23.32	3486.66	−6967.31	41.14	0.00
	LG-3	69.79 (69.44–70.14)	0.444 (0.435–0.453)	1.500 (1.475–1.525)	23.69	3439.89	−6873.77	134.68	0.00
Adjusted-RC	VBG-3	81.50 (80.0–83.10)	0.170 (0.165–0.175)	−1.363 (−1.383––1.343)	16.89	3413.10	−6820.20	0.00	1.00
	VBG-2	75.52 (74.02–77.12)	0.215 (0.206–0.224)	-	14.50	3352.04	−6700.08	120.11	0.00
	GG-3	73.16 (72.51–73.81)	0.308 (0.299–0.317)	1.020 (0.999–1.039)	18.60	3395.95	−6785.90	34.30	0.00
	LG-3	69.53 (69.39–70.19)	0.454 (0.434–0.455)	2.040 (2.017–2.067)	19.71	3354.06	−6702.12	118.08	0.00
Adjusted-DC	VBG-3	81.87 (80.37–83.47)	0.168 (0.163–0.173)	−1.384 (−1.404––1.364)	16.96	3454.00	−6902.00	0.00	1.00
	VBG-2	75.59 (74.09–77.19)	0.213 (0.204–0.222)	-	14.50	3386.07	−6768.15	133.85	0.00
	GG-3	73.36 (72.71–74.01)	0.305 (0.296–0.314)	1.030 (1.011–1.051)	18.65	3436.19	−6866.39	35.61	0.00
	LG-3	69.70 (69.35–70.05)	0.449 (0.439–0.460)	2.063 (2.040–2.090)	19.77	3392.80	−6779.60	122.40	0.00

4. Discussion

We found sexual dimorphism by size between sexes, with females (TLmax = 84 cm) being larger than males (TLmax = 58 cm). Similar differences in sizes between sexes were described for *N. entemedor* by Villavicencio-Garayzar [24] off the west coast of Baja California Sur (Bahía Magdalena) and for other closely-related species, such as *N. brasiliensis* [52], *Torpedo californica* [53], *T. marmorata* [54] and *T. nobiliana* [55]. This suggests a selective advantage for larger-sized females; as has been mentioned, the larger size in female elasmobranchs facilitates the accommodation and nourishment of embryos [56]. Similarly, the differences in estimated maximum ages between sexes (6 years for males, 14 years for females) suggests both sexes are not equally long-lived. Furthermore, elasmobranch sexual segregation has been documented related to sex differences in body size, which possibly confers differences in attributes such as predation risk and nutritional requirements [57]. Therefore, the absence of males TL > 59 cm (and >6 years old) in our sample could be explained by differential mortality (or differential longevity) or by sexual segregation. In addition, gear selectivity can be ruled out as a factor for the observed differences because the same gear was used during the full study; thus, differences in size and age by sex seem to be characteristic of the species.

In recent decades, the need to perform age validation has been stressed [31,58]. Villavicencio-Garayzar [24] verified the annual deposition through marginal increments for

N. entemedor off the west coast of Baja California, concluding that annual band formation occurs during June. Moreover, annual deposition has been verified through marginal increment analysis for another Torpediniformes species, *Torpedo marmorata* [54], as well as for other batoid species such as *Dipturus trachyderma* [59], *Bathyraja parmifera* [60] and *Urotrygon rogersi* [7]. Accounting for this, annual deposition of a pair of bands was assumed in our study; however, we suggest being cautious since MIR did not show monthly significant differences. A pattern was observed both in MIR and edge analyses which suggests that annual band deposition concludes in April.

Although the VBG-3 model in the unadjusted age data set showed the maximum −log-likelihood value among the three data sets, the length at birth described by this model, as estimated from the intersection of the growth curve with the length axis (L_0 = 23.0 cm TL), was the farthest from the length of birth described for the species in the southern Gulf of California (14.5 cm LT; [28]). Considering that the rest of the parameters varied slightly between data sets, according to statistical results (–log-likelihood, *AIC* and W_i) and biological interpretation (L_∞, *k* and L_0 values), we considered the VBG-3 based on the adjusted age to date of capture (age Adjustment 2) to be the growth model that best describes individual growth of *N. entemedor*. The multimodel inference approach allows the analysis of an alternative growth hypothesis for *N. entemedor* and avoids the risk of using an inappropriate model *a priori* [61].

The estimated ages for females in our study, 2–14 years, were similar to ages, 1–15 years, estimated for females by Villavicencio-Garayzar [24] off the west coast of Peninsula Baja California, Mexico. Nonetheless, the ages found for males in our study were 1–6 years (LT = 41–59 cm), which differed from males aged 1–11 years (LT = 24–67 cm) obtained by Villavicencio-Garayzar [24]. This could be due to differences in longevity between study areas related to differences in environmental conditions (e.g., temperature; [62]) or differences in the spatial distribution of older males since sexual segregation among elasmobranchs have been documented related to sex differences in body size [57].

The L_∞ estimated for females in our study (TL = 81.87 cm) was similar to L_∞ = 82.6 cm TL estimated for females by Villavicencio-Garayzar [24] and close to the maximum observed length (TL = 84 cm). On the other hand, the growth coefficient estimated for females in our study, k = 0.17 year^{-1}, was lower than k = 0.30 year^{-1} obtained for females by Villavicencio-Garayzar [24]. In this regard, it has been documented that a limited representation of the sizes in the sample, particularly of small and/or large individuals, can bias parameter estimates using the VBG [63]. However, in our case, the back-calculated lengths at previous ages provided a good representation of younger organisms, and this was observed in the closeness of our L_0 estimation to the observed birth length for *N. entemedor*. Compared with closely-related ray species, our estimation was similar to k = 0.18 year^{-1} found for combined sexes of *T. marmorata* [54] but proved higher than k = 0.07 year^{-1} obtained for *T. californica* females [53].

5. Conclusions

Acknowledging the difficulty of sampling young individuals, we suggest the use of back-calculation estimations where information associated with early stages is limited. Furthermore, we conclude that age adjustment is a useful practice. Although age has traditionally been analyzed on a yearly basis (as a discrete variable), our study indicated that adjusting age to biological features, such as birth date, catch date and the periodicity of growth band deposition, may result in a more precise description of individual growth. Finally, we conclude that *N. entemedor* is a moderate body size elasmobranch species with moderate longevity and fast growth, which is a life history pattern typical of species that grow quickly to overcome mortality in the early life stages.

Author Contributions: Conceptualization, P.M.-Z., P.A.M.-F., A.F.N. and V.H.C.-E.; methodology, P.M.-Z., P.A.M.-F., A.F.N., F.N.M.-B., E.M.-B. and X.A.P.-P.; validation, P.M.-Z., P.A.M.-F. and A.F.N.; formal analysis, P.M.-Z., P.A.M.-F., A.F.N., F.N.M.-B., E.M.-B. and X.A.P.-P.; investigation, P.M.-Z., P.A.M.-F., V.H.C.-E., A.F.N., F.N.M.-B., E.M.-B. and X.A.P.-P.; resources, V.H.C.-E.; data curation, P.M.-Z., P.A.M.-F., A.F.N. and F.N.M.-B.; writing—original draft preparation, P.M.-Z., P.A.M.-F., A.F.N., F.N.M.-B., V.H.C.-E., E.M.-B. and X.A.P.-P. writing—review and editing, P.M.-Z., P.A.M.-F., A.F.N., F.N.M.-B., V.H.C.-E., E.M.-B. and X.A.P.-P.; visualization, P.M.-Z., P.A.M.-F., A.F.N., F.N.M.-B., V.H.C.-E., E.M.-B. and X.A.P.-P.; supervision, P.A.M.-F., A.F.N., F.N.M.-B., E.M.-B. and V.H.C.-E.; project administration, V.H.C.-E.; funding acquisition, V.H.C.-E. All authors have read and agreed to the published version of the manuscript.

Funding: This research was funded by the Consejo Nacional de Ciencia y Tecnología de México (SEP-CONACyT/CB-2012/180894) and Instituto Politécnico Nacional, Centro Interdisciplinario de Ciencias Marinas (IPN-SIP/20210127). P.M.-Z. was funded by a scholarship from CONACyT (contract number 422410). A.F.N. and P.A.M.-F. were funded by a postdoctoral fellowship from SEP-CONACyT project (SEP-CONACyT/CB-2012/180894). VHCE and FNMB were supported by programs at the Instituto Politécnico Nacional: Estímulo al Desempeño de los Investigadores (EDI) and Comisión de Operación y Fomento de Actividades Académicas (COFAA). E.M.B. and VHCE are fellowships from Sistema Nacional de Investigadores (CONACyT-SNI).

Institutional Review Board Statement: Ethical review and approval was not required for the animal study because the specimens used within this analysis came from coastal artisanal fishermen from La Paz Bay, Mexico, who have official permits for the activity (CONAPESCA-103053993316-1). We used the animals captured by the fishermen once they are brought to their fishing grounds, at that time the specimens are dead, and we then took advantage for the collection of biological samples. We did not participate in fishing operations, therefore, at no time did we handle live specimens, and thus, the study complies with the ethical guidelines supported by our institution.

Data Availability Statement: Not applicable.

Acknowledgments: The authors thank Juan Higuera and the students of the "Demografía de los batoideos costeros más abundantes en el Pacífico mexicano centro-norte" project for their collaboration in fieldwork and specimen processing. Thanks to Armando Hernández for the final edition of the likelihood contours graphs.

Conflicts of Interest: The authors declare no conflict of interest.

References

1. Holden, M.J. Elasmobranchs. In *Fish Population Dynamics*; Gulland, J.A., Ed.; John Wiley & Sons: London, UK, 1977; pp. 187–215.
2. Stevens, J.D. Variable Resilience to Fishing Pressure in Two Sharks: The Significance of Different Ecological and Life History Parameters. In *Life in the Slow Lane: Ecology and Conservation of Long-Lived Marine Animals*; Musick, J.A., Ed.; American Fisheries Society symposium: Bethesda, MD, USA, 1999; pp. 11–15. Available online: http://hdl.handle.net/102.100.100/214311?index=1 (accessed on 4 November 2021).
3. Cortés, E.; Arocha, F.; Beerkircher, L.; Carvalho, F.; Domingo, A.; Heupel, M.; Holtzhausen, H.; Santos, M.N.; Ribera, M.; Simpfendorfer, C. Ecological Risk Assessment of Pelagic Sharks Caught in Atlantic Pelagic Longline Fisheries. *Aquat. Living Resour.* **2010**, *23*, 25–34. [CrossRef]
4. Coelho, R.; Alpizar-Jara, R.; Erzini, K. Demography of a Deep-Sea Lantern Shark (*Etmopterus spinax*) Caught in Trawl Fisheries of the Northeastern Atlantic: Application of Leslie Matrices with Incorporated Uncertainties. *Deep Sea Res. Part II Top. Stud. Oceanogr.* **2015**, *115*, 64–72. [CrossRef]
5. Walker, P. Sensitive Skates or Resilient Rays? Spatial and Temporal Shifts in Ray Species Composition in the Central and North-Western North Sea between 1930 and the Present Day. *ICES J. Mar. Sci.* **1998**, *55*, 392–402. [CrossRef]
6. Simpfendorfer, C.A. Mortality Estimates and Demographic Analysis for the Australian Sharpnose Shark, *Rhizoprionodon Taylori*, from Northern Australia. *Fish. Bull.* **1999**, *97*, 978–986. [CrossRef]
7. Mejía-Falla, P.A.; Cortés, E.; Navia, A.F.; Zapata, F.A. Age and Growth of the Round Stingray *Urotrygon rogersi*, a Particularly Fast-Growing and Short-Lived Elasmobranch. *PLoS ONE* **2014**, *9*, e96077. [CrossRef] [PubMed]
8. Cailliet, G.; Mollet, H.; Pittenger, G.; Bedford, D.; Natanson, L. Growth and Demography of the Pacific Angle Shark (*Squatina californica*), Based upon Tag Returns off California. *Mar. Freshw. Res.* **1992**, *43*, 1313–1330. [CrossRef]
9. Mollet, H.F.; Cailliet, G.M. Comparative Population Demography of Elasmobranchs Using Life History Tables, Leslie Matrices and Stage-Based Matrix Models. *Mar. Freshw. Res.* **2002**, *53*, 503–515. [CrossRef]
10. Cortés, E. Chondrichthyan Demographic Modelling: An Essay on Its Use, Abuse and Future. *Mar. Freshw. Res.* **2007**, *58*, 4–6. [CrossRef]

11. Cortés, E.; Brooks, E.N.; Gedamke, T. Population dynamics, demography, and stock assessment. In *Biology of Sharks and Their Relatives*; Carrier, J.C., Musick, J.A., Heithaus, M.R., Eds.; CRC Press, Taylor & Francis Group: Boca Raton, FL, USA, 2012; pp. 453–485; ISBN 084931514X.

12. Pollerspöck, J.; Straube, N. Bibliography database of living/fossil sharks, rays and chimaeras (Chondrichthyes: Elasmobranchii, Holocephali) Papers of the Year 2018. Available online: https://shark-references.com/ (accessed on 9 December 2021).

13. Cailliet, G.M.; Goldman, K.J. Age determination and validation in chondrichthyan fishes. In *Biology of Sharks and Their Relatives*; Carrier, J., Musick, J.A., Heithaus, M.R., Eds.; CRC Press: Boca Raton, FL, USA, 2004; pp. 399–439; ISBN 9780203491317.

14. Goldman, K.; Cailliet, G.; Andrews, A.; Natanson, L. Assessing the age and growth of chondrichthyan fishes. In *Biology of Sharks and Their Relatives*; Carrier, J.C., Musick, J.A., Heithaus, M.R., Eds.; CRC Press: Boca Raton, FL, USA, 2012; pp. 423–451.

15. Cortés, E. Incorporating Uncertainty into Demographic Modeling: Application to Shark Populations and Their Conservation. *Conserv. Biol.* **2002**, *16*, 1048–1062. [CrossRef]

16. Dulvy, N.K.; Reynolds, J.D. Predicting Extinction Vulnerability in Skates. *Conserv. Biol.* **2002**, *16*, 440–450. [CrossRef]

17. Dulvy, N.K.; Reynolds, J.D. Life history, population dynamics and extinction risks in chondrichthyans. In *Biology of Sharks and Their Relatives II: Biodiversity, Adaptive Physiology, and Conservation*; Carrier, J., Musick, J.A., Heithaus, M.R., Eds.; CRC Press, Taylor & Francis Group: Boca Raton, FL, USA, 2010; pp. 639–679.

18. Cortés, E. Perspectives on the Intrinsic Rate of Population Growth. *Methods Ecol. Evol.* **2016**, *7*, 1136–1145. [CrossRef]

19. Bizzarro, J.J.; Smith, W.D.; Hueter, R.E.; Tyminski, J.; Márquez-farias, J.F.; Castillo-Géniz, J.L.; Cailliet, G.M.; Villavicencio Garayzar, C.J. *El Estado Actual de Los Tiburones y Rayas Sujetos a Explotación Comercial En El Golfo de California: Una Investigación Aplicada al Mejoramiento de Su Manejo Pesquero y Conservación*; Moss Landing Marine Laboratories Tech. Pub. 2009–02; 2007. Available online: https://aquadocs.org/bitstream/handle/1834/20226/MLML_Tech_Pub_09_02.pdf?sequence=1&isAllowed=y (accessed on 9 December 2021).

20. Bizzarro, J.J.; Smith, W.D.; Márquez-Farías, J.F.; Tyminski, J.; Hueter, R.E. Temporal Variation in the Artisanal Elasmobranch Fishery of Sonora, Mexico. *Fish. Res.* **2009**, *97*, 103–117. [CrossRef]

21. Smith, W.D.; Bizzarro, J.J.J.; Cailliet, G.M. The Artisanal Elasmobranch Fishery on the East Coast of Baja California, Mexico: Characteristics and Management Considerations. *Ciencias Marinas*. **2009**, *35*, 209–236. [CrossRef]

22. Ramírez-Amaro, S.R.; Cartamil, D.; Galvan-Magaña, F.; Gonzalez-Barba, G.; Graham, J.B.; Carrera-Fernandez, M.; Escobar-Sanchez, O.; Sosa-Nishizaki, O.; Rochin-Alamillo, A. The artisanal elasmobranch fishery of the Pacific coast of Baja California Sur, Mexico, management implications. *Sci. Mar.* **2013**, *77*, 473–487. [CrossRef]

23. Cartamil, D.; Santana-Morales, O.; Escobedo-Olvera, M.; Kacev, D.; Castillo-Geniz, L.; Graham, J.B.; Rubin, R.D.; Sosa-Nishizaki, O. The Artisanal Elasmobranch Fishery of the Pacific Coast of Baja California, Mexico. *Fish. Res.* **2011**, *108*, 393–403. [CrossRef]

24. Villavicencio-Garayzar, C.J. Taxonomia, Abundancia Estacional, Edad y Crecimiento y Biologia Reproductiva de *Narcine entemedor* Jordan y Starks (Chondrichthyes; Narcinidae), En Bahia Almejas, B.C.S, Mexico. Ph.D. Dissertation, Universidad Autónoma de Nuevo León, Nuevo Leon 64460, Mexico, 2000; 138p.

25. Márquez-Farías, J.F. The Artisanal Ray Fishery in the Gulf of California: Development, Fisheries Research and Management Issues. The IUCN/SSC Shark Specialist Group. *Shark News* **2002**, *14*, 12–13.

26. González-González, L.; Mejía-Falla, P.A.; Navia, A.F.; de la Cruz-Agüero, G.; Ehemann, N.R.; Peterson, M.S.; Cruz-Escalona, V.H. The Espiritu Santo Island as a Critical Area for Conserving Batoid Assemblage Species within the Gulf of California. *Environ. Biol. Fishes* **2021**, *104*, 1359–1379. [CrossRef]

27. Valadez-González, C. Distribución, Abundancia y Alimentación de Las Rayas Bentónicas de La Costa de Jalisco y Colima, México. Ph.D. Dissertation, Centro Interdisciplinario de Ciencias Marinas, La Paz, Mexico, 2007; p. 119.

28. Burgos-Vázquez, M.I.; Mejía-Falla, P.A.; Cruz-Escalona, V.H.; Brown-Peterson, N.J. Reproductive Strategy of the Giant Electric Ray in the Southern Gulf of California. *Mar. Coast. Fish.* **2017**, *9*, 577–596. [CrossRef]

29. Goldman, K.K. Age and growth of elasmobranch fishes. In *Management Techniques for Elasmobranch Fisheries, FAO Fisheries Tech. Pap., 474*; Musick, J.A., Bonfil, R., Eds.; Food and Agriculture Organization of the United Nations: Rome, Italy, 2005; pp. 76–102; ISBN 92-5-105403-7.

30. Zar, J. *Biostatistical Analysis*; Pearson Prentice-Hall: Hoboken, NJ, USA, 2010; ISBN 9788578110796.

31. Campana, S.E. Accuracy, Precision and Quality Control in Age Determination, Including a Review of the Use and Abuse of Age Validation Methods. *J. Fish Biol.* **2001**, *59*, 197–242. [CrossRef]

32. Cailliet, G.M. Elasmobranch age determination and verification: An updated review. In *Elasmobranchs as Living Resources: Advances in the Biology, Ecology, Systematics and the Status of the Fisheries*; NOAA Technical Report NMFS; Pratt, H.L., Gruber, S.H., Taniuchi, T., Eds.; National Marine Fisheries Service, NOAA: Washington, DC, USA, 1990; pp. 157–165.

33. Beamish, R.J.; Fournier, D.A. A Method for Comparing the Precision of a Set of Age Determinations. *Can. J. Fish. Aquat. Sci.* **1981**, *38*, 982–983. [CrossRef]

34. Chang, W.Y.B. A Statistical Method for Evaluating the Reproducibility of Age Determination. *Can. J. Fish. Aquat. Sci.* **1982**, *39*, 1208–1210. [CrossRef]

35. Campana, S.E.; Annand, M.C.; McMillan, J.I. Graphical and Statistical Methods for Determining the Consistency of Age Determinations. *Trans. Am. Fish. Soc.* **1995**, *124*, 131–138. [CrossRef]

36. Bowker, A.H. A Test for Symmetry in Contingency Tables. *J. Am. Stat. Assoc.* **1948**, *43*, 572–574. [CrossRef]

37. Kusher, D.I.; Smith, S.E.; Cailliet, G.M. Validated Age and Growth of the Leopard Shark, *Triakis semifasciata*, with Comments on Reproduction. *Environ. Biol. Fishes* **1992**, *35*, 187–203. [CrossRef]

38. Branstetter, S. Age, Growth and Reproductive Biology of the Silky Shark, *Carcharhinus falciformis*, and the Scalloped Hammerhead, *Sphyrna lewini*, from the Northwestern Gulf of Mexico. *Environ. Biol. Fishes* **1987**, *19*, 161–173. [CrossRef]

39. Metochis, C.P.; Carmona-Antoñanzas, G.; Kousteni, V.; Damalas, D.; Megalofonou, P. Population Structure and Aspects of the Reproductive Biology of the Blackmouth Catshark, *Galeus Melastomus* Rafinesque, 1810 (Chondrichthyes: Scyliorhinidae) Caught Accidentally off the Greek Coasts. *J. Mar. Biol. Assoc. UK* **2018**, *98*, 909–925. [CrossRef]

40. Francis, R.I.C.C. Back-Calculation of Fish Length: A Critical Review. *J. Fish Biol.* **1990**, *36*, 883–902. [CrossRef]

41. Hile, R. Age and Growth of the Rock Bass, *Ambloplites rupestris* (Rafinesque), in Nebish Lake, Wisconsin. *Trans. Wis. Acad. Sci. Arts Lett.* **1941**, *33*, 189–337.

42. Burnham, K.P.; Anderson, D.R. *Model Selection and Multimodel Inference: A Practical Information Theoretic Approach*, 2nd ed.; Springer: New York, NY, USA, 2002; ISBN 0387953647.

43. von Bertalanffy, L. A Quantitative Theory of Organic Growth (Inquiries on Growth Laws II). *Hum. Biol.* **1938**, *10*, 181–213.

44. Fabens, A.J. Properties and Fitting of the Von Bertalanffy Growth Curve. *Growth* **1965**, *29*, 265–289.

45. Ricker, W.E. Computation and Interpretation of Biological Statistics of Fish Populations. *Bull. Fish. Res. Board Can.* **1975**, *191*, 1–382. [CrossRef]

46. Wang, Y.; Liu, Q. Comparison of Akaike Information Criterion (AIC) and Bayesian Information Criterion (BIC) in Selection of Stock-Recruitment Relationships. *Fish. Res.* **2006**, *77*, 220–225. [CrossRef]

47. García-Borbón, J.A.; Morales-Bojórquez, E.; Aguirre-Villaseñor, H. Long-Term Changes in the Fraction of Mature Brown Shrimp *Farfantepenaeus californiensis* (Holmes, 1900) Females and Their Impact on Length at First Maturity. *J. Shellfish. Res.* **2018**, *37*, 1103–1111. [CrossRef]

48. Hilborn, R.; Mangel, M. *The Ecological Detective. Confronting Models with Data*, 1st ed.; Princeton University Press: Princeton, NJ, USA, 1997.

49. Haddon, M. *Modelling and Quantitative Methods in Fisheries*; Statistics/Biology; Chapman and Hall: Boca Raton, FL, USA, 2001; ISBN 9781584881773.

50. Cerdenares-Ladrón de Guevara, G.; Morales-Bojórquez, E.; Rodriguez-Sánchez, R. Age and Growth of the Sailfish *Istiophorus Platypterus* (Istiophoridae) in the Gulf of Tehuantepec, Mexico. *Mar. Biol. Res.* **2011**, *7*, 488–499. [CrossRef]

51. Luquin-Covarrubias, M.A.; Morales-Bojórquez, E.; González-Peláez, S.S.; Hidalgo-De-La-Toba, J.Á.; Lluch-Cota, D.B. Modeling of Growth Depensation of Geoduck Clam *Panopea Globosa* Based on a Multimodel Inference Approach. *J. Shellfish. Res.* **2016**, *35*, 379–387. [CrossRef]

52. Rolim, F.; Caltabellotta, F.; Rotundo, M.; Vaske-Júnior, T. Sexual Dimorphism Based on Body Proportions and Ontogenetic Changes in the Brazilian Electric Ray *Narcine Brasiliensis* (von Olfers, 1831) (Chondrichthyes: Narcinidae). *Afr. J. Mar. Sci.* **2015**, *37*, 167–176. [CrossRef]

53. Neer, J.A.; Cailliet, G.M. Aspects of the Life History of the Pacific Electric Ray, *Torpedo californica* (Ayres). *Copeia* **2001**, *2001*, 842–847. [CrossRef]

54. Omer, V.D.; Basusta, N. Age and Growth Characteristics of Marbled Electric Ray *Torpedo marmorata* (Risso, 1810) Inhabiting Iskenderun Bay, North-Eastern Mediterranean Sea. *Turk. J. Fish. Aquat. Sci.* **2013**, *13*, 541–549. [CrossRef]

55. Kaya, G.; Basusta, N. A Study on Age and Growth of Juvenile and Semi Adult *Torpedo nobiliana* Bonaparte, 1835 Inhabiting Iskenderun Bay, Northeastern Mediterranean Sea. *Acta Biol. Turc.* **2016**, *29*, 143–149.

56. Hamlett, W.C. Reproductive Biology and Phylogeny of Chondrichthyes: Sharks, Batoids and Chimaeras. In *Reproductive Biology and Phylogeny*; Hamlett, W.C., Ed.; Science Publishers, Inc.: Enfield, NH, USA, 2005; ISBN 9781578082711.

57. Wearmouth, V.J.; Sims, D.W. Sexual Segregation in Marine Fish, Reptiles, Birds and Mammals Behaviour Patterns, Mechanisms and Conservation Implications. *Adv. Mar. Biol.* **2008**, *54*, 107–170. [CrossRef] [PubMed]

58. Cailliet, G.; Martin, D.; Kusher, P.; Wolf, B.; Welde, A. Techniques for Enhancing Vertebral Bands in Age Estimation of California Elasmobranchs. In *Proceedings of an International Workshop on Age Determination of Oceanic Pelagic Fishes: Tunas, Billfishes, and Sharks, NOAA Tech. Rep. NMFS 8*; Prince, E.D., Pulos, L.M., Eds.; National Marine Fisheries Service, NOAA: Miami, FL, USA, 1983; pp. 157–165.

59. Licandeo, R.; Cerna, F.; Céspedes, R. Age, Growth, and Reproduction of the Roughskin Skate, *Dipturus trachyderma*, from the Southeastern Pacific. *ICES J. Mar. Sci.* **2007**, *64*, 141–148. [CrossRef]

60. Matta, M.E.; Gunderson, D.R. Age, Growth, Maturity, and Mortality of the Alaska Skate, *Bathyraja parmifera*, in the Eastern Bering Sea. *Environ. Biol. Fishes* **2007**, *80*, 309–323. [CrossRef]

61. Smart, J.J.; Chin, A.; Tobin, A.J.; Simpfendorfer, C.A. Multimodel Approaches in Shark and Ray Growth Studies: Strengths, Weaknesses and the Future. *Fish Fish.* **2016**, *17*, 955–971. [CrossRef]

62. Keil, G.; Cummings, E.; de Magalhães, J.P. Being Cool: How Body Temperature Influences Ageing and Longevity. *Biogerontology* **2015**, *16*, 383–397. [CrossRef] [PubMed]

63. Cailliet, G.; Tanaka, S. Recommendations for research needed to better understand the age and growth of elasmobranchs. In *Elasmobranchs as Living Resources: Advances in the Biology, Ecology, Systematics, and the Status of the Fisheries*; Pratt, W.S., Jr., Gruber, S.H., Taniuchi, T., Eds.; National Marine Fisheries Service, NOAA: Washington, DC, USA, 1990; pp. 505–507.

Article

Length at Maturity, Sex Ratio, and Proportions of Maturity of the Giant Electric Ray, *Narcine entemedor*, in Its Septentrional Distribution

Xchel Aurora Pérez-Palafox [1], Enrique Morales-Bojórquez [2], Hugo Aguirre-Villaseñor [3] and Víctor Hugo Cruz-Escalona [1,*]

[1] Centro Interdisciplinario de Ciencias Marinas, Instituto Politécnico Nacional, Av. Instituto Politécnico Nacional S/N, Col. Playa Palo de Sta. Rita, La Paz 23096, Baja California Sur, Mexico; xapp39@gmail.com
[2] Centro de Investigaciones Biológicas del Noroeste, Av. Instituto Politécnico Nacional 195, Col. Playa Palo de Santa Rita, La Paz 23096, Baja California Sur, Mexico; emorales@cibnor.mx
[3] Instituto Nacional de Pesca y Acuacultura. Centro Regional de Investigación Acuícola y Pesquera, Calzada Sábalo-Cerritos S/N, Mazatlán 82112, Sinaloa, Mexico; aguirre_hugo@hotmail.com
* Correspondence: vescalon@ipn.mx; Tel.: +52-1-612-141-8669

Simple Summary: The size at which 50 percent of a fish population reaches sexual maturity is an important parameter of life history and is useful for setting conservation goals and fishing efforts. Based on 305 individuals in a population of giant electric rays, *Narcine entemedor*, collected in artisanal fisheries in the Bahía de La Paz, Mexico in its northern distribution over a 2-year period, females were larger than males, but males dominated the sex ratio. Total length at maturity for females was 55.87 cm with mature females present all year; there was no apparent seasonality in the reproductive pattern. Using these data sets, there appeared to be continuous annual reproductive activity.

Abstract: The size at which a certain fraction of a fish population reaches sexual maturity is an important parameter of life history. The estimation of this parameter based on logistic or sigmoid models could provide different ogives and values of length at maturity, which must be analyzed and considered as a basic feature of biological reproduction for the species. A total of 305 individuals of *Narcine entemedor* (*N. entemedor*) were obtained from artisanal fisheries in the Bahía de La Paz, Mexico. For the organisms sampled, sexes were determined and total length (TL) in cm was measured from October 2013 to December 2015. The results indicated that the females were larger, ranging from 48.5 cm to 84 cm TL, while males varied from 41.5 cm to 58.5 cm TL. The sex ratio was dominated by males ranging from 45–55 cm TL, while females were more abundant from 60 to 85 cm TL. Mature females were present all year long, exhibiting a continuous annual reproductive cycle. The length at maturity data were described by the Gompertz model with value of 55.87 cm TL. The comparison between models, and the model selection between them, showed that the Gompertz model had maximum likelihood and smaller Akaike information criterion, indicating that this model was a better fit to the maturity proportion data of *N. entemedor*.

Keywords: maturity; reproductive peak; sigmoid model; length structure

Citation: Pérez-Palafox, X.A.; Morales-Bojórquez, E.; Aguirre-Villaseñor, H.; Cruz-Escalona, V.H. Length at Maturity, Sex Ratio, and Proportions of Maturity of the Giant Electric Ray, *Narcine entemedor*, in Its Septentrional Distribution. *Animals* **2022**, *12*, 120. https://doi.org/10.3390/ani12010120

Academic Editors: Martina Francesca Marongiu and James Albert

Received: 11 November 2021
Accepted: 29 December 2021
Published: 5 January 2022

Publisher's Note: MDPI stays neutral with regard to jurisdictional claims in published maps and institutional affiliations.

1. Introduction

The size at which a certain fraction of the fish population reaches sexual maturity is an important parameter of life history [1]. This information is relevant for demographic analysis, stock assessment, and providing information for fishery control rules, such as establishment of minimum legal length and closed fishing seasons [2,3]. In such analyses, one can achieve biological reference points, defined as metrics of stock statuses, such as fishing mortality values and biomass level [4,5].

The data used to estimate an appropriate size at sexual maturity should provide information on two aspects: (1) an observed proportion of physiologically mature individuals, meaning organisms capable of producing viable gametes, and (2) the proportion of these that are actually producing eggs at a given time [6]. Thus, the logistic model is most commonly used to describe the relation between body size and sexual maturity [1,7]. The plot of this model represents a proportion of mature females in each size class, consequently an S-shaped relationship with an asymptote approaching 1.0 for the largest sizes is commonly estimated. However, in some iteroparous species, not all females are physiologically mature during the reproductive season. In several cases reproduction occurs through batch fecundity (e.g., anchovies, sardines) and, for these species, the values of length at maturity are highly variable and the asymptote denoting the proportion of maturity will differ from 1.0 [6].

There are many methods for estimating the length at maturity, often referred to as the length at which 50% of the organisms are mature (L_{50}). These include the models of Gompertz [8], Lysack [9], and more recently White et al. [10]. In many studies, the criteria for selecting a model are often arbitrary [11,12]. Therefore, the estimation of L_{50} parameters and their precision in these models are based solely on a single average model [13]. To have a better and more robust approach, the model selection based on information theory and maximum likelihood theory is a relatively new paradigm in which several models are compared to each other, evaluating the support of the observed data with respect to each model [13,14]. Studies in reproductive biology using multimodel inference to estimate L_{50} are scarcely reported in the literature, and this is mainly in teleost fishes, such as bigeye tuna (*Thunnus obesus*) [15] and herring (*Opistonema libertate*) [3]. In elasmobranchs, particularly from Mexican waters, only the shark, *Rhizoprionodon terraenovae* [12], and the bat ray (*Myliobatis californica*) [16] have been analyzed using the approach previously described.

According to the data observed, the estimations of L_{50} based on logistic or sigmoid models could provide different ogives, such that the choice of the candidate model influences the expected L_{50}. In addition, its values could be biased whether the model used does not fitthe data set adequately, affecting the biological interpretation of the length at which 50% of the organisms are mature. This situation is crucial when the species exhibits viviparity, such as in several species of elasmobranchs. Biologically, the knowledge of L_{50} does not indicate a maternity condition for all those females reaching this length. According to Walker [17] (pp. 81–127), the ogives of L_{50} and the maturity condition are independent; particularly the L_{50} could provide the beginning of the maternity condition. Therefore, the accuracy of L_{50} is relevant because it would be indicating the females that are recruited to the reproductive stock. However, the adequate quantification of the number of births is necessary for the estimation of the size at maternity, which must be larger than the size at maturity. Hence, the first step is the estimation of an L_{50} value that is sufficiently informative for simultaneously understanding the length at which 50% of the females are mature, as well as the beginning of their potential size at maternity. Consequently, the multimodel inference approach for estimating L_{50} is a statistical procedure useful for this purpose.

According to this biological background, the giant electric ray, *Narcine entemedor* (Jordan and Starks, 1895), is identified as a viviparous species with a continuous annual reproductive cycle and limited histotrophy as a reproductive mode, exhibiting embryonic diapause [18]. The population is distributed from Bahía Magdalena, on the west coast of Baja California Sur to Peru, including the Gulf of California and Galapagos Islands [19,20]. This species is incidentally captured by artisanal fisheries in the Eastern Tropical Pacific and is bycatch from fisheries that target higher-valued teleost or crustaceans [21–23]. Given this incidental feature, the species has been poorly studied and there is limited biological information in the region. The information is mainly associated with descriptions about its reproductive biology, age, and growth, along with its food and feeding patterns; therefore, key population features based on quantitative analysis are necessary to understand the demography of this species. Thus, in this study, we reanalyzed the length at maturity for

Narcine entemedor using a multimodel inference approach based on candidate models with different shapes, number of parameters, and biological assumptions.

2. Materials and Methods

2.1. Collection of Samples

A total of 305 individuals were obtained from artisanal fisheries in the Bahía de La Paz, which is located in the Gulf of California, Mexico between 24°07′ and 24°21′ latitude north and 110°17′ and 110°40′ longitude west. The individuals collected are very common and abundant in the Gulf of California; the species is not protected throughout its range, and is a very well-known commercial species. Additionally, all applicable international, national, and/or institutional guidelines for the care and use of animals were followed. In this study, experimental use of organisms was not required. Sex was determined and total length was measured (TL, cm) for all individuals sampled from October 2013 through December 2015. The maturity data of *N. entemedor* were taken from Burgos-Vázquez et al. [18].

2.2. Criteria for Evaluating Maturity

Maturity in *N. entemedor* individuals was defined as immature (0) or mature (1), with macroscopic characteristics using the criteria proposed by Burgos-Vázquez et al. [18]. For females, the total length and degree of vitellogenesis of the ovarian follicles in the ovary, as well as the anterior oviduct and uterus condition, were considered. Females that presented ovaries with translucent ovarian follicles ≤5 cm and abundant ovarian stroma, slight differentiation between the anterior oviducts and the uterus, and uteri between 0.2 cm and 1.2 cm wide without eggs or embryos were considered as immature. Females that presented ovaries with yellow ovarian follicles ≥6 cm, a uterus that was well differentiated from the anterior oviducts, with widths ≥1.3 cm, with or without eggs or embryos were considered as mature. Based on this microscopic evidence, the macroscopic criteria for defining the binomial classification (0,1) were validated such that the macroscopic and microscopic condition of the ovaries and uterus showed matches [18]. Consequently, the uncertainty associated to the binomial classification describing the observed length at maturity from the macroscopic characteristics of *N. entemedor* is negligible.

2.3. Sex Ratio

Sex ratio was calculated monthly. The sex ratios were compared using a chi-squared (X^2) test, assuming that the sex ratio was 0.5. The null hypothesis was rejected if the X^2 estimated value was greater than 3.84 ($\alpha < 0.05$, df = 1) [24,25]. Additionally, the sex ratio was also represented for each 5 cm (TL) length class.

2.4. Length at Maturity

Length at maturity of females was estimated using a binomial code (immature = 0 and mature = 1), the data were modeled into two length-at-maturity models (Table 1). *Pi* was the estimated proportion of mature fish in size class *i*, *exp* refers to the exponent which is the number of times a number is multiplied by itself, TL_i was the total length of size class *i*, γ was the rate parameter related to the speed of size change from non-reproductive to reproductive status, L_{50} was the length at which 50% of the organisms were mature, ε was the maximum proportion of maturity reached, L_{95} was the length at which 95% of the organisms are mature, and μ was the amplitude of the maturity ogive. The WHI equation was modified, expressing it as a three-parameter function for modeling changes in the proportion maturity; thus, the ε parameter varied as follows: $0 \leq \varepsilon \leq 1$, which allowed for the maximum fraction of mature females to be less than 1 [6].

Table 1. Candidate length-at-maturity models used to estimate L_{50} for *Narcine entemedor*.

Model	Abbreviation	Function	Source
Gompertz	GOM	$P_i = exp^{-exp^{-\gamma(TL_i - L_{50})}}$	[8]
White	WHI	$P_i = \dfrac{\varepsilon}{1 + exp^{[-\ln(19)\frac{(TL_i - L_{50})}{(L_{95} - L_{50})}]}}$	[10]

The objective function for estimating the parameters in the candidate length-at-maturity models were fitted by minimizing the negative log-likelihood $(-\ln \mathcal{L})$ [26]:

$$-\ln \mathcal{L} = -\sum_{i=1}^{n}\left[m_i \times \ln\left(\frac{P_i}{1 - P_i}\right) + P_i \times \ln(k) \right]$$

where n_i was the number of individuals in size class i, m_i was the number of mature fish in size class i, and the quantity $\kappa = \begin{pmatrix} n_i \\ m_i \end{pmatrix}$ was defined as the binomial coefficient and was computed as $\kappa = \frac{n_i!}{m_i! \times (n_i - m_i)!}$. Given that these models exhibited a correlation between parameters, estimates of confidence intervals (CI) in each model were obtained using the likelihood contour method [27]. A chi-squared distribution with df = 2 was used, such that values that were equal to or less than 5.99 were accepted within the CI [24]. The chi-squared estimator was [28]:

$$CI = 2[-\ln \mathcal{L}(\theta_{est}) - (-\ln \mathcal{L}(\theta_i))] \leq \chi^2_{df, 1-\alpha}$$

where $-\ln \mathcal{L}(\theta_{est})$ was the negative log-likelihood of the most likely value of θ_i, $-\ln \mathcal{L}(\theta_i)$ was the negative log-likelihood based on hypotheses of the value of θ_i, $\chi^2_{1-\alpha}$ was the value of the chi-squared distribution with a confidence level of 1-α = 0.05 and df = 2 [28]. Model performance was evaluated using Akaike's information criterion (AIC), where the best model was the one with the lowest AIC value [29,30].

3. Results

In total, 260 females and 45 males were collected from October 2013 to December 2015 in the Bahía de La Paz, Mexico. Females ranged in size from 48.5 cm to 84 cm TL, males ranged from 41.5 cm to 58.5 cm TL. Thus, the females were larger than males in the biological samples during the study period. The sex ratio of *Narcine entemedor* showed that there was a dominance of males in the range of 45–55 cm TL; conversely, the females were more abundant from 60 cm to 85 cm TL (Figure 1). The monthly sex ratio showed a dominance of females and an absence of males was observed during January, April, and June. However, during July–September, the presence of males increased (Figure 2). Nonetheless, the sex ratio assessed from the X^2 test ($p < 0.05$) showed that only during three months the sex ratio was 1:1. These months were March (X^2 = 1.80, df = 1), September (X^2 = 0.75, df = 1), and November (X^2 = 1.80, df = 1) (Table 2).

All males analyzed in the present study were mature. Of the total number of females analyzed, 17.7% were immature. The proportion of maturity, expressed as the relationship between immature and mature females, showed that the larger females of 55 cm TL were mature, and the dominance of mature females was observed from 65 cm TL. An overlap between immature and mature females was identified for individuals smaller than 70 cm TL (Figure 3). The monthly proportions of maturity showed that mature females were present all year round, with the first change in proportions of immature females observed from January to April, with high values during January–February, and low proportions during March–April. A second change in the proportion of immature females occurred with a decrease from May to September, and the third change was an increase in proportions of immature females observed from October to December (Figure 4). These results suggested that there was no seasonality in the reproductive pattern for *N. entemedor*, given than the females were mature from 55 cm TL and in high proportions throughout the year.

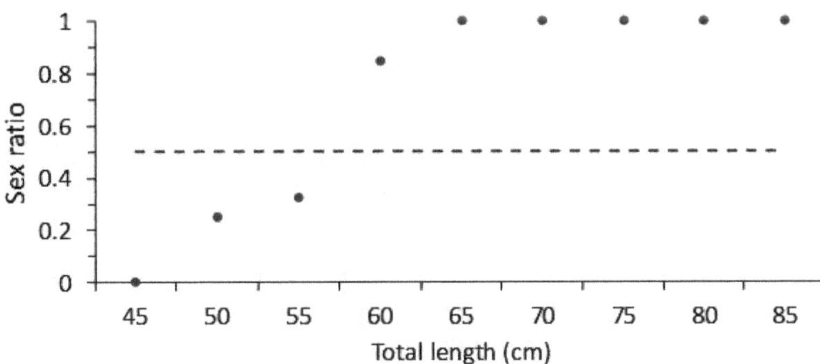

Figure 1. Sex ratio female:male (F:M) of giant electric rays, *Narcine entemedor*, by size intervals.

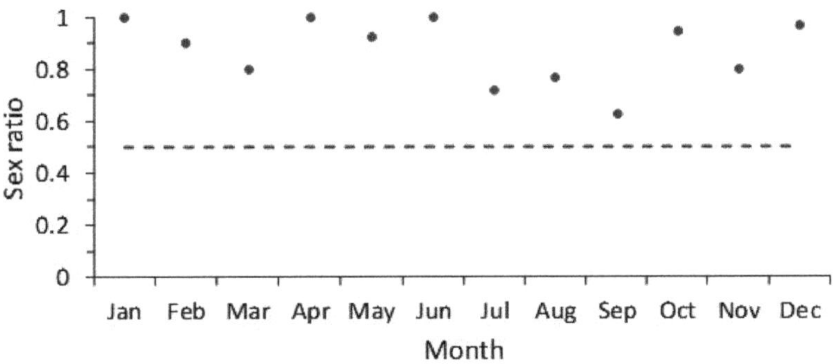

Figure 2. Monthly sex ratio (F:M) of giant electric rays, *Narcine entemedor*.

Table 2. Monthly values of chi-squared (X^2) test estimated for *Narcine entemedor* from the Bahía de La Paz, Baja California Sur, Mexico.

	Female	Male	X^2
January	21	0	10.50
February	37	4	13.28
March	8	2	1.80
April	8	0	4.00
May	35	3	13.47
June	17	0	8.50
July	33	13	4.35
August	33	10	6.15
September	15	9	0.75
October	16	1	6.62
November	8	2	1.80
December	29	1	13.07

The estimates of length at maturity and parameters for each model are shown in Table 3. The L_{50} value estimated through GOM and WHI showed a difference of approximately 2 cm, where the GOM exhibited a smaller value. The parameterization of WHI indicated that the asymptotic value expressed from ε was 1, indicating that the females from 60.64 cm TL progressively increased their maturity proportions at length until reaching the total length of 85 cm, although the asymptote was promptly described by both models from 65 cm. The comparison between models and the model selection between them showed that the GOM model had the maximum likelihood (73.6) and smaller AIC, indicating that this model was a better fit to the maturity proportion data of *N. entemedor*. A partial overlap between trajectories estimated that mature proportions for the two models were observed. The trajectories computed for both models showed that the GOM underestimated the maturity proportions at length for smaller lengths (55 cm TL) (Figure 5).

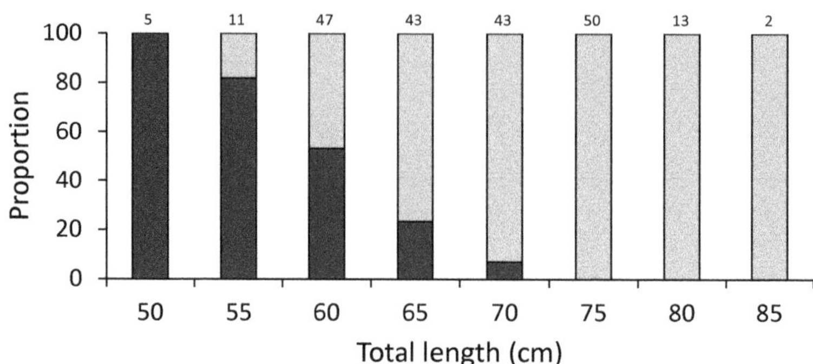

Figure 3. Proportion at length of immature (dark gray bars) and mature (light gray bars) females of *Narcine entemedor* throughout the study period (*n* = 214). The sample size is denoted above each bar.

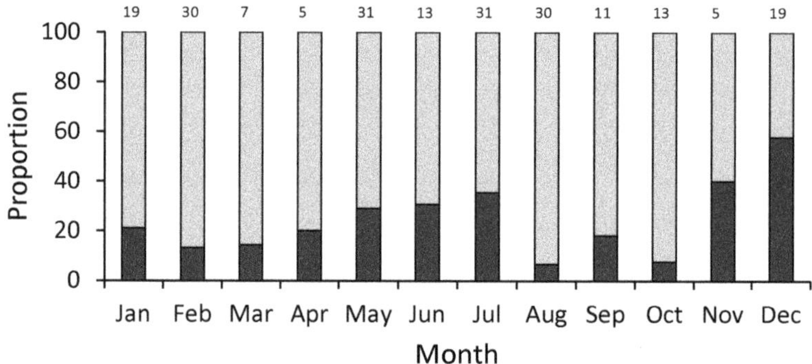

Figure 4. Monthly proportions of immature (dark gray bars) and mature (light gray bars) females of *Narcine entemedor* throughout the study period (*n* = 214). The sample size is denoted above each bar.

Table 3. Parameters (in bold) and confidence intervals (in parenthesis) estimated from negative ln-likelihood contours ($p < 0.05$). L_{50} is the length at which 50% of the organisms were mature, L_{95} is the length at which 95% of the organisms are mature.

Model	L_{50}	γ	L_{95}	ε	$-\ln\mathcal{L}$	AIC
GOM	**58.50** (56.2–60)	**0.231** (0.16–0.31)			73.585	151.16
WHI	**60.64** (58.6–62.2)		**70.37** (67.6–74.4)	1	72.970	151.94

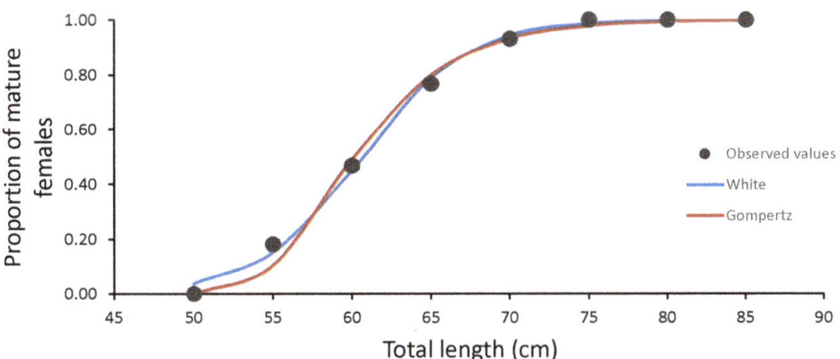

Figure 5. Sigmoid and logistic models (lines) fitted to the length at maturity data (points) for female giant electric rays, *Narcine entemedor*.

4. Discussion

4.1. Length at Maturity

This study reanalyzed the length at maturity data for *N. entemedor*, mainly because an interesting feature was observed in previous studies. This feature was the symmetry of the model reported and its performance for fitting observed data, which was apparently well distributed around the model. Our results showed a lack of symmetry in the observed data fitted to the L_{50} model. Commonly, the observations could be expected, assuming that the cumulative distribution function was symmetric; however, the data set provided more information close to the asymptotic value, therefore the functional form of both models and observed data were important for estimating L_{50}. In this study, two models were statistically compared (GOM and WHI), avoiding an analysis based on mathematical expressions yielding similar estimates of L_{50}, such as was documented by Oviedo-Pérez et al. [12] and García-Rodríguez et al. [16]. The estimates obtained from GOM indicated that the L_{50} value for this species was similar to the length reported by Burgos-Vázquez et al. [18], while the WHI provided a larger L_{50} value. This comparison suggested that the data set was distributed around the model, covering all the size classes. Whether this condition was observed or not could have caused misspecifications in the models, providing bias in L_{50} estimates with evident poor fit [31–33].

Estimates of length at maturity were different between the two models used. The values associated with the Akaike information criterion indicated values of 151.16 (GOM) and 151.94 (WHI); consequently, the GOM was the best model selected using the maximum likelihood values estimated [30]. In this study, the WHI was implemented for the final estimation of L_{50}, the main assumption was that the reproductive event in *Narcine entemedor* was a nonlinear process related to its total length, assuming that not all mature females had reproductive activity at the same time; thus, the maximum proportion of maturity reached will be different to an asymptotic value of 1 [6] (pp. 81–127, [17]). However, this assumption was not satisfied for this species. Conversely, the proportion of maturity observed in *N. entemedor* was sufficiently informative for an asymptote equal to 1, this was clearly influenced for females larger than the 65 cm size class (TL), indicating that it was a coincident with a continuous annual reproductive cycle and the absence of a reproductive peak for mature females.

4.2. Features of the Reproductive Biology Affecting L_{50}

This analysis was supported by biological information obtained from commercial artisanal fisheries, where *N. entemedor* was not a target species. Consequently, the data were limited, nonetheless there was the biological information necessary for analyzing the basic features of reproduction for the giant electric ray. We observed that the sex ratio of *Narcine entemedor* was dominated by males with a total length less than 55 cm, while the females with a total length above 65 cm were more abundant. Females were dominant across all months observed through the annual cycle. According to Villavicencio-Garayzar [19], the males of this species are scarce; thus, the annual sex ratio estimated during 1992 was ~11:1. This change in sex ratio could be attributed to differences in the growth pattern by sex. Smith et al. [34] reported for *Hypanus dipterura*, that the age structure of the population was sexually dimorphic, in that males had a longevity of 19 years while that of the females was 28 years. A similar age structure was found for *Himatura astra*, in which the males had 19 age classes and the females had 30 age classes [35]. A similar pattern was described for *Platyrhina sinensis*, with a maximum age reported for males of 5 years and for females the respective age was 12 years [36].

The length structure of *Narcine entemedor* was also different by sex, with females being larger than males. Similarly, for *Hypanus sayi* (*H. sayi*), *Torpedo torpedo* (*T. torpedo*), and *Torpedo marmorata* (*T. marmorata*) the females attained a larger size than the males [37,38]. For *H. sayi*, the largest female had a 72.9 cm disc width (DW) and the largest male had a 52.1 cm DW [37], while for *T. torpedo*, the total length reported for females was 47.7 cm TL and for males it was 44.5 cm TL. For *T. marmorata*, the difference was also 20 cm TL (55.3 cm TL for females and 36.4 cm TL for males). Additionally, in *T. torpedo*, the sex ratio did not differ among size groups, but in T. marmorata, the presence of females exceeded males at sizes >34.1 cm TL [38]. According to Rolim et al. [39], the length structure of *Narcine brasiliensis* is different between males (from 23.6 cm to 38.0 cm LT) and females (from 23.7 cm to 47.0 cm LT). The dominance of females measured from sex ratio for the populations of rays previously referred to were always higher than males, with females attaining larger sizes.

According to Koob and Callard [40], the reproductive cycles in elasmobranchs can be classified into three types, made up of distinct species assemblages: (1) continuous breeders, (2) seasonal breeders, and (3) punctuated breeders. *Narcine entemedor* is classified as an organism with lecithotrophic viviparity. Lecithotrophy is a developmental pattern in which yolk, produced by the maternal liver and sequestered in the yolk sac, provides embryonic nutrition [17]. However, Burgos-Vázquez et al. [18] suggested that the giant electric ray presented limited histotrophy as a reproductive mode and has a continuous annual reproductive cycle; one peak of ovulation occurs between July and September, but two peaks of parturition occur (minor peak in January–February and major peak in August–September). In the Bahía de La Paz, the reproductive period of *Narcine entemedor* was not temporally defined because the presence of a reproductive peak for mature females was not observed from the mature/immature ratio. This feature could be associated to its northernmost distribution zone, where the environmental conditions (e.g., food, temperature) have an influence on its population structure and reproductive strategy.

Species with lecithotrophic oviparity, in which a continuous annual reproductive cycle has been reported, are relatively commonly. These include *Raja clavata* (Gulf of Gabés) with an absence of a reproductive peak [41]. This species attains maturity at a younger age off the Strait of Sicily, meanwhile in North Wales the maturity is commonly observed at an older age [42,43]. Conversely, *Leucoraja naevus* (*L. naevus*) has a reproductive peak in Southern European waters (January–May), Celtic Seas (February), and the North Sea (September–December) [44]. Additionally, both species exhibit a latitudinal gradient in size structure: *Raja clavate* (*R. clavate*) was larger in the Celtic Seas (98 cm TL) than in the North Sea (92 cm TL). For *L. naevus*, the maximum total length was 69 cm in the Celtic Seas and 62 cm TL in the North Sea [45].

The length at maturity of *N. entemedor* off the southwest coast of the Baja California Peninsula (Bahía Magdalena) was reported to vary between 62 cm and 63 cm TL. This length interval represented approximately 68% of its maximum total length [19]. Meanwhile, Burgos-Vázquez et al. [18] estimated a value of 58.5 cm TL (CI = 51.7–65.4 cm TL) in the Bahía de la Paz (Gulf of California) using a logistic regression model. In this study, the length at maturity was less than that previously reported by Villavicencio-Garayzar [19]. This value was supported by a sigmoid model (GOM) with a different trajectory in comparison to WHI. The advantage of the GOM was that it had a more flexible form; it had a rapid inflexion point in the first length classes, showing a slower approach to the maximum fraction of mature females (asymptotic value). Thus, the comparison among the length-at-maturity models more frequently used and reported in the literature showed that the GOM fitted to the data was better than the logistic model.

For several batoid populations, changes in length structure and L_{50} estimates have been found, mainly through latitudinal gradients. For *N. entemedor*, differences were found in the Equatorial zone, where this species attains larger sizes (110 cm TL) and $L_{50} = 70$ cm TL [46]. This study reported lower values, with the maximum length being 84 cm TL and $L_{50} = 55.8$ cm TL. Moreover, *Raja clavata* distributed in the Atlantic Ocean from Iceland to southern Africa [47] has differences in L_{50} estimates. According to McCully et al. [45], the females inhabiting the North Sea showed the values $L_{50} = 73.7$ cm and 77.1 cm TL, versus those from the Black Sea varying between $L_{50} = 66.7$ cm and 74.6 cm TL. Similar results were reported for *Leucoraja naevus* distributed from Norway to Morocco and Tunisia, including in the Mediterranean Sea [48]. McCully et al. [45] found significant statistical differences between estimates from the North Sea ($L_{50} = 53.6$ cm TL) and the Celtic Sea ($L_{50} = 59.8$ cm TL).

5. Conclusions

In conclusion, considering that *Narcine entemedor* is distributed from the northwest Mexican Pacific to Peru, the population in this study inhabited the northernmost limit for the species. Therefore, reproductive biology values were different from populations elsewhere. This included sex ratio, proportions of maturity, and length at maturity over a year's period. This species did not appear to have a reproductive peak and had a continuous annual reproductive cycle. The estimates of L_{50} for this species showed that a sigmoid model (GOM) was better than the logistic model.

Author Contributions: Conceptualization, X.A.P.-P., H.A.-V. and E.M.-B.; methodology, X.A.P.-P., H.A.-V. and E.M.-B.; validation, X.A.P.-P. and E.M.-B.; formal analysis, X.A.P.-P. and E.M.-B.; investigation, X.A.P.-P., E.M.-B. and V.H.C.-E.; resources, V.H.C.-E.; data curation, X.A.P.-P. and E.M.-B.; writing—original draft preparation, X.A.P.-P., E.M.-B. and V.H.C.-E.; writing—review and editing, X.A.P.-P., E.M.-B. and V.H.C.-E.; supervision, E.M.-B.; project administration, V.H.C.-E.; funding acquisition, V.H.C.-E. All authors have read and agreed to the published version of the manuscript.

Funding: This research was funded by the Consejo Nacional de Ciencia y Tecnología de México (SEP-CONACyT/CB-2012/180894) and Instituto Politécnico Nacional, Centro Interdisciplinario de Ciencias Marinas (IPN-SIP/20210127). XAPP was a recipient of the Consejo Nacional de Ciencia y Tecnología México (CONACYT) postgraduate fellowship (contract number 290124). VHCE was supported by programs at the Instituto Politécnico Nacional: Estímulo al Desempeño de los Investigadores (EDI) and Comisión de Operación y Fomento de Actividades Académicas (COFAA). E.M.B., H.A.V. and V.H.C.E. are fellows of Sistema Nacional de Investigadores (CONACyT-SNI).

Institutional Review Board Statement: Ethical review and approval was not required for the animal study because the specimens used within this analysis came from coastal artisanal fishermen from La Paz Bay, Mexico, who have legal licenses for the commercial harvest (CONAPESCA-103053993316-1). We used the animals captured by the fishermen once they are brought to their fishing grounds, at that time the specimens are dead, and we then took advantage for the collection of biological samples. We did not participate in fishing operations, therefore, at no time did we handle live specimens, and thus, the study complies with the ethical guidelines supported by our institution.

Data Availability Statement: Not applicable.

Acknowledgments: Thanks to technicians at the Centro Interdisciplinario de Ciencias Marinas of the Instituto Politécnico Nacional, México and students and researchers who assisted with field and laboratory work. The authors thank Juan Higuera for his collaboration in fieldwork and specimen processing. Gregor M. Cailliet, Moss Landing Marine Laboratories (MLML), and Mark S. Peterson (University of Southern Mississippi (USM)) provided useful comments that improved the manuscript.

Conflicts of Interest: The authors declare no conflict of interest.

References

1. Chen, Y.; Paloheimo, J.E. Estimating fish length and age at 50% maturity using a logistic type model. *Aquat. Sci.* **1994**, *56*, 206–219. [CrossRef]
2. Cerdenares-Ladrón de Guevara, G.; Morales-Bojórquez, E.; Rodriguez-Sánchez, R. Age and growth of the sailfish *Istiophorus platypterus* (Istiophoridae) in the Gulf of Tehuantepec, Mexico. *Mar. Biol. Res.* **2011**, *7*, 488–499. [CrossRef]
3. Jacob-Cervantes, M.L.; Aguirre-Villaseñor, H. Inferencia multimodelo y selección de modelos aplicados a la determinación de L_{50} para la sardina crinuda *Opisthonema libertate* del sur del golfo de California. *Cienc. Pesq.* **2014**, *22*, 61–68.
4. Caddy, J.F.; Mahon, R. *Reference Points for Fisheries Management*; FAO Fisheries Technical Paper. No. 347; FAO: Rome, Italy, 1995.
5. Midway, S.R.; Scharf, F.S. Histological analysis reveals larger size at maturity for southern flounder with implications for biological reference points. *Mar. Coast. Fish. Dyn. Manag. Ecosyst. Sci.* **2012**, *4*, 628–638. [CrossRef]
6. Restrepo, V.R.; Watson, R.A. An approach to modeling crustacean egg-bearing fractions as a function of size and season. *Can. J. Fish. Aquat. Sci.* **1991**, *48*, 1431–1436. [CrossRef]
7. Roa, R.; Ernst, B.; Tapia, F. Estimation of size at sexual maturity: An evaluation of analytical and resampling procedures. *Fish. Bull.* **1999**, *97*, 570–580.
8. Gompertz, B. On the nature of the function expressive of the law of human mortality, and on a new mode of determining the value of life contingencies. *Philos. Trans. R. Soc. Lond.* **1825**, *115*, 513–583. [CrossRef]
9. Lysack, W. *1979 Lake Winnipeg Fish Stock Assessment Program*; Manitoba Department of Natural Resources; Fisheries Branch: Winnipeg, MB, Canada, 1980.
10. White, T.; Hall, G.; Potter, C. Size and age compositions and reproductive biology of the nervous shark *Carcharhinus cautus* in a large subtropical embayment, including an analysis of growth during pre- and postnatal life. *Mar. Biol.* **2002**, 1153–1164. [CrossRef]
11. Flores, L.; Ernst, B.; Parma, A.M. Growth pattern of the sea urchin, *Loxechinus albus* (Molina, 1782) in Southern Chile: Evaluation of growth models. *Mar. Biol.* **2010**, *157*, 967–977. [CrossRef]
12. Oviedo-Pérez, J.L.; Zea-De la Cruz, H.; Aguirre-Villaseñor, H.; Meiners-Mandujano, C.; Jiménez-Badillo, L.; González-Ocarranza, L. Talla de madurez sexual del tiburón *Rhizoprionodon terraenovae* en Veracruz, México. *Cienc. Pesq.* **2014**, *22*, 37–45.
13. Katsanevakis, S.; Maravelias, C.D. Modelling fish growth: Multi-model inference as a better alternative to a priori using von Bertalanffy equation. *Fish Fish.* **2008**, *9*, 178–187. [CrossRef]
14. Morales-Bojórquez, E.; Aguirre-Villaseñor, H.; Cerdenares-Ladrón De Guevara, G. Confrontación de hipótesis múltiples en pesquerías, teoría de la información y selección de modelos. *Cienc. Pesq.* **2014**, *22*, 9–10.
15. Zhu, G.P.; Dai, X.J.; Song, L.M.; Xu, L.X. Size at sexual maturity of bigeye tuna *Thunnus obesus* (Perciformes: Scombridae) in the tropical waters: A comparative analysis. *Turk. J. Fish. Aquat. Sci.* **2011**, *11*, 149–156. [CrossRef]
16. García-Rodríguez, A.; Hernández-Herrera, A.; Galván-Magaña, F.; Ceballos-Vázquez, B.P.; Pelamatti, T.; Tovar-Ávila, J. Estimation of the size at sexual maturity of the bat ray (*Myliobatis californica*) in Northwestern Mexico through a multi-model inference. *Fish. Res.* **2020**, *231*, 105712. [CrossRef]
17. Hamlett, W.C. *Reproductive Biology and Phylogeny of Chondrichthyes: Sharks, Batoids and Chimaeras*; Hamlett, W.C., Ed.; Reproductive Biology and Phylogeny; Science Publishers, Inc.: Enfield, NH, USA, 2005; ISBN 9781578082711.
18. Burgos-Vázquez, M.I.; Mejía-Falla, P.A.; Cruz-Escalona, V.H.; Brown-Peterson, N.J. Reproductive strategy of the giant electric ray in the Southern Gulf of California. *Mar. Coast. Fish.* **2017**, *9*, 577–596. [CrossRef]
19. Villavicencio-Garayzar, C.J. Taxonomía, Abundancia Estacional, Edad y Crecimiento y Biología Reproductiva de *Narcine entemedor* Jordan y Starks (Chondrichthyes; Narcinidae), en Bahía Almejas, B.C.S, México. Ph.D. Thesis, Universidad Autónoma de Nuevo Leon, San Nicolás de los Garza, México, 2000.
20. Robertson, D.R.; Allen, G. Peces Costeros del Pacífico Oriental Tropical: Un Sistema de Información en Línea. Available online: https://biogeodb.stri.si.edu/sftep/es/pages (accessed on 8 November 2021).
21. Clarke, T.M.; Espinoza, M.; Ahrens, R.; Wehrtmann, I.S. Elasmobranch bycatch associated with the shrimp trawl fishery off the Pacific Coast of Costa Rica, Central America. *Fish. Bull.* **2016**, *114*, 1–17. [CrossRef]
22. Márquez-Farías, J.F. The artisanal ray fishery in the Gulf of California: Development, fisheries research and management issues. *IUCN/SSC Shark Spec. Group* **2002**, *14*, 12–13.
23. Ramírez-Amaro, S.R.; Cartamil, D.; Galvan-Magaña, F.; Gonzalez-Barba, G.; Graham, J.B.; Carrera-Fernandez, M.; Escobar-Sanchez, O.; Sosa-Nishizaki, O.; Rochin-Alamillo, A. The artisanal elasmobranch fishery of the Pacific Coast of Baja California, Mexico. *Sci. Mar.* **2013**, *77*, 473–487. [CrossRef]
24. Zar, J. *Biostatistical Analysis*; Pearson Prentice-Hall: Hoboken, NJ, USA, 2010; ISBN 9788578110796.
25. Cerdenares-Ladrón de Guevara, G.; Morales-Bojórquez, E.; Rodríguez-Jaramillo, C.; Hernández-Herrera, A.; Abitia-Cárdenas, A. Seasonal reproduction of sailfish *Istiophorus platypterus* from the Southeast Mexican Pacific. *Mar. Biol. Res.* **2013**, *9*, 407–420. [CrossRef]
26. Brouwer, S.L.; Griffiths, M.H. Reproductive biology of carpenter seabream (*Argyrozona argyrozona*) (Pisces: Sparidae) in a marine protected area. *Fish. Bull.* **2005**, *103*, 258–269.
27. Luquin-Covarrubias, M.A.; Morales-Bojórquez, E.; González-Peláez, S.S.; Hidalgo-De-La-Toba, J.Á.; Lluch-Cota, D.B. Modeling of growth depensation of geoduck clam *Panopea globosa* based on a multimodel inference approach. *J. Shellfish. Res.* **2016**, *35*, 379–387. [CrossRef]

28. Haddon, M. *Modelling and Quantitative Methods in Fisheries*; Statistics/Biology; Chapman and Hall: Boca Raton, FL, USA, 2001; ISBN 9781584881773.

29. Burnham, K.P.; Anderson, D.R. *Model Selection and Multimodel Inference: A Practical Information Theoretic Approach*, 2nd ed.; Springer: New York, NY, USA, 2002; ISBN 0387953647.

30. Okamura, H.; Punt, A.E.; Semba, Y.; Ichinokawa, M. Marginal increment analysis: A new statistical approach of testing for temporal periodicity in fish age verification. *J. Fish Biol.* **2013**, *82*, 1239–1249. [CrossRef] [PubMed]

31. Remya Mohan, S.; Harikrishnan, M.; Sherly Williams, E. Reproductive biology of a gobiid fish *Oxyurichthys tentacularis* (Valenciennes, 1837) inhabiting ashtamudi lake, S. India. *J. Appl. Ichthyol.* **2018**, *34*, 1099–1107. [CrossRef]

32. Chelapurath Radhakrishnan, R.; Kuttanelloor, R.; Balakrishna, M.K. Reproductive biology of the endemic cyprinid fish *Hypselobarbus thomassi* (Day, 1874) from Kallada River in the Western Ghats, India. *J. Appl. Ichthyol.* **2020**, *36*, 604–612. [CrossRef]

33. Chang, H.-Y.; Sun, C.-L.; Yeh, S.-Z.; Chang, Y.-J.; Su, N.-J.; DiNardo, G. Reproductive biology of female striped marlin *Kajikia audax* in the Western Pacific Ocean. *J. Fish Biol.* **2018**, *92*, 105–130. [CrossRef] [PubMed]

34. Smith, W.D.; Cailliet, G.M.; Melendez, E.M. Maturity and growth characteristics of a commercially exploited stingray, *Dasyatis dipterura*. *Mar. Freshw. Res.* **2007**, *58*, 54–66. [CrossRef]

35. Jacobsen, I.P.; Bennett, M.B. Life history of the blackspotted whipray *Himantura astra*. *J. Fish Biol.* **2011**, *78*, 1249–1268. [CrossRef] [PubMed]

36. Kume, G.; Furumitsu, K.; Yamaguchi, A. Age, growth and age at sexual maturity of fan ray *Platyrhina sinensis* (Batoidea: Platyrhinidae) in Ariake Bay, Japan. *Fish. Sci.* **2008**, *74*, 736–742. [CrossRef]

37. Snelson, F.F., Jr.; Williams-Hooper, S.E.; Schmid, T.H. Biology of the bluntnose stingray, *Dasyatis sayi*, in Florida Coastal Lagoons. *Bull. Mar. Sci.* **1989**, *45*, 15–25.

38. Consalvo, I.; Scacco, U.; Romanelli, M.; Vacchi, M. Comparative study on the reproductive biology of *Torpedo torpedo* (Linnaeus, 1758) and *T. Marmorata* (Risso, 1810) in the Central Mediterranean Sea. *Sci. Mar.* **2007**, *71*, 213–222. [CrossRef]

39. Rolim, F.A.; Rotundo, M.M.; Vaske-Júnior, T. Notes on the reproductive biology of the brazilian electric ray *Narcine brasiliensis* (Elasmobranchii: Narcinidae). *J. Fish Biol.* **2016**, *89*, 1105–1111. [CrossRef] [PubMed]

40. Koob, T.J.; Callard, I.P. Reproductive endocrinology of female elasmobranchs: Lessons from the Little Skate (*Raja erinacea*) and Spiny Dogfish (*Squalus acanthias*). *J. Exp. Zool.* **1999**, *284*, 557–574. [CrossRef]

41. Kadri, H.; Marouani, S.; Saïdi, B.; Bradai, M.N.; Bouaïn, A.; Morize, E. Age, growth, sexual maturity and reproduction of the thornback ray, *Raja clavata* (L.), of the Gulf of Gabès (South-Central Mediterranean Sea). *Mar. Biol. Res.* **2014**, *10*, 416–425. [CrossRef]

42. Cannizzaro, L.; Garofalo, G.; Levi, D.; Rizzo, P.; Gancitano, S. *Raja clavata* (Linneo,1758) Nel Canale Di Sicilia: Crescita, distribuzione e abbondanza. *Biol. Mar. Mediterr.* **1995**, *2*, 257–262.

43. Whittamore, J.M.; McCarthy, I.D. The population biology of the thornback ray, *Raja clavata* in Caernarfon Bay, North Wales. *J. Mar. Biol. Assoc. U.K.* **2005**, *85*, 1089–1094. [CrossRef]

44. Maia, C.; Erzini, K.; Serra-Pereira, B.; Figueiredo, I. Reproductive biology of cuckoo ray *Leucoraja naevus*. *J. Fish Biol.* **2012**, *81*, 1285–1296. [CrossRef]

45. McCully, S.R.; Scott, F.; Ellis, J.R. Lengths at maturity and conversion factors for skates (Rajidae) around the British Isles, with an analysis of data in the literature. *ICES J. Mar. Sci.* **2012**, *69*, 1812–1822. [CrossRef]

46. Palma-Chávez, J.J. Biología Reproductiva de Las Rayas: Mariposa *Gymnura marmorata* (Cooper, 1863) y Torpedo *Narcine entemedor* (Jordan and Starks, 1895) Desembarcados En Santa Rosa-Salinas. Master's Thesis, Universidad Laica Eloy Alfaro de Manabí, Manta, Ecuador, 2014.

47. Ellis, J. *Raja clavata. The IUCN Red List of Threatened Species 2016*: E.T39399A103110667; 2016; Available online: http://dx.doi.org/10.2305/IUCN.UK.2016-3.RLTS.T39399A103110667.en (accessed on 8 November 2021).

48. Ellis, J.; Dulvy, N.; Walls, R. *Leucoraja naevus. The IUCN Red List of Threatened Species 2015*: E.T161626A48949434; 2015. Available online: http://dx.doi.org/10.2305/IUCN.UK.2015-1.RLTS.T161626A48949434.en (accessed on 8 November 2021).

Article

Integrated Taxonomy Revealed Genetic Differences in Morphologically Similar and Non-Sympatric *Scoliodon macrorhynchos* and *S. laticaudus*

Kean Chong Lim [1,2], William T. White [3], Amy Y. H. Then [4,*], Gavin J. P. Naylor [5], Sirachai Arunrugstichai [6] and Kar-Hoe Loh [1,*]

[1] Institute of Ocean and Earth Sciences, Universiti Malaya, Kuala Lumpur 50603, Malaysia; keanchonglim@gmail.com
[2] Institute of Advanced Studies, Universiti Malaya, Kuala Lumpur 50603, Malaysia
[3] CSIRO National Research Collections Australia, Australia National Fish Collection, Hobart, TAS 7001, Australia; william.white@csiro.au
[4] Institute of Biological Sciences, Universiti Malaya, Kuala Lumpur 50603, Malaysia
[5] Florida Museum of Natural History, Dickinson Hall, Gainesville, FL 32601, USA; gjpnaylor@gmail.com
[6] Aow Thai Marine Ecology Centre, Bangkok 10100, Thailand; carcharodon.shinalodon@gmail.com
* Correspondence: amy_then@um.edu.my (A.Y.H.T.); khloh@um.edu.my (K.-H.L.)

Citation: Lim, K.C.; White, W.T.; Then, A.Y.H.; Naylor, G.J.P.; Arunrugstichai, S.; Loh, K.-H. Integrated Taxonomy Revealed Genetic Differences in Morphologically Similar and Non-Sympatric *Scoliodon macrorhynchos* and *S. laticaudus*. *Animals* 2022, 12, 681. https://doi.org/10.3390/ani12060681

Academic Editor: Martina Francesca Marongiu

Received: 27 January 2022
Accepted: 5 March 2022
Published: 8 March 2022

Publisher's Note: MDPI stays neutral with regard to jurisdictional claims in published maps and institutional affiliations.

Simple Summary: In this study, the species identities of similar-looking coastal spadenose sharks from different areas were clarified by adding new molecular markers and more individual body measurements, including animals from the Malaysian Peninsula that had not been examined previously. Collective evidence showed that there are two genetically distinct species that do not overlap in their spatial occurrence. The Malacca Strait acts as a boundary delineating the distribution range of the Pacific spadenose shark *Scoliodon macrorhynchos* to the east and, of the Northern Indian Ocean, *S. laticaudus* to the west. In addition, the need to determine the species status of *Scoliodon* animals from Indonesian waters was identified. The present study reinforced the need to rely on comprehensive genetic information in addition to external characteristics to assess the species identities and distribution range for small sharks and rays that have apparent contiguous coastal distribution with limited dispersal abilities.

Abstract: Previous examination of the mitochondrial *NADH2* gene and morphological characteristics led to the resurrection of *Scoliodon macrorhynchos* as a second valid species in the genus, in addition to *S. laticaudus*. This study applied an integrated taxonomic approach to revisit the classification of the genus *Scoliodon* based on new materials from the Malaysian Peninsula, Malaysian Borneo and Eastern Bay of Bengal. Mitochondrial DNA data suggested the possibility of three species of *Scoliodon* in the Indo-West Pacific, while the nuclear DNA data showed partially concordant results with a monophyletic clade of *S. macrorhynchos* and paraphyletic clades of *S. laticaudus* and *S. cf. laticaudus* from the Malacca Strait. Morphological, meristic and dental characteristics overlapped between the three putative species. Collective molecular and morphological evidence suggested that the differences that exist among the non-sympatric species of *Scoliodon* are consistent with isolation by distance, and *Scoliodon macrorhynchos* remains as a valid species, while *S. cf. laticaudus* is assigned as *S. laticaudus*. The Malacca Strait acts as a spatial delineator in separating the Pacific *S. macrorhynchos* (including South China Sea) from the Northern Indian Ocean *S. laticaudus*. Future taxonomic work should focus on clarifying the taxonomic status of *Scoliodon* from the Indonesian waters.

Keywords: spadenose sharks; integrated taxonomy; synonymy; Indo-West Pacific; morphometrics; genetics; distribution range

1. Introduction

The genus *Scoliodon* was proposed by Müller and Henle [1] for *S. laticaudus* Müller and Henle [2]. Within the family Carcharhinidae, this genus is distinguished from other genera by its clasper and cranial morphology and very shallowly concave post-ventral caudal margin [3]. The genus *Scoliodon* is morphologically similar to hammerhead sharks (family Sphyrnidae) in a number of proportional body measurements but is placed in Carcharhinidae, as it does not have the laterally expanded head that is characteristic of hammerheads [4]. The genus sits within the subfamily Scoliodontinae and differs from the other genera within the subfamily, i.e., *Rhizoprionodon* and *Loxodon*, by having a greatly depressed, trowel-shaped head, broader and more triangular pectoral fins and a more posteriorly located first dorsal fin (free rear tip about over mid-bases of the caudal fin) [5].

Scoliodon has long been considered to be a monotypic genus until White et al. [5] resurrected *S. macrorhynchos* [6] as a second species within the genus. *Scoliodon laticaudus* is common along the insular shelf extending from the Northern Indian Ocean to Northeastern Africa [7]. *Scoliodon macrorhynchos* is known from Southeast Asia from Taiwan and China to Indonesia and Sarawak, Malaysia [5]. A possible third species was also reported from the Bay of Bengal by White et al. [5] and Naylor et al. [8] based on *NADH2* sequence data. These authors suggested that *Carcharhias* (*Physodon*) *muelleri* Müller and Henle [2], described from Bengal may be an available name for this species, but in the absence of specimens, this species was not formally resurrected.

The spadenose shark is one of the smallest carcharhinid species, attaining a maximum total length of 74 cm [9], occurring in shallow muddy and sandy bottom habitats [10]. Nearshore elasmobranchs generally have limited dispersal capabilities [4]. For instance, the bambooshark *Chiloscyllium punctatum* [11] and the stingray *Neotrygon* species [12], both of which are small, show regional population subdivisions with limited genetic mixing throughout the Indo-West Pacific. When geographic barriers and the lack of suitable contiguous habitats are combined with a proclivity not to disperse, allopatric speciation becomes more likely. These factors influenced the redescription of *S. macrorhynchos* from the Eastern South China Sea and the suggestion that *S. muelleri* from the Bay of Bengal might also be a distinct species [5].

White et al. [5] found that *S. macrorhynchos* and *S. laticaudus* showed high intraspecific variations from morphometric data (as high as 5.2% in some head and snout measurements) but low interspecific variations; only a limited number of morphometric measurements differed between the two species, with partly overlapping ranges. For the molecular analysis, the interspecific genetic distance of the *NADH dehydrogenase subunit 2* (*NADH2*) gene between *S. macrorhynchos* and *S. laticaudus* was about 3%. This degree of divergence falls at the borderline of "intra" versus "inter"-specific genetic variations in sharks and rays. *Mobula kuhlii* and *M. eregoodoo* were viewed as one species based on their close genetic distance (interspecific distance < 1.5%) but viewed as distinct species based on morphological data [13]. *Hypanus berthalutzae* was viewed as a distinct species from other closely related *Hypanus* species based on genetics (interspecific distance 0.82–3.11%), morphology, and ecological niche modeling data [14]. These examples highlight the challenge of distinguishing similar-looking but potentially distinct species, such as those in the genus *Scoliodon*.

Reliance on mitochondrial DNA (mtDNA) alone in elucidating phylogenetic relationships among closely related species has been called into question. Reviews by Galtier et al. [15] and Balloux [16] presented some of the limitations associated with reliance on mitochondrial data. The concerns raised arose from limited cases of non-maternally transmitted mtDNA that may call into question the assumption of reduced within-individual diversity [17–19], non-neutral evolution through selection [20–22], and the nonconstant mutation rate in mtDNA [23–25]. While these concerns may not necessarily be applicable in the representation of within-species history for *Scoliodon*, the genetic basis for delineating *S. macrorhynchos* as a separate species from *S. laticaudus* [5] merits a critical review using more representative sampling with additional markers.

In this study, both nuclear and mtDNA markers were used in addition to morphological data sample specimens across known geographical range of *Scoliodon* to clarify the phylogenetic relationships for the group. We included specimens of *Scoliodon* from the Malacca Strait, the west coast of Peninsula Malaysia, that had not been previously examined. The fine-scale contemporary distribution range of the *Scoliodon* genus, especially in the Indo-Malaya region, and knowledge gaps were discussed.

2. Materials and Methods

2.1. Sample Collection

Specimens of *Scoliodon* were acquired at fish landing sites located in the Malacca Strait on the west coast of Peninsular Malaysia, i.e., Hutan Melintang (3°52′13.6″ N 100°55′39.3″ E), Sungai Besar (3°40′15.2″ N 100°58′52.3″ E), and Pasir Penambang (3°21′03.9″ N 101°15′07.0″ E), henceforth labeled as *S.* cf. *laticaudus* and *S. macrorhynchos* from two landing sites in Sarawak in Malaysian Borneo, i.e., Kuching (1°34′04.7″ N, 110°22′45.8″ E) and Mukah (2°53′50.6″ N, 112°05′45.6″ E). Tissue samples were taken from a random subset of specimens (10 each from Malacca Strait and from Sarawak) and stored in 95% alcohol prior to molecular analyses, while the whole specimens were fixed using 10% formalin and store in 70% alcohol. A subset of specimens, 21 from Malacca Strait and 13 from Sarawak, was preserved whole and retained for subsequent morphological analysis by one of us (KCL). Eleven whole specimens of *S.* cf. *laticaudus* were also collected from the Ranong fishing port in Thailand, Eastern Bay of Bengal, during recent surveys of that landing site [26].

2.2. Molecular Analyses

Two mitochondrial DNA (*cytochrome oxidase subunit 1 (COI)* and *NADH dehydrogenase subunit 2 (NADH2)* regions) were used in molecular species identification and seven nuclear genes following Aschliman et al. [27] DNA (*actin-like protein (ACT)*, *kelch repeat and BTB domain-containing protein 2 (KBTBD2)*, *prospero homeobox protein 1 (PROX1)*, *recombination activating gene 1 (RAG1)*, *recombination activating gene 2 (RAG2)*, *sec1 family domain-containing protein 2 (SCFD2)*, and *transducer of ERBB2.1 (TOB1)* region) were used to verify the taxonomic assignment using mitochondrial DNA. DNA was extracted using 10% Chelex resin incubated for two minutes at 60 °C, followed by 25 min at 103 °C (modified from Hyde et al. [28]). Extracted DNA was subjected to Polymerase Chain Reaction (PCR) to amplify all targeted DNA markers. PCR were carried out either using iTaq™ Plus DNA Polymerase (iNtRON Biotechnology, INC., Seongnam-si, Korea) or MyTaq™ Red Mix (Bioline, London, UK) in 20 μL of reaction mix containing 2 μL of 10x PCR buffer; 0.5 μL of dNTP mixture (2.5 mM each); 1 μL of 10-pmol primer (both primers); 1.25 unit of Taq DNA polymerase; 1 μL of 50-pg–1.0-μg DNA templates; and top up with molecular-grade water or 10 μL of MyTaq™ Red Mix premix (mixture of 10x PCR buffer, dNTPs, and Taq polymerase); 1 μL of 10-pmol primer (both primers); 1 μL of 50-pg–1.0-μg DNA templates; and top up with molecular-grade water, respectively. The PCR cycles for mitochondrial DNA comprised of 2-min initial denaturation at 94 °C, followed by 30 cycles of 20 s at 94 °C, 20 s at 44 °C (*COI*) or 52 °C (*NADH2*), and 1 min at 72 °C and, subsequently, a final extension of 5 min at 72 °C. The PCR cycles for nuclear DNA comprised 3-min initial denaturation at 95 °C, followed by 35 cycles of 15 s at 95 °C, 15 s at 52–60 °C, and 1 min at 72 °C and, subsequently, a final extension of 5 min at 72 °C. Touchdown PCR with annealing temperature that decreased 0.3 °C/cycle from 68 °C to 58 °C was performed on *PROX1* due to the amplification of nonspecific DNA at all tested temperatures between 45 and 60 °C. The primer sets used for all the targeted regions are listed in Table 1. All PCR products were examined using 1% agarose in TAE buffer prior to the Sanger sequencing service at Apical Scientific Sdn Bhd (Selangor, Malaysia).

Table 1. Primers used in this study and their references.

Marker	Forward Primer (5′–3′)	Reverse Primer (5′–3′)	References
COI	FishF2—TCG ACT AAT CAT AAA GAT ATC GGC AC	FishR2—ACT TCA GGG TGA CCG AAG AAT CAG AA	Ward et al. [29]
NADH2	ILEM—AAG GAG CAG TTT GAT AGA GT	ASNM—AAC GCT TAG CTG TTA ATT AA	Naylor et al. [30]
ACT	ACT-F—GCT TTC ATC TCC TTC GGC AGT TTG	ACT-R—CCA CTG GTA ATT GGG ATA CTT GGC	Design based on GN's sequence of sample GN1680
KBTBD2	KBT-F—CTC AGT ATC TAT CTT CAG TCC TTG GC	KBT-R—GCT CTT ACA CAG GGA TCA GAG TAG C	Design based on GN's sequence of sample GN1680
PROX1	PRO1-F—AAT TCT TCA AGG GAA AGT GCC CAA G	PRO1-R—CAG ACT GCT CCG ACG AGT TTT TG	Design based on GN's sequence of sample GN1680
RAG1	RAG1-F—CTT ATT CAA ACC ATC AAC AAC ACA ACA	RAG1-R—CTG CAT GAC TGC TTC CAA CTC ATC	Design based on GN's sequence of sample GN1680
RAG2	RAG2-F—TCA GAA TCA AAC AGC CTC ATT TAC C	RAG2-R—TTA ATT TCA TTG GAC CAT TCT GGG G	Design based on GN's sequence of sample GN1680
SCFD2	SCFD-F—AGG TGA AAG CGG TAT TTG TGG TG	SCFD-R—TGA GCT GCA GAA CTT CAA ACA TAG	Design based on GN's sequence of sample GN1680
TOB1	TOB1-F—ATA TGA AGG TCA CTG GTA TCC AGA C	TOB1-R—GAA AAC AAA CTC CTT GGC ATT GGG A	Design based on GN's sequence of sample GN1680

2.3. Phylogenetic Analysis

Sequences were reviewed manually using BioEdit [31], aligned using ClustalX [32], and finally, trimmed using BioEdit [31]. They were all submitted to the NCBI GenBank database, with the accession numbers provided in Supplementary Table S1. The following analyses applied to individual marker, as well as grouped markers by mitochondrial DNA and nuclear DNA. The aligned sequences were subjected to the best model search based on Akaike Information Criterion (AIC) and Bayesian Information Criterion (BIC) for Maximum Likelihood (ML) and Bayesian Inference (BI) analyses, respectively, using Kakusan v 3 [33], as shown in Supplementary Table S2. The generated files were subsequently used for phylogenetic tree construction using Treefinder for ML [34] and MrBayes for BI [35]. The ML analyses were performed with 1000 bootstrap replicates. The Bayesian analyses were initiated with a random starting tree and two parallel runs, each of which consisted of running four chains of Markov chain Monte Carlo (MCMC) iterations for 2,000,000 generations (sampled every 100th generation for each chain). The convergence and burn-in from "sump" commands in MrBayes were used to evaluate likelihood values for post-analysis trees and parameters. Five thousand trees generated were discarded as burn-in (where the likelihood values were stabilized prior before the burn-in), and the remaining trees after burn-in were used to calculate the posterior probabilities using the "sumt" command.

The finalized ML and BI phylogenetic trees were processed via Figtree v 1.3.1 (http://tree.bio.ed.ac.uk/software/figtree/, accessed on 1 October 2014). For mitochondrial DNA, sequences of closely related species *Loxodon macrorhinus* and *Lamiopsis tephrodes* were used as outgroups. Sequences of *Sphyrna lewini* and *Rhizoprionodon acutus*, on the other hand, were used as the outgroup for nuclear DNA, as the sequences for *Loxodon* and *Lamiopsis* were not available. As such, the sequence of *S. lewini* was added to mitochondrial DNA analyses to facilitate the comparison between mitochondrial and nuclear DNA. Some other sequences available in the National Center for Biotechnology Information (NCBI) GenBank and Barcode of Life Data (BOLD) systems were also used in the tree construction for comparison (Supplementary Table S3). Uncorrected p-distance was calculated using PAUP* 40b10 software [36] to evaluate the genetic divergence among the sampled *Scoliodon* species by sampling areas.

We tested species delimitation using a multispecies coalescent analysis implemented in ASTRAL 5.7.7 [37,38] and BPP 4.3 [39–41]. In the ASTRAL analysis, two hundred gene trees were searched under ML + rapid bootstrap for each of the genes using raxmlGUI 1.5 beta [42]. All generated gene trees were combined manually as input into ASTRAL to generate a ASTRAL species tree and normalized quartet score. The normalized quartet score refers to the proportion of gene trees that matched with the species tree; a higher score indicates greater agreement between gene trees and species tree. In the BPP analysis, we performed an unguided species delimitation analysis (A11) to test if the *Scoliodon* species can be assigned as a single species. We set multiple theta (population size) and tau (divergence time) combinations using the inverse gamma prior to IG (2, X), with X being 0.1, 0.01, and 0.001. Each analysis was performed twice to confirm the stability of the results.

2.4. Morphological and Meristic Data

Measurement terminology followed Compagno [3,4,43], who assigned names and abbreviations to measurements often indicated by descriptive phrases (example: snout to upper caudal origin = precaudal length = PRC). Dentitional terms generally followed Compagno [3,43,44]. Vertebral terminology, method of counting, and vertebral ratios followed Springer and Garrick [45] and Compagno [3,43,44].

A total of 83 morphometric measurements were obtained from 74 *Scoliodon* specimens from a range of locations encompassing a large proportion of the geographic range of the three 'species' types: *S. laticaudus*, *S.* cf. *laticaudus*, and *S. macrorhynchos* (Figure 1). A total of 8 specimens of *S. laticaudus* (India); 32 specimens of *S.* cf. *laticaudus* (including the *S. muelleri* holotype from 'Bengal', Malacca Strait, and the Ranong fishing port in the Andaman Sea); and 34 specimens of *S. macrorhynchos* (Hong Kong, Indonesia, Borneo, and Taiwan) were measured in full (Table 2). Vertebral counts were taken from radiographs of 13 specimens of *S.* cf. *laticaudus* and 13 specimens of *S. macrorhynchos*. Counts were obtained separately for the trunk (monospondylous), precaudal (monospondylous and diplospondylous to the origin of upper lobe of caudal fin), and caudal (centra of the caudal fin) vertebrae. Tooth row counts were taken in situ or from excised jaws of 7 specimens of *S. laticaudus*, 5 specimens of *S.* cf. *laticaudus*, and 8 specimens of *S. macrorhynchos*.

Figure 1. Lateral view of *Scoliodon* 'species'. (**a**) *S. macrorhynchos* IPPS WWPLAL#1 (adult male 426 mm TL, fresh), (**b**) *S.* cf. *laticaudus* CSIRO H 8401-09 (adult male 394 mm TL), and (**c**) *S. laticaudus* MNHN 1123 (female 524 mm TL, preserved).

Table 2. Ranges of proportional dimensions as percentages of the total length for the three 'species' of *Scoliodon*.

	S. laticaudus		*S.* cf. *laticaudus*		*S. macrorhynchos*	
	n = 8		*n* = 32		*n* = 34	
	Min.	Max.	Min.	Max.	Min.	Max.
Total length (mm)	169	524	239	490	227	562
Precaudal length	75.6	78.0	75.3	78.0	73.6	78.0
Pre-second dorsal length	62.6	65.4	62.9	66.7	61.5	66.5
Pre-first dorsal length	35.1	38.8	33.0	37.7	33.0	38.1
Head length	21.5	29.1	21.5	26.3	21.3	25.6
Head length (horiz)	21.0	28.6	21.0	25.1	20.9	25.0
Pre-branchial length	17.1	23.5	17.1	20.4	16.5	20.7
Pre-branchial length (horiz)	16.6	22.6	16.5	19.8	16.0	19.5
Preorbital length	8.9	12.6	8.9	11.7	8.5	11.6
Preorbital length (horiz)	8.1	11.3	7.9	10.8	7.0	10.7
Preoral length	7.1	11.1	7.1	10.4	7.2	10.4
Pre-narial length	6.6	9.1	6.6	8.7	6.2	8.4
Pre-narial length (horiz)	5.9	8.2	5.6	8.1	4.8	7.8
Pre-pectoral length	22.1	26.4	21.5	26.6	20.1	26.2
Pre-pelvic length	43.9	48.4	43.9	50.2	43.8	49.2
Snout–vent length	45.9	49.2	45.9	51.4	45.4	50.6
Preanal length	56.7	59.9	56.7	62.0	54.8	60.4
Interdorsal space	16.1	21.7	17.9	21.7	17.9	22.2
Dorsal-caudal space	7.2	9.3	7.2	9.9	7.2	9.4
Pectoral–pelvic space	16.7	19.7	16.9	20.7	16.8	21.6
Pelvic–anal space	5.2	9.0	5.6	11.1	4.8	8.7
Anal–caudal space	6.4	9.1	6.4	8.8	6.4	9.1
Eye length	1.5	2.2	1.6	2.5	1.3	2.4
Eye height	1.3	2.5	1.3	2.5	1.5	2.2
Interorbital space	7.4	11.2	7.4	9.8	7.5	10.3
Nostril width	1.4	2.0	1.5	2.3	1.4	2.3
Internarial space	4.9	6.9	4.9	6.7	4.9	6.5
Anterior nasal flap length	0.2	0.6	0.2	0.5	0.2	0.6
Mouth length	4.5	5.6	4.1	4.9	3.5	5.2
Mouth width	6.0	7.6	5.3	7.6	5.7	7.6
Upper labial furrow length	0.2	0.6	0.1	0.5	0.1	0.5
Lower labial furrow length	0.8	1.2	0.2	1.5	0.3	1.4
First gill slit height	2.3	3.1	2.3	4.1	2.2	4.0
Second gill slit height	2.3	3.6	2.1	2.6	2.2	3.2
Third gill slit height	2.4	3.8	2.2	4.7	2.3	4.4
Fourth gill slit height	2.4	3.7	2.0	2.8	2.4	3.3
Fifth gill slit height	2.2	3.2	2.1	3.3	2.3	3.1
Intergill length	4.6	5.9	4.6	5.4	4.5	6.4

Table 2. *Cont.*

| | S. laticaudus | | S. cf. laticaudus | | S. macrorhynchos | |
| | n = 8 | | n = 32 | | n = 34 | |
	Min.	Max.	Min.	Max.	Min.	Max.
Head height	6.1	10.2	7.7	9.9	7.0	10.6
Trunk height	7.9	10.8	8.3	10.8	7.8	13.1
Abdomen height	7.5	11.2	10.0	11.4	9.4	13.9
Tail height	6.3	10.2	7.0	9.4	7.5	11.3
Caudal peduncle height	3.9	4.5	3.8	4.5	4.0	5.0
Head width	7.3	9.4	6.9	9.9	7.9	10.8
Trunk width	6.4	8.5	6.5	8.8	6.2	11.8
Abdomen width	5.2	7.1	4.9	6.9	5.2	8.9
Tail width	4.1	5.6	4.2	5.6	4.6	6.5
Caudal peduncle width	1.9	2.7	2.3	3.5	2.2	3.7
Pectoral length	10.2	12.1	9.8	11.6	9.9	11.7
Pectoral anterior margin	9.5	12.1	9.4	11.5	9.2	11.9
Pectoral base	4.5	6.6	5.2	6.4	4.8	6.6
Pectoral height	7.8	10.3	7.4	10.3	7.5	10.1
Pectoral inner margin	5.2	6.2	4.6	6.4	4.3	6.2
Pectoral posterior margin	6.3	10.6	6.8	12.5	6.8	9.8
Pelvic length	7.3	8.7	7.1	8.9	6.9	8.3
Pelvic anterior margin	4.7	5.4	4.3	6.0	4.3	5.6
Pelvic base	4.7	5.6	4.3	6.3	4.3	6.1
Pelvic height	3.2	4.3	2.3	4.4	2.7	4.2
Pelvic inner margin length	2.2	3.7	2.1	3.9	2.2	3.5
Pelvic posterior margin length	3.4	5.3	3.4	5.3	3.8	5.1
Clasper outer length	6.0	9.0	4.5	10.2	4.0	10.0
Clasper inner length	8.4	11.8	6.4	12.4	6.5	12.1
Clasper base width	1.0	1.4	0.6	1.7	0.6	1.4
First dorsal length	13.3	15.6	13.3	15.7	12.9	15.5
First dorsal anterior margin	11.1	13.5	11.8	14.3	11.2	14.6
First dorsal base	8.9	10.9	8.9	11.4	8.8	11.0
First dorsal height	6.6	8.6	5.8	8.8	6.5	9.0
First dorsal inner margin	3.8	5.1	3.9	5.3	3.5	4.9
First dorsal posterior margin	6.7	9.2	5.7	9.0	6.2	8.9
Second dorsal length	7.5	9.3	7.3	9.1	6.9	8.6
Second dorsal Anterior margin	4.1	5.5	3.4	5.5	3.4	5.0
Second dorsal base	4.0	4.8	3.2	4.9	3.2	4.8
Second dorsal height	1.7	2.2	1.2	2.4	1.3	2.0
Second dorsal inner margin	3.2	4.7	3.8	5.1	3.3	4.8
Second dorsal posterior margin	3.8	5.3	3.9	4.9	3.6	4.7
Anal length	11.4	13.5	9.6	13.7	10.8	14.1
Anal anterior margin	5.1	6.7	4.1	7.0	4.9	7.8

Table 2. *Cont.*

	S. laticaudus		S. cf. laticaudus		S. macrorhynchos	
	n = 8		n = 32		n = 34	
	Min.	Max.	Min.	Max.	Min.	Max.
Anal base	8.0	10.3	6.1	10.3	7.2	11.2
Anal height	2.8	3.7	2.2	3.6	2.6	3.8
Anal Inner margin	3.0	3.9	3.0	3.9	2.8	4.1
Anal posterior margin	6.6	8.4	5.8	8.8	6.5	8.9
Dorsal caudal margin	22.0	24.9	21.6	24.6	21.9	25.6
Pre-ventral caudal margin	8.5	10.2	7.8	10.7	8.0	10.5
Lower post-ventral caudal margin	3.4	4.7	2.9	5.0	2.9	4.8
Upper post-ventral caudal margin	9.5	11.5	8.9	11.0	9.1	12.3
Caudal fork width	5.4	7.5	5.4	6.8	5.9	7.1
Caudal fork length	7.8	9.7	8.0	9.8	7.8	9.8
Subterminal caudal margin	3.9	5.6	3.1	4.7	3.1	5.3
Subterminal caudal width	2.6	3.4	2.6	3.5	2.7	3.4
Terminal caudal margin	4.5	7.4	4.8	6.8	4.9	7.3
Terminal caudal lobe	7.6	8.9	6.8	8.6	7.2	9.3
Second dorsal origin	4.6	6.9	3.0	6.9	5.2	9.1
Second dorsal insertion	0.5	2.2	0.6	2.0	0.6	2.7
Mid-base first dorsal fin to pectoral insertion	10.9	12.7	10.5	13.4	11.0	14.6
Mid-base first dorsal fin to pelvic origin	4.4	6.2	4.6	7.9	4.4	7.6
First dorsal insertion to pelvic mid-base	2.8	3.9	2.8	5.4	1.9	5.1
Pelvic mid-base to second dorsal origin	12.9	18.1	13.6	18.1	13.5	19.0

2.5. Multivariate Analyses

Morphometric measurements, as % total length (TL), were subjected to nonmetric multidimensional scaling (MDS) ordination (Primer v 7.0 package, Quest Research Limited, Auckland, New Zealand) to determine whether significant differences between putative species exist or whether intraspecific variations of a single species is a factor. One-way Analyses of Similarity (ANOSIM) were employed to test whether morphometric measurements differed significantly among size classes. Similarity Percentages (SIMPER) were employed when a pairwise ANOSIM result was significant at $p < 0.05$ to determine what characters contributed most to the observed differences. To determine if significant differences between size classes exist, samples were allocated to one of four arbitrary size classes: (1) <249 mm TL, (2) 250–299 mm TL, (3) 300–399 mm TL, and (4) >400 mm TL. Morphometric measurements were analyzed without transformation since the preliminary analyses revealed that the stress levels were acceptable (i.e., <0.3) for MDS analyses (see Clarke and Gorley [46]). Several measurements, associated with the clasper and trunk and abdomen heights and widths, were not available for measurement for all individuals, so these characters were excluded from the MDS analysis.

2.6. Museum Holdings

Collection details for the 74 *Scoliodon* specimens examined are provided in Supplementary Data S1. Specimens are referred to by the following prefixes for their registration numbers: BMNH, British Museum of Natural History, London; IPPS, Sarawak Fisheries Research Institute, Bintawa, Malaysia; CSIRO, Australian National Fish Collection, Hobart;

RMNH, Rikjsmuseum van Natuurlkjke Histoire, Leiden; and MNHN, Museum National d'Histoire Naturelle, Paris, France.

3. Results

3.1. Molecular Analysis

Using the *NADH2* and *COI* mitochondrial DNA sequences (Figure 2 and Supplementary Figure S1a,b), three monophyletic groups with moderate-to-full support bootstrap values (ML 58.3—100/BI 68—100) were identified based on sampling locations, i.e., *Scoliodon laticaudus* from the Indian Ocean (based on samples from the west coast of India), *Scoliodon macrorhynchos* from South China Sea (Kuching and Mukah, both localities in Sarawak, which were grouped with samples from China and Taiwan), and a possible third species from the Malacca Strait, tentatively labeled as *S.* cf. *laticaudus*, were grouped with samples from Bangladesh, Myanmar, and Thailand. The uncorrected p-distances among these three monophyletic groups ranged from 0.61 to 3.06% for *COI*, 2.98 to 4.23% for *NADH2*, and 2.12 to 3.19% for the combined mitochondrial DNA (Table 3).

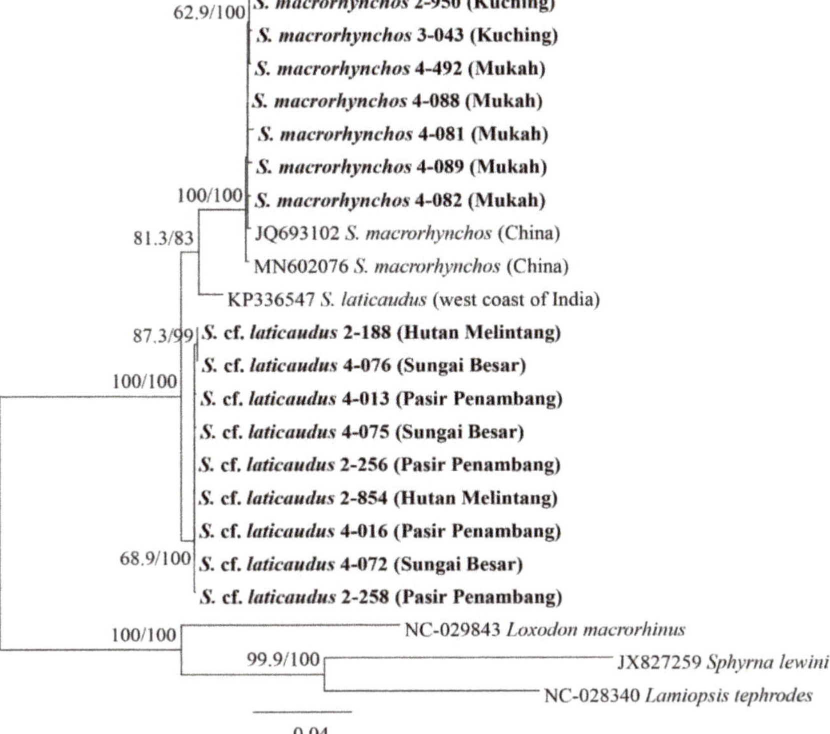

Figure 2. NADH2COI gene mid-point rooting phylogenetic relationships of *Scoliodon* 'species' (phylogram). The bootstrap values (ML/BI) are shown at branches. Sequence names in bold are from the present study.

Table 3. Genetic distance range (mean, in percent) among monophyletic groups in mitochondrial DNA and nuclear DNA phylogenetic trees. Slat—*Scoliodon laticaudus*, Scflat—*S.* cf. *laticaudus*, and Smac—*S. macrorhynchos*.

	Slat-Scflat	Slat-Smac	Scflat-Smac
COI	0.82 (0.61–1.53)	2.35 (1.99–2.75)	2.29 (2.14–3.06)
NADH2	3.05 (2.98–3.27)	3.06 (2.98–3.26)	3.64 (3.46–4.23)
Mitochondrial	2.16 (2.12–2.18)	2.82 (2.71–2.89)	3.05 (2.95–3.18)
ACT	0.10 (0.00–0.25)	0.50 (0.50–0.50)	0.50 (0.25–0.74)
KBTBD2	0.00 (0.00–0.00)	0.22 (0.22–0.22)	0.22 (0.22–0.22)
PROX1	0.00 (0.00–0.00)	0.02 (0.00–0.11)	0.02 (0.00–0.11)
RAG1	0.12 (0.12–0.12)	0.12 (0.12–0.12)	0.02 (0.00–0.12)
RAG2	0.54 (0.45–0.61)	0.91 (0.91–0.91)	0.58 (0.45–0.61)
SCFD2	0.13 (0.00–0.21)	0.21 (0.21–0.21)	0.17 (0.00–0.42)
TOB1	0.00 (0.00–0.00)	0.00 (0.00–0.00)	0.00 (0.00–0.00)
Nuclear	0.12 (0.10–0.14)	0.25 (0.25–0.25)	0.19 (0.16–0.21)

The estimated trees for *Scoliodon* species using nuclear DNA (Figures 2 and 3 and Supplementary Figure S1) showed partial agreement with those using mitochondrial DNA. Three out of five individual nuclear DNA gene trees indicated monophyly of the *Scoliodon* genus (*Prox1, RAG1,* and *TOB1*) (Supplementary Figure S1c–i). Topologies of concatenated nuclear DNA estimated tree showed two monophyletic groups, *S. macrorhynchos* and *S. laticaudus–S.* cf. *laticaudus* groups (Figure 3). The uncorrected p-distance for nuclear DNA among the three monophyletic groups identified from mitochondrial DNA ranged from 0 to 0.91% (mean 0.2%) (Table 3).

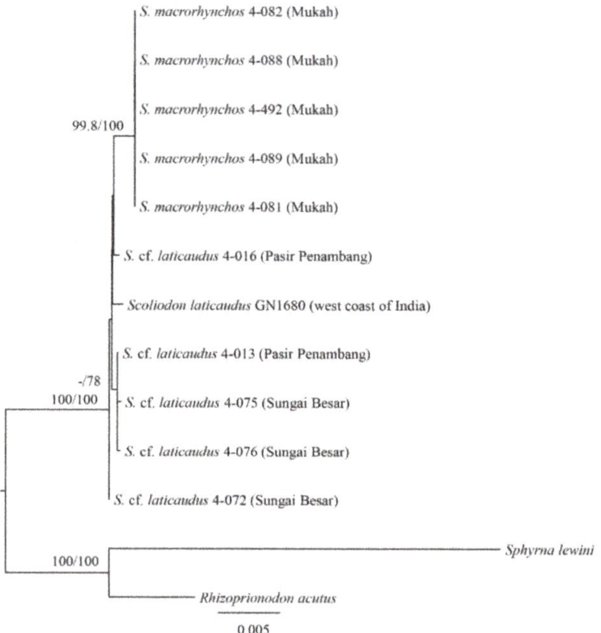

Figure 3. Nuclear gene mid-point rooting phylogenetic relationships of *Scoliodon* 'species' (phylogram). The bootstrap values (ML/BI) are shown at branches.

The species tree estimated in ASTRAL for both mitochondrial and nuclear DNA were topologically congruent with their respective gene trees and had a normalized quartet score of 0.81 and 0.61, respectively (Figure 4). The BPP run supported both estimations from traditional phylogenetic analyses (Bayesian and ML) and ASTRAL. Specifically, the BPP run on mtDNA supported the separation of *Scoliodon* into three separate species with a probability of 1 under any combination of the theta and tau priors. The BPP run on nuclear DNA, on the other hand, varied depending on the theta and tau prior settings; settings of theta at 0.1 regardless of tau prior supported the monospecificity of *Scoliodon* (probability > 0.99), theta at 0.001 in combinations with tau at 0.01 and at 0.001 supported separation into three species (probability 0.65–0.88), and other settings in between supported the combination of *S. laticaudus* and *S.* cf. *laticaudus* as a separate group from *S. macrorhynchos* (probability 0.51–0.61).

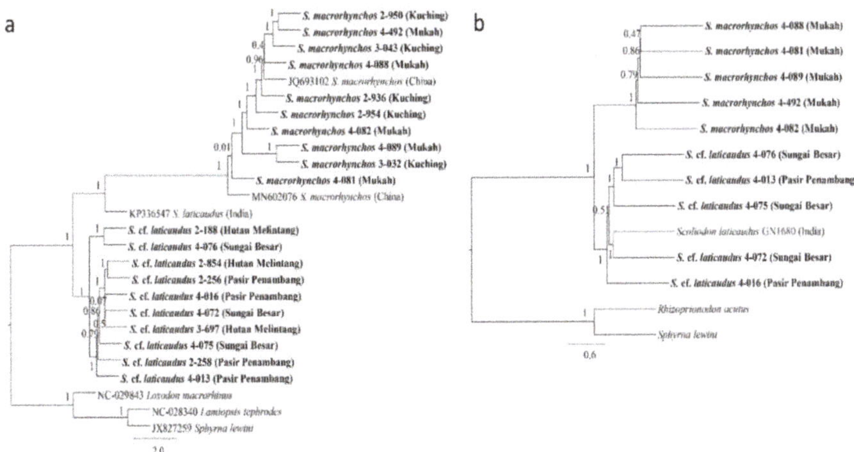

Figure 4. ASTRAL species tree of *Scoliodon* species for (**a**) mtDNA and (**b**) nuclear DNA.

3.2. Morphology and Meristics

No nonoverlapping morphometric ranges were found between the three putative 'species' of *Scoliodon*. Likewise, vertebral counts strongly overlapped between the three 'species'. No dental morphological differences were detected between the three *Scoliodon* 'species'.

The MDS analysis of the measured *Scoliodon* specimens showed considerable overlaps among the three 'species' (Figure 5a). Measurements of the limited *S. laticaudus* samples were highly variable but generally fell within the two overlapping clusters of *S. macrorhynchos* and *S.* cf. *laticaudus* animals. ANOSIM showed that the 'species' were significantly different overall ($p < 0.01$) although the global R2 value was very low (0.24). Similarly, pairwise comparisons between the three 'species' were also significantly different ($p < 0.01$) but with low R2 values (0.18–0.42).

When the same ordination plot was coded by size class ($1 \leq 250$ mm TL, $2 = 250$–299 mm TL, $3 = 300$–399 mm TL, and $4 \geq 400$ mm TL), the samples for each size class formed only partially overlapping groups, with the smallest specimens to the left of the plot and the largest to the right (Figure 5b). ANOSIM showed that the size classes significantly different overall ($p < 0.01$), and with a higher global R2 value (0.54). All pairwise comparisons of size classes were also significantly different ($p < 0.01$), with generally higher R2 values (0.3–0.96). The measurements shown by SIMPER to be the most responsible for the differences between the size classes were pre-anal length, pectoral–pelvic space, pre-pectoral length, pre-pelvic length, head length, and pre-first dorsal length.

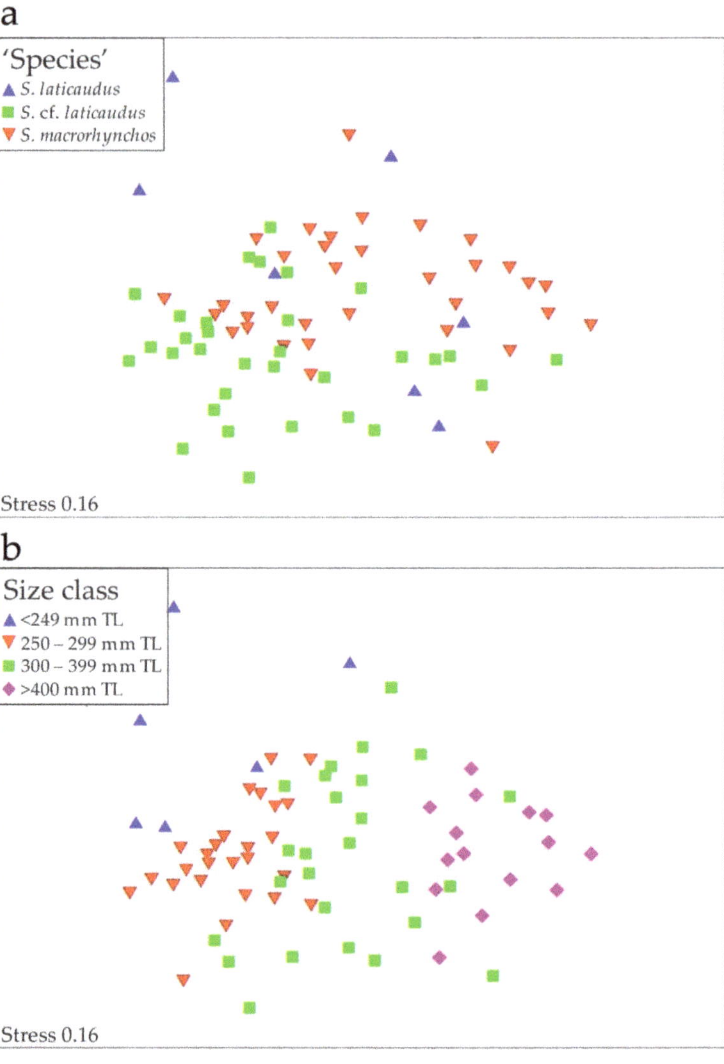

Figure 5. Nonmetric multidimensional scaling (MDS) ordination of *Scoliodon* 'species' morphometric percentages (% TL): (**a**) coded by species and (**b**) coded by size class.

4. Discussion

Based on a combination of nuclear and mitochondrial markers, the evidence supports the split proposed by White et al. [5]. Evidence from mtDNA suggests genetic isolation among the three 'species' types; *S. laticaudus* from India is a separate species from *S. macrorhynchos* from Sarawak, Malaysian Borneo that appears to cluster with samples from China and Taiwan. Evidence from the pooled nuclear markers group *S.* cf. *laticaudus* (from Malacca Strait) with *S. laticaudus*. Both molecular and morphological data presented suggest that any differences that exist among the species of *Scoliodon* are consistent with isolation by distance. We found no evidence of sympatry among any of the three 'species'. Presently, we cautiously recommend that *S.* cf. *laticaudus* of the Malacca Strait be assigned as *S. laticaudus*. These results and the updated distributional range of the *Scoliodon* species are discussed below.

4.1. Taxonomic Conclusions and Recommendations

The decision to resurrect *S. macrorhynchos* as distinct from *S. laticaudus* was primarily based on the *NADH2* sequence data obtained in White et al. [5]. Recent studies have highlighted that the use of single mitochondrial markers alone to distinguish between species can be questionable, especially in light of discordant species trees using mitochondrial and nuclear DNA (for example, *Chimaera ogilbyi* in Finucci et al. [47], freshwater snail genus *Cipangopaludina* in Hirano et al. [48], and terrapins (family Emydidae) in Wiens et al. [49]). In the case of *Scoliodon*, there is considerable concordance between mitochondrial and nuclear signals to support the conclusion of White et al. [5], i.e., the resurrection of *S. macrorhynchos* as a valid species and separate from *S. laticaudus* from India.

Phylogenetic and species trees using combined mitochondrial markers group *S. macrorhynchos* from Sarawak Borneo and from China together, but the same cannot be said for nuclear markers due to the nonavailability of China sequences. Both mitochondrial and nuclear phylogenetic trees mostly support *S. macrorhynchos* from Sarawak Borneo as separate from *S.* cf. *laticaudus* from the Malacca Strait. The discordance between mitochondrial and nuclear signals arises regarding the relationship of *S.* cf. *laticaudus* and *S. laticaudus*. Ambiguity in individual nuclear signals underscores the need to use multiple genes to infer species relationship, and concatenated nuclear signals provisionally group *Scoliodon* individuals from the Malacca Strait as *S. laticaudus*. In addition to congruence between mitochondrial and nuclear data, congruence between molecular and morphological characteristics has also been employed to delimit species (e.g., Finucci et al. [47] and Petean et al. [14]). For *Scoliodon*, White et al. [5] documented only mean differences in several morphometric characteristics but with ranges partially overlapping, i.e., head length, pre-pectoral length, lower labial furrow length, and second dorsal fin origin to anal fin origin. The more comprehensive morphological data presented in this study did not find any nonoverlapping morphological differences in the *Scoliodon* specimens examined. However, given the high intraspecific variability in measurements from *S. laticaudus* (Figure 5a), measurements from additional individuals across a broad distribution range are important to clarify the morphological distinctions between *S. laticaudus* and *S. macrorhynchos*.

The available molecular evidence delimits the Malacca Strait as the easternmost boundary for the range of *S. laticaudus*, thereby extending the distribution of the species based on the most recent International Union for Conservation of Nature (IUCN) assessment [10]. The Malay Peninsula appears to serve as a contemporary physical barrier between the two species. This pattern has been seen for a number of coastal-associated species with limited dispersal abilities, such as bamboosharks [11], guitarfishes [50], groupers [51], sea snails [52], and a number of mangrove species [53]. The molecular differences between morphologically similar but non-sympatric *S. macrorhynchos* and *S. laticaudus* suggest a relatively recent divergence due to geographical isolation with limited mixing that drove allopatric speciation, which is feasible given the complexity of the past geological history of the Sundaland region [54]. Further population genetic studies to corroborate this will help shed light on the evolutionary history and biogeography of the species.

Another important aspect to investigate for *Scoliodon* is the population genetic structure. *Scoliodon* is one of the top landed sharks in terms of both abundance and biomass in surveyed areas within Malaysia [55,56]. A strong coastal affiliation [7] and limited dispersal due to small size are traits that likely promote genetic differentiation and, thus, increase their vulnerability to localized fishing impacts. A similar pattern of a fine-scale population structure has been revealed for a similar small-sized benthic coastal shark, *Chiloscyllium punctatum*, that is subject to high fishing pressure in the Southeast Asian region [11,57]. Further investigation into the genetic structure of *Scoliodon* in Southeast Asia and Indian waters is warranted given the high fishing pressure exerted [58].

4.2. Geographic Range

Distributional ranges for species are often based on a combination of literature sources and expert opinions; therefore, validating some occurrences can be difficult. Since *Scoliodon* is herein confirmed with two valid species, notwithstanding the possibility of another in the Bay of Bengal, it is important to critically investigate the full distributional range for *S. laticaudus* and *S. macrorhynchos*. The identity of *Scoliodon* at locations without genetic sequences is putatively assigned as either *S.* cf. *laticaudus* or *S.* cf. *macrorhynchos* using the Malay Peninsula as the genus distribution break. The resulting distributional range is displayed in Figure 6, with questionable occurrences noted. Investigation of the range is discussed below in an east to west direction.

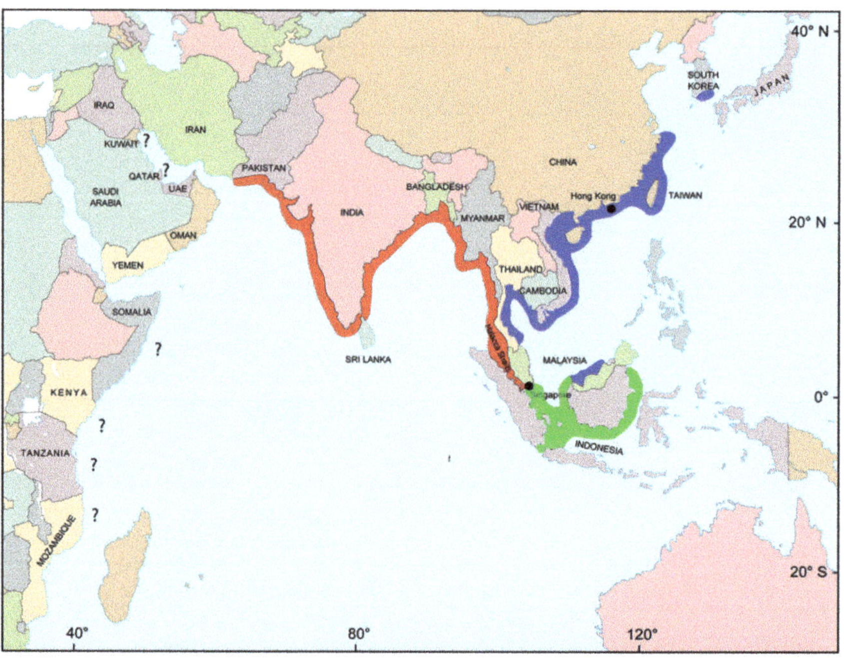

Figure 6. Map of the Indo-West Pacific region showing the refined range of *Scoliodon* species based on the materials examined and a critical examination of the literature. Dubious range locations are highlighted with a question mark. Red = *S. laticaudus*, blue = *S. macrorhynchos*, and green = *Scoliodon* sp. (verification needed).

Off Japan, *S.* cf. *macrorhynchos* has been recorded as a rare occurrence from Kochi Prefecture [59] (as *S. sorrakowah*). Although listed as occurring off the Pacific coast of Southern Japan by Nakaya [60] and Nakabo [61], it is noticeably absent from checklists of coastal fishes in prefectures on the Pacific coast of Southern Japan, e.g., Mie [62], Kagoshima [63], and Nagasaki [64]. Furthermore, nine specimens of *Scoliodon* deposited in Japanese collections with geographic data were caught in either China, Taiwan, or Vietnam (via http://science-net.kahaku.go.jp/, accessed on 28 February 2022). The distribution off Southern Japan appears to be erroneous and should not be included in the range of this species. It has not been previously recorded from South Korea, but Cho et al. [65] reported on a single specimen collected from a Yeosu fish market, Busan in 1995 identified as *S. laticaudus* and supposedly caught from the South Sea of Korea. Off China, Wang [66] noted that *S. macrorhynchos* was abundant off Wenzhou in Southern Zhejiang Province in late spring and early summer but rarely caught in the northern part of the province. Zhu et al. [67] also recorded *S. macrorhynchos* (identified as *S. laticaudus*) from Zhejiang Province but throughout much of the year. Lam and Sadovy de Micheson [68] found that *Scoliodon*, identified as *S. laticaudus*, was the most abundant shark species present during comprehensive market surveys off the Fujian, Hainan, and Guangdong Provinces of China, as well as off Hong Kong. Likewise, Ebert et al. [69] noted that this species was very abundant in fisheries catches around Taiwan.

Naylor et al. [8] provided numerous *NADH2* sequences from specimens caught off Vietnam recorded during local ichthyofaunal surveys. Orlov [70] listed *Scoliodon* spp. as one of the pelagic predators found in marine waters off Cambodia, which likely refers wholly or in part to *S.* cf. *macrorhynchos*. Deechum [71] and Springer [72] included records of *Scoliodon* (identified as *S. laticaudus*) from the Gulf of Thailand. No *Scoliodon* individuals were recorded during comprehensive ichthyofaunal surveys along the east coast of Peninsular Malaysia ([56] Lim et al., unpublished data) but are caught in high abundance in the waters of the west coast of Peninsular Malaysia. As verified by Compagno et al. [73], *Scoliodon* was largely absent in the Philippines. A recent listing of this species in the Philippines elasmobranch identification guide by Alava et al. [74] was likely based on an old record of misidentified *Loxodon* or *Rhizoprionodon*. In Malaysian Borneo, none were recorded from off Sabah from multiple fish surveys, but *S. macrorhynchos* is caught in high abundance off Sarawak ([75] Lim et al., unpublished data, and Manjaji-Matsumoto pers. comm.). *Scoliodon* was not recorded in shark catches off Bintan Island in the Riau Archipelago of Indonesia just to the southeast of Singapore [76].

In Indonesia, *S.* cf. *macrorhynchos* appears to be restricted to Kalimantan [75] and around the river outflows of Eastern Sumatra that flow into the Malacca Strait [77]. It has not been recorded in the literature from West Sumatra or from recent landing site surveys (Fahmi, pers. comm.). Although Bleeker [6] described *S. macrorhynchos* from a juvenile specimen from off Batavia (= Jakarta), which would have likely been caught locally, it has not been recorded off Java in surveys over the last half a century (e.g., Widodo et al. [78] and Widodo and Mahiswara [79]). Springer [72] also listed a specimen deposited at the Smithsonian Institute (USNM 72479) from Batavia (= Jakarta, West Java). This specimen was collected by Owen Bryant and William Palmer in 1909 during a natural history specimen collection trip [80]. Despite being the most abundant species found in recent surveys of the Muara Baru fishing port in Jakarta [81], these were caught in South Kalimantan and only landed in Jakarta. Due to the lack of accurate baseline information, it is not possible to determine whether *Scoliodon* has been extirpated from Javan waters due to overexploitation.

Arunrugstichai et al. [26] recorded *S. laticaudus* as one of the most abundant shark species landed off the Andaman Coast of Thailand. Psomadakis et al. [82] stated that this species is found in coastal waters and lower reaches of the rivers in Myanmar. Jit et al. [83] recorded it as the most abundant shark species based on surveys of two landing centers in Bangladesh, i.e., Chittagong and Cox's Bazar. *Scoliodon laticaudus* is abundant off the Indian coastline, with verified records from all coastal states (from east to west): Andaman and Nicobar archipelago [84,85], West Bengal [86], Orissa [87,88], Andhra Pradesh [89], Tamil Nadu [90], Kerala [91], Karnataka [92,93], Goa [94], Maharashtra [95,96], and Gujarat [97]. *Scoliodon laticaudus* has not been recorded from the Indian union territory of Lakshadweep (formerly Laccadive Archipelago) nor further south in the Maldives or Chagos Archipelago. The presence of *S. laticaudus* off Sri Lanka is less clear. Some checklists have included this species from Sri Lankan waters, e.g., Misra [98] (as *Scoliodon sorrakowah*), Mendis [99] (as *Carcharias laticaudus*), and De Silva [100]. However, recent surveys of 15 fish markets around Sri Lanka recorded no *Scoliodon* [101]. Likewise, Moron et al. [102] did not include this species as present off the west coast of Sri Lanka. Given that *Scoliodon* is usually found in abundance where it occurs, its absence is notable in these studies. Thus, it may be absent from Sri Lankan waters or restricted to only the northern part of Sri Lanka around Palk Bay and the Gulf of Mannar, where it is known to be abundant on the respective Indian coastlines. Off Pakistan, *S.* cf. *laticaudus* was recorded from the coasts of the Sindh Province (Misra [103] as *S. sorrakowah*) and a single specimen recorded during port surveys at Jiwani in Westernmost Balochistan Province, close to the Iranian border [104].

The range of *Scoliodon* has recently included the Persian Gulf and parts of East Africa [7,9]. Bishop [105] and Sivasubramanian and Ibrahim [106] recorded it from off Kuwait and Qatar, respectively, but more recent comprehensive surveys of these locations, as well as of Bahrain and the United Arab Emirates, did not record any *S. laticaudus* in fisheries landings [107–109]. Amojil et al. [110] included this species as only possibly occurring in the Persian Gulf due to the lack of verifiable records. *Scoliodon* cf. *laticaudus* was not recorded during comprehensive surveys of fish landing sites in Oman [111,112]. It was also not recorded from catches of Russian trawlers operating off the entire Yemen coast (including Socotra Island) between 1985 and 1990 [113] or in a recent comprehensive survey of the fish fauna of Socotra Islands [114].

Scoliodon cf. *laticaudus* was included as part of the marine fauna of Somalia [115] and reported as rare in the Somali shark fishery [116]. Although included in a species catalog of Kenya [117], surveys of catches in small-scale fisheries off Kenya over the last decade have not recorded any individuals of this species ([118] B. Kiilu, pers. comm.). Compagno [4] included Tanzania in the range for *S.* cf. *laticaudus* and also included it as present in Mozambique [119]. However, this species has not been recorded from fishery bycatches in recent years in either Mozambique or Tanzania (A. Marshall, S. Pierce, C. Rohner, and D. Ebert, pers. comm.). The presence of *Scoliodon* in the fauna of East Africa from Somalia to Mozambique is dubious. Where *S. laticaudus* is found, they are typically caught in high numbers and common in coastal waters. It is more likely that they are misidentifications of similar species, e.g., *Rhizoprionodon acutus*, which was previously referred to as *Scoliodon walbeehmi* throughout the Indo-West Pacific before being synonymized. Thus, the East Africa distribution of *S. laticaudus* is treated as dubious.

The present distribution delineation is mostly consistent with the recently published IUCN assessment for *S. laticaudus* [10] and *S. macrorhynchos* [120]. In a largely contiguous coastline distribution of *Scoliodon* (Figure 6), we noted two contemporary spatial 'breaks', i.e., along the east coast of the Malaysian Peninsula and off the Sabah coastline of North-eastern Borneo. These breaks could be due to the presence of unsuitable bottom habitats for the species (Manjaji-Matsumoto, pers. comm.) and also reflect the complex evolutionary history of the Sundaland region. Notably, the presence and taxonomic status of *Scoliodon* in the Indonesian region, especially along Eastern Sumatra and along the Kalimantan coastline (Figure 6), needs to be investigated using an integrative approach, i.e., molecular and morphological analyses. It was hypothesized that animals along Eastern Sumatra are *S. laticaudus*, while those in Kalimantan waters are *S. macrorhynchos*, with the Karimata Strait acting as a physical and/or genetic barrier—this is consistent with evidence presented for the genetic structure seen for *C. punctatum* [11].

5. Conclusions

Collective evidence from mitochondrial DNA, nuclear DNA, and morphological analyses clearly supports the previous resurrection of *S. macrorhynchos* as distinct species from *S. laticaudus*. Genetic distinctiveness between the two species is likely a product of isolation by distance with the Malaysian Peninsula acting as a physical barrier. The identity of *Scoliodon* from Indonesian waters remained unverified and should be the focus for future taxonomic studies. Both *Scoliodon* species are currently classified as "near threatened" in the IUCN Red List. With the new evidence from this study, we recommend updating the distribution information of these species and investigating the taxonomic status of *Scoliodon* animals from Indonesian coastal waters.

Supplementary Materials: The following supporting information can be downloaded at https://www.mdpi.com/article/10.3390/ani12060681/s1: Table S1: Genetic samples used in this study with locality data and GenBank accession numbers for each of the mitochondrial and nuclear markers. Table S2: Best model selected for maximum likelihood and Bayesian inference analysis according to each marker and the combined markers. Table S3: NCBI GenBank and Barcode of Life Data (BOLD) Systems accession number of the reference sequences used in the analyses. Data S1: Collection data for all specimens of *Scoliodon* examined in this study. Figure S1a: *COI* gene phylogenetic relationships of *Scoliodon* species (phylogram). The bootstrap values (ML/BI) are shown at branches. Sequence names in bold were from the present study. Figure S1b: *NADH2* gene phylogenetic relationships of *Scoliodon* species (phylogram). The bootstrap values (ML/BI) are shown at branches. Sequence names in bold were from the present study. Figure S1c: *ACT* phylogenetic relationships of *Scoliodon* species (phylogram). The bootstrap values (ML/BI) are shown at branches. Figure S1d: *KBTBD2* phylogenetic relationships of *Scoliodon* species (phylogram). The bootstrap values (ML/BI) are shown at branches. Figure S1e: *PROX1* phylogenetic relationships of *Scoliodon* species (phylogram). The bootstrap values (ML/BI) are shown at branches. Figure S1f: *RAG1* phylogenetic relationships of *Scoliodon* species (phylogram). The bootstrap values (ML/BI) are shown at branches. Figure S1g: *RAG2* phylogenetic relationships of *Scoliodon* species (phylogram). The bootstrap values (ML/BI) are shown at branches. Figure S1h: *SCFD1* phylogenetic relationships of *Scoliodon* species (phylogram). The bootstrap values (ML/BI) are shown at branches. Figure S1i: *TOB1* phylogenetic relationships of *Scoliodon* species (phylogram). The bootstrap values (ML/BI) are shown at branches.

Author Contributions: Designed the study: K.C.L., W.T.W., A.Y.H.T. and K.-H.L. Performed the field work: K.C.L., K.-H.L. and S.A. Conducted the statistical analysis of the data: K.C.L. and W.T.W. Conducted the molecular analyses: K.C.L. and G.J.P.N. Captured morphological and meristic data: W.T.W. and S.A. Wrote the manuscript: K.C.L., W.T.W. and A.Y.H.T. All authors have read and agreed to the published version of the manuscript.

Funding: Financial support for the lead author K.C.L. was provided by the University Malaya Research Grant RP018C-16SUS and the University Malaya Research Fund Assistance BK018-2015; for K.-H.L., funding for rental the facility instruments of molecular analyses in the IOES was provided by the World Wide Fund-Malaysia PV049-2019, the APC was provided by Third Institute of Oceanography, State Oceanic Administration, China IF004-2022 and the UM Research Grant RU009H-2020; for W.T.W., by the CSIRO National Research Collections Australia; and for G.J.P.N., funding for some of the molecular analyses was provided through a US National Science Foundation Division of Environmental Biology Award (#1132229). Financial support to collect specimens was provided by the Australian Centre for Agricultural Research (Indonesian projects FIS/2000/006 and FIS/2003/037), National Science Foundation (Borneo projects DEB 0103640 and DEB 0542846), Murdoch University (internal funding for Taiwan collection trip), CSIRO, and the University of Hong Kong (discretionary funding for guest lecture and collection trip in Hong Kong), Centre for Biodiversity in Peninsular Thailand, and Trocadero group funding (to S.A. for Thailand collection).

Institutional Review Board Statement: The specimens of the spadenose sharks used in this study were all taken from fish market surveys of Southeast Asia between 2001 and 2016. All specimens were dead prior to the surveys being undertaken. The fishing port surveys in Thailand were conducted by SA as part of his Master's Thesis program through the Prince of Songkhla University, and permission to collect data was granted in accordance with the Thailand Department of Fisheries. The fishing port surveys in Malaysia were conducted by KCL as part of his PhD Thesis program through the University of Malaya. No permission to collect data from Peninsular Malaysia was needed at the time of the study, and permission to collect data from Sarawak waters was granted by the Fisheries Research Institute Sarawak, Department of Fisheries Malaysia. All comparative materials used in this study were deposited in museum collections around the world and were borrowed with official loan documentation from these collections. No live animals were collected or killed during this study. All applicable international, national, and/or institutional guidelines for the care and use of animals were followed by the authors.

Data Availability Statement: Ranges of the morphological data obtained in this study are provided in Table 2. The raw morphological data generated during this current study are available from the corresponding author on reasonable request. All sequences used in this study have been deposited in GenBank, and the related accession numbers are provided in the related figure and text sections.

Acknowledgments: Thanks to J. Caira (University of Connecticut); K. Jensen (University of Kansas); P. Last, G. Yearsley, and J. Stevens (CSIRO); Mabel Manjaji-Matsumoto (Universiti Malaysia Sabah); Annie Lim (Fisheries Biosecurity Centre Sarawak); Fahmi (Indonesian Institute of Sciences); V. Lam and S. Lea (University of Hong Kong); and Dharmadi (Research Centre for Capture Fisheries, Jakarta) for their valuable work in the field. We would also like to acknowledge J. Pogonoski (CSIRO) for capturing the meristic data, H. O'Neill (CSIRO) for providing editorial comments, A. Graham (CSIRO) for providing collection information and registering specimens, and L. Conboy (CSIRO) for image preparation. We would also like to thank the following museum staff for allowing accessing and assisting with specimen examination: M. van Oijen and R. de Ruiter at the Rijksmuseum van Natuurlijke Histoire (RMNH) in Leiden; R. Causse, B. Séret, G. Duhamel, and P. Pruvost at the

Animals **2022**, *12*, 681

Muséum national d'Histoire naturelle (MNHN) in Paris; and P. Campbell and J. Maclaine at the British Museum of Natural History (BMNH) in London.

Conflicts of Interest: The authors declare that they have no conflicts of interest.

References

1. Müller, J.; Henle, F.G.J. Gattungen der Haifische und Rochen nach einer von ihm mit Hrn. Henle unternommenen gemeinschaftlichen Arbeit über die Naturgeschichte der Knorpelfische. *Ber. Königlichen Preuss. Akad. Wiss. Berl.* **1837**, *1837*, 111–118.
2. Müller, J.; Henle, F.G.J. *Systematische Beschreibung der Plagiostomen*; Veit und Comp: Berlin, Germany, 1938–1941; (pp. 1–28 published in 1838, 29–102 in 1839, 103–200 in 1841).
3. Compagno, L.J.V. *Sharks of the Order Carcharhiniformes*; The Blackburn Press: Caldwell, NJ, USA, 1988.
4. Compagno, L.J.V. *FAO Species Catalogue. Volume 4, Sharks of the World. An Annotated and Illustrated Catalogue of Shark Species Known to Date*; FAO Fisheries Synopsis No 125; FAO: Rome, Italy, 1984.
5. White, W.T.; Last, P.R.; Naylor, G.J.P. *Scoliodon macrorhynchos* (Bleeker, 1852), a second species of spadenose shark from the Western Pacific (Carcharhiniformes: Carcharhinidae). In *Descriptions of New Australian Chondrichthyans, CSIRO Marine and Atmospheric Research Paper 32*; Last, P.R., White, W.T., Pogonoski, J.J., Eds.; CSIRO: Hobart, Australia, 2010; pp. 61–76.
6. Bleeker, P. Bijdrage tot de kennis der Plagiostomen van den Indischen Archipel. *Verh. Batav. Genoots. Kuns.* **1852**, *24*, 1–92.
7. Simpfendorfer, C. *Scoliodon laticaudus*. The IUCN Red List of Threatened Species. 2009. Available online: https://doi.org/10.2305/IUCN.UK.2009-2.RLTS.T39383A10188364.en (accessed on 14 February 2020).
8. Naylor, G.J.P.; Caira, J.N.; Jensen, K.; Rosana, K.A.M.; White, W.T.; Last, P.R. A DNA sequence-based approach to the identification of shark and ray species and its implications for global elasmobranch diversity and parasitology. *Bull. Am. Nat. Hist. Mus.* **2012**, *367*, 1–263. [CrossRef]
9. Ebert, D.A.; Fowler, S.; Compagno, L. *Sharks of the World: A Fully Illustrated Guide*; Wild Nature Press: Plymouth, UK, 2013.
10. Dulvy, N.K.; Simpfendorfer, C.; Akhilesh, K.V.; Derrick, D.; Elhassan, I.; Fernando, D.; Haque, A.B.; Jabado, R.W.; Maung, A.; Valinassab, T.; et al. *Scoliodon laticaudus*. The IUCN Red List of Threatened Species. 2021. Available online: https://doi.org/10.2305/IUCN.UK.2021-2.RLTS.T169234201A173436322.en (accessed on 14 December 2021).
11. Lim, K.C.; Then, A.Y.; Wee, A.K.S.; Sade, A.; Rumpet, R.; Loh, K.-H. Brown banded bamboo shark (*Chiloscyllium punctatum*) shows high genetic diversity and differentiation in Malaysian waters. *Sci. Rep.* **2021**, *11*, 14874. [CrossRef] [PubMed]
12. Puckridge, M.; Last, P.R.; White, W.T.; Andreakis, N. Phylogeography of the Indo-West Pacific maskrays (Dasyatidae, Neotrygon): A complex example of chondrichthyan radiation in the Cenozoic. *Ecol. Evol.* **2013**, *3*, 217–232. [CrossRef]
13. White, W.T.; Corrigan, S.; Yang, L.; Henderson, A.C.; Bazinet, A.L.; Swofford, D.L.; Naylor, G.J. Phylogeny of the manta and devilrays (Chondrichthyes: Mobulidae), with an updated taxonomic arrangement for the family. *Zool. J. Linnean. Soc.* **2018**, *182*, 50–75. [CrossRef]
14. Petean, F.F.; Naylor, G.J.; Lima, S.M. Integrative taxonomy identifies a new stingray species of the genus *Hypanus* Rafinesque, 1818 (Dasyatidae, Myliobatiformes), from the Tropical Southwestern Atlantic. *J. Fish. Biol.* **2020**, *97*, 1120–1142. [CrossRef]
15. Galtier, N.; Nabholz, B.; Glémin, S.; Hurst, G.D.D. Mitochondrial DNA as a marker of molecular diversity: A reappraisal. *Mol. Ecol.* **2009**, *18*, 4541–4550. [CrossRef]
16. Balloux, F. The worm in the fruit of the mitochondrial DNA tree. *Heredity* **2010**, *104*, 419–420. [CrossRef]
17. Awadalla, P.; Eyre-Walker, A.; Maynard-Smith, J. Linkage disequilibrium and recombination in hominid mitochondrial DNA. *Science* **1999**, *286*, 2524–2525. [CrossRef]
18. Eyre-Walker, A.; Smith, N.H.; Maynard-Smith, J. How clonal are human mitochondria? *Proc. R. Soc. B Biol. Sci.* **1999**, *266*, 477–483. [CrossRef] [PubMed]
19. Hagelberg, E.; Goldman, N.; Lió, P.; Whelan, S.; Schiefenhöel, W.; Clegg, J.B.; Bowden, D.K. Evidence for mitochondrial DNA recombination in a human population of island Melanesia. *Proc. R. Soc. B Biol. Sci.* **1999**, *266*, 485–492. [CrossRef]
20. Rand, D.M. The units of selection on mitochondrial DNA. *Annu. Rev. Ecol. Syst.* **2001**, *32*, 415–448. [CrossRef]
21. Bazin, E.; Glémin, S.; Galtier, N. Population size does not influence mitochondrial genetic diversity in animals. *Science* **2006**, *312*, 570–572. [CrossRef] [PubMed]
22. Castoe, T.A.; de Koning, A.J.; Kim, H.M.; Gu, W.; Noonan, B.P.; Naylor, G.; Jiang, Z.J.; Parkinson, C.L.; Pollock, D.D. Evidence for an ancient adaptive episode of convergent molecular evolution. *Proc. Natl. Acad. Sci. USA* **2009**, *106*, 8986–8991. [CrossRef] [PubMed]
23. Xu, W.; Jameson, D.; Tang, B.; Higgs, P.G. The relationship between the rate of molecular evolution and the rate of genome rearrangement in animal mitochondrial genomes. *J. Mol. Evol.* **2006**, *63*, 375–392. [CrossRef] [PubMed]
24. Nabholz, B.; Glémin, S.; Galtier, N. Strong variations of mitochondrial mutation rate across mammals—The longevity hypothesis. *Mol. Biol. Evol.* **2008**, *25*, 120–130. [CrossRef]
25. Nabholz, B.; Glémin, S.; Galtier, N. The erratic mitochondrial clock: Variations of mutation rate, not population size, affect mtDNA diversity across mammals and birds. *BMC Evol. Biol.* **2009**, *9*, 54. [CrossRef]
26. Arunrugstichai, S.; True, J.D.; White, W.T. Catch composition and aspects of the biology of sharks caught by Thai commercial fisheries in the Andaman Sea. *J. Fish. Biol.* **2018**, *92*, 1487–1504. [CrossRef]

27. Aschliman, N.C.; Cleason, K.M.; McEachran, J.D. Phylogeny of batoidea. In *Biology of Sharks and Their Relatives,* 2nd ed.; Carrier, J.C., Musick, J.A., Heithaus, M.R., Eds.; CRC Press: Boca Raton, FL, USA, 2012; pp. 57–95.

28. Hyde, J.R.; Lynn, E.; Humphreys, R., Jr.; Musyl, M.; West, A.P.; Vetter, R. Shipboard identification of fish eggs and larvae by multiplex PCR, and a description of fertilized eggs of blue marlin, short bill spearfish, and wahoo. *Mar. Ecol. Prog. Ser.* **2005**, *286*, 269–277. [CrossRef]

29. Ward, R.D.; Holmes, B.H.; White, W.T.; Last, P.R. DNA barcoding Australasian chondrichthyans: Results and potential uses in conservation. *Mar. Freshwat. Res.* **2008**, *59*, 57–71. [CrossRef]

30. Naylor, G.J.P.; Ryburn, J.A.; Fedrigo, O.; Lopez, J.A. Phylogenetic relationships among the major lineages of modern elasmobranchs. In *Reproductive Biology and Phylogeny: Sharks, Skates, Stingrays, and Chimaeras*; Hamlett, W.C., Jamieson, B.G.M., Eds.; Science Publishers Inc.: Plymouth, UK, 2005; pp. 1–25.

31. Hall, T.A. BioEdit: A user-friendly biological sequence alignment editor and analysis program for Windows 95/98/NT. *Nucleic Acids Symp. Ser.* **1999**, *41*, 95–98.

32. Thompson, J.D.; Gibson, T.J.; Plewniak, F.; Jeanmougin, F.; Higgins, D.G. The CLUSTAL_X windows interface: Flexible strategies for multiple sequence alignment aided by quality analysis tools. *Nucleic Acids Res.* **1997**, *25*, 4876–4882. [CrossRef] [PubMed]

33. Tanabe, A.S. Kakusan: A computer program to automate the selection of a nucleotide substitution model and the configuration of a mixed model on multilocus data. *Mol. Ecol. Notes* **2007**, *7*, 962–964. [CrossRef]

34. Jobb, G.; von Haeseler, A.; Strimmer, K. Treefinder: A powerful graphical analysis environment for molecular phylogenetics. *BMC Evol. Biol.* **2004**, *4*, 18. [CrossRef]

35. Huelsenbeck, J.P.; Ronquist, F. MrBayes: Bayesian Inference of Phylogenetic Trees. *Bioinformatics* **2001**, *17*, 754–755. [CrossRef]

36. Swofford, D.L. *PAUP*: Phylogenetic Analysis Using Parsimony (* and Other Methods), Version 4*; Sinauer Associates: Sunderland, MA, USA, 2002.

37. Zhang, C.; Rabiee, M.; Sayyari, E.; Mirarab, S. ASTRAL-III: Polynomial time species tree reconstruction from partially resolved gene trees. *BMC Bioinform.* **2018**, *19*, 153. [CrossRef]

38. Rabiee, M.; Sayyari, E.; Mirarab, S. Multi-Allele Species Reconstruction Using ASTRAL. *Mol. Phylogenet. Evol.* **2019**, *130*, 286–296. [CrossRef]

39. Yang, Z.; Rannala, B. Unguided species delimitation using DNA sequence data from multiple loci. *Mol. Biol. Evol.* **2014**, *31*, 3125–3135. [CrossRef]

40. Rannala, B.; Yang, Z. Efficient Bayesian species tree inference under the multispecies coalescent. *Syst. Biol.* **2017**, *66*, 823–842. [CrossRef]

41. Flouri, T.; Jiao, X.; Rannala, B.; Yang, Z. Species tree inference with BPP using genomic sequences and the multispecies coalescent. *Mol. Biol. Evol.* **2018**, *35*, 2585–2593. [CrossRef] [PubMed]

42. Silvestro, D.; Michalak, I. raxmlGUI: A graphical front-end for RAxML. *Org. Divers. Evol.* **2012**, *12*, 335–337. [CrossRef]

43. Compagno, L.J.V. *Sharks of the World: An Annotated and Illustrated Catalogue of Shark Species Known to Date, Volume 2. Bullhead, Mackerel and Carpet Sharks (Heterodontiformes, Lamniformes and Orectolobiformes)*; FAO: Rome, Italy, 2001.

44. Compagno, L.J.V. Carcharhinoid Sharks: Morphology, Systematics and Phylogeny. Ph.D. Thesis, Stanford University, Stanford, CA, USA, 1979.

45. Springer, V.G.; Garrick, J.A.F. A survey of vertebral numbers in sharks. *Proc. U. S. Natl. Mus.* **1964**, *116*, 73–96. [CrossRef]

46. Clarke, K.R.; Gorley, R.N. *PRIMER v6 User Manual/Tutorial*; Primer-E Ltd.: Plymouth, UK, 2006.

47. Finucci, B.; White, W.T.; Kemper, J.M.; Naylor, G.J. Redescription of *Chimaera ogilbyi* (Chimaeriformes; Chimaeridae) from the Indo-Australian region. *Zootaxa* **2018**, *4375*, 191–210. [CrossRef] [PubMed]

48. Hirano, T.; Saito, T.; Tsunamoto, Y.; Koseki, J.; Ye, B.; Miura, O.; Suyama, Y.; Chiba, S. Enigmatic incongruence between mtDNA and nDNA revealed by multi-locus phylogenomic analyses in freshwater snails. *Sci. Rep.* **2019**, *9*, 6223. [CrossRef] [PubMed]

49. Wiens, J.J.; Kuczynski, C.A.; Stephens, P.R. Discordant mitochondrial and nuclear gene phylogenies in emydid turtles: Implications for speciation and conservation. *Biol. J. Linn. Soc.* **2010**, *99*, 445–461. [CrossRef]

50. Giles, J.L.; Riginos, C.; Naylor, G.J.; Dharmadi; Ovenden, J.R. Genetic and phenotypic diversity in the wedgefish *Rhynchobatus australiae*, a threatened ray of high value in the shark fin trade. *Mar. Ecol. Prog. Ser.* **2016**, *548*, 165–180. [CrossRef]

51. Ma, K.Y.; van Herverden, L.; Newman, S.J.; Brumen, M.L.; Choat, J.H.; Chu, K.H.; de Mitcheson, Y.S. Contrasting population genetic structure in three aggregating groupers (Percoidei: Epinephelidae) in the Indo-West Pacific: The importance of reproductive mode. *BMC Evol. Biol.* **2018**, *18*, 180. [CrossRef]

52. Reid, D.G.; Lal, K.; Mackenzie-Dodds, J.; Kaligis, F.; Littlewood, D.T.J.; Williams, S.T. Comparative phylogeography and species boundaries in *Echinolittorina* snails in the central Indo-West Pacific. *J. Biogeogr.* **2006**, *33*, 990–1006. [CrossRef]

53. Mantiquilla, J.A.; Shiao, M.S.; Shih, H.C.; Chen, W.H.; Chiang, Y.C. A review on the genetic structure of ecologically and economically important mangrove species in the Indo-West Pacific. *Ecol. Genet. Genom.* **2021**, *18*, 100078. [CrossRef]

54. Hall, R. Cenozoic geological and plate tectonic evolution of SE Asia and the SW Pacific: Computer-based reconstructions, model and animations. *J. Asian Earth Sci.* **2002**, *20*, 353–431. [CrossRef]

55. Ahmad, A.; Abdul Haris Hilmi, A.A.; Ismail, I. *Implementation of the National Plan of Action for Conservation and Management of Shark Resources in Malaysia (Malaysia NPOA-Shark)*; Terminal Report; SEAFDEC/MFRDMD: Kuala Terengganu, Malaysia, 2015.

56. Arai, T.; Azri, A. Diversity, occurrence and conservation of sharks in the southern South China Sea. *PLoS ONE* **2019**, *14*, e0213864. [CrossRef] [PubMed]

57. Fahmi; Tibbetts, I.R.; Bennett, M.B.; Ali, A.; Krajangdara, T.; Dudgeon, C.L. Population structure of the brown-banded bamboo shark, *Chiloscyllium punctatum* and its relation to fisheries management in the Indo-Malay region. *Fish. Res.* **2021**, *240*, 105972. [CrossRef]

58. FAO. *FAO Yearbook. Fishery and Aquaculture Statistics 2016*; FAO: Rome, Italy, 2018; 104p.

59. Kamohara, T. Revised catalogue of fishes of Kochi Prefecture, Japan. *Rep. USA Mar. Biol. Stat.* **1964**, *11*, 1–99.

60. Nakaya, K. Carcharhinidae. In *The Fishes of the Japanese Archipelago*; Masuda, H., Amaoka, K., Araga, C., Uyeno, T., Yoshino, T., Eds.; Tokai University Press: Tokyo, Japan, 1984; pp. 5–6.

61. Nakabo, T. *Fishes of Japan with Pictorial Keys to the Species*, 3rd ed.; Tokai University Press: Hadano, Japan, 2013.

62. Okada, Y.; Mori, K. Descriptions and figures of marine fishes obtained at Mie Prefecture, the middle of Honshu, Japan. *J. Fac. Fish. Pref. Univ. Mie* **1958**, *3*, 1–39.

63. Kamohara, T. List of fishes from Amami-Oshima and adjacent regions, Kagoshima Prefecture, Japan. *Rep. USA Mar. Biol. Stat.* **1957**, *4*, 1–65.

64. Shinohara, G.; Matsuura, K.; Shirai, S. Fishes of Tachibana Bay, Nagasaki, Japan. *Mem. Nat. Sci. Mus. Tokyo* **1998**, *30*, 105–138.

65. Cho, H.G.; Kweon, S.M.; Kim, B.J. New record of the spadenose shark, *Scoliodon laticaudus* (Carcharhiniformes: Carcharhinidae) from South Sea, Korea. *Korean J. Ichthyol.* **2014**, *26*, 336–339.

66. Wang, K.F. Preliminary notes on the fishes of Chekiang (Elasmobranches). *Contr. Biol. Lab. Sci. Soc. China Zool. Ser.* **1933**, *9*, 87–117.

67. Zhu, J.F.; Dai, X.J.; Li, Y. Preliminary study on biological characteristics of spadenose shark, *Scoliodon laticaudus*, caught from coastal waters of Zhejiang Province. *J. Shanghai Fish. Inst.* **2008**, *17*, 635–639.

68. Lam, V.Y.Y.; Sadovy de Micheson, Y. The sharks of South East Asia—Unknown, unmonitored and unmanaged. *Fish Fish.* **2010**, *12*, 51–74. [CrossRef]

69. Ebert, D.A.; White, W.T.; Ho, H.C.; Last, P.R.; Nakaya, K.; Séret, B.; Straube, N.; Naylor, G.J.P.; de Carvalho, M.R. An annotated checklist of the chondrichthyans of Taiwan. *Zootaxa* **2013**, *3752*, 279–386. [CrossRef] [PubMed]

70. Orlov, A.M. Brief review of the marine ichthyofauna of Cambodia. *J. Ichthyol.* **1995**, *35*, 81–87.

71. Deechum, W. Species Compositions and Some Biological Aspects of Sharks and Rays from the Gulf of Thailand and Andaman Landing Sites. Ph.D. Thesis, Prince of Songkla University, Songkla, Thailand, 2009.

72. Springer, V.G. A revision of the carcharhinid shark genera *Scoliodon, Loxodon*, and *Rhizoprionodon*. *Proc. U. S. Nat. Mus.* **1964**, *115*, 559–632. [CrossRef]

73. Compagno, L.J.V.; Last, P.R.; Stevens, J.D.; Alava, M.N.R. Checklist of Philippine chondrichthyes. *CSIRO Mar. Lab. Rep.* **2005**, *243*, 103.

74. Alava, M.N.R.; Gaudiano, J.P.A.; Utzurrum, J.T.; Capuli, E.E.; Aquino, M.T.R.; Luchvez-Maypa, M.M.A.; Santos, M.D. Pating Ka Ba? In *An Identification Guide to Sharks, Batoids and Chimaeras of the Philippines*; Department of Agriculture Bureau of Fisheries and Aquatic Resources—National Fisheries Research and Development Institute and the Marine Wildlife Watch of the Philippines: Metro Manila, Philippines, 2014; 200p.

75. Last, P.R.; White, W.T.; Caira, J.N.; Jensen, K.; Lim, A.P.K.; Manjaji-Matsumoto, B.M.; Naylor, G.J.P.; Pogonoski, J.J.; Stevens, J.D.; Yearsley, G.K. *Sharks and Rays of Borneo*; CSIRO Publishing: Melbourne, Australia, 2010.

76. Emiliya, P.A.; Putra, R.D. *Identifikasi jenis hiu Hasil Tangkapan Nelayan di Pulau Bintan Provinsi Kepulauan Riau [Identification of the Type Shark Fishermen Catch in Bintan Island Riau Islands Province]*; Project Report; Universitas Maritim Raja Ali Haji: Kota Tanjung Pinang, Indonesia, 2017.

77. Teshima, K.; Ahmad, M.; Mizue, K. Studies on sharks—XIV. Reproduction in the Telok Anson shark collected from Perak River, Malaysia. *Jpn. J. Ichthyol.* **1978**, *25*, 181–189.

78. Widodo, J.; Pralampita, W.A.; Chodriyah, U. Length-weight relationships and condition factors of sharks landed from the Indian Ocean south of Java, Bali, and Lombok, Indonesia. In Proceedings of the First Annual Meeting on Artisanal Shark and Rays Fisheries in East Indonesia: Their Socio-Economic and Fishery Characteristics and relationship to Australian Resources, Perth, Australia, 4–5 April 2002.

79. Widodo, A.A.; Mahiswara, M. Sumberdaya ikan cucut (hiu) yang tertangkap nelayan di perairan Laut Jawa [The shark resource caught by fishermen in Java Sea]. *J. Iktiol. Indones.* **2007**, *7*, 15–21.

80. Bean, B.A.; Weed, A.C. Notes on a collection of fishes from Java, made by Owen Bryant & William Palmer in 1909. *Proc. U. S. Nat. Mus.* **1912**, *42*, 587–611.

81. White, W.T. Catch composition and reproductive biology of whaler sharks (Carcharhiniformes: Carcharhinidae) caught by fisheries in Indonesia. *J. Fish Biol.* **2007**, *71*, 1512–1540. [CrossRef]

82. Psomadakis, P.N.; Thein, H.; Russell, B.C.; Tun, M.T. *Field Identification Guide to the Living Marine Resources of Myanmar, FAO Species Identification Guide for Fishery Purposes*; FAO: Rome, Italy, 2019.

83. Jit, R.B.; Alam, M.F.; Rhaman, M.G.; Singha, N.K.; Akhtar, A. Landing trends, species composition and percentage composition of sharks and rays in Chittagong and Cox's Bazar, Bangladesh. *Int. J. Adv. Res. Biol. Sci.* **2014**, *1*, 81–93.

84. Kumar, R.R.; Venu, S.; Akhilesh, K.V.; Bineesh, K.K.; Rajan, P.T. First report of four deep-sea chondrichthyans (Elasmobranchii and Holocephali) from Andaman waters, India with an updated checklist from the region. *Acta Ichthyol. Piscat.* **2018**, *48*, 289–301. [CrossRef]

85. Tyabi, Z.; Jabado, R.W.; Sutaria, D. New records of sharks (Elasmobranchii) from the Andaman and Nicobar Archipelago in India with notes on current checklists. *Biodivers. Data J.* **2018**, *6*, e28593. [CrossRef] [PubMed]
86. Sen, S.; Dash, G.; Mukherjee, I. An overview of elasmobranch fisheries of West Bengal in 2018. *Mar. Fish Infor. Serv. Tech. Ext. Ser.* **2018**, *238*, 18–22.
87. Barman, R.P.; Mishra, S.S.; Kar, S.; Mukherjee, P.; Saren, S.C. Marine and estuarine fish fauna of Orissa. *Rec. Zool. Surv. India Occas. Pap.* **2007**, *260*, 1–186.
88. Talwar, P.K. A contribution to the taxonomy of *Rhizoprionodon oligolinx* Springer 1964: An important component of the shark fishery of Orissa, India. *Indian J. Fish.* **1974**, *21*, 604–607.
89. Rao, S.C.V.S. Scientific, common and local names of commercially important edible marine fin and shell fishes of Andhra Pradesh. *Mar. Fish. Infor. Serv. Tech. Ext. Ser.* **1991**, *108*, 1–10.
90. Joshi, K.K.; Sreeram, M.P.; Zacharia, P.U.; Abdussamad, E.M.; Varghese, M.; Habeeb Mohammed, O.M.M.J.; Jayabalan, K.; Kanthan, K.P.; Kannan, K.; Sreekumar, K.M.; et al. Check list of fishes of the Gulf of Mannar ecosystem, Tamil Nadu, India. *J. Mar. Biol. Assoc. India* **2016**, *58*, 34–54. [CrossRef]
91. Bineesh, K.K.; Gopalakrishnan, A.; Akhilesh, K.V.; Sajeela, K.A.; Abdussamad, E.M.; Pillai, N.G.K.; Basheer, V.S.; Jena, J.K.; Ward, R.D. DNA barcoding reveals species composition of sharks and rays in the Indian commercial fishery. *Mitochondrial DNA A* **2016**, *28*, 458–472. [CrossRef]
92. Kulkarni, G.N.; Shanbhogue, S.L.; Udupa, K.S. Length-weight relationship of *Scoliodon laticaudus* Muller and Henle and *Carcharhinus limbatus* (Muller and Henle), from Dakshina Kannada coast. *Indian J. Fish.* **1988**, *35*, 300–301.
93. Veena, S.; Thomas, S.; Raje, S.G.; Durgekar, R. Case of leucism in the spadenose shark, *Scoliodon laticaudus* (Muller & Henle, 1838) from Mangalore, Karnataka. *Indian J. Fish.* **2011**, *58*, 109–112.
94. Pillai, P.P.; Parakkal, B. Pelagic sharks in the Indian seas their exploitation, trade, management and conservation. *CMFRI Spec. Publ.* **2000**, *70*, 1–95.
95. Nair, R.V.; Appukuttan, K.K.; Rajapandian, M.E. On the systematics and identity of four pelagic sharks of the family Carcharhinidae from Indian region. *Indian J. Fish.* **1974**, *21*, 220–232.
96. Mathew, C.J.; Devaraj, M. The biology and population dynamics of the spadenose shark *Scoliodon laticaudus* in the coastal waters of Maharastra State, India. *Indian J. Fish.* **1997**, *44*, 11–27.
97. Fofandi, M.D.; Zala, M.; Koya, M. Observations on selected biological aspects of the spadenose shark (*Scoliodon laticaudus* Müller & Henle, 1838), landed along Saurashtra coast. *Indian J. Fish.* **2013**, *60*, 51–54.
98. Misra, K.S. A check list of the fishes of India, Burma & Ceylon. Part I. Elasmobranchii and Holocephali. *Rec. Indian Mus.* **1947**, *45*, 1–46.
99. Mendis, A.S. Fishes of Ceylon: A catalogue, key & bibliography. *Fish. Res. Stat. Bull.* **1954**, *2*, 1–222.
100. De Silva, R.I. Taxonomy and status of the sharks and rays of Sri Lanka. *Fauna Sri Lanka* **2006**, *2006*, 294–301.
101. Fernando, D.; Bown, R.M.K.; Tanna, A.; Gobiraj, R.; Ralicki, H.; Jockusch, E.L.; Ebert, D.A.; Jensen, K.; Caira, J.N. New insights into the identities of the elasmobranch fauna of Sri Lanka. *Zootaxa* **2019**, *4585*, 201–238. [CrossRef]
102. Moron, J.; Bertrand, B.; Last, P. A check-list of sharks and rays of western Sri Lanka. *J. Mar. Biol. Assoc. India* **1998**, *40*, 142–157.
103. Misra, K.S. An aid to the identification of the common commercial fishes of India and Pakistan. *Rec. Indian Mus.* **1962**, *57*, 1–320.
104. Gore, M.; Waqas, U.; Khan, M.M.; Ahmad, E.; Baloch, A.S.; Baloch, A.R. A first account of the elasmobranch fishery of Balochistan, south-west Pakistan. *West. Indian Ocean. J. Mar. Sci.* **2019**, *18*, 95–105. [CrossRef]
105. Bishop, J.M. History and current checklist of Kuwait's ichthyofauna. *J. Arid. Environ.* **2003**, *54*, 237–256. [CrossRef]
106. Sivasubramanian, K.; Ibrahim, M.A. *Common Fishes of Qatar. Scientific Atlas of Qatar 1*; Doha Modern Printing Press: Doha, Qatar, 1982.
107. Moore, A.B.M.; McCarthy, I.D.; Carvalho, G.R.; Peirce, R. Species, sex, size and male maturity composition of previously unreported elasmobranch landings in Kuwait, Qatar and Abu Dhabi Emirate. *J. Fish Biol.* **2012**, *80*, 1619–1642. [CrossRef] [PubMed]
108. Moore, A.B.M.; Peirce, R. Composition of elasmobranch landings in Bahrain. *Afr. J. Mar. Sci.* **2013**, *35*, 593–596. [CrossRef]
109. Jabado, R.W.; Al Ghais, S.M.; Hamza, W.; Shivji, M.S.; Henderson, A.C. Shark diversity in the Arabian/Persian Gulf higher than previously thought: Insights based on species composition of shark landings in the United Arab Emirates. *Mar. Biodiv.* **2015**, *45*, 719–731. [CrossRef]
110. Almojil, D.K.; Moore, A.B.M.; White, W.T. *Sharks and Rays of the Arabian/Persian Gulf*; MBG (INT) Ltd.: London, UK, 2015.
111. Henderson, A.C.; McIlwain, J.L.; Al-Oufi, H.S.; Al-Sheili, S. The Sultanate of Oman shark fishery: Species composition, seasonality and diversity. *Fish. Res.* **2007**, *86*, 159–168. [CrossRef]
112. Henderson, A.C.; McIlwain, J.L.; Al-Oufi, H.S.; Al-Sheile, S.; Al-Abri, N. Size distributions and sex ratios of sharks caught by Oman's artisanal fishery. *Afr. J. Mar. Sci.* **2009**, *31*, 233–239. [CrossRef]
113. Al Sakaff, H.; Esseen, M. Occurrence and distribution of fish species off Yemen (Gulf of Aden and Arabian Sea). *Naga ICLARM Q.* **1999**, *22*, 43–47.
114. Zajonz, U.; Lavergne, E.; Bogorodsky, S.V.; Saeed, F.N.; Aideed, M.S.; Krupp, F. Coastal fish diversity of the Socotra Archipelago, Yemen. *Zootaxa* **2019**, *4636*, 1–108. [CrossRef]
115. Sommer, C.; Schneider, W.; Poutiers, J.M. *Living Marine Resources of Somalia. FAO Species Identification Field Guide for Fishery Purposes*; FAO: Rome, Italy, 1996.

116. Marshall, N.T. The Somali shark fishery in the Gulf of Aden and the Western Indian Ocean. In *Trade in Sharks and Shark Products in the Western Indian and Southeast Atlantic Oceans*; Marshall, N.T., Barnes, R., Eds.; TRAFFIC East/Southern Africa: Nairobi, Kenya, 1997; pp. 24–30.
117. Anam, R.; Mostarda, E. *Field Identification Guide to the Living Marine Resources of Kenya. FAO Species Identification Field Guide for Fishery Purposes*; FAO: Rome, Italy, 2012.
118. Kiilu, B.K.; Ndegwa, S. *Shark Bycatch—Small Scale Tuna Fishery Interactions along the Kenyan Coast*; IOTC-2013-WPEB09-13; Indian Ocean Tuna Commission (IOTC): Victoria, Seychelles, 2013.
119. Fischer, W.; Sousa, I.; Silva, C.; de Freitas, A.; Poutiers, J.M.; Schneider, W.; Borges, T.C.; Feral, J.P.; Massinga, A. *Fichas FAO de Identificaçao de Espécies para Actividades de Pesca. Guia de Campo Das Espécies Comerciais Marinhas e de águas Salobras de Moçambique*; FAO: Rome, Italy, 1990.
120. Rigby, C.L.; Bin Ali, A.; Bineesh, K.K.; Chen, X.; Derrick, D.; Dharmadi Ebert, D.A.; Fahmi Fernando, D.; Gautama, D.A.; Haque, A.B.; Herman, K.; et al. *Scoliodon macrorhynchos*. The IUCN Red List of Threatened Species. 2020, p. e.T169233669A169233911. Available online: https://doi.org/10.2305/IUCN.UK.2020-3.RLTS.T169233669A169233911.en (accessed on 13 December 2021).

Communication

What Is in Your Shark Fin Soup? Probably an Endangered Shark Species and a Bit of Mercury

Christina Pei Pei Choy [1] **and Benjamin J. Wainwright** [2,*]

[1] Independent Researcher, Singapore 521112, Singapore; christinachoy01@gmail.com
[2] Yale-NUS College, National University of Singapore, Singapore 138527, Singapore
* Correspondence: ben.wainwright@yale-nus.edu.sg

Simple Summary: Shark fin soup is consumed by many Asian communities throughout the world and is one of the main drivers of the demand for shark fin. The demand for shark products has seen shark populations decline by as much as 70%. The fins found in soups break down into a fibrous mass meaning that identifying the species of shark that a fin came from is impossible by visual methods. Here, we use molecular techniques to identify the species of sharks found in bowls of soup collected in Singapore. We identified a number of endangered species in the surveyed soups, and many of these species have been shown to contain high levels of mercury, a potent neurotoxin. It is highly likely that consumers of shark fin soup are consuming levels of mercury that are above safe allowable limits, and at the same time are contributing to the massive declines in global shark populations.

Abstract: Shark fin soup, consumed by Asian communities throughout the world, is one of the principal drivers of the demand of shark fins. This near USD 1 billion global industry has contributed to a shark population declines of up to 70%. In an effort to arrest these declines, the trade in several species of sharks is regulated under the auspices of the Convention on International Trade in Endangered Species of Wild Fauna and Flora (CITES). Despite this legal framework, the dried fins of trade-regulated sharks are frequently sold in markets and consumed in shark fin soup. Shark fins found in soups break down into a fibrous mass of ceratotrichia, meaning that identifying the species of sharks in the soup becomes impossible by visual methods. In this paper, we use DNA barcoding to identify the species of sharks found in bowls of shark fin soup collected in Singapore. The most common species identified in our samples was the blue shark (*Prionace glauca*), a species listed as Near Threatened on the International Union for Conservation of Nature (IUCN) Red List with a decreasing population, on which scientific data suggests catch limits should be imposed. We identified four other shark species that are listed on CITES Appendix II, and in total ten species that are assessed as Critically Endangered, Endangered or Vulnerable under the IUCN Red List of Threatened Species. Globally, the blue shark has been shown to contain levels of mercury that frequently exceed safe dose limits. Given the prevalence of this species in the examined soups and the global nature of the fin trade, it is extremely likely that consumers of shark fin soup will be exposed to unsafe levels of this neurotoxin.

Keywords: CITES; IUCN; conservation; Singapore; DNA barcoding; mercury

Citation: Choy, C.P.P.; Wainwright, B.J. What Is in Your Shark Fin Soup? Probably an Endangered Shark Species and a Bit of Mercury. *Animals* **2022**, *12*, 802. https://doi.org/10.3390/ani12070802

Academic Editor: Martina Francesca Marongiu

Received: 14 February 2022
Accepted: 16 March 2022
Published: 22 March 2022

Publisher's Note: MDPI stays neutral with regard to jurisdictional claims in published maps and institutional affiliations.

1. Introduction

Shark fin soup is considered a delicacy served in many Asian communities throughout the world [1,2]; it is also highly prized in traditional Chinese Medicine [3–5] where it is thought to help to alleviate a host of ailments and have beneficial properties throughout the body. The fishing industry that supplies the fins for this dish is arguably one of the principle drivers of shark overexploitation [6,7], with declines of 71% reported for oceanic sharks since the 1970s [8]. These declines are largely attributed to the increased fishing efforts required to meet growing market demands, with predictions suggesting that shark consumption and further declines will accelerate if regulations are not effectively enforced [7,8]. This overexploitation has led to the inclusion of several shark species under the Convention on International Trade in Endangered Species of Wild Fauna and Flora (CITES) Appendix II. This list includes species not necessarily threatened with extinction, but in which trade must be controlled in order to avoid utilization incompatible with their survival (CITES, 1994). Unfortunately, work from around the globe regularly shows that fins from CITES-listed sharks are traded within, and throughout many countries [7,9,10]. This trade is made possible and is extremely difficult to prevent because once a fin is removed from a carcass and processed for sale, the majority of the diagnostic features that can be used in visual identification are lost, then becoming nearly impossible to identify the species of shark to which a fin belonged without molecular methods [9,11,12]. The practice of finning occurs at sea; the high value fins are removed and the lower value carcass is discarded at sea [13]. This maximizes the number of fins that can be collected and minimizes the storage space occupied by the lower value carcass.

The consumption of seafood can lead to the inadvertent ingestion of mercury by humans where it is associated with a number of adverse health risks [14]. Mercury (Hg) poisoning has been implicated in central nervous system and brain damage, infant death, and can retard fetal cognitive development when mothers consume mercury-containing seafood [15,16]. The burning of fossil fuels is the primary source of atmospheric mercury [17], which then dissolves in the oceans, where Hg concentrations in surface waters have increased by a factor of three since the industrial revolution [18–20]. In marine systems, bacteria transform mercury into its organic form, methylmercury (MeHg), where it has the potential to accumulate and biomagnify in large upper trophic level predators, such as sharks [20–23]. Unsurprisingly, shark fins consumed in soups frequently exceed safe mercury concentrations, with some studies showing all examined samples significantly above the established maximum limits for mercury consumption [22].

Shark fin soup is primarily made up of collagenous protein fibers, or ceratotrichia that are found on the inside of the fin. The consumption of this dish and other shark products offers a potential pathway that can expose humans to elevated levels of mercury. Studies show that children and adults who consume shark products once a week are exposed to three times more mercury than what is recommended by the United States Environmental Protection Agency (U.S. EPA) [24]. While the majority of studies generally report mercury and levels of other toxic elements in muscle tissue, the liver or from vertebrae where they readily accumulate [20,24,25], it is important to note that fins can contain higher levels of mercury and other toxic elements, such as lead and cadmium, in comparison to muscle and other non-fin tissues [23,26]. Additionally, work performed in Hong Kong and China shows that the total amount of mercury found in shark fins regularly exceeded the prescribed Hong Kong and China legal limit of 0.5 ppm [22], and the 1 ppm legal limit used for predatory fish in Singapore [27].

Different species of sharks contain varying levels of toxic elements; for example, the mean percentage of mercury found in silky sharks is nearly twice that found in the scalloped hammerhead, and there is a general trend of higher levels of mercury in coastal sharks when compared to oceanic species [22]. If shark products are legally required to carry correct and comprehensive labelling that indicated the species and where it was caught, a consumer then has the ability to select products that come from species of sharks acknowledged to contain lower levels of mercury, purchase products caught from a sustainable fishery, or from a species that is not endangered.

In this paper, we take a mini-DNA barcoding approach to identify the shark species found in shark fin soups purchased in Singapore. The fins used in soup are already processed, meaning that any DNA is likely degraded, with this DNA becoming further degraded by heat in the cooking process. Consequently, amplifying the full 650 bp cytochrome c oxidase 1 (COI) region becomes difficult, if not impossible. To overcome these challenges, mini-DNA barcoding techniques that use a small fragment of the original gene have been developed [28–30]. Using this smaller fragment, we expect to identify many endangered and CITES-listed species. It is also highly probable that many of these fins will come from species that have previously been shown to contain high levels of mercury and other aquatic toxins or toxic elements.

2. Materials and Methods

We sequenced 92 samples collected from bowls of shark fin soup purchased at various locations throughout Singapore. Where present, intact fins were removed. When no obvious fins were present, we collected ceratotrichia (Figure 1). Fins and ceratotrichia were rinsed in sterile deionized water, cooled on ice to prevent further DNA degradation and then placed in individual sterile falcon tubes, transported to the laboratory on ice and stored at $-80\ ^\circ$C until DNA extraction was performed.

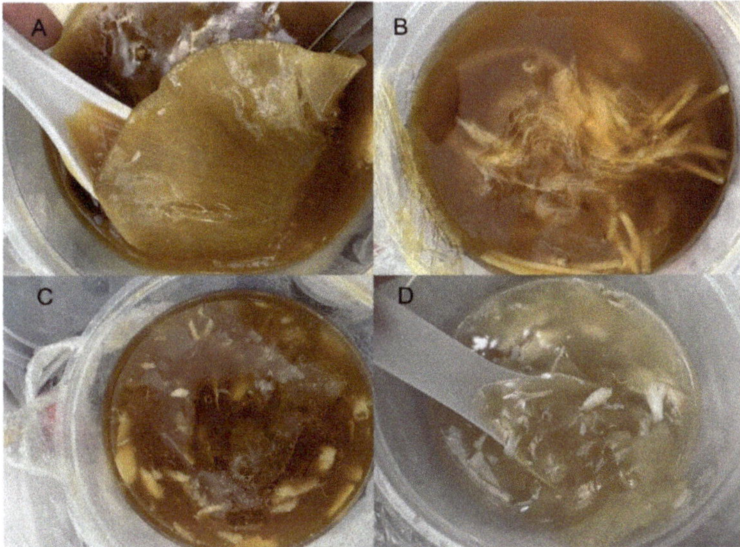

Figure 1. Examples of the products purchased in this work. (**A**) = soup 1, (**B**) = soup 2, (**C**) = soup 3 and (**D**) = soup 4. See Table 1 for the species identified in each.

DNA was extracted using a blood and tissue kit (Qiagen) following the manufacturer's instructions. Where more than one sample was collected from each soup, a separate extraction was performed for each. We initially attempted to amplify an approximate 300 bp fragment of the cytochrome c oxidase subunit I (COI) gene using the mlCOIintF (5′-GGW ACW GGW TGA ACW GTW TAY CCY CC-3′) and LoboR1 (5′-TAA ACY TCW GGR TGW CCR AAR AAY CA-3′) primers [31,32] using the following Polymerase chain reaction (PCR) cycling conditions 94 °C for 60 s, 5 cycles of 94 °C for 30 s, 48 °C for 120 s, 72 °C for 60 s, and 35 cycles of 94 °C for 30 s, 54 °C for 120 s, 72 °C for 60 s and 72 °C for 5 min. Each reaction contained 1.0 µL of MgCl2 (2.5 mM), 1.0 µL of each primer at 10 µM, 1 µL BSA (20 mg/mL), 2 µL of DNA template, 12.5 µL GoTaq mastermix green (Promega), and PCR grade water to 25 µL. We then attempted to amplify any reaction that failed under the conditions described above with the following primer pair that amplifies an approximate 150 bp fragment of the COI gene. M-13 tailed forward primer VF2_tl (5′-GTA AAA CGA CGG CCA GTC AAC CAA CCA CAA AGA CAT TGG CAC-3′) and reverse primer Shark150R (5′ -AAG ATT ACA AAA GCG TGG GC-3′) [30]. Reactions were performed in 25 µL volumes, each reaction contained 12.5 µL GoTaq mastermix green (Promega), 1 µL forward primer (10 µM), 1 µL reverse primer (10 µM), 8.5 µL PCR grade water, and 2 µL of undiluted DNA template. PCR thermal cycling conditions followed an initial denaturation period of 2 min at 94 °C, followed by 35 cycles at 94 °C for 1 min, 52 °C for 1 min, 72 °C for 1 min, and a final extension period of 10 min at 72 °C.

Table 1. Details of the species of sharks in shark fin soups collected in Singapore, their common names and conservation status according to the International Union for Conservation of Nature (IUCN) Red List of Threatened Species and Convention on International Trade in Endangered Species of Wild Fauna and Flora (CITES) listings.

	Price USD	Number of Samples Taken	Number of Species Identified	Species Identified	Common Name	CITES Listing	IUCN Red List Status
Soup 1	15.66	5	2	*Prionace glauca*	Blue shark	–	NT
				Carcharhinus falciformis	Silky shark	II	VU
Soup 2	53.49	10	2	*Alopias pelagicus*	Pelagic thresher	II	EN
				Sphyrna lewini	Scalloped hammerhead	II	CR
Soup 3	20.59	5	3	*Prionace glauca* *	Blue shark	–	NT
				Hemigaleus microstoma *	Sicklefin weasel shark	–	VU
				Carcharias taurus *	Sand tiger shark	–	CR
Soup 4	9.11	5	1	*Galeorhinus galeus*	School shark	–	CR
Soup 5	11.28	5	2	*Prionace glauca*	Blue shark	–	NT
				Rhizoprionodon acutus	Milk shark	–	VU
Soup 6	29.93	5	3	*Mustelus antarcticus*	Gummy shark	–	LC
				Galeorhinus galeus	School shark	–	CR
				Squalus spp	–	–	–
Soup 7	45.63	5	3	*Hemigaleus microstoma*	Sicklefin weasel shark	–	VU
				Prionace glauca	Blue shark	–	NT
				Carcharias taurus	Sand tiger shark	–	CR
Soup 8	36.62	6	1	*Prionace glauca*	Blue shark	–	NT
Soup 9	22.87	6	2	*Mustelus henlei*	Brown smooth-hound	–	LC
				Mustelus spp *	–	–	–
Soup 10	30.88	12	2	*Prionace glauca*	Blue shark	–	NT
				Alopias superciliosus	Bigeye thresher	II	VU
Soup 11	14.15	5	1	*Loxodon macrorhinus*	Sliteye shark	–	NT
Soup 12	33.09	10	2	*Prionace glauca*	Blue shark	–	NT
				Callorhinchus callorynchus	American Elephantfish	–	VU
Soup 13	41.07	6	1	*Prionace glauca*	Blue shark	–	NT
Soup 14	31.51	7	2	*Galeorhinus galeus* *	School shark	–	CR
				Mustelus spp *	–	–	–

* Indicates 150 bp fragment used for identification; all other identifications were made using an approximate 300 bp fragment.

PCR products were cleaned and Sanger sequenced by Macrogen (Seoul, Korea). Geneious v2020.2.4 (http://www.geneious.com; accessed 15 March 2022 [33] was used to view sequence data. We used The Barcode of Life Data System (BOLD, https://www.boldsystems.org; accessed 15 March 2022) and the Nucleotide BLAST (BLASTn) function in Genbank (http://www.ncbi.nlm.nih.gov; accessed 15 March 2022) to make species identifications from DNA data. Species identifications were considered positive, and unambiguous if BOLD indicated the ID was solid with no closely allied congeneric species currently known, and the same species was then identified as the top match in BLAST.

3. Results

We sequenced 92 samples collected from 14 bowls of soup and successfully identified 14 species of sharks at the genus or species level and 1 chimaera (Table 1). Of these, four species are listed on CITES Appendix II and tenare determined to be Critically Endangered (CR), Endangered (EN) or Vulnerable (VU) under the International Union for Conservation of Nature (IUCN) Red List of Threatened Species (Table 1). Six samples did not amplify or produce a usable sequence. If the same species was identified more than once in the same bowl of soup, we considered it the same individual.

Prionace glauca, the blue shark, was found in the most bowls of soup (8 of 14), *Galeorhinus galeus*, the school shark, in three and *Hemigaleus microstoma*, *Mustelus antarcticus*, and *Mustelus* spp. each found in two bowls; the rest only occurred once.

4. Discussion

Similar to other work that used DNA barcoding to identify the species of sharks involved in the fin trade [1,10,34–37], this study showed that a number of CITES Appendix II listed sharks, along with several other species of shark that are listed as threatened (Critically Endangered, Endangered, or Vulnerable) under the IUCN Red List of Threatened Species, are available to consumers in Singapore, and in this case directly for public consumption in the form of shark fin soup. From a conservation and fisheries management perspective, knowing what species are involved in the trade is essential, particularly if they are endangered [12,38,39]. Not knowing what species of sharks are involved in the trade makes designing successful management strategies and establishing correct CITES and IUCN designations very difficult. Additionally, setting appropriate catch quotas to ensure legality of products and sustainability of wild populations becomes very challenging, if not impossible when accurate identifications cannot be made [36].

Given the acknowledged role that biodiversity plays in promoting ecosystem stability and functioning [40–42], along with the structuring influence of large predators [40] and other apex predators can have on marine ecosystems [43], the declines in shark populations—some of which have been reduced by more than 70%—is troublesome [8], particularly as the shark product and fin trade involves many endangered species [44] and appears to continue largely unabated and unenforced.

The continuing global consumption of shark fins and products suggests consumers are either unconcerned by shark population declines, are unaware that their actions may be contributing to these declines, or do not know they could be eating endangered species. If the precarious conservation status of many sharks is not enough to discourage the consumption of shark fins and products, the high levels of elements such as mercury should be concerning to consumers, particularly as they have an established and unambiguous track record of causing severe disorders and adverse medical conditions in humans. Evidence from fins collected and analyzed throughout the world indicates that the presence of toxic elements in sharks is a global phenomenon, not restricted to specific bodies of water. Studies from the Atlantic, Indian, Pacific, Caribbean, South China Sea and Australian waters all find high levels of lead, arsenic and mercury in shark products, frequently exceeding safe advisory levels [20–24,26,45–47], and many of these tainted products enter human food chains, especially shark fins, via soups.

The most commonly encountered shark in this work was the blue shark. This shark was found in 8 of 14 bowls, and though not listed under CITES and classified as Near Threatened under the IUCN Red List of Threatened Species, blue sharks are one of the most commonly encountered sharks in the global fin trade and are traded extensively throughout Hong Kong and Singapore [9,12,48,49]. Scientific evidence suggests that this species is overexploited and should have its catch regulated to avoid population crashes [44,50]. Blue sharks collected from the Atlantic and Australian waters have been documented to contain high levels of mercury and selenium in muscle and liver tissue [25,51]. Dried blue shark fins and blue shark fins collected directly from bowls of soup in Hong Kong have levels of mercury frequently exceeding Hong Kong, European Commission and United States regulatory body maximum permissible levels [22]. While we have not determined the concentrations of mercury, or other toxic elements present in our samples collected from Singapore, it seems reasonable to suggest that the levels of mercury and other metals present in these samples will be comparable to other regions, more so given the global nature of the fin trade and the incidence of reports documenting high levels of these contaminants in blue sharks from the planet's oceans. It is a similar story with the other sharks we identified in this paper, especially the silky and hammerhead sharks.

Anecdotal evidence collected from visiting retail establishments selling dried shark products suggests that all species of hammerhead shark fins are highly regarded and command a premium price in Singapore, which appears to be corroborated by our data. The most expensive bowl of soup was the only one that contained fins from the CITES Appendix II listed scalloped hammerhead shark (*Sphyrna lewini*). The size of the fin is also a key determinant in pricing, with larger fins commanding a higher price. While it is impossible to determine the size of the fin from the collected ceratotrichia, this bowl also contained fins from the pelagic thresher shark (*Alopias pelagicus*), a species of shark renowned for its large caudal and distinctive pectoral fins, which could also help to account for the higher price of this bowl. Soups in this study ranged in price from a minimum of USD 9.11 to a maximum of USD 53.49.

Over a quarter (29%) of the sharks we identified are listed on CITES Appendix II, and 10 of the 14 bowls contained at least one species of shark that is categorized under the IUCN Red List of Threatened Species as Critically Endangered, Endangered, or Vulnerable, making it extremely likely consumers in Singapore are eating species of sharks that are threatened with extinction and in some cases trade regulated. Additionally, and for the reasons discussed earlier, there is a high chance that consumers could be unknowingly ingesting toxic elements, such as mercury, arsenic, selenium and others, at concentrations that exceed maximum recommended levels. Mandatory, effective labelling of foods that contain shark fins and other shark products, detailing the particular species and where it was caught, would allow the consumer to make an informed choice on whether it was safe to eat and if the species was endangered or came from a potentially sustainably managed stock [11].

5. Conclusions

Overall, our work shows that endangered species of sharks continue to be consumed in Singapore, and it is also extremely probable that consumers of shark fin soup are exposing themselves to unsafe levels of mercury. If shark fishing continues unabated, it is very likely they will be completely removed from marine ecosystems in the not-too-distant future. Sustainable shark fishing is possible, and the better labelling of products with the species names and geographic origin would offer a way to prevent the overexploitation of species that are already endangered. Additionally, if products were labelled in a more comprehensive and rigorous fashion, the consumer would be able to make a more informed choice and only purchase fins and other shark products from fisheries that are acknowledged to be sustainable. Better labelling could also be important from a human health perspective; for example, if certain species of shark are known to accumulate mercury at higher concentrations, then these could then be avoided.

Author Contributions: Conceptualization, C.P.P.C., B.J.W.; formal analysis, C.P.P.C., B.J.W.; investigation, C.P.P.C., B.J.W.; resources, B.J.W.; data curation, C.P.P.C., B.J.W.; writing—original draft preparation, C.P.P.C., B.J.W.; writing—review and editing C.P.P.C., B.J.W.; funding acquisition, B.J.W. All authors have read and agreed to the published version of the manuscript.

Funding: This research was funded by financial support for laboratory research was provided by the Yale-NUS College Start-up Fund (A-0007210-00-00). Christina Pei Pei Choy was supported by a grant from the Shark Conservation Fund.

Institutional Review Board Statement: Not applicable.

Informed Consent Statement: Not applicable.

Data Availability Statement: Not applicable.

Conflicts of Interest: The authors declare no conflict of interest.

References

1. O'Bryhim, J.R.; Parsons, E.C.M.; Lance, S.L. Forensic Species Identification of Elasmobranch Products Sold in Costa Rican Markets. *Fish. Res.* **2017**, *186*, 144–150. [CrossRef]
2. Ip, Y.C.A.; Chang, J.J.M.; Lim, K.K.P.; Jaafar, Z.; Wainwright, B.J.; Huang, D. Seeing through Sedimented Waters: Environmental DNA Reduces the Phantom Diversity of Sharks and Rays in Turbid Marine Habitats. *BMC Ecol. Evol.* **2021**, *21*, 166. [CrossRef] [PubMed]
3. Cardeñosa, D.; Fields, A.T.; Babcock, E.; Shea, S.K.H.; Feldheim, K.A.; Kraft, D.W.; Hutchinson, M.; Herrera, M.A.; Caballero, S.; Chapman, D.D. Indo-Pacific Origins of Silky Shark Fins in Major Shark Fin Markets Highlights Supply Chains and Management Bodies Key for Conservation. *Conserv. Lett.* **2021**, *14*, e12780. [CrossRef]
4. Mondo, K.; Broc Glover, W.; Murch, S.J.; Liu, G.; Cai, Y.; Davis, D.A.; Mash, D.C. Environmental Neurotoxins β-N-Methylamino-l-Alanine (BMAA) and Mercury in Shark Cartilage Dietary Supplements. *Food Chem. Toxicol.* **2014**, *70*, 26–32. [CrossRef]
5. Hammerschlag, N.; Davis, D.A.; Mondo, K.; Seely, M.S.; Murch, S.J.; Glover, W.B.; Divoll, T.; Evers, D.C.; Mash, D.C. Cyanobacterial Neurotoxin BMAA and Mercury in Sharks. *Toxins* **2016**, *8*, 238. [CrossRef]
6. Fields, A.T.; Abercrombie, D.L.; Eng, R.; Feldheim, K.; Chapman, D.D. A Novel Mini-DNA Barcoding Assay to Identify Processed Fins from Internationally Protected Shark Species. *PLoS ONE* **2015**, *10*, e0114844. [CrossRef]
7. Hobbs, C.A.D.; Potts, R.W.A.; Walsh, M.B.; Usher, J.; Griffiths, A.M. Using DNA Barcoding to Investigate Patterns of Species Utilisation in UK Shark Products Reveals Threatened Species on Sale. *Sci. Rep.* **2019**, *9*, 1028. [CrossRef]
8. Pacoureau, N.; Rigby, C.L.; Kyne, P.M.; Sherley, R.B.; Winker, H.; Carlson, J.K.; Fordham, S.V.; Barreto, R.; Fernando, D.; Francis, M.P.; et al. Half a Century of Global Decline in Oceanic Sharks and Rays. *Nature* **2021**, *589*, 567–571. [CrossRef]
9. Liu, C.J.N.; Neo, S.; Rengifo, N.M.; French, I.; Chiang, S.; Ooi, M.; Heng, J.M.; Soon, N.; Yeo, J.Y.; Bungum, H.Z.; et al. Sharks in Hot Soup: DNA Barcoding of Shark Species Traded in Singapore. *Fish. Res.* **2021**, *241*, 105994. [CrossRef]
10. Marchetti, P.; Mottola, A.; Piredda, R.; Ciccarese, G.; Di Pinto, A. Determining the Authenticity of Shark Meat Products by DNA Sequencing. *Foods* **2020**, *9*, 1194. [CrossRef]
11. Choo, M.Y.; Choy, C.P.P.; Ip, Y.C.A.; Rao, M.; Huang, D. Diversity and Origins of Giant Guitarfish and Wedgefish Products in Singapore. *Aquat. Conserv. Mar. Freshw. Ecosyst.* **2020**, *31*, 1636–1649. [CrossRef]
12. Wainwright, B.J.; Ip, Y.C.A.; Neo, M.L.; Chang, J.J.M.; Gan, C.Z.; Clark-Shen, N.; Huang, D.; Rao, M. DNA Barcoding of Traded Shark Fins, Meat and Mobulid Gill Plates in Singapore Uncovers Numerous Threatened Species. *Conserv. Genet.* **2018**, *19*, 1393–1399. [CrossRef]
13. Clarke, S.; Milner-Gulland, E.J.; Bjørndal, T. Social, Economic, and Regulatory Drivers of the Shark Fin Trade. *Mar. Resour. Econ.* **2007**, *22*, 305–327. [CrossRef]
14. Baishaw, S.; Edwards, J.; Daughtry, B.; Ross, K. Mercury in Seafood: Mechanisms of Accumulation and Consequences for Consumer Health. *Rev. Environ. Health* **2007**, *22*, 91–114. [CrossRef] [PubMed]
15. Driscoll, C.T.; Mason, R.P.; Chan, H.M.; Jacob, D.J.; Pirrone, N. Mercury as a Global Pollutant: Sources, Pathways, and Effects. *Environ. Sci. Technol.* **2013**, *47*, 4967–4983. [CrossRef] [PubMed]
16. Park, J.-D.; Zheng, W. Human Exposure and Health Effects of Inorganic and Elemental Mercury. *J. Prev. Med. Public Health* **2012**, *45*, 344–352. [CrossRef] [PubMed]
17. Pacyna, E.G.; Pacyna, J.M.; Sundseth, K.; Munthe, J.; Kindbom, K.; Wilson, S.; Steenhuisen, F.; Maxson, P. Global Emission of Mercury to the Atmosphere from Anthropogenic Sources in 2005 and Projections to 2020. *Atmos. Environ.* **2010**, *44*, 2487–2499. [CrossRef]
18. Lamborg, C.H.; Hammerschmidt, C.R.; Bowman, K.L.; Swarr, G.J.; Munson, K.M.; Ohnemus, D.C.; Lam, P.J.; Heimbürger, L.-E.; Rijkenberg, M.J.A.; Saito, M.A. A Global Ocean Inventory of Anthropogenic Mercury Based on Water Column Measurements. *Nature* **2014**, *512*, 65–68. [CrossRef]
19. Mason, R.P.; Choi, A.L.; Fitzgerald, W.F.; Hammerschmidt, C.R.; Lamborg, C.H.; Soerensen, A.L.; Sunderland, E.M. Mercury Biogeochemical Cycling in the Ocean and Policy Implications. *Environ. Res.* **2012**, *119*, 101–117. [CrossRef]
20. Matulik, A.G.; Kerstetter, D.W.; Hammerschlag, N.; Divoll, T.; Hammerschmidt, C.R.; Evers, D.C. Bioaccumulation and Biomagnification of Mercury and Methylmercury in Four Sympatric Coastal Sharks in a Protected Subtropical Lagoon. *Mar. Pollut. Bull.* **2017**, *116*, 357–364. [CrossRef] [PubMed]
21. Biton-Porsmoguer, S.; Bǎnaru, D.; Boudouresque, C.F.; Dekeyser, I.; Bouchoucha, M.; Marco-Miralles, F.; Lebreton, B.; Guillou, G.; Harmelin-Vivien, M. Mercury in Blue Shark (*Prionace glauca*) and Shortfin Mako (*Isurus oxyrinchus*) from North-Eastern Atlantic: Implication for Fishery Management. *Mar. Pollut. Bull.* **2018**, *127*, 131–138. [CrossRef]
22. Garcia Barcia, L.; Argiro, J.; Babcock, E.A.; Cai, Y.; Shea, S.K.H.; Chapman, D.D. Mercury and Arsenic in Processed Fins from Nine of the Most Traded Shark Species in the Hong Kong and China Dried Seafood Markets: The Potential Health Risks of Shark Fin Soup. *Mar. Pollut. Bull.* **2020**, *157*, 111281. [CrossRef] [PubMed]
23. Mohammed, A.; Mohammed, T. Mercury, Arsenic, Cadmium and Lead in Two Commercial Shark Species (*Sphyrna lewini* and *Caraharinus porosus*) in Trinidad and Tobago. *Mar. Pollut. Bull.* **2017**, *119*, 214–218. [CrossRef]
24. Tiktak, G.P.; Butcher, D.; Lawrence, P.J.; Norrey, J.; Bradley, L.; Shaw, K.; Preziosi, R.; Megson, D. Are Concentrations of Pollutants in Sharks, Rays and Skates (Elasmobranchii) a Cause for Concern? A Systematic Review. *Mar. Pollut. Bull.* **2020**, *160*, 111701. [CrossRef]

25. Vas, P.; Stevens, J.D.; Bonwick, G.A.; Tizini, O.A. Cd, Mn, and Zn Concentrations in Vertebrae of Blue Shark and Shortfin Mako in Australian Coastal Waters. *Mar. Pollut. Bull.* **1990**, *21*, 203–206. [CrossRef]

26. Boldrocchi, G.; Spanu, D.; Mazzoni, M.; Omar, M.; Baneschi, I.; Boschi, C.; Zinzula, L.; Bettinetti, R.; Monticelli, D. Bioaccumulation and Biomagnification in Elasmobranchs: A Concurrent Assessment of Trophic Transfer of Trace Elements in 12 Species from the Indian Ocean. *Mar. Pollut. Bull.* **2021**, *172*, 112853. [CrossRef]

27. Singapore Food Agency (SFA) Heavy-Metals-in-Food. Pdf. Available online: https://www.sfa.gov.sg/docs/default-source/default-document-library/heavy-metals-in-food.pdf (accessed on 15 March 2022).

28. Hajibabaei, M.; Smith, M.A.; Janzen, D.H.; Rodriguez, J.J.; Whitfield, J.B.; Hebert, P.D.N. A Minimalist Barcode Can Identify a Specimen Whose DNA Is Degraded. *Mol. Ecol. Notes* **2006**, *6*, 959–964. [CrossRef]

29. Meusnier, I.; Singer, G.A.; Landry, J.-F.; Hickey, D.A.; Hebert, P.D.; Hajibabaei, M. A Universal DNA Mini-Barcode for Biodiversity Analysis. *BMC Genom.* **2008**, *9*, 214. [CrossRef]

30. Cardeñosa, D.; Fields, A.; Abercrombie, D.; Feldheim, K.; Shea, S.K.H.; Chapman, D.D. A Multiplex PCR Mini-Barcode Assay to Identify Processed Shark Products in the Global Trade. *PLoS ONE* **2017**, *12*, e0185368. [CrossRef] [PubMed]

31. Leray, M.; Yang, J.Y.; Meyer, C.P.; Mills, S.C.; Agudelo, N.; Ranwez, V.; Boehm, J.T.; Machida, R.J. A New Versatile Primer Set Targeting a Short Fragment of the Mitochondrial COI Region for Metabarcoding Metazoan Diversity: Application for Characterizing Coral Reef Fish Gut Contents. *Front. Zool.* **2013**, *10*, 34. [CrossRef]

32. Lobo, J.; Costa, P.M.; Teixeira, M.A.; Ferreira, M.S.; Costa, M.H.; Costa, F.O. Enhanced Primers for Amplification of DNA Barcodes from a Broad Range of Marine Metazoans. *BMC Ecol.* **2013**, *13*, 34. [CrossRef]

33. Kearse, M.; Moir, R.; Wilson, A.; Stones-Havas, S.; Cheung, M.; Sturrock, S.; Buxton, S.; Cooper, A.; Markowitz, S.; Duran, C.; et al. Geneious Basic: An Integrated and Extendable Desktop Software Platform for the Organization and Analysis of Sequence Data. *Bioinformatics* **2012**, *28*, 1647–1649. [CrossRef] [PubMed]

34. Appleyard, S.A.; White, W.T.; Vieira, S.; Sabub, B. Artisanal Shark Fishing in Milne Bay Province, Papua New Guinea: Biomass Estimation from Genetically Identified Shark and Ray Fins. *Sci. Rep.* **2018**, *8*, 6693. [CrossRef] [PubMed]

35. Hernández, S.; Gallardo-Escárate, C.; Álvarez-Borrego, J.; González, M.T.; Haye, P.A. A Multidisciplinary Approach to Identify Pelagic Shark Fins by Molecular, Morphometric and Digital Correlation Data. *Hidrobiológica* **2010**, *20*, 71–80.

36. Holmes, B.H.; Steinke, D.; Ward, R.D. Identification of Shark and Ray Fins Using DNA Barcoding. *Fish. Res.* **2009**, *95*, 280–288. [CrossRef]

37. Ward, R.D.; Holmes, B.H.; White, W.T.; Last, P.R. DNA Barcoding Australasian Chondrichthyans: Results and Potential Uses in Conservation. *Mar. Freshw. Res.* **2008**, *59*, 57. [CrossRef]

38. Neo, S.; Kibat, C.; Wainwright, B.J. Seafood Mislabelling in Singapore. *Food Control* **2022**, *135*, 108821. [CrossRef]

39. Rehman, A.; Jafar, S.; Ashraf Raja, N.; Mahar, J. Use of DNA Barcoding to Control the IllegalWildlife Trade: A CITES Case Report FromPakistan. *J. Bioresour. Manag.* **2015**, *2*, 3. [CrossRef]

40. Cochrane, S.K.J.; Andersen, J.H.; Berg, T.; Blanchet, H.; Borja, A.; Carstensen, J.; Elliott, M.; Hummel, H.; Niquil, N.; Renaud, P.E. What Is Marine Biodiversity? Towards Common Concepts and Their Implications for Assessing Biodiversity Status. *Front. Mar. Sci.* **2016**, *3*, 248. [CrossRef]

41. Loreau, M. Biodiversity and Ecosystem Functioning: Recent Theoretical Advances. *Oikos* **2000**, *91*, 3–17. [CrossRef]

42. Yachi, S.; Loreau, M. Biodiversity and Ecosystem Productivity in a Fluctuating Environment: The Insurance Hypothesis. *Proc. Natl. Acad. Sci. USA* **1999**, *96*, 1463–1468. [CrossRef] [PubMed]

43. Hammerschlag, N.; Schmitz, O.J.; Flecker, A.S.; Lafferty, K.D.; Sih, A.; Atwood, T.B.; Gallagher, A.J.; Irschick, D.J.; Skubel, R.; Cooke, S.J. Ecosystem Function and Services of Aquatic Predators in the Anthropocene. *Trends Ecol. Evol.* **2019**, *34*, 369–383. [CrossRef]

44. French, I.; Wainwright, B.J. DNA Barcoding Identifies Endangered Sharks in Pet Food Sold in Singapore. *Front. Mar. Sci.* **2022**, *9*, 9. [CrossRef]

45. Rodríguez-Gutiérrez, J.; Galván-Magaña, F.; Jacobo-Estrada, T.; Arreola-Mendoza, L.; Sujitha, S.B.; Jonathan, M.P. Mercury–Selenium Concentrations in Silky Sharks (*Carcharhinus falciformis*) and Their Toxicological Concerns in the Southern Mexican Pacific. *Mar. Pollut. Bull.* **2020**, *153*, 111011. [CrossRef] [PubMed]

46. Kim, S.W.; Han, S.J.; Kim, Y.; Jun, J.W.; Giri, S.S.; Chi, C.; Yun, S.; Kim, H.J.; Kim, S.G.; Kang, J.W.; et al. Heavy Metal Accumulation in and Food Safety of Shark Meat from Jeju Island, Republic of Korea. *PLoS ONE* **2019**, *14*, e0212410. [CrossRef] [PubMed]

47. Le Bourg, B.; Kiszka, J.J.; Bustamante, P.; Heithaus, M.R.; Jaquemet, S.; Humber, F. Effect of Body Length, Trophic Position and Habitat Use on Mercury Concentrations of Sharks from Contrasted Ecosystems in the Southwestern Indian Ocean. *Environ. Res.* **2019**, *169*, 387–395. [CrossRef] [PubMed]

48. Fields, A.T.; Fischer, G.A.; Shea, S.K.H.; Zhang, H.; Feldheim, K.A.; Chapman, D.D. DNA Zip-coding: Identifying the Source Populations Supplying the International Trade of a Critically Endangered Coastal Shark. *Anim. Conserv.* **2020**, *23*, 670–678. [CrossRef]

49. Fields, A.T.; Fischer, G.A.; Shea, S.K.H.; Zhang, H.; Abercrombie, D.L.; Feldheim, K.A.; Babcock, E.A.; Chapman, D.D. Species Composition of the International Shark Fin Trade Assessed through a Retail-Market Survey in Hong Kong: Shark Fin Trade. *Conserv. Biol.* **2018**, *32*, 376–389. [CrossRef]

50. Simpfendorfer, C.A.; Dulvy, N.K. Bright Spots of Sustainable Shark Fishing. *Curr. Biol.* **2017**, *27*, R97–R98. [CrossRef] [PubMed]
51. Branco, V.; Vale, C.; Canário, J.; Dos Santos, M.N. Mercury and Selenium in Blue Shark (*Prionace glauca*, L. 1758) and Swordfish (*Xiphias gladius*, L. 1758) from Two Areas of the Atlantic Ocean. *Environ. Pollut.* **2007**, *150*, 373–380. [CrossRef]

Article

How Well Do 'Catch-Only' Assessment Models Capture Catch Time Series Start Years and Default Life History Prior Values? A Preliminary Stock Assessment of the South Atlantic Ocean Blue Shark Using a Catch-Based Model

Richard Kindong [1,2,3,4,5], Feng Wu [1,2,3,4,5,*], Siquan Tian [1,2,3,4,5,*] and Ousmane Sarr [1]

1. College of Marine Sciences, Shanghai Ocean University, Shanghai 201306, China; kindong@shou.edu.cn (R.K.); ousmanesarr218@gmail.com (O.S.)
2. National Engineering Research Center for Oceanic Fisheries, Shanghai Ocean University, Shanghai 201306, China
3. Key Laboratory of Sustainable Exploitation of Oceanic Fisheries Resources, Ministry of Education, Shanghai 201306, China
4. Key Laboratory of Oceanic Fisheries Exploitation, Ministry of Agriculture, Shanghai 201306, China
5. Scientific Observing and Experimental Station of Oceanic Fishery Resources, Ministry of Agriculture, Shanghai 201306, China
* Correspondence: fwu@shou.edu.cn (F.W.); sqtian@shou.edu.cn (S.T.)

Citation: Kindong, R.; Wu, F.; Tian, S.; Sarr, O. How Well Do 'Catch-Only' Assessment Models Capture Catch Time Series Start Years and Default Life History Prior Values? A Preliminary Stock Assessment of the South Atlantic Ocean Blue Shark Using a Catch-Based Model. *Animals* **2022**, *12*, 1386. https://doi.org/10.3390/ani12111386

Academic Editor: Martina Francesca Marongiu

Received: 13 April 2022
Accepted: 26 May 2022
Published: 27 May 2022

Publisher's Note: MDPI stays neutral with regard to jurisdictional claims in published maps and institutional affiliations.

Simple Summary: Blue shark species are at the top of the list of captured bycatch sharks in most tuna and tuna-like fisheries. As a consequence, their populations have been declining due to overfishing; thus, there is a need for the assessment of their stocks to better understand blue sharks' stock status. Most bycatch species lack sufficient data for traditional stock assessment models to be implemented. Blue sharks in the South Atlantic have been assessed in the past using a state-space production model. Given the development of new assessment models and the use of up-to-date data, their stock status was evaluated using a new state-space production model (CMSY++). We used different catch time series, abundance indices and priors to measure the intrinsic growth rate r to evaluate their influence on the outputs of CMSY++. We identified from many scenarios that the blue shark stock in the South Atlantic may be witnessing overfishing and is being overfished.

Abstract: CMSY++, an improved version of the CMSY approach developed from Catch-MSY which uses a Bayesian implementation of a modified Schaefer model and can predict stock status and exploitation, was used in the present study. Evaluating relative performance is vital in situations when dealing with fisheries with different catch time series start years and biological prior information. To identify the influences of data inputs on CMSY++ outputs, this paper evaluated the use of a nominal reported catch and a reconstructed catch dataset of the South Atlantic blue shark alongside different priors of the blue shark's productivity/resilience (r) coupled with different indices of abundance. Results from the present study showed that different catch time series start years did not have a significant influence on the estimation of the biomass and fishing reference points reported by CMSY++. However, uninformative priors of r affected the output results of the model. The developed model runs with varying and joint abundance indices showed conflicting results, as classification rates in the final year changed with respect to the type of index used. However, the model runs indicated that South Atlantic blue shark stock could be overfished (B2020/Bmsy = 0.623 to 1.15) and that overfishing could be occurring (F2020/Fmsy = 0.818 to 1.78). This result is consistent with the results from a previous assessment using a state-space surplus production model applied for the same stock in 2015. Though some potential could be observed when using CMSY++, the results from this model ought to be taken with caution. Additionally, the continuous development of prior information useful for this model would help strengthen its performance.

Keywords: overfishing; management; CMSY++; bycatch species; ICCAT; BSP

---

1. Introduction

The blue shark *Prionace glauca* is a highly migratory pelagic shark species with a circumglobal distribution found throughout all oceans in tropical, subtropical, and temperate waters [1–3]. It is particularly vulnerable as bycatch in longline fisheries and has a near-threatened status according to the International Union for Conservation of Nature and Natural Resources (IUCN) [4,5]. Blue sharks are usually captured from the surface to a depth of at least 600 m [1,6]. The blue shark is a top shark bycatch species for many commercial fisheries, especially those targeting tunas and tuna-like species in the Atlantic, Pacific, and Indian oceans [7–11]. Because large numbers of blue sharks are caught by various fisheries, the species' stock status has become an issue of great concern for regional fishery management organizations (RFMOs), such as the International Commission for the Conservation of Atlantic Tunas (ICCAT), the Western and Central Pacific Fisheries Commission, and the Indian Ocean Tuna Commission. However, the IUCN reports that the population status of this species has been declining globally [5].

The continuous removal of blue sharks as well as other bycatch species, such as rays, marine mammals, and other sharks, may seriously alter the marine ecosystem structure [12]. Due to this growing concern, tuna RFMOs have made it their prime objective to use various approaches to effectively manage global marine resources to ensure balance in the marine ecosystem. Blue shark population trends have been obtained for populations in the three main oceans. This information was obtained from different assessment results from the North Atlantic [7], South Atlantic [13], North Pacific [10], South Pacific [8], and Indian oceans [14]. These assessments reported that different stocks were sustainably harvested; however, further assessment works need to be performed with much-updated data given that catches of blue sharks continue to increase.

The last stock assessment for the South Atlantic blue shark population was carried out in 2015 [7,13], and no assessment with updated data has been completed since. It is essential to track this stock's population trend given the constant increase in catches, and also to test the effectiveness of newly developed assessment models on stocks such as the blue shark. The author of [13] used a state-space Bayesian Surplus Production (BSP) model to assess the South Atlantic blue shark stock. Ref. [7] also used a BSP approach for blue shark stock assessment in the Atlantic to assess the status of the same stock, but conflicting results were observed from the two models. However, these reports paved way for the implementation of other assessment methods for testing and comparison between models.

The data available for blue sharks in this region can be classified as moderate. Newly developed data-poor/-moderate assessment models such as the CMSY++ can be applied to evaluate the stock status of this threatened bycatch species. CMSY++ is an advanced state-space Bayesian method for stock assessment that estimates fishery reference points (maximum sustainable yield (MSY), Fmsy, Bmsy) as well as status or relative stock size (B/Bmsy) and fishing pressure or exploitation (F/Fmsy) from catch and (optionally) abundance data, a prior for resilience or productivity r, and broad priors for the ratio of biomass to unfished biomass (B/k) at the beginning, an intermediate year, and the end of the time series [15]. CMSY++ can incorporate, in addition to the catch time series, a wide variety of additional data and supplementary information in a rigorous Bayesian context that tends to reduce the dependency on prior information as much as possible, while remaining robust and thus usable in data-limited/-moderate situations [15,16].

This updated version of CMSY has recently been used by [16,17] to estimate the biomass and exploitation levels of some of the world's commercially exploited species. These reports indicate the effectiveness of CMSY++ to evaluate species' stock status, especially when combined with informative priors and some indices of abundance data. Some of the species analyzed using CMSY++ were cod, sole, European anchovy, yellowfin tuna, sardinella (round and Madeiran), and many others, as indicated in these reports [15–17]. These reports highlight the effectiveness of this model in providing basic reference points, particularly for stocks with no data available. Furthermore, the flexibility in the implementation of CMSY++ makes it interesting, as its performance can be improved whenever more data are available to develop priors from external sources [16]. This method can, at the same time, be applied either in a data-poor (catch-only) or a data-moderate (catch and catch-per-unit-of-effort (CPUE)) situation, making it suitable to assess fishery stocks worldwide, which will contribute to a much-needed better understanding of the world's fisheries [18]. However, given that most assessment models are easily influenced by various input data, particularly data-poor and data-moderate models, it is also necessary to evaluate the performance of the CMSY++ model. Hence, the significance of evaluating the performance of the CMSY++ model, a newly developed state-space Bayesian method, using fishery data with different catch time series start years, CPUE indices, and biological prior information.

The last stock assessment of the South Atlantic blue shark indicated that any future increase in fishing mortality could cause the stock to be overfished and/or experience overfishing due to some unsustainable harvests witnessed in past years [7,13]. The recent updates on the blue shark in the Atlantic by the IUCN indicate that populations continue to decline, probably due to an increase in fishing mortality [4]. Therefore, up-to-date assessment studies using the best possible up-to-date data are necessary to better understand their population trend. This study used a recently developed stock assessment approach, CMSY++, to evaluate the South Atlantic Ocean blue shark population using available catch and abundance indices data. Catch data available from the ICCAT Task I nominal catch database and reconstructed catch presented in the ICCAT 2015 blue shark data preparatory meeting were used to test the influence of different catch time series start years on the outputs of the stock assessment model. Additionally, priors of the intrinsic growth rate r presented in the FishBase database and that were used in the last blue shark stock assessment were evaluated to identify their effects on the outputs of the assessment model. Furthermore, the present study also evaluates the effect on the model's output of different abundance indices tested as additional runs in the model.

2. Materials and Methods

2.1. Study Area and Data Source

Blue sharks in the Atlantic Ocean are divided into two populations (north and south). This study focused on the South Atlantic blue shark population. Blue shark data used in this analysis were mainly catch data (Task I's reported catch and reconstructed catches developed for blue sharks, ICCAT) and indices of abundance available for some countries. The catch data used in the present work were obtained from the 2015 blue shark data preparatory meeting report [19], which made estimations for many fleets and nations based on the best available information. This report presented a reconstructed catch dataset of the 2015 blue shark stock assessment [13,20]—different from the ICCAT Task I nominal catch data (T1NC) for the blue shark—using a ratio-based method. The general approach that was used to fill in some missing catches in the historical reconstructed catch dataset was the average catch between two adjacent years to capture the localized tendency. Figure 1 presents the catch time series for the reconstructed catch (catch series starts from 1971) used in the 2015 assessment and the nominal catch data (starting from 1991 (8 tons negligible)) for the South Atlantic Ocean blue shark. For this assessment study, we focused on both catch time series to evaluate the influence of this different catch time series information on the final assessment results. The blue shark data preparatory meeting also presented relative indices of abundance developed from the standardized catch-per-unit-of-effort (CPUE) time series available for some countries [19].

The abundance indices considered for this assessment were based on CPUE indices from longline fishery data for Uruguay (URG), Brazil (BRS), Japan (JPN), Spain (ESP), Taiwan, China/Chinese Taipei (CHTP), and combined CPUE (J_C). The same abundance indices were used in this study as those used in the 2015 blue shark assessment.

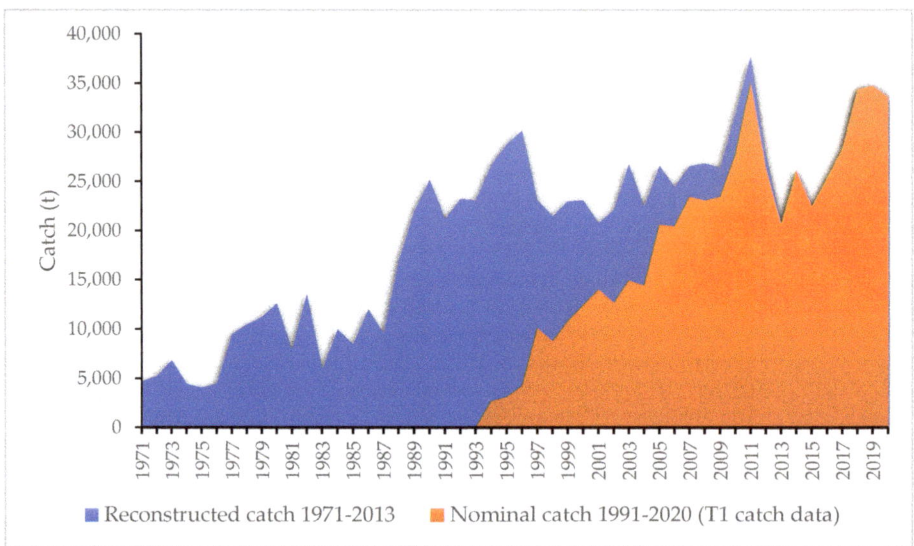

Figure 1. Historical reconstructed catches and nominal catch (T1NC) for the South Atlantic blue shark population. Catch reconstructed from 1971 to 2013; total used catch from 1971 to 2020. Nominal reported catch (T1NC) data from 1991 to 2020.

The biological information needed from this assessment was also obtained from the 2015 blue shark assessment report to ensure a sort of comparison of the present results to the previous ones. The bounds for the intrinsic growth rate (r) were obtained from FishBase [21] and the 2015 blue shark assessment report [13,20]. On one hand, the intrinsic growth rate bounds of (0.045–0.3) from r = 0.21 used in the 2015 blue shark assessment [20] were used, and on the other hand, the intrinsic growth rate r = 0.06 (0.045–0.10) was obtained from FishBase [21].

2.2. CMSY++ (CMSY and BSM)

The method CMSY++ is an innovative state-space Bayesian method that comprises two solid analytical methodological parts, both based on a modified Schaefer surplus production model: (1) the CMSY part, the method that treats catch-only data, and (2) the other part of the method, BSM (Bayesian Schaefer Model) that requires additional abundance data [15]. Furthermore, typical production models use catch time series data and indices of abundance data to estimate productivity. In its place, the CMSY++ method uses a catch dataset and a prior for resilience or productivity (r), abundance data (optionally), and broad priors for the ratio of biomass to unfished biomass (B/k) at the beginning, an intermediate year, and at the end of the time series. These inputs by CMSY++ are used to estimate biomass status or relative stock size (B/Bmsy), fishing pressure or exploitation rate (F/Fmsy), and related fishery biological reference points (MSY, Fmsy, Bmsy). The BSM applied in CMSY++, compared to other implementations of surplus production models, focuses on informative priors and also accepts short and fragmented years of abundance data [15]. In doing so, the CMSY++ provides an alternative assessment tool for situations where CPUE indices are not available or potentially unreliable.

The CMSY++ method is an improvement of the CMSY method presented in [22], which is an improvement of the Catch-MSY method of [23]. The main technical differences between CMSY++ and CMSY are the implementation of a fully Bayesian approach with MCMC (Markov chain Monte Carlo) modelling even when only catch data are available (CMSY analysis), and the prediction of default biomass priors from catch using an AI (Artificial Intelligence) neural network [15]. Additionally, in the previous version of CMSY, priors of r and k (carrying capacity) were uniformly distributed, whereas in the new CMSY++ an introduction of multivariate normal priors for r and k in log space is applied for both the CMSY and BSM methods. Hence, this allows the easy estimation of the 'best' r–k pair in CMSY and faster run times [15]. It is worth noting that CMSY++ addresses the overestimation of productivity at very low stock sizes (a general shortcoming of production models) by implementing a linear decline in surplus production when biomass falls below 1/4k.

2.3. Setting Ranges of Prior Parameters to Be Explored

Biological information is vital for properly informing the priors of CMSY++. The catch time series were derived from the 2015 blue shark data preparatory meeting report and updated blue shark's T1NC data (up to 2020). Relative abundance indices for blue sharks consisted of standardized catch-per-unit efforts (CPUEs) for Japanese, Brazilian, Uruguayan, Spanish, Taiwanese, and Chinese longline fisheries.

The smallest and largest values of r obtained from FishBase ((0.045–0.1) [21] and the value of r used in the 2015 blue shark assessment (r = 0.21; ICCAT/SCRS/2015/014) were used to set the bounds of r (0.045–0.3) explored in CMSY++. In addition to this, CMSY++ was also run using the default approach, in which the resilience value available on FishBase was used to define the range of r. For the blue shark, resilience was estimated to be low using r = 0.21 [20] (Table 1), reflecting what we know of blue sharks: they are a slow-growing, late-maturing species that can produce many offspring (4–135 pups; FishBase) [21].

Table 1. Ranges of different categories of the intrinsic growth rate or resilience (r) and the depletion rate or biomass relative to the unfished stock (B/k).

Resilience/Intrinsic Growth Rate (r)	Prior r Range
High	0.6–1.5
Medium	0.2–0.8
Low	0.05–0.5
Very low	0.015–0.1
Depletion rate	**Prior relative biomass (B/k) range**
Very strong depletion	0.01–0.2
Strong depletion	0.01–0.4
Medium depletion	0.2–0.6
Low depletion	0.4–0.8
Nearly unexploited	0.75–1.0

Regarding the range of depletion rates (B/k) at the start of the time series (1971), the stock is believed to be experiencing a low to medium depletion state, as can be seen in Figure 1 and the categories expressed in Table 1. Therefore, a wider initial depletion rate (B/k) of 0.4–0.8 was defined (Table 1 and specified in Table 2) based on ranges of depletion and resilience rates as stated in Froese et al. [15]. A larger range for the intermediate depletion rate was set to 0.2–0.9 for the year 1995 (reconstructed catch data) and 2011 for nominal catch data, to give the model more freedom. In order not to overly constrain the estimated stock trajectory, a wider range, between 0.1 and 0.7, was given as the depletion rate for the final year (2020).

Table 2. Prior values tested using CMSY++ (14 runs). Each run is represented by different catch per unit effort (CPUE) types (BSH_ATS: blue shark Atlantic South with reconstructed catch dataset from 1971; BSH_ATS_n: blue shark Atlantic South with nominal reported catch dataset from 1992; UR: Uruguay; BR: Brazil; JP: Japan; ESP: Spain; CHTP: Taiwan, China; and JCPUE: combined CPUE.

Run	CPUE	Start Year	End Year	r.Low	r.Hi	stb. Low	stb. Hi	intb. Yr	intb. Low	intb. Hi	endb. Low	endb. Hi	Btype	force. Cmsy	Process Error
1	BSH_ATS	1971	2020	0.045	0.3	0.4	0.8	1995	0.2	0.9	0.1	0.7	None	T	0.05
2	BSH_ATS_CPUE_UR	1971	2020	0.045	0.3	0.4	0.8	1995	0.2	0.9	0.1	0.7	CPUE	F	0.05
3	BSH_ATS_CPUE_BR	1971	2020	0.045	0.3	0.4	0.8	1995	0.2	0.9	0.1	0.7	CPUE	F	0.05
4	BSH_ATS_CPUE_JP	1971	2020	0.045	0.3	0.4	0.8	1995	0.2	0.9	0.1	0.7	CPUE	F	0.05
5	BSH_ATS_CPUE_ESP	1971	2020	0.045	0.3	0.4	0.8	1995	0.2	0.9	0.1	0.7	CPUE	F	0.05
6	BSH_ATS_CPUE_CHTP	1971	2020	0.045	0.3	0.4	0.8	1995	0.2	0.9	0.1	0.7	CPUE	F	0.05
7	BSH_ATS_JCPUE	1971	2020	0.045	0.3	0.4	0.8	1995	0.2	0.9	0.1	0.7	CPUE	F	0.05
8	BSH_ATS_n	1992	2020	0.045	0.3	0.4	0.8	2011	0.2	0.9	0.1	0.7	None	T	0.05
9	BSH_ATS_n_CPUE_UR	1992	2020	0.045	0.3	0.4	0.8	2011	0.2	0.9	0.1	0.7	CPUE	F	0.05
10	BSH_ATS_n_CPUE_BR	1992	2020	0.045	0.3	0.4	0.8	2011	0.2	0.9	0.1	0.7	CPUE	F	0.05
11	BSH_ATS_n_CPUE_JP	1992	2020	0.045	0.3	0.4	0.8	2011	0.2	0.9	0.1	0.7	CPUE	F	0.05
12	BSH_ATS_n_CPUE_ESP	1992	2020	0.045	0.1	0.4	0.8	2011	0.2	0.9	0.1	0.7	CPUE	F	0.05
13	BSH_ATS_n_CPUE_CHTP	1992	2020	0.045	0.1	0.4	0.8	2011	0.2	0.9	0.1	0.7	CPUE	F	0.05
14	BSH_ATS_n_JCPUE	1992	2020	0.045	0.1	0.4	0.8	2011	0.2	0.9	0.1	0.7	CPUE	F	0.05

Note: r.Low/Hi: the prior range of intrinsic growth rate for the species; stb.Low/Hi: the prior biomass range relative to the unexploited biomass (B/k) at the beginning of the catch time series; intb.Yr: a year in the time series for an intermediate biomass level; intb.Low/Hi: the estimated intermediate relative biomass range; endb.Low/Hi: the prior relative biomass (B/k) range at the end of the catch time series; btype: the type of information in the bt column of the catch file.

Further model configuration and scenarios involved the choice of variances for the catch data (observation errors), CPUEs, and process errors. Process errors enable the population dynamics to deviate from the exact values given by the model, while still conforming to the assumptions of the model on average. The incorporation of process errors is useful for two reasons: (1) when the model is trying to fit an abundance index, process errors can reduce bias arising from lack of fit in a deterministic SRA whenever dynamics are poorly explained by catch history alone, and (2) with or without an abundance index (or other auxiliary information), the stochastic portion is necessary to obtain plausible uncertainty intervals in the final estimates [24,25].

2.4. Scenarios

A total of fourteen (14) scenario runs were performed; these scenarios were chosen from combinations of two input sources: catch time series (reconstructed versus T1NC reported catches) and different indices of abundance (Table 2). Seven runs comprised reconstructed catches from 1971 to 2020 and the other seven catches were from 1992 to 2020. Among the 14 runs, 1 run that included reconstructed catches (1971–2020), all input CPUE indices (standardized catch-per-unit effort (CPUE) for Japan, Brazil, Uruguay, Spain, and Taiwan, longline fisheries) and prior mean values was developed as a base case (BSH_ATS_JCPUE, Table 1: Run 14). Two scenarios were tested without any indices of abundance (only catch data, CMSY analysis) and twelve scenarios with catch and different indices of abundance (CMSY and BSM), with the aim to evaluate the sensitivity of the model to different assumptions regarding the changes in input data. The process error was set to 0.05 for all scenarios. The coefficients of variations (CVs) for catches and CPUEs were set to 0.2 and 0.15, respectively. The process error and CVs fixed in the present study were the same as those used in the Bayesian state-space surplus production model of the 2015 South Atlantic blue shark assessment [13,20]. Kobe plots were also presented. For the present study, we investigated the effect of time series start year, including catch time series starting in 1971 or 1992 (Table 2).

3. Results

Effect of the Starting Year of the Catch Time Series and Choice of the Prior of r

The two options used to define the prior of r did change significantly, especially on runs with catch and CPUE data for both cases (Tables 3 and 4). For the first case, when the prior of r obtained from FishBase was used, the r for all fourteen runs did not change significantly (Table 3) whereas, for the second case (r = 0.21), significant changes were observed between runs with only catch data and runs comprising catch and CPUE data (Table 4). The values obtained in the second case for all runs in Table 4 fell within the range presented in the 2015 blue shark stock assessment. For further analysis, the prior of r from FishBase was dropped given that this prior of r did not specify from which blue shark population (north or south) the value r = 0.06 was obtained. This study focused on the prior of r for the South Atlantic blue shark as used in the 2015 blue shark assessment (Table 4) so as to facilitate comparing the final results.

Table 3. A summary of the results from CMSY++ (catch-only and BSM) runs including catch-per-unit effort (CPUE) indices from Uruguay (UR), Brazil (BR), Japan (JP), Spain (ESP), Taiwan, China (CHTP), and combined CPUE (J_CPUE). Intrinsic growth rate r of 0.06 (0.045–0.10) obtained from FishBase [21]. Runs 1 to 7 CPUE indices denoted n_CPUE for Uruguay (UR), Brazil (BR), Japan (JP), Spain (ESP), Taiwan, China (CHTP), and combined CPUE (J_CPUE) represent runs used with nominal reported catches from 1992. r represents resilience or intrinsic growth rate; K—maximum stock size or carrying capacity; MSY— maximum sustainable yield; B—biomass level; F—fishing mortality rate; Bmsy—biomass at MSY level; Fmsy—fishing mortality at MSY level; B/Bmsy—biomass relative to Bmsy; F/Fmsy—fishing mortality relative to Fmsy.

Run	CPUE	Start Year	End Year	r	K	MSY	B	Bmsy	B/Bmsy	F	Fmsy	F/Fmsy
1	BSH_ATS_n	1992	2020	0.062	2452	37.9	1645	1226	1.2	0.021	0.031	0.716
2	BSH_ATS_n_CPUE_UR	1992	2020	0.073	1312	23.9	660	656	1	0.051	0.036	1.42
3	BSH_ATS_n_CPUE_BR	1992	2020	0.072	1484	27	869	742	1.18	0.039	0.036	1.08
4	BSH_ATS_n_CPUE_JP	1992	2020	0.064	1932	30.7	924	966	0.976	0.037	0.032	1.15
5	BSH_ATS_n_CPUE_ESP	1992	2020	0.074	1281	23.7	659	641	1.04	0.051	0.036	1.39
6	BSH_ATS_n_CPUE_CHTP	1992	2020	0.068	1673	28.7	959	836	1.15	0.035	0.034	1.04
7	BSH_ATS_n_JCPUE	1992	2020	0.072	1528	27.7	910	764	1.2	0.037	0.036	1.03
8	BSH_ATS	1971	2020	0.061	2516	38.4	1592	1258	1.16	0.021	0.031	0.753
9	BSH_ATS_CPUE_UR	1971	2020	0.074	1276	24	597	638	0.942	0.057	0.037	1.5
10	BSH_ATS_CPUE_BR	1971	2020	0.071	1607	28.5	911	803	1.14	0.037	0.035	1.05
11	BSH_ATS_CPUE_JP	1971	2020	0.069	1728	29.7	512	864	0.594	0.067	0.035	1.94
12	BSH_ATS_CPUE_ESP	1971	2020	0.074	1235	23.1	559	618	0.913	0.061	0.037	1.62
13	BSH_ATS_CPUE_CHTP	1971	2020	0.067	1675	28.3	872	838	1.04	0.039	0.033	1.16
14	BSH_ATS_JCPUE	1971	2020	0.075	1319	24.9	716	660	1.1	0.047	0.037	1.26

Table 4. A summary of the results from CMSY++ (catch-only and BSM) runs including catch-per-unit effort (CPUE) indices from Uruguay (UR), Brazil (BR), Japan (JP), Spain (ESP), Taiwan, China (CHTP), and combined CPUE (J_CPUE). Intrinsic growth rate bounds of (0.045–0.3) from r = 0.21 used in the 2015 blue shark assessment [20]. Runs 1 to 7 CPUE indices denote n_CPUE for Uruguay (UR), Brazil (BR), Japan (JP), Spain (ESP), Taiwan, China (CHTP), and combined CPUE (J_CPUE) with runs used with nominal reported catches from 1992. r represents resilience or intrinsic growth rate; K—maximum stock size or carrying capacity; MSY—maximum sustainable Yield; B—biomass level; F—fishing mortality rate; Bmsy—biomass at MSY level; Fmsy—fishing mortality at MSY level; B/Bmsy—biomass relative to Bmsy; F/Fmsy—fishing mortality relative to Fmsy.

Run	CPUE	Start Year	End Year	r	K	MSY	Bmsy	B	B/Bmsy	F	Fmsy	F/Fmsy
1	BSH_ATS_n	1992	2020	0.074	2027	37.5	1014	1325	1.16	0.026	0.037	0.823
2	BSH_ATS_n_CPUE_UR	1992	2020	0.257	368	24.4	184	178	0.97	0.191	0.128	1.45
3	BSH_ATS_n_CPUE_BR	1992	2020	0.142	815	29	408	494	1.22	0.068	0.071	0.972
4	BSH_ATS_n_CPUE_JP	1992	2020	0.075	1663	31.3	832	795	0.951	0.043	0.038	1.15
5	BSH_ATS_n_CPUE_ESP	1992	2020	0.223	415	23.8	207	202	0.974	0.171	0.111	1.47
6	BSH_ATS_n_CPUE_CHTP	1992	2020	0.11	1046	28.7	523	605	1.17	0.056	0.055	1.02
7	BSH_ATS_n_JCPUE	1992	2020	0.159	723	28.7	361	438	1.21	0.077	0.079	1.01
8	BSH_ATS	1971	2020	0.072	2065	37.3	1032	1345	1.15	0.025	0.036	0.818
9	BSH_ATS_CPUE_UR	1971	2020	0.227	441	25.3	220	193	0.903	0.177	0.114	1.51
10	BSH_ATS_CPUE_BR	1971	2020	0.118	1036	30.5	518	613	1.2	0.055	0.06	0.944
11	BSH_ATS_CPUE_JP	1971	2020	0.119	1049	30.8	525	322	0.623	0.105	0.057	1.78
12	BSH_ATS_CPUE_ESP	1971	2020	0.237	413	24.6	206	160	0.8	0.212	0.118	1.75
13	BSH_ATS_CPUE_CHTP	1971	2020	0.126	907	28.4	453	509	1.12	0.067	0.063	1.07
14	BSH_ATS_JCPUE	1971	2020	0.154	724	27.9	362	409	1.15	0.084	0.077	1.07

As seen in Tables 3 and 4, setting the start year to 1971 or 1992 had no great influence on r when running CMSY++ with only catch data (CMSY) and no CPUE indices data (Runs 1 and 8). A slight increase in K was observed with the longer time series of catch data from 2027kt to 2065kt (Table 4). Similarly, we found that the start year of the catch time series had a negligible impact on the results for different runs when including CPUE in the catch data, except for the runs that included the Japanese CPUE (Runs 4 and 11: Table 4). The Japanese CPUE run with a short times series (Run 4: Table 4) had similar r values as runs with catch-only data. The runs with the Japanese CPUE examined across different start years had similar results, and presented extreme values for B/Bmsy (0.623, lowest value: Run 11, Table 4) and F/Fmsy (1.78, highest value: Run 11, Table 4). This difference observed in start years for Japanese CPUE and different runs with other CPUE indices may be attributed to differences in CPUE time series. The Japanese CPUE started in 1971, contrary to other CPUE indices, so might have influenced the results for different catch start years. The lack of influence of the start year observed for the other CPUE indices may be attributed to the catch rate before 1971 being lower than the current catch rate (Figure 1), and therefore having little impact on the estimates of K. Consequently, this low influence of start year observed for most runs advises the selection of the full dataset available; thereby, we adjusted our final model configuration to include the start year of 1971.

The Japanese, Spanish and Uruguayan BSMs indicated that biomass had dropped below Bmsy by 2020; meanwhile, CMSY (catch-only) for these three indices indicated biomass above BMSY. Spain's BSM indicated biomass dropping below Bmsy by 2012, while this occurred from 2018 to 2020 for Uruguay, and from 1992 to 2020 for Japan, almost dropping to half of Bmsy, indicating a proxy of reduced recruitment (Figure 2, Row 2). Brazil, Taiwan, China and joint CPUE BSM estimates were in more accordance with the CMSY (catch-only) stock status (biomass) estimate than the other three indices and showed decreasing biomass, but were still higher than Bmsy by 2020 (Figure 2, Row 2). Further results indicated that the B/Bmsy values estimated by CMSY (catch-only) were more

in line with the BSMs from Brazil, Taiwan and China, and joint CPUE with B/Bmsy > 1
indicated that the stock was not overfished (Runs 8, 10, 13–14: Table 3). Contrary to that, the
Japanese, Spanish and Uruguayan BSM estimates indicated that the stock was overfished
with B/Bmsy < 1.

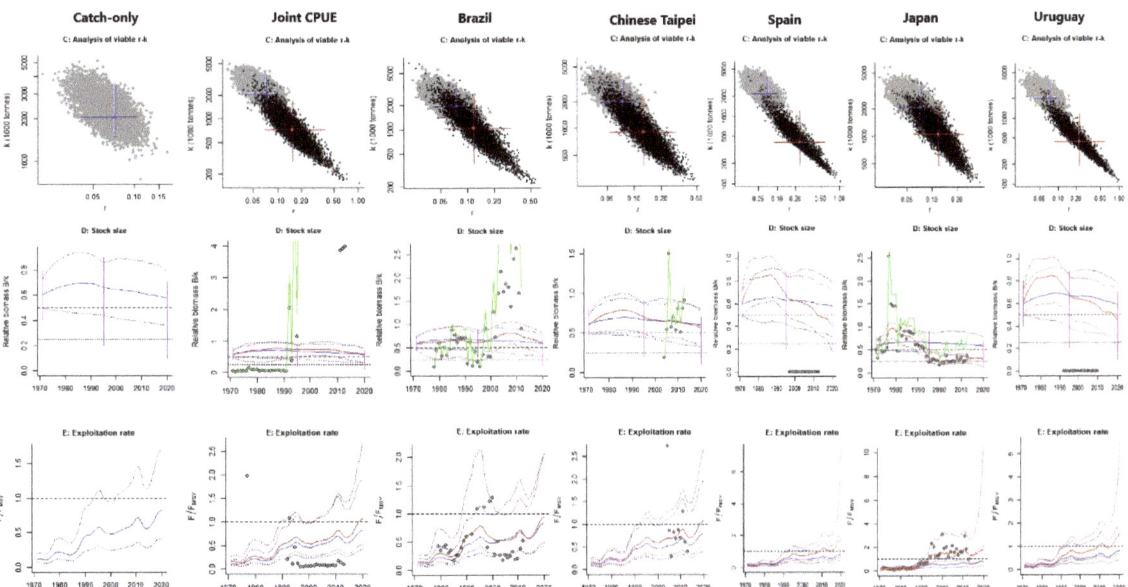

Figure 2. The output of the CMSY++ (CMSY and BSM models) using catch-only data (column 1) and
the catch-per-unit effort (CPUE) indices (from columns 2 to 7) of the joint CPUEs of Brazil, Taiwan,
China, Spain, Japan, and Uruguay. **Row 1** displays the r/k pairs found by CMSY catch-only and the
BSM when using the full catch time series (from 1971 to 2020) with the dark grey points indicating
possible r–k pairs found by the CMSY model and black dots indicating possible r–k pairs found by the
BSM model. The blue crosses indicate the most probable r–k pairs found by the CMSY (catch-only)
and the red crosses indicate the most probable r–k pairs found by the BSM and their 95% confidence
limits. **Row 2** shows the estimated biomass relative to K (red), i.e., the CPUE data, scaled to the BSM
estimate of Bmsy = 0.5 k, and the biomass trajectory estimated by CMSY in blue. Dotted blue and red
lines indicate the 2.5th and 97.5th percentiles. The dots indicate the CPUE data scaled and corrected
by BSM, and the green line indicates the uncorrected CPUE. Vertical purple lines indicate the prior
biomass ranges. Horizontal dashed and dotted black lines indicate the 0.5 and 0.25 biomasses (BMSY),
respectively. **Row 3** indicates the exploitation rate estimated from the BSM (red) and the CMSY (blue).
The black horizontal–dotted line indicates where F/Fmsy = 1.

The Japanese BSM indicated that overfishing has been occurring throughout the last
three decades. CMSY outputs indicated that exploitation rates begun relatively low, started
rising in the mid-1980s, declined around 2010, and rose exponentially from 2016 to 2020, but
remained below overfishing limits (Figure 1, Row 3). Apart from CMSY (catch-only) and the
Brazilian BSM, the outputs from the other BSMs indicated that the stock was experiencing
overfishing (F/Fmsy > 1) (Table 3; Figure 1, Row 3). Stock trajectories produced by each
tested run were also evaluated. The CMSY (catch-only) run indicated a probability of 63.5%
of the stock in the last year falling in the green area. When the abundance indices were
put together (joint CPUE), BSM indicated that the probability of the stock falling in the
orange area in the last year was 28.3%, and there was an 86% probability of this same stock
falling in the red area in 2020 for the Japanese BSM (Figure 3). Furthermore, probabilities

from the BSMs of Uruguay (61.5%) and Spain (72.6%) also indicated stock in the red area; the Chinese Taipei BSM indicated stock in the orange area (26.5% probability); and the Brazilian BSM showed that the stock in the last year fell in the green area (54.3%).

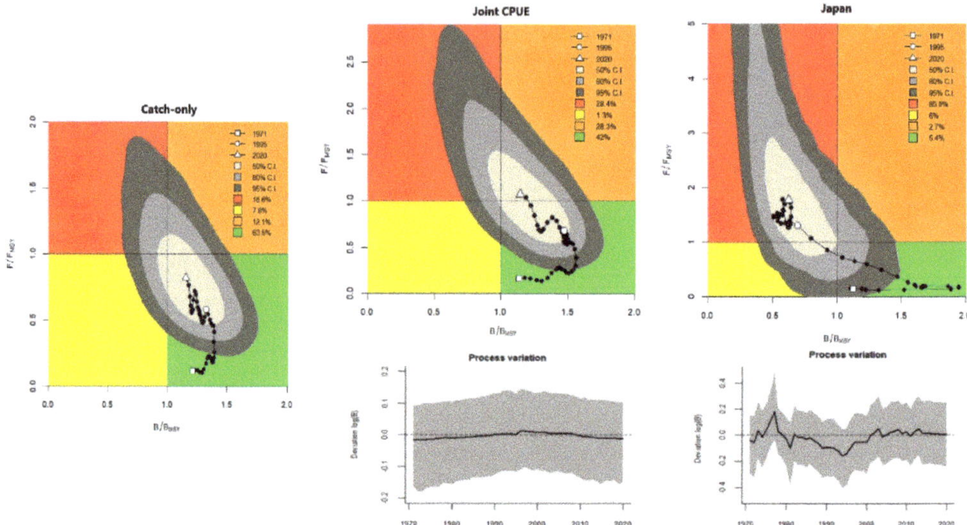

Figure 3. Kobe plots show the synchronized changes in exploitation (F/Fmsy) and the relative biomass (B/Bmsy). The orange area indicates healthy stock sizes that are about to be depleted by overfishing. The red area indicates that the stock is overfished and is undergoing overfishing. The yellow area indicates reduced fishing pressure on stocks recovering from still too low biomass levels. The green area indicates sustainable fishing pressure and a healthy stock size capable of producing high yields close to maximum sustainable yield (MSY). The lower panel figures represent process error deviations indicating changes in biomass diverging from the Schaefer model expectations (thick bold vs. dashed lines).

The process variation figures show the deviation between deterministic expectation (surplus production minus catch) and stochastic realization (after adding process errors); a strong deviation of the bold curve from the dashed line indicates that changes in biomass diverge from the Schaefer model expectations, perhaps due to the CPUE not properly describing the abundance or the priors being mis-specified (Froese et al., 2021). Slight deviations could be observed for the Japanese CPUE run and an insignificant change was observed when the CPUEs were combined.

4. Discussion

The present study shows that the start year catch time series had an insignificant influence on the final results, whereas the assignment of intrinsic growth (r) greatly affected the outputs of the model if not carefully selected. This study tested two sources of priors of r, and significant changes in the posterior r and the reference points determined by different runs for both cases could be observed. The abundance indices presented in this study also influenced the final results of the CMSY++ model, as results from the CMSY (catch-only) model generally differed from those presented by the BSMs. When CMSY++ ran with catch-only data it indicated a healthy stock status for the South Atlantic blue shark. However, when abundance indices were added to the catch data, most runs indicated a decline in biomass and an increase in overfishing levels for the stock. Though in our investigation we saw that the outputs given by the CMSY (catch-only) method indicated a healthy stock level, these outputs also showed exponential increases in fishing efforts in recent years, which may lead to overfishing and consequently an overfished state if the present level of fishing continues. Three BSM runs have already indicated that the South Atlantic blue shark is overfished and witnessing overfishing.

The fishery status of the South Atlantic blue shark was evaluated using different inputs of informative data applied in the CMSY++ approach. We limited the use of default settings as inputs in different runs since this is not advisable when using catch-only approaches [16]. Thus, prior estimates on r and biomass depletion rates were defined based on available knowledge of the stock. In this study, we tested runs using priors of r obtained from FishBase [21] and the prior used in the 2015 blue shark assessment report [7]. We found that priors of r in both cases for all runs indicated a healthy stock status with increasing fishing pressure when using the CMSY (catch-only) method, and almost all runs by the BSM indicated the overfishing and overfished status of the stock. We also observed a significant difference in the estimated posterior of r for both cases, especially for runs having catch and abundance indices; this change eventually influenced the reference points used in defining stock status (Table 4). When r = 0.21 [7] was used, our posterior r values for all runs tested (0.118–0.237) fell within the estimated range of values obtained in the ICCAT blue shark assessment report [7]. This range of resilience indicating a low to medium level of stock exploitation [15] shows increasing fishing pressure on South Atlantic blue sharks. Therefore, we note the importance of reliable priors when using CMSY++ to obtain good estimates of depletion in the final year, since this model gave results close to past assessment results when we applied a prior of r that closely depicted the South Atlantic blue shark stock.

This study also showed that outputs from CMSY++ can be influenced when different CPUE indices are combined with catch data. Among all abundance indices used, only the BSM run of the Brazil CPUE index had a similar output outcome to the catch-only method run, with the BSMs of the other indices indicating varying results from the CMSY runs. For example, the CMSY (catch-only) and the Brazilian BSM runs both indicated that the blue shark stock was in a healthy state, while the other runs either stated that the stock was currently either witnessing overfishing or was overfished and witnessing overfishing at the same time. The Japanese, Uruguayan and Spanish BSM runs indicated that overfishing may be occurring and that the stock may also be overfished.

Compared to the assessment methods used in the 2015 South Atlantic blue shark assessment meeting, our results correlate more closely with the runs from the state-space production model in JAGS than with the Bayesian Surplus Production (BSP) model. Though model runs in the state-space production model used in the 2015 assessment meeting had combined CPUE indices with three different process errors (0.05, 0.01 and 0), most runs indicated that the stock could be overfished (B2013/Bmsy = 0.78 to 1.29 against B2020/Bmsy = 0.623 to 1.15) and that overfishing could be occurring (F2013/Fmsy = 0.54 to 1.19 against F2020/Fmsy = 0.818 to 1.78). The scenarios with the BSP model indicated that the stock was not overfished (B2013/Bmsy = 1.96 to 2.03) and that overfishing was not occurring (F2013/Fmsy = 0.01 to 0.11). The BSP results greatly differed from the present study, which may be due to the configuration and assumptions of the models used and also maybe the shorter catch time series used in the BSP. The conclusions from the 2015 assessment meeting were that the estimates obtained with the state-space BSP were generally less optimistic, given that their outputs changed and had more pessimistic results, especially when process errors were not included [7]. Process variations may indicate changes in biomass that differ from Schaefer model expectations; this may be due to strong environmental variation, CPUEs not properly describing the abundance, or the priors used being mis-specified [15]. Our study showed minimal changes in process deviations for all model runs, indicating that our results may have better outcomes.

As indicated in [15,16], catch-only methods, including CMSY++, could have broad application in achieving fisheries' sustainable development goals at national as well as regional fishery management levels, given their performances and their flexible usage. However, discrepancies in results may arise when using catch-only models without good knowledge of priors, as was observed in the present study when some prior information was obtained from FishBase (in our study, prior of r). Some past studies also support our results, stating that discrepancies may arise when using catch-only methods for evaluations when using life–history meta-analyses from platforms such as FishBase [16,26,27]. As seen in the present study, the default setting of r (from FishBase, Table 3) resulted in posterior resilience, indicating a very low depletion biomass state of the stock; this was a distorted picture of the evolution of the stock, given that the blue shark population in the South Atlantic is currently declining, thus corresponding to medium to high biomass exploitation levels [4,15].

5. Conclusions

When using informative priors of initial depletion, especially the prior of r combined with a well-defined index of abundance, CMSY++ model performance improved, depicting clearer biomass trends of blue shark stock. Besides catch time series start years, abundance indices and biological prior information, the potential performance of stock assessment models may also be affected by factors such as gear efficiency [26]. Although the usage of CMSY++ in the present study showed various limitations when comparing outputs to the BSP model used in 2015, we still think some positive and probably fairly robust information can be obtained from the analysis that could be helpful to further guide the implementation of CMSY++ to other stocks, and also to present relevant information on the stock status of blue sharks in the South Atlantic Ocean.

Author Contributions: Conceptualization, R.K.; formal analysis, F.W., R.K.; investigation, S.T.; resources, S.T.; data curation, R.K., O.S.; writing—original draft preparation, R.K.; writing—review and editing R.K., F.W.; funding acquisition, F.W., S.T. All authors have read and agreed to the published version of the manuscript.

Funding: This research was funded by financial support for the project on the Survey and Monitoring-Evaluation of Global Fishery Resources sponsored by the Ministry of Agriculture and Rural Affairs (21-0109-02), by the National Key R&D Programs of China (2019YFD0901404) and by the National Programme on Global Change and Air–sea Interaction (GASI-01-EIND-YD01aut/02aut).

Institutional Review Board Statement: Not applicable.

Informed Consent Statement: Not applicable.

Data Availability Statement: Data used in this study is publicly available from the ICCAT database.

Acknowledgments: We would like to acknowledge our institutional colleagues for their input in this manuscript.

Conflicts of Interest: The authors declare no conflict of interest.

References

1. Compagno, L.J.V. *FAO Species Catalogue. Vol. 4. Sharks of the World. An Annotated and Illustrated Catalogue of Shark Species Known to Date. Part 2—Carcharhiniformes*; FAO Fisheries Synopsis; FAO: Rome, Italy, 1984; Volume 125, pp. 251–655.
2. Last, P.R.; Stevens, J.D. *Sharks and Rays of Australia*, 2nd ed.; CSIRO: Melbourne, Australia, 2009.
3. Ebert, D.A.; Fowler, S.; Compagno, L. *Sharks of the World. A Fully Illustrated Guide*; Wild Nature Press: Plymouth, UK, 2013.
4. Rigby, C.L.; Barreto, R.; Carlson, J.; Fernando, D.; Fordham, S.; Francis, M.P.; Herman, K.; Jabado, R.W.; Liu, K.M.; Marshall, A.; et al. *Prionace glauca. The IUCN Red List of Threatened Species*; IUCN: London, UK, 2019; p. e.T39381A2915850. [CrossRef]
5. IUCN. The IUCN Red List of Threatened Species. Version 2019-3. 2019. Available online: www.iucnredlist.org (accessed on 10 December 2019).
6. Nakano, H.; Stevens, J. The Biology and Ecology of the Blue Shark. In *Prionace glauca. Sharks of the Open Ocean: Biology, Fisheries and Conservation*; Camhi, M.D., Pikitch, E.K., Babcock, E.A., Eds.; Blackwell Publishing: Oxford, UK, 2008; pp. 141–151.
7. ICCAT. Report of the 2015 ICCAT Blue Shark Stock Assessment Session. In Proceedings of the International Commission for the Conservation of Atlantic Tunas (ICCAT), Lisbon, Portugal, 27–31 July 2015.
8. Takeuchi, Y.; Tremblay-Boyer, L.; Pilling, G.M.; Hampton, J. Assessment of Blue Shark in the Southwestern Pacific. In Proceedings of the Western Central Pacific Fisheries Commission. Scientific Committee Twelfth Regular Session, Bali, Indonesia, 3–11 August 2016; WCPFC-SC12-2016/SA-WP-08 Rev 1.

9. Indian Ocean Tuna Commission (IOTC). Stock Assessment of Blue Shark in the Indian Ocean. In Proceedings of the IOTC—17th Working Party on Ecosystems & Bycatch (Assessment), Victoria, Seychelles, 6–10 September 2021; IOTC-2021-WPEB17(AS)-15_rev1.

10. International Scientific Committee for Tuna and Tuna-like species in the north Pacific Ocean (ISC). Stock Assessment and Future Projections of Blue Shark in the North Pacific Ocean through 2015. Report of the Shark Working Group. In Proceedings of the Western Central Pacific Fisheries Commission Scientific Committee Thirteenth Regular Session, Rarotonga, Cook Islands, 9–17 August 2017; WCPFC-SC13-2017/SA-WP-10.

11. Xia, M.; Carruthers, T.; Kindong, R.; Dai, L.; Geng, Z.; Dai, X.; Wu, F. Quantifying bycatch risk factors for the Chinese distant water fishery. *ICES J. Mar. Sci.* **2021**, fsab210. [CrossRef]

12. Myers, R.A.; Baum, J.K.; Shepherd, T.D.; Powers, S.P.; Peterson, C.H. Cascading Effects of the loss of apex predatory sharks from a coastal ocean. *Science* **2007**, *315*, 1846–1850. [CrossRef] [PubMed]

13. Carvalho, F.; Winker, H. *Stock Assessment of South Atlantic Blue Shark (Prionace glauca) through 2013*; SCRS/2015/153; Collective Volumes of Scientific Papers; ICCAT: Madrid, Spain, 2015.

14. Rice, J. Stock Assessment Blue Shark (Prionace glauca) in the Indian Ocean Using Stock Synthesis. In Proceedings of the IOTC Secrtetariat, Indian Ocean Tuna Commission, Victoria, Seychelles, 17–22 October 2017; IOTC-2017-WPEB13-33.

15. Froese, R.; Demirel, N.; Coro, G.; Winker, H. *User Guide for CMSY++*; GEOMAR: Kiel, Germany, 2021; p. 17. Available online: http://oceanrep.geomar.de/52147/ (accessed on 5 April 2022).

16. Sharma, R.; Winker, H.; Levontin, P.; Kell, L.; Ovando, D.; Palomares, M.L.D.; Pinto, C.; Ye, Y. Assessing the Potential of Catch-Only Models to Inform on the State of Global Fisheries and the UN's SDGs. *Sustainability* **2021**, *13*, 6101. [CrossRef]

17. Palomares, M.L.D.; Baxter, S.; Bailly, N.; Chu, E.; Derrick, B.; Frias-Donaghey, M.; Simon-Luc, N.; Emmalai, P.; Rebecca, S.; Jessika, W.; et al. Estimating the biomass of commercially exploited fisheries stocks left in the ocean. *Fish. Cent. Res. Rep.* **2021**, *29*, 74.

18. Pauly, D.; Cheung, W.W.L.; Sumaila, U.R. "What Are Global Fisheries Studies?". In *Marine and Freshwater Miscellanea*; Pauly, D., Ruiz-Leotaud, V., Eds.; Fisheries Centre Research Reports; Institute for the Oceans and Fisheries, University of British Columbia: Vancouver, BC, Canada, 2018; Volume 26, pp. 33–37.

19. ICCAT. *2015 BLUE SHARK DATA PREPARATORY MEETING*; Document SCRS/2015/012; ICCA: Madrid, Spain, 2015.

20. ICCAT. *2015 BLUE SHARK STOCK ASSESSMENT MEETING*; Document SCRS/2015/014; ICCA: Madrid, Spain, 2015.

21. Froese, R.; Pauly, D. (Eds.) FishBase. World Wide Web Electronic Publication. 2022. Available online: www.fishbase.org (accessed on 5 April 2022).

22. Froese, R.; Demirel, N.; Gianpaolo, C.; Kleisner, K.M.; Winker, H. Estimating fisheries reference points from catch and resilience. *Fish Fish.* **2017**, *18*, 506–526. [CrossRef]

23. Martell, S.; Froese, R. A simple method for estimating MSY from catch and resilience. *Fish Fish.* **2013**, *14*, 504–514. [CrossRef]

24. Walters, C.J.; Martell, S.J.; Korman, J. A stochastic approach to stock reduction analysis. *Can. J. Fish. Aquat. Sci.* **2006**, *63*, 212–223. [CrossRef]

25. Thorson, J.T.; Rudd, M.B.; Winker, H. The case for estimating recruitment variation in data-moderate and data-poor age-structured models. *Fish. Res.* **2019**, *217*, 87–97. [CrossRef]

26. Rosenberg, A.A.; Fogarty, M.J.; Cooper, A.B.; Dickey-Collas, M.; Fulton, E.A.; Gutiérrez, N.L.; Hyde, K.J.W.; Kleisner, K.M.; Kristiansen, T.; Longo, C.; et al. Developing New Approaches to Global Stock Status Assessment and Fishery Production Potential of the Seas. In *FAO Fisheries and Aquaculture Circular No. 1086*; FAO: Rome, Italy, 2014; p. 175.

27. Froese, R.; Winker, H.; Coro, G.; Demirel, N.; Tsikliras, A.C.; Dimarchopoulou, D.; Scarcella, G.; Palomares, M.L.D.; Dureuil, M.; Pauly, D. Estimating stock status from relative abundance and resilience. *ICES J. Mar. Sci.* **2019**, *77*, 527–538. [CrossRef]

Article

New Occurrences of the Tiger Shark (*Galeocerdo cuvier*) (Carcharhinidae) off the Coast of Rio de Janeiro, Southeastern Brazil: Seasonality Indications

Izar Aximoff [1,*], Rodrigo Cumplido [2], Marcelo Tardelli Rodrigues [3], Ubirajara Gonçalves de Melo [4], Eduardo Barros Fagundes Netto [5], Sérgio Ricardo Santos [6,7] and Rachel Ann Hauser-Davis [8,*]

[1] Laboratório de Radioecologia e Mudanças Globais, Núcleo de Fotografia Científica Ambiental, Instituto de Biologia Roberto Alcântara Gomes, Universidade do Estado do Rio de Janeiro (UERJ), Rua São Francisco Xavier, n° 524, PHLC Subsolo, Maracanã, Rio de Janeiro 20550-900, RJ, Brazil
[2] Programa de Pós-Graduação em Oceanografia (PPG-OCN), Centro de Tecnologia e Ciências, Faculdade de Oceanografia, Universidade do Estado do Rio de Janeiro (UERJ), Rua São Francisco Xavier, n° 524, Maracanã, Rio de Janeiro 20550-000, RJ, Brazil
[3] Laboratório de Ecotoxicologia e Microbiologia Ambiental (LEMAM), Instituto Federal de Educação, Ciência e Tecnologia Fluminense (IFF), Campus Cabo Frio, Estrada Cabo Frio-Búzios, s/n°, Baía Formosa, Cabo Frio 28909-971, RJ, Brazil
[4] Programa Associado de Pós-Graduação em Biotecnologia Marinha (PPGBM), Instituto de Estudos do Mar Almirante Paulo Moreira (IEAPM), Universidade Federal Fluminense (UFF), Rua Kioto, n° 253, Praia dos Anjos, Arraial do Cabo 28930-000, RJ, Brazil
[5] Divisão de Oceanografia Biológica, Departamento de Oceanografia, Instituto de Estudos do Mar Almirante Paulo Moreira (IEAPM), Rua Kioto, n° 253, Praia dos Anjos, Arraial do Cabo 28930-000, RJ, Brazil
[6] Laboratório de Biologia e Tecnologia Pesqueira (BioTecPesca), Universidade Federal do Rio de Janeiro (UFRJ), Rio de Janeiro 21941-590, RJ, Brazil
[7] Instituto Museu Aquário Marinho do Rio de Janeiro (IMAM-AquaRio), Rio de Janeiro 20220-360, RJ, Brazil
[8] Laboratório de Avaliação e Promoção da Saúde Ambiental, Instituto Oswaldo Cruz, Fiocruz, Avenida Brasil, n° 4.365, Manguinhos, Rio de Janeiro 21040-360, RJ, Brazil
* Correspondence: izar.aximoff@gmail.com (I.A.); rachel.hauser.davis@gmail.com (R.A.H.-D.)

Citation: Aximoff, I.; Cumplido, R.; Rodrigues, M.T.; de Melo, U.G.; Fagundes Netto, E.B.; Santos, S.R.; Hauser-Davis, R.A. New Occurrences of the Tiger Shark (*Galeocerdo cuvier*) (Carcharhinidae) off the Coast of Rio de Janeiro, Southeastern Brazil: Seasonality Indications. *Animals* **2022**, *12*, 2774. https://doi.org/10.3390/ani12202774

Academic Editor: Martina Francesca Marongiu

Received: 22 August 2022
Accepted: 10 October 2022
Published: 14 October 2022

Publisher's Note: MDPI stays neutral with regard to jurisdictional claims in published maps and institutional affiliations.

Simple Summary: There is a lack of detailed information on the capture pressures and records regarding the tiger shark *Galeocerdo cuvier* (Péron & Lesueur, 1822) (Carcharhinidae) for the state of Rio de Janeiro, Brazil. This study aimed to expand the tiger shark record database and to improve upon future conservation and management strategies in this area. A total of 23 new records were obtained, increasing the number of tiger shark records off the coast of the state of Rio de Janeiro approximately 5-fold. Possible seasonality patterns concerning tiger shark sizes were noted, indicating the need to consider the coast of Rio de Janeiro as an especially relevant area for at least part of the life history of tiger sharks.

Abstract: The tiger shark *Galeocerdo cuvier* (Péron & Lesueur, 1822) (Carcharhinidae) is classified as near-threatened along the Brazilian coast, in line with its global categorization. Although Rio de Janeiro, located in southeastern Brazil, is internationally identified as a priority shark conservation area, many shark species, including tiger sharks, are landed by both industrial and artisanal fisheries in this state. However, there is a lack of detailed information on the species capture pressures and records for the state of Rio de Janeiro. Therefore, the aims of this study were to expand the tiger shark record database and to improve upon future conservation and management strategies. Tiger shark records from four coastal Rio de Janeiro regions were obtained by direct observation. The information obtained from fishery colonies/associations, environmental guards, researchers, and scientific articles, totaling 23 records, resulted in an approximately 5-fold increase in the number of tiger shark records off the coast of the state of Rio de Janeiro. A possible seasonality pattern concerning the size of the captured/observed animals was noted, emphasizing the need to consider the coast of Rio de Janeiro as an especially relevant area for at least part of the life history of tiger sharks.

Keywords: elasmobranchs; geographic distribution; citizen science; southeastern Brazil

1. Introduction

Elasmobranchs generally play important roles in shallow-water ecosystems in reefs, bays, and estuaries in tropical and temperate oceans [1]. However, despite their environmental importance, the global abundance of oceanic sharks and rays has declined by 71% since 1970 due to an 18-fold increase in relative fishing pressure [2]. In addition, other pressures associated with climate change and chemical contamination of the oceans are also significant negative stressors to this taxonomic group [3,4]. In the long term, these pressures jeopardize populations, a problem compounded by the fact that sharks and rays are a long-lived and a late-maturing species, thus increasing the risk of global extinction to the point that three-fourths of all elasmobranch species are threatened with extinction [2].

One of the priority areas concerning global elasmobranch conservation is the Atlantic coast of South America, which presents the highest concentration of threatened species [5] and where scientific data and resources are limited, with the complexity of local fisheries often being overlooked [6]. In Brazil, the monitoring of fisheries ceased from 2011, which had landed more than 20 tons of cartilaginous fish in the previous decade [7]. It is estimated that Brazil currently ranks among the countries with the highest elasmobranch capture rates worldwide [8]. While battling finning has exposed the global fin market to public scrutiny and control agencies [9,10], the emphasis given to Brazil is validated by the increased elasmobranch meat consumption and non-fin products noted in recent decades [11]. Despite the commercial importance of elasmobranchs, the reconstruction of fishery production data indicates a considerable drop in shark landings, from a peak in the early 1980s declining in the following years to the current estimates of around 9.000 tons/year [12].

The pressure and lack of information associated with the overfishing of species traded without consumers knowing they are buying and consuming shark meat have put these animals at significant risk in Brazil [13]. Currently, one-third of Brazilian shark species are threatened [14], although the true status of many species is unknown due to lack of information [13]. This is true for the tiger shark *Galeocerdo cuvier* (Péron & Lesueur, 1822) (Carcharhinidae), which, although one of the least representative species in northeast [15], southeast [16], and south [17] Brazil, has been classified as near threatened along the Brazilian coast, in line with its global categorization [14,18]. Australia has, in fact, recently identified declines in this species abundance [19]. In southeastern Brazil, although the state of Rio de Janeiro is internationally identified as a priority shark conservation area [20], it is responsible for 2.5% of the total number of sharks caught per year in Brazil [21].

Among the several shark species caught by artisanal fisher colonies in Rio de Janeiro, the tiger shark has always been one of the least recorded [16,22,23]. The tiger shark is a large iconic predator that can reach 500 cm in length and weigh around 900 kg [24]. It exhibits a circumglobal distribution, inhabiting tropical and temperate waters in all oceans [25], and, similarly to many other sharks from an ecosystem point of view, it plays an essential role as a trophic regulator of ecosystem interaction networks [26]. Tiger sharks exhibit a very varied diet, composed of a wide variety of marine and terrestrial organisms, in addition to carcasses [27–29]. The species is ovoviviparous and exhibits a high fecundity, producing between 10 and 82 neonates per posture [30], although it is late to mature, reaching maturity between seven (males) and eight (females) years old [31]. It is important to note that Brazilian coastal records indicate rapid tiger shark growth when compared to other regions worldwide, evidenced by a smaller first maturation size in the northeast coast. ($L_{50}♀$ = 210 cm, $L_{50}♂$ = 237 cm) when compared to the North Atlantic population ($L_{50}♀$ = 320 cm, $L_{50}♂$ = 310 cm) [24,31,32].

Due to a lack of detailed information, capture pressures, and records along the Rio de Janeiro coast, the aim of this study was to expand the tiger shark record database and to improve upon future conservation and management strategies.

2. Materials and Methods

2.1. Study Area

The coast of Rio de Janeiro comprises 1.160 km, covering a total of 33 municipalities, with its northern limit in the municipality of São Francisco de Itabapoana (21°18′ S) and southern limit in Paraty (23°21′ S). These municipalities represent 40.1% of the state's territory, where 83% of the state's population lives. This area exhibits significant economic importance, accounting for 96% of the national offshore oil production and 77% of the national gas production [33]. The coast is divided into four regions (Norte Fluminense, Baixada Litorânea, Metropolitana, and Baía da Ilha Grande) (Figure 1), where a total of 28 artisanal fisher colonies are distributed [34].

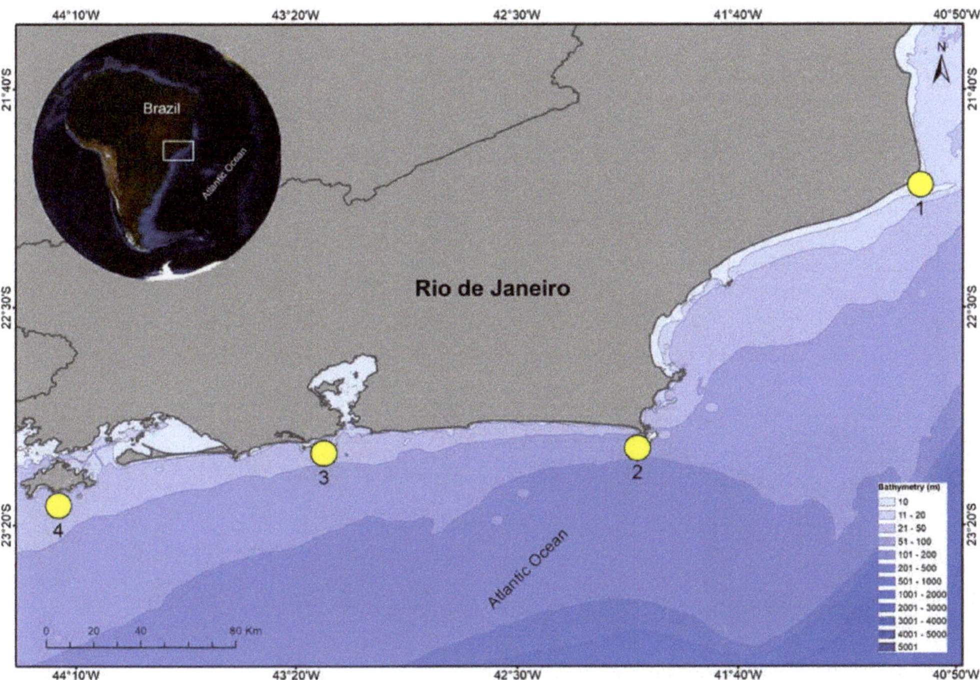

Figure 1. Map indicating the four information sampling regions on *Galeocerdo cuvier* occurrences in the state of Rio de Janeiro, Southeastern Brazil. 1—Norte Fluminense region, 2—Baixada Litorânea region, 3—Metropolitan region and 4—Ilha Grande Bay region.

All *Galeocerdo cuvier* records obtained off the coast of the state of Rio de Janeiro in this assessment are exhibited in the map depicted in Figure 2.

Table 1. Morphometric *Galeocerdo cuvier* data for the records obtained off the coast of the state of Rio de Janeiro. Caption: TL = Total Length; TW = Total Weight, * = live animal.** juvenile animal.

Study ID	Month/Year	Sex	TL (cm)	TW (kg)	Source
1	Not informed/1985	-	161	20	Santos et al. (2022a)
2	Not informed/1990	-	170	-	Santos et al. (2022a)
3	Not informed/2000	-	338	220	Current study
4	January/2002	-	300	-	Current study
5	January/2003 *	-	300	-	Current study
6	Not informed/2003	-	393	360	Santos et al. (2022a)
7	Not informed/2004	F	80	-	Silva-Junior et al. (2008)
8	Not informed/2005	-	-	-	Silva-Junior et al. (2008)
9	July/2013	-	150	40	Current study
10	August/2013	F	186	103	Current study
11	February/2016	-	280	220	Current study
12	December/2016	-	250	350	Current study
13	April/2018	F	89	-	Araujo et al. (2020)
14	July/2018	F	160	23	Current study
15	Not informed/2018	-	243	76	Current study
16	February/2019	M	250	200	Current study
17	March/2019	M	335	418	Miranda et al. (2021)
18	September/2019	F	150	-	Araujo et al. (2020)
19	February/2020	-	350	400	Current study
20	December/2021	F	350	300	Current study
21	January/2022 *	-	250	200	Current study
22	June/2022	F	180	75	Current study
23	July/2022 **	F	-	-	Current study

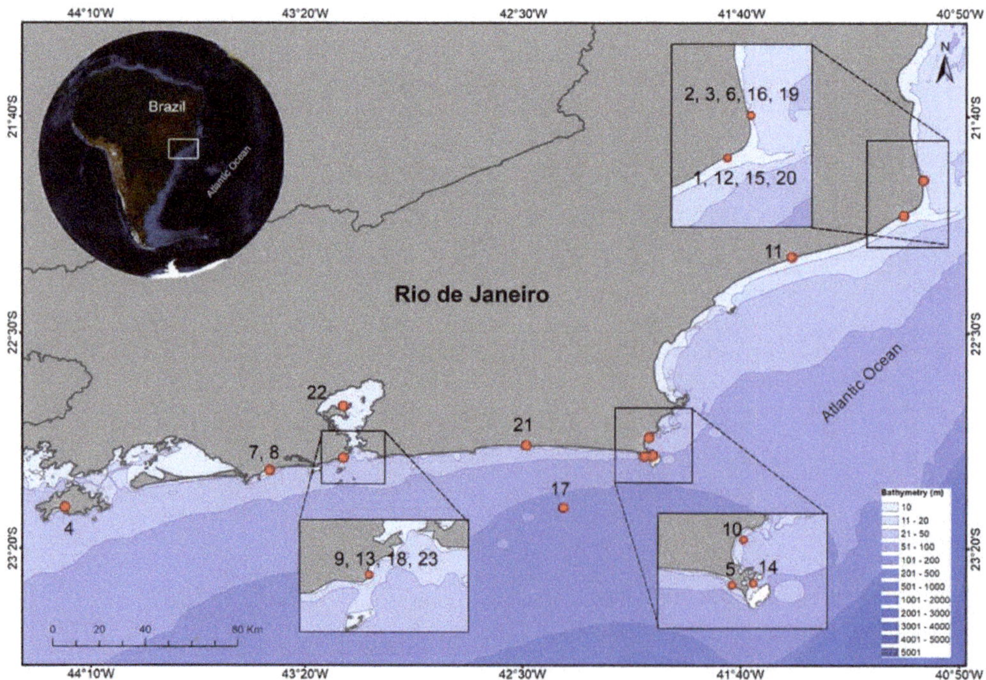

Figure 2. Map of the distribution of *Galeocerdo cuvier* records obtained off the coast of the state of Rio de Janeiro. Each study is identified by the code displayed in Table 1.

2.2. Data Sampling

Tiger shark records from the four coastal Rio de Janeiro regions were obtained by direct observation and from fisher colonies/associations, environmental guards, researchers, and published reports. The consulted fisher colonies were the Z13 colony in the metropolitan region of the city of Rio de Janeiro; the Z1 colony in the city of Gargaú, São Francisco de Itabapoana; the Z19 colony in the city of Campos dos Goytacazes (Northern Rio de Janeiro); and the Z17 colony in the city of Angrados Reis (Ilha Grande Bay region). The Tamoios Fishing Association in Cabo Frio (Baixadas Litorâneas region) was also consulted. Consultations were carried out with the fisher colonies/associations and with the environmental guards of each city over the last four years, obtaining reports and some photographic/video records. Lastly, voucher specimens from ichthyological collections of research centers, museums, and universities were recovered from the online databases FishNet 2—www.fishnet2.net (accessed on 12 July 2022), Information System for the Brazilian Biodiversity (SiBBr)—www.ala-bie.sibbr.gov.br (accessed on 12 July 2022) and Rede SpeciesLink—www.specieslink.net (accessed on 12 July 2022) and direct contact with the curators.

3. Results

A total of 23 tiger shark records were obtained, with 4 other records identified in articles published in the last 40 years (between 1982 and 2022) along the coast of the state of Rio de Janeiro (Table 1). The coastal areas with the most records were Rio de Janeiro (N = 6), followed by São João da Barra (N = 5), Campos dos Goytacazes (N = 4), Arraial do Cabo (N = 3), Cabo Frio, Saquarema, Quissamã, and Angra dos Reis (N = 1 each). The northern region of the state exhibited the highest number of records (N = 10), followed by the metropolitan lowland regions (N = 6), coastal region (N = 5), and Ilha Grande Bay (N = 1). Two previous records were recovered from voucher specimens deposited at the National Museu (MNRJ) and NUPEM (UFRJ) ichthyological collections. Two neonate specimens, a male and a female, were recovered from the Macaé Municipal Fish Market on 12 December 2017 (NPM 5772). Another individual tiger shark, captured along the Rio de Janeiro coast, was deposited under the catalog number MNRJ 10576.

Four additional records for the city of Rio de Janeiro were obtained from scientific articles by [16,22]. At least six abstracts presented between 1993 and 2018 at scientific events were identified (i.e., Oceanography Weekland Meetings of the Brazilian Society for the Study of Elasmobranchs, among others) mentioning the presence of the species on the coast of Rio de Janeiro, but we did not analyze these data in compliance with journal rules. Only two records of live animals were obtained, whilst the remaining records were from animals that were accidentally caught in artisanal fishing nets. There were four records from artisanal fishers from the northern coast of Rio de Janeiro, with data on catch size and/or weight, photographic evidence, or photographic evidence only [35].

Some morphometric parameters could be obtained from the new records on tiger shark occurrences in this study. The Total Length (TL) of the specimens ranged from 80 to 339.3 cm, with Total Weight (TW) ranging from 23 to 418 kg. Only eight individuals were sexed, comprising two males and six females. The females ranged from 80 cm to 350 cm TL and males from 250 to 335 cm TL. Regarding weight, females ranged from 23 to 300 kg and males, from 200 to 418 kg. The five individuals with the shortest TL (equal to or below 150 cm) were recorded in the city of Rio de Janeiro, and four of these records were obtained between March and September, during the autumn and winter seasons. Another four individuals smaller than 180 cm TL were recorded in the same period in the northern regions of Rio de Janeiro and its coastal lowlands. Another nine individuals (above 200 cm) were recorded between December and March, during the summer season. Several records for juveniles and neonates were obtained, suggesting seasonality patterns and the possibility that the coast of Rio de Janeiro may be a strategic breeding and nesting site for tiger sharks.

Figure 3 displays some of the tiger shark records obtained for the coast of the state of Rio de Janeiro in the present study.

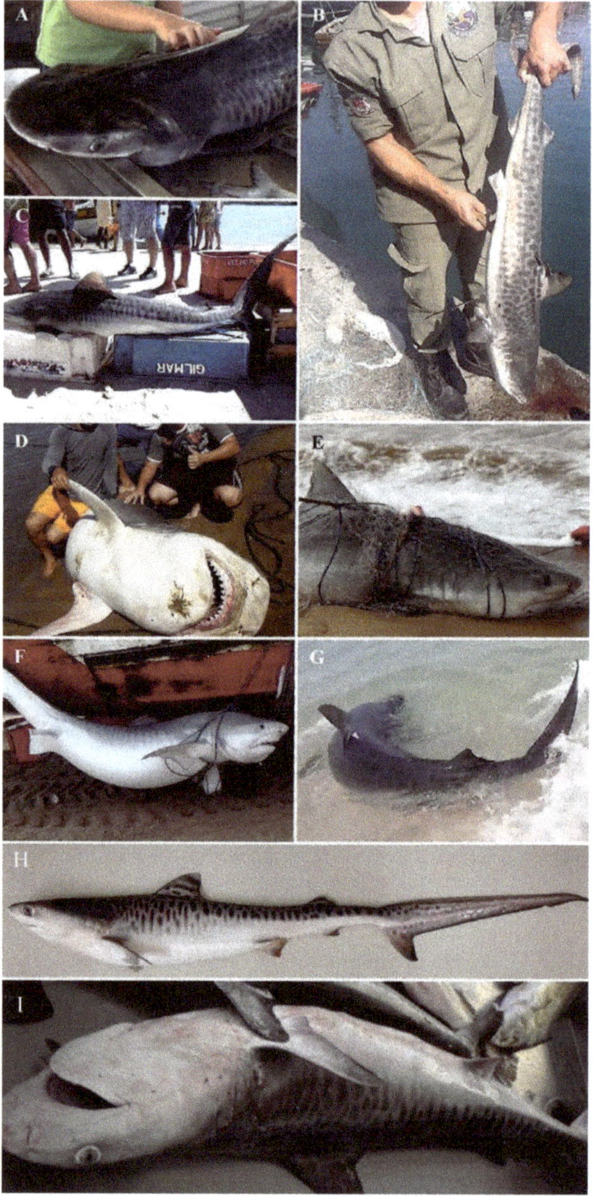

Figure 3. Some of the tiger shark records obtained for the coast of the state of Rio de Janeiro. Photos: (**A**)—Copacabana Beach (2013), (**B**)—Arraial do Cabo (2018), (**C**)—Cabo Frio (2013), (**D**)—Campos dos Goytacazes (2016), (**E**)—São João da Barra (2020), (**F**)—Campos dos Goytacazes (2021), (**G**)—Saquarema (2022), (**H**)—Copacabana Beach (2018), (**I**)—Copacabana Beach (2022).

4. Discussion

Our results increase the number of records of tiger sharks off the coast of the state of Rio de Janeiro approximately 5-fold and provide information on a possible seasonality pattern concerning the size of the captured/observed animals. Prior to our study, there was some mention of specimens without further details for the cities of Atafona [36], Angra dos Reis, and Rio de Janeiro [37,38]. More recently, four new records for the northern Rio de Janeiro coast were recovered from local fisher knowledge from the last 50 years [35]. Only Araujo et al. [16] presented detailed records of two individuals captured in 2018 and 2019 on Copacabana beach by the Z13 colony, in the metropolitan region of Rio de Janeiro. Although the tiger shark has been reported as among the least abundant shark species on the coast of different Brazilian states, including Rio de Janeiro, we believe that the number of tiger shark records may be higher than the records obtained herein, as we analyzed data from the last 40 years for only about 7% of the 28 colonies and fisher associations in the state. Studies on zooarchaeology in sambaqui shell mounds have identified traces of tiger sharks off the coast of the state on islands near the cities of Angra dos Reis, Cabo Frio, and Guanabara Bay [39–41]. Future studies should expand the space–time scale and the number of towns and fishing colonies to be consulted.

The high number of records obtained in northern Rio de Janeiro and the coastal Baixada regions may be due to the presence of the Paraíba do Sul River estuary and a local upwelling system, respectively, together with the fact that tiger sharks display the ability to enter estuarine and freshwater environments to feed. The section near the mouth of the Paraíba do Sul River has, in fact, been identified as presenting the greatest fish richness and diversity during the flooding season when compared to other sections of the same river [42]. However, even though there were no recorded captures of tiger sharks at the mouth of the river, the Paraíba do Sul River plume, which extends to Barra do Açú, results in an increased biomass caused by the greater discharges from the continent, attracting opportunistic species such as tiger sharks. The species may also occur in this region chasing *bonito* and squid shoals that occur during the summer. The upwelling phenomenon of the coastal lowland region, an oceanographic characteristic where deep waters rise continuously resulting in a considerable abundance of nektonic species, makes this one of the most productive fishing areas in the state [43]. The only two live animal observations swimming close to the coast in shallow waters were obtained in this region in 2003 and 2022, both associated with fish schools (Rodrigues and Cumplido pers. obs.).

However, the only record for the Ilha Grande Bay region comprises a 300 cm TL specimen captured 20 years ago on the south side of Ilha Grande. In this region, the consulted young local fisherman did not mention any recent records, and his grandfather reported having caught tiger sharks over 50 years ago. The oldest existing report of a tiger shark in Rio de Janeiro cites the capture of a large specimen in that same region in 1956 [37]. The absence of tiger shark reports among younger fishers has also been identified in the northeast of Brazilian [44]. This suggests that the high industrial fishing pressure caused by the high number of vessels in Ilha Grande Bay (Aximoff pers. obs.) may have caused a decline in the tiger shark population, therefore being responsible for the lack of more recent records. However, we also do not rule out artisanal fishing pressures, as identified in northern Rio de Janeiro [35].

In general, the tiger shark is among the least abundant shark species along the coast of different Brazilian states, such as Recife, Pernambuco [15], Rio de Janeiro itself [16], and near Mel Island in the state of Paraná [17]. This is possibly due to tiger sharks not being the main object of fishing unlike other shark species. This may be taking place in Rio de Janeiro, where few records of tiger shark captures are noted when compared to more abundant species such as sharpnose sharks (*Rhizoprionodon porosus* and *R. lalandii*), angel sharks (*Squatina guggenheim*), and scalloped hammerhead sharks (*Sphyrna lewini*) [16,23]. According to the data from the latest report by the Rio de Janeiro Fisheries Institute Foundation, these same three species presented the highest biomass captured in the state in 2016, at almost 10 tons [45]. According to [35], artisanal fishers in northern Rio de Janeiro consider that the tiger shark used to more common and has suffered a decline in catches over the last few decades.

The records obtained at Copacabana beach, in the metropolitan region of the state, were of two juveniles and one neonate measuring only 89 cm in TL, as these animals are born ranging from about 50 to 70 cm TL [37]. Ref. [16] also identified juveniles of other shark species in Copacabana, suggesting this site as a strategic breeding and nesting site for several species. We suggest that Copacabana beach is strongly influenced by Guanabara Bay, which is considered as one of the most important and productive estuaries in Brazil [46], still exhibiting significant marine and estuarine biodiversity but with significant observed losses for elasmobranch populations [47].

Another study, carried out at Recreio dos Bandeirantes, about 30 km from Copacabana Beach, obtained records of juvenile sharks of several species, indicating that the coast of the metropolitan region of Rio de Janeiro is vital for the reproduction of several elasmo-branchs, as indicated by [22], with Guanabara Bay being indicated as a nursery for the butterfly stingray *Gymnura altavela* [48], categorized as Endangered by the International Union for Conservation of Nature. According to [49], some of the criteria employed to identify a nursery area are a high frequency of sharks and the repeated use of the habitat over the years. Two other areas on the coast of the state of São Paulo, southeastern Brazil, are also considered nurseries, thus demonstrating the need for specific conservation measures [40,50].

Regarding the recording period, juveniles have been identified in coastal waters during the autumn and winter seasons, similar to records obtained in the state of Recife [51]. Given that juvenile tiger sharks exhibit high growth rates [24], sexual maturity is reached when these animals reach about 200 cm [37]. At this stage, they then move to deeper waters [52], although they still regularly move to coastal areas to forage [53]. In the Gulf of Mexico, sub-adult and adult tiger sharks achieve significantly higher movement rates than juveniles and inhabit deeper habitats, particularly during the fall and winter seasons [52]. It is considered that the same behavior recorded at the Gulf of Mexico may take place at Guanabara Bay, with tiger sharks that have already surpassed the juvenile stage migrating to new areas to the north and south coasts of the state. In any case, the presence of juvenile tiger sharks demonstrates the importance of the coast of the state of Rio de Janeiro for the species, probably as a feeding and growth area or potentially for reproduction, indicated by the presence of specimens close to the sizes reported for neonates.

The records retrieved herein may comprise the beginning of a baseline for future environmental impact assessments, such as concerning the effects of oil spills and the movements and distribution of tiger sharks off the coast of Rio de Janeiro, which currently accounts for 83% of the offshore oil production in the country [54]. Telemetry monitoring over time throughout the coast of the state is, therefore, paramount to verify the hypotheses in various studies, such as in the study carried out in the oil area of the Gulf of Mexico, where Ajemian et al. [52] investigated the habitat use and the movement patterns of 56 monitored tiger sharks. Telemetry has, in fact, been proven a useful tool to identify variable patterns of the space used by tiger sharks, including residence and migration periods [55].

Finally, the survey carried out here recorded 23 tiger shark occurrences along the coast of Rio de Janeiro with special emphasis on juveniles (eight records), further emphasizing the need to consider the coast of Rio de Janeiro as an especially relevant area for at least part of tiger shark life history. The preservation of Rio de Janeiro's marine ecosystems significantly depends on the maintenance of its natural top predators, many of which have been considered threatened by the fishing intensity applied to their populations throughout the 20th century [14,35]. Moreover, Guanabara Bay is heavily contaminated by several chemical contaminants due to the inflow of significant amounts of untreated sewage and runoff from landfills, industries, and shipyards [4,56]. In fact, several elasmobranch species captured in or near this area have been reported as containing extremely high levels of both inorganic [57–60] and organic contaminants ([61], Hauser-Davis, pers. Obs.). This, in turn, comprises a significant threat to their reproduction, development, and survival, especially as many contaminants in the aforementioned studies have been detected in both elasmobranch reproductive and sensory organs.

Furthermore, tiger sharks exhibited a lower reproductive output than expected from the literature for specimens captured along the coast of Hawai'i [62], making juvenile survival and the protection of nursery, feeding, and growing areas potentially critical for conservation strategies, even more so as chemical contamination is of significant concern in juvenile stages [63]. It is also important to note that although pollution has recently been identified as a priority for conservation efforts, especially for elasmobranchs [64,65], data are still scarce for many species, posing a significant knowledge gap for elasmobranch conservation. Additionally, a database such as the one produced herein is the first step toward the design of new research on a species that, although not yet categorized as threatened, is considered key to food web maintenance. This is even more vital in the case of the coast of Rio de Janeiro, where other large predators, many of them also large sharks, similarly face clear declines in their populations.

5. Conclusions

This study has expanded tiger shark records 5-fold for Rio de Janeiro, Brazil, reporting 23 new records for this area and providing a baseline for future environmental impact assessments regarding this species. Potential seasonality patterns were noted, and several new records comprised juveniles, which indicates the need to consider the coast of Rio de Janeiro as an especially relevant area for at least part of this species life history and for future conservation strategies.

Author Contributions: U.G.d.M.: Conceptualization, Data curation, Formal analysis, Investigation, Methodology, Validation, Visualization, Writing—original draft, and Writing—review and editing. E.B.F.N.: Conceptualization, Data curation, Formal analysis, Investigation, Methodology, Validation, Visualization, Writing—original draft, and Writing—review and editing. I.A.: Data curation, Formal analysis, Investigation, Methodology, Project administration, Validation, Visualization, Writing—original draft, and Writing—review and editing. R.C.: Data curation, Formal analysis, Funding acquisition, Investigation, Methodology Validation, Visualization, Writing—original draft, and Writing—review and editing. M.T.R.: Data curation, Formal analysis, Investigation, Methodology, Validation, Validation, Visualization, Writing—original draft, and Writing—review and editing. S.R.S.: Data curation, Formal analysis, Investigation, Methodology, Supervision, Validation, Visualization, Writing—original draft, and Writing—review and editing. R.A.H.-D.: Data curation, Investigation, Visualization, Writing—original draft, and Writing—review and editing. All authors have read and agreed to the published version of the manuscript.

Funding: R.A.H.-D acknowledges the Carlos Chagas Filho Foundation for Research Support of the State of Rio de Janeiro (FAPERJ) through a Jovem Cientista do Nosso Estado 2021–2024 grant (process number E-26/201.270/202) and an ARC grant (process number E-26/21.460/2019), and the Brazilian National Council of Scientific and Technological Development (CNPq) through a productivity grant. The implementation of the Projeto Pesquisa Marinha e Pesqueira is a compensatory measure established by the Conduct Adjustment Agreement under the responsibility of the PRIO company, conducted by the Federal Public Ministry—MPF/RJ.

Institutional Review Board Statement: Not applicable.

Informed Consent Statement: Not applicable.

Data Availability Statement: Not applicable.

Acknowledgments: The authors would like to thank institutions and professionals who provided the photos: Environmental Guard of Arraial do Cabo (Figure 3 Photo B), Charles Guimarães Neves (Figure 3, Photo C), Matheus Manhães (Figure 3, Photo D), colonies and fisher associations (Figure 3, Photos A, E, F, G). Thanks are also due to Sérgio C. Moreira for kindly preparing the maps used in this report. The authors would also like to thank Fábio Di Dario (NPM-NUPEM/UFRJ) who provided access to museum specimens and associated data.

Conflicts of Interest: The authors declare no conflict of interest.

References

1. Hammerschlag, N.; Schmitz, O.J.; Flecker, A.S.; Lafferty, K.D.; Sih, A.; Atwood, T.B.; Gallagher, A.J.; Irschick, D.J.; Skubel, R.; Cooke, S.J. Ecosystem function and services of aquatic predators in the anthropocene. *Trends Ecol. Evol.* **2019**, *34*, 369–383. [CrossRef]
2. Pacoureau, N.; Rigby, C.L.; Kyne, P.M.; Sherley, R.B.; Winker, H.; Carlson, J.K.; Dulvy, N.K. Half a century of global decline in oceanic sharks and rays. *Nature* **2021**, *589*, 567–571. [CrossRef]
3. Simpfendorfer, C.A.; Dulvy, N.K. Bright spots of sustainable shark fishing. *Curr. Biol.* **2017**, *27*, 97–98. [CrossRef]
4. Hauser-Davis, R.A.; Amorim-Lopes, C.; Araujo, N.L.F.; Rebouças, M.; Gomes, R.A.; Rocha, R.C.C.; Dos Santos, L.N. On mobulid rays and metals: Metal content for the first Mobulamobular record for the state of Rio de Janeiro, Brazil and a review on metal ecotoxicology assessments for the Manta and Mobula genera. *Mar. Pollut. Bull.* **2021**, *168*, 112472. [CrossRef]
5. Dulvy, N.K.; Fowler, S.L.; Musick, J.A.; Cavanagh, R.D.; Kyne, P.M.; Harrison, L.R.; Pollock, C.M. Extinction risk and conservation of the world's sharks and rays. *Elife* **2014**, *3*, e00590. [CrossRef]
6. Booth, H.; Squires, D.; Milner-Gulland, E.J. The neglected complexities of shark fisheries, and priorities for holistic risk-based management. *Ocean. Coast. Manag.* **2019**, *182*, 104994. [CrossRef]
7. Dent, F.; Clarke, S. *State of the Global Market for Shark Products*; FAO Fisheries and Aquaculture Technical Paper 590; FAO: Rome, Italy, 2015.
8. FAO. *The State of World Fisheries and Aquaculture: Sustainability in Action*; Food and Agriculture Organization of the United Nations: Rome, Italy, 2020.
9. Ferrette, B.L.S.; Domingues, R.R.; Ussami, L.H.F.; Moraes, L.; Magalhães, C.O.; Amorim, A.F.; Hilsdorf, A.W.S.; Oliveira, C.; Foresti, F.; Mendonça, F.F. DNA-based species identification of shark finning seizures in Southwest Atlantic: Implications for wildlife trade surveillance and law enforcement. *Biodivers. Conserv.* **2019**, *28*, 4007–4025. [CrossRef]
10. Ferretti, F.; Jacoby, D.M.P.; Pfleger, M.O.; White, T.D.; Dent, F.; Micheli, F.; Rosenberg, A.A.; Crowder, L.B.; Block, B.A. Shark fin trade bans and sustainable shark fisheries. *Conserv. Lett.* **2020**, *13*, e12708. [CrossRef]
11. Pincinato, R.B.M.; Gasalla, M.A.; Garlock, T.; Anderson, J.L. Market incentives for shark fisheries. *Mar. Policy* **2022**, *139*, 105031. [CrossRef]
12. Freire, K.M.F.; Almeida, Z.S.; Amador, J.R.E.T.; Aragão, J.A.; Araújo, A.R.R.; Ávila-da-Silva, A.O.; Bentes, B.; Carneiro, M.H.; Chiquieri, J.; Fernandes, C.A.F.; et al. Reconstruction of marine commercial landings for the Brazilian industrial and artisanal fisheries from 1950 to 2015. *Front. Mar. Sci.* **2021**, *8*, 659110. [CrossRef]
13. Cruz, M.; Szynwelski, B.E.; Ochotorena de Freitas, T.R. Biodiversity on sale: The shark meat market threatens elasmobranchs in Brazil. *Aquat. Conserv. Mar. Freshw. Ecosyst.* **2021**, *31*, 3437–3450. [CrossRef]
14. ICMBio. *Livro Vermelho Da Fauna Brasileira Ameaçada de Extinção: Volume VI—Peixes*; Instituto Chico Mendes de Conservação da Biodiversidade/Ministério do Meio Ambiente: Brasília, Brazil, 2018. Available online: https://www.Icmbio.gov.br/portal/images/stories/comunicacao/publicacoes/publicacoesdiversas/livro_vermelho_2018_vol6.pdf (accessed on 12 July 2022).
15. Hazin, F.H.V.; Wanderley Júnior, J.A.D.M.; Mattos, S.M.G.D. Distribuição e abundância relativa de tubarões no litoral do Estado de Pernambuco, Brasil. *Arquivos de Ciências do Mar* **2002**, *33*, 33–42.
16. Araujo, N.L.F.; Lopes, C.A.; Brito, V.B.; Santos, L.N.D.; Barbosa Filho, M.L.V.; Amaral, C.R.L.D.; Hauser-Davis, R.A. Artisanally landed elasmobranchs along the coast of Rio de Janeiro, Brazil. *Bol. Do Laboratório De Hidrobiol.* **2020**, *30*, 33–53. [CrossRef]
17. Rupp, A.; Bornatowski, H. Food web model to assess the fishing impacts and ecological role of elasmobranchs in a coastal ecosystem of Southern Brazil. *Environ. Biol. Fishes* **2021**, *104*, 905–921. [CrossRef]
18. IUCN. International Union for Conservation of Nature Red List. The IUCN Red List of Threatened Species 2018. Available online: https://www.iucnredlist.org/species/39378/2913541 (accessed on 12 July 2022).
19. Brown, C.J.; Roff, G. Life-history traits inform population trends assessing the conservation status of a declining tiger shark population. *Biol. Conserv.* **2019**, *239*, 108230. [CrossRef]
20. Lucifora, L.O.; García, V.B.; Menni, R.C.; Worm, B. Spatial patterns in the diversity of sharks, rays, and chimaeras (Chondrichthyes) in the Southwest Atlantic. *Biodivers. Conserv.* **2012**, *2*, 407–419. [CrossRef]
21. Ministério da Pesca e Aquicultura. *Pesca Artesanal*; Ministério da Pesca e Aquicultura: Brasília, Brazil, 2014. Available online: http://www.mpa.gov.br/pesca/artesanal (accessed on 12 July 2022).
22. Silva-Júnior, L.C.D.; Andrade, A.C.D.; Vianna, M. Caracterização de uma pescaria de pequena escala em uma área de importância ecológica para elasmobrânquios, no Recreio dos Bandeirantes, Rio de Janeiro. *Arq. De Ciências Do Mar* **2008**, *41*, 47–57.
23. Tomás, A.R.G.; Gomes, U.L.; Ferreira, B.P. Distribuição temporal dos elasmobrânquios na pesca de pequena escala de Barra de Guaratiba, Rio De Janeiro, Brasil. *Bol. Do Inst. De Pesca* **2018**, *36*, 317–324.
24. Afonso, A.S.; Hazin, F.H.V.; Barreto, R.R.; Santana, F.M.; Lessa, R.P. Extraordinary growth in tiger sharks *Galeocerdo cuvier* from the South Atlantic Ocean. *J. Fish Biol.* **2012**, *81*, 2080–2085. [CrossRef] [PubMed]
25. Ebert, D.A.; Fowler, S.; Compagno, L. *Sharks of the World: A Fully Illustrated Guide*; Wild Nature Press: Plymouth, UK, 2013; 528p.
26. Ferreira, L.C.; Thums, M.; Heithaus, M.R.; Barnett, A.; Abrantes, K.G.; Holmes, B.J.; Zamora, L.M.; Frisch, A.J.; Pepperell, J.G.; Burkholder, D.; et al. The trophic role of a large marine predator, the tiger shark *Galeocerdo Cuvier*. *Sci. Rep.* **2017**, *7*, 7641. [CrossRef]
27. Bornatowski, H.; Wedekin, L.L.; Heithaus, M.R.; Marcondes, M.C.C.; Rossi-Santos, M.R. Shark scavenging and predation on cetaceans at Abrolhos Bank, eastern Brazil. *J. Mar. Biol. Assoc. U. K.* **2012**, *92*, 1767–1772. [CrossRef]

28. Dicken, M.L.; Hussey, N.E.; Christiansen, H.N.; Smale, M.J.; Nkabi, N.; Cliff, G.; Wintner, S.P. Diet and trophic ecology of the tiger shark (*Galeocerdo cuvier*) from South African waters. *PLoS ONE* **2017**, *12*, e0177897. [CrossRef] [PubMed]

29. Miranda, F.R.; Santos, P.M.; Rodrigues, M.T.; Cumplido, R.; Kersul, M.G. Consumption of a maned sloth (*Bradypus torquatus* Illiger, 1811) by a tiger shark (*Galeocerdo cuvier* Péron & LeSueur, 1822) in southeastern Brazil. *Notas Sobre Mamíferos Sudam.* **2021**, *3*, 1–7.

30. Randall, J.E. Review of the biology of the tiger shark (*Galeocerdo cuvier*). *Aust. J. Mar. Freshw. Res.* **1992**, *43*, 21–31. [CrossRef]

31. Branstetter, S.; Musick, J.A.; Colvocoresses, J.A. A comparison of the age and growth of the tiger shark, *Galeocerdo cuvieri*, from off Virginia and from the Northwestern Gulf of Mexico. *Fish. Bull.* **1987**, *85*, 269–279.

32. Trindade-Santos, I.; Freire, K.M.F. Analysis of reproductive patterns of fishes from three Large Marine Ecosystems. *Front. Mar. Sci.* **2015**, *2*, 38. [CrossRef]

33. INEA. 2021. Available online: http://www.inea.rj.gov.br/biodiversidade-territorio/gerenciamento-costeiro/ (accessed on 12 July 2022).

34. FIPERJ. *Entidades do Setor. Colônia*; Fundação Instituto de Pesca do Estado do Rio de Janeiro: Rio de Janeiro, Brazil, 2019. Available online: http://www.fiperj.rj.gov.br/index.php/entidade (accessed on 12 July 2022).

35. Santos, S.R.; Macedo, M.L.C.; Maciel, T.R.; Souza, G.B.G.; Almeida, L.S.; Gadig, O.B.F.; Vianna, M. A tale that never loses in the telling: Considerations for the shifting ethnobaseline based on artisanal fisher records from the southwestern Atlantic. *Ethnobiol. Conserv.* **2022**, *11*, 3.

36. Rosas, F.C.W.; Capistrano, L.C.; Di Beneditto, A.P.; Ramos, R. *Hydruga leptonyx* recovered from the stmoach of a tiger shark captured off the Rio de Janeiro coast, Brazil. *Mammalia* **1992**, *56*, 153–155.

37. Gomes, U.L.; Signori, C.N.; Gadig, O.B.F.; Santos, U.R.S. *Guia para Identificação de Tubarões e Raias do Rio de Janeiro*, 1st ed.; Technical Books Editora: Rio de Janeiro, Brasil, 2010; 234p.

38. Gomes, U.L.; Santos, H.R.S.; Gadig, O.B.F.; Signorini, C.N.; Vicente, M.M. Guia para identificação dos tubarões, raias e quimeras do Estado Rio de Janeiro (Chondrichthyes: Elasmobranchii e Holocephali). *Revista Nordestina de Biologia* **2019**, *27*, 171–368.

39. Lopes, M.S.; Bertucci, T.C.P.; Rapagnã, L.; Tubino, R.D.A.; Monteiro-Neto, C.; Tomas, A.R.G.; Aguilera Socorro, O. The path towards endangered species: Prehistoric fisheries in Southeastern Brazil. *PLoS ONE* **2016**, *11*, e0154476. [CrossRef]

40. Lopes, E.Q.; dos Santos, B.D.P.; de Faria, I.C.; de Lima, T.G.; de Lucca, D.S.Q. Registros de ocorrências de tubarões tigres (*Galeocerdo cuvier*, Peron e Lesueur, 1822) no litoral de Peruíbe-São Paulo-Brasil, APA-CIP e Unidades de Conservação do Mosaico Jureia Itatins-SP. *Braz. J. Anim. Environ. Res.* **2020**, *3*, 4270–4282. [CrossRef]

41. Lopes, M.S.; Grouard, S.; Gaspar, M.D.; Sabadini-Santos, E.; Bailon, S.; Aguilera, O. Middle Holocene marine and land-tetrapod biodiversity recovered from Galeão shell mound, Guanabara Bay, Brasil. *Quat. Int.* **2022**, *610*, 80–96. [CrossRef]

42. Teixeira, T.P.; Pinto, B.C.; Terra, B.D.F.; Estiliano, E.O.; Gracia, D.; Araújo, F.G. Diversidade das assembléias de peixes nas quatro unidades geográficas do rio Paraíba do Sul. *Iheringia: Série Zoologia* **2005**, *95*, 347–357. [CrossRef]

43. Valentin, J.L. The Cabo Frio Upwelling System, Brazil. In *Coastal Marine Ecosystems of Latin America*; Seeliger, U., Kjerfve, B., Eds.; Springer: Berlin/Heidelberg, Germany, 2001; pp. 97–105.

44. Leduc, A.O.; De Carvalho, F.H.; Hussey, N.E.; Reis-Filho, J.A.; Longo, G.O.; Lopes, P.F. Local ecological knowledge to assist conservation status assessments in data poor contexts: A case study with the threatened sharks of the Brazilian Northeast. *Biodivers. Conserv.* **2021**, *30*, 819–845. [CrossRef]

45. FIPERJ. Relatório. 2016. Available online: http://www.fiperj.rj.gov.br/fiperj_imagens/arquivos/revistarelatorios2015.pdf (accessed on 12 July 2022).

46. Da Silva, D.R.; Paranhos, R.; Vianna, M. Spatial patterns of distribution and the influence of seasonal and abiotic factors on demersal ichthyofauna in an estuarine tropical bay. *J. Fish Biol.* **2016**, *89*, 821–846. [CrossRef] [PubMed]

47. Leite, C.V.T.; Lima, A.P.; Maciel, T.R.; Santos, S.R.B.; Vianna, M. A baía de Guanabara é um ambiente importante para a conservação neotropical? Uma abordagem ictiológica. *Diversidade e Gestão* **2018**, *2*, 76–89.

48. Silva, F.G.; Vianna, M. Diet and reproductive aspects of the endangered butterfly ray *Gymnura altavela* raising the discussion of a possible nursery area in a highly impacted environment. *Braz. J. Oceanogr.* **2018**, *66*, 315–324. [CrossRef]

49. Heupel, M.R.; Carlson, J.K.; Simpfendorfer, C.A. Shark nursery areas: Concepts, definition, characterization and assumptions. *Mar. Ecol. Prog. Ser.* **2007**, *337*, 287–297. [CrossRef]

50. Santos, P.R.; Balanin, S.; Gadig, O.B.; Garrone-Neto, D. The historical and contemporary knowledge on the elasmobranchs of Cananeia and adjacent waters, a coastal marine hotspot of southeastern Brazil. *Reg. Stud. Mar. Sci.* **2022**, *51*, 102224. [CrossRef]

51. Afonso, A.S.; Andrade, H.A.; Hazin, F.H. Structure and dynamics of the shark assemblage off Recife, northeastern Brazil. *PLoS ONE* **2014**, *9*, e102369. [CrossRef] [PubMed]

52. Ajemian, M.J.; Drymon, J.M.; Hammerschlag, N.; Wells, R.J.D.; Street, G.; Falterman, B.; Mckinney, J.A.; Drigers, W.B., III; Hoffmayer, E.R.; Fischer, C.; et al. Movement patterns and habitat use of tiger sharks (*Galeocerdo cuvier*) across ontogeny in the Gulf of Mexico. *PLoS ONE* **2020**, *15*, e0234868. [CrossRef] [PubMed]

53. Heithaus, M.; Dill, L.; Marshall, G.; Buhleier, B. Habitat use and foraging behavior of tiger sharks (*Galeocerdo cuvier*) in a seagrass ecosystem. *Mar. Biol.* **2002**, *140*, 237–248.

54. EBC. 2021. Available online: https://agenciabrasil.ebc.com.br/economia/noticia/2021-08/em-11-anos-arrecadacao-de-royalties-de-petroleo-no-rio-subiu-225 (accessed on 12 July 2022).

55. Holland, K.N.; Anderson, J.M.; Coffey, D.M.; Holmes, B.J.; Meyer, C.G.; Royer, M.A. A Perspective on Future tiger shark Research. *Front. Mar. Sci.* **2019**, *6*, 37. [CrossRef]

56. Fistarol, G.O.; Coutinho, F.H.; Moreira, A.P.B.; Venas, T.; Ca´novas, A.; de Paula, S.E.M.; Coutinho, R.; de Moura, R.L.; Valentin, J.L.; Tenenbaum, D.R.; et al. Environmental and sanitary conditions of Guanabara Bay, Rio de Janeiro. *Front. Microbiol.* **2015**, *6*, 1232. [CrossRef]

57. Amorim-Lopes, C.; Willmer, I.Q.; Araujo, N.L.; de SPereira, L.H.S.; Monteiro, F.; Rocha, R.C.; Saint'Pierre, T.D.; dos Santos, L.N.; Siciliano, S.; Vianna, M.; et al. Mercury screening in highly consumed sharpnose sharks (*Rhizoprionodon lalandii* and *R. porosus*) caught artisanally in southeastern Brazil. *Elem. Sci. Anthr.* **2020**, *8*, 22. [CrossRef]

58. Hauser-Davis, R.A.; Pereira, C.F.; Pinto, F.; Torres, J.P.M.; Malm, O.; Vianna, M. Mercury contamination in the recently described Brazilian white-tail dogfish *Squalus albicaudus* (Squalidae, Chondrichthyes). *Chemosphere* **2020**, *250*, 126228. [CrossRef]

59. Maciel, O.L.C.; Willmer Isabel, Q.; Saintpierre, T.D.; Machado, W.T.V.; Siciliano, S.; Hauser-Davis, R.A. Arsenic contamination in widely consumed Caribbean sharpnose sharks in southeastern Brazil: Baseline data and concerns regarding fisheries resources. *Mar. Pollut. Bull.* **2021**, *172*, 112905. [CrossRef]

60. Willmer, I.Q.; Wosnick, N.; Rocha, R.C.C.; Saintpierre, T.D.; Vianna, M.; Hauser-Davis, R.A. First report on metal and metalloid contamination of Ampullae of Lorenzini in sharks: A case study employing the Brazilian Sharpnose Shark Rhizoprionodonlalandii from Southeastern Brazil as an ecotoxicological model. *Mar. Pollut. Bull.* **2022**, *179*, 113671. [CrossRef]

61. Paiva, L.; Vannuci-Silva, M.; Correa, B.; Santos-Neto, E.; Vianna, M.; Lailson-Brito, J.L. Additional pressure to a threatened species: High persistent organic pollutant concentrations in the tropical estuarine batoid *Gymnuraaltavela*. *Bull. Environ. Contam. Toxicol.* **2021**, *107*, 37–44. [CrossRef]

62. Whitney, N.M.; Crow, G.L. Reproductive biology of the tiger shark (*Galeocerdo cuvier*) in Hawaii. *Mar. Biol.* **2007**, *151*, 63–70. [CrossRef]

63. Lyons, K.; Adams, D.H. Maternal offloading of organochlorine contaminants in the yolk-sac placental scalloped hammerhead shark (*Sphyrna lewini*). *Ecotoxicology* **2015**, *24*, 553–562. [CrossRef] [PubMed]

64. Consales, G.; Marsili, L. Assessment of the conservation status of Chondrichthyans: Underestimation of the pollution threat. *Eur. Zool. J.* **2021**, *88*, 165–180. [CrossRef]

65. Cooke, S.J.; Bergman, J.N.; Madliger, C.L.; Cramp, R.L.; Beardall, J.; Burness, G.; Chown, S.L. One hundred research questions in conservation physiology for generating actionable evidence to inform conservation policy and practice. *Conserv. Physiol.* **2021**, *9*, coab009. [CrossRef] [PubMed]

Article

Demographic Analysis of Shortfin Mako Shark (*Isurus oxyrinchus*) in the South Pacific Ocean

Hoang Huy Huynh [1,2,3], Chun-Yi Hung [1] and Wen-Pei Tsai [2,*]

[1] Institute of Aquatic Science and Technology, National Kaohsiung University of Science and Technology, Kaohsiung 81157, Taiwan
[2] Department of Fisheries Production and Management, National Kaohsiung University of Science and Technology, Kaohsiung 81157, Taiwan
[3] Division of Fisheries Ecology and Aquatic Resources, Research Institute for Aquaculture No. 2, Ho Chi Minh 710000, Vietnam
* Correspondence: wptsai@nkust.edu.tw

Simple Summary: Statistical of baseline and species-specific data are largely unavailable for several elasmobranchs (e.g., sharks, skates, and rays). Populations of many shark species such as shortfin mako sharks (*Isurus oxyrinchus*) have declined, mainly because of overexploitation and fishing pressure-related vulnerability. Mako shark conservation and management have become major concerns. The available catch, fishing effort, and biological composition data of mako sharks, particularly in the South Pacific Ocean, are insufficient, making stock assessments difficult. We applied demographic analysis to increase the current understanding of the mako shark stock status despite insufficient fishery data. Our results revealed that mako shark populations would only be able to sustain a lower fishing mortality (20% of natural mortality) in the study area. The findings strongly suggest that more conservative management measures should be implemented to ensure sustainable utilization of the mako shark stock. The current models may be applicable to other shark species and taxa with limited data availability for preserving their ecological balance and establishing management and conservation measures for them.

Citation: Huynh, H.H.; Hung, C.-Y.; Tsai, W.-P. Demographic Analysis of Shortfin Mako Shark (*Isurus oxyrinchus*) in the South Pacific Ocean. *Animals* **2022**, *12*, 3229. https://doi.org/10.3390/ani12223229

Academic Editor: Martina Francesca Marongiu

Received: 22 September 2022
Accepted: 14 November 2022
Published: 21 November 2022

Publisher's Note: MDPI stays neutral with regard to jurisdictional claims in published maps and institutional affiliations.

Abstract: The shortfin mako shark (*Isurus oxyrinchus*) demonstrates low productivity and is thus relatively sensitive to fishing. Natural mortality (M) and fishing mortality (F) data are critical to determine their population dynamics. However, catch and fishing effort data are unavailable for this species in the South Pacific Ocean, making stock assessments difficult. Demographic quantitative methods aid in analyzing species with limited data availability. We used a two-sex stage-structured matrix population model to examine the demographic stock status of mako sharks. However, data-limited models to determine fishery management strategies have limitations. We performed Monte Carlo simulations to evaluate the effects of uncertainty on the estimated mako shark population growth rate. Under unfished conditions, the simulations demonstrated that the mako sharks showed a higher finite population growth rate in the 2-year reproductive cycle compared to the 3-year reproductive cycle. Protecting immature mako sharks led to a higher population growth rate than protecting mature mako sharks. According to the sex-specific data, protecting immature male and female sharks led to a higher population growth rate than protecting mature male and female sharks. In conclusion, sex-specific management measures can facilitate the sustainable mako shark conservation and management.

Keywords: demographic analysis; data-limited; Monte Carlo simulation; shortfin mako shark; two-sex stage-based matrix model; uncertainties

1. Introduction

Globally, the scientific evaluation of many overexploited marine species, particularly sharks, is lacking, which has prevented their sustainable exploitation and effective conservation [1–3]. In recent years, various regional fisheries management organizations such as the International Commission for the Conservation of Atlantic Tunas (ICCAT), the Indian Ocean Tuna Commission (IOTC), and the Western and Central Pacific Fisheries Commission (WCPFC) have conducted shark stock assessments and have consequently placed restrictions on the commercial retention of various shark species including the shortfin mako shark (hereafter, referred to as "mako shark"). However, the Inter-American Tropical Tuna Commission (IATTC) is yet to report its shark stock assessment results. The management measures implemented by the aforementioned organizations indicate the urgent need for an improved understanding of shark stock status.

The mako shark (*Isurus oxyrinchus* Rafinesque, 1810, from the family Lamnidae) is fast-swimming, and is a vital component of the pelagic shark community and is widely distributed in tropical and temperate waters with a temperature of >16 °C [4,5]. In 2019, it was classified as Endangered (EN) on the International Union for Conservation of Nature (IUCN) Red List [6] and was included in the Convention on International Trade in Endangered Species of Wild Fauna and Flora (CITES) Appendix II [7]. This species is known for its long life, low fecundity [8,9], late maturity [10,11], and extended reproductive cycle (2–3 years) [9,10,12]. Consequently, it is extremely susceptible to overexploitation. Given its high susceptibility and low productivity, the mako shark has been identified as one of the most vulnerable species in the Atlantic, Indian, and Pacific Oceans, as per an ecological risk assessment [13–15].

Shark conservation status demonstrates considerable variations across regions, and shark population trends are unpredictable. Several recent studies have focused on mako shark stock status in various regions worldwide including in the Atlantic Ocean [16], northwestern Pacific Ocean [17–20], and Indian Ocean [21,22]. However, apart from two recent preliminary mako shark stock assessment reports in the Southwest Pacific [23,24], its population status remains poorly known, and in terms of biological information, only three investigations have been conducted to date [25–27]. Obtaining the long-term catch and effort data of fish is difficult, and the use of biomass dynamics models (e.g., Bayesian biomass dynamics model, virtual population analysis, or surplus production model), which have been used to assess the bigeye thresher shark (*Alopias superciliosus*) [28] and mako shark [17], can be complicated. Similarly, age-structured assessment models (e.g., stock synthesis) such as that used for bigeye tuna (*Thunnus obesus*) [29] generally require steepness values of stock–recruitment relationships. The traditional Leslie matrix method (based on an age and stage–based model), along with deterministic and stochastic approaches, have been used to assess the stock status of pelagic thresher sharks (*Alopias pelagicus*) [30,31], the silky shark (*Carcharhinus falciformis*) [32], the mako shark [19,20], and the blue shark (*Prionace glauca*) [33]. Data-limited management methods are increasingly being used to reflect the state of fishery areas [34–36], and demographic models may be used to provide valuable insights into the management recommendations for these species.

In fisheries, demographic models have several advantages over conventional stock assessment models. Traditional stock assessment approaches, surplus production models, or age-structured population models require a large amount of data (e.g., catch, effort, and abundance indices). In contrast to conventional models, demographic models require only data-limited life history parameters (i.e., survival rate, age at maturity, longevity, litter size, and number of embryos) [37,38]. Many elasmobranch populations in the Pacific and Atlantic Oceans have been studied using these models [16,19,30,39–43]. In addition, these models are more applicable for long-living, slow-growing shark species [42] such as the mako shark. They can compensate for data imbalance and hence serve as a valuable assessment tool until sufficient fisheries data for traditional stock assessments are available or until alternative stock assessment methods are developed [37,38]. Moreover, marine fish growth patterns are sex-specific and dimorphic, as in the mako shark [9,17–19,44,45], boarfish (*Capros aper*) [46], electric knifefish (*Gymnorhamphichthys rondoni*) [47], and European sea bass (*Dicentrarchus labrax* L.) [48]. Two-sex models are applied when species (e.g., the mako shark) exhibit considerable sexual dimorphism [45,49–51].

Acquiring species-specific biological information for stock assessment is critical to ensure that shark populations are exploited sustainably. For this purpose, by using Monte Carlo simulations, we constructed a stochastic framework accounting for uncertainty in crucial parameters; this is the standard methodology currently used to assess shark populations [19,38,45,52,53]. The objectives of the current study were to **(i)** develop and use a two-sex stage-based matrix model with Monte Carlo simulations to assess the uncertainty in the life history parameters of mako sharks; **(ii)** estimate their finite rate of population growth (λ) and the intrinsic rate of population increase (r); **(iii)** generate various scenarios to assess the influence of their fishing mortality (F) using many approaches to assess natural mortality (M); and **(iv)** compare their demography in other oceans. In the current study, we, for the first time, provide information applicable to mako shark population management and conservation in the South Pacific Ocean by using a demographic methodology; our methods may be broadly applicable to other data-limited shark species as well as to other taxa.

2. Materials and Methods

2.1. Study Area and Biological Parameters

This study was conducted in waters in the South Pacific Ocean, extending from $0°0'0''$ to $60°0'0''$ S and from $130°0'0''$ E to $70°0'0''$ W (Figure 1), and we assumed the existence of a unit stock for mako sharks to simplify the development of the matrix model. The values of the biological parameters for mako sharks in the South Pacific Ocean have been estimated by Chan [25], Bishop et al. [26], and Cerna and Licandeo [27]. However, Cerna and Licandeo [27] provided parameters with a wider range and estimates with higher accuracy. Therefore, these parameters were used as the foundation for the current demographic study (Table 1).

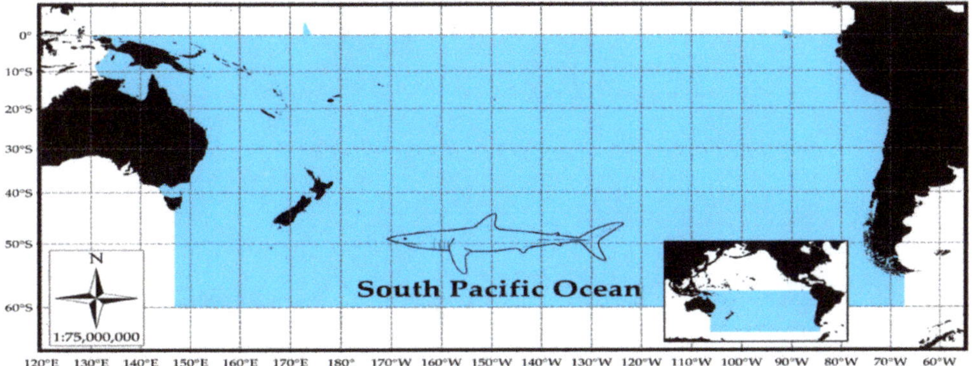

Figure 1. Map of the study area. The blue area depicts the South Pacific Ocean's location on the world map.

Table 1. Biological parameters of mako sharks used in this study.

	L_∞ (cm)	K (Year^{-1})	t_0 (Years)	L_0 (cm)	Reference
Male	296.604	0.087	−3.579	79.3	[52]
Female	325.293	0.076	−3.182	70.0	[52]

L_∞: asymptotic length, k: growth coefficient, t_0: age at length 0, L_0: length at birth.

2.2. Mako Shark Life History

The reproductive biology of mako sharks in the study area was provided by Bishop et al. [26]. The life phases and duration of mako sharks were established by Tribuzio and Kruse [42], who examined their life history; the developmental stages of mako sharks include the neonate, juvenile, subadult, and adult stages. Mature female mako sharks can be either pregnant (adult pregnant) or not pregnant (adult resting); they typically alternate between the two states. A female in the resting stage must return to the pregnant stage after 1 year, and a pregnant female may return to the pregnancy stage or move to the resting stage [54]. Gestation lasts 15–18 months [55,56]; in other words, pregnancy in a female adult mako shark lasts for approximately 2 years.

In the present study, the life history of female mako sharks was divided into six stages: neonate (0–1 year), juvenile (1–17 years), subadult (17–21 years), pregnant adult (1 year), parturient adult (1 year), and resting adult (1 year); only in case of a 3-year reproductive cycle. Moreover, the life history of male mako sharks was divided into four stages: neonate (0–1 year), juvenile (1–6 years), subadult (6–10 years), and adults (>10 years; Figure 2 and Table 2). According to the recommendations of Duffy and Francis [57], female sharks have a 3-year reproductive cycle with a 2-year gestation period and a 1-year resting period; however, some of them may have a 2-year reproductive cycle [10]. Consequently, we included both the 2- and 3-year reproductive cycles in our analysis.

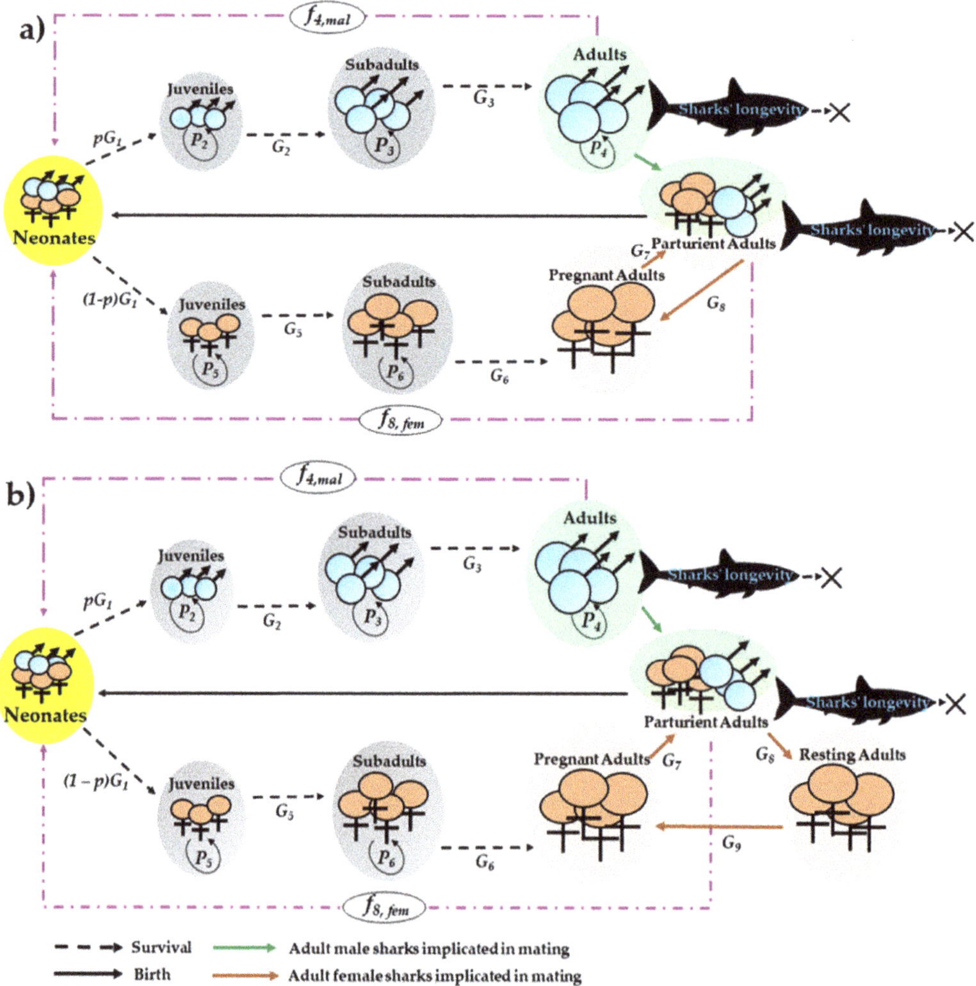

Figure 2. Stage-based matrix model for 2-year (**a**) and 3-year (**b**) reproductive cycles.

Table 2. Life history stages of mako sharks in the South Pacific Ocean.

Sex	Stage-Class	Approximate Ages (Years)	Expected Stage Duration (Years)
Male	Neonates	0–1	1
	Juveniles	1–6	5
	Subadults	$6–a_{mat}$	1–4
	Adults	$a_{mat}–a_{max}$	12–24
Female	Neonates	0–1	1
	Juveniles	1–17	16
	Subadults	$17–a_{mat}$	2–4
	Adults	$a_{mat}–a_{max}$	15–26
Included	Pregnant adults	>21	1
	Parturient adults	>21	1
	Resting adults *	>21	1

a_{mat}: 7–10 years for males and 19–21 years for females, a_{max}: 22–31 years for males and 36–45 years for females, * Only related to the 3-year reproductive cycle.

2.3. Model Development

We used a stage-structured model as a component of an ungenerous technique to estimate allowed injuries because it has been applied to the majority of threatened species with limited age-specific data. Management strategies are more likely to be related to life cycle stages than to age classes. The demographics of the mako shark in the South Pacific Ocean were therefore examined using a stage-structured Leslie matrix population projection model [58]. This is the fundamental equation to estimate the population's stage structure at any time t:

$$N_{t+1} = A_t N_t \tag{1}$$

where N_t is the vector of the number of sharks in each stage class at time t, and A_t is the life-history projection matrix comprising survival and fecundity for each stage at time t [37,59].

Our model, nevertheless, assumed that harvesting occurred before natural mortality (M) and reproduction. As the highly migratory mako shark is found over a wide range of area throughout the South Pacific Ocean, any regulatory theory based on competition for resources in a space-limited environment is more suitable than a hypothesis. Thus, the dynamics of density dependence on parameters such as survival, fecundity, and growth were ignored in this study. A stage-structured model based on maturity or breeding conditions involves the concept of knife edge (i.e., step-like) changes from one stage to another. Hence, knife-edge maturity was applied in this model.

Stage-based projection models are typically used based on their ability to analytically solve A_t to derive significant demographic variables and determine their sensitivity to the parameter estimations. Here, we established two-sex models in which the abundance of both male and female sharks influences fecundity [19,45,49] in the 2-year reproductive cycle and the 3-year reproductive cycle with a 1-year resting period; these are shown in Equations (2) and (3), respectively:

$$A_t = \begin{bmatrix} 0 & 0 & 0 & f_4^{mal} & 0 & 0 & 0 & G_8^{fem} \times f_8^{fem} \\ \rho G_1 & p_2^{mal} & 0 & 0 & 0 & 0 & 0 & 0 \\ 0 & G_2^{mal} & p_3^{mal} & 0 & 0 & 0 & 0 & 0 \\ 0 & 0 & G_3^{mal} & p_4^{mal} & 0 & 0 & 0 & 0 \\ (1-\rho)G_1 & 0 & 0 & 0 & p_5^{fem} & 0 & 0 & 0 \\ 0 & 0 & 0 & 0 & G_5^{fem} & p_6^{fem} & 0 & 0 \\ 0 & 0 & 0 & 0 & 0 & G_6^{fem} & 0 & G_8^{fem} \\ 0 & 0 & 0 & 0 & 0 & 0 & G_7^{fem} & 0 \end{bmatrix} \tag{2}$$

$$
A_t = \begin{bmatrix}
0 & 0 & 0 & f_4^{mal} & 0 & 0 & 0 & G_8^{fem} \times f_8^{fem} & 0 \\
\rho G_1 & p_2^{mal} & 0 & 0 & 0 & 0 & 0 & 0 & 0 \\
0 & G_2^{mal} & p_3^{mal} & 0 & 0 & 0 & 0 & 0 & 0 \\
0 & 0 & G_3^{mal} & p_4^{mal} & 0 & 0 & 0 & 0 & 0 \\
(1-\rho)G_1 & 0 & 0 & 0 & p_5^{fem} & 0 & 0 & 0 & 0 \\
0 & 0 & 0 & 0 & G_5^{fem} & p_6^{fem} & 0 & 0 & 0 \\
0 & 0 & 0 & 0 & 0 & G_6^{fem} & 0 & 0 & G_9^{fem} \\
0 & 0 & 0 & 0 & 0 & 0 & G_7^{fem} & 0 & 0 \\
0 & 0 & 0 & 0 & 0 & 0 & 0 & G_8^{fem} & 0
\end{bmatrix} \tag{3}
$$

Here, the male and female parameters of matrix A_t are indicated using the subscripts *mal* and *fem*, respectively, and ρ denotes the sex ratio at birth (set at 0.5 to achieve an equal offspring sex ratio) [10]. In the post-breeding survey, the fertility coefficient f_i measures the stage-specific per-capita fecundity, which can be calculated as follows: $f_i = fi \times \sigma_i$ [59]. Caswell's [59] modified harmonized mean birth function was used to analyze the influence of the sex ratio on the mating success, as shown in Equations (4) and (5), respectively:

$$
f_4^{mal}(n) = \frac{kn_f}{n_f + n_m} \tag{4}
$$

$$
f_8^{fem} = \frac{kn_m}{n_f + n_m} \tag{5}
$$

where k denotes the litter size and n_m and n_f denote the reproductive densities of male and female sharks, respectively. Considering the lack of evidence on the mating systems of mako sharks, monogamous mating and an equal litter size for both sexes were assumed in Equations (4) and (5)—independent of size or even other stage characteristics.

$G_{i,s}$ is the product of the probability of an individual in stage i surviving ($\sigma_{i,s}$) and the probability of shifting to another stage ($\gamma_{i,s}$) for each sex: $G_{i,s} = \sigma_{i,s} \times \gamma_{i,s}$; moreover, $p_{i,s}$ is the probability of an individual surviving and remaining in its current stage: $p_{i,s} = \sigma_{i,s} \times (1 - \gamma_{i,s})$ [54,60]. For a single stage, $\sigma_{i,s}$ is denoted as $\sigma_{i,s} = e^{-(M_{i,s}+F_{i,s})}$, here, $F_{i,s}$ is the F at stage i and is set as the average over-age of age-specific F ($F_{a,s}$), and $M_{i,s}$ is the M at stage i, which is set as the average over-age of age-specific natural mortality. $\gamma_{i,s}$ is calculated as follows:

$$
\gamma_{i,s} = \frac{(\sigma_{i,s}/\lambda_{init})^{T_{i,s}} - (\sigma_{i,s}/\lambda_{init})^{T_{i,s}-1}}{(\sigma_{i,s}/\lambda_{init})^{T_{i,s}} - 1} = \frac{(\sigma_{i,s})^{T_{i,s}} - (\sigma_{i,s})^{T_{i,s}-1}}{(\sigma_{i,s})^{T_{i,s}} - 1} \tag{6}
$$

where λ_{init} is the initial population growth rate (=1), and $T_{i,s}$ is the duration of each sex and stage. An iterative approach was used to estimate the matrix parameters using Equation (6). The finite population growth rate (λ) can be obtained by solving the following Equation:

$$
|A - \lambda I = 0| \tag{7}
$$

where I is the identify matrix [59]. λ for each sex can be calculated by separating A_t into the matrices for male and female sharks, as illustrated in Equations (2) and (3). Moreover, the intrinsic rate of population growth (r) may be related to λ obtained through the matrix projection [37]:

$$
\lambda = e^r \tag{8}
$$

The net productive rate R_0 is the mean number of pups produced by which a newborn individual is replaced by the end of its life. Generation time (T) [61] is defined as the time required for the population to increase by a factor of R_0 and can be calculated as follows:

$$
T = \frac{\ln R_0}{\ln \lambda} = \frac{\ln R_0}{r} \tag{9}
$$

Here, when the finite rate of population growth $\lambda = 1$, the population is stationary; when $\lambda > 1$, overpopulation has occurred (this is when harvesting may be considered an alternative to maintaining a stable population); and when $\lambda < 1$, the population is declining.

The relative proportion of individuals in each stage remains consistent throughout time after obtaining the stable stage distribution (SSD) vector w, which is a right eigenvector and one of the two demographic traits that can be derived from A_t. The reproductive value (RV) vector v is a left eigenvector, which denotes the proportion of offspring that individuals have yet to give birth at a given age [62]. RV depends on the number of individuals surviving to maturity and reproduction. Thus, individuals in the earliest stages should have the lowest RVs because they must survive and reach maturity before reproducing. An eigen analysis can be used to calculate the w and v of A_t. For any $\{n \times n\}$ matrix, up to n scalar values $(\lambda_{1 \dots n})$ and n-associated right and left vectors so that:

$$Aw = \lambda w, \ vA^T = \lambda v \tag{10}$$

where A^T is the transpose of A_t; and w and v are the right and left eigenvectors of A_t, respectively.

2.4. Mortality Estimation

Mortality is a crucial parameter for understanding the dynamics of any population, and it should be estimated for mako sharks. Calculation of the natural mortality (M) of shark species is difficult [63,64]. The magnitude of M is closely related to population productivity, environmental parameters, exploitation rates, management variables (e.g., maximum sustainable yield), and biological reference points [63,65] and it is frequently the main source of uncertainty in population dynamics modeling and stock assessment.

In this study, mako shark data and methods for the direct estimation of their M were unavailable. Therefore, we estimated M by using indirect methods. Sex-specific M (M_s) was assumed to remain constant in the subsequent deterministic analyses, which were based on the median obtained of the 14 methods listed in Table 3.

Table 3. Method and formulas used to estimate M_s of mako sharks in the South Pacific Ocean.

ID	Method	Equation
M1	Campana et al. [66]	$M_s = -\dfrac{\ln(0.01)}{a_{\max}}$
M2	Hoeing [67]	$M_s = 0.941 - 0.873 \ln(a_{\max})$
M3	Then et al. [68]	$M_s = 4.899 a_{\max}^{-0.916}$
M3	Then et al. [68]	$\ln(M_s) = 1.717 - 1.01 \times \ln(a_{\max})$
M5	Hamel [69]	$M_s = \dfrac{4.374}{a_{\max}}$
M6	Zhang and Megrey [70]	$M_s = \dfrac{3k}{e^{k0.38 a_{\max}} - 1}$
M7	Jensen [71]	$M_s = \dfrac{1.65}{a_{mat}}$
M8	Hisano et al. [41]	$M_s = \dfrac{1.6}{a_{mat} - t_0}$
M9	Frisk et al. [72]	$M_s = \dfrac{1}{0.44 a_{mat} + 1.87}$
M10	Cubillos et al. [73]	$M_s = 4.31 \left[t_0 - \dfrac{\ln(0.05)}{k} \right]^{-1.01}$
M11	Jensen [71]	$M_s = 1.6k$
M12	Frisk et al. [72]	$\ln(M_s) = 0.42 \ln(k) - 0.83$
M13	Hamel [69]	$M_s = 1.753k$
M14	Then et al. [68]	$M_s = 4.118 k^{0.73} L_\infty^{-0.33}$

a_{max} = longevity, set at 22–31 years for male sharks and 36–45 years for female sharks, a_{mat} = age at maturity, set at 7–10 years for male sharks and 19–21 for female sharks, L_∞, t_0 and k = growth parameters of the Von Bertalanffy Growth Equation (VBGE) data (Table 1).

2.5. Demographic Methods Accounting for Uncertainty

As with many shark species, estimating the rates of vital parameters is difficult due to limited data and vulnerability to uncertainty, so accounting for the effects of uncertainty by integrating stochasticity in demographic models is essential [74]. In the current study, Monte Carlo simulations were used [52,75]. This study accounted for uncertainty by using a simulation methodology similar to that used by [52], where statistical distribution functions based on published data were defined for each life history and biological parameter. Because detailed historical catch and effort data for mako sharks in the South Pacific Ocean are rare, each significant parameter was calculated with (Z = % of considered M) and without ($Z = M + F$) for the 2- and 3-year reproductive cycles; here, Z is the total mortality rate of a population.

Fecundity is a crucial component in demographic analysis. In several studies, the majority of observations lies between 8 and 18, and the mean litter size for the studied species was determined to be 12.5 [8,10,76–78]; moreover, the triangular distribution including the lower and upper limits and the assignment of a most likely value between these limits was assumed.

In the case of M_s uncertainty, the mean M_s values and their standard deviation (SD) were obtained in each scenario using the 14 indirect techniques for each sex and stage indicated in Table 3; these scenarios were used to generate a lognormal distribution. A lognormal error structure for M_s calculation ensures that the converted estimates are positive and between 0 and 1.

We assumed the lower and upper bounds of the uniform probability distribution approach in the case of age at maturity uncertainty to be 7–10 in male sharks and 19–21 in female sharks.

In the presence of longevity uncertainty, theoretical longevity estimates were based on Equations (11) and (12) for male sharks [79,80] and females [80,81], respectively:

$$T_{\max} = 7\ln\left(\frac{2}{k}\right) [79], \; T_{\max} = t_0 - \frac{\ln(0.05)}{k} \tag{11}$$

$$T_{\max} = \frac{1}{k}\ln\left[\frac{(L_\infty - L_0)}{L_\infty(1 - x)}\right] \tag{12}$$

where k, L_∞ and L_0 are the growth parameters of the VBGE [27]; and x is the proportion of L_∞ reached at $T_{\max}$, which is commonly assumed to be 0.95 [80,82]; the estimator with an estimated $T_{\max}$ using $x = 0.95$ is hereafter referred to as L_∞ 95. These theoretical estimated values were used to define the upper and lower bounds of a discrete uniform distribution of longevity. The uncertainties involved in fecundity, M_s, age at maturity, and longevity were evaluated using corresponding assumed distributions and are displayed in Table 4.

Table 4. Uncertainty used in the stochastic simulations.

Sources of Uncertainty	Male	Female	Assumed Distribution
Fecundity	Triangle (8, 12.5, 18)	Triangle (8, 12.5, 18)	Triangular distribution
Natural mortality	ln (mean, SD) *	ln (mean, SD) *	Lognormal
Age at maturity	7–10 years	19–21 years	Uniform
Longevity	22–31 years	36–45 years	Uniform

* Mean and (SD) obtained for M_s across 14 approaches for each sex and stage used to establish a lognormal distribution.

2.6. Size Limit Measures

Estimating the current fishing levels is difficult mainly because accurate harvest data are lacking; this hinders quantitative stock assessments for mako sharks in the study area. As the mako shark demonstrates considerable sex and size segregation on the area scale [9,83], along with segregation between juveniles and other stages [84], setting size limits for mako sharks may be simple. Consequently, we conducted simulations to analyze the existing status and to evaluate various management strategies by generating different F values based on a range of methodologies in the theoretical estimations of M_s.

To investigate stock status, 11 scenarios of harvest strategy were established in this study. For both the 2- and 3-year reproductive cycle models, five scenarios for harvest strategies and six scenarios for protective measures were as follows:

- Scenario 1 (S1): Under natural conditions, F was set to 0 for all ages;
- Scenario 2 (S2): F was set to 80% of M_s by stage;
- Scenario 3 (S3): F was set to 60% of M_s by stage;
- Scenario 4 (S4): F was set to 40% of M_s by stage;
- Scenarios 5 (S5): F was set to 20% of M_s by stage;
- Scenario 6 (S6): With the protection of immature sharks, F was set to 80% of M_s by stage; F values of neonates, juveniles, and subadults of both sexes were set to 0;
- Scenario 7 (S7): With the protection of mature sharks, F was set to 80% of M_s by stage, and F values of adults of both sexes were set to 0;
- Scenario 8 (S8): With the protection of immature male sharks, F was set to 80% of M_s by stage, F values of male neonates, juveniles, and subadults were set to 0;
- Scenarios 9 (S9): With the protection of mature male sharks, F was set to 80% of M_s by stage, F values of male adults was set to 0;
- Scenario 10 (S10): With the protection of immature female sharks, F was set to 80% of M_s by stage; F values of female neonates, juveniles, and subadults were set to 0;
- Scenario 11 (S11): With the protection of mature female sharks, F was set to 80% of M_s by stage; F value of female adults was set to 0.

The parameter values were calculated as the means of 10,000 replicates, with 95% confidence intervals (CIs) being the 2.5th and 97.5th percentiles. λ and r were obtained using Monte Carlo simulations accounting for parameter uncertainty for each scenario. All the aforementioned demographic and simulation analyses were performed using PopTools [85], and programmed using R [86].

3. Results

3.1. M_s Estimation

In all circumstances, male sharks had higher estimated M_s values than female sharks (Figure 3)—with the median M_s being 0.156 and 0.126 year^{-1}, respectively. M_s values obtained using the 14 estimators in Table 3 for male and female sharks are illustrated in Table S1. As shown in Figure 3, compared with other estimators, M3 and M14 demonstrated significant differences in male sharks, and M3 and M9 exhibited significant differences in the female sharks.

3.2. Demographic Analysis for the 2- and 3-Year Reproductive Cycle Models

Given the uncertainty in the mortality estimates and life history parameter availability, a range of scenarios, all containing stochasticity (noted through the Monte Carlo simulation), were examined. According to the most optimistic input parameters, the deterministic model of unfished conditions indicated an increasing population with mean and 95% CI for our 2- and 3-year reproductive cycle models, with sex-specificity evident in all scenarios (Tables S2 and S3).

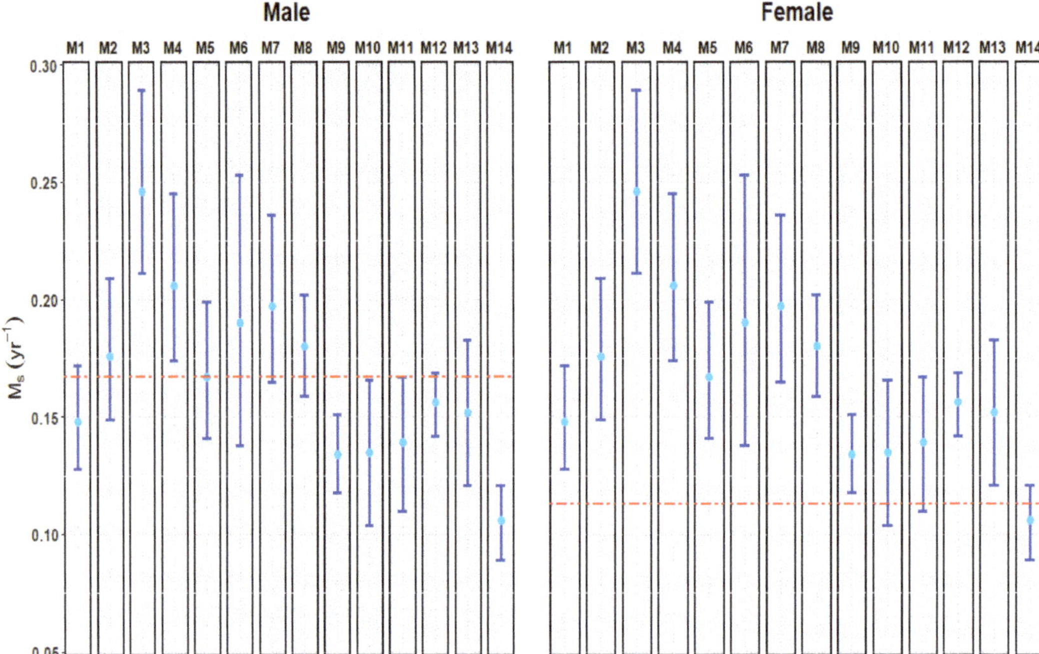

Figure 3. M_s of male and female mako sharks in the South Pacific Ocean. The lower and upper values are delineated by vertical navy blue lines, and each azure dot reflects the median values of each estimator, respectively. The red dashed line in each graph represents the mean M_s of the 14 estimators.

Under unfished conditions, the stage-structured model based 2- and 3-year reproductive cycle models exhibited equivalent population growth with $\lambda > 1$ (1.098 and 1.063 $year^{-1}$, respectively); however, the net reproduction rates, R_0, differed slightly (3.630 and 2.472, respectively) because the duration spent in the adult stages differed (Table S2). The λ estimates for the 2-year reproductive cycle were higher than those for the 3-year reproductive cycle. In addition, without F, the mean mako shark generation time, T, (95% CI) was anticipated to be 13.468 (10.874–16.518) and 14.260 (11.491–17.584) years in the 2- and 3-year reproductive cycle models, respectively. When F was incorporated into the models (Scenarios 2–5), all the 2- and 3-year reproductive cycle models tended to provide a reduced value of λ. Our results further indicate that under Scenario 5 (0.2 M_s; λ of >1) and the given harvest conditions (F = 20% of M), a population of mako sharks would be stable (Tables S2 and S3).

In summary, the aforementioned findings confirmed that the 3-year reproductive cycle model estimated a lower population growth rate compared with the 2-year reproductive cycle, particularly in female sharks. Moreover, the estimated elements of A_t for both reproductive cycle models under unfished conditions are presented in Table S4.

3.3. SSD and RV

The stable stage distribution (SSD) and the reproductive value (RV) of our 2- and 3-year reproductive cycle models demonstrated some similarities and differences. In general, the SSDs were extremely similar across the five harvest strategy scenarios, and the RVs varied across the scenarios. Although the SSDs shifted mainly to juveniles in both male and female sharks in both reproductive cycle models, the general pattern of stage distributions remained unchanged (Figure 4).

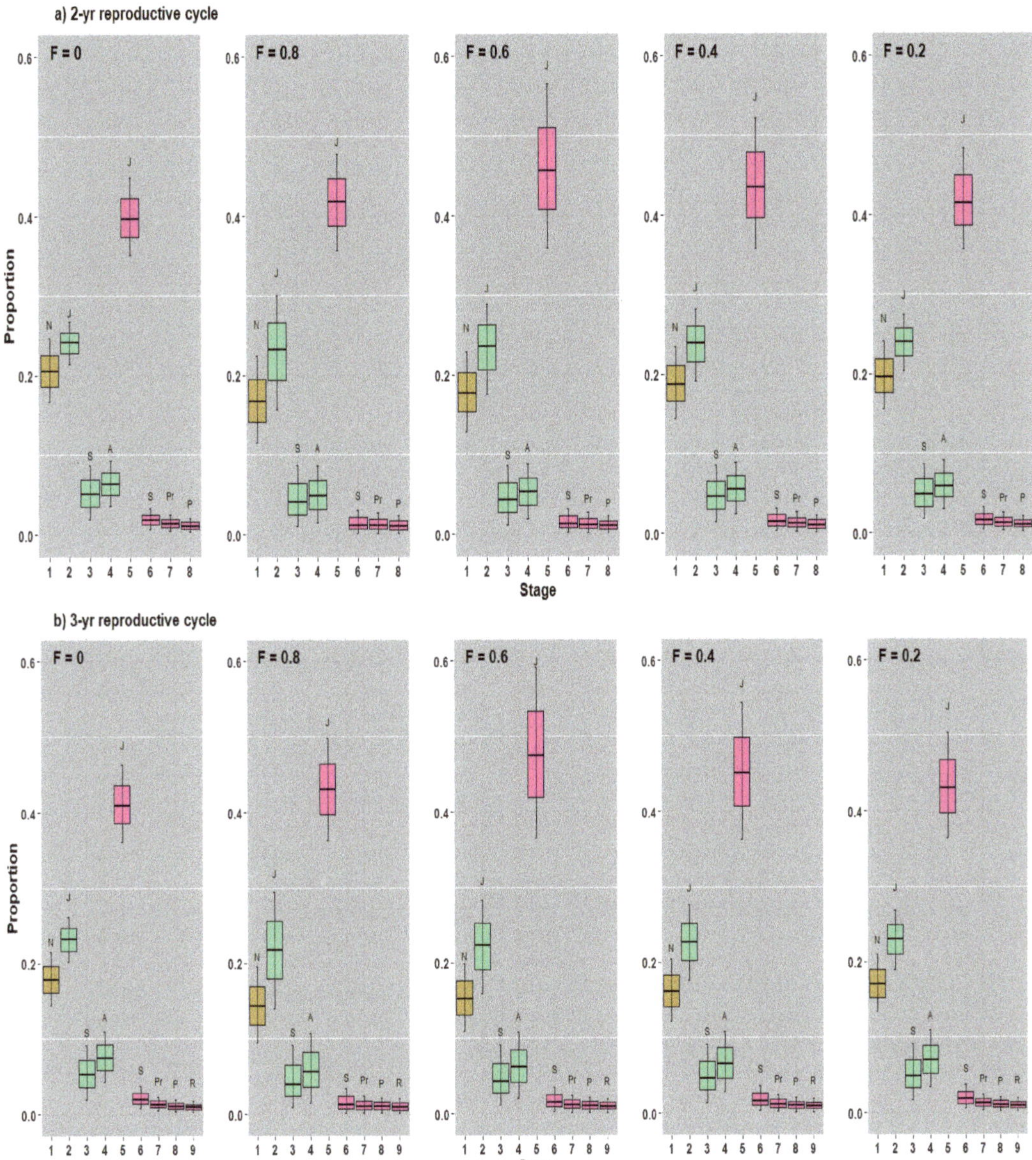

Figure 4. SSD values at various levels at F = 0, 0.8, 0.6, 0.4, and 0.2. (**a**,**b**) SSD values in the 2- and 3-year reproductive cycle models, respectively (where N = neonates, J = juveniles, S = subadults, A = adults, Pr = pregnant adults, P = parturient adults, R = resting adults). Both sexes (male [in aquamarine] and female [in hotpink]) are shown.

For all scenarios investigated, RVs tended to increase in the 2-year reproductive cycle model, peaking in adult male and adult parturient female sharks (Figure 5a). In our 3-year reproductive cycle model, RVs also tended to increase, peaking in adult male and adult parturient female sharks, but they decreased in adult resting female sharks (Figure 5b).

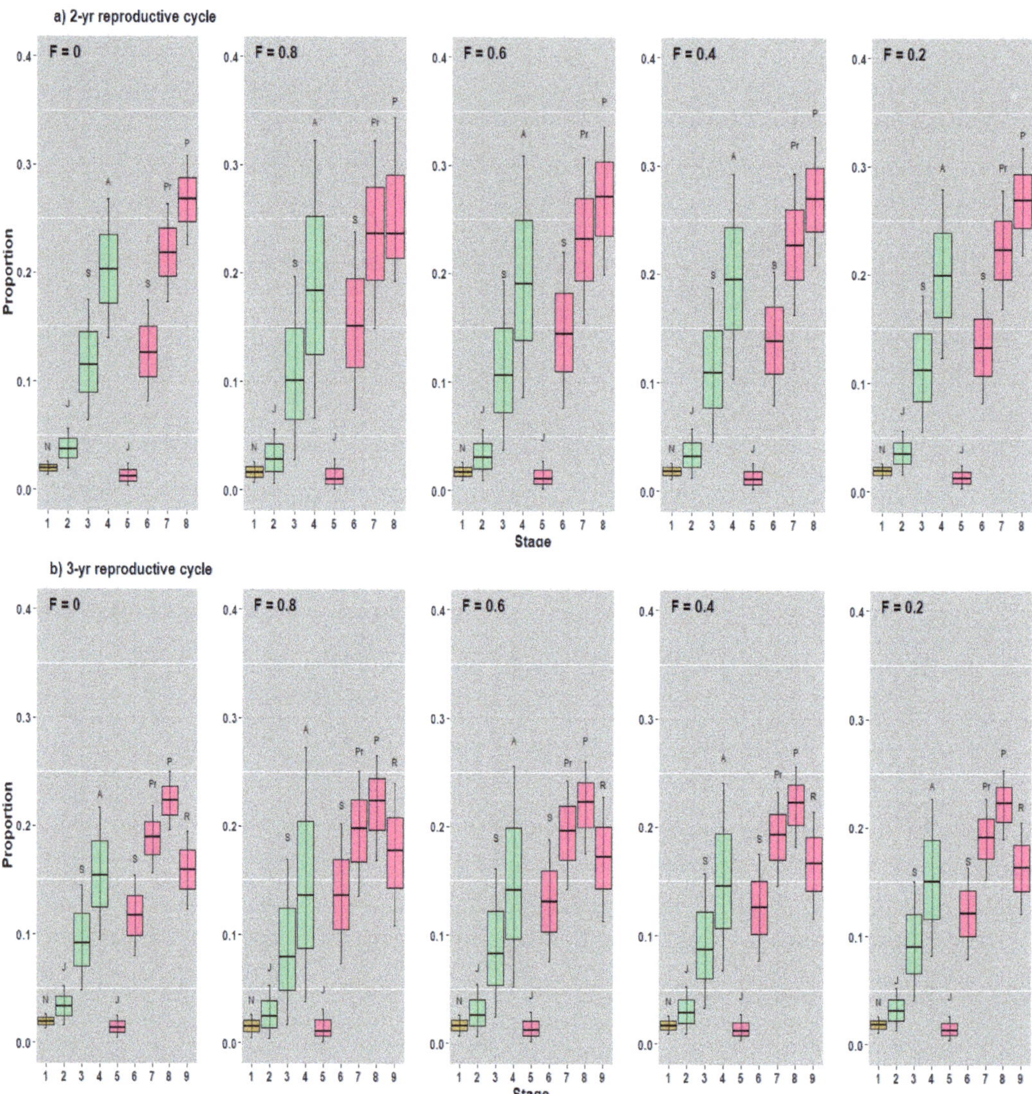

Figure 5. RV values at various levels at F = 0, 0.8, 0.6, 0.4, and 0.2. (**a,b**): RVs in the 2- and 3-year reproductive cycle models, respectively (where N = neonates, J = juveniles, S = subadults, A = adults, Pr = pregnant adults, P = parturient adults, R = resting adults). Both sexes (male [in aquamarine] and female [in hotpink]) are shown.

Moreover, the neonates in the stage-based model were expressed as a percentage of the population, which was approximately 20% (Figure 4). Although the RV for all stages probably exhibited a nearly bell-shaped curve indicating changes between stages, the RVs for neonates, juveniles, and subadults were low, and almost all RV values were observed for adult male sharks, adult pregnant females, and adult parturient female sharks for the 2- and 3-year reproductive cycle models, respectively, and adult resting female stages (only in a 3-year reproductive cycle) (Figure 5).

3.4. Size Limits Measures

In general, the λ values for mako sharks fluctuated considerably across the different management measure scenarios evaluated in the demographic models. Compared with the 3-year reproductive cycle model, the 2-year reproductive cycle model estimated higher management measurements including the population growth rates (Figures 6 and 7). Furthermore, because of the parameter uncertainty, the λ values of both sexes, males and females, exhibited significant variation in most of the scenarios.

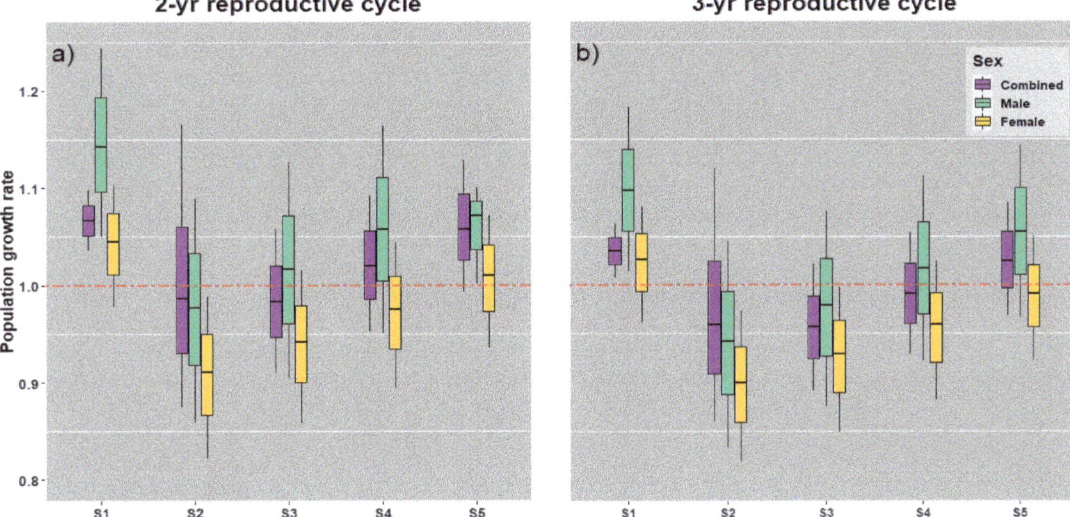

Figure 6. Box plots for the mako shark population growth rate in various harvesting scenarios with 2-year (**a**) and 3-year (**b**) reproduction cycle models. The red dashed line represents the stable level of population growth.

The simulation results demonstrated that the stock status under 20% fishing pressure (Scenario 5) led to a stable population growth rate in our 2- and 3-year reproductive cycle models (Figure 6a,b). However, λ was the lowest in female sharks (0.991 year^{-1} in Scenario 5 in the 3-year reproductive cycle model), which was lower than the threshold λ of 1. Notably, under the F put into these models (Scenarios 2 and 3) in both the reproductive cycle models, all λ values were <1, particularly for female sharks in Scenarios 2–4.

In both the reproductive cycle models, the population increases were noted when immature mako sharks were protected (Scenario 6; Figure 7a,b). Both the reproductive cycles predicted a higher population growth rate when immature male and female sharks were protected than when mature male and female sharks were protected (Figure 7). However, under these measures, the female stocks demonstrated a significant decline when immature male sharks were protected (Scenario 8; Figure 7a,b), particularly when λ was relatively lower than 1. Even when mature male sharks were protected in Scenario 9 (Figure 7), the population growth rate of the female stock was possibly <1. Variance in the aforementioned scenarios was due to the influence of uncertainty in age at maturity and the longevity of male and female sharks in the South Pacific Ocean.

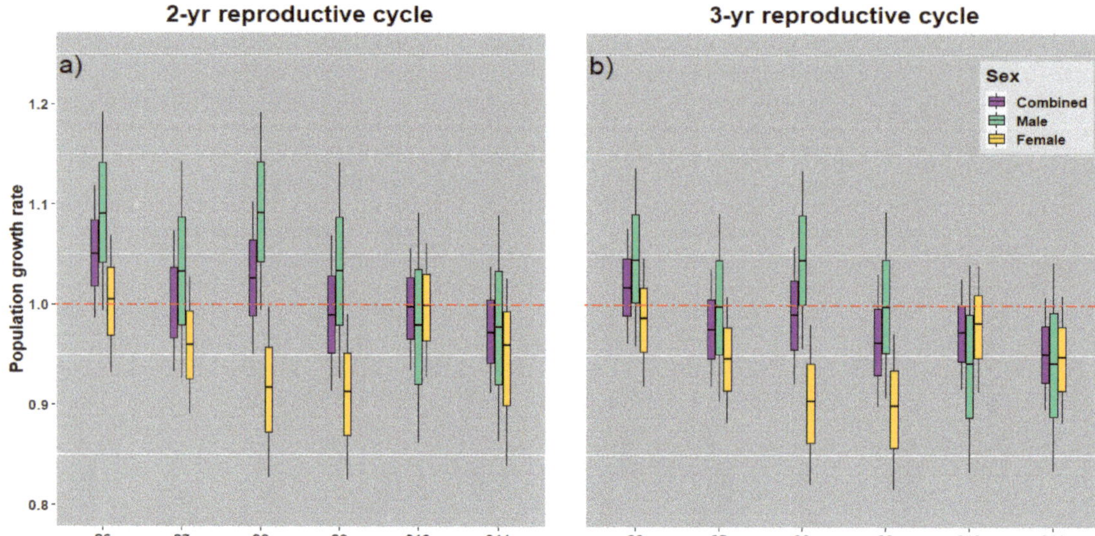

Figure 7. Box plots for mako shark population growth rate in various stock protection scenarios with 2-year (**a**) and 3-year (**b**) reproductive cycle models. The red dashed line represents the stable level of population growth.

4. Discussion

4.1. M_s

In this study, despite limited data availability, we estimated the M_s of mako sharks in the South Pacific Ocean by using indirect (empirical) approaches. Several indirect approaches have been proposed for estimating the M of fish including some proposed for species with limited data availability such as that reported by Kenchington [64] as well as some newly developed methods. Estimating M—an essential parameter in population dynamics—is difficult [87]. Therefore, we employed 14 previously reported empirical estimators along with a detailed assessment of uncertainty with the related parameters. Although Brodziak et al. [63] and Hoenig et al. [87] recommended comparing indirect estimates of M with the field-derived measurements of M, direct estimates of M are unavailable due to the lack of mako shark capture and effort data. In the current study, in the South Pacific Ocean, male sharks had a higher M_s than female sharks (0.106–0.246 vs. 0.069–0.166 year^{-1}), similar to the data obtained at other locations: Takeuchi et al. [16] reported that based on the average mortality estimation methodology, the M_s of mako sharks in the Atlantic Ocean was 0.122–0.168; moreover, the M_s of male and female mako sharks in the Northwest Pacific Ocean was estimated to be 0.093–0.200 and 0.077–0.242, respectively, according to Tsai et al. [18] and 0.119–0.141 and 0.091–0.124 according to Tsai et al. [19].

Substantial variation in the M values is related to the indirect approach used; these approaches differ in terms of the number of parameters used, their uncertainty, and their theoretical foundation. For instance, two life history parameters were considered in M14 (k and L_∞) [68] and M8 (age at maturity and t_0) [41]. M12 [72] and M13 [69] considered only the somatic growth rate k; finally, M1 [66] and M2 [67] considered only the longevity parameter. Empirical methods are often perceived to be less reliable than direct methods such as mark-recapture [88,89] and telemetry [90,91]; moreover, the measurement of life history parameters can be erroneous [69,92].

4.2. Demographic Analysis for Mako Shark Stock Assessment

Demographic analysis of long-living, highly migratory species such as mako sharks has limitations. Nevertheless, this study is the first to assess mako shark stock in the South Pacific Ocean through demographic analysis and to forecast how mako shark populations respond to alternative management strategies. Demographic analysis of shark populations can facilitate the effective management of their stocks and the assessment of their population dynamics. This may be used to estimate the most fundamental statistical parameters related to shark stock fluctuation and management: r and λ. Moreover, we constructed a two-sex stage-structured matrix population model accurately reflecting mako shark life cycles to examine the influence of biological parameters on the population growth rate.

According to our demographic estimates obtained using the 2- and 3-year reproductive cycle models, mako sharks are one of the least productive shark species. In the Ecological Risk Assessment (ERA) ranking for longline gear, mako sharks have been allocated the highest vulnerability level (No. 1) [93]. In the models without F, the mean population growth rates in this study ($\lambda = 1.098$ and 1.063 year^{-1} for the 2- and 3-year reproductive cycle models, respectively) were moderately higher than those derived from previous demographic analyses [20,94,95]; moreover, they differed considerably from those reported by Cortés [52] (Table 5). The reasons underlying these differences are diverse, with the most typical being the use of different models (single-sex vs. two-sex, age-structured vs. stage-structured, Bayesian surplus production model) and life history parameters (e.g., M_s and population growth rate) in the aforementioned studies; moreover, the findings for mako sharks in the South Pacific Ocean may not be applicable to those in other oceans, since the shark populations in the different oceans have varying traits.

Conventional demographic analysis assumes that male sharks do not control population growth; most studies have used the data of only female sharks. Two-sex models are, however, required to analyze the ecological and evolutionary implications of sexual dimorphism [96]; this is because single-sex models provide unsatisfactory results if demographic parameters such as size at age differ across sexes [49,50,97]. Consequently, male and female dynamics may differ, necessitating the incorporation of the data of both sexes in population models to understand population dynamics. The current findings demonstrate that when sex is considered in the model, the population growth rates are higher compared to those reported previously [19,21,45]. However, Yokoi et al. [53] reported values greater than this study's demographic parameters because of the age at maturity and longevity differences across the studied oceans. Therefore, this study confirmed that age at maturity and longevity were significantly and negatively correlated with the population growth rate of sharks [30,52,79].

The demographic method can be used to establish the prior distribution of important parameters in stock assessment (e.g., r), especially for parameters required in Bayesian stock assessment [98]. These population growth rate values can be used as the prior distribution or input variables for the Bayesian surplus production model (BSPM) under the stock assessment framework and ERA productivity index. For instance, Tsai et al. [99] estimated bigeye thresher shark population development by using a Bayesian population model; moreover, the vulnerability of pelagic sharks and mako sharks has also been assessed using ERA [100,101]. The present findings may aid in performing similar assessments. The aforementioned BSPM, for example, has been used to calculate the stock sizes of mako sharks in the South and North Atlantic Oceans. Thus, the high population growth rate estimated here can improve the future stock estimates of mako sharks in the South Pacific Ocean, at least until sufficient fishery data become available.

Table 5. Comparison of mako shark population growth rates collected from different studies.

Area	Model/Method	Sex	a_{mat}	a_{max}	λ (yr^{-1})	Reference
Northwestern Atlantic	Age-structured matrix population model	F	6, 7, 8	17	1.141 [a] (1.098–1.181) [t]	[52]
Atlantic	Age-structured matrix population model	F	6, 7, 8	17	* 1.076 [a]	[94]
North Atlantic	Bayesian surplus production model	F	18	32	* 1.060–1.061 [y]	[95]
South Atlantic	Bayesian surplus production model	F	18	32	* 1.060–1.064 [y]	[95]
North Pacific	Two-sex stage structured matrix population model	C	-	-	1.078 [b(2yr)], 1.051 [b(3yr)]	[19]
		M	11–15	24–31		
		F	18–21	31–41		
North Pacific	Two-sex stage-structured matrix population model with polyandrous and polygynous	C	-	-	1.010–1.082 [c,y]	[45]
		M	11–15	24–31		
		F	18–21	31–41		
Global	Two-sex age-structured matrix population model	C	-	-	* 1.107 [d] (1.007–1.374) [x]	[53]
		M	3	23		
		F	17	38		
Northwestern Pacific	Stage-based models	F	19–22	42	1.059 [e] (1.023–1.098) [z]	[20]
Indian Ocean	Two-sex age-structured matrix population model	C	-	-	* 1.120 [f] (1.062–1.141) [x]	[21]
		M	7	23–29		
		F	15	32–38		
South Pacific	Two-sex stage-structured matrix population model	C	-	-	1.098 [g (2yr)] (1.036–1.165) [t] 1.063 [g (3yr)] (1.008–1.119) [t]	This study
		M	7–10	22–31		
		F	19–21	36–45		

C: Combined, M: Male, F: Female, a_{mat}: age at maturity, a_{max}: longevity, [2yr and 3yr]: 2- and 3-year reproductive cycle models, respectively, [x] Minimum to maximum population growth rates, [y] Population growth rate estimated in several scenarios, [z] 95% CI, [t] 95% CI, calculated as the 2.5th and 97.5th percentiles, * r of the matrix population model, converted to, λ[a] Estimated using six empirical methods: (1) Hoenig [67], (2) Pauly [102], (3) Chen and Watanabe [103], (4) Peterson and Wroblewski [104], and (5) and (6) Jensen [71]. [b] Estimated using life history invariants of (1) Hoenig [67], (2) Campana et al. [66], (3) and (4) Jensen [71], and (5) Petersen and Wroblewski [104]. [c] Estimated using the methods of (1) Hoenig [67], (2) Jensen [71], (3) Peterson and Wroblewski [104], and (4) Campana et al. [66]. [d] Estimated using several theoretical models: (1) Chen and Watanabe [103], (2) Jensen [71], (3) Pauly [102], (4) Hoenig [67], and (5) Petersen and Wroblewski [104]. [e] Estimated by randomly selecting from the following four methods: (1) Hoenig [67], (2) and (3) Jensen [71], and (4) Petersen and Wroblewski [104]. [f] Estimated by randomly selecting from the following methods: (1) Petersen and Wroblewski [104] and (2) Hoenig [67]. [g] Estimated by randomly selecting from the following 14 methods: (1) Hoeing [67], (2) Campana et al. [66], (3–5) Then et al. [68], (6–7) Hamel [69], (8) Zhang and Megrey [70], (9–10) Jensen [71], (11–12) Frisk et al. [72], (13) Hisano et al. [41], and (14) Cubillos et al. [73].

Variations in biology, morphology, behavior, and reproductive functions between male and female sharks in the ocean [105,106] can lead to sex differences in ecological and demographic parameters [107–110]. Moreover, life history parameters such as longevity and mortality demonstrate sex differences. Therefore, a two-sex stage-based population model, rather than the conventional single-sex assessment model, was considered appropriate for the current study. Moreover, accessing multiple paternity as a mating strategy because of region-specific movement behavior suggested that mako sharks have significantly more opportunities to meet mating partners than previously believed, potentially indicating the existence of polyandry [45], contributing to improved demographic models. In addition, when formulating management measures, a species' possible sexual dimorphism must be carefully evaluated to determine whether more accurate estimates of decline risk can be derived.

4.3. Conservation Problems and Management Strategies for Mako Sharks

The current demographic analysis results demonstrate that fishing negatively affects the mako shark population, and for the first time, provided data for implementing mako shark management and conservation in the South Pacific Ocean using a demographic method. Studies have indicated that mako shark communities in other regions experience considerable changes: the Mediterranean Sea stock has collapsed and is at the risk of extinction [6], the North Atlantic stock remains overfished [111], and the status of the South Atlantic stock remains undetermined. The most recent comprehensive studies on the North Pacific and Indian Ocean stocks have indicated that the populations were neither overfished nor stable [22,112].

The current conclusions are based on a two-sex stage-structured matrix population model, and stochastic approaches for parameters with uncertainty may provide a relatively realistic perspective of the population dynamics of this species. Our results revealed that mako shark populations would only be able to sustain a relatively low fishing mortality (20% of natural mortality) in the study area. Little research has focused on quantitative stock assessment or fishery status indicators. Most researchers have used a few remarkable methods for including M_s into their demographic models. Several slow-growing, long-living shark species have demonstrated substantially lower λ [52]. When we incorporated F into our demographic analysis, the population growth rates of mako sharks decreased. This result also showed that any added anthropogenic source of mortality will likely result in its population decline.

In this study, the effects of protection (Figure 7) differed according to the sex-specific hypothesis. In both reproductive cycle models, the protection of immature mako sharks of both sexes resulted in a higher population growth rate than the protection of mature sharks. Shark demographic studies have reported that the population growth rate is the most sensitive to the immaturity stage; it is particularly high in juvenile females [30,31,38,79]. The differential availability of shark sexes in fisheries possibly arises from sex segregation in the population [83]; moreover, the mako shark demonstrates substantial sex and size segregation as well as segregation between juveniles and other developmental stages. Therefore, protection of this species using an optimal management approach dependent on the varying sex-specific hypothesis can be the most efficient measure for mako shark conservation. However, as ocean boundaries between male and female sharks have not been documented in detail, the potential role of sexual segregation in the overexploitation of shark populations remains unclear [83], particularly in the current study, considering that the knowledge on the South Pacific mako shark population is limited. A population with a clear structure and a 2-year reproductive cycle was noted to remain stable, whereas that with a longer (3-year) reproductive cycle was predicted to diminish.

4.4. Life History Parameter Uncertainty and Their Limitation

This study considered not only the estimated parameters but also the CIs to reflect and account for these uncertainties. For this, we applied Monte Carlo simulations for the survival and fecundity parameters input in the Leslie matrices. The results of the demographic analysis using stage-based matrix models in this study should be interpreted with caution because the analysis was based on some restricted assumptions. The biological and life history parameter data of South Pacific mako sharks are possibly influenced by bias. Our analysis was based only on the relevant biological parameters derived for mako sharks in the Southeastern Pacific Ocean (Cerna and Licandeo [27]). Moreover, the life history data reported by Bishop et al. [26] in the Southwestern Pacific Ocean were also used in this study. More robust growth curves for improved stock assessments and demographic analyses are warranted.

Stage-based models have been modified based on the life history parameters of different species and have been efficiently applied in several studies including those on the sandbar shark (*Carcharhinus plumbeus*) [54], spiny dogfish (*Squalus suckleyi*) [42], mako shark [19,20,45], silky shark (*Carcharhinus falciformis*) [32], and pelagic thresher shark (*Alopias pelagicus*) [31]. In this study, rather than adopting an age-based approach, we used a stage-based matrix, and simulations revealed that a 2-year reproductive cycle predicted more optimistic results than a 3-year reproductive cycle. Hence, the existence of a resting stage significantly affects population growth rates. However, because of the large size and wide oceanic distribution of mature female sharks as well as high fishing activity, determining the reproductive and life history parameters of mako sharks is difficult [10,83,113].

Age at maturity is a crucial parameter potentially influencing the demographic analysis results [52,59,75]. In this study, we used a stochastic methodology to account for annual fluctuations in age at maturity in the models and improved the quality of the results. Female maturation age has been reported to have the greatest effect on the mako shark population growth rate [53]. This study demonstrated the importance of considering the sex differences in age at maturity when estimating the population growth rates, with extreme differences in life history parameters—corroborating the results of previous studies [19,32,52,114,115]. Fecundity, one of the most important productivity indicators, is uncertain in most cases; this relationship has not been confirmed for the litter size and the size of female mako sharks [8–10,12,76–78]. Therefore, rather than lognormal or uniform assumed distributions, we used a triangular distribution in this study to determine the fecundity parameters and assign the most likely value between the limits. Further research into the reproductive traits of this species is required to effectively assess its stock. Future studies should focus on collecting more reliable litter size data through onboard observation to reduce uncertainty.

The unavailability of longevity data was also a potential source of bias influencing our demographic analysis results. Obtaining biological data including information on the longevity of a species is difficult, and the lack of these data has hindered the demographic analysis of shark populations [39]. Almost all sharks are bycatch species with large body sizes, small populations, high susceptibility, and widespread distribution. The oldest fish may be caught in some cases. As obtaining shark longevity data through direct age verification is difficult, longevity is typically calculated through indirect analysis. The current study incorporated stochastic methodology to determine the longevity by using four equations reported previously [79–81]. This approach may reflect the actual conditions more closely. Simultaneously, the periodicity of band pair deposition may vary at the different life stages of mako sharks [116], leading to age estimation uncertainties. Furthermore, because the observed maximum age reported in the South Pacific Ocean was smaller than that in other oceans, longevity may have been underestimated. Nevertheless, we did not use the observed longevity in this study.

Compensatory density dependence is vital for fisheries management because they can offset the loss of individuals [117,118]. Although the concept of compensation is simple, it is one of the most contentious topics in population dynamics [118–121]. These responses may be smaller in shark species [19,32,54], and they are particularly unlikely to influence mako sharks due to their extremely susceptible life cycle characteristics (i.e., long lifespan, low fecundity, late maturity, and an extended reproductive cycle), all of which make them more susceptible to the overexploitation of high fishing pressure. Although a sustainable harvest is possible in the absence of compensating density dependency in theory, it is ecologically impractical [118]. Nevertheless, in this study, when F was introduced and the demographic model did not consider density dependency or compensation, mako shark stock status demonstrated a decline (Figure 6). This result is similar to those of Tribuzio and Kruse [42], Tsai et al. [18,31,32], Coelho et al. [115], and Chang and Liu [20]. Therefore, even though we ignored the compensation of density reliance in the South Pacific mako shark population, the current results could be considered as acceptable. Demographic models provide certain theoretical and empirical results that facilitate future advancements in compensating process measurements and understanding.

The estimations of stage-based models in this study assumed that mako sharks in the South Pacific form a unit stock. According to Heist et al. [122] and Schrey and Heist [123], a unit stock of mako sharks was confirmed in the eastern Pacific Ocean across the equator. The data derived using molecular genetic techniques have recently revealed at least three genetic populations of mako sharks in the Pacific Ocean including the North Pacific, Southeast Pacific, and Southwestern Pacific Ocean, with little movement of fish between the populations [124]. Moreover, any changes in the life history parameters in demographics influence the population dynamics [125], with the focus being on the population growth rate. Life history parameters can be input into species-specific demographic models based on unit stock. However, the South Pacific Ocean mako shark stock structure is unclear, mostly due to the lack of relevant data. Clarifying whether mako sharks in the entire South Pacific Ocean are a unit stock [124] is necessary to establish a completely separate demographic model for each stock and to gain a further understanding of the population dynamics of the species.

Additionally, because of the parameters used, the accuracy of the information, and the large variability degree of uncertainty, it is impossible to recommend a single value, demonstrating that the indirect estimation of M is less accurate [64]. Notably, Zhou et al. [126] indicated concerns regarding the estimation of M for elasmobranchs (including sharks) by using estimators generated mostly from teleost data. Therefore, in this study, we emphasized the uncertainty of M estimates for stock assessment and management [63], particularly in demographic models. Actual variations in M may be much higher than assumed in this study. Thus, a robust method for estimating M by using tagging research findings and catch and effort data and thus calculating the total mortality rate is required.

5. Conclusions

Even though mako sharks are vulnerable to overexploitation, management strategies have not been established for this species thus far. Here, we examined mako shark demographics in the South Pacific Ocean by using limited data on their life history. Nevertheless, our sex-specific demographic models demonstrated that management decisions in the South Pacific Ocean mako shark population are required to ensure the sustainability of mako shark stocks, and that the population growth rate is higher when immature mako sharks are protected than when mature mako sharks are protected. Better estimates of natural and fishing mortality are required to understand the impact of commercial fisheries in the South Pacific Ocean's shortfin mako shark population. It is also suggested that our approach can be utilized as an assessment tool for shark species with insufficient catch and effort data.

Supplementary Materials: The following supporting information can be downloaded at: https://www.mdpi.com/article/10.3390/ani12223229/s1, Table S1: Ms (year^{-1}) estimated using 14 indirect approaches; Table S2: Estimated demographics in various scenarios based on a 2-year reproductive cycle; Table S3: Estimated demographics in various scenarios based on a 3-year reproductive cycle; Table S4: At for the 2- and 3-year reproductive cycles under unfished conditions.

Author Contributions: H.H.H.: Data curation, formal analysis, investigation, visualization, writing—original draft. C.-Y.H.: Formal analysis, investigation, methodology, visualization. W.-P.T.: Methodology, software, supervision, validation, writing—review & editing. All authors have read and agreed to the published version of the manuscript.

Funding: This study was financially supported by the Ministry of Science and Technology, Taiwan (MOST) under Contract No. MOST 111-2313-B-992-001.

Institutional Review Board Statement: Not applicable.

Informed Consent Statement: Not applicable.

Data Availability Statement: Not applicable.

Acknowledgments: We thank three anonymous reviewers for their helpful comments, which greatly improved this manuscript.

References

1. Dulvy, N.K.; Fowler, S.L.; Musick, J.A.; Cavanagh, R.D.; Kyne, P.M.; Harrison, L.R.; Carlson, J.K.; Davidson, L.N.K.; Fordham, S.V.; Francis, M.P.; et al. Extinction risk and conservation of the world's sharks and rays. *eLife* **2014**, *3*, e00590. [CrossRef]
2. Costello, C.; Ovando, D.; Clavelle, T.; Strauss, C.K.; Hilborn, R.; Melnychuk, M.C.; Branch, T.A.; Gaines, S.D.; Szuwalski, C.S.; Cabral, R.B.; et al. Global fishery prospects under contrasting management regimes. *Proc. Natl. Acad. Sci. USA* **2016**, *113*, 5125–5129. [CrossRef]
3. Simpfendorfer, C.A.; Dulvy, N.K. Bright spots of sustainable shark fishing. *Curr. Biol.* **2017**, *27*, R97–R98. [CrossRef]
4. Compagno, L.J.V. *Sharks of the World: An Annotated and Illustrated Catalogue of Shark Species Known to Date. Bullhead, Mackerel and Carpet Sharks (Heterodontiformes, Lamniformes and Orectolobiformes)*; Food and Agriculture Organization of the United Nations: Rome, Italy, 2001; p. 269.
5. Compagno, L.J.V.; Dando, M.; Fowler, S. *A Field Guide to the Sharks of the World*; Princeton University Press: Princeton, NJ, USA, 2005.
6. Rigby, C.L.; Barreto, R.; Carlson, J.; Fernando, D.; Fordham, S.; Francis, M.P.; Jabado, R.W.; Liu, K.-M.; Marshall, A.; Pacoureau, N.; et al. *Isurus oxyrinchus*. The IUCN Red List of Threatened Species. 2019. Available online: https://www.iucnredlist.org/species/39341/2903170 (accessed on 21 February 2021).
7. CITES (Convention on International Trade in Endangered Species of Wild Fauna and Flora). Updates on Decisions Made on Proposals to Amend Appendices I and II at CoP18. Decisions Made on Proposals to Amend Appendices I and II. 2019. Available online: https://cites.org/eng/updates_decisions_cop18_species_proposals (accessed on 21 February 2022).
8. Stevens, J.D. Observations on reproduction in the Shortfin Mako, *Isurus oxyrinchus*. *Copeia* **1983**, *1*, 126–130. [CrossRef]
9. Semba, Y.; Aoki, I.; Yokawa, K. Size at maturity and reproductive traits of shortfin mako, *Isurus oxyrinchus*, in the western and central North Pacific. *Mar. Freshw. Res.* **2011**, *62*, 20–29. [CrossRef]
10. Mollet, H.F.; Cliff, G.; Pratt, H.L.; Stevens, J.D., Jr. Reproductive biology of the female shortfin mako, *Isurus oxyrinchus*, Rafinesque, 1810, with comments on the embryonic development of lamnoids. *Fish. Bull.* **2000**, *98*, 299–318.
11. Francis, M.P.; Duffy, C. Length at maturity in three pelagic sharks (*Lamna nasus, Isurus oxyrinchus, and Prionace glauca*) from New Zealand. *Fish. Bull.* **2005**, *103*, 489–500.
12. Joung, S.-J.; Hsu, H.-H. Reproduction and embryonic development of the Shortfin Mako, *Isurus oxyrinchus*, Rafinesque, 1810, in the northwestern Pacific. *Zool. Stud.* **2005**, *44*, 487–496.
13. Murua, H.; Coelho, R.; Santos, M.N.; Arrizabalaga, H.; Yokawa, K.; Romanov, E.; Zhu, J.F.; Kim, Z.G.; Bach, P.; Chavance, P.; et al. Preliminary Ecological Risk Assessment (ERA) for shark species caught in fisheries managed by the Indian Ocean Tuna Commission (IOTC). *IOTC* **2012**, *13*, 16.
14. Cortés, E.; Domingo, A.; Miller, P.; Forselledo, R.; Mas, F.; Arocha, F.; Campana, S.; Coelho, R.; Da Silva, C.; Holtzhausen, H.; et al. Expanded ecological risk assessment of pelagic sharks caught in Atlantic pelagic longline fisheries. *Collect. Vol. Sci. Pap. ICCAT* **2015**, *71*, 2637–2688.
15. Griffiths, S.; Duffy, L.; Aires-da-Silva, A. A preliminar ecological risk assessment of the large-scale tuna longline fishery in the Eastern Pacific Ocean using productivity-susceptibility analysis. In Proceedings of the 8th Meeting of the Scientific Advisory Committee of the IATTC, La Jolla, CA, USA, 8–12 May 2017.
16. Takeuchi, Y.; Senba, Y.; Nakano, H. Demographic analysis on Atlantic blue and shortfin mako sharks. *Collect. Vol. Sci. Pap. ICCAT* **2005**, *58*, 1157–1165.
17. Chang, J.-H.; Liu, K.-M. Stock assessment of the Shortfin Mako shark, *Isurus oxyrinchus*, in the Northwest Pacific Ocean using per recruit and virtual population analyses. *Fish. Res.* **2009**, *98*, 92–101. [CrossRef]
18. Tsai, W.-P.; Sun, C.-L.; Wang, S.-P.; Liu, K.-M. Evaluating the impacts of uncertainty on the estimation of biological reference points for the shortfin mako shark, *Isurus oxyrinchus*, in the Northwest Pacific Ocean. *Mar. Freshw. Res.* **2011**, *62*, 1383–1394. [CrossRef]
19. Tsai, W.-P.; Sun, C.-L.; Punt, A.E.; Liu, K.-M. Demographic analysis of the shortfin mako shark, *Isurus oxyrinchus*, in the Northwestern Pacific using a two-sex stage-based matrix model. *ICES J. Mar. Sci.* **2014**, *71*, 1604–1618. [CrossRef]
20. Chang, J.-H.; Liu, K.-M. Demographic Analysis of the Shortfin Mako Shark, *Isurus oxyrinchus*, in the western North Pacific Using Stage-Based Models. *J. Fish. Soc. Taiwan* **2018**, *45*, 153–172.
21. Semba, Y.; Yokoi, H.; Kai, M. Estimate of intrinsic rate of natural increase (*r*) of shortfin mako, *Isurus oxyrinchus*, based on life history parameters from Indian Ocean. In Proceedings of the IOTC—15th Working Party on Ecosystem and Bycatch, La Saline Les Bains, Reunion Island, France, 3–7 September 2019; IOTC: Victoria, Seychelles, 2019. WPEB15-20.
22. Wu, X.-H.; Liu, S.Y.V.; Wang, S.-P.; Tsai, W.-P. Distribution patterns and relative abundance of shortfin mako shark caught by the Taiwanese large-scale longline fishery in the Indian Ocean. *Reg. Stud. Mar. Sci.* **2021**, *44*, 101691. [CrossRef]
23. Large, K.; Neubauer, P.; Brouwer, S.; Kai, M. *Stock Assessment of Southwest Pacific Shortfin Mako Shark*; WCPFC: Colonia, Micronesia, 2022; (WCPFC-SC18-2022/SA-IP-13).

24. Large, K.; Neubauer, P.; Brouwer, S. *Input Data for the 2022 South Pacific Shortfin Mako Shark Stock Assessment*; WCPFC: Colonia, Micronesia, 2022; (WCPFC-SC18-2022/SA-WP-02).
25. Chan, R.W.K. Biological Studies on Sharks Caught Off the Coast of New South Wales. Ph.D. Thesis, University of New South Wales, Sydney, Australia, 2001; p. 323.
26. Bishop, S.D.H.; Francis, M.P.; Duffy, C.; Montgomery, J.C. Age, growth, maturity, longevity and natural mortality of the shortfin mako shark, *Isurus oxyrinchus*, in New Zealand waters. *Mar. Freshw. Res.* **2006**, *57*, 143–154. [CrossRef]
27. Cerna, F.; Licandeo, R. Age and growth of the shortfin mako, *Isurus oxyrinchus*, in the southeastern Pacific off Chile. *Mar. Freshw. Res.* **2009**, *60*, 394–403. [CrossRef]
28. Tsai, W.-P.; Liu, K.-M.; Chang, Y.-J. Evaluation of Biological Reference Points for Conservation and Management of the Bigeye Thresher Shark, *Alopias superciliosus*, in the Northwest Pacific. *Sustainability* **2020**, *12*, 8646. [CrossRef]
29. Zhu, J.; Chen, Y.; Dai, X.; Harley, S.J.; Hoyle, S.D.; Maunder, M.N.; Aires-da Silva, A.M. Implications of uncertainty in the spawner–recruitment relationship for fisheries management: An illustration using bigeye tuna, *Thunnus obesus*, in the eastern Pacific ocean. *Fish. Res.* **2012**, *119*, 89–93. [CrossRef]
30. Tsai, W.-P.; Liu, K.-M.; Joung, S.J. Demographic analysis of the pelagic thresher shark, *Alopias pelagicus*, in the Northwestern Pacific using a stochastic stage-based model. *Mar. Freshw. Res.* **2010**, *61*, 1056–1066. [CrossRef]
31. Tsai, W.-P.; Huang, C.-H. Data–limited approach to the management and conservation of the pelagic thresher shark in the Northwest Pacific. *Conserv. Sci. Pract.* **2022**, *4*, e12682. [CrossRef]
32. Tsai, W.-P.; Wang, Y.-J.; Yamaguchi, A. Demographic analyses of the data limited silky shark population in the Indian Ocean using a two-sex stochastic matrix framework. *J. Mar. Sci. Technol.* **2019**, *27*, 55–63.
33. Geng, Z.; Wang, Y.; Kingdong, R.; Zhu, J.; Dai, X. Demographic and harvest analysis for blue shark, *Prionace glauca*, in the Indian Ocean. *Reg. Stud. Mar. Sci.* **2021**, *41*, 101583. [CrossRef]
34. Dowling, N.A.; Dichmont, C.M.; Haddon, M.; Smith, D.C.; Smith, A.D.M.; Sainsbury, K. Guidelines for developing formal harvest strategies for data-poor species and fisheries. *Fish. Res.* **2015**, *171*, 130–140. [CrossRef]
35. Dowling, N.A.; Smith, A.D.M.; Smith, D.C.; Parma, A.M.; Dichmont, C.M.; Sainsbury, K.; Wilson, J.R.; Dougherty, D.T.; Cpoe, J.M. Generic solutions for data-limited fishery assessments are not so simple. *Fish Fish.* **2019**, *20*, 174–188. [CrossRef]
36. Chrysafi, A.; Kuparinen, A. Assessing abundance of populations with limited data: Lessons learned from data-poor fisheries stock assessment. *Environ. Rev.* **2016**, *24*, 25–38. [CrossRef]
37. Simpfendorfer, C.A. Demographic models: Life tables, matrix models and rebound potential. In *Management Techniques for Elasmobranch Fisheries*; Musick, J.A., Bonfil, R., Eds.; Fisheries Technical Paper (474); FAO: Rome, Italy, 2005; pp. 143–144.
38. Smart, J.J.; Chin, A.; Tobin, A.; White, W.; Kumasi, B.; Simpfendorger, C.A. Stochastic demographic analyses of the silvertip shark, *Carcharhinus albimarginatus*, and the common blacktip shark, *Carcharhinus limbatus*, from the Indo-Pacific. *Fish. Res.* **2017**, *191*, 95–107. [CrossRef]
39. Chen, P.; Yuan, W. Demographic analysis based on the growth parameter of sharks. *Fish. Res.* **2006**, *78*, 374–379. [CrossRef]
40. Romine, J.G.; Musick, J.A.; Burgess, G.H. Demographic analyses of the dusky shark, *Carcharhinus obscurus*, in the Northwest Atlantic incorporating hooking mortality estimates and revised reproductive parameters. *Environ. Biol. Fishes* **2009**, *84*, 277–289. [CrossRef]
41. Hisano, M.; Connolly, S.R.; Robbins, W.D. Population growth rates of reef sharks with and without fishing on the great barrier reef: Robust estimation with multiple models. *PLoS ONE* **2011**, *6*, e25028. [CrossRef]
42. Tribuzio, C.A.; Kruse, G.H. Demographic and risk analyses of spiny dogfish, *Squalus suckleyi*, in the Gulf of Alaska using age and stage-based population models. *Mar. Freshw. Res.* **2011**, *62*, 1395–1406. [CrossRef]
43. Cortés, E. Estimates of maximum population growth rate and steepness for blue sharks in the north and south Atlantic ocean. *Collect. Vol. Sci. Pap. ICCAT* **2016**, *72*, 1180–1185.
44. Semba, Y.; Nakano, H.; Aoki, I. Age and growth analysis of the shortfin mako, *Isurus oxyrinchus*, in the western and central North Pacific Ocean. *Environ. Biol. Fishes* **2009**, *84*, 377–391. [CrossRef]
45. Tsai, W.-P.; Liu, K.-M.; Punt, A.E.; Sun, C.-L. Assessing the potential biases of ignoring sexual dimorphism and mating mechanism in using a single-sex demographic model: The shortfin mako shark as a case study. *ICES J. Mar. Sci.* **2015**, *72*, 793–803. [CrossRef]
46. Hüssy, K.; Coad, J.O.; Farrell, E.D.; Clausen, L.W.; Clarke, M.W. Sexual dimorphism in size, age, maturation, and growth characteristics of boarfish, *Capros aper*, in the Northeast Atlantic. *ICES J. Mar. Sci.* **2012**, *69*, 1729–1735. [CrossRef]
47. Garcia, E.Q.; Zuanon, J. Sexual dimorphism in the electric knifefish, *Gymnorhamphichthys rondoni*, Rhamphichthyidae: Gymnotiformes. *Acta Amazon.* **2019**, *49*, 213–220. [CrossRef]
48. Faggion, S.; Vandeputte, M.; Vergnet, A.; Clota, F.; Blanc, M.; Sanchez, P.; Ruelle, F.; Allal, F. Sex dimorphism in European sea bass, *Dicentrarchus labrax* L.: New insights into sexrelated growth patterns during very early life stages. *PLoS ONE* **2021**, *16*, e0239791. [CrossRef]
49. Caswell, H.; Weeks, D. Two-sex models: Chaos, extinction, and other dynamic consequences of sex. *Am. Nat.* **1986**, *128*, 41–47. [CrossRef]
50. Lindström, J.; Kokkom, H. Sexual reproduction and population dynamics: The role of polygyny and demographic sex differences. *Proc. Biol. Sci.* **1998**, *265*, 483–488. [CrossRef]
51. Gerber, L.R.; White, E.R. Two-sex matrix models in assessing population viability: When do male dynamics matter? *J. Appl. Ecol.* **2014**, *51*, 270–278. [CrossRef]

52. Cortés, E. Incorporating uncertainty into demographic modeling: Application to shark populations and their conservation. *Conserv. Biol.* **2002**, *16*, 1048–1062. [CrossRef]
53. Yokoi, H.; Ijima, H.; Ohshimo, S.; Yokawa, K. Impact of biology knowledge on the conservation and management of large pelagic sharks. *Sci. Rep.* **2017**, *7*, 10619. [CrossRef]
54. Brewster-Geisz, K.K.; Miller, T.J. Management of the sandbar shark, *Carcharhinus plumbeus*, implications of a stage-based model. *Fish. Bull.* **2000**, *98*, 236–249.
55. Gilmore, G.R. Reproductive biology of lamnoid sharks. *Environ. Biol. Fishes* **1990**, *38*, 95–114. [CrossRef]
56. Snelson, F.F.; Roman, B.L.; Burgess, G.H. The reproductive biology of pelagic elasmobranchs. In *Sharks of the Open Ocean: Biology, Fisheries and Conservation*; Camhi, M.D., Pikitch, E.K., Babcock, E.A., Eds.; Blackwell Publishing Ltd.: Oxford, UK, 2008; pp. 24–54.
57. Duffy, C.; Francis, M.P. Evidence of summer parturition in shortfin mako, *Isurus oxyrinchus*, sharks from New Zealand waters. *N. Z. J. Mar. Freshw. Res.* **2001**, *35*, 319–324. [CrossRef]
58. Caswell, H. *Matrix Population Models: Construction, Analysis and Interpretation*; Sinauer Associates: Sunderland, MA, USA, 2002.
59. Caswell, H. *Matrix Population Models. Construction, Analysis, and Interpretation*, 2nd ed.; Sinauer Associates, Inc.: Sunderland, MA, USA, 2001; Volume 22, 772p.
60. Frisk, M.G.; Miller, T.J.; Fogarty, M.J. The population dynamics of little skate, *Leucoraja erinacea*, winter skate, *Leucoraja cellata*, and barndoor skate, *Dipturus laevis*: Predicting exploitation limits using matrix analyses. *ICES J. Mar. Sci.* **2002**, *59*, 576–586. [CrossRef]
61. Coale, A.J. *The Growth and Structure of Human Populations: A Mathematical Investigation*; Princeton University Press: Princeton, NJ, USA, 1972.
62. Gotelli, N.J. *A Primer of Ecology*; Sinauer Associates: Sunderland, MA, USA, 1995; p. 206.
63. Brodziak, J.; Ianelli, J.; Lorenzen, K.; Methot, R.D. *Estimating Natural Mortality in Stock Assessment Applications*; NOAA Technical Memorandum NMFS-F/SPO-119; U.S. Department of Commerce: Washington, DC, USA, 2011; p. 38.
64. Kenchington, T.J. Natural mortality estimators for informationlimited fisheries. *Fish Fish.* **2014**, *15*, 533–562. [CrossRef]
65. Punt, A.E.; Castillo-Jordán, C.; Hamel, O.S.; Cope, J.M.; Maunder, M.N.; Ianelli, J.N. Consequences of error in natural mortality and its estimation in stock assessment models. *Fish. Res.* **2021**, *233*, 105759. [CrossRef]
66. Campana, S.; Joyce, W.; Marks, L.; Harley, S. *Analytical Assessment of the Porbeagle Shark, Lamna nasus, Population in the Northwest Atlantic, with Estimates of Long-Term Sustainable Yield*; Canadian Stock Assessment Research Document 2001/067; Canadian Science Advisory Secretariat: Ottawa, ON, Canada, 2001.
67. Hoenig, J.M. Empirical use of longevity data to estimate mortality rates. *Fish. Bull.* **1983**, *82*, 898–903.
68. Then, A.Y.; Hoenig, J.M.; Hall, N.G.; Hewitt, D.A. Evaluating the predictive performance of empirical estimators of natural mortality rate using information on over 200 fish species. *ICES J. Mar. Sci.* **2015**, *72*, 82–92. [CrossRef]
69. Hamel, O.S. A method for calculating a meta-analytical prior for the natural mortality rate using multiple life history correlates. *ICES J. Mar. Sci.* **2015**, *72*, 62–69. [CrossRef]
70. Zhang, C.I.; Megrey, B.A. A revised Alverson and Carney model for estimating the instantaneous rate of natural mortality. *Trans. Am. Fish. Soc.* **2006**, *135*, 620–633. [CrossRef]
71. Jensen, A.L. Beverton and Holt life history invariants result from optimal trade-off of reproduction and survival. *Can. J. Fish. Aquat. Sci.* **1996**, *53*, 820–822. [CrossRef]
72. Frisk, M.G.; Miller, T.J.; Fogarty, M.J. Estimation and analysis of biological parameters in elasmobranch fishes: A comparative life history study. *Can. J. Fish. Aquat. Sci.* **2001**, *58*, 969–981. [CrossRef]
73. Cubillos, L.A.; Alarcon, R.; Brante, A. Empirical estimates of natural mortality for the Chilean hake, *Merluccius gayi*, evaluation of precision. *Fish. Res.* **1999**, *42*, 147–153. [CrossRef]
74. Caswell, H.; Brault, S.; Read, A.J.; Smith, T.D. Harbor porpoise and fisheries: An uncertainty analysis of incidental mortality. *Ecol. Appl.* **1998**, *8*, 1226–1238. [CrossRef]
75. Cortés, E. A stochastic stage-based population model of the sandbar shark in the western North Atlantic. *Am. Fish. Soc. Symp.* **1999**, *23*, 115–136.
76. Branstetter, S. Biological notes on the sharks of the north central Gulf of Mexico. *Contrib. Mar. Sci.* **1981**, *24*, 13–34.
77. Stevens, J.D. The Biology and Ecology of the shortfin mako shark, *Isurus oxyrinchus*. In *Sharks of the Open Ocean: Biology, Fisheries and Conservation*; Camhi, M.D., Pikitch, E.K., Babcock, E.A., Eds.; Blackwell: Oxford, UK, 2008; pp. 87–91.
78. Taniuchi, T. Some biological aspects of sharks caught by floating longlines-3. Reproduction. *Rep. Jpn. Group Elasmobranch Stud.* **1997**, *33*, 6–13.
79. Mollet, H.F.; Cailliet, G.M. Comparative population demography of elasmobranches using life history tables, Leslie matrices and stage-based matrix models. *Mar. Freshw. Res.* **2002**, *53*, 503–516. [CrossRef]
80. Taylor, C.C. Cod growth and temperature. *ICES J. Mar. Sci.* **1958**, *23*, 366–370. [CrossRef]
81. Gompertz, B. On the nature of the function expressive of the law of human mortality and on a new mode of determining life contingencies. *Philos. Trans. R. Soc. Lond.* **1825**, *115*, 513–585.
82. Ricker, W.E. Growth rates and models. In *Fish Physiology*; Hoar, W.S., Randall, D.J., Brett, J.R., Eds.; Academic Press: New York, NY, USA, 1979; Volume 8, pp. 677–743.
83. Mucientes, G.R.; Queiroz, N.; Sousa, L.L.; Tarroso, P.; Sims, D.W. Sexual segregation of pelagic sharks and the potential threat from fisheries. *Biol. Lett.* **2009**, *5*, 156–159. [CrossRef]

84. Nakano, H.; Nagasawa, K. Distribution of pelagic elasmobranchs caught by salmon research gillnets in the North Pacific. *Fish. Sci.* **1996**, *62*, 860–865. [CrossRef]

85. Hood, G. *PopTools. Pest Animal Control Co-Operative Research Center*; CSIRO: Canberra, ACT, Austraila, 2004.

86. R Development Core Team. *R: A Language and Environment for Statistical Computing*; R Foundation for Statistical Computing: Vienna, Austria, 2021.

87. Hoenig, J.M.; Then, A.Y.-H.; Babcock, E.A.; Hall, N.G.; Hewitt, D.A.; Hesp, S.A. The logic of comparative life history studies for estimating key parameters, with a focus on natural mortality rate. *ICES J. Mar. Sci.* **2016**, *73*, 2453–2467. [CrossRef]

88. Brooks, E.N.; Pollock, K.H.; Hoenig, J.M.; Hern, W.S. Estimation of fishing and natural mortality from tagging studies on fisheries with two user groups. *Can. J. Fish. Aquat. Sci.* **1998**, *55*, 2001–2010. [CrossRef]

89. Hewitt, D.A.; Lambert, D.M.; Hoenig, J.M.; Lipcius, R.N.; Bunnell, D.B.; Miller, T.J. Direct and indirect estimates of natural mortality for Chesapeake Bay blue crab. *Trans. Am. Fish. Soc.* **2007**, *136*, 1030–1040. [CrossRef]

90. Hightower, J.E.; Jackson, J.R.; Pollock, K.H. Use of telemetry methods to estimate natural and fishing mortality of striped bass in Lake Gaston, North Carolina. *Trans. Am. Fish. Soc.* **2001**, *130*, 557–567. [CrossRef]

91. Heupel, M.R.; Simpfendorfer, C.A. Estimation of mortality of juvenile blacktip sharks, *Carcharhinus limbatus*, within a nursery area using telemetry data. *Can. J. Fish. Aquat. Sci.* **2002**, *59*, 624–632. [CrossRef]

92. Rudd, M.B.; Thorson, J.T.; Sagarese, S.R.; Kuparinen, A. Ensemble models for data-poor assessment: Accounting for uncertainty in life-history information. *ICES J. Mar. Sci.* **2019**, *76*, 870–883. [CrossRef]

93. Murua, H.; Santiago, J.; Coelho, R.; Zudaire, I.; Neves, C.; Rosa, D.; Semba, Y.; Geng, Z.; Bach, P.; Arrizabalaga, H.; et al. Updated Ecological Risk Assessment (ERA) for Shark Species Caught in Fisheries Managed by the Indian Ocean Tuna Commission (IOTC). IOTC–2018–SC21–14_Rev_1. Available online: https://www.fao.org/3/bj492e/bj492e.pdf (accessed on 1 March 2021).

94. Cortés, E.; Arocha, F.; Beerkircher, L.; Carvalho, F.; Domingo, A.; Heupel, M.; Holtzhausen, H.; Santos, M.N.; Ribera, M.; Simpfendorfer, C. Ecological risk assessment of pelagic sharks caught in Atlantic pelagic longline fisheries. *Aquat. Living Resour.* **2010**, *23*, 25–34. [CrossRef]

95. ICCAT. 2012 Shortfin mako stock assessment and ecological risk assessment meeting. *Collect. Vol. Sci. Pap. ICCAT* **2013**, *69*, 1427–1570.

96. Miller, T.E.X.; Inouye, B.D. Confronting two-sex demographic models with data. *Ecology* **2011**, *92*, 2141–2151. [CrossRef]

97. Kokko, H.; Rankin, D.J. Lonely hearts or sex in the city? density-dependent effects in mating systems. *Philos. Trans. R. Soc. B Biol. Sci.* **2006**, *361*, 319–334. [CrossRef]

98. McAllister, M.K.; Pikitch, E.K.; Babcock, E.A. Using demographic methods to construct Bayesian priors for the intrinsic rate of increase in the Schaefer model and implications for stock rebuilding. *Can. J. Fish. Aquat. Sci.* **2001**, *58*, 1871–1890. [CrossRef]

99. Tsai, W.-P.; Chang, Y.-Y.; Liu, K.-M. Development and testing of a Bayesian population model for the bigeye thresher shark, *Alopias superciliosus*, in an area subset of the Western North Pacific. *Fish. Manag. Ecol.* **2019**, *26*, 269–294. [CrossRef]

100. Duarte, H.O.; Droguet, E.L.; Moura, M.C. Quantitative ecological risk assessment of shortfin mako shark, *Isurus oxyrinchus*: Proposed model and application example. *Appl. Ecol. Environ. Res.* **2018**, *16*, 3691–3709. [CrossRef]

101. Liu, K.-M.; Wu, C.-B.; Joung, S.-J.; Tsai, W.-P.; Su, K.-Y. Multi-Model Approach on Growth Estimation and Association With Life History Trait for Elasmobranchs. *Front. Mar. Sci.* **2021**, *8*, 591692. [CrossRef]

102. Pauly, D. On the interrelationships between natural mortality, growth parameters, and mean environmental temperature in 175 fish stocks. *ICES J. Mar. Sci.* **1980**, *39*, 175–192. [CrossRef]

103. Chen, S.; Watanabe, S. Age dependence of natural mortality coefficient in fish population dynamics. *Nippon Suisan Gakk.* **1989**, *55*, 205–208. [CrossRef]

104. Peterson, I.; Wroblewski, J.S. Mortality rate of fishes in the pelagic ecosystem. *Can. J. Fish. Aquat. Sci.* **1984**, *41*, 1117–1120. [CrossRef]

105. Ruckstuhl, K.E.; Cultton-Brock, T.H. Sexual segregation and the ecology of the two sexes. In *Sexual Segregation in Vertebrates*; Ruckstuhl, K.E., Neuhaus, P., Eds.; Cambridge University Press: Cambride, UK, 2006; pp. 3–8.

106. Breed, M.D.; Moore, J. *Animal Behavior*, 2nd ed.; Elsevier: San Diego, CA, USA, 2015.

107. Kraus, C.; Eberle, M.; Kappeler, P.M. The costs of risky male behaviour: Sex differences in seasonal survival in a small sexually monomorphic primate. *Proc. R. Soc. B Biol. Sci.* **2008**, *275*, 1635–1644. [CrossRef]

108. Oro, D.; Torres, R.; Rodríguez, C.; Drummond, H. Climatic influence on demographic parameters of a tropical seabird varies with age and sex. *Ecology* **2010**, *91*, 1205–1214. [CrossRef]

109. Jenouvrier, S.; Holland, M.; Stroeve, J.; Barbraud, C.; Weimerskirch, H.; Serreze, M.; Caswell, H. Effects of climate change on an emperor penguin population: Analysis of coupled demographic and climate models. *Glob. Chang. Biol.* **2012**, *18*, 2756–2770. [CrossRef]

110. Vaughn, D.; Turnross, O.R.; Carrington, E. Sex-specific temperature dependence of foraging and growth of intertidal snails. *Mar. Biol.* **2014**, *161*, 75–87. [CrossRef]

111. ICCAT. Report of the 2019 shortfin mako shark stock assessment update meeting. In Proceedings of the SMA SHK SA Intersessional Meeting, Madrid, Spain, 20–24 May 2019; Volume 76, pp. 1–77.

112. de Bruyn, P. *Report of the 2017 ICCAT Shortfin Mako Stock Assessment Meeting*; International Commission for the Conservation of Atlantic Tunas: Madrid, Spain, 2018; Volume 74, pp. 1465–1561.

113. Casey, J.G.; Kohler, N.E. Tagging studies on the shortfin mako shark, *Isurus oxyrinchus*, in the Western North Atlantic. *Aust. J. Mar. Freshw. Res.* **1992**, *43*, 45–60. [CrossRef]
114. Frisk, M.G.; Miller, T.J.; Dulvy, N.K. Life histories and vulnerability to exploitation of elasmobranchs: Inferences from elasticity, perturbation and phylogenetic analyses. *J. Northwest Atl. Fish. Sci.* **2005**, *35*, 27–45. [CrossRef]
115. Coelho, R.; Alpizar-Jara, R.; Erzini, K. Demography of a deep-sea lantern shark, *Etmopterus spinax*, caught in trawl fisheries of the northeastern Atlantic: Application of Leslie matrices with incorporated uncertainties. *Deep-Sea Res. II Top. Stud. Oceanogr.* **2015**, *115*, 64–72. [CrossRef]
116. Kinney, M.; Wells, R.; Kohin, S. Oxytetracycline age validation of an adult shortfin mako shark, *Isurus oxyrinchus*, after 6 years at liberty. *J. Fish Biol.* **2016**, *89*, 1828–1833. [CrossRef]
117. Fogarty, M.J.; Rosenberg, A.A.; Sissenwine, M.P. Fisheries risk assessment: A case study of Georges Bank haddock. *Environ. Sci. Technol.* **1992**, *26*, 440–447. [CrossRef]
118. Rose, K.A.; Cowam, J.H.; Winemiller, K.O.; Myers, R.A.; Hilborn, R. Compensatory density dependence in fish populations: Importance, controversy, understanding and prognosis. *Fish Fish.* **2002**, *2*, 293–327. [CrossRef]
119. Fletcher, R.I.; Deriso, R.B. Fishing in dangerous waters: Remarks on a controversial appeal to spawner-recruit theory for long-term impact assessment. *Am. Fish. Soc. Monogr.* **1988**, *4*, 232–243.
120. Barnthouse, L.W.; Klauda, R.J.; Vaughan, D.S. Introduction to the monograph. In *Science, Law, and Hudson River Power Plants: A Case Study in Environmental Impact Assessment*; U.S. Department of Energy Office of Scientific and Technical Information: Washington, DC, USA, 1988; Volume 4, pp. 1–8.
121. Macaluso, J. Red snapper season offered. In *The Advocate*; The Advocate: Baton Rouge, LA, USA, 1999.
122. Heist, E.J.; Musick, J.A.; Graves, J.E. Genetic population structure of shortfin mako, *Isurus oxyrinchus*, inferred from restriction fragment length polymorphism analysis of mitochondrial DNA. *Can. J. Fish. Aquat. Sci.* **1996**, *53*, 583–588. [CrossRef]
123. Schery, A.W.; Heist, E.J. Microsatellite markers for the shortfin mako and cross-species amplification in lamniformes. *Conserv. Genet.* **2002**, *3*, 459–461. [CrossRef]
124. Francis, M.P.; Shivji, M.S.; Duffy, C.A.J.; Rogers, P.J.; Byrne, M.E.; Wetherbee, B.M.; Tindale, S.C.; Lyon, W.S.; Meyres, M.M. Oceanic nomad or coastal resident? Behavioural switching in the shortfin mako shark, *Isurus oxyrinchus*. *Mar. Biol.* **2019**, *166*, 5. [CrossRef]
125. Sæther, B.E.; Coulson, T.; Grøtan, V.; Engen, S.; Altwegg, R.; Armitage, K.B.; Barbraud, C.; Becker, P.H.; Blumstein, D.T.; Dobson, F.S.; et al. How Life History Influences Population Dynamics in Fluctuating Environments. *Am. Nat.* **2013**, *182*, 743–759. [CrossRef]
126. Zhou, S.; Deng, R.A.; Dunn, M.R.; Hoyle, S.D.; Lei, Y.; Williams, A.J. Evaluating methods for estimating shark natural mortality rate and management reference points using life-history parameters. *Fish Fish.* **2021**, *23*, 462–477. [CrossRef]

Article

Influence of Environmental Factors on Prey Discrimination of Bait-Attracted White Sharks from Gansbaai, South Africa

Francesca Romana Reinero [1,*], Emilio Sperone [2], Gianni Giglio [2], Antonio Pacifico [1,3], Makenna Mahrer [4] and Primo Micarelli [1,5]

[1] Sharks Studies Center-Scientific Institute, 58024 Massa Marittima, Italy
[2] Department of Biology, Ecology and Earth Sciences, University of Calabria, 87036 Rende, Italy
[3] Department of Political Science and CEFOP-LUISS, LUISS Guido Carli University, 00197 Rome, Italy
[4] W. M. Keck Science Department, Claremont McKenna College, Claremont, CA 91711, USA
[5] Department of Physical Sciences, Earth and Environment, University of Siena, 53100 Siena, Italy
* Correspondence: fr.reinero@gmail.com; Tel.: +39-32-9685-2552

Simple Summary: Predator–prey interactions can be influenced by environmental factors and depend on the sensory capabilities of a predator and its prey. Environmental factors influence the prey discrimination of white sharks in Gansbaai during their passive prey predatory behavior. Tide range is the most important factor that influences the white sharks' prey choice, followed by underwater visibility, water temperature, and sea conditions. With high tide, better underwater visibility, cooler water temperature, and better sea conditions, sharks choose the energetical richer prey, the seal-shaped decoy, instead of the tuna bait. This study confirms the importance of visual ability in mature and immature white sharks with dietary shifts in their feeding pattern.

Abstract: The influence of environmental factors on prey discrimination of bait-attracted white sharks was studied over a six-year period (2008–2013) at Dyer Island Nature Reserve (Gansbaai, South Africa). Across 240 bait-attracted feeding events observed in this period, both immature and mature white sharks were attracted by the seal-shaped decoy rather than the tuna bait, except for the years 2008 and 2011. Tide ranges, underwater visibility, water temperature, and sea conditions were, in decreasing order, the factors which drove white sharks to select the seal-shaped decoy. High tide lowered the minimum depth from which sharks could approach seals close to the shore, while extended visibility helped the sharks in making predatory choices towards the more energy-rich prey source, the odorless seal-shaped decoy. On the contrary, warmer water is associated with an increase in phytoplankton that reduces underwater visibility and increases the diversity of teleosts including tuna—a known prey of white sharks—driving the sharks to favor the tuna bait. Overall, sea conditions were almost always slightly rough, ensuring a good average underwater visibility. Recommendations for future research work at this site are presented.

Keywords: elasmobranch; environmental effects; predatory behavior; prey choice; sensory ecology

Citation: Reinero, F.R.; Sperone, E.; Giglio, G.; Pacifico, A.; Mahrer, M.; Micarelli, P. Influence of Environmental Factors on Prey Discrimination of Bait-Attracted White Sharks from Gansbaai, South Africa. *Animals* **2022**, *12*, 3276. https://doi.org/10.3390/ani12233276

Academic Editor: Martina Francesca Marongiu

Received: 27 October 2022
Accepted: 22 November 2022
Published: 24 November 2022

1. Introduction

It is relatively simple to study the white shark (WS) *Carcharodon carcharias* (Linnaeus, 1758) predatory behavior, thanks to the ease with which it can be observed from the surface attacking and feeding on pinniped colonies close to rocky islands, where white sharks congregate [1]. Predator–prey interactions can be influenced by environmental factors and depend on the sensory capabilities of a predator and its prey [2]. Accordingly, these factors can drive the former's ability to detect the prey, the latter's ability to avoid the attack, behavioral patterns, and predator activity peaks within marine ecosystems [3]. Studying environmental factors that can affect predator–prey interactions and predation frequencies is also important, considering the populations of many predators are declining globally [4]. Although several works describe the surface predatory behavior [5–9] and the passive prey discrimination of WSs when bait-attracted [10–12], studies pertaining to the influence of environmental factors on predator–prey interactions are rare [2,13] and have never specifically focused on bait attraction.

Abiotic factors, such as tide ranges, light levels, underwater visibility, water temperature, and sea conditions, are thought to influence the sharks' predatory behavior directly through their effects on physiology and sensory ecology, limiting or improving their access to prey [14].

Vision plays an important role in feeding patterns [10,15]. Under normal conditions, seals' visual acuity can detect WSs from the surface. However, light levels could help sharks to hide from their prey when hunting from the bottom, giving them the possibility of ambushing from the depths to the surface under low light conditions [5,12,16]. Furthermore, Huveeners et al. [17] stated that, at Neptune Island (Australia), sharks hunt near the surface swimming with the sun directly behind them, which hides them in shadow while illuminating the prey ahead. Follows et al. [13], at Seal Island (South Africa), stated that out of 1.476 recorded natural predation events, the white shark attack frequencies and seal capture success rates were higher during periods of high lunar illumination, when seal silhouettes along the surface were easier to discern.

At the same time, olfactory bulbs of WSs comprise 18% of their total brain mass, suggesting the importance of this sense for feeding activity [18]. Winds, which have a direct effect on sea conditions, can propel chemical stimuli, such as blood, excreta, or other organic fractions, that can enable sharks to locate their prey [2]. In addition, northerly winds that predominate during the South African winter storms force seals to return to Seal Island, swimming against the waves and current and producing sounds that aid in prey detection [2].

Tidal height can also influence predation success, since high tide reduces both the minimum depth from which sharks can approach seals and their haul-out area by forcing pinnipeds to go into the water, increasing white shark prey availability [2].

Water temperature, on the other hand, is considered one of the most important abiotic environmental factors that influences the distribution, movements, foraging, and reproductive strategies of sharks, playing a significant role in determining seasonal prey distribution and abundance [14]. Warmer water, which can result in blooms of diatoms, for example, is associated with an increase in abundance and diversity of teleosts and chondrichthyans that are in turn WSs prey [14]. Since it is widely known that WSs are endothermic and able to regulate their internal body temperature, maintaining metabolic rates though specialized circulatory mechanisms [19], the abiotic factor of water temperature for this species has a greater influence on prey distribution and abundance rather than on a shark's physiological water preference.

The present paper looks more deeply into the previous work of Micarelli et al. [10] at Gansbaai (South Africa), where the authors investigated the surface predatory behavior and the passive prey discrimination of bait-attracted WSs between 2008 and 2013. They observed 240 predator–prey interactions and highlighted that both immature and mature WSs were attracted to the seal-shaped decoy rather than the tuna bait, except during the years 2008 and 2011. Authors ran a non-parametric statistical test, Cochran's Q test, displaying that the sharks' predatory choice was not random (rejecting the null) and citing a difference in the effectiveness of sharks' prey preferences. This finding was likely due to the dietary shift that occurs in WSs 200–340 cm in total length (TL), as well as the sharks' ability to visually locate and select the highest energy prey source. The latter point was also confirmed in immature sharks.

The aim of this study was to assess if, between 2008 and 2013, environmental factors, such as tide ranges, light levels, underwater visibility, water temperature, and sea conditions, could have influenced the choice of passive prey (tuna bait and seal-shaped decoy) by bait-attracted WSs from Gansbaai, and, in particular: (i) prove that the sharks' prey preferences are not random but affected by potential not directly observed and measured environmental factors varying over the time; (ii) highlight the main environmental factors causing the WSs to prefer the seal-shaped decoy instead of the tuna bait; and (iii) evaluate major environmental factors affecting predatory behavior. These aspects aim to confirm the influence of environmental factors on the feeding behavior of WSs.

2. Materials and Methods

In Gansbaai (South Africa), at the Dyer Island Nature Reserve, a large WSs population with a prevalence of immature individuals is present and can be observed thanks to the support of local ecotourism operators authorized to reach the field observation sites [1]. Dyer Island Nature Reserve is located 7.5 km south-east of Gansbaai (34°41′ S; 19°24′ E) and includes two islands: Dyer Island, a low-profile island ca. 1.5 km long and 0.5 km wide, and Geyser Rock, a small islet ca. 0.5 km long and 180 m wide, characterized by different seabird colonies and a colony of Cape fur seals *Arctocephalus pusillus pusillus* (Schreber, 1775), respectively, (Figure 1).

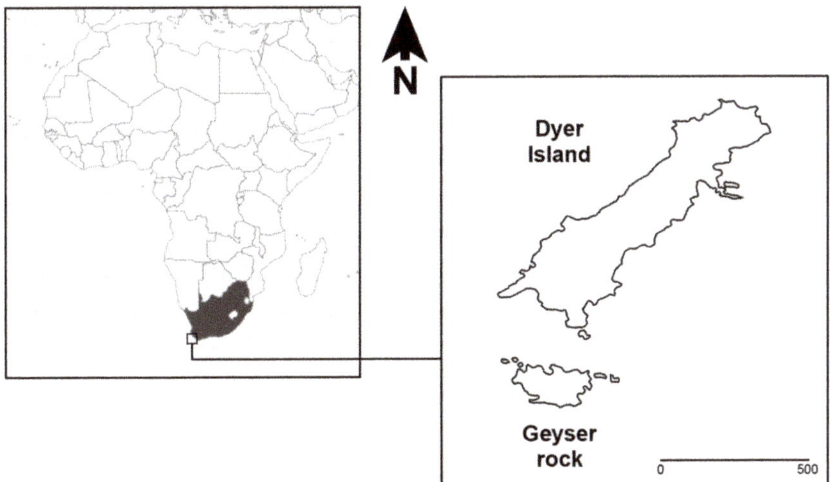

Figure 1. Dyer Island and Geyser Rock Nature Reserve.

The methodology follows and goes further than the one used by Micarelli et al. [10]. The study period, from 2008 to 2013 in the autumn season between March and May, required a total of 247 h of effort (on average 41 h/y). Environmental observations and data collection were performed from the 13 m "Barracuda" boat owned by the "Unlimited Shark diving", anchored at 100–150 m off Dyer Island. An anti-shark cage was fixed to the side of the boat for the entire duration of the observations. To attract sharks to the boats, olfactory stimulants (chum) composed of sea water, cod liver oil (*Gadus* sp.), tuna blood, and small pieces of fish were used, according to the methods described by Laroche et al. [20], Ferreira and Ferreira [21], Strong et al. [22], and Sperone et al. [23].

Passive target preys chosen to attract sharks for passive prey discrimination and predatory observations were an odorless seal-shaped decoy and odorous buoy floating baits of tuna pieces (tuna bait), both of similar size (70 cm long and 32 cm wide juvenile Cape fur seal decoy and 60 cm in diameter floating tuna baits), in line with the research protocol adopted by Sperone et al. [10,15]. Both passive target preys were positioned at the bow 10 m from each other with the aim of testing that the odorless seal-shaped decoy remained isolated from the odorous tuna bait.

Sharks were individually identified, and environmental data were collected by the same Sharks Studies Center—Scientific Institute (Massa Marittima, Italy) research team for the duration of the study. Sometimes a specimen was observed several times throughout the day and all exhibited behaviors were recorded. Sharks' TL and sex were estimated according to the known size of the cage length and the pelvic fin area observations (males if claspers were seen; females if the lack of claspers was verified and their pelvic fin area was filmed from the cage), respectively [1]. All other specimens were categorized as unknown sex. White sharks' size at sexual maturity was estimated, according to Hewitt et al. [24] and Micarelli et al. [1]: TL $\geq$ 450 cm for mature females and TL $\geq$ 350 cm for mature males. Sex was estimated by a maximum of three observers at a time from the cage, while both passive prey discrimination of bait-attracted WSs and TL were estimated by the same operator from the lower or upper deck of the boat.

During the observations, the following environmental factors were recorded by the same operator every three hours from arrival at the sampling area until the end of daily observations:

(1) Sea conditions were divided according to the Douglas scale of wave height [25]:

    (a) "Calm", which included glassy and rippled (0–10 cm wave height).
    (b) "Slightly rough", which included smooth, slight, and moderate (11–250 cm wave height).
    (c) "Rough", which included rough, very rough, high, very high, and phenomenal (>250 cm wave height).

(2) Light levels were expressed in oktas, a unit of measurement that indicates the cloudiness of the sky, estimated in terms of how many eighths of it are obscured by clouds [26]. Measurement intervals used to assess the sky coverage were as follows:

    (a) 0–2 oktas corresponded to clear sky.
    (b) 3–5 oktas corresponded to partly cloudy sky.
    (c) 6–8 oktas corresponded to a totally covered sky.

(3) Tide ranges were divided into (a) low and (b) high classifications, obtained from the Windguru Database (www.windguru.cz, accessed on 1 June 2015) for the "Kleinbaai, Marine Dynamics" area.
(4) Water temperature, expressed in degrees Celsius (°C), was obtained by the boat's instruments.
(5) Underwater visibility, expressed in meters (m), was calculated by operator with the Secchi disc.

The empirical methodology aims to prove that: (i) both passive preys (seal-shaped decoy and tuna bait) do not show spatial autocorrelation; (ii) and the sharks' passive prey discrimination is affected by potential environmental factors. (i) To measure the overall spatial autocorrelation between passive preys, Moran's test has been assessed. In our study, the spatial autocorrelation is multi-dimensional—accounting for multiple data objects in the analysis—and then we need to test if the observations are not independent. More precisely, according to Moran's test, we are able to measure how one object is similar to the others surrounding it. The possible results are equal to: -1, indicating perfect clustering of dissimilar values (perfect dispersion); 0, denoting no autocorrelation (perfect randomness); and $+1$, indicating perfect clustering of similar values (no perfect dispersion). Thus, we construct a hypothesis testing where the null is in the data that are randomly distributed, while the alternative is that the data are more spatially clustered. Calculations for Moran's test are based on a weighted matrix, with units i and j, where $i \neq j$. Similarities between units are computed as the product of the differences between y_i and y_j, with y_l denoting the variable of interest and $l = i, j$. In this context, y_l denotes the tuna bait and the seal-shaped decoy for all different units. The formula to compute the test statistic is:

$$z_{test} = \frac{1}{s^2} \times \frac{\sum_i \sum_j (y_i - \bar{y})(y_j - \bar{y})}{\sum_i \sum_j w_{ij}}$$

where $s^2 = \frac{1}{n}\sum_i (y_i - \bar{y})^2$ denotes the sample variance, w_{ij} the weighted matrix, and z_{test} refers to the test statistic approximatively distributed as a normal standard with a sufficiently large sample ($n \geq 30$). (ii) In this study, five main factors assessed during 240 interactions, spanning the period 2008–2013 are considered. They consist of two not directly observed variables and three not directly measured factors split, respectively, into: (i) light levels, measured in oktas; (ii) water temperature, measured in degrees Celsius (°C); (iii) underwater visibility, measured in meters (m); (iv) sea conditions, classified as calm, slightly rough, and rough; and (v) tide, classified as high or low.

The last two elements have been evaluated through proxy discrete variables. More precisely, an ordinal variable was used to assess sea conditions, assuming values 1 (whether the sea is calm), 2 (whether the sea is slightly rough), and 3 (whether the sea is rough). Tide was computed through a dummy variable equal to 1 (low tide) or 0 (high tide).

The water temperature has been grouped in three classes to analytically evaluate its effects on the prey preference and potential interactions with the other covariates. The related discrete variable was then equal to 1 for the class 11.0–13.9 °C; 2 for the class 14.0–16.9 °C; and 3 for the class 17.0–19.9 °C.

The variable of interest denotes sharks' prey preferences and was measured by means of a dummy variable of either 0, where the sharks preferred the bait, or 1, where the choice corresponded to the seal-shaped decoy.

The statistical methodology consists of a three-step approach called the Three-Step System Multivariate Classification (TSMC). This method combines a first-step non-parametric statistical test for matching heterogeneous groups of variables with a supervised machine learning (ML) technique for consistently estimating all parameters of interest in a multivariate non-linear context.

The first step evaluates a permutational multivariate analysis of variance (PERMANOVA) that can: (i) evaluate potential heterogeneity among units; (ii) investigate heterogeneous effects among factors and outcomes (prey preferences); and (iii) deals with potential endogeneity issues. Let X_{ik} be a $N * K$ matrix, with $i = 1, \ldots, N$ and $k = 1, \ldots, K$ denoting units and factors, respectively. We estimate a PERMANOVA between the elements 'years' ($t = 1, \ldots, T$) and 'effects' (k) across each sample-observed unit. The null hypothesis stands for homogeneity among groups (the dispersion is equivalent for all groups), while a rejection of the null highlights that the spread of any two variables investigated in the experiment is different between the groups. PERMANOVA was estimated between the years and environmental effects across each sample-observed unit. The variables are labelled as so: (i) 'oktas' denoting light levels; (ii) 'sea' referring to sea conditions; (iii) 'tide' referring to tide ranges; (iv) 'visibility' denoting underwater visibility; and (v) 'temp.' describing water temperature grouped across classes. The variable of interest is defined as 'prey' and accounts for sharks' prey preferences (tuna bait or seal-shaped decoy). The main assumptions held are: (i) variables in the dataset are exchangeable under the null; (ii) exchangeable variables, such as sites, observations, and factors, are independent; and (iii) exchangeable variables have similar multivariate dispersion (each unit of observation has a similar degree of multivariate scatter).

Given that every environmental factor observed over time significantly affects the outcome, the second step estimates a logit regression analysis to predict the categorical dependent variable (prey preferences) through a set of covariates. It is one of the most popular supervised ML algorithms because it can provide probabilities and classify non-homogeneous data using continuous and discrete datasets (as in this study). Let P_i be the probability that $y_i = 1$ (preference about seal shape), conditional to the information set Ω_i, which contains exogenous and quantitative variables, a model for binary response that serves to model this conditional probability. Notably, P_i corresponds to the binary response model thought of as modelling the conditional expectation and can be expressed as:

$$P_i \equiv \Pr(y_i = 1 | \Omega_i) = E(y_i | \Omega_i) \tag{1}$$

In that context, we cannot use the linear model because a linear regression on X fails to impose the condition $0 \leq E(y_i | \Omega_i) \leq 1$. To solve (1), we need to impose the constraint by a proper functional form equal to:

$$P_i \equiv \Pr(y_i | \Omega_i) = F(X_i \beta) \tag{2}$$

where X_i is the matrix defined in the first step—stacked for k—containing all the covariates, the $K * 1$ vector β_k is an unknown parameter estimated to predict y_i, and $F()$ is a logistic function with the form:

$$P_i \equiv \frac{e^{X_i\beta}}{1 + e^{X_i\beta}} \tag{3}$$

The logit model can be written as:

$$log\left(\frac{P_i}{1 - P_i}\right) = X_i\beta \tag{4}$$

which means that the logarithm of the odd ratio is equal to $X_i\beta$.

The last step consists of computing the sample's marginal effects. The main aim is to assess, in order of importance, the major predictors affecting the prey preferences across units over time. Let the logit model be expressed by Equation (4), changes in the values of X_i affecting $E(y_i|\Omega_i)$ in a non-linear fashion can be computed by:

$$\frac{\partial P_i}{\partial x_{ik}} = \frac{\partial F(X_i\beta)}{\partial x_{ik}} = f(X_i\beta) * \beta_k \tag{5}$$

For most transformation functions used, $f(X_i\beta)$ reaches its maximum at $X_i\beta = 0$, meaning that the effect of a change in x'_{ik} on P_i is at maximum when $P_i = 0.5$ and reduced when P_i is close to 0 or 1. Given (potential) unobserved interactions among factors, a White's correction for the presence of heteroskedasticity in the calculation of marginal effects standard errors is applied, obtaining larger significance and lower AIC (Akaike information criterion (AIC) is usually used to match similar regression functions and choose the best one).

3. Results

3.1. Moran's Test

The Moran's test statistic obtained was equal to -3.93 and, in absolute value with two-side alternative hypothesis, was bigger than the critical value $z_{\frac{\alpha}{2}} = z_{0.025} = 1.96$, with $\alpha = 5\%$ as default. Thus, the hypothesis testing is significant and, letting the statistic be negative, denotes perfect dispersion between the tuna bait and the seal-shaped decoy.

3.2. PERMANOVA Test

The first step, the PERMANOVA Test (Table 1), proves that the sharks' prey preferences are not random but are affected by potential that is not directly observed and measured environmental factors varying over time.

Table 1. PERMANOVA analysis between the elements 'years' and 'effects' across units. Here, the labels stand for 'degrees of freedom' (df), 'sum of squared dissimilarities' (SS), 'Pseudo F statistic' (Pseudo-F), and 'associated p-value' (Pr ($>$F)). The significance levels are: (**) significance at 5%; and (***) significance at 1%.

Source	df	SS	Pseudo-F	Pr ($>$F)
Prey	1	9.8800×10^{-7}	73.2246	0.001 ***
Oktas	1	4.2400×10^{-7}	4.8126	0.031 **
Sea	1	3.6560×10^{-6}	61.5374	0.001 ***
Tide	1	2.8770×10^{-6}	82.6926	0.001 ***
Visibility	1	4.0000×10^{-8}	45.3016	0.500
Temp.	1	7.4930×10^{-6}	77.1439	0.001 ***
Residual	233	1.5005×10^{-7}		
Total	239	3.6743×10^{-7}	1	

Table 1 displays the estimation outputs, where the significance level is $\alpha = 5\%$ (as default). Almost all predictors are significant (except 'visibility'), displaying a p-value close to zero and lower than α, and the residuals are null, highlighting the robustness and validity of the results. Looking closely at the estimates, two findings should be dealt with carefully: (i) according to the predictor 'visibility', it would not be significant across units over time even if its related sum of squared dissimilarities and Pseudo-F statistic were null and highly large, respectively; (ii) the predictor 'oktas' displays, on average, larger p-values and lower Pseudo-F than the other significant ones. These findings highlight that multicollinearity problems would matter. For instance, the factor 'visibility' could be strongly correlated with one or more predictors within the model, rejecting the assumption in (ii) (exchangeable variables, such as sites, observations, and factors, are independent). To deal with this, we ran a new PERMANOVA test dropping 'oktas', the factor displaying lower Pseudo F statistic, and obtained the expected result (Table 2). The variable 'visibility' becomes significant with lower SS and higher Pseudo-F, compared to before. According to the other predictors, the estimated results improve as well. Thus, the estimates are now valid and unbiased.

Table 2. PERMANOVA analysis between the elements 'years' and 'effects' across units, dealing with multicollinearity problems. Here, the labels stand for 'degrees of freedom' (df), 'sum of squared dissimilarities' (SS), 'Pseudo F statistic' (Pseudo-F), and 'associated p-value' (Pr (>F)). The significance level is: (***) significance at 1%.

Source	df	SS	Pseudo-F	Pr (>F)
Prey	1	7.4628×10^{-8}	84.1295	0.001 ***
Sea	1	4.3409×10^{-7}	65.3846	0.001 ***
Tide	1	4.1240×10^{-7}	87.9824	0.000 ***
Visibility	1	5.0832×10^{-8}	73.3625	0.000 ***
Temperature	1	8.1254×10^{-7}	81.9472	0.000 ***
Residual	233	3.8756×10^{-7}		
Total	238	4.9857×10^{-7}	1	

The five main considerations are, in order: (i) there is a relevant heterogeneity among environmental effects. Thus, they are time-varying and change significantly year by year; (ii) heterogeneity also matters across units, and thus these effects heterogeneously affect sharks' prey preferences over time; (iii) non-linear relationships between predictors and outcomes need to be quantified through appropriate econometric algorithms; (iv) the environmental factors all have a strong effect on WSs' prey preferences; (v) potential environmental problems can be addressed in a particular time. For instance, we are interested to investigate why WSs would prefer the tuna bait rather than the seal-shaped decoy in 2008 and 2011. The last two points are addressed in the second and third step, respectively.

3.3. LOGIT Regression Model

The second step, the LOGIT regression model (Table 3), highlights the main environmental factors causing the WSs to prefer the seal-shaped decoy instead of the tuna bait.

Table 3 displays the estimation outputs, where the significance level is $\alpha = 5\%$ (as a default). The five main findings are, in order: (i) all the results displayed in the first step are significant; (ii) all predictors, except 'temperature', positively affect the variable of interest 'prey'. For instance, better sea conditions increase the seal-shaped decoy preferences across sightings. The same accounts for high tide and better underwater visibility. On the contrary, higher water temperature increases the tuna bait preferences across sightings; (iii) according to the magnitude of the model, the factor that most affects the outcome is 'tide' (2.10), followed by 'visibility' (0.83), 'temperature' (−0.66), and 'sea' (0.65). The constant has not been considered, displaying an effect on the outcome, with all predictors being fixed ($X_i = 0$); (iv) 'visibility' and 'oktas' predictors are strongly negatively correlated (−43%) and show a significant linear relationship (Figure 2). Thus, by dropping 'oktas' within the system, any (potential) collinearity problems are dealt with, and we obtain better predicted outcomes; (v) the variable 'year' is also significant and positive, highlighting the presence of relevant endogeneity issues affecting sharks' predatory behavior and heterogeneous effects given the unexpected environmental effects.

Table 3. Logit regression model accounting for all the five environmental factors across units over time. Here, 'Coefficients' refers to the factors within the model; 'Estimate' denotes b_k (the estimated β_k); 'SE' stands for standard error; 'z-value' denotes the test statistic obtained for each predictor; and 'Pr ($>|z|$)' refers to the associated p-value in a two-sided hypothesis test (where the null accounts for non-significance). The significance levels are: (**) significance at 5%; and (***) significance at 1%.

Coefficients	Estimate	SE	z-Value	Pr ($>|z|$)
Costant	−1049.7508	269.6977	−3.892	9.93×10^{-5} ***
Sea	0.6342	0.3426	2.040	0.041335 **
Tide	2.0951	0.3536	5.925	3.12×10^{-9} ***
Visibility	0.8361	0.2743	3.048	0.002303 ***
Temperature	−0.6630	0.3250	−1.910	0.041689 **
Year	0.5203	0.1343	3.874	0.000107 ***

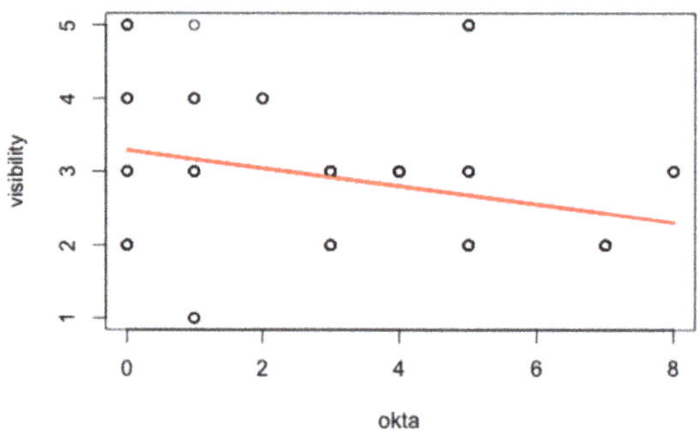

Figure 2. Scatter plot by regressing 'okta' (independent variable) on 'visibility' (dependent variable). The estimation output is negative and significant, displaying a consistent relationship between the variables. The scatter plot shows a large component of outliers affecting the output due to omitted variables and unobserved heterogeneity (endogeneity issues) dealt with the logistic function. The significance level used in the analysis is $\alpha = 5\%$ (as default).

3.4. Sample Marginal Effects

The third and final step, the sample marginal effects (Table 4), evaluates major environmental factors affecting feeding behavior.

Table 4. Sample marginal effects for each observation unit, given n observations, are accounted for. Here, 'Coefficients' refers to the factors within the model; 'dF/dx' denotes the partial derivatives displaying the marginal effects of the predictors (x'_{ik}) on y_i ('prey'); 'SE' stands for standard error; 'z-value' denotes the test statistic obtained for each predictor; and 'Pr ($>|z|$)' refers to the associated p-value in a two-sided hypothesis test (where the null accounts for non-significance). The significance level is: (***) significance at 1%.

Coefficients	dF/dx	SE	z-Value	Pr ($>|z|$)
Sea	0.492973	0.034583	3.5558	0.0003768 ***
Tide	0.610744	0.058782	6.9875	2.798×10^{-12} ***
Visibility	0.557756	0.035013	4.5056	6.618×10^{-6} ***
Temperature	−0.512029	0.037561	−2.9826	0.0028581 ***
Year	0.157797	0.020775	4.2261	2.377×10^{-5} ***

Table 4 displays the estimation outputs, where the significance level is $\alpha = 5\%$ (as default). Three relevant findings are addressed: (i) the main factors affecting the outcome correspond to the ones found in the logistic function (second step). More precisely, in order of importance, we have: (1) 'tide', (2) 'visibility', (3) 'temperature', and (4) 'sea'; (ii) the marginal effect values appear sensible. For instance, a one-unit change associated with an observation increases the probability of preferring the seal-shaped decoy by 61% with high 'tide', 55% with better 'visibility', 51% with less water temperature ('temp.'), and 49% with better 'sea' conditions; (iii) the factor 'year', though it affects the outcome with lower associated probability, is significant and must be included in the analysis when studying environmental problems.

Finally, we look more deeply into these sample marginal effects, referring to the average individual effects to investigate the same effects in every period (Table 5). We focus on 2008 and 2011, representing the time periods in which white sharks preferred the tuna bait rather than the seal-shaped decoy [10]. In 2011, according to the total sightings, tide events were exclusively characterized by low tide and the highest water temperature was registered. The same results were found in 2013, with two key differences: (i) in 2011, the total sightings were more than double (63 vs. 29), highlighting the highest sample marginal effects according to the predictor 'tide'; (ii) the water temperature (the third most significant outcome-affecting factor) was, on average, balanced in 2013 compared to the other time periods, whereas in 2011 the temperature held at 18 °C. In 2008, the most sightings were characterized by low tide (25 on 40). Similar effects were observed in 2009, showing the same visibility and water temperature. However, in 2009, tide ranges were similar (12 observations occurred during low tide and 10 during high tide out of 22 total sightings), showing that, once again, 'tide' denotes the most significant/important factor in terms of affecting sharks' predatory behavior.

Table 5. Summary and descriptive table for the average individual (total sightings) over time. Here, 'Tot. Sigh.' stands for total sightings, and 'W. Temp.', 'U. Visibility', and 'Sea Cond.' refer, in average $\pm$ SD, to water temperature (°C), underwater visibility (m), and sea conditions, respectively. The ranges would correspond to the minimum and maximum values displayed in the table.

Years	Tot Sigh.	Low Tide	High Tide	W. Temp. (Mean $\pm$ SD)	U. Visibility (Mean $\pm$ SD)	Sea Cond. (Mean $\pm$ SD)
2008	40	25	15	13.25 $\pm$ 1.83	2.55 $\pm$ 0.33	2.45 $\pm$ 0.35
2009	22	12	10	14 $\pm$ 1.08	2.55 $\pm$ 0.33	2.45 $\pm$ 0.035
2010	54	19	35	15 $\pm$ 0.08	2.83 $\pm$ 0.05	2.17 $\pm$ 0.07
2011	63	63	0	18 $\pm$ 2.93	2.79 $\pm$ 0.09	1.94 $\pm$ 0.16
2012	32	16	16	15.2 $\pm$ 0.12	3 $\pm$ 0.12	2 $\pm$ 0.1
2013	29	29	0	15 $\pm$ 0.08	3.55 $\pm$ 0.67	1.59 $\pm$ 0.51

4. Discussion

This study demonstrates that prey discrimination of bait-attracted WSs in Gansbaai is affected by environmental factors that drive the sharks to detect and choose prey depending on the conditions. Passive preys (tuna bait and seal-shaped decoy) do not show spatial autocorrelation, as demonstrated by the Moran's Test, confirming that the odor from the tuna bait did not serve as chum within the 10 m range of the experiment. In Gansbaai, as stated by Micarelli et al. [10] from 2008–2013, the majority of WSs transient population, immatures included, were attracted by the seal-shaped decoy rather than by the tuna bait, except in the two years of 2008 and 2011. As observed by the first PERMANOVA test (Table 1), 'visibility' was not significant and 'oktas' displayed larger p-values and lower Pseudo F than other variables due to a multicollinearity problem, since underwater visibility is strongly correlated (-43%) with the light levels, as stated by the linear regression model in Figure 2. For that reason, light levels were omitted in the second PERMANOVA test (Table 2). As suggested by Strong [12], one of these aspects is a consequence of the other: low light levels (represented by high oktas values) reduce the surface underwater visibility that disadvantages the seal's ability to detect the WS coming from below.

We observed a relevant heterogeneity among environmental effects that undergoes large changes year by year and affect sharks' prey choice over the time. According to the sample marginal effects, the factors influencing the prey discrimination and the choice of the seal-shaped decoy for a one-unit change associated with the observation (Table 4) are, in decreasing order, tide ranges (61%), underwater visibility (55%), water temperature (51%), and sea conditions (49%).

Tide range is the most important environmental factor that influences prey choice. At Seal Island in South Africa, Hammerschlag et al. [2] stated that attack frequency to Cape fur seals increased significantly during high tides within 400 m of the island and over a depth range of 5–31 m where seals swim over shallow reefs. During high tide, the minimum depth from which sharks could approach seals was closer to shore, increasing the availability of prey and the possibility of predation. Additionally, at the Farallon Islands in California, WSs' attacks increased with high tide that reduced the haul-out area for elephant seals, forcing them into the water and increasing predation events [11,16,27]. In fact, during high tide events, the number of Cape fur seals around our boat anchored at 100–150 m from Dyer Island increased, inducing sharks to use their visual acuity both to attack them and our seal-shaped decoy. For that reason, we propose that high tide could induce WSs from Gansbaai to turn their attention towards seal-shaped decoys rather than towards the tuna bait, consequently obtaining more caloric sources over low-energy ones, according to the theory of optimal foraging [28,29].

It is true that most of the years (2008, 2009, 2011, and 2013) were characterized by low tide, but it is also necessary to observe other environmental parameters that influence prey choice, such as underwater visibility and water temperature.

Underwater visibility is the second environmental factor that influences the sharks' discrimination of prey and, on average, was good during all the years (at a visibility range registered from 1 to 5 m during all the years, minimum average visibility was 2.55 m in 2008 and 2009; maximum average visibility was 3.55 m in 2013). Higher average visibility and, consequently, low oktas values seemed to facilitate the attack on the seal-shaped decoy in Gansbaai. Our findings do not support the Seal Island results of Hammerschlag et al. [2] that observed a higher predatory success of WSs on Cape fur seals during low light levels and low underwater visibility. This inconsistency could be linked to two reasons: (i) stalking and ambush with low visibility are important for successful prey capture in many fishes, as well as sharks, including *C. carcharias* [12,30–32]. However, a depth range of 26–30 m around Seal Island is optimal for sharks to remain undetected and stalk seals from below with low visibility and enough vertical distance to build the perfect moment for attack and surface strike, as observed by Hammerschlag et al. [2]; on the other hand, around the Dyer Island Nature Reserve and in the Shark Alley channel that divides Dyer Island from Geyser Rock, the maximum depth is ca. 8–10 m and the vertical distance is not enough to surprise seals by exploiting the low visibility. For that reason, most of the attacks to passive preys at Gansbaai were horizontal [15] and in the presence of a good average underwater visibility. Sometimes, as observed by Huveeners et al. [17] in Neptune Island (South Australia), sharks hunting near the surface orient the sun directly behind them during approach, directly illuminating prey and increasing their detectability. (ii) As demonstrated by Micarelli et al. [10], vision plays an important role in mature and immature sharks with a dietary shift in their feeding pattern, and a good average visibility could help them to more consistently secure higher-energy prey sources, such as our shaped decoy.

Water temperature is the third environmental factor that influences prey selection and plays an important role in the physiology of ectotherms, regulating their internal functions, distribution, foraging, reproduction, prey distribution, and abundance [14]. Unlike most sharks and rays, the white shark is an endothermic animal within the family Lamnidae, and water temperature has an influence on its prey distribution rather than on the shark's physiology [19]. Warmer water resulting in blooms of diatoms is associated with an increase in the number and diversity of teleosts, which are prey of white sharks [14]. Different tuna species, used as passive prey to attract white sharks, are present in South African waters and around Gansbaai, such as Albacore or Longfin tuna *Thunnus alalunga* (Bonnaterre, 1788), Yellowfin tuna *Thunnus albacares* (Bonnaterre, 1788), Big-Eye tuna *Thunnus obesus* (Lowe, 1839), and Skipjack tuna *Katswonus pelamis* (Linnaeus, 1758) [33]. All of these species, except the Albacore, are mostly found off of the Southern Cape Coast of South Africa from spring to summer (www.deep-sea-fishing.co.za, accessed on 15 April 2022) when waters are warmer; their higher availability and abundance during these seasons and during our observation period in which the water was warmer likely influenced the tuna bait choice. Although WSs generally prefer to feed on juvenile Cape fur seals that leave Geyser Rock to forage offshore for the first time during the autumn and winter season, warmer water induces phytoplankton blooms that reduce the underwater visibility, hindering the shark's ability to detect the odorless seal-shaped decoy and encouraging the tuna bait choice.

According to what has been said, warmer water temperatures registered in 2011 (18 °C on average) were associated with only low tide events (63), one of the lowest average underwater visibilities (2.79 m), and the highest number of interactions (63) that could have influenced the choice of the tuna bait. Additionally, in 2008 sharks preferred the tuna bait, even if the average water temperatures were the lowest registered (13.25 °C), but most of the interactions were characterized by low tide events (25 out of 40 total interactions) and by the lowest average underwater visibility (2.55 m), forcing the sharks to choose the tuna bait. During the other years (2009, 2010, 2012, and 2013) sharks preferred the seal-shaped decoy. In 2009, tide ranges were similar between high and low (10 vs. 12, respectively). The average water temperature in 2009 was also slightly higher than 2008 (14 °C vs. 13.25 °C, respectively) despite having the lowest average underwater visibility (2.55 m), justifying the choice, since both total sightings were the lowest registered during the entire study (22) and tide range remained the main influencing factor. In 2010, mostly high tide events were recorded (35 out of 54 total interactions) with a cool average water temperature (15 °C) and good average underwater visibility (2.83 m), justifying the seal-shaped decoy choice. In 2012, the situation was like 2009, since the number of low and high tide events were identical (16); the average water temperature was cool (15.2 °C), the underwater average visibility was good (3 m), and total sightings were few (32). In 2013, high tide events were not observed, the average water temperature was cool (15 °C) and total sightings were few (29), but average underwater visibility was the best out of all the years (3.55 m), probably justifying the seal-shaped decoy choice.

Ultimately, sea condition is the variable that affects prey selection the least. In any case, sea condition was, on average, slightly rough during all the years, except for 2013 and 2011 where it was calm. Wind intensity has a direct effect on the creation of rough conditions, and, therefore, could propel pinnipeds' chemical stimuli (e.g., excreta, blood, organic compounds) from the Dyer Island Nature Reserve to the open sea, helping the sharks to locate the area where the seals are concentrated, as observed in the study of Strong [12] and Hammerschlag et al. [2] for Seal Island. At the same time, pinnipeds are forced to swim against the current with rough sea, producing sounds and losing their ability to maintain subsurface vigilance, which gives the sharks greater possibilities to strike [12]. However, it is also true that we used an odorless seal-shaped decoy to attract the sharks. A calm or slightly rough sea also has a direct effect on underwater visibility that, in our study, is the second most important environmental factor to influence prey discrimination, helping the sharks in Gansbaai to be able to visually detect the seal decoy more successfully. Notably, 2011 saw an average calm sea condition (1.94), but the other environmental factors, such as tide ranges events, underwater visibility, and water temperature, had a greater influence on the choice of the prey, which, in this case, was tuna. On the other hand, we registered the roughest average sea condition in 2008 (2.45 in terms of mean), alongside the lowest underwater visibility, the lowest water temperature, and a higher number of low tide events, justifying the tuna bait selection.

5. Conclusions

Environmental factors influence the WSs' sensory ecology to choose between passive prey types during interactions. The choice of the seal-shaped decoy is strictly related to high tide events, good underwater visibility, cooler water temperature, and calm or slightly rough sea. The last three conditions highlight the importance of the mature and immature WSs' visual ability in detecting the prey. WSs are initially attracted by the olfactory trace emitted by the tuna bait, but they subsequently shift their focus to the more calorically rich seal prey, helped by good environmental conditions. High tide, on the other hand, gives the sharks more chances to get closer to the island and reduces the prey haul-out area, increasing the possibility of predation on seals and the seal decoys alike. Some environmental factors were not taken into consideration during the study, such as wind direction, currents, and distance from the island, and could, therefore, be considered as topics of future inquiry at this site.

Author Contributions: Conceptualization, F.R.R., P.M. and E.S.; methodology, F.R.R., P.M. and A.P.; software, G.G. and A.P.; validation, P.M., F.R.R., E.S., A.P. and M.M.; formal analysis, A.P., P.M. and F.R.R.; investigation, P.M., F.R.R. and E.S.; data curation, F.R.R., P.M. and A.P.; writing—original draft preparation, F.R.R., M.M. and P.M.; writing—review and editing, F.R.R., P.M., E.S., A.P., G.G. and M.M.; funding acquisition, P.M. and E.S. All authors have read and agreed to the published version of the manuscript.

Funding: This research received no external funding.

Institutional Review Board Statement: The study and experimentation protocols were reviewed and approved in accordance 166 with the Directive 2010/63/EU.

Informed Consent Statement: Not applicable.

Data Availability Statement: https://www.researchgate.net/project/Great-White-Shark-Carcharodon-carcharias-Behaviour-Ecology-and-cotoxicology/update/61312b952897145fbd6df39c accessed on 10 April 2022.

Acknowledgments: We are grateful to the CSS team members that carried out 11 expeditions for their indirect financial support of this research, and thanks are also due to Shark Diving Unlimited for the logistical assistance and all other field assistance with data collection. All research protocols were approved by the University of Calabria and Siena.

Conflicts of Interest: The authors declare no conflict of interest.

References

1. Micarelli, P.; Bonsignori, D.; Compagno, L.J.V.; Pacifico, A.; Romano, C.; Reinero, F.R. Analysis of sightings of white sharks in Gansbaai (South Africa). *Eur. Zool. J.* **2021**, *88*, 363–374. [CrossRef]
2. Hammerschlag, N.; Martin, R.A.; Fallows, C. Effects of environmental conditions on predatory-prey interactions between white sharks (Carcharodon carcharias) and Cape fur seals (Arctocephalus pusillus pusillus) at Seal Island, South Africa. *Environ. Biol. Fishes* **2006**, *76*, 341–350. [CrossRef]
3. Heithaus, M.R. Predator-prey interactions. In *Biology of Sharks and Their Relatives*, 1st ed.; CRC Press: Boca Raton, FL, USA, 2004; pp. 488–512.
4. Estes, J.A.; Terborgh, J.; Brashares, J.S.; Power, M.E.; Berger, J.; Bond, W.J.; Carpenter, S.R.; Essington, T.E.; Holt, R.D.; Jackson, J.B.; et al. Trophic downgrading of planet earth. *Science* **2011**, *333*, 301–306. [CrossRef] [PubMed]
5. Martin, R.A.; Hammerschlag, N.; Collier, R.S.; Fallows, C. Predatory behaviour of White Sharks (Carcharodon carcharias) at Seal Island, South Africa. *J. Mar. Biol. Ass. UK* **2005**, *85*, 1121–1135. [CrossRef]
6. Martin, R.A.; Rossmo, D.K.; Hammerschlag, N. Hunting patterns and geographic profiling of white shark predation. *J. Zool.* **2009**, *279*, 111–118. [CrossRef]
7. Klimley, A.P.; Pyle, P.; Anderson, S.D. The behavior of white sharks and their pinniped prey during predatory attacks. In *Great White Shark: The Biology of Carcharodon Charcharias*; Klimley, A.P., Ainley, D.G., Eds.; Academic Press: San Diego, CA, USA, 1996; pp. 175–192.
8. Klimley, A.P.; Le Boeuf, B.J.; Cantara, K.M.; Richert, J.E.; Davis, S.F.; Van Sommeran, S.; Kelly, J.T. The hunting strategy of white sharks (Carcharodon carcharias) near a seal colony. *Mar. Biol.* **2001**, *138*, 617–636. [CrossRef]
9. Klimley, A.P. The predatory behavior of the white shark. *Am. Sci.* **1994**, *82*, 122–133.

10. Micarelli, M.; Chieppa, F.; Pacifico, A.; Rabboni, E.; Reinero, F.R. Passive Prey Discrimination in Surface Predatory Behaviour of Bait-Attracted White Sharks from Gansbaai, South Africa. *Animals* **2021**, *11*, 2583. [CrossRef] [PubMed]

11. Anderson, S.D.; Henderson, R.P.; Pyle, P. White shark reaction to unbaited decoys. In *Great White Shark: The Biology of Carcharodon Carcharias*; Klimley, A.P., Ainley, D.G., Eds.; Academic Press: San Diego, CA, USA, 1996; pp. 223–228.

12. Strong, W.R. Shape discrimination and visual predatory tactics in white sharks. In *Great White Shark: The Biology of Carcharodon Carcharias*; Klimley, A.P., Ainley, D.G., Eds.; Academic Press: San Diego, CA, USA, 1996; pp. 229–240.

13. Fallows, C.; Fallows, M.; Hammerschlag, N. Effects of lunar phase on predator-prey interactions between white sharks (Carcharodon carcharias) and Cape fur seals (Arctocephalus pusillus pusillus). *Env. Biol. Fishes* **2016**, *99*, 805–812. [CrossRef]

14. Schlaff, A.M.; Heupel, M.R.; Simpfendorfer, C.A. Influence of environmental factors on shark and ray movement, behaviour and habitat use: A review. *Rev. Fish. Biol. Fish.* **2014**, *24*, 1089–1103. [CrossRef]

15. Sperone, E.; Micarelli, P.; Andreotti, S.; Spinetti, S.; Andreani, A.; Serena, F.; Brunelli, E.; Tripepi, S. Social interactions among bait-attracted white sharks at Dyer Island (South Africa). *Mar. Biol. Res.* **2012**, *6*, 408–414. [CrossRef]

16. Pyle, P.; Anderson, A.; Klimley, P.; Henderson, R.P. Environmental effects on white shark occurrence and behavior at the South Farallon Islands, California. In *Great White Shark: The Biology of Carcharodon Carcharias*; Klimley, A.P., Ainley, D.G., Eds.; Academic Press: San Diego, CA, USA, 1996; pp. 281–291.

17. Huveneers, C.; Holman, D.; Robbins, R.; Fox, A.; Endler, J.A.; Taylor, A.H. White sharks exploit the sun during predatory approaches. *Am. Nat.* **2015**, *185*, 562–570. [CrossRef] [PubMed]

18. Demski, L.S.; Northcutt, R.G. The brain and cranial nerves of the white shark: An evolutionary perspective. In *Great White Shark: The Biology of Carcharodon Carcharias*; Klimley, A.P., Ainley, D.G., Eds.; Academic Press: San Diego, CA, USA, 1996; pp. 121–130.

19. Bernal, D.; Carlson, J.K.; Goldman, K.J.; Lowe, C.J. Energetic, metabolism, and endothermy in sharks and rays. In *Biology of Sharks and Their Relatives*, 2nd ed.; CRC Press: Boca Raton, FL, USA, 2012; pp. 211–237.

20. Laroche, R.; Kock, A.A.; Dill, L.M.; Oosthuizen, W. Effects of provisioning ecotourism activity on the behaviour of white sharks, Carcharodon carcharias. *Mar. Ecol. Progr. Ser.* **2007**, *338*, 199–209. [CrossRef]

21. Ferreira, C.A.; Ferreiera, T.P. Population dynamics of white sharks in South Africa. In *Great White Shark: The Biology of Carcharodon Carcharias*; Klimley, A.P., Ainley, D.G., Eds.; Academic Press: San Diego, CA, USA, 1996; pp. 381–391.

22. Strong, W.R.; Murphy, R.C.; Bruce, B.D.; Nelson, D.R. Movements and associated observations of bait-attracted white sharks, Carcharodon carcharias. A preliminary report. *Aust. J. Mar. Freshw. Res.* **1992**, *43*, 13–20. [CrossRef]

23. Sperone, E.; Micarelli, P.; Andreotti, S.; Brandmayr, P.; Bernabò, I.; Brunelli, E.; Tripepi, S. Surface behaviour of bait-attracted white sharks at Dyer Island (South Africa). *Mar. Biol. Res.* **2012**, *8*, 982–991. [CrossRef]

24. Hewitt, A.M.; Kock, A.A.; Booth, A.J.; Griffiths, C.L. Trends in sightings and population structure of white sharks, Carcharodon carcharias, at Seal Island, False Bay, South Africa, and the emigration of subadult female sharks approaching maturity. *Environ. Biol. Fishes* **2017**, *101*, 39–54. [CrossRef]

25. Alfahmi, F.; Hakim, O.S.; Dewi, R.C.; Khaerima, A. Utilization of data mining classification tecniques to identify the effect of Madden-Julian Oscillation on increasing sea wave hight over East Java Waters. *Earth Environ. Sci.* **2019**, *399*, 012062.

26. Rees, W.G. *Physical Principles of Remote Sensing*, 3rd ed; Cambridge University Press: Cambridge, UK, 2001; pp. 67–70.

27. Anderson, S.D.; Klimley, A.P.; Pyle, P.; Henderson, R.P. Tidal Height and White Shark Predation at the Farallon Islands, California. In *Great White Shark: The Biology of Carcharodon Carcharias*; Klimley, A.P., Ainley, D.G., Eds.; Academic Press: San Diego, CA, USA, 1996; pp. 275–279.

28. Gerking, S.D. *Feeding Ecology of Fish*; Elsevier: Amsterdam, The Netherlands, 2014; p. 416.

29. Helfamn, G.; Collette, B.B.; Facey, D.E.; Bowen, B.W. *The Diversity of Fishes: Biology, Evolution, and Ecology*; John Wiley and Sons: Hoboken, NJ, USA, 2009; p. 736.

30. Ebert, D.A. Diet of the sevengill shark Notorynchus cepedianus in the temperate coastal waters of southern Africa. *S. Afr. J. Mar. Sci.* **1991**, *11*, 565–572. [CrossRef]

31. Tricas, T.C.; McCosker, J.E. Predator behaviour of the white shark (Carcharodon carcharias), with notes on its biology. *Proc. Calif. Acad. Sci.* **1984**, *43*, 221–238.

32. Goldman, K.J.; Anderson, S.D. Space utilization and swimming depth of white sharks, Carcharodon carcharias, at the South Farallon Islands, Central California. *Environ. Biol. Fishes* **1999**, *56*, 351–364. [CrossRef]

33. Van Der Elst, R. *A Guide to the Common Sea Fishes of South Africa*, 3rd ed.; Struik, C., Ed.; FAO: Cape Town, South Africa, 1997; pp. 57–72.

Article

Courtship and Reproduction of the Whitetip Reef Shark *Triaenodon obesus* (Carcharhiniformes: Carcharhinidae) in an Ex Situ Environment, with a Description of the Late Embryonic Developmental Stage

Sérgio Ricardo Santos [1,2,*], Veronica Takatsuka [1], Shayra P. Bonatelli [3], Nicole L. L. Amaral [3], Matheus F. Goés [1] and Rafael F. Valle [1]

[1] Instituto Museu Aquário Marinho do Rio de Janeiro—IMAM/AquaRio, Praça Muhammad Ali, Gambôa, Rio de Janeiro 20220-360, Brazil

[2] Laboratório de Biologia e Tecnologia Pesqueira—BioTecPesca, Universidade Federal do Rio de Janeiro (UFRJ), Av. Carlos Chagas Filho, 373—Sala A1-083, Cidade Universitária, Rio de Janeiro 21941-902, Brazil

[3] RADIUS—Centro Especializado em Diagnóstico por Imagem Veterinário, R. José Siqueira, 156—Sala 2, Dom Bosco, Itajaí 88307-310, Brazil

* Correspondence: srbs.ufrj@gmail.com

Citation: Santos, S.R.; Takatsuka, V.; Bonatelli, S.P.; Amaral, N.L.L.; Goés, M.F.; Valle, R.F. Courtship and Reproduction of the Whitetip Reef Shark *Triaenodon obesus* (Carcharhiniformes: Carcharhinidae) in an Ex Situ Environment, with a Description of the Late Embryonic Developmental Stage. *Animals* 2022, *12*, 3291. https://doi.org/10.3390/ani12233291

Academic Editor: Martina Francesca Marongiu

Received: 30 October 2022
Accepted: 22 November 2022
Published: 25 November 2022

Publisher's Note: MDPI stays neutral with regard to jurisdictional claims in published maps and institutional affiliations.

Simple Summary: The reproduction of key reef species is still largely unknown due to difficulties in documenting all elements and steps involved. Sharks are particularly affected by this scarcity of information due to being long-lived species, and witnessing courtship, gestation, and birth is still mostly limited to fortuitous encounters by divers or specimens captured by fishers. Still scarcely described in the literature, our study reports the successful reproduction of *Triaenodon obesus* in an ex situ environment, which offers an opportunity to observe all steps of the reproduction in detail. Furthermore, we offer the first description of the late embryonic developmental stage based on ultrasound imagery.

Abstract: Elasmobranchs represent a group of species under considerable anthropic pressure because of the scale of industrial and artisanal fisheries and the loss of essential areas for nursery and feeding, which are causing substantial population losses around the world. Reproduction in an ex situ environment enables a healthy population to be built and maintained in networks of public aquariums, increasing our knowledge of elasmobranch reproductive biology and offering the opportunity for reintroductions in areas where native populations have been removed. The study reports two successful pregnancies of the whitetip reef shark *Triaenodon obesus*, considered a vulnerable species by the International Union for the Conservation of Nature. Copulation and gestation data are provided, including ultrasound recordings of the late stage of embryo development. Ultrasonography was performed with the GE Logiq and convex transducer and revealed a fetus with defined fins and organogenesis, with definition of eyes, gills, liver, a heart with individualized chambers, partially defined kidneys, and a well-defined spiral intestine. A cartilaginous skeleton forming a posterior acoustic shadow was detailed, as well as a moving fetus with a biparietal diameter of 6.47 cm and a heart rate of 62 Beats Per Minute on spectral Doppler. This is the first successful reproduction of *T. obesus* in an aquarium in Brazil.

Keywords: ultrasound; elasmobranch; gestational; aquarium; conservation

1. Introduction

Declines in elasmobranch populations related to diverse and continuous anthropic impacts have been widely reported in the ecosystems of several parts of the world [1–4]. These losses represent only a fraction of the actual damage suffered by populations in these ecosystems, where elasmobranchs represent important components of the food web [5–8]. Fishery impacts are also underestimated, given the scarcity of data for most marine ecosystems, particularly in tropical waters and off the coasts of developing countries, with numerous unmonitored fishing communities associated with small-scale fisheries [9]. The growing popular appeal of the conservation of marine environments, the public demand to establish the extent of environmental changes, and the worsening of the conservation status of several flagship species of elasmobranchs have promoted the growth of sustainable management plans in the last decades. A compromise between governments and non-governmental organizations (NGOs) was established based on commitments assumed since the Rio 1992 convention [10,11], yet the environmental policies and the species protection efforts implemented in the following decades have still been unable to revert or mitigate the population declines of several species. Special interest has been devoted to the conservation of coral reefs and their inhabitants, considering the fragility of these ecosystems in the face of climate change and a wide variety of human impacts, with the whitetip reef shark *Triaenodon obesus* (Rüppell, 1837) being a known representative.

Despite being a coastal demersal species with a limited dispersal capacity and home range, *T. obesus* is the most widely distributed reef shark species in the world, found from the Red Sea to Micronesia, as well as from the coast of Mexico to Panama in the Eastern Pacific, and its presence is suspected even in the southwestern Atlantic [12–14]. This wide distribution, however, does not mean that the species is not threatened. The continuous pressure exerted by human impacts on the ecosystems they inhabit, whether through direct activities such as fishing [15] or through environmental degradation and loss of habitat [16,17], exposes populations of this species to conditions hostile to their survival. The situation is worsening in the context of climate change and its effects on marine ecosystems, within which coral reefs are especially impacted areas [16,18]. As a direct consequence of this situation, the decline seen in the last three generations of whitetip reef shark populations led to the classification of their conservation status as 'vulnerable' by the Red List of the International Union for Conservation of Nature (IUCN) [19].

Triaenodon obesus, despite its wide distribution, is assessed as vulnerable due to population decline and habitat loss in many areas of its range [20,21], but gaps in knowledge persist. The whitetip reef shark is a placentotrophic viviparous species, with a matrotrophic nutrition of the fetus [22,23]. That is, *T. obesus* is a live-bearing shark with the development of the embryo sustained by nutrients produced by the mother and delivered by a placental connection, a feature only identified within five families of Carcharhiniformes (Carcharhinidae, Sphyrnidae, Hemigaleidae, Leptochariidae, and Triakidae) [22]. Recently, ultrasound studies on elasmobranchs in captivity were published, focusing mainly on determining gestation and general anatomy [24–27]. However, there is a scarcity of studies focused on the description of the ultrasonographic aspects of fetal organs and their organogenesis in sharks. Since *T. obesus* is considered vulnerable, the detailed monitoring of pregnancy is of paramount importance for the preservation of the species.

While the high cost of field studies hinders the in situ observation of the species, alternative sources are available. The limited home range, average size of adult specimens, and ease of access and exchange among public aquariums have made the species a common sight in public exhibitions around the world [28]. The need for data that support the handling of *T. obesus* under human care and the knowledge about the biology of the species that can be acquired in public aquariums and research centers, especially by monitoring aspects that are difficult to track in its natural environment, such as reproduction, make ex situ studies fundamental in reversing its population decline. The present study describes the reproductive success at the Rio de Janeiro Marine Aquarium (AquaRio) in 2022 with the birth of a male and a female specimen of whitetip reef shark, as well as presenting continuous monitoring of the courtship, gestation, parturition, and postpartum processes. Due to the scarcity of ultrasound data of pregnancy for the species, the objective of the work also included a description of the fetal structures in the final stage of gestation.

2. Materials and Methods

The Rio de Janeiro Marine Aquarium's breeding stock consists of more than 2600 specimens of 306 species of marine and freshwater organisms, with an emphasis on 67 individuals of 11 species of Elasmobranchii. The whitetip reef shark *T. obesus* was originally represented by five specimens (three females, two males) from the coast of Indonesia that were transported to Brazil in August 2017. At the time, all specimens were smaller than 90 cm TL (total length), which identified all individuals as juveniles, considering the size at first maturation reported in the literature ($\male$= 104 cm TL, $\female$= 105 cm TL) [29]. After a period of quarantine, the sharks were introduced into the 3.5 million-liter marine exhibition tank, with a maximum depth of 7 m, where they have been kept for the past four years. The oceanic tank is equipped with a mechanical filtration system comprising 20 sand filters, 9 skimmers (Altamar 25c) with ozone injection (O3R and Altamar, 40 g/h), and a degasser (1200 gal/h). Reverse bottom biological filtration is complemented by weekly siphoning. Temperature control is achieved via a heat exchanger. The diet of whitetip reef shark adults is offered daily and includes *Euthynnus alletteratus* (Rafinesque, 1810), *Coryphaena hippurus* Linnaeus, 1758, *Sardinella brasiliensis* (Steindachner, 1879), *Cynoscion* spp., *Anchoa* spp., *Pseudupeneus maculatus* (Bloch, 1793), and *Dorytheutis* spp. Feeding is unrestricted three times a week at four points in the oceanic tank where specimens of *T. obesus* have been conditioned to go.

Two observation strategies were used on our protocol: The first one comprises two programmed daily visits on the entire perimeter of the oceanic tank performed by a veterinarian or an aquarist. The ethogram records any deviation in behavior from the known baseline for each elasmobranch. The second source of observations derives from opportunistic records (i.e., video, photographs, or reports) from environmental educators, a group comprised of a dozen professionals that are continuously present around the oceanic tank for ten hours a day. When isolated, the pregnant female is closely monitored throughout the day by a team of aquarists and veterinarians.

Ultrasound Examination

The ultrasound examination of the first parturient whitetip reef shark female was performed using a Logiq e model (General Electric®, Boston, MA, USA) equipped with a convex transducer ranging in frequency from 3 to 5 MHz. Latex gloves coated with a generous amount of gel were used to protect the transducer from possible damage caused by contact with salt water and shark skin and to promote adequate transmission of the ultrasound wave. The animal was physically restrained using the tonic immobility maneuver and positioned in dorsal decubitus throughout the examination, according to the structure being analyzed and with the patient's cooperation. A state also known as animal hypnosis, it induces loss of muscle tone and equilibrium while the shark remains unresponsive to major stimulation. It allows safer and less stressful procedures and, unlike chemical anesthetics, allows immediate recovery and minimal disruption to respiration [30,31]. An ultrasound scan of the entire coelomic cavity, the diagnosis of pregnancy, and the detailing of the fetus were performed. Anatomical descriptions follow Crow and Brock [32] and Mylniczenko [23].

3. Results

3.1. Courtship, Copulation, and Pregnancy

The presence of bite scars on the female's fins was noticed in the first days of June 2021, and the first courtship with copulation was recorded on 2 June 2021. The event lasted for 3 min 55 s from the moment of the bite until the removal of the clasper and separation of participants (Video S1). The courtship began with the male's pursuit of the female until the male was anchored by biting the female's left pectoral fin, starting a spiral descent to the bottom of the tank with slow or no swimming. Still in the water column, the male inserted its right clasper into the female's cloaca, remaining attached until the end of copulation. The position of the couple varied throughout the process, with the snout keeping in contact with the bottom and the body angle varying between 90° and 45° with the substratum. The swimming of both stopped, and the male moved its pelvic region toward the female. Copulation was closely observed by the second male of the squad and by another female, and physical contact with the mating couple was recorded, removing them from their initial position in the water column and both coming to lay their bodies directly over the substrate. Although they suffered this interference, the second male did not bite the female, nor was there any attempt to insert its clasper. Upon completion of copulation, the male and female did not immediately separate. The clasper remained inserted in the cloaca while both began to swim in opposite directions, forcing the withdrawal of the clasper and the couple's separation.

The monitoring of the behavior of the specimens of *T. obesus* in the oceanic tank also allowed observation of the female's escape strategies from unaccepted copulations. Several events were observed in the ocean tank, two of which were recorded on video and are described here (Video S2). The beginning of courtship followed the expected course and started with the chase by the male. The first strategy was to escape by fast swimming, avoiding the direct approach to the exposed pectoral fin and the necessary bite to anchor. If anchoring was successful, the female moved erratically and swam at different speeds while the male stopped swimming. Finally, the female made use of the substrate's complexity to release the male's bite, initiating a quick escape soon afterwards before the male could restart its approach.

3.2. Internal Morphology

Upon ultrasound examination, it was possible to detail a fetus with a well-defined brain, tail, and pectoral and dorsal fins, with well-defined gills and coelomic cavity (Figures 1, 2a–d, and 3a–d). The fetal liver was characterized by a hypoechoic, homogeneous parenchyma with regular contours, also showing the gallbladder with anechogenic content (Figure 2b). The kidneys were partially defined and presented a slightly heterogeneous hypoechogenic parenchyma. The intestinal spiral with the characteristic spiral pattern was evident, as was a defined parietal stratification (Figure 3a). The cartilaginous skeleton was hyperechoic and formed a moderate posterior acoustic shadow artifact (Figure 3b). It was possible to detail the cranial conformation, define the orbital fossa housing the anechoic eyeball with a hyperechoic capsule (Figure 2c), and evaluate the biparietal diameter, which measured approximately 6.47 cm (Figure 3c).

The vertebrae of the spine presented themselves as hyperechoic, individualized structures, homogeneously spaced, with an anechoic central area, possibly the spinal canal (Figure 3b). In the cross-section, the spinous processes of the vertebrae were also detailed. The gills were defined as parallel slits of alternating echogenicity (hyperechogenic and anechogenic) (Figure 2c), which, on color Doppler mapping, showed an evident signal, indicating the presence of fluid passage. The heart presented individualized chambers and an evident blood flow signal when we used the color Doppler tool (Figure 2d). With spectral Doppler, it was possible to measure the heart rate, which presented a value of 62 beats per minute at the time of the examination (Figure 3d). Considering that the average gestation period observed by Schaller [24] is 387 days and knowing that the neonate described above was born 12 days after the ultrasound examination, we calculated that it was approximately 375 days old. Therefore, the description of organogenesis by this method characterized the final third of the pregnancy.

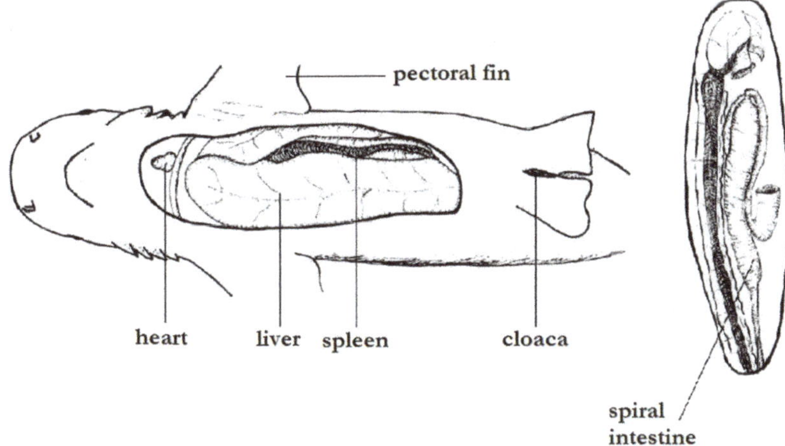

Figure 1. Basic internal anatomy of a carcharhinid shark (*Carcharhinus melanopterus*), showing the location of principal organs. Adapted from Crow and Brock [32].

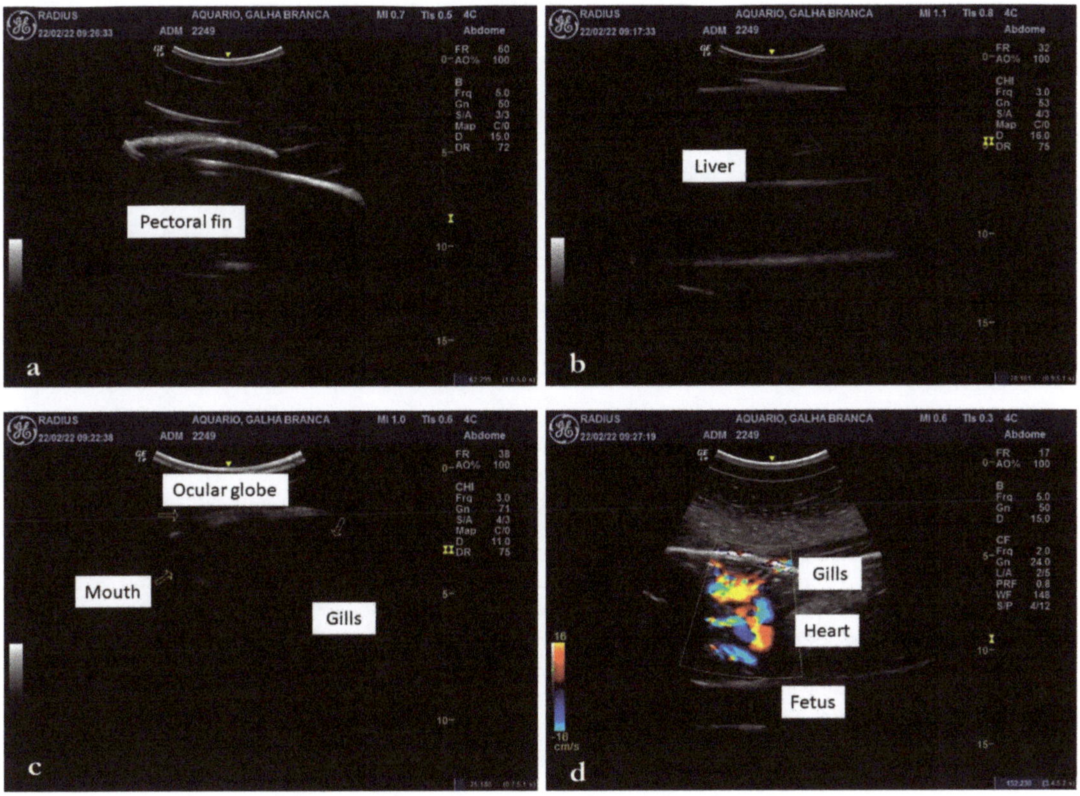

Figure 2. Ultrasonography of a female *Triaenodon obesus* (Rüppell, 1837) in late pregnancy: fetal orientation (brain/tail) and pectoral fins (**a**), hypoechogenic liver (**b**), eyes and gills (**c**), and heart with individualized chambers (**d**).

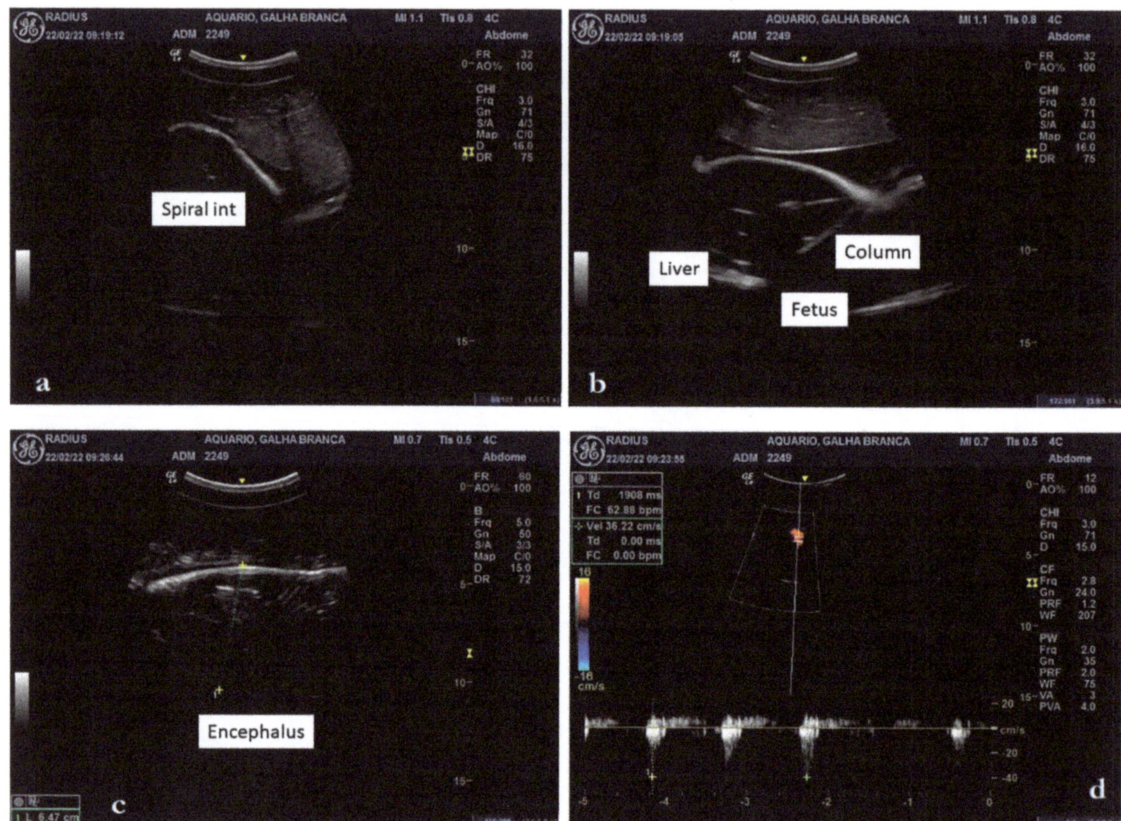

Figure 3. Ultrasonography of a female *Triaenodon obesus* (Rüppell, 1837) in late pregnancy: intestinal spiral and parietal stratification (**a**), cartilaginous skeleton, forming a posterior acoustic shadow artifact (**b**), biparietal diameter of 6.47 cm (**c**), and cardiac blood flow of 62 beats per minute (**d**).

3.3. Birth and Feeding

A *T. obesus* female of 145.0 cm TL and 20.1 kg TW (total weight) successfully completed her reproductive process with the birth on 7 March 2022 of a male of 58.0 cm TL and 1.15 kg TW. A second female of 136.4 cm TL and 19.0 kg TW gave birth to a female neonate of 56.4 cm TL and 1.0 kg TW on 7 May 2022. This was the first occurrence of the reproduction of a whitetip reef shark in an aquarium in Brazil. Both offspring were active, displaying constant swimming even though they rejected food for the first three and seven days, respectively. The first neonate fed for the first time on the third day, consuming 15 g of tilapia (*Oreochromis* sp.). The diet in the first trimester consisted of Brazilian sardine *Sardinella brasiliensis*, tilapia, manjubas *Anchoa* spp., shrimp *Penaeus* spp., and squid *Doryteuthis* spp. Over the initial three months, teleosts composed 76% of the diet, followed by shrimp (21%) and squid (3%). As for the second neonate born in 2022, feeding only started after seven days, and any food offered before then was rejected. The items offered in the diet were the same as those of the male, with tilapia accepted first, while in the following feedings, the male consumed mostly manjubas and sardines. The pregnant female isolated during the quarantine showed a marked reduction in her feeding during gestation, fasting for up to ten days, with erratic intervals in the following days until giving birth. Following veterinary recommendation and internal postpartum protocol, the recovering female was medicated via intramuscular injection with Ceftazidime 30 mg/kg and sodium methylprednisolone succinate 2 mg/kg, every 72 h, totaling five applications, due to a slight increase of leukocytes. Afterwards, she returned to her normal appetite and was transferred to the ocean tank.

4. Discussion

4.1. Courtship, Copulation, and Pregnancy

These records, made between 2021 and 2022, allowed us to evaluate the process of courtship with copulation and the gestation of the whitetip reef shark in an ex situ environment. The observations were compared with events already described in the literature and coincided with the copulations recorded previously. Whitney et al. [20] described the estimated period in the natural environment for the mating season and birth, based on photo identification made off the coast of Kona, Hawaii (USA). The authors pointed out a possible overlap of the birth period with the mating period, estimating that gestation spans one year. In the case reported here, the records of mating scars were made in June 2021, while births occurred between March and May, close to but shorter than the period reported for the species by Whitney et al. [20]. Comparing the mating record obtained in 2021 and the date of the first birth in 2022, the gestation period for the female born ex situ would have been 278 days, or just over 9 months. However, if the filmed copulation and the birth of the second neonate are considered, the gestation period would have been 339 days, which differs little from that observed for the species. Schaller [24] reported 5 births of *T. obesus* at the Steinhart Aquarium (San Francisco, CA, USA) between 2001 and 2004 and observed a longer gestation period, between 355 and 422 days, based on the first mating scar and the date of birth. Based on specimens captured off Indonesia by a commercial fishing fleet, White [33] observed two pregnant females of sizes close to our females (158.1 and 140.6 cm TL), with four formed embryos gestated by the first and three embryos found in the second, and estimated the total gestation time to be greater than six months.

Whitney et al. [34] observed three courtship events in the Coco Islands (Costa Rica), two of which involved only one male and one female. The beginning was recorded at mid-water, with the descent to the substrate provoked after the bite on the female's pectoral fin and, consequently, the inhibition of swimming. The courtship concluded with copulation in only one of the events. The courtship attempts we observed in our oceanic tank were all initiated in mid-water. Similar to the observations made by Whitney et al. [34], most attempts did not result in copulation. Incomplete copulation occurred owing to the female's detachment from the pectoral fin bite and the male and female's inability to keep swimming in parallel and in close enough contact to allow a second bite to restart the descent to the bottom. Despite being in an ex situ environment, the female's nonacceptance and escape strategies observed at AquaRio confirmed previous observations made in a natural environment [34] that indicated the success of copulation being largely dependent on acceptance from the female. Pratt Jr. and Carrier [35], in a review of reproductive behavior in elasmobranchs, noted that biting is apparently a universal characteristic of the group, although for larger species some degree of cooperation or acceptance by the female is necessary for actual copulation to occur. Copulation events observed or recorded within other elasmobranch species from our aquarium (*Ginglymostoma cirratum* and *Stegostoma tigrinum*) support this finding, especially for an event involving a female of *S. tigrinum*. In this observed event, after a tail bite, the female moved to the substrate and exposed its belly, even though the male did not conclude the copulation. In the case of *T. obesus*, the nonacceptance of courtship was associated with accelerated swimming, even after the pectoral bite, and erratic movement, which made it difficult for the male to chase. The cloaca was protected by the female's body movement and the structural complexity of the habitat was used to enact the escape or disarm the bite.

Tricas and Le Feuvre [36] described two mating events for the species from observations on Molokini Island, Hawaii (USA), in a shallow (7 m) protected reef area. The description began with the male anchored to the female by biting the pectoral fin, the male's right clasper inserted into the cloaca, and the bodies positioned with the head downward and the rest of the body suspended in the water column at an angle of 45°. The two copulation events observed in our oceanic tank coincided with the events described by Tricas and Le Feuvre [36], occurring shortly after the spiral descent. Whitney et al. [34] expanded on the findings of Tricas and Le Feuvre [36], having observed a larger set of courtships, copulations, and the functioning of the siphon bags. Whitney et al. [34] reported the presence and behavior of other males that did not participate directly in copulation but remained present and sometimes participated in the initial moment of copulation, biting the female's pectoral fin to provoke the descent to the bottom, but separating soon after without copulating. In the two courtships with copulation recorded by us, the second male was observed circling the main couple, not having participated with any bite on the pectoral fin. In the first event, in 2021, the male's physical contact caused the couple to leave their suspended and inclined position in the water column, but there was no insertion of the clasper by the second male, who moved away from the couple shortly afterwards. The most recent event did not present any interference from the second male other than constantly circling the couple during mating.

Studies on changes in behavior and feeding during pregnancy and in the period before birth are scarce in the literature for most species [37,38]. The whitetip reef shark is characterized by greater activity at night, remaining at rest for most of the day [39,40]. After monitoring the activity of adults by recording swimming acceleration and the intensity of breathing in neonates, Whitney et al. [39] pointed out that the latter assumed the pattern observed for adults soon after birth. The circumstances of the two neonates born in quarantine and in the ocean tank agreed with the observation made by Whitney et al. [39], with the same circadian pattern as adults being verified in neonates. Even so, the female born in quarantine and observed in more detail demonstrated more active diurnal swimming in the first day after birth before assuming the nocturnal pattern similar to that described for the species in the following days.

The feeding during the pregnancy of the first female of *T. obesus* was intermittent, being generally characterized by the acceptance of food every two days. It is important to note that the transfer process from the oceanic tank to the isolation in the quarantine tank was followed by a period of ten days in which the pregnant female remained within the expected activity pattern for an adult, but without accepting any food. Although we cannot assume this to be a feeding pattern of a pregnant whitetip reef shark, the disturbance of the transfer may have provoked enough stress to cause an interruption in the feeding of a pregnant female, despite it being a routine activity that does not cause the same reaction in nonpregnant females.

Given that it was the first pregnancy for the two females in question, in addition to the known relative sizes of the pregnant female and the embryo in *T. obesus*, the occurrence of only one offspring each could have been caused by the restricted space for their development. However, based on a visual census conducted in 21 reefs of the Great Barrier Reef (Australia) and records for 76 females, Robbins et al. [41] and Robbins [1] identified pregnancies of 1 to 4 offspring with an average of 2.07 neonates and showed no significant correlation with the size of the pregnant female or her age, with this being considered a particularly low level of fecundity compared to other species of Carcharhinidae.

4.2. Internal Morphology

Many of the studies devoted to the embryonic development of sharks have been carried out using oviparous species, given the ease of obtaining fertilized eggs to monitor the different stages of development [42]. In viviparous or ovoviviparous species, monitoring is more complex owing to restrictions on obtaining pregnant females found in the natural environment [43–46]. Research heavily relies on accidental captures by fishing of pregnant females at different stages of embryonic development to compose a sequence of embryonic stages [38,39]. Pregnancy monitoring in an ex situ environment offers the opportunity to monitor the same females from mating to birth, recording behavioral and physiological changes throughout the reproductive cycle [24]. The monitoring, however, depends on routine exams that allow the identification of pregnant females at the beginning of embryonic development. Maintenance of the usual behavioral pattern and the minor morphological alteration of the female during the initial and intermediate stages of gestation did not call attention to our whitetip reef shark until the last months before birth. While ultrasound exams are dedicated to the diagnosis and treatment of individuals suspected to be ill, monitoring by routine scheduled exams is necessary for viviparous species of elasmobranchs to enable the identification of pregnant females and recording of the early stages of embryo development.

Sharks of the Triakidae (*Mustelus* spp.) and Carcharhinidae (*Rhizoprionodon* spp.), placentotrophic species with annual reproduction, have concomitant pre-ovulatory vitellogenic follicles and well-developed embryos [45,46]. Although the ultrasound examination did not allow the visualization of follicles for the female at the end of gestation, a new courtship and attempt at copulation by the males started a few days after birth and the female's return to the ocean tank. Considering that the average gestation period observed by Schaller [24] is 387 days and that the first neonate described here was born 12 days after the ultrasound examination, we assumed that the female was pregnant for approximately 1 year. Therefore, the description of organogenesis via this method characterized the final third of pregnancy. This study offers the first images of the development of a healthy embryo in the late gestation period of *T. obesus*.

Viviparous placental development has been recorded only in Selachii of the order Carcharhiniformes, in the families Leptochariidae, Triakidae, Hemigaleidae, and Carcharhinidae [22,47]. The literature describing the developmental stages for placental species is scarcer than that available for oviparous lecitotrophic species [42] or viviparous aplacental species with oophagia, histotrophy, and adelphophagy [48,49]. Although the stages for *T. obesus* have not yet been described, the whitetip reef shark embryo described herein, based on an ultrasound taken 12 days before birth, showed full structural development, with developed fins and skeleton, as well as internal organs without apparent anomalies. Behavior, feeding patterns, and routine veterinary examinations in the months following birth confirmed that both newborns were fully developed, healthy individuals.

4.3. Birth and Feeding

The ex situ reproduction of a shark species classified as vulnerable, such as *T. obesus* [19], highlights the need to integrate public aquariums in the consolidation of a viable ex situ population, according to criteria already outlined by Buckley et al. [50], and an essential step for this is the reproductive success and genetic variability of this population. Schaller [24] described four litters of *T. obesus* obtained between 2001 and 2003 in work conducted at the Steinhart Aquarium (USA), providing precise data on courtship, pregnancy, and parturition for the species. Portnoy et al. [51] described the birth of a stillborn whitetip reef shark at the BioPark Aquarium (USA), with no apparent reason for death or signs of trauma or malformation. It is important to note that Portnoy et al. [51] also pointed out that all parthenogenesis cases reported for sharks generated females, which initially suggested that the female pup born at AquaRio could have been a parthenogenetic event. Still, photographic records and recordings obtained in 2021 made it possible to follow courtship and copulation events among the *T. obesus* group, supporting the suspicion that the births were the product of sexual reproduction. The appearance of bite scars in June 2021 and the presence of a male pup also suggested that parthenogenesis did not take place. Furthermore, a DNA analysis is going to be conducted to establish if the same male was responsible for both shark pups. The diet of adult whitetip reef sharks consists mainly of teleosts (Holocentridae, Lutjanidae, Pomacentridae, Scaridae, Acanthuridae, Balistidae, Mullidae), which can comprise 90% of their food, although mollusks and crustaceans have also been reported [29,52]. The food preference displayed by the neonates reflected the described diet for adults, consisting mainly of teleosts, but with a relevant inclusion of penaeid shrimps.

5. Conclusions

The study compared the data available for the reproduction of *T. obesus* in the literature, based mostly on in situ observations, and the case reported by Schaller [24] for births registered at the Steinhart Aquarium (USA), making it possible to corroborate the courtship stages already reported, as well as to identify female escape strategies and contribute data on gestation duration and the initial development of neonates and their food preference. Furthermore, ultrasonography is a noninvasive examination technique and it allowed us to detail fetal organogenesis and study the viability of the fetus in whitetip reef sharks, with this being the first ultrasonographic description of fetal organogenesis in the species and the first report of reproduction of the species in a Brazilian aquarium.

Completing the reproductive cycle of endangered elasmobranchs in an ex situ environment is another relevant step that may lead to an effective plan for the preservation of these species. While natural populations are under pressure from habitat loss and overfishing, a healthy population kept safe from these impacts will enable a species reintroduction plan to be implemented after minimum living conditions are restored for the species in their original distribution. This strategy is even more vital when it comes to coral reefs, highly productive ecosystems with rich fauna and complex food webs. Existing literature reports several threats already affecting these communities, within which *T. obesus* is an essential predator. Reproductive success will offer hope for Brazilian species of Carcharhinidae, as many share characteristics with the whitetip reef shark, such as being viviparous placentotrophs. The lessons learned from whitetip reef sharks under human care may assist in the development of protocols that help with the proper handling of other carcharinid sharks in ex situ conditions, with the possibility of increasing the chances of successful reproduction of endangered species, such as the highly threatened sandbar shark *Carcharhinus plumbeus* (Nardo, 1827) and the small-tail shark *Carcharhinus porosus* (Ranzani, 1839).

Supplementary Materials: The following supporting information can be downloaded at: https://www.mdpi.com/article/10.3390/ani12233291/s1, Video S1: Courtship and copulation of *Triaenodon obesus* (Rüppell, 1837), recorded at the Rio de Janeiro Marine Aquarium. Video S2: Failed attempt at courtship of *Triaenodon obesus* (Rüppell, 1837), recorded at the Rio de Janeiro Marine Aquarium.

Author Contributions: Conceptualization, S.R.S., V.T. and M.F.G.; methodology, V.T., S.P.B. and M.F.G.; validation, S.R.S., V.T., S.P.B. and M.F.G.; investigation, S.R.S., V.T., S.P.B., N.L.L.A. and M.F.G.; resources, R.F.V.; data curation, S.R.S. and V.T.; writing—original draft preparation, S.R.S., V.T., S.P.B., N.L.L.A., M.F.G. and R.F.V.; writing—review and editing, S.R.S.; visualization, S.R.S.; supervision, M.F.G. and R.F.V.; project administration, R.F.V.; funding acquisition, R.F.V. All authors have read and agreed to the published version of the manuscript.

Funding: This research was funded by Seasub 7 S.A. through the partnership with IMAM-AquaRio.

Institutional Review Board Statement: Not applicable.

Informed Consent Statement: Not applicable, as this research did not involve humans.

Data Availability Statement: Not applicable.

Acknowledgments: The authors would like to thank the combined efforts of aquarists and veterinarians that help preserve the lives of thousands of fish, crustaceans, and mollusks of different species, among other groups, under a strict daily routine. Their ethological observations and constant surveillance helped to identify the first signs of courtship and pregnancy and build a detailed profile of the behavioral evolution of both pregnant females and neonates. We also thank both anonymous reviewers for their careful reading of our manuscript and their insightful remarks and suggestions.

Conflicts of Interest: The authors declare no conflict of interest.

References

1. Robbins, W.D. Abundance, Demography and Population Structure of the Grey Reef Shark (*Carcharhinus amblyrhynchos*) and the White Tip Reef Shark (*Triaenodon obesus*) (Fam. Carcharhinidae). Ph.D. Thesis, James Cook University, Douglas, Australia, 2006.
2. Ferretti, F.; Worm, B.; Britten, G.L.; Heithaus, M.R.; Lotze, H.K. Patterns and ecosystem consequences of shark declines in the ocean. *Ecol. Lett.* **2010**, *13*, 1055–1071. [CrossRef] [PubMed]
3. Pistevos, J.C.A.; Nagelkerken, I.; Rossi, T.; Olmos, M.; Connell, S.D. Ocean acidification and global warming impair shark hunting behavior and growth. *Sci. Rep.* **2015**, *5*, 16293. [CrossRef] [PubMed]
4. Pacoureau, N.; Rigby, C.L.; Kyne, P.M.; Sherley, R.B.; Winker, H.; Carlson, J.K.; Fordham, S.V.; Barreto, R.; Fernando, D.; Francis, M.P.; et al. Half a century of global decline in oceanic sharks and rays. *Nature* **2021**, *587*, 567–571. [CrossRef] [PubMed]
5. Ruppert, J.L.W.; Travers, M.J.; Smith, L.L.; Fortin, M.-J.; Meekan, M.G. Caught in the Middle: Combined Impacts of Shark Removal and Coral Loss on the Fish Communities of Coral Reefs. *PLoS ONE* **2013**, *8*, e74648. [CrossRef] [PubMed]
6. Heithaus, M.R.; Alcoverro, T.; Arthur, R.; Burkholder, D.A.; Coates, K.A.; Christianen, M.J.A.; Kelkar, N.; Manuel, S.A.; Wirsing, A.J. Seagrasses in the age of sea turtle conservation and shark overfishing. *Front. Mar. Sci.* **2014**, *1*, 28. [CrossRef]
7. Roff, G.; Doropoulos, C.; Rogers, A.; Bozec, Y.-M.; Krueck, N.C.; Aurellado, E.; Priest, M.; Birrell, C.; Mumby, P.J. The ecological role of sharks on coral reefs. *Trends Ecol. Evol.* **2016**, *31*, 395–407. [CrossRef] [PubMed]
8. Ruppert, J.L.W.; Fortin, M.-J.; Meekan, M.G. The ecological role of sharks on coral reefs: Response to Roff et al. *Trends Ecol. Evol.* **2016**, *31*, 586–587. [CrossRef]
9. Pauly, D.; Zeller, D. Accurate catches and the sustainability of coral reef fisheries. *Curr. Opin. Environ. Sustain.* **2014**, *7*, 44–51. [CrossRef]
10. Barange, M. Ecosystem science and the sustainable management of marine resources: From Rio to Johannesburg. *Front. Ecol. Environ.* **2003**, *1*, 190–196. [CrossRef]
11. Humphreys, J.; Clark, R.W.E. A critical history of marine protected areas. In *Marine Protected Areas: Science, Policy and Management*, 1st ed.; Humphreys, J., Clark, R.W.E., Eds.; Elsevier Ltd.: Amsterdam, The Netherlands, 2020; pp. 1–12.
12. Whitney, N.M.; Robbins, W.D.; Schultz, J.K.; Bowen, B.W.; Holland, K.N. Oceanic dispersal in a sedentary reef shark (*Triaenodon obesus*): Genetic evidence for extensive connectivity without a pelagic larval stage. *J. Biogeogr.* **2012**, *39*, 1144–1156. [CrossRef]
13. Bornatowski, H.; Loose, R.; Sampaio, C.L.S.; Gadig, O.B.F.; Carvalho-Filho, A.; Domingues, R.R. Human introduction or natural dispersion? Atlantic Ocean occurrence of the Indo-Pacific whitetip reef shark *Triaenodon obesus*. *J. Fish Biol.* **2018**, *92*, 537–542. [CrossRef] [PubMed]
14. López-Garro, A.; Zanella, I.; Golfín-Duarte, G.; Pérez-Montero, M. Residencia del tiburón punta blanca de arrecife, *Triaenodon obesus*, (Carcharhiniformes: Carcharhinidae) en las bahías Chatam y Wafer del Parque Nacional Isla del Coco, Costa Rica. *Rev. Biol. Trop.* **2020**, *68*, S330–S339. [CrossRef]
15. Dillon, E.M.; McCauley, D.J.; Morales-Saldaña, J.M.; Leonard, N.D.; Zhao, J. Fossil dermal denticles reveal the preexploitation baseline of a Caribbean coral reef shark community. *Proc. Natl. Acad. Sci. USA* **2021**, *118*, e2017735118. [CrossRef]
16. Graham, N.A.J.; Wilson, S.K.; Jennings, S.; Polunin, N.V.C.; Robinson, J.; Bijoux, J.P.; Daw, T.M. Lag effects in the impacts of mass coral bleaching on coral reef fish, fisheries, and ecosystems. *Conserv. Biol.* **2007**, *21*, 1291–1300. [CrossRef]
17. Heron, S.F.; Maynard, J.A.; van Hooidonk, R.; Eakin, C.M. Warming trends and bleaching stress of the world's coral reefs 1985–2012. *Sci. Rep.* **2016**, *6*, 38402. [CrossRef] [PubMed]
18. Cornwall, C.E.; Comeau, S.; Kornder, N.A.; Perry, C.T.; van Hooidonk, R.; DeCarlo, T.M.; Pratchett, M.S.; Anderson, K.D.; Browne, N.; Carpenter, R.; et al. Global declines in coral reef calcium carbonate production under ocean acidification and warming. *Proc. Natl. Acad. Sci. USA* **2021**, *118*, e2015265118. [CrossRef]
19. Simpfendorfer, C.; Yuneni, R.R.; Tanay, D.; Seyha, L.; Haque, A.B.; Bineesh, K.K.; Bin Ali, A.; Gautama, D.A.; Maung, A.; Sianipar, A.; et al. *Triaenodon obesus*. IUCN Red List. Threat. Species **2020**, 2020, e.T39384A173436715. [CrossRef]
20. Whitney, N.M.; Pyle, R.L.; Holland, K.N.; Barez, J.T. Movements, reproductive seasonality, and fisheries interactions in the whitetip reef shark (*Triaenodon obesus*) from community-contributed photographs. *Environ. Biol. Fishes* **2012**, *93*, 121–136. [CrossRef]
21. Ebert, D.A.; Fowler, S.; Compagno, L. *Sharks of the World: A Fully Illustrated Guide*, 1st ed.; Wild Nature Press: Plymouth, UK, 2013.
22. Buddle, A.L.; Van Dyke, J.U.; Thompson, M.B.; Simpfendorfer, C.A.; Whittington, C.M. Evolution of placentotrophy: Using viviparous sharks as a model to understand vertebrate placental evolution. *Mar. Freshw. Res.* **2018**, *70*, 908–924. [CrossRef]
23. Mylniczenko, N.D. A1: Anatomy and taxonomy. In *Clinical Guide to Fish Medicine*, 1st ed.; Hadfield, C.A., Clayton, L.A., Eds.; John Willey & Sons, Inc.: Hoboken, NJ, USA, 2021; pp. 3–34.
24. Schaller, P. Husbandry and reproduction of whitetip reef sharks *Triaenodon obesus* at Steinhart Aquarium, San Francisco. *Int. Zoo Yearb.* **2006**, *40*, 232–240. [CrossRef]
25. Anderson, B.; Belcher, C.; Slack, J.A.; Gelsleichter, J. Evaluation of the use of portable ultrasonography to determine pregnancy status and fecundity in bonnethead shark *Sphyrna tiburo*. *J. Fish Biol.* **2018**, *93*, 1163–1170. [CrossRef] [PubMed]
26. Carrier, J.C.; Murru, F.L.; Walsh, M.T.; Pratt, H.L., Jr. Assessing reproductive potential and gestation in nurse sharks (*Ginglymostoma cirratum*) using ultrasonography and endoscopy: An example of bridging the gap between field research and captive studies. *Zoo Biol.* **2003**, *22*, 179–187. [CrossRef]
27. Grant, K.R.; Campbell, T.W.; Silver, T.I.; Olea-Popelka, F.J. Validation of an ultrasound-guided technique to establish a liver-to-coelom ratio and a comparative analysis of the ratios among acclimated and recently wild-caught southern stingrays, *Dasyatis americana*. *Zoo Biol.* **2012**, *32*, 104–111. [CrossRef] [PubMed]

28. Janse, M.; Zimmerman, B.; Geerlings, L.; Brown, C.; Nagelkerke, L.A.J. Sustainable species management of the elasmobranch populations within European aquariums: A conservation challenge. *J. Zoo Aquar. Res.* **2017**, *5*, 172–181. [CrossRef]
29. Compagno, L.J.V. FAO Species Catalogue Vol. 4. Sharks of the World. An annotated and illustrated catalogue of shark species known to date. Part 2. Carcharhiniformes. *FAO Fish. Synop.* **1984**, *125*, 251–655.
30. Henningsen, A.D. Tonic immobility in 12 elasmobranchs: Use as an aid in captive husbandry. *Zoo Biol.* **1994**, *13*, 325–332. [CrossRef]
31. Kessel, S.T.; Hussey, N.E. Tonic immobility as an anaesthetic for elasmobranchs during surgical implantation procedures. *Can. J. Fish. Aquat. Sci.* **2015**, *72*, 1287–1291. [CrossRef]
32. Crow, G.L.; Brock, J.A. Chapter 30: Necropsy methods and procedures for elasmobranchs. In *The Elasmobranch Husbandry Manual: Captive Care of Sharks, Rays and Their Relatives*, 1st ed.; Smith, M., Warmolts, D., Thoney, D., Hueter, R., Eds.; Ohio Biological Survey, Inc.: Columbus, OH, USA, 2004; pp. 467–472.
33. White, W.T. Catch composition and reproductive biology of whaler sharks (Carcharhiniformes: Carcharhinidae) caught by fisheries in Indonesia. *J. Fish Biol.* **2007**, *71*, 1512–1540. [CrossRef]
34. Whitney, N.M.; Pratt, H.L., Jr.; Carrier, J.C. Group courtship, mating behaviour and siphon sac function in the whitetip reef shark, *Triaenodon obesus*. *Anim. Behav.* **2004**, *68*, 1435–1442. [CrossRef]
35. Pratt, H.L., Jr.; Carrier, J.C. A review of elasmobranch reproductive behavior with a case study on the nurse shark, *Ginglymostoma cirratum*. *Environ. Biol. Fishes* **2001**, *60*, 157–188. [CrossRef]
36. Tricas, T.C.; Le Feuvre, E.M. Mating in the reef white-tip shark *Triaenodon obesus*. *Mar. Biol.* **1985**, *84*, 233–237. [CrossRef]
37. Ebert, D.A.; Ebert, T.B. Reproduction, diet and habitat use of leopard sharks *Triakis semifasciata* (Girard), in Humbolt Bay, California, USA. *Mar. Freshw. Res.* **2005**, *56*, 1089–1098. [CrossRef]
38. Rangel, B.S.; Hammerschlag, N.; Sulikowski, J.A.; Moreira, R.G. Dietary and reproductive biomarkers in a generalist apex predator reveal differences in nutritional ecology across life stages. *Mar. Ecol. Prog. Ser.* **2021**, *664*, 149–163. [CrossRef]
39. Whitney, N.M.; Papastamatiou, Y.P.; Holland, K.N.; Lowe, C.G. Use of an acceleration data logger to measure diel activity patterns in captive whitetip reef sharks, *Triaenodon obesus*. *Aquat. Living Resour.* **2007**, *20*, 299–305. [CrossRef]
40. Mejía-Falla, P.A.; Navia, A.F.; Lozano, R.; Tóbon-López, A.; Narváez, K.; Muñoz-Osorio, L.A.; Mejía-Ladino, L.M.; López-García, J. Uso de hábitat de *Triaenodon obesus* (Carcharhiniformes: Carcharhinidae), *Rhincodon typus* (Orectolobiformes: Rhincodontidae) y *Manta birostris* (Myliobatiformes: Myliobatidae) en el Parque Nacional Gorgona, Pacífico Colombiano. *Rev. Biol. Trop.* **2014**, *62*, 329–342. [CrossRef]
41. Robbins, W.D.; Hisano, M.; Connolly, S.R.; Choat, J.H. Ongoing collapse of coral-reef shark populations. *Curr. Biol.* **2006**, *16*, 2314–2319. [CrossRef] [PubMed]
42. Onimaru, K.; Motone, F.; Kiyatake, I.; Nishida, K.; Kuraku, S. A staging table for the embryonic development of the brownbanded bamboo shark (*Chiloscyllium punctatum*). *Dev. Dyn.* **2018**, *247*, 712–723. [CrossRef]
43. Guallart, J.; Vicent, J.J. Changes in composition during embryo development of the gulper shark, *Centrophorus granulosus* (Elasmobranchii, Centrophoridae): An assessment of maternal-embryonic nutritional relationship. *Environ. Biol. Fishes* **2001**, *61*, 135–150. [CrossRef]
44. Sato, K.; Nakamura, M.; Tomita, T.; Toda, M.; Miyamoto, K.; Nozu, R. How great white sharks nourish their embryos to a large size: Evidence of lipid histotrophy in lamnoid shark reproduction. *Biol. Open* **2016**, *5*, 1211–1215. [CrossRef]
45. Castro, J.I. Biology of the blacktip shark, *Carcharhinus limbatus*, off the southeastern United States. *Bull. Mar. Sci.* **1996**, *59*, 508–522.
46. Macías-Cuyare, M.; Tavares, R.; Oddone, M.C. Reproductive biology and placentotrophic embryonic development of the smalleye smooth-hound shark, *Mustelus higmani*, from the south-eastern Caribbean. *J. Mar. Biol. Assoc. U. K.* **2020**, *100*, 1337–1347. [CrossRef]
47. López, J.A.; Ryburn, J.A.; Fedrigo, O.; Naylor, G.J.P. Phylogeny of sharks of the family Triakidae (Carcharhiniformes) and its implications for the evolution of carcharhiniform placental viviparity. *Mol. Phylogenet. Evol.* **2006**, *40*, 50–60. [CrossRef] [PubMed]
48. Gilmore, R.G.; Dodrill, J.W.; Linley, P.A. Reproduction and embryonic development of the sand tiger shark, *Odontaspis taurus* (Rafinesque). *Fish. Bull.* **1983**, *81*, 201–225.
49. Joung, S.-J.; Hsu, H.-H. Reproduction and Embryonic Development of the Shortfin Mako, *Isurus oxyrinchus* Rafinesque, 1810, in the Northwestern Pacific. *Zool. Stud.* **2005**, *44*, 487–496.
50. Buckley, K.A.; Crook, D.A.; Pillans, R.D.; Smith, L.; Kyne, P.M. Sustainability of threatened species displayed in public aquaria, with a case study of Australian sharks and rays. *Rev. Fish Biol. Fish.* **2018**, *28*, 137–151. [CrossRef]
51. Portnoy, D.S.; Hollenbeck, C.M.; Johnston, J.S.; Casman, H.M.; Gold, J.R. Parthenogenesis in a whitetip reef shark *Triaenodon obesus* involves a reduction in ploidy. *J. Fish Biol.* **2014**, *85*, 502–508. [CrossRef]
52. Frisch, A.J.; Ireland, M.; Rizzari, J.R.; Lönnstedt, O.M.; Magnenat, K.A.; Mirbach, C.E.; Hobbs, J.-P.A. Reassessing the trophic role of reef sharks as apex predators on coral reefs. *Coral Reefs* **2016**, *35*, 459–472. [CrossRef]

Article

Growth Characteristics of Long-Nosed Skate *Dipturus oxyrinchus* (Linnaeus, 1758) Inhabiting the Northeastern Mediterranean Sea

Nuri Başusta [1,*] and Fatih Volkan Ozel [2]

[1] Faculty of Fisheries, Firat University, Elazig 23119, Turkey
[2] Graduate School of Natural and Applied Sciences, Firat University, Elazig 23119, Turkey
* Correspondence: nbasusta@firat.edu.tr; Tel.: +90-5445184995

Simple Summary: Skates and rays generally have low fecundity, delayed maturation age and slow growth rates. These life history traits make them very vulnerable to commercial fisheries' activities, although they are not usually target species. Understanding age and growth is important for stock assessment of the long-nose skate. We believe that determining these can be the first step toward correct management actions. Little is known about the growth characteristics of the long-nose skate in the Northeastern Mediterranean Sea. Our study aimed at comparing three different growth models (the von Bertalanffy, the Robertson (logistic), and Gompertz) and also the absolute and relative growth characteristics of this species. Our results show that the relationship between the age of the long-nose skate and its total length is adequately explained by the Robertson (logistic) growth model, with the Gompertz growth model being the second best.

Citation: Başusta, N.; Ozel, F.V. Growth Characteristics of Long-Nosed Skate *Dipturus oxyrinchus* (Linnaeus, 1758) Inhabiting the Northeastern Mediterranean Sea. *Animals* **2022**, *12*, 3443. https://doi.org/10.3390/ani12233443

Academic Editor: Martina Francesca Marongiu

Received: 29 October 2022
Accepted: 2 December 2022
Published: 6 December 2022

Publisher's Note: MDPI stays neutral with regard to jurisdictional claims in published maps and institutional affiliations.

Abstract: This study aims to determine the age and growth characteristics of *Dipturus oxyrinchus* living in the Northeastern Mediterranean Sea and to present data that can provide a comparison with previous studies on the same subject. A total of 255 long-nose skates at a total length of 12.2–93.5 cm and weight of 8.34–3828 g were collected as non-target species from a commercial fishing boat. The male−female ratio was determined as 1:1.27. Using the von Bertalanffy equation and the Gompertz or logistic growth models, the growth parameters of *Dipturus oxyrinchus* were estimated as L∞ = 154.0, K = 0.064, t_0 = −1.622; L∞ = 104.0, K = 0.35, I = 4.99; L∞ = 128.40, K = 0.19, I = 4.39 for all individuals, respectively. Maximum absolute growth was calculated as 9.33 cm at 5–6 years of age. Maximum relative growth at 1–2 years of age was estimated as 36.39%. Both absolute and relative growth were minimal in the 11–12 age group. The highest condition factor value was estimated as 0.416 in the 8-year-old group. As a result, the growth data of long-nose skates were obtained for the first time in the Northeastern Mediterranean Sea.

Keywords: age; growth; *Dipturus oxyrinchus*; long-nosed skate; Northeastern Mediterranean Sea

1. Introduction

The long-nosed skate, *Dipturus oxyrinchus* (L. 1758), is a demersal species that lives at depths of 70−1230 m on sandy or muddy substrates and, rarely, on rocky and pebbly grounds; it is usually found at depths of 200−500 m [1–6]. The long-nosed skate ranges from Norway to Senegal, from the Northeast Atlantic to the Faroe Islands, Skagerrak (the strait connecting the Baltic Sea to the North Sea), the Canary Islands, the Madeira Islands and the Mediterranean Sea [3] and has little commercial value [4]. Sizes between 60 and 100 cm are common, but the largest recorded individual was 150 cm [1]. *Dipturus oxyrinchus* is globally recognized as a Near Threatened (NT) species by the International Union for Conservation of Nature (IUCN) [7]. Long-nosed skates have been studied satisfactorily by researchers during recent years in other parts of the Mediterranean Sea in terms of age, growth, distribution, systematics, length–weight relationships (LWR) and feeding habits. However, no data on growth parameters are available for this species in the Northeastern Mediterranean Sea [8–12]. The presence of juveniles and mature males of the long-nosed skate was previously reported by Başusta and Başusta [13] from the same region. Griffiths et al. [14] concluded that Mediterranean long-nosed skates may have been genetically isolated from other stocks (e.g., Atlantic). This study aimed to determine the growth characteristics of long-nosed skates living in the Northeastern Mediterranean Sea and compare them with the data reported in other studies conducted in the same region.

2. Materials and Methods

2.1. Collecting of Samples

The long-nosed skate individuals were caught by a commercial trawler (F/V NIHAT BABA/31-A-1463) in Iskenderun Bay (Figure 1; 36°29′200″ N; 35°05′973″ E–36°07′052″ N; 35°17′936″ E–36°07′148″ N; 35°17′978″ E–36°13′720″ N; 35°22′998″ E–36°13′650″ N; 35°23′032″ N–36°16′622″ E; 35°18′509″ N) between May 2015 and June 2016. The samples were collected monthly. The bottom trawl gear used was equipped with a 42 mm stretched-mesh size net at the cod-end. Each hauling lasted three hours, with a trawling speed of 2.2–2.9 knots. Approximate sampling depths ranged between 100–150 m, 150–200 m and 200–400 m. After fishing, all samples were transported with ice to the laboratory of the Fisheries Faculty, Firat University. Total length (L) was measured as a straight line from the tip of the rostrum to the end of the tail to the nearest mm and body mass (W) was weighed with a 1 g accuracy for each individual.

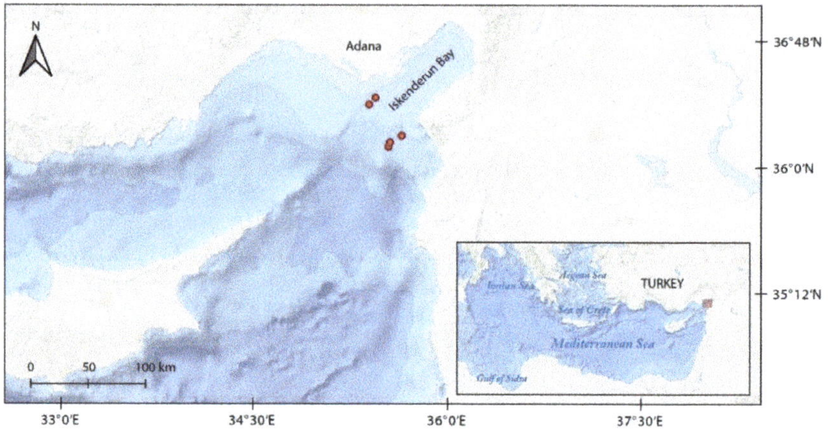

Figure 1. Red dots indicate out of Iskenderun Bay, Turkey, where *Dipturus oxyrinchus* specimens were collected.

2.2. Processing of Vertebral Centra

A section of 10–12 vertebral centra was removed from the widest portion of the body of 255 long-nosed skate specimens (143 females and 112 males) and subsequently labeled, frozen and stored until further processing. Vertebrae were later thawed and cleaned of excess tissue, rinsed in tap water and then stored in a 70% ethanol solution. Three random vertebrae from each sample were removed from the ethanol and air-dried [15,16]. Smaller centra of less than 5 mm were fixed to a clear glass slide using resin (Crystol bond 509™) and were sanded with a Dremel™ tool to replicate a sagittal cut [17–19]. Vertebral sections (0.6 mm thick) were taken using a Ray Tech™ (Littleton, CO, USA) gam saw for large centra >5 mm in diameter [20]. Vertebral cross-sections were mounted on microscope slides using clear resin (Cytoseal 60; Fisher Scientific, Pittsburgh, PA, USA) [21].

2.3. Age Assessment and Verification

Vertebral cross-sections were examined under a Leica S8 APO™ (Singapore) microscope using LAS software (Version 4.8.0, Leica Microsystems Limited, Heerbrugg, Switzerland). One growth band was defined as an opaque and translucent band pair that traversed the intermedialia and clearly extended into the corpus calcareum (Figure 2) [22–24].

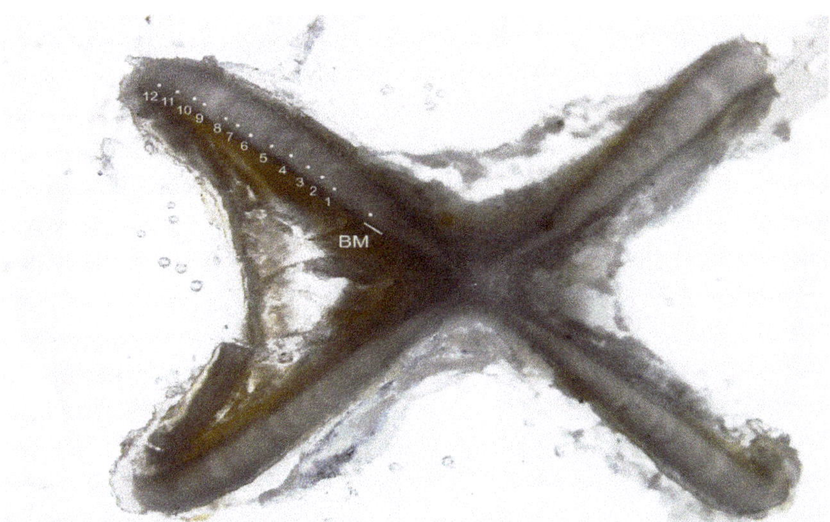

Figure 2. A vertebral cross-section of an estimated 12-year-old *Dipturus oxyrinchus* (total length = 93.5 cm, female) (BM, Birth Mark). White dots indicate opaque bands.

Each vertebral cross section was examined by two readers (reader 1 = NB and reader 2 = FVO). Reader 1 made two nonconsecutive band-counts of sampled vertebral cross-sections without prior information of the long-nosed skate's length or former counts. Reader 2 made two consecutive counts from 50 randomly selected vertebrae sections. Vertebral cross-sections that had an instability of more than two years between each reading were eliminated from further analyses. Count reproducibility was compared by the percent agreement (%PA) and coefficient of variation (%CV) [25], as well as the index of average percent error between the readers (%IAPE) [26]. All were determined using the following mathematical equations:

$$\text{PA} = \frac{No.\ agreed}{No.\ read} \times 100 \tag{1}$$

$$\text{CV}_j = 100 \times \frac{\sqrt{\sum_{i=1}^{R} \frac{(X_{ij}-X_j)^2}{R-1}}}{x_j} \tag{2}$$

$$\text{IAPE} = \frac{1}{R} \sum_{i=1}^{R} \frac{|x_{ij} - x_j|}{x_j} \times 100 \tag{3}$$

where R is the number of readings; X_{ij} is the count from the *j*th fish at the *i*th reading and X_j is the mean age calculated for the *j*th fish from *i* readings.

Pair-wise age-reader comparisons were independently generated by the two readers by making nonconsecutive band counts from a random sample of 50 vertebral sections [27]. All statistical tests were performed with R software version 4.1 [28], with a significance level set at 5%.

2.4. Total Length–Weight Relationship and Growth Modeling

The total length–weight relationship parameters of long-nosed skate were estimated according to the equation given below [29]:

$$W = aL^b \tag{4}$$

where L = total length (cm); W = body mass (g); a is a constant of proportionality and b is the allometric factor. The deviance of the estimated b values for long-nosed skate from the hypothetical value of 3 (i.e., isometric growth) was tested by a t-test at the 0.01 significance level [29,30].

The absolute length growth (ALG) and the relative length growth (RLG) of long-nosed skate were calculated as:

$$\text{ALG} = (Lt + \Delta t - Lt)/\Delta t \tag{5}$$

$$\text{RLG} = (Lt + \Delta t - Lt/Lt) \times 100 \tag{6}$$

where L_t is total length at the start of the time interval and $L_{t+\Delta t}$ is total length at the end of the time interval (Δt) [30,31].

The values of condition factor were obtained with the formula:

$$Kn = \left(\frac{W}{Lb}\right) \times 100 \tag{7}$$

where W, L and b are as defined above [32].

The observed length-at-age data of the long-nosed skate were used as the dependent variable and the age as the independent variable, with the three most-used models to describe the growth of fish. The first model employed was the von Bertalanffy [33] growth equation (VBGM). The VBGM was formulated by Beverton and Holt [34] as:

$$L_t = L_\infty \left(1 - e^{-K(t-t_0)}\right), \tag{8}$$

where L_t is the expected total length at age t years; L_∞ is the asymptotic total length; K is the growth coefficient or curvature parameter indicating the rate at which long-nosed skates grow toward their L_∞ and t_0 is the theoretical age at zero total length.

The second growth model used was that of Gompertz [35], an S-shaped growth model (GGM) [36–38]:

$$L_t = L_\infty e^{-e^{-G(t-t_i)}}, \tag{9}$$

where t, L_t and L_∞ are the same as in the VBGM; t_i is the age at the inflection point of the growth curve, i.e., the age at which the absolute growth rate starts to decrease and G is the instantaneous growth rate coefficient at age t_i, where growth becomes asymmetrical.

The Robertson (logistic) model was also used as the last growth model. While t, t_i, L_t and L_∞ are the same as in the previous models, K is a parameter that affects the rate of exponential growth:

$$Lt = \frac{L_\infty}{(1 + e(-K(t - t_i)))} \tag{10}$$

Akaike's information criterion (AIC) was used to compare the different growth models. A smaller value of the AIC indicates that the observed data are closer to the fitted model [29,39]. The AIC is defined as −2 times the maximum value of the log likelihood ($\hat{L}$) plus 2 times the number of parameters (p) in the model including the estimated variance [40]:

$$\text{AIC} = -2\ln(\hat{L}) + 2p \tag{11}$$

3. Results

3.1. Sample Composition, Sex and Vertebral Analysis

A total of 255 long-nosed skates ranging from 12.2 to 93.5 cm in total length and 8.34–3828 g in weight were collected as bycatch or discard species from a commercial fishing vessel between May 2015 and June 2016 (Table 1). The sex ratio (M/F) was determined as 1:1.27. The ratio of females to males was not statistically different from the expected 1:1 ratio between sexes ($p > 0.05$). The size frequency based on total length and age group is presented in Figure 3.

Table 1. Descriptive statistics for *Dipturus oxyrinchus* inhabiting the Northeastern Mediterranean Sea.

Group	n	Total Weight (g)		Total Length (TL, cm)	
		Mean SE	Min–Max	Mean SE	Min–Max
Female	143	1000.28 ± 902.78	8.50–3828.00	57.38 ± 22.05	12.20–93.50
Male	112	825.44 ± 666.55	8.34–2234.00	53.97 ± 19.24	14.60–80.10

SE: Standard Error; Min: Minimum; Max: Maximum.

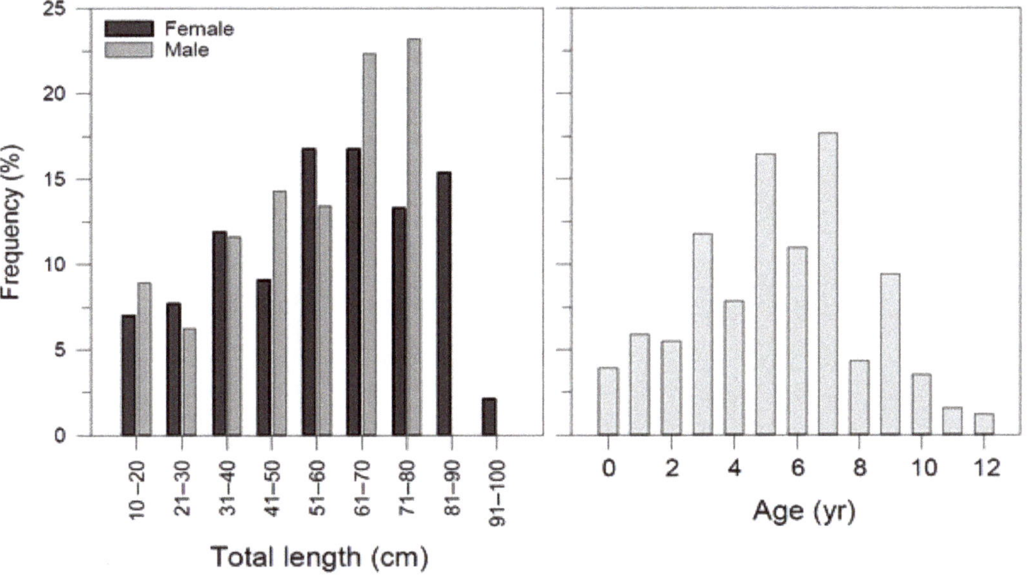

Figure 3. Frequency distribution of *Dipturus oxyrinchus* inhabiting the Northeastern Mediterranean Sea, according to their total length and age.

3.2. Age Estimation, Reading Precision and Age Bias

In this study, estimated ages ranged from 0 to 12 for females and from 0 to 9 for males. Age estimation for males and females by two independent readers did not show a considerable variation (Figure 4a,b). The highest PA with lowest IAPE and CV were found for males than female (Table 2).

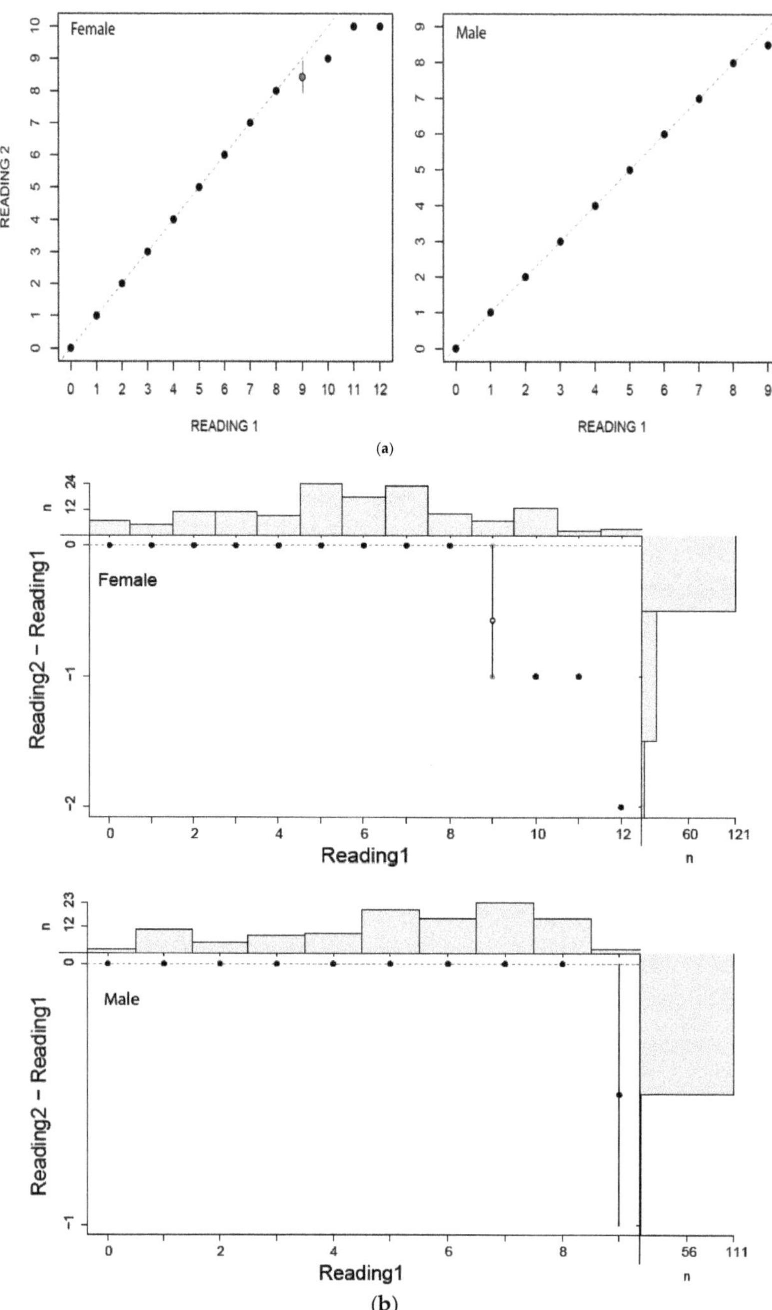

(a)

(b)

Figure 4. (**a**) Age bias graphs for *Dipturus oxyrinchus* inhabiting the Northeastern Mediterranean Sea. (**b**) Age bias plots for two readers for aging *Dipturus oxyrinchus* inhabiting the Northeastern Mediterranean Sea. Plots illustrate reference reader's age estimates on the *x*-axis; the mean difference (circles) and distribution of the differences between corresponding ages (vertical lines) are represented on the *y*-axis.

Table 2. Summary statistics for the coefficient of variation (CV), percentage of agreement (%PA) and index of average percentage error (IAPE) to determine the precision of age readings of *Dipturus oxyrinchus* inhabiting the Northeastern Mediterranean Sea.

Group	Readers	n	R	CV	PA
Overall	Reader 1 vs. Reader 2	255	2	0.75	90.98
Female	Reader 1 vs. Reader 2	143	2	1.27	84.62
Male	Reader 1 vs. Reader 2	112	2	0.74	99.11

n = sample size and R = number of readings.

3.3. Length–Weight Relationships, Growth Patterns and Condition Factor

3.3.1. Length–Weight Relationships

The total length–weight relationships of *Dipturus oxyrinchus* for female, male and overall are presented in Figure 5. According to these results, positive allometric growth ($b > 3$) was demonstrated for the overall category. Regression results showed that the length and weight of the long-nosed skate was predicted significantly (r = 0.987, r^2 = 0.974, F1,242 = 9092.525 $p < 0.001$) in all sexes. It is possible to state that 97% of the increase is due to that of the size of *Dipturus oxyrinchus* in all individuals for the present study. In addition, when *t*-test results related to the significance of the regression coefficients were analyzed (*t*-test = 95.355 $p < 0.01$), the proportion of the individual's length to weight was found to be important.

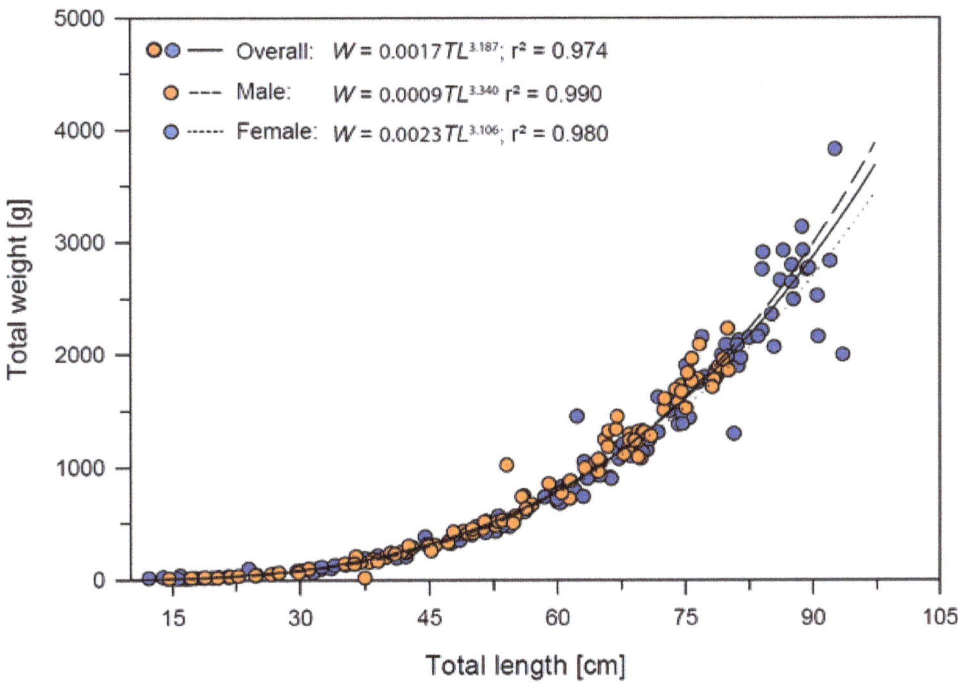

Figure 5. Total length–weight relationships of *Dipturus oxyrinchus* inhabiting the Northeastern Mediterranean Sea.

3.3.2. Growth Characteristics

The growth parameters were estimated for overall and each sex separately from the VBGM, the Robertson (logistic) and Gompertz growth model using nonlinear regression analysis, as presented in Table 3 and Figure 6. The relationship between total length and age was adequately described by the Robertson (logistic) growth model followed by the Gompertz growth model. The von Bertalanffy model performed relatively weakly compared to other growth models based on AIC values.

Table 3. The growth parameters for *Dipturus oxyrinchus* inhabiting the Northeastern Mediterranean Sea derived from different growth models.

Group	Growth Models	L_∞ (cm)	K (year^{-1})	t_0 (year)	I (year)	AIC
Overall	von Bertalanffy	154.01	0.06	−1.62		1401.60
	Logistic	104.14	0.35		4.99	1350.69
	Gompertz	128.40	0.19		4.39	1363.98
Female	von Bertalanffy	152.55	0.06	−1.62		788.90
	Logistic	103.54	0.36		4.91	751.83
	Gompertz	124.73	0.19		4.19	761.12
Male	von Bertalanffy	161.73	0.06	−1.63		788.90
	Logistic	103.23	0.35		4.98	595.02
	Gompertz	141.44	0.17		5.02	598.61

L_∞ = the asymptotic length (cm); K = the growth rate (year^{-1}); t_0 = the time when $L = 0$ (year); I = the age at the inflection point; and AIC = Akaike Information Criterion.

3.3.3. The Relative and Absolute Growth Rates and Condition Factor

The maximum absolute growth was estimated as 9.33 cm with 5–6 years of age. The maximum relative growth was calculated as 36.39% with 1–2 years of age. Both in absolute and relative growth were observed as minimum in the 11–12 age group (Table 4). Average condition factor value of population was estimated as 0.363. The highest and lowest condition factor values were estimated as 0.416 in age group 8 and 0.308 in age group 2, respectively (Table 5).

Table 4. The relative and absolute growth rates of for *Dipturus oxyrinchus* inhabiting the Northeastern Mediterranean Sea.

	Absolute Growth Rate			Relative Growth Rate		
Age Group	Overall	Female	Male	Overall	Female	Male
0–1	4.94	5.14	5.22	32.59	33.72	35.17
1–2	7.31	6.74	7.73	36.39	33.06	38.5
2–3	8.86	8.32	9.56	32.3	30.7	34.39
3–4	8.71	9.95	9.13	24.01	28.07	24.42
4–5	9.09	11.53	9.67	20.2	25.4	20.8
5–6	9.33	7.7	9.33	17.26	13.53	16.62
6–7	7.54	6.28	5.48	11.9	9.72	8.36
7–8	5.2	5.39	4.82	7.33	7.6	6.79
8–9	4.23	4.97	3.1	5.56	6.52	4.09
9–10	6.67	5.78	-	8.3	7.11	-
10–11	2.94	2.94	-	3.38	3.38	-
11–12	2.73	2.73	-	3.03	3.03	-

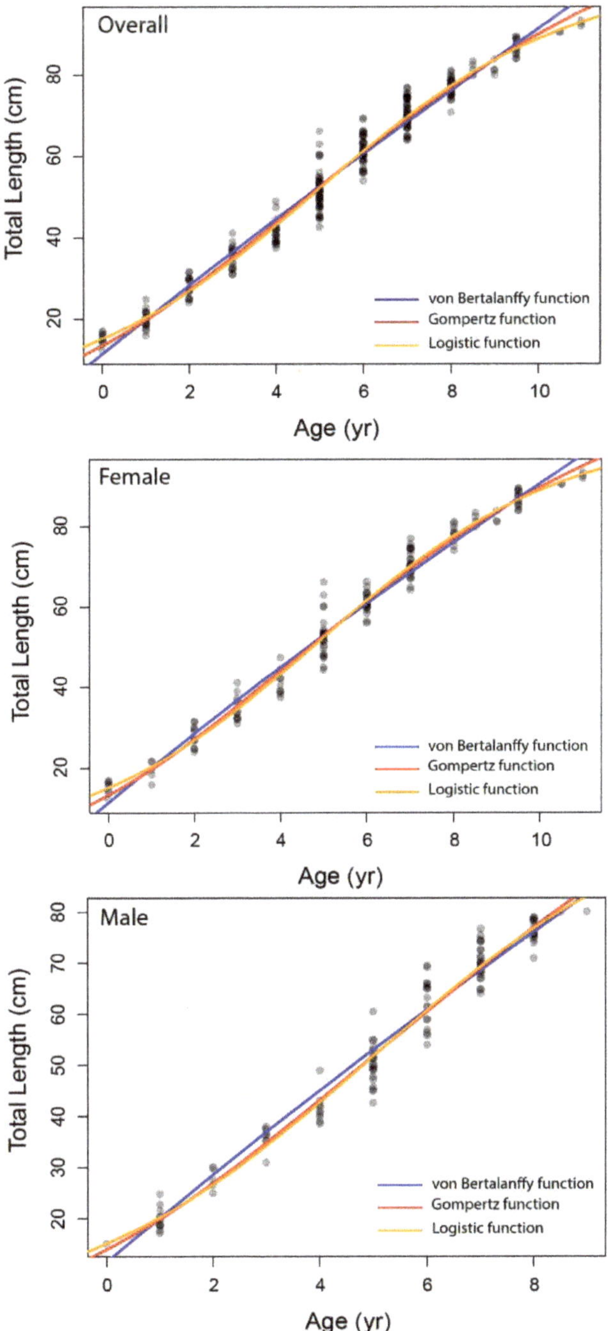

Figure 6. The different growth models fitted to the data of *Dipturus oxyrinchus* inhabiting the Northeastern Mediterranean Sea.

Table 5. The condition factor for *Dipturus oxyrinchus* inhabiting the Northeastern Mediterranean Sea.

	ALL		FEMALE		MALE	
AGE	*N*	**K**	*N*	**K**	*N*	**K**
0	10	0.409	8	0.454	2	0.273
1	15	0.317	4	0.221	11	0.279
2	14	0.308	9	0.377	5	0.267
3	30	0.311	17	0.298	13	0.303
4	20	0.335	5	0.25	15	0.349
5	42	0.36	24	0.268	18	0.38
6	28	0.375	18	0.32	10	0.416
7	45	0.383	24	0.365	21	0.389
8	11	0.416	3	0.404	8	0.418
9	24	0.386	15	0.389	9	0.386
10	9	0.405	9	0.411	-	-
11	4	0.351	4	0.322	-	-
12	3	0.362	3	0.362	-	-
		0.363 ± 0.037		0.342 ± 0.060		0.346 ± 0.060

4. Discussion

The overall length ranges recorded in our study were found to be smaller than those reported by Alkusairy and Saad [11] for the same species in Syrian waters (34.1–100 cm for females and 34.5–81.6 cm for males). Yigin and Işmen [8] reported a total length of 14.9–100 cm in females in Saros Bay (the North Aegean Sea), while Bellodi et al. [12] reported these values as 15.2–86.5 cm in males and 10.4–117.5 cm in females in Sardinian waters. Finally, Kadri et al. [9] reported 16.5–105 cm in females and 15.5–95 cm in males in the Gulf of Gabès (Southern Tunisia, Central Mediterranean). These values are very close to the values we found in our study. The Robertson (logistic) growth model estimates indicated that the long-nosed skate showed sexual dimorphism with females larger than males. These observations are consistent with other studies for *Dipturus oxyrinchus* in the Mediterranean Sea (e.g., Yigin and Işmen [8] in the North Aegean Sea, Kadri et al. [9] in Southern Tunisia and Bellodi et al. in Sardinian waters [12]). Sexual dimorphism appears to be a common feature for the Rajidae (e.g., blonde ray *Raja brachyura* [41] and thornback ray *Raja clavata* [42]). The percentage of females and males for all samples was 56.07% and 43.93%, respectively. This was not statistically different from the expected 1:1 ratio between the genders. All genders were equally distributed confirming the pattern proposed by Yigin and Işmen [8], Kadri et al. [9] and Bellodi et al. [12]. Growth bands were highly legible and visible in cross-sections, with an easily recognizable birthmark. No staining technique was used to determine the age of *Dipturus oxyrinchus*, with a maximum age of 12 noted in females and 9 in males. Differences in age determination after age 9 for both sexes are related to the small number of individuals being sampled. The age estimation process ensured a high level of count repeatability among readers (IAPE = 0.53%; %CV = 0.75; %PA = 90.98) and no signs sensitive to bias were detected among readers. These precision values are acceptable [43]. The b parameter of the length–weight relationship of *Dipturus oxyrinchus* showed positive allometric growth for both genders in the current study. The estimated *b* values for *Dipturus oxyrinchus* by region are shown in Table 6, which are very close to our study's findings. Other b values were reported as 3.539 for the south coasts of Portugal by Borges et al. [44] and 3.40 for North Aegean Sea by Filiz and Bilge [45]. These values differ from those of our study, which may be due to the small sample size of the fish or the fact that samples were made in different seasons. Relative and absolute growth rates and condition factor for *Dipturus oxyrinchus* were calculated for the first time in our study and therefore no comparison with other studies could be made. Absolute growth rates indicate actual growth between two years (ages) in terms of weight or length. Absolute growth rate decreases with age (t) and provides information about the years (ages) when growth is highest. The way the absolute growth rate is calculated depends strongly on the size the fish has reached. For comparison purposes, the relative growth rate may be more useful. This is used to determine age-related growth rate in natural populations [30].

Table 6. Length–weight relationship values for *Dipturus oxyrinchus* from different regions.

	Sexes	n	$L_{min-max}$ (cm)	$W_{min-max}$ (g)	a	b	r^2	Researchers
South coasts, Portugal	Σ	8	30.2–55.4	88.0–702.7	0.00048	3.539	0.99	Borges vd. [44]
North Aegean Sea	Σ	8	17.9–62.2	10.44–850.48	0.0007	3.40	0.99	Filiz and Bilge [45]
South of Sicily and South of Malta	Σ	-	23.0–124.0	-	0.00128	3.250	-	Geraci et al. [46]
Northwestern Mediterranean Sea	Σ	2	27.4–33.6	58–84	0.0025	3	-	Barría et al. [47]
Gulf of Saros, Turkey	Σ	118	10.0–63.2	9.0–4056.0	0.00423	3.291	0.998	Işmen et al. [48]
Sardinian waters, Italy	♀	531	10.9–115.5	-	0.0012	3.2498	0.98	Bellodi et al. [12]
	♂	448	14.7–101.5	-	0.0009	3.327	0.99	
Gulf of Saros, Turkey	Σ	179	14.9–100.0	8.0–4047.0	0.00083	3.35	0.996	
	♀	89	14.9–100.0	8.0–4047.0	0.00077	3.37	0.997	Yigin and Işmen [8]
	♂	90	15.2–86.5	8.0–2510.0	0.00088	3.34	0.996	
Mersin Bay, Northeastern Mediterranean Turkey	Σ	255	12.20–93.50	8.34–3828.0	0.0017	3.187	0.974	
	♀	143	12.20–93.50	8.50–3828.0	0.0023	3.106	0.990	In this study
	♂	112	14.60–80.10	8.34–2234.0	0.0009	3.340	0.980	

n: sample size, $L_{min-max}$: minimum-maximum length, $W_{min-max}$: minimum-maximum weight, a: intercept, b: slope of the equation, r^2: coefficient of determination.

Bellodi et al. [12] emphasized that the Gompertz function provided the best fit among the four growth models examined. Additionally, Liu et al. [49] stated that multiple model applications should be tested in elasmobranch age and growth studies. They also indicated that the Robertson (logistic) and Gompertz models provide the best fit for small-sized demersal skates/rays living in deep water. In our study, according to the AIC values, the logistic and Gompertz models were found to be more appropriate in describing the growth parameters of *Dipturus oxyrinchus*. These results agree with Liu et al. [49] and Bellodi et al. [12]. The logistic parameters determined in our study show that females attain a slightly larger asymptotic TL∞ (103.54 cm) than males (103.23 cm). In addition, the K values of the long-nosed skate were found to be similar for both genders. These growth rates appear to be similar with other skate species of similar size in the Mediterranean Sea. Bellodi et al. [12] suggested that the best growth model was the Gompertz model and estimated L∞ as 127.55 cm for all genders. This result is very close to the asymptotic value (L∞ = 128.40 cm) calculated with the Gompertz model for all individuals in our study. Yigin and Işmen [8] reported that, for both genders, the L∞ was 256.46 cm and the K was 0.04, the t_0 value was −1.17 year and the maximum age was 9 years. The above authors [8] found individuals aged up to 9 years and a total length of up to 100 cm in their study. The largest individual captured in our study (93.5 cm) was smaller than that reported by the above researchers, despite being older. This shows that there may be a mistake in reading the older age rings. This leads to an overcalculation of the asymptotic length. For females, Kadri et al. [9] reported the L∞ as 123.9 cm, the K as 0.08, the t_0 as −1.26 and the maximum age as 25; for males, the L∞ as 102.1 cm, the K as 0.12, the t_0 as −1.18 and the maximum age as 22. These researchers found that the oldest individual was 105 cm in total length and 25 years old. Again, these values were very high compared to those in our study, which therefore indicated to us that mistakes might have been made in the reading of the age rings of the older individuals. These errors may have caused an underestimation of the L∞ value. Considering that the largest *Dipturus oxyrinchus* caught in nature is 150 cm in TL, the L∞ value should not be lower than this value.

5. Conclusions

Our work ensured basic growth parameters and the best fit among the three growth models for the long-nosed skate and that the Robertson (logistic) growth model was the best model to describe the species growth. Results of the research showed that the long-nosed skate has life history features similar to other Rajidae species in the Mediterranean Sea and are long-lived and slow growing. The present study has provided the first analysis of growth characteristics and data for *Dipturus oxyrinchus* for conservative management plans in the Northeastern Mediterranean Sea.

Author Contributions: Conceptualization, N.B.; methodology, N.B.; software, F.V.O.; validation, F.V.O.; formal analysis, N.B.; investigation, F.V.O.; resources, F.V.O.; data curation, F.V.O.; writing—original draft preparation, N.B.; writing—review and editing, N.B.; visualization, N.B.; supervision, N.B.; project administration, N.B.; funding acquisition, N.B. All authors have read and agreed to the published version of the manuscript.

Funding: This research was funded by the Scientific Research Projects Coordination Unit of Firat University, grant number: SUF.15.04.

Institutional Review Board Statement: The animal study protocol was approved by the Animal Experimentation Ethics Committee of Firat University, Turkey (protocol code 54/109 and 27.05.2015).

Informed Consent Statement: Not applicable.

Data Availability Statement: Data associated with this research are available and can be obtained by contacting the corresponding author.

Acknowledgments: Specimen collection was permitted through the General Directorate of Fisheries and Aquaculture of the Turkish Ministry of Agriculture and Forestry (permit no: 26.05.2015–2065-42337). A part of this work was presented as Abstract at 11–13 May 2017, at the Ecology International Symposium in Kayseri, Turkey.

Conflicts of Interest: The authors declare no conflict of interest.

References

1. Serena, F. *Field Identification Guide to the Sharks and Rays of the Mediterranean and Black Sea*; Food and Agriculture Organization of The United Nation: Rome, Italy, 2005; p. 97, ISSN 1020-6868.
2. Ebert, D.A.; Stehmann, M.F.W. Sharks, Batoids, and Chimaeras of the North Atlantic. *FAO Species Cat. Fish. Purp.* **2013**, *7*, 1–523.
3. Stehmann, M.F.W.; Bürkel, D.L. Rajidae. In *Fishes of the North-Western Atlantic and the Mediterranean*; Whitehead, P.J.P., Bauchot, M.L., Hureau, J.C., Nielsen, J., Tortonese, E., Eds.; UNESCO: Paris, France, 1984; Volume I, pp. 163–196.
4. Golani, D.; Öztürk, B.; Başusta, N. *Fishes of the Eastern Mediterranean*, 1st ed.; Turkish Marine Research Foundation: Istanbul, Turkey, 2006; p. 259.
5. Weigmann, S. Annotated checklist of the living sharks, batoids and chimaeras (Chondrichthyes) of the world, with a focus on biogeographical diversity. *J. Fish Biol.* **2016**, *88*, 837–1037. [CrossRef] [PubMed]
6. Last, P.R.; Séret, B.; Stehmann, M.F.W.; Weigmann, S. Skates, Family Rajidae. In *Rays of the World*; Last, P.R., White, W.T., de Carvalho, M.R., Séret, B., Stehmann, M.F.W., Naylor, G.J.P., Eds.; CSIRO Publishing: Melbourne, Australia, 2016; pp. 204–363.
7. Ellis, J.; Abella, A.; Serena, F.; Stehmann, M.F.W.; Walls, R. Dipturus oxyrinchus. In *The IUCN Red List of Threatened Species*; IUCN Global Species Programme Red List Unit: UK, 2015; p. e.T63100A48908629. Available online: https://www.iucnredlist.org/species/63100/48908629 (accessed on 7 November 2022).
8. Yigin, C.; Işmen, A. Age, growth, reproduction and feed of long-nosed skate, *Dipturus oxyrinchus* (Linnaeus 1758) in Saros Bay, the north Aegean Sea. *J. Appl. Ichthyol.* **2010**, *26*, 913–919. [CrossRef]
9. Kadri, H.; Marouani, S.; Bradai, M.N.; Bouaïn, A.; Morize, E. Age, growth, longevity, mortality and reproductive biology of *Dipturus oxyrinchus*, (Chondrichthyes: Rajidae) off the Gulf of Gabès (Southern Tunisia, central Mediterranean). *J. Mar. Biol. Assoc. U. K.* **2014**, *95*, 569–577. [CrossRef]
10. Eronat, E.G.T.; Özaydın, O. Length-weight relationship of cartilaginous fish species from Central Aegean Sea (Izmir Bay and Sığacık Bay). *Ege J. Fish Aqua. Sci.* **2014**, *31*, 119–125. [CrossRef]
11. Alkusairy, H.H.; Saad, A.A. Some morphological and biological aspects of long-nosed skate, *Dipturus oxyrinchus* (Elasmobranchii: Rajiformes: Rajidae), in Syrian marine waters (Eastern Mediterranean). *Acta Ichthyol. Piscat.* **2017**, *47*, 371–383. [CrossRef]
12. Bellodi, A.; Porcu, C.; Cannas, R.; Cau, A.; Marongiu, M.F.; Mulas, A.; Vittori, S.; Follesa, M.C. Life-history traits of the long-nosed skate *Dipturus oxyrinchus*. *J. Fish Biol.* **2017**, *90*, 867–888. [CrossRef]
13. Başusta, N.; Başusta, A. Occurence of adult male and juveniles of *Dipturus Oxyrinchus* from North-Eastern Mediterranean Sea. *Ecol. Life Sci.* **2019**, *14*, 40–42. [CrossRef]
14. Griffiths, A.M.; Sims, D.W.; Johnson, A.; Lynghammar, A.; McHugh, M.; Bakken, T.; Genner, M.J. Levels of connectivity between longnose skate (*Dipturus oxyrinchus*) in the Mediterranean Sea and the north-eastern Atlantic Ocean. *Conserv. Genet.* **2011**, *12*, 577–582. [CrossRef]

15. Martin, L.K.; Cailliet, G.M. Age and growth determination of the bat ray, *Myliobatis california* Gill, in central California. *Copeia* **1988**, *3*, 762–773. [CrossRef]

16. Duman, O.V.; Başusta, N. Age and growth characteristics of marbled electric ray *Torpedo marmorata* (Risso, 1810) inhabiting Iskenderun Bay, North-eastern Mediterranean Sea. *Turk. J. Fish. Aquat. Sci.* **2013**, *13*, 551–559. [CrossRef]

17. Başusta, N.; Demirhan, S.A.; Çiçek, E.; Başusta, A.; Kuleli, T. Age and growth of the common guitarfish, *Rhinobatos rhinobatos* (Linnaeus, 1758), in Iskenderun Bay (northeastern Mediterranean, Turkey). *J. Mar. Biol. Assoc. U. K.* **2008**, *88*, 837–842. [CrossRef]

18. Başusta, N.; Aslan, E. Age and growth of bull ray *Aetomylaeus bovinus* (Chondrichthyes: Myliobatidae) from the northeastern Mediterranean coast of Turkey. *Cah. Biol. Mar.* **2018**, *59*, 107–114. [CrossRef]

19. Girgin, H.; Başusta, N. Testing staining techniques to determine age and growth of *Dasyatis pastinaca* (Linnaeus, 1758) captured in Iskenderun Bay, northeastern Mediterranean. *J. Appl. Ichthyol.* **2016**, *32*, 595–601. [CrossRef]

20. Başusta, N.; Sulikowski, J.A. The oldest estimated age for roughtail stingray (*Dasyatis centroura*; Mitchill, 1815) from the Mediterranean Sea. *J. Appl. Ichthyol.* **2012**, *28*, 641–642. [CrossRef]

21. Başusta, N.; Başusta, A.; Çiçek, E.; Cicia, A.M.; Sulikowski, J.A. First Estimates of Age and Growth of the Lusitanian Cownose Ray (*Rhinoptera marginata*) from the Mediterranean Sea. *J. Mar. Sci. Eng.* **2022**, *10*, 685. [CrossRef]

22. Sulikowski, J.A.; Morin, M.D.; Suk, S.H.; Howell, W.H. Age and growth estimates of the winter skate (*Leucoraja ocellata*) in the western gulf of Maine. *Fish. Bull.* **2003**, *101*, 405–413.

23. Sulikowski, J.A.; Kneebone, J.; Elzey, S.; Jurek, J.; Danley, P.D.; Howell, W.H.; Tsang, P.C.W. Age and growth estimates of the thorny skate (*Amblyraja radiata*) in the western Gulf of Maine. *Fish. Bull.* **2005**, *103*, 161–168.

24. Goldman, K.J.; Cailliet, G.M.; Andrews, A.H.; Natanson, L.J. Assessing the age and growth of chondrichthyan fishes. In *Biology of Sharks and Their Relatives*, 2nd ed.; Carrier, J.C., Musick, J.A., Heithaus, M.R., Eds.; CRC Press, Taylor & Francis: Boca Raton, FL, USA, 2012; pp. 423–451.

25. Chang, W.Y.B. A statistical method for evaluating the reproducibility of age determination. *Can. J. Fish. Aquat. Sci.* **1982**, *39*, 1208–1210. [CrossRef]

26. Beamish, R.J.; Fournier, D.A. A method for comparing the precision of a set of age determinations. *Can. J. Fish. Aquat. Sci.* **1981**, *38*, 982–983. [CrossRef]

27. Natanson, L.J.; Mello, J.J.; Campana, S.E. Validated age and growth of the porbeagle shark, *Lamna nasus*, in the western North Atlantic Ocean. *Fish. Bull.* **2002**, *100*, 266–278.

28. R Core Team. *R: A Language and Environment for Statistical Computing*; Version 4.1; R Foundation for Statistical Computing: Vienna, Austria, 2017.

29. Quinn, T.J., II; Deriso, R.B. *Quantitative Fish Dynamics*; Oxford University Press: New York, NY, USA, 1999.

30. Tıraşın, E.M. Balık populasyonlarının büyüme parametrelerinin araştırılması. *Turk. J. Zool.* **1993**, *17*, 29–82. (In Turkish)

31. Richer, W.E. *Computation and Interpretation of Biological Statistics of Fish Populations*; Bullettin of the Fisheries Research Board Canada: Ottowa, ON, Canada, 1975; Volume 191, p. 382.

32. Le Cren, C.D. The length-weight relationship and seasonal cycle in gonad weight and condition in Perch, *Perca fluviatilis*. *J. Anim. Ecol.* **1951**, *20*, 201–219. [CrossRef]

33. von Bertalanffy, L. A quantitative theory of organic growth (inquiries on growth laws. II). *Hum. Biol.* **1938**, *10*, 181–213.

34. Beverton, R.J.H.; Holt, S.J. *On the Dynamics of Exploited Fish Populations*; Fisheries Investigations, Series 2; Springer Science+Business Media, B.V.: London, UK, 1957.

35. Gompertz, B. On the nature of the function expressive of the law of human mortality and on a new mode of determining the value of life contingencies. *Philos. Trans. R. Soc. Lond.* **1825**, *115*, 515–585.

36. Ricker, W.E. Growth rates and Models. In *Bioenergetics and Growth*; Hoar, W.S., Randal, D.J., Brett, J.R., Eds.; Academic Press: Cambridge, MA, USA, 1979; Fish ISSN:1307-3311 Physiology; Volume VIII, pp. 677–743.

37. Smart, J.J.; Chin, A.; Tobin, A.J.; Simpfendorfer, C.A. Multimodel approaches in 420 shark and ray growth studies: Strengths, weaknesses and the future. *Fish Fish.* **2016**, *17*, 955–971. [CrossRef]

38. Tjørve, K.M.C.; Tjørve, E. The use of Gompertz models in growth analyses, and new Gompertz-model approach: An addition to the Unified-Richards family. *PLoS ONE* **2017**, *12*, e0178691. [CrossRef]

39. Başusta, N.; Başusta, A.; Tıraşın, E.M.; Sulikowski, J.A. Age and growth of the blackchin guitarfish *Glaucostegus cemiculus* (Geoffroy Saint-Hilaire, 1817) from Iskenderun Bay (Northeastern Mediterranean). *J. Appl. Ichthyol.* **2019**, *36*, 880–887. [CrossRef]

40. Akaike, H. Information theory as an extension of the maximum likelihood principle. In Proceedings of the Second International Symposium on Information Theory, Armenia, USSR, 2–8 September 1971; Petrov, B.N., Csaki, F., Eds.; Akadémiai Kiadó: Budapest, Hungary, 1973; pp. 267–281.

41. Porcu, C.; Bellodi, A.; Cannas, R.; Marongiu, M.F.; Mulas, A.; Follesa, M.C. Life-history traits of a commercial ray, *Raja brachyura* from the central western Mediterranean Sea. *Medit. Mar. Sci.* **2015**, *16*, 90–102. [CrossRef]

42. Gallagher, M.J.; Nolan, C.P.; Jeal, F. Age, growth and maturity of the commercial ray species from the Irish Sea. *J. Northwest Atl. Fish. Sci.* **2005**, *35*, 47–66. [CrossRef]

43. Campana, S.E. Accuracy, precision and quality control in age determination, including of the use and abuse of age validation methods. *J. Fish Biol.* **2001**, *59*, 197–242. [CrossRef]

44. Borges, T.C.; Olim, S.; Erzini, K. Weight-length relationship for fish species discarded in commercial fisheries of the Algarve (southern Portugal). *J. Appl. Ichthyol.* **2003**, *19*, 394–396. [CrossRef]

45. Filiz, H.; Bilge, G. Length-weight relationships of 24 fish species from the North Aegean Sea, Turkey. *J. Appl. Ichthyol.* **2004**, *20*, 431–432. [CrossRef]
46. Geraci, M.L.; Ragonese, S.; Scannella, D.; Falsone, F.; Gancitano, V.; Mifsud, J.; Gambin, M.; Said, A.; Vitale, S. Batoid abundances, spatial distribution, and life history traits in the Strait of Sicily (Central Mediterranean Sea): Bridging a knowledge gap through three decades of survey. *Animals* **2021**, *11*, 2189. [CrossRef] [PubMed]
47. Barría, C.; Navarro, J.; Coll, M.; Fernancez-Arcaya, U.; Sáez-Liante, R. Morphological parameters of abundant and threatened chondrichthyans of the northwestern Mediterranean Sea. *J. Appl. Ichthyol.* **2015**, *31*, 114–119. [CrossRef]
48. Işmen, A.; Özen, Ö.; Altınağaç, U.; Özekinci, U.; Ayaz, A. Weight–length relationships of 63 fish species in Saros Bay, Turkey. *J. Appl. Ichthyol.* **2007**, *23*, 707–708. [CrossRef]
49. Liu, K.M.; Wu, C.B.; Joung, S.J.; Tsai, W.P.; Su, K.Y. Multi-model approach on growth estimation and association with life history trait for elasmobranchs. *Front. Mar. Sci.* **2021**, *8*, 591692. [CrossRef]

Article

Long-Term Changes in the Distribution and Abundance of Nine Deep-Water Skates (Arhynchobatidae: Rajiformes: Chondrichthyes) in the Northwestern Pacific

Alexei M. Orlov [1,2,3,4,5,*] and Igor V. Volvenko [6]

1 Laboratory of Oceanic Ichthyofauna, Shirshov Institute of Oceanology, Russian Academy of Sciences, 117218 Moscow, Russia
2 Laboratory of Behavior of Lower Vertebrates, A.N. Severtsov Institute of Ecology and Evolution, Russian Academy of Sciences, 119071 Moscow, Russia
3 Department of Ichthyology and Hydrobiology, Tomsk State University, 634050 Tomsk, Russia
4 Department of Ichthyology, Dagestan State University, 367000 Makhachkala, Russia
5 Laboratory of Marine Biology, Caspian Institute of Biological Resources, Dagestan Federal Research Center, Russian Academy of Sciences, 367000 Makhachkala, Russia
6 Laboratory for Modeling of Biological Processes, Pacific Branch of the Russian Federal Research Institute of Fisheries and Oceanography, 690091 Vladivostok, Russia
* Correspondence: orlov@vniro.ru or orlov.am@ocean.ru

Simple Summary: Changes in the spatial distribution and abundance of common skate species of the genus *Bathyraja* in the Russian waters of the Northwestern Pacific, where their fishery is considered promising, but still not developed, are traced. Over the past six decades, the boundaries of their ranges have slowly fluctuated near some average annual position and abundance trends have coincided with previously identified ecosystem rearrangements under the influence of climatic and oceanological changes. It will be useful to compare the results obtained with abundance trends in the near future, when fishing for these species is expected to intensify here, as well as with other areas where fishing pressure for these species already exists.

Abstract: Based on the analysis of long-term data from bottom trawl surveys (1977–2021), changes in the spatial distribution, position of the boundaries of the ranges and the catch rates of the nine most common deep-sea skates of the genus *Bathyraja* in the Russian waters of the Northwestern Pacific (*B. violacea, B. aleutica, B. matsubarai, B. maculata, B. bergi, B. taranetzi, B. minispinosa, B. interrupta,* and *B. isotrachys*) are considered. During the surveyed period, significant changes in the spatial distribution were observed, which are probably due to both subjective reasons (changes in the number of trawling stations, surveyed depths, etc.) and climatic changes. No monotonous displacement of the northern or southern boundaries of the range or its center in one direction was observed in any area of any species during the entire observation period. At the same time, shifts in the boundaries of the ranges of different species in different areas occurred for different decades, i.e., the boundaries of the ranges slowly fluctuated or "pulsed" near some average annual position. In general, from the 1970s to the 1980s, the number of skates grew; from the 1980s to the 1990s, it decreased; from the 1990s to the 2000s, it fluctuated at the achieved level; from the 2000s to the 2010s, it grew again; and from the 2010s to the 2020s, it decreased again. These trends coincide with previously identified ecosystem rearrangements under the influence of climatic and oceanological changes. The identification of links between changes in spatial distribution, range boundaries and catch rates with climatic and oceanological factors require separate additional studies.

Keywords: spatial distribution; catch dynamics; latitudinal range; western Bering Sea; Sea of Okhotsk; northwestern Sea of Japan; Sakhalin Island; Pacific waters; Kuril Islands; Kamchatka

Citation: Orlov, A.M.; Volvenko, I.V. Long-Term Changes in the Distribution and Abundance of Nine Deep-Water Skates (Arhynchobatidae: Rajiformes: Chondrichthyes) in the Northwestern Pacific. *Animals* **2022**, *12*, 3485. https://doi.org/10.3390/ani12243485

Academic Editor: Martina Francesca Marongiu

Received: 23 October 2022
Accepted: 6 December 2022
Published: 9 December 2022

Publisher's Note: MDPI stays neutral with regard to jurisdictional claims in published maps and institutional affiliations.

1. Introduction

Representatives of the order Rajiformes are widely distributed in the North Pacific Ocean. They play an important role in marine ecosystems [1], being mainly top predators consuming many valuable commercial invertebrates and fish [2–9].

Skates have biological features common to most cartilaginous fish (slow growth rates, late sexual maturation and low reproductive rates), which makes their stocks extremely vulnerable to fishing [10]. However, they are commercially important fishing targets in many countries [11], particularly in Southeast Asia, where skate wings are used for consumption [12]. The flesh of skates is rich in nutrients and contains an almost complete set of essential amino acids. Their livers, being rich in vitamin A, though less so than those of sharks, can be used for the extraction of oil for veterinary and medical purposes. Skates have strong and thick skin that may be used for the manufacture of many leather articles. The flesh of skates is suitable for processing to fish meat jelly, which is used in the preparation of various Japanese national dishes [12–15].

The biomass of skates within Russian waters of the northwestern Pacific amounted to 677 thousand tons [16], and in some areas they account for up to 10% of the total biomass of groundfish [17]. Their fishing in Russian waters has long been considered promising [18,19], but has not yet been developed. In the Far Eastern Fisheries basin of Russia in the 1990s, only 11.4–11.7 thousand tons were recommended for harvesting; and in the 2000s, 7.2–11.9; in the 2010s, 11.2–14.0; and in the 2020s, 11.2–11.3 thousand tons (first within estimated total allowable catch—TAC, then within possible catch—PC or recommended catch—RC). Skates were not separated by species, but their actual total catch never reached the recommended values, since these fish got on board Russian fishing vessels as bycatch only, were not retained and utilized and were subsequently discarded [20–22]. The situation has changed in recent years after some Russian companies started to export skate wings to China and the British Virgin Islands at prices between 0.9 and 1.3 USD per kg, while they also may be profitably marketed in Japan and Korea [23,24].

According to long-term monitoring data [25,26], at least 28 species of skates are regularly found within the Russian waters and adjacent areas of the North Pacific, and all of them have commercial value [27,28]. To date, there is published information about the general features of the spatial distribution and dynamics of catches of the most common and some rare species in the Pacific waters off the North Kuril Islands and southeastern Kamchatka [17,29–31]. Recently, data on the spatial distribution and dynamics of catches of Alaska skate *Bathyraja parmifera*, Okhotsk skate *Bathyraja violacea* and Aleutian skate *Bathyraja aleutica* throughout the North Pacific have also been described [32–34]. Additionally, there is information on the distribution and biology of the bottom skate *Bathyraja bergi* in the Russian waters of the Sea of Japan [35]. Nevertheless, the features of spatial distribution, dynamics of skate catches and their long-term changes have not been considered in detail for Russian waters of the northwestern Pacific. Meanwhile, insufficient information prevents the evaluation of correct values of the TAC, PC or RC differentiated by species and area, and the subsequent rational exploitation of skates stocks in the Russian waters of the northwestern Pacific, which during the period of increasing intensity of their fishing may negatively affect the condition of populations of these fish [36].

The goal of this paper is to identify changes in the spatial distribution and catch rates of the nine most common skates within the Russian Exclusive Economic Zone (EEZ) of the Northwestern Pacific (the northwestern Sea of Japan and the Sea of Okhotsk, western Bering Sea and Pacific waters off the Kuril Islands and Eastern Kamchatka) deep-water skates of the genus *Batyraja* based on the analysis of long-term data of bottom trawl surveys.

2. Material and Methods

Material was obtained from database [37] supplemented with data from the recent bottom trawl surveys conducted by the Pacific Branch of the Russian Federal Research Institute of Fisheries and Oceanography through 2021. We selected from it trawl stations located within the Russian EEZ (there were 39,420 of them, see Figure 1) and calculated the occurrence of all skate species over the research period (1977–2021). Species encountered less than 100 times are not included in the list. *B. parmifera* and *Bathyraja smirnovi* were also excluded from it, since until recently, information about the distribution of these two morphologically similar species was very contradictory [27,32,38–44]; as a result, research teams confused them with each other. Thus, nine of the most common species were identified, which became the objects of this study (Table 1).

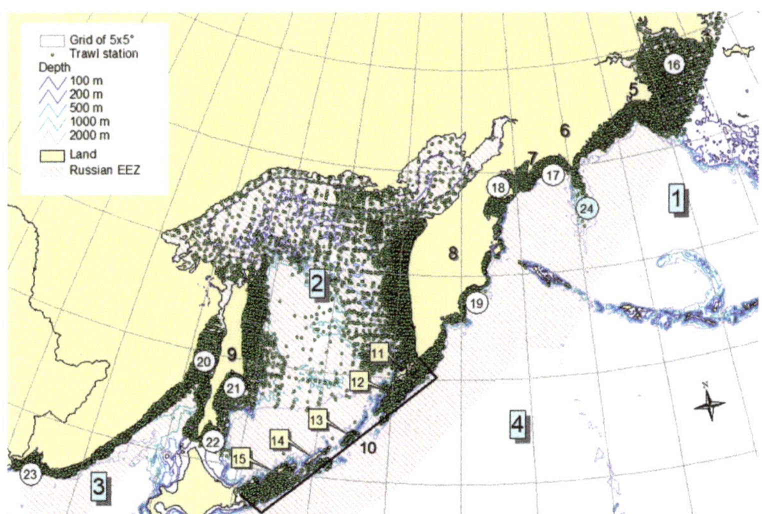

Figure 1. Location of trawl stations in the surveyed area. The numbers indicate the geographical names mentioned in the text: 1 — Bering Sea, 2 — Sea of Okhotsk, 3 — Sea of Japan, 4 — Pacific Ocean, 5 — Navarin Cape, 6 — Koryak coast, 7 — Goven Cape, 8 — Kamchatka, 9 — Sakhalin, 10 — Kuril Islands, 11 — Paramushir Isl., 12 — Onekotan Isl., 13 — Simushir Isl., 14 — Urup Isl., 15 — Iturup Isl., 16 — Gulf of Anadyr, 17 — Olyutor Bay, 18 — Karagin Bay, 19 — Kronotsk Bay, 20 — Tatar Strait, 21 — Terpeniya Bay, 22 — Aniva Bay, 23 — Peter the Great Bay, 24 — Shirshov Underwater Ridge.

Table 1. Number of records of nine skate species within the Russian EEZ, 1977–2021.

Species	Common Name	Bering Sea	Sea of Okhotsk	Sea of Japan	Pacific Ocean	All Areas Combined
Bathyraja violacea (Suvorov, 1935)	Okhotsk skate	578	1340	4 *	835	2757
Bathyraja aleutica (Gilbert, 1896)	Aleutian skate	628	260	0	265	1153
Bathyraja matsubarai (Ishiyama, 1952)	Dusky-purple skate	323	290	0	244	857
Bathyraja maculata (Ishiyama et Ishihara, 1977)	Whiteblotched skate	146	177	0	177	500
Bathyraja bergi (Dolganov, 1983)	Bottom skate	0	16	254	33	303
Bathyraja taranetzi (Dolganov, 1983)	Mud skate	125	4 *	0	168	297
Bathyraja minispinosa (Ishiyama et Ishihara, 1977)	Whitebrow skate	155	46	0	36	237
Bathyraja interrupta (Gill et Townsend, 1897)	Sandpaper skate	176	0	0	0	176
Bathyraja isotrachys (Günther, 1877)	Challenger's skate	0	77	0	29	106
All species combined		2131	2210	258	1787	6386

Note: Species are sorted in descending order of total occurrence. Areas with records of particular species <15 (marked with *) excluded from the analysis.

Identification of skates to species level during past surveys by research teams was conducted using available guides [45–47]. During recent surveys, scientists for this purpose additionally used publications [48,49]. Scientific (Latin) and common names of skates used in this paper are provided according to the most authoritative sources [50,51].

To analyze long-term changes in their distribution and abundance, we divided the time scale into six decades: (1) 1970s, (2) 1980s, (3) 1990s, (4) 2000s, (5) 2010s and (6) 2020s. The sample sizes thus obtained are shown in Table 2.

Table 2. Description of the data used for the analysis of spatial distribution and relative abundance of nine skate species in the Russian waters of the northwestern Pacific, 1977–2021.

Parameter	Area	Decade						Total
		1	2	3	4	5	6	
Survey years	Bering Sea	1977–1979	1980–1989	1990–1999	2000–2008	2010–2019	2020–2021	1977–2021
	Sea of Okhotsk	1977–1979	1980–1989	1990–1999	2000–2009	2010–2019	2020–2021	1977–2021
	Sea of Japan	1978	1981–1989	1990–1999	2000–2009	2010–2019	-	1978–2019
	Pacific Ocean	1977–1979	1980–1989	1990–1998	2000–2008	2012–2019	2021	1977–2021
	Total	1977–1979	1980–1989	1990–1999	2000–2009	2010–2019	2020–2021	1977–2021
Surveyed depths, m	Bering Sea	10–1150	14–1400	20–1000	12–773	10–968	18–957	10–1400
	Sea of Okhotsk	30–1350	10–2000	10–1100	5–1100	9–981	11–459	5–2000
	Sea of Japan	23–255	14–740	5–700	5–940	5–926	-	5–940
	Pacific Ocean	20–1500	10–1860	30–1530	23–1260	20–1054	70–480	10–1860
	Total	10–1500	10–2000	5–1530	5–1260	5–1054	11–957	5–2000
Stations at depths $\geq$400 m, %	Bering Sea	11.4	15.4	7.0	6.5	11.8	24.3	11.5
	Sea of Okhotsk	66.5	21.4	13.2	10.8	14.6	1.7	17.4
	Sea of Japan	0.0	13.3	7.6	12.1	14.7	0.0	12.4
	Pacific Ocean	11.8	27.5	17.2	17.0	26.8	5.9	20.8
	Total	21.4	19.2	10.2	11.8	14.5	8.0	15.1
Number of stations	Bering Sea	852	2942	1760	1343	1617	235	8749
	Sea of Okhotsk	358	5007	1321	2267	2931	583	12,467
	Sea of Japan	56	3212	1986	3011	3977	0	12,242
	Pacific Ocean	677	2329	982	1656	250	68	5962
	Total	1943	13,490	6049	8277	8775	886	39,420
Surveyed area, km^2	Bering Sea	70.7	297.3	282.0	82.7	103.5	9.0	845.2
	Sea of Okhotsk	35.3	429.9	119.7	191.1	107.2	17.5	900.7
	Sea of Japan	2.9	274.8	100.0	160.0	84.1	0.0	621.8
	Pacific Ocean	71.2	295.6	236.8	753.6	9.5	2.4	1369.1
	Total	180.1	1297.5	738.5	1187.5	304.3	28.9	3736.8

Note: Surveyed area was calculated as the sum of areas covered by trawl hauls. Area of a trawl haul was calculated by multiplying trawl horizontal opening by hauling distance.

For each species in each decade and area, we found the maximum (max), minimum (min) and average (avg) latitudes of all catches to determine the northern and southern boundaries of the range, and its center in decimal degrees. The abundance of the species at each capture point, as well as the average for decades and areas, was estimated by catch per unit of effort (CPUE). For unit of effort, we took the area of a trawl haul (in square kilometers), which was calculated by multiplying the trawl horizontal opening by hauling distance, i.e., the unit of measure of CPUE is ind./km^2.

In addition to the decadal changes in the four variables, which are described in detail in this paper, we estimated their total trends for all time using four simple regression models: (1) Linear: $y = a + b \cdot x$, (2) Logarithmic: $y = a + b \cdot ln(x)$, (3) Exponential: $y = exp(a + b \cdot x)$ and (4) Multiplicative: $y = a\, x^b$, where x is the decade number and y is the variable being approximated. In each particular case the best model was chosen by the minimum of the residual and p-value, and the maximum of the correlation coefficient (r) modulo. We hoped that the r sign would help classify trends as positive or negative. However, the variations in the variables turned out to be not monotonous, poorly approximated by smooth lines. Accordingly, the r values in most cases turned out to be non-significant at the 95% confidence level. All graphs, their equations, and r and p values are given in the Supplement, and the results are shown in the last columns of Tables 3 and 4.

Table 3. Trends in shifts of the northern (max) and southern (min) boundaries of a range and its center (avg) of nine skate species in the western Bering Sea (B), Sea of Okhotsk (O), northwestern Sea of Japan (J), and Pacific Ocean (P): "+"—northward shift, positive correlation (green); "–"— southward shift, negative correlation (brown); "0"— no shift (yellow); "?"— unknown, no records of species; "ns"— non-significant correlation (yellow).

Species	Latitudinal Boundary	Area	Shift from One Decade to Another					Correlation
			1–2	2–3	3–4	4–5	5–6	
Bathyraja violacea	max	B	+	−	+	+	−	+
		O	+	−	+	+	−	ns
		P	+	−	+	−	−	−
	avg	B	+	−	+	+	+	ns
		O	+	+	+	−	+	+
		P	−	−	−	+	+	ns
	min	B	+	−	−	+	+	ns
		O	−	+	−	+	−	ns
		P	−	+	−	−	+	ns
Bathyraja aleutica	max	B	+	−	−	+	−	ns
		O	+	−	−	+	?	ns
		P	+	+	−	+	−	ns
	avg	B	−	−	−	+	−	ns
		O	+	+	−	+	?	+
		P	+	−	−	+	−	ns
	min	B	−	+	+	−	0	ns
		O	−	+	+	+	?	ns
		P	−	+	−	−	+	ns
Bathyraja matsubarai	max	B	?	−	+	+	−	ns
		O	?	−	−	+	?	ns
		P	?	−	+	+	−	ns
	avg	B	?	−	+	−	−	ns
		O	?	+	−	+	?	ns
		P	?	−	−	+	+	ns
	min	B	?	+	+	−	+	ns
		O	?	+	+	−	?	+
		P	?	+	−	+	+	+
Bathyraja maculata	max	B	?	−	+	−	−	ns
		O	?	−	+	+	−	ns
		P	+	−	−	+	−	ns
	avg	B	?	−	+	+	−	ns
		O	?	+	−	+	+	+
		P	+	−	−	+	−	ns
	min	B	?	+	+	−	+	ns
		O	?	+	−	−	+	ns
		P	−	+	−	+	+	ns
Bathyraja bergi	max	O	?	?	?	?	?	ns
		P	?	?	?	?	?	ns
		J	?	+	+	+	?	ns
	avg	O	?	?	?	?	?	ns
		P	?	?	?	?	?	ns
		J	?	−	+	−	?	ns
	min	O	?	?	?	?	?	ns
		P	?	?	?	?	?	ns
		J	?	−	−	−	?	ns
	max	P	?	−	−	+	−	ns
	avg	B	?	?	?	−	+	ns
		P	?	−	−	+	+	ns
	min	B	?	?	?	−	+	ns
		P	?	+	−	−	+	ns
Bathyraja minispinosa	max	B	?	?	?	+	−	ns
		O	?	−	−	+	?	ns
		P	?	−	+	+	−	ns
	avg	B	?	?	?	+	−	ns
		O	?	+	−	+	?	ns
		P	?	−	−	+	+	ns
	min	B	?	?	?	−	0	ns
		O	?	−	−	−	?	ns
		P	?	−	−	+	+	ns

Table 3. *Cont.*

Species	Latitudinal Boundary	Area	Shift from One Decade to Another					Correlation
			1–2	2–3	3–4	4–5	5–6	
Bathyraja interrupta	max	B	+	+	?	?	?	ns
	avg	B	+	+	?	?	?	ns
	min	B	−	+	?	?	?	ns
Bathyraja isotrachys	max	O	?	?	?	?	?	ns
		P	?	+	−	−	+	ns
	avg	O	?	?	?	?	?	ns
		P	?	+	−	+	+	ns
	min	O	?	?	?	?	?	ns
		P	?	+	−	+	+	ns

Note: Correlation coefficients (r), p-values, regression equations and plots are presented in the Supplementary Figures S1–S27.

Table 4. Trends in CPUE changes of nine skate species in the western Bering Sea (B), Sea of Okhotsk (O), northwestern Sea of Japan (J), Pacific Ocean (P), and all areas combined (T): "+"—increasing, positive correlation (green); "–"—decreasing, negative correlation (brown); "0"—no changes (yellow); "ns"—non-significant correlation (yellow).

Species	Area	From One Decade to Another					Correlation
		1–2	2–3	3–4	4–5	5–6	
Bathyraja violacea	B	+	+	+	+	−	+
	O	−	−	−	+	−	ns
	P	+	−	−	+	−	ns
	T	+	−	−	+	+	ns
Bathyraja aleutica	B	+	−	+	+	+	+
	O	+	+	+	+	−	ns
	P	+	−	−	+	+	ns
	T	+	−	+	+	+	+
Bathyraja matsubarai	B	+	−	+	+	+	+
	O	+	+	−	+	−	ns
	P	+	−	−	+	−	ns
	T	+	−	+	+	+	+
Bathyraja maculata	B	+	+	+	+	−	+
	O	+	−	+	+	−	ns
	P	+	−	−	+	+	ns
	T	+	−	−	+	+	ns
Bathyraja bergi	O	+	−	0	+	−	ns
	P	+	−	0	+	−	ns
	J	+	+	−	+	−	ns
	T	+	+	−	+	−	ns
Rhinoraja taranetzi	B	+	−	+	+	−	ns
	P	+	−	+	+	−	+
	T	+	−	+	+	−	+
Bathyraja minispinosa	B	+	−	+	+	−	ns
	O	+	+	−	+	−	ns
	P	+	+	−	+	−	ns
	T	+	+	−	+	−	+
Bathyraja interrupta	B	+	−	−	+	−	ns
Bathyraja isotrachys	O	+	−	0	+	−	ns
	P	+	−	+	+	−	+
	T	+	−	+	+	−	+

Note: Correlation coefficients (r), p-values, regression equations and plots are presented in the Supplementary Figures S28–S36.

3. Results

3.1. Okhotsk Skate

Spatial distribution (Figure 2). During the research period, the spatial distribution of the Okhotsk skate has changed significantly. In the first decade, it was caught only in the waters off the central and northern Kuril Islands and southern Kamchatka. The maximum catches were taken off the southwestern Kamchatka in the Sea of Okhotsk and off the central part of the East Kamchatka coast in the Pacific Ocean. In the 1980s, this species was found most widely within the surveyed area, i.e., from the Southern Kuril Islands in the south to Navarin Cape in the western Bering Sea in the north. In the Sea of Okhotsk, it was observed almost everywhere except in the southern part of the sea, and the maximum catches were recorded off both coasts of Kamchatka and the northern Kuril Islands. In the next decade, the distribution area of the Okhotsk skate was reduced. It was recorded only in the western Bering Sea, off the northern Kuril Islands, western and southeastern Kamchatka. Maximum catches were recorded in the Sea of Okhotsk off the central part of the West Kamchatka coast. In the 2000s, the distribution area of the Okhotsk skate expanded again. It began to occur off the eastern Sakhalin, in the northern Sea of Okhotsk, and off the southern and central Kuril Islands, but ceased to occur near the East Kamchatka coast. The maximum catches during this period were recorded off the southern and northern Kuril Islands and in the western Bering Sea. The spatial distribution of the Okhotsk skate in the 2010s did not fundamentally differ from that in the 2000s. However, the maximum catches shifted to the waters of the western and southeastern coasts of Kamchatka, and also remained off the northern Kuril Islands. At the beginning of the current decade, the Okhotsk skate disappeared from catches in the northern Sea of Okhotsk and off the eastern Sakhalin, but it was still observed in the western Bering Sea, off the northern Kuril Islands and western Kamchatka with maximum catches in the waters off the northern Kuril Island.

Boundaries of the ranges (Figure 3, Supplementary Figures S1–S3). In general, the distribution center of the Okhotsk skate shifted significantly to the north during the research period. At the same time, the trends in different areas differed significantly. Whereas the northern distribution boundaries in the western Bering Sea and Sea of Okhotsk have shifted northward, in the Pacific waters of the Kuril Islands and eastern Kamchatka, they shifted south. Similar trends in orientation, but not so pronounced, were noted in relation to distribution centers. The southern borders of the ranges showed slightly different dynamics. Whereas in the Sea of Okhotsk and Pacific waters the borders shifted northward, in the Bering Sea, they showed a weak but opposite trend.

Catch dynamics (Figure 4 and Supplementary Figure S28). In general, during the research period, the value of catches per unit effort of Okhotsk skate increased by almost 3 times. At the same time, an increase in catches was noted in the western Bering Sea and Pacific waters (especially sharp, starting from the fourth decade). In the Sea of Okhotsk, on the contrary, a decrease in the catches rates (especially at the beginning of the research period) was observed.

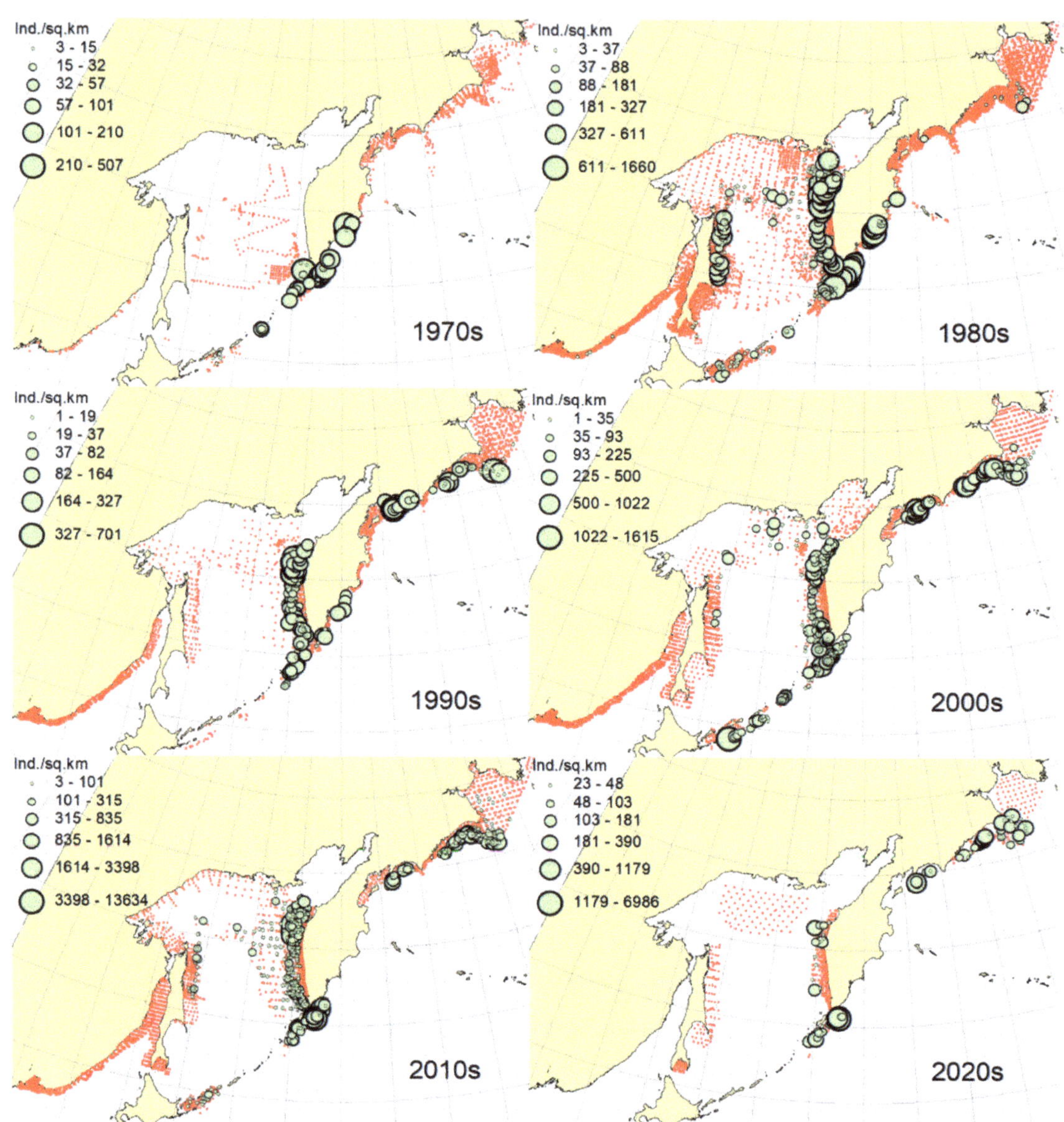

Figure 2. Decadal changes in the spatial distribution of the Okhotsk skate *Bathyraja violacea* in the Northwestern Pacific. Here and on Figures 2–9, red dots indicate stations where this species was not recorded.

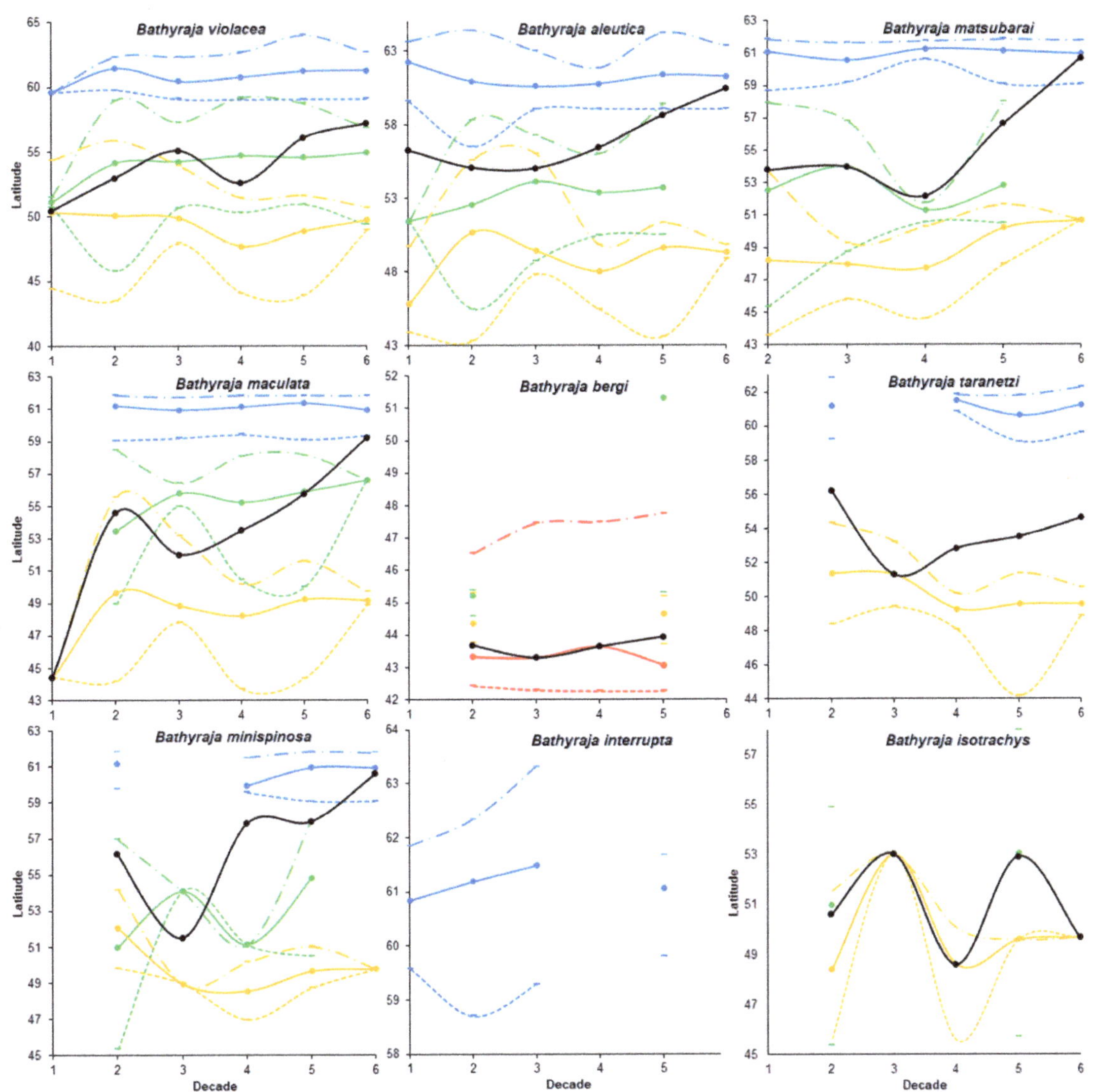

Figure 3. Decadal changes in the position of the range of nine skate species in the Northwestern Pacific. The Bering Sea is in blue, Sea of Okhotsk in green, Sea of Japan in red, and Pacific Ocean in yellow, and all areas combined are in black. The northern boundaries of the ranges are shown by dash-dotted lines, southern boundaries by dashed lines, and middle of the ranges by solid lines with circles.

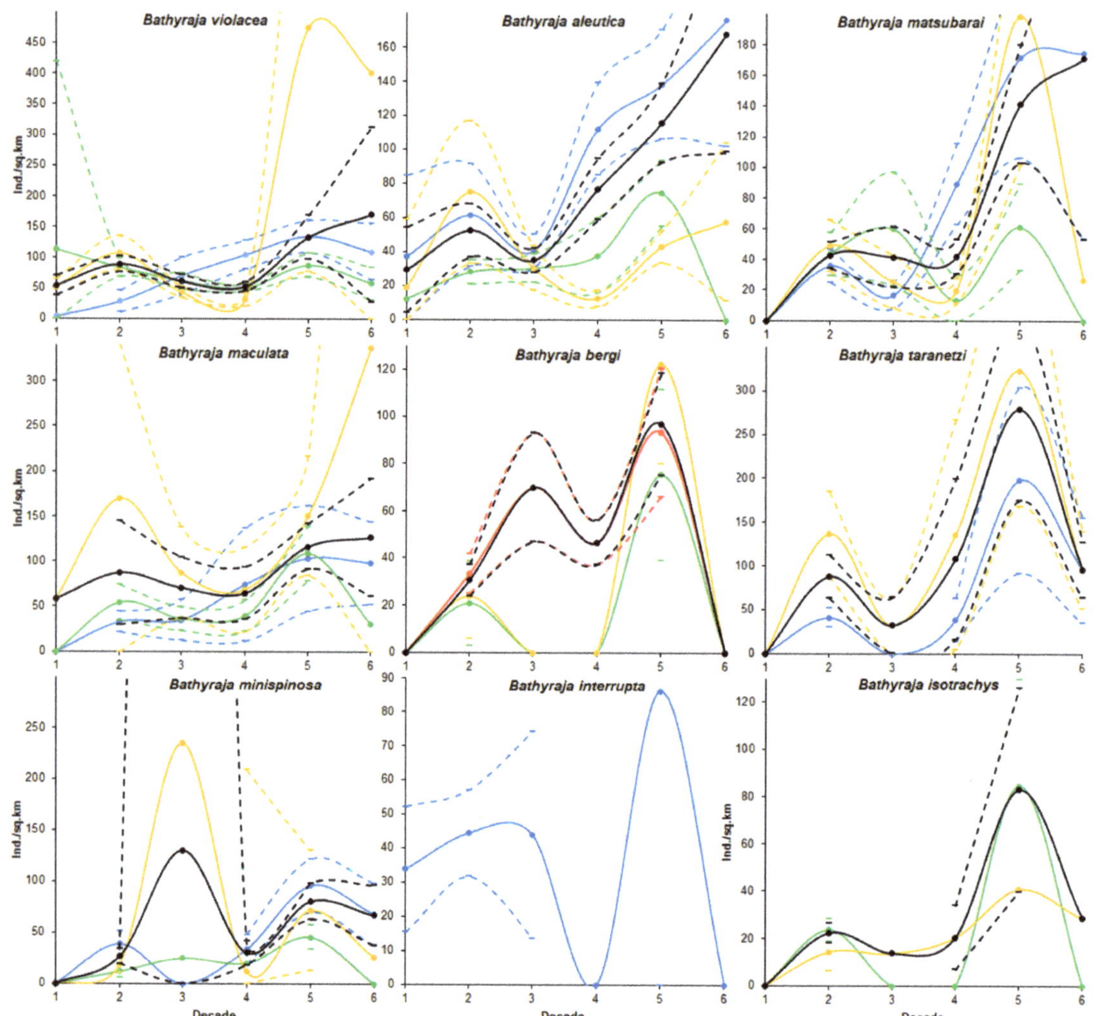

Figure 4. Decadal changes in the CPUE of nine skate species in the Northwestern Pacific. The Bering Sea is in blue, Sea of Okhotsk in green, Sea of Japan in red, and Pacific Ocean in yellow, and all areas combined in black. The solid lines with circles indicate the average CPUEs, and the dotted lines indicate the 95% confidence intervals.

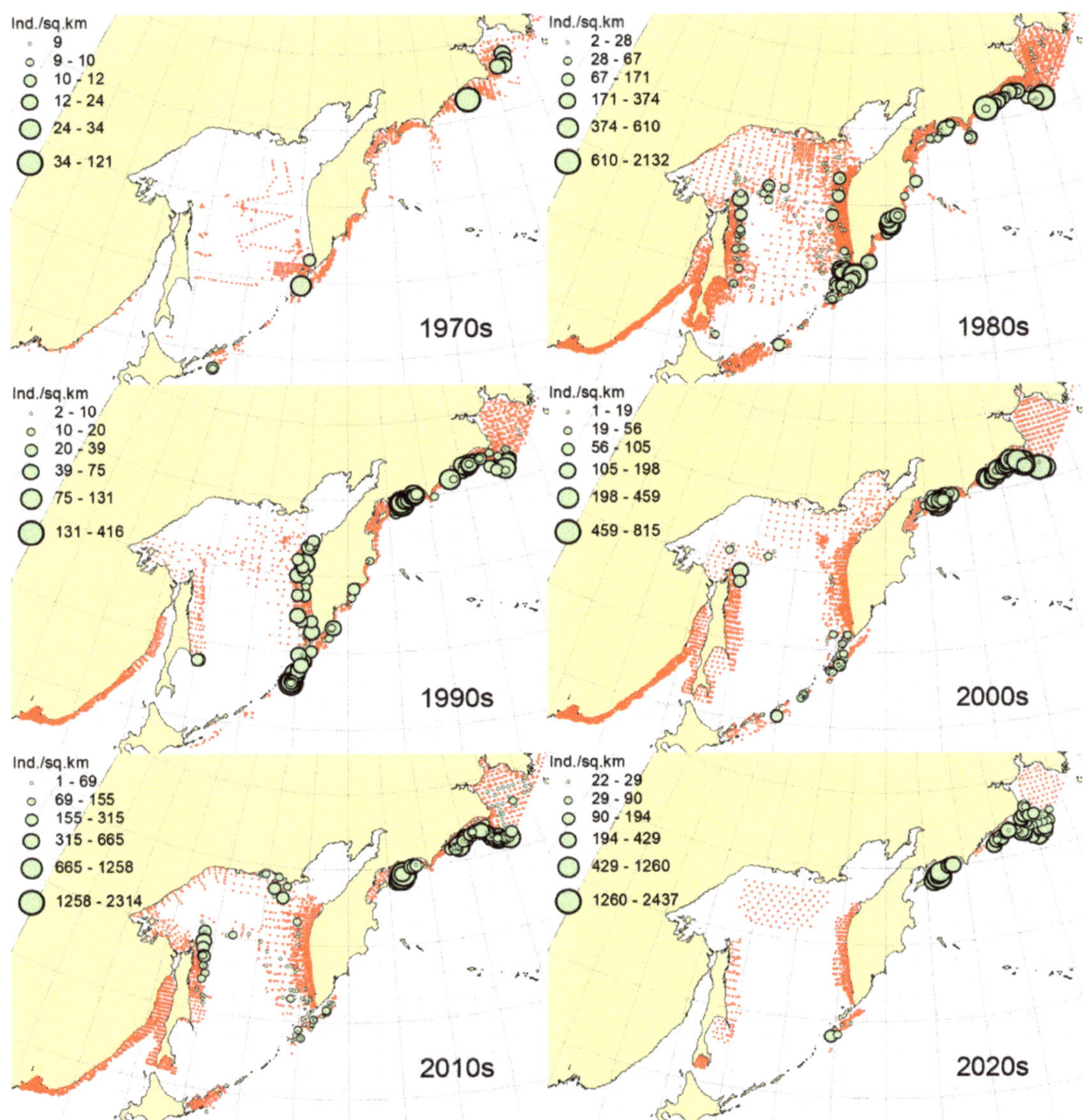

Figure 5. Decadal changes in the spatial distribution of the Aleutian skate *Bathyraja aleutica* in the Northwestern Pacific. Designations as in Figure 1.

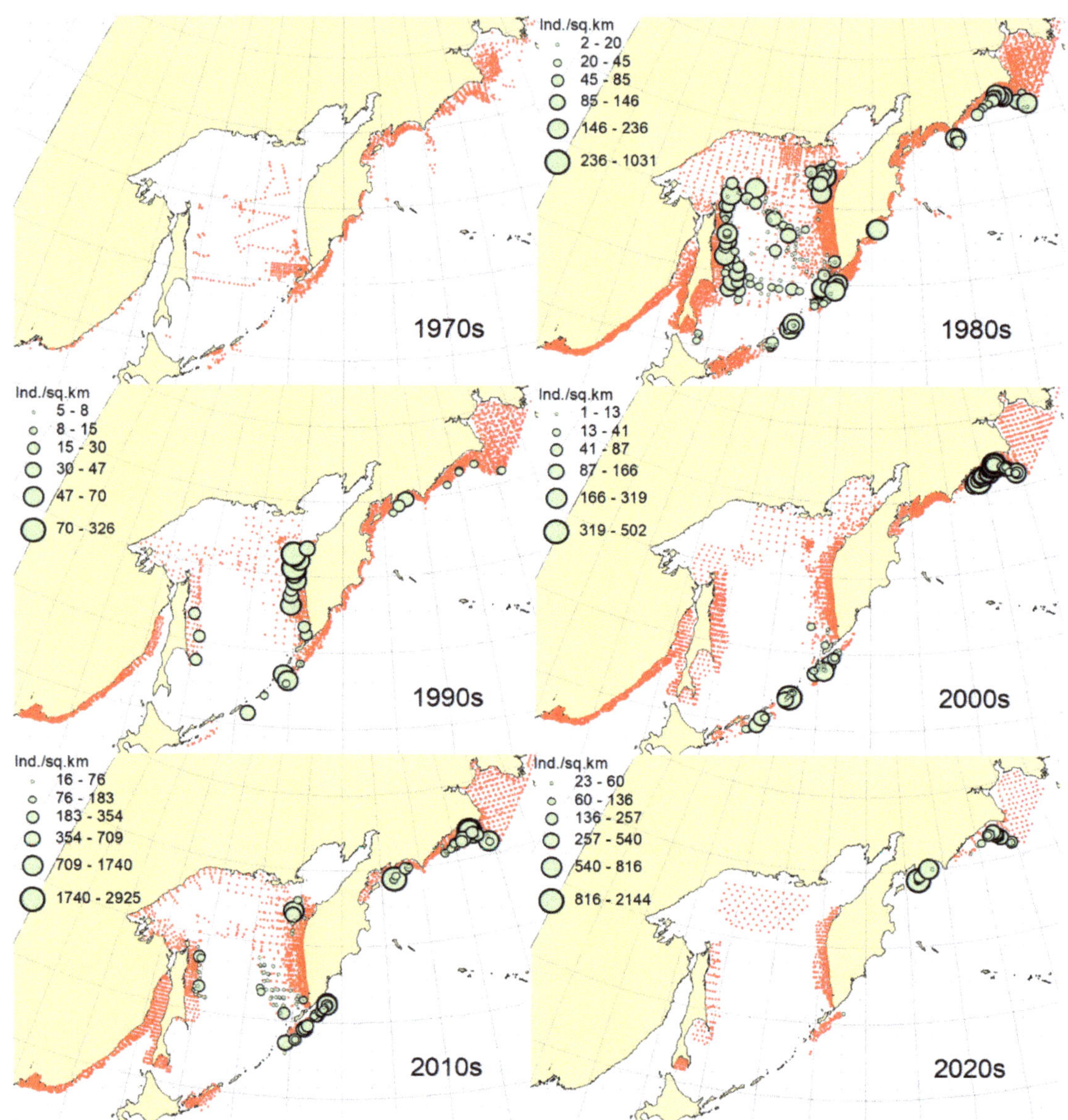

Figure 6. Decadal changes in the spatial distribution of the dusky-purple skate *Bathyraja matsubarai* in the Northwestern Pacific. Designations as in Figure 1.

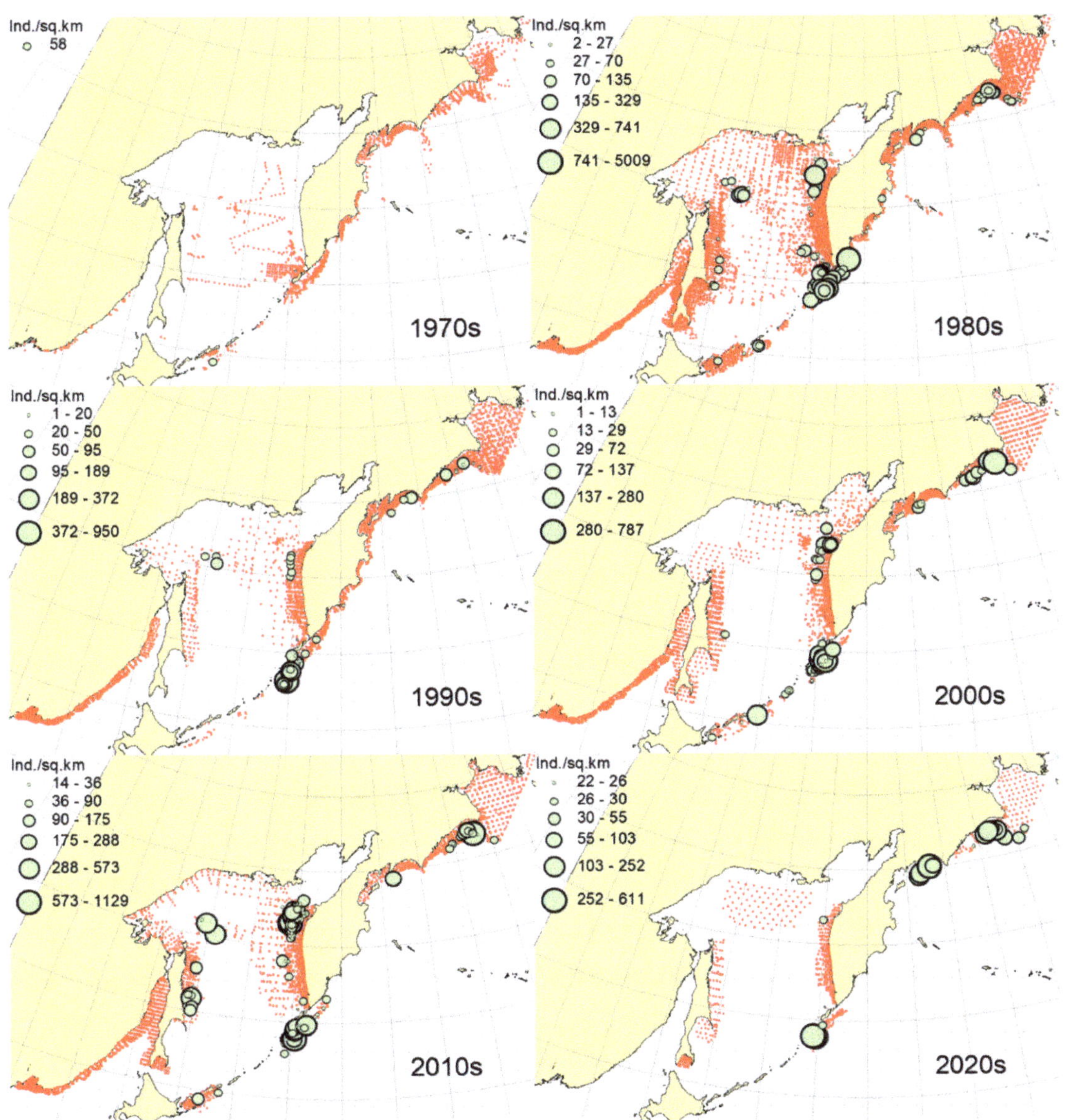

Figure 7. Decadal changes in the spatial distribution of the whiteblotched skate *Bathyraja maculata* in the Northwestern Pacific. Designations as in Figure 1.

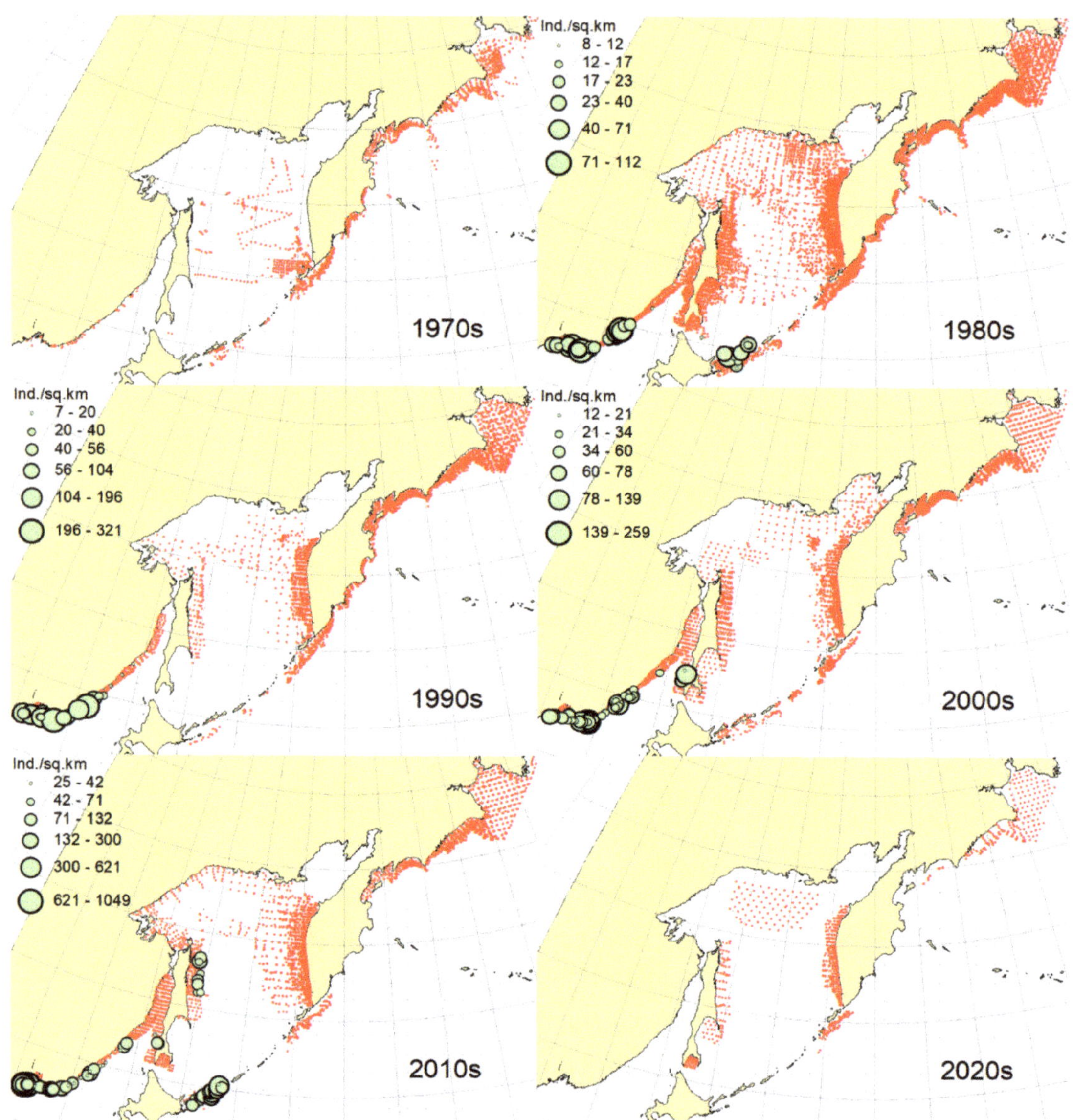

Figure 8. Decadal changes in the spatial distribution of the bottom skate *Bathyraja bergi* in the Northwestern Pacific. Designations as in Figure 1.

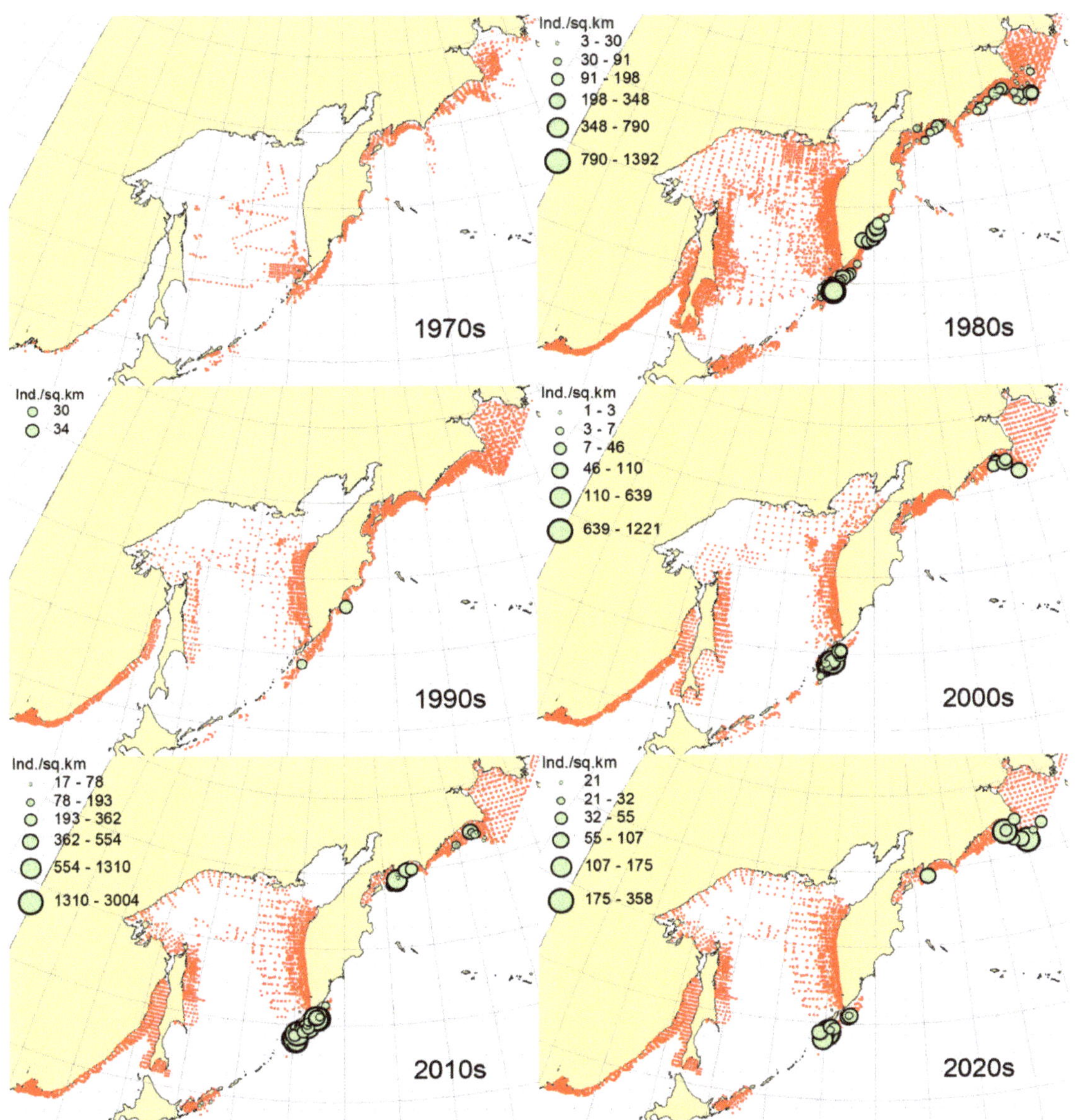

Figure 9. Decadal changes in the spatial distribution of the mud skate *Bathyraja taranetzi* in the Northwestern Pacific. Designations as in Figure 1.

3.2. Aleutian Skate

Spatial distribution (Figure 5). In the 1970s, the Aleutian skate was sporadically observed in catches off the southern and northern Kuril Islands, off southwestern Kamchatka and in the western Bering Sea. Its most frequent captures were recorded in the Gulf of Anadyr. The maximum catches were recorded in the central part of the Koryak coast and in the Pacific waters of the northern Kuril Islands. During the next decade, this species was most widely distributed within the study area. Its captures were recorded from the southern Kuril Islands in the south to the northern part of the Gulf of Anadyr in the north.

In the Sea of Okhotsk, it was found almost everywhere except in the southwestern part of the sea. The maximum catches were in the Pacific waters of the northern Kuril Islands, off the central parts of the East Kamchatka and Koryak coasts and off the Navarin Cape. In the third decade, catches of the Aleutian skate were recorded mainly along the western and eastern coasts of Kamchatka, in the Pacific waters of the northern Kuril Islands and in the western Bering Sea. Several catches were recorded off the Eastern Sakhalin to the east of Terpeniya Bay. The maximum catches were recorded in the Olyutor Bay of the Bering Sea and in the Pacific waters of the northern Kuril Islands. In the 2000s, the distribution of the Aleutian skate changed significantly. Catches from Eastern Sakhalin have shifted far to the north. Off the western Kamchatka and northern Kuril Islands, their values have significantly decreased, and the captures themselves have shifted south to the Simushir Island. In the western Bering Sea, the distribution pattern has practically not changed, but the value of catches has increased significantly. In the next decade, the distribution of this species changed again. The occurrence has increased near northeastern Sakhalin, in the northern Sea of Okhotsk and off western Kamchatka. Individual catches have been recorded in the Pacific waters of the southern Kuril Islands. In the western Bering Sea, the catches in the Gulf of Anadyr have not only increased, but they have also been recorded almost within the entire Gulf of Anadyr. For several decades, the maximum catches were recorded in the Olyutor Bay, in the central part of the Koryak coast and off the Navarin Cape. At the beginning of the current decade, the Aleutian skate was found mainly in the western Bering Sea and also in a few catches in the Pacific waters of the northern Kuril Islands.

Boundaries of the ranges (Figure 3, Supplementary Figures S4–S6). In general, the research area was characterized by a shift of the distribution center in the north direction. At the same time, the northern border of the range has shifted significantly to the north in the Sea of Okhotsk. In the Bering Sea and Pacific waters, on the contrary, it has shifted to the south (in the latter area it is noticeably stronger). The center of the range has shifted to the north in the Sea of Okhotsk and Pacific waters and to a lesser extent to the south in the Bering Sea. The southern distribution boundaries in all three areas shifted to the north (most noticeably in the last two areas).

Catch dynamics (Figure 4, Supplementary Figure S29). During the research period, the value of the average catch per unit effort for all areas increased several times. At the same time, catches of the Aleutian skate in the western Bering Sea demonstrated the maximum growth. To a lesser extent, the increase in catches was typical for Pacific waters. In the Sea of Okhotsk, despite a sharp decrease in the number of catches per unit effort in recent years, the general trend showed a certain increase in them as a whole for the entire period of research.

3.3. Dusky-Purple Skate

Spatial distribution (Figure 6). In the first decade, there were no captures of the dusky-purple skate within the surveyed area. In the 1980s, it was most widely distributed, occurring from the southern Kuril Islands in the south to m. Navarin in the Bering Sea in the north. In the Sea of Okhotsk, it was found almost everywhere, with the exception of a deep-water basin in the southwestern part of the sea. The maximum catches in this period were typical for the waters of Eastern Sakhalin, central and northern Kuril Islands, eastern Kamchatka and the western Bering Sea. In the 1990s, the area of occurrence and number of catches of this species decreased significantly. The main catches during this period were observed in the waters of Western Kamchatka. The second most common area of occurrence was the western Bering Sea. Occasional catches were also recorded off Eastern Sakhalin and in the Pacific waters of the central Kuril Islands. In the next decade, the range of the dusky-purple skate decreased even more. It has practically disappeared from catches in the waters of Western Kamchatka and Eastern Sakhalin but has become more numerous in the western Bering Sea and Pacific waters of the Kuril Islands from Paramushir Island to Iturup Island. In the 2010s, the area of distribution of this species increased markedly.

It reappeared in the waters of Eastern Sakhalin and Western Kamchatka. In the Pacific waters of the Kuril Islands, catches shifted northward; in the western Bering Sea, the occurrence expanded westward to the Karagin Bay. The maximum catches were recorded in the western Bering Sea. In the early years of the current decade, the dusky-purple skates were recorded in the western Bering Sea only and also caught once in the Pacific waters of southeastern Kamchatka. The maximum catches were recorded in the Olyutor Bay of the Bering Sea.

Boundaries of the ranges (Figure 3, Supplementary Figures S7–S9). In general, the period of research was characterized by a significant shift of the distribution center to the north. At the same time, the position of the northern border of the range in the Bering Sea has not changed, while in the Sea of Okhotsk and Pacific waters it has shifted to the south. The center of distribution in the Bering Sea also did not undergo displacement; while in the Sea of Okhotsk, it shifted somewhat southward; and in Pacific waters, on the contrary, to the north. The southern boundary of the range in all three areas shifted northward (to a lesser extent in the western Bering Sea).

Catch dynamics (Figure 4, Supplementary Figure S30). In general, for the research period, catches per unit effort increased many times over the course of six decades. At the same time, the maximum increase in catches was recorded in the Pacific waters and western Bering Sea. In the former area, after a sharp increase in the catch rates in the fifth decade, an equally sharp decline has been noted in recent years. In the Sea of Okhotsk during the research periods, there was an alternation of ups and downs of the catch rates, as a result of which the overall trend turned out to be weakly positive.

3.4. Whiteblotched Skate

Spatial distribution (Figure 7). In the 1970s, the whiteblotched skate in the research area was marked by a single capture in the waters of the southern Kuril Islands to the east of the Iturup Island. In the second decade, it was found in catches from southern Kuril Islands in the south to Navarin Cape in the Bering Sea in the north. In the Sea of Okhotsk, it was recorded near Eastern Sakhalin; in the northwestern part of the sea, off Western Kamchatka; and the northern Kuril Islands. The maximum catches were typical for the waters of the northern Kuril Islands and southeastern Kamchatka. In the 1990s, the range of the whiteblotched skate significantly decreased. It disappeared from catches in the waters of Eastern Sakhalin, southwestern Kamchatka and the Sea of Okhotsk waters off the northern Kuril Islands. The main area of occurrence and maximum catches remained the Pacific waters of the northern Kuril Islands. The second most important area was the western Bering Sea. The pattern of catch distribution of this species has not fundamentally changed in the next decade, except for individual catches off Eastern Sakhalin and in the Pacific waters of the southern and central Kuril Islands. The main area of occurrence and maximum catches were still the Pacific waters of the northern Kuril Islands. The value of catches in the waters of Western Kamchatka and the western Bering Sea has increased markedly in comparison with the previous decade. A similar distribution pattern of whiteblotched skate catches was typical for the 2010s. The only difference between these two adjacent decades was the appearance of this species in catches off Eastern Sakhalin and the northwestern Sea of Okhotsk. In addition to the Pacific waters of the northern Kuril Islands and the western Bering Sea, the maximum catches were also recorded in the waters of Western Kamchatka. In the current decade, the western Bering Sea has remained the main area of occurrence and maximum catches of whiteblotched skate. It was also recorded by single capture off the northwestern Kamchatka. Catches in the Pacific waters of the northern Kuril Islands have become occasional, although they have retained their high value.

Boundaries of the ranges (Figure 3, Supplementary Figures S10–S12). In general, in the research period, the center of the whiteblotched skate range showed a trend of displacement in the north direction. At the same time, the northern boundaries of its range in the Bering and Okhotsk Seas have hardly changed their position, and in the Pacific waters have noticeably shifted to the north (especially in the second decade). The center of the range in the Bering Sea had not undergone a significant shift, while in the other two areas it noticeably shifted northward. A similar pattern was observed with respect to the southern borders of the whiteblotched skate range.

Catch dynamics (Figure 4 and Supplementary Figure S31). In general, a noticeable increase in catches per unit effort (more than 2 times) can be noted for the study period. The catches of whiteblotched skates in Pacific waters demonstrate a particularly sharp increase, where after a considerable decrease in their value in the third and fourth decades, there was a sharp increase. In the Bering Sea during the entire period, catches showed a progressive increase. In the Sea of Okhotsk, despite the decrease in the value of catches in recent years, the general trends showed noticeable growth.

3.5. Bottom Skate

Spatial distribution (Figure 8). In the 1970s and 2020s, bottom skate was not recorded in catches within the study area. In the 1980s, the main area of its occurrence and maximum catches were the northwestern Sea of Japan. The second most important area was the waters around the southern Kuril Islands. Occasional catches were recorded in the southern part of the Tatar Strait (mainland) and Sea of Okhotsk to the southeast of the Aniva Bay. In the next decade, the northwestern Sea of Japan retained its importance as the main area of occurrence and maximum catches of bottom skate. The catches in the mainland part of the Tatar Strait began to be recorded somewhat northward in comparison with the previous decade. In the 2000s, the pattern of bottom skate distribution did not fundamentally change, but catches began to occur in the eastern part of the Tatar Strait off southwestern Sakhalin. In the next decade, the range of the bottom skate significantly expanded. It reappeared in the waters of the southern Kuril Islands, and also began to occur off northeastern Sakhalin. The maximum catches during this period were recorded in the waters of the Peter the Great Bay of the Sea of Japan.

Boundaries of the ranges (Figure 3, Supplementary Figures S13–S15). During decades two to five, the center of the range shifted slightly to the north on average. In the Sea of Japan, the northern border of the range has shifted noticeably to the north. The center of distribution and the southern border of the range, on the contrary, have shifted slightly southward. Data on other areas (the Sea of Okhotsk and Pacific waters) are not sufficient to judge the displacement of the boundaries of the range.

Catch dynamics (Figure 4, Supplementary Figure S32). The available data allow us to correctly judge only the dynamics of bottom skate catches in the Sea of Japan. During the period under review, from the second to the fifth decade, the catches per unit effort showed an increase with some decrease in the fourth decade. In general, the catches of this species were characterized by a trend of increasing catches both in the Sea of Japan and other areas.

3.6. Mud Skate

Spatial distribution (Figure 9). In the 1970s, the mud skate was not observed in the catches. During the second decade, this species was recorded in catches from the northern Kuril Islands in the south to the southern Gulf of Anadyr in the north. The maximum catches were recorded in the Pacific waters of the northern Kuril Islands and the bays of Eastern Kamchatka. In the 1990s, the mud skate was observed by occasional catches in the Pacific waters of the northern Kuril Islands and southeastern Kamchatka. In the next decade, the Pacific waters of the northern Kuril Islands again became the main area of occurrence and maximum catches. The second most important area was the western Bering Sea to the east of Navarin Cape. In the 2010s, the pattern of the distribution of catches in comparison with the previous period practically did not change. The Pacific waters of the northern Kuril Islands remained the main area of occurrence and maximum catches. However, in the western Bering Sea, large catches were recorded in the northern part of the Karagin Bay and Olyutor Bay. In recent years, the character of the spatial distribution has hardly changed, but in the western Bering Sea, the main catches have shifted to the area of Navarin Cape.

Boundaries of the ranges (Figure 3, Supplementary Figures S16–S18). In general, the center of the range of the mud skate showed a southward shift from the second to the third decade, after which it shifted to the north. At the same time, the northern borders and distribution centers in the Bering Sea and Pacific waters showed trends of displacement to the south. Additionally, the trend of displacement of the southern boundary of the range in Pacific waters had a southward direction, while in the Bering Sea it was directed in the opposite direction.

Catch dynamics (Figure 4, Supplementary Figure S33). Catches per unit effort of mud skate both in the Bering Sea and Pacific waters showed similar dynamics, i.e., increasing in the second, fourth and fifth decades and a fall in the third and sixth decades. Nevertheless, trends showed an increase in the value of catches per unit effort during the entire research period, both for each area and for all areas combined.

3.7. Whitebrow Skate

Spatial distribution (Figure 10). In the first decade, catches of the whitebrow skate in the research area were not registered. In the 1980s, the area of maximum occurrence and catches was the western Bering Sea from the central part of the Koryak coast to Navarin Cape. The second most important area was the Kronotsk Bay. Single captures of this species have been recorded in the Olyutor Bay, off northwestern and southwestern Kamchatka, the central part of the Sea of Okhotsk and the Pacific waters of the northern Kuril Islands. In the 1990s, the whitebrow skate was recorded only in a single catch in the Pacific waters of the northern Kuril Islands to the east of Onekotan Island. In the 2020s, the pattern of its distribution was close to that of the 1980s, but the maximum occurrence and catch rates were characteristic of the Olyutor and Karagin Bays of the Bering Sea. In the 2010s, the range of the whitebrow skate has significantly expanded, and the number of catches has increased markedly. In addition to the western Bering Sea, which remained the main area of occurrence and maximum catches, this species began to be frequently observed along the entire Western Kamchatka and in the Pacific waters of the northern Kuril Islands. Also, its single capture was registered to the northeast of Sakhalin. In recent years, the whitebrow skate had been recorded only in the western Bering Sea from the northern part of the Karagin Bay to Navarin Cape.

Boundaries of the ranges (Figure 3, Supplementary Figures S19–S21). From the second to the third decade, as a whole, a significant southward shift of the center of the range of the whitebrow skate was observed, after which it consistently and significantly shifted to the north. The northern boundaries of the distribution in all areas showed a tendency to shift in a southerly direction (well expressed in Pacific waters and less noticeable in other areas). The center of the range shifted slightly to the south in the Bering Sea and more noticeably in Pacific waters, while in the Sea of Okhotsk it showed the opposite tendency. Similar trends were noted in relation to the southern borders of the range of the whitebrow skate.

Catch dynamics (Figure 4, Supplementary Figure S34). The dynamics of catches per unit effort of the whitebrow skate as a whole for all areas of its occurrence showed a series of their growth (2nd and 4th decades) and fall (3rd and 5th decades). Nevertheless, all areas combined were characterized by positive trends in the increase in catches per unit effort, i.e., the strongest in the Pacific waters and less pronounced in the Bering and Okhotsk Seas.

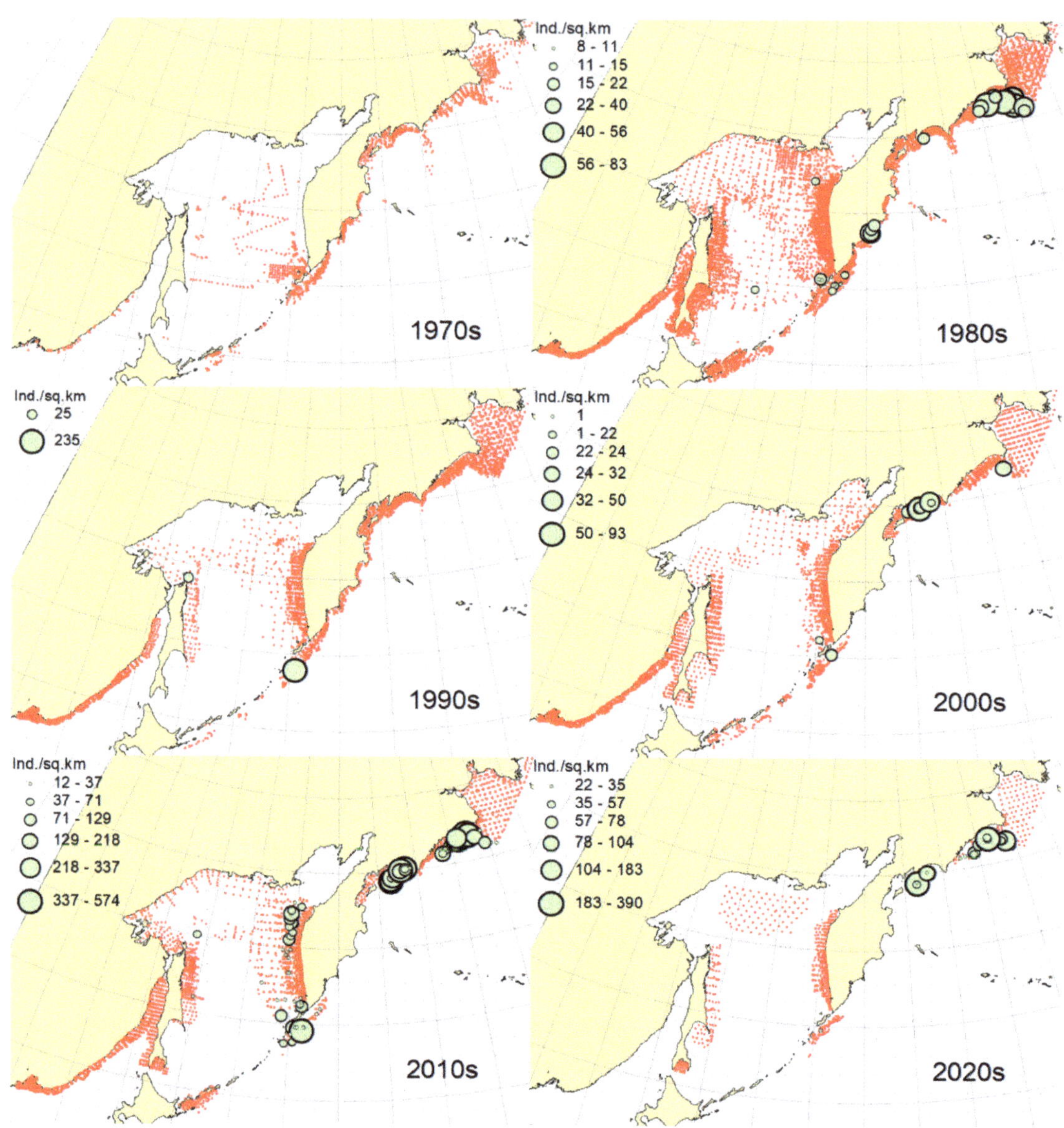

Figure 10. Decadal changes in the spatial distribution of the whitebrow skate *Bathyraja minispinosa* in the Northwestern Pacific. Designations as in Figure 1.

3.8. Sandpaper Skate

Spatial distribution (Figure 11). During the entire period of research, the sandpaper skate was observed in the western Bering Sea only. In the fourth and sixth decades, catches of this species in the research area were not recorded. In the 1970s, it was found in catches in the western Bering Sea from Goven Cape to Navarin Cape, and the maximum catches were recorded in the central part of the Koryak coast. In the 1980s, its number and distribution area in the western Bering Sea increased markedly. Its catches began to be observed south of Olyutor Bay in the waters of the Shirshov underwater ridge, as well as in the southern part of the Gulf of Anadyr. In the 1990s, the distribution pattern of the sandpaper skate in the western Bering Sea did not undergo significant changes, except for the extension of the range northward to the central part of the Gulf of Anadyr and absence of catches in the Olyutor Bay and waters of the western part of the Koryak coast. In the 2010s, the sandpaper skate was recorded in only a few catches off the Navarin Cape.

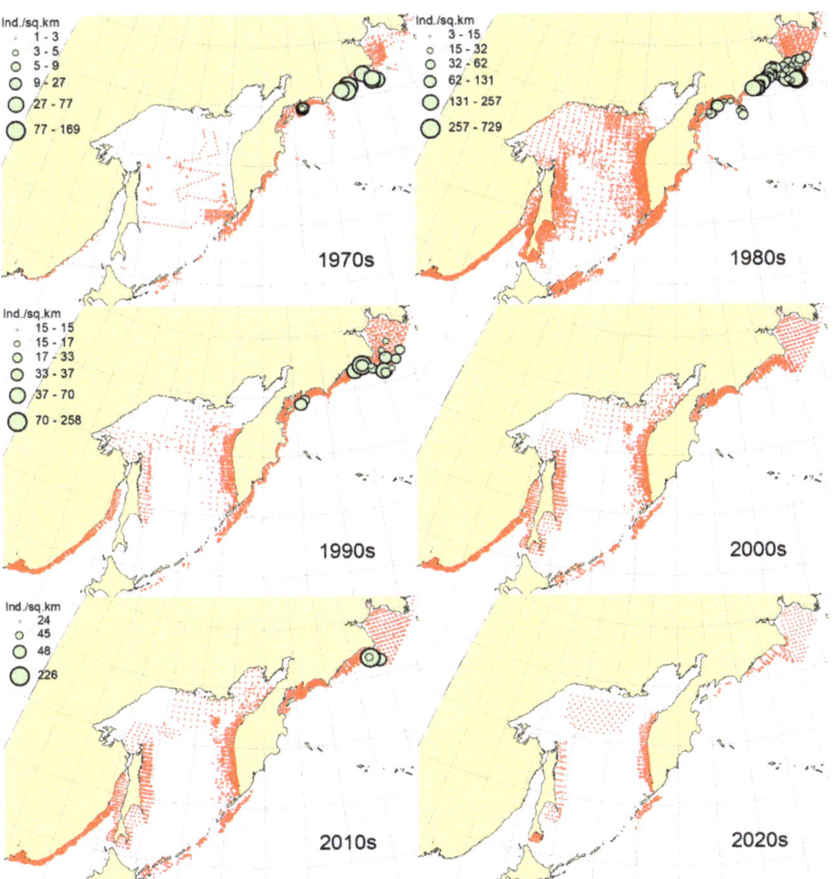

Figure 11. Decadal changes in the spatial distribution of the sandpaper skate *Bathyraja interrupta* in the Northwestern Pacific. Designations as in Figure 1.

Boundaries of the ranges (Figure 3, Supplementary Figures S22–S24). The fragmentary nature of the data and the short time series of observations do not allow us to judge with certainty about the dynamics of the position of sandpaper skate range boundaries. In general, we can only note a trend towards a southward shift of the northern border and a trend towards a northward shift of the center of the range and its southern border.

Catch dynamics (Figure 4 and Supplementary Figure S35). It is difficult to judge the dynamics of sandpaper skate catches in the western Bering Sea due to the lack of a continuous series of data. Nevertheless, the trend of catch per unit effort during the study period demonstrated their sharp decline.

3.9. Challenger's Skate

Spatial distribution (Figure 12). In the first decade, there were no captures of the Challenger's skate within the surveyed area. Occasional catches in the waters of southeastern Kamchatka and the Pacific waters of the northern Kuril Islands were recorded, respectively, in the 1990s and the current decade. This species was most widely distributed in the study area in the 1980s. During this period, it was most numerous in the Sea of Okhotsk, occurring mainly in the central part from southeastern and northeastern Sakhalin to southwestern Kamchatka. In Pacific waters, this species has been recorded in catches from Urup Island to Avacha Bay in Kamchatka. The maximum catches were recorded in the central part of the Eastern Sakhalin, southwestern Kamchatka and in the central part of the Sea of Okhotsk. The 2000s were characterized by the capture of the Challenger's skate exclusively in the Pacific waters of the Kuril Islands from Urup Island to Paramushir Island with maximum catches eastward to Onekotan Island. In the 2010s, catches of this species in the Pacific waters of the Kuril Islands were occasional, and its greatest occurrence was observed off Eastern Sakhalin and Western Kamchatka. The maximum catches were recorded in the northern Sea of Okhotsk off the coast of northwestern Kamchatka.

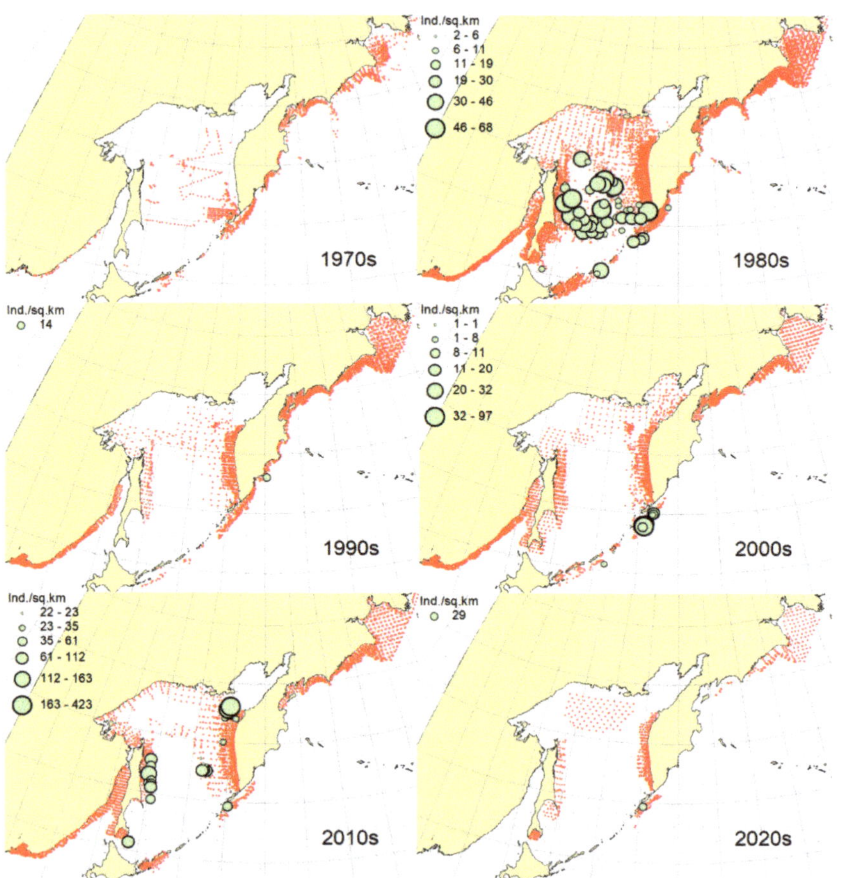

Figure 12. Decadal changes in the spatial distribution of the Challenger's skate *Bathyraja isotrachys* in the Northwestern Pacific. Designations as in Figure 1.

Boundaries of the ranges (Figure 3, Supplementary Figures S25–S27). In general, the center of the range of the Challenger's skate was characterized by a series of shifts to the north (2nd and 4th decades) and south (3rd and 5th decades). At the same time, the most complete data are available only for Pacific waters. The northern border of the range in this area showed a well-marked trend of southward displacement; the center of the range also shifted to the south (but less sharply), and the southern border of the range, on the contrary, shifted to the north.

Catch dynamics (Figure 4, Supplementary Figure S36). Generalized data for all areas showed that Challenger's skate catches per unit effort increased during the study period, with the exception of recent years. Nevertheless, in general, they showed a positive trend. In the Pacific waters and Sea of Okhotsk (despite the limited data in the latter area), catches showed very similar long-term dynamics in increasing catches during the entire study period.

4. Discussion

4.1. Spatial Distribution

An analysis of long-term changes in the spatial distribution of nine studied species showed that in the 1970s only three species were recorded in catches: Okhotsk, Aleutian and sandpaper skates. We see the reason for this situation as follows. Two species (white-blotched and whitebrow) were first described only in the late 1970s [52] while bottom and mud skates were described several years later [53]. However, two species described much earlier, Challenger's skate at the end of the 19th century [54] and dusky-purple skate in the middle of the 20th century [55], were also not recorded in catches. The reason for this, in our opinion, was the lack of field guides to skates, which were developed and published only in the early 1980s [46]. This did not allow species identification during surveys in the 1970s, and therefore all rays caught were identified only to the genus level. Additionally, data for the first decade include survey results for only one year for the Sea of Japan (1978) and three years (1977–1979) for the remaining areas. Therefore, they should hardly be considered as representative ones.

The vast majority of species differed in the widest distribution within the surveyed water area in the 1980s. The most likely reason for this is the specifics of surveys. During this period, research funding was at its maximum. The research fleet consisted of dozens of vessels, including medium- and large-tonnage ones, capable of bottom trawling to a depth of about 2 km [37,56,57]. This period was characterized by the maximum number of stations, surveyed depths and areas. All this, in our opinion, led to the most complete account of skates within the study area.

In the 1990s, the surveyed area, number of trawl stations (more than 2 times) and range of surveyed depth significantly decreased. This was due to the collapse of the former USSR and economic crisis that followed when research funding was significantly reduced. In the 2000s, with the improvement of the economic situation in Russia, the amount of funding for scientific research increased slightly, which made it possible to significantly increase the number of stations and expand surveyed areas. However, the range of surveyed depths noticeably narrowed, which affected the occurrence of skates in catches. In the 2010s, the number of hauls compared with the previous period hardly changed. However, the range of surveyed depths became even narrower, which resulted in a large degree of underestimation of skates in the study area. Additionally, it should be noted that in the fourth and fifth decades, the total surveys were not carried out in the waters of Eastern Kamchatka, or in the fifth decade in the Pacific waters of the central Kuril Islands. The data for the last (sixth) decade cannot be considered representative since they were obtained for only two years (2020–2021).

Thus, the nature of the data presented by us on the spatial distribution of nine species of deep-sea skates within the Russian EEZ in the northwestern Pacific is largely due to some artifacts. However, there are no other such detailed and long-term observations. The value of the presented data lies in the fact that, despite their scarcity, they still make it possible to present the nature of the spatial distribution and its long-term changes in nine species of skates in the study area, to determine the areas of their main concentrations based on the catches and assess the fishing potential.

Until recently, the features of the spatial distribution of the Okhotsk skate were described in general terms only for the Pacific waters of the northern Kuril Islands and southeastern Kamchatka [17,29]. The distribution of this species throughout the North Pacific, including the waters of Russia, is reviewed in a recent publication [33]. Despite the fact that the same value (ind./km^2) was used to characterize the distribution of catches in our and the mentioned work, the data on the long-term dynamics of the distribution turned out to be difficult to compare, since different time intervals were used and the visual representation of the features of the distribution of species varies greatly.

Data on the spatial distribution of the Aleutian, whiteblotched, whitebrow, mud, dusky-purple and Challenger's skates have so far been limited only to information on the occurrence and general features of the localization of catches and their magnitude in the Pacific waters of the northern Kuril Islands and southeastern Kamchatka, obtained in the 1990s [17,29–31]. Data on the distribution of the sandpaper skate in the northwestern Pacific have not yet been presented in the literature. Therefore, this article significantly expands the understanding of the occurrence, size of the catches, features of the spatial distribution and its long-term changes of the nine most common species of deep-sea skates within the Russian EEZ

4.2. Boundaries of the Ranges

Nowhere in any species is there a monotonous shift of the northern or southern border of the range or its center for the entire observation period in one direction—only to the south or only to the north. The shifts alternate in a different order. At the same time, positive and negative shifts in different species in different water bodies occur in different decades (Table 3). It can be said that the ranges slowly fluctuate or pulsate near some average long-term positions.

We tried to evaluate the results of the six-decade shifts of all changes using regression analysis. Despite the fact that this analysis revealed some trends in the shifts of the boundaries and centers of ranges in all nine species studied (see Supplementary Materials), statistically significant changes are observed only in a small number of cases (last column in Table 3). Thus, the northern boundary of the range of the Okhotsk skate has significantly shifted to the north in the Bering Sea and, on the contrary, to the south in Pacific waters. The center of the range of this species in the Sea of Okhotsk has shifted to the north. The distribution centers of the Aleutian and whiteblotched skates in the Sea of Okhotsk have significantly shifted to the north. The southern boundaries of the dusky-purple skate range in the Sea of Okhotsk and Pacific waters have also shifted northward. The reasons for such shifts in ranges are still unclear, although, most likely, they are due to climate change. However, finding links between changes in range and climate (oceanological conditions) requires additional research and will probably be the subject of a separate future publication. Only then will it be possible to understand whether the observed ranges' fluctuations are random or associated with some external factors.

4.3. Catch Dynamics

Until recently, the catch dynamics of the Okhotsk skate were considered [17] for the Pacific waters of the northern Kuril Islands and southeastern Kamchatka, and only for a time period limited to eight years from 1993 to 2000. The dynamics of this species' catches in various areas of the North Pacific, including the waters of Russia, are also presented in a recent publication [33]. If in our and the mentioned works the trends of catches per effort of this species in the western part of the Bering Sea coincide, then in the Sea of Okhotsk and Pacific waters they show the opposite pattern. The reason for such differences, in addition to different primary data, may be a different calculation method.

Unfortunately, the above-mentioned publications do not provide a statistical analysis of the calculated trends.

Data on the dynamics of catches of the Aleutian, whiteblotched, whitebrow, mud and dusky-purple skates have so far been limited only to information on changes in the magnitude of catches in the Pacific waters of the northern Kuril Islands and southeastern Kamchatka, obtained in 1993–2000 [17,31]. Data on long-term changes in the catches of the sandpaper skate in the northwestern Pacific have not yet been presented in literature. Therefore, the data presented by us on the long-term dynamics of catches of the nine most common species of deep-sea skates within the Russian EEZ will help in the future to better understand the reasons that determine the fluctuations in the abundance of not only skates of the genus Bathyraja, but also, possibly, other deep-sea fish species, which today remain poorly understood from various points of view [58].

Despite the fact that the analysis performed made it possible to identify certain trends in the dynamics of catches in all nine species studied during the study period, statistically significant estimates were obtained only in a small number of cases (Table 4). Thus, for seven of the nine studied species, statistically significant trends in the increase in catches were revealed. For the Okhotsk and whiteblotched skates, a positive trend was noted in the Bering Sea, for the Aleutian and dusky-purple rays, in the Bering Sea and in general throughout the study area. For mud and Challenger's skates, a similar trend was found in Pacific waters and in the study area as a whole; for whitebrow skate, only in Pacific waters. The reasons for such dynamics of catches are still unclear, but, as in the case of the displacement of the range boundaries, they are probably associated with climate change. However, establishing links between catch dynamics and climate change requires additional data and research. Perhaps this will be the subject of a separate publication in the future.

However, judging by Table 4—where pluses over minuses absolutely predominate in the third and sixth columns (30 vs. 1 and 31 vs. 0, respectively); in the fourth and seventh column, minuses predominate over pluses (22 vs. 9 and 8 vs. 23, respectively); and in the fifth, their number is the same (14)—it can be concluded that in general, from the 1970s to 1980s, the number of skates increased; from the 1980s to 1990s, it decreased; from the 1990s to 2000s, it fluctuated at the achieved level; from the 2000s to 2010s, it increased again, and; from the 2010s to 2020s, it again decreased. These trends coincide with previously identified regime shifts associated with rearrangements in the abundance of the most common species under the influence of climatic and oceanological factors [59–63]: the 1980s are considered the era of high bio- and fish productivity in the northwestern Pacific, 1991–1995—the lowest, the second half of the 1990s—a period of recovery, and 2000s—of high productivity again. This correlation seems plausible in connection with the absence of targeted fishing for skates in the study area noted in the Introduction, because in its absence, the abundance of populations is regulated by natural causes.

The decline in skate abundance from 2010s to 2020s may be associated both with the beginning of their target fishing and an extremely small sample over the last decade—only two years 2020 and 2021 (see Table 2). This will only become clear as more information becomes available. Theoretically, in the prevailing decade, abundance changes (the predominance of pluses or minuses) could be associated with sample sizes, since their changes (see Table 1) coincide with the prevailing abundance dynamics described in the previous paragraph (see Table 4). However, it is difficult to come up with a rational interpretation of such a relationship, since the sample size affects the accuracy of the estimate of the mean, but not its value.

It is also noteworthy that among the long-term trends in the abundance of skates for all decades, which were estimated by regression equations (Supplement), only positive ones are statistically significant. All negative trends are non-significant (see the last column in Table 4). It can be said that for six decades not a single species has been significantly affected by either fishing or climate change—not even Okhotsk skate, which has a record number of four minuses in the Sea of Okhotsk (2nd line in Table 4).

5. Conclusions

In this paper, for the first time, we analyzed all unique available data (from bottom trawl surveys 1977–2021) on changes in the spatial distribution and abundance of the nine most common skate species in the Russian waters of the Northwestern Pacific, where their fishery is considered promising, but still not developed. We hope that it will be useful to compare the results obtained with developments in the near future, when there is expected to be a sharp increase in fishing for these species, as well as with other areas where there is already fishing pressure on these species.

Supplementary Materials: The following supporting information can be downloaded at: https://www.mdpi.com/article/10.3390/ani12243485/s1, Figures S1–S27: Trends of shifts of the northern and southern boundaries of range and its center of nine skate species in the Northwestern Pacific; Figures S28–S36: Trends of CPUE changes of nine skate species in the Northwestern Pacific.

Author Contributions: A.M.O. and I.V.V. equally contributed to the preparation of the paper. All authors have read and agreed to the published version of the manuscript.

Funding: Preparation of this paper was funded by the Ministry of Science and Higher Education, Russian Federation (grant No. 13.1902.21.0012, contract No. 075-15-2020-796).

Institutional Review Board Statement: Not applicable.

Informed Consent Statement: Not applicable.

Data Availability Statement: Not applicable.

Acknowledgments: The authors are grateful to their numerous colleagues who collected material at sea. They also thank two anonymous reviewers for valuable comments and suggestions that allowed for improvement of the manuscript.

Conflicts of Interest: The authors declare no conflict of interest.

References

1. Dolganov, V.N. Skates of the Far Eastern Seas of the Soviet Union. Ph.D. Thesis, Institute of Marine Biology, Far Eastern Scientific Center of the USSR Academy of Sciences, Vladivostok, Russia, 1987; pp. 1–24.
2. Mito, K. Food Relationships among Benthic Fish Populations in the Bering Sea on the *Theragra chalcogramma* Fishing Grounds in October and November of M. Master's Thesis, Hokkaido University Graduate School, Hakodate, Japan, 1974; pp. 1–135.
3. Brodeur, R.D.; Livingston, P.A. Food habits and diet overlap of various Eastern Bering Sea fishes. *U.S. Dep. Commer. NOAA Tech. Memo. NMFS F/NWC* **1988**, *127*, 1–76.
4. Livingston, P.A.; deReynier, Y. Groundfish food habits and predation on commercially important prey species in the Eastern Bering Sea from 1990 to 1992. *U.S. Dep. Commer. NOAA/NMFS AFSC Proc. Rep.* **1996**, *96-04*, 12–14.
5. Orlov, A.M. The diets and feeding habits of some deep-water benthic skates (Rajidae) in the Pacific waters off the northern Kuril Islands and southeastern Kamchatka. *Alsk. Fish. Res. Bull.* **1998**, *5*, 1–17.
6. Orlov, A.M. Feeding habits of some deep-benthic skates (Rajidae) in the western Bering Sea. In Proceedings of the MTS/IEEE Oceans an Ocean Odyssey, Honolulu, HI, USA, 5–8 November 2001; Volume 2, pp. 842–855. [CrossRef]
7. Orlov, A.M. Diets, feeding habits, and trophic relations of six deep-benthic skates (Rajidae) in the western Bering Sea. *Aqua. J. Ichthyol. Aquat. Biol.* **2003**, *7*, 45–60.
8. Chuchukalo, V.I.; Lapko, V.V.; Kuznetsova, N.A.; Slabinsky, A.M.; Napazakov, V.V.; Nadtochy, V.A.; Koblikov, V.N.; Pushchina, O.I. Feeding of groundfishes on the shelf and continental slope of the northern Sea of Okhotsk. *Izv. TINRO* **1999**, *126*, 24–57.
9. Chuchukalo, V.I.; Napazakov, V.V. Feeding and trophic state of the mass species of skates (Rajidae) from the western part of the Bering Sea. *Izv. TINRO* **2002**, *130*, 422–428.
10. Dulvy, N.K.; Reynolds, J.D. Predicting extinction vulnerability in skates. *Conserv. Biol.* **2002**, *16*, 440–450. [CrossRef]
11. Orlov, A.M. Groundfish resources of the northern North Pacific continental slope: From science to sustainable fishery. In Proceedings of the Seventh North Pacific Rim Fisheries Conference, Busan, Republic of Korea, 18–20 May 2005; Alaska Pacific University: Anchorage, AK, USA, 2005; pp. 139–150.
12. Ishihara, H. The skates and rays of the western North Pacific: An overview of their fisheries, utilization, and classification. *U.S. Dep. Commer. NOAA Tech. Rep. NMFS* **1990**, *90*, 485–497.
13. Nasedkina, E.A. Characteristics of flesh of skates. *Rybn. Khoz.* **1969**, *6*, 59–60.
14. Baidalova, G.F.; Dubnitskaya, G.M.; Nechaev, A.P.; Severinenko, S.M. Chemical composition and properties of thorny skate liver oil depending on the fishing season. *Rybn. Khoz.* **1984**, *10*, 60–62.

15. Ershov, A.M.; Petrov, B.F.; Korchunov, V.V.; Nikolaenko, O.A. Problems of using of thorny skate for human food purposes. In *Prospects of Development of Russian Fisheries Complex—XXI Century, Proceedings of the Abstracts of the Scientific-Practical Conference, Moscow, Russia, 27–28 June 2002*; VNIRO Publishing: Moscow, Russia, 2002; pp. 101–102.
16. Dolganov, V.N. Reserves of the skates in Far Eastern seas of Russia and their prospective fishery. *Izv. TINRO* **1999**, *126*, 650–652.
17. Orlov, A.; Tokranov, A.; Fatykhov, R. Common deep-benthic skates (Rajidae) of the northwestern Pacific: Basic ecological and biological features. *Cybium* **2006**, *30* (Suppl. S4), 49–65.
18. Suvorov, E. Use of sharks and skates. *Za Rybn. Ind. Sev.* **1933**, *7*, 28–29.
19. Dvinin, Y.F.; Konstantinova, L.L.; Kuz'mina, V.I. Techno-chemical characteristics of some skates and chimaeras. *Tr. PINRO* **1981**, *47*, 29–51.
20. Antonov, N.P.; Klovatch, N.V.; Orlov, A.M.; Datsky, A.V.; Lepskaya, V.A.; Kuznetsov, V.V.; Yarzhombek, A.A.; Abramov, A.A.; Alekseev, D.O.; Moiseev, S.I.; et al. Fishing in the Russian Far East fishery basin in 2013. *Tr. VNIRO* **2013**, *160*, 133–211.
21. Badaev, O.Z. Estimation of bycatch and discards of longline fish in Far East seas. *Probl. Fish.* **2018**, *19*, 58–72.
22. Zolotov, A.O. Modern specialized fishery of sea fish in the western Bering Sea. *Izv. TINRO* **2021**, *201*, 76–101. [CrossRef]
23. Orlov, A.M. Bycatch reduction of multispecies bottom trawl fisheries: Some approaches to solve the problem in the Northwest Pacific. *Am. Fish. Soc. Symp.* **2007**, *49*, 241–258.
24. Orlov, A.M. Some approaches to reducing non-commercial bycatch of bottom trawl fisheries in the western Bering Sea. *Asian Fish. Sci.* **2011**, *24*, 397–412. [CrossRef]
25. Volvenko, I.V.; Gebruk, A.V.; Katugin, O.N.; Ogorodnikova, A.A.; Vinogradov, G.M.; Maznikova, O.A.; Orlov, A.M. Commercial value of trawl macrofauna of the North Pacific and adjacent seas. *Environ. Rev.* **2020**, *28*, 269–283. [CrossRef]
26. Orlov, A.M.; Tokranov, A.M. Checklist of deep-sea fishes of the Russian northwestern Pacific Ocean found at depths below 1000 m. *Prog. Oceanogr.* **2019**, *176*, 102143. [CrossRef]
27. Grigorov, I.V.; Orlov, A.M. Species diversity and conservation status of cartilaginous fishes (Chondrichthyes) of Russian waters. *J. Ichthyol.* **2013**, *53*, 923–936. [CrossRef]
28. Volvenko, I.V.; Orlov, A.M.; Gebruk, A.V.; Katugin, O.N.; Vinogradov, G.M.; Maznikova, O.A. Species richness and taxonomic composition of trawl macrofauna of the North Pacific and its adjacent seas. *Sci. Rep.* **2018**, *8*, 16604. [CrossRef] [PubMed]
29. Orlov, A.M. Trophic relationships of commercial fishes in the Pacific waters off southeastern Kamchatka and the northern Kuril Islands. In *Ecosystem Approaches for Fisheries Management*; Alaska Sea Grant College Program: Fairbanks, AK, USA, 1999; pp. 231–263.
30. Orlov, A.M.; Tokranov, A.M. New Data on Two Rare Skate Species of the Genus *Bathyraja. J. Ichthyol.* **2005**, *45*, 444–450.
31. Orlov, A.M.; Tokranov, A.M. Reanalysis of long-term surveys on the ecology and biology of mud skate (*Rhinoraja taranetzi* Dolganov, 1985) in the northwestern Pacific (1993–2002). *J. Appl. Ichthyol.* **2010**, *26*, 861–871. [CrossRef]
32. Grigorov, I.V.; Orlov, A.M.; Baitalyuk, A.A. Spatial distribution, size composition, feeding habits, and dynamics of abundance of Alaska skate *Bathyraja parmifera* in the North Pacific. *J. Ichthyol.* **2015**, *55*, 644–663. [CrossRef]
33. Grigorov, I.V.; Baitalyuk, A.A.; Orlov, A.M. Spatial distribution, size composition, and dynamics of catches of the Okhotsk skate *Bathyraja violacea* in the North Pacific Ocean. *J. Ichthyol.* **2017**, *57*, 706–720. [CrossRef]
34. Grigorov, I.V.; Kivva, K.K.; Orlov, A.M. The Aleutians and Beyond: Distribution, Size Composition, and Catch Dynamics of the Aleutian Skate *Bathyraja aleutica* across the North Pacific. *Animals* 2022, *accepted*.
35. Panchenko, V.V.; Pushchina, O.I.; Boiko, M.I.; Kalchugin, P.V. Distribution and some biological features of bottom skate *Bathyraja bergi* in the Russian waters of the Sea of Japan. *J. Ichthyol.* **2017**, *57*, 560–568. [CrossRef]
36. Orlov, A.M. Substantiation of commercial size of Far Eastern skates (fam. Rajidae) by example of widespread species from the Bering Sea. *Tr. VNIRO* **2006**, *146*, 252–264.
37. Volvenko, I.V. The new large database of the Russian bottom trawl surveys in the Far Eastern seas and the North Pacific Ocean in 1977. *Int. J. Environ. Monit. Anal.* **2014**, *2*, 302–312. [CrossRef]
38. Orlov, A.M.; Smirnov, A.A. New data on sexual dimorphism and reproductive biology of Alaska Skate *Bathyraja parmifera* from the Northwestern Pacific Ocean. *J. Ichthyol.* **2011**, *51*, 590–603. [CrossRef]
39. Dyldin, Y.V. Annotated checklist of the sharks, batoids and chimaeras (Chondrichthyes: Elasmobranchii, Holocephali) from waters of Russia and adjacent areas. *Bull. Seto Mar. Biol. Lab.* **2015**, *43*, 40–91. [CrossRef]
40. Dyldin, Y.V.; Orlov, A.M.; Velikanov, A.Y.; Makeev, S.S.; Romanov, V.I.; Hanel, L. An annotated list of the marine and brackish-water ichthyofauna of Aniva Bay (Sea of Okhotsk, Sakhalin Island): Petromyzontidae–Agonidae Families. *J. Ichthyol.* **2018**, *58*, 473–501. [CrossRef]
41. Dyldin, Y.V.; Orlov, A.M.; Hanel, L.; Romanov, V.I.; Fricke, R.; Vasil'eva, E.D. Ichthyofauna of the Fresh and Brackish Waters of Russia and Adjacent Areas: Annotated List with Taxonomic Comments. Families Petromyzontidae–Pristigasteridae. *J. Ichthyol.* **2022**, *62*, 385–414. [CrossRef]
42. Dyldin, Y.V.; Orlov, A.M. An annotated list of cartilaginous fishes (Chondrichthyes: Elasmobranchii, Holocephali) of the coastal waters of Sakhalin Island and the adjacent southern part of the Sea of Okhotsk. *J. Ichthyol.* **2018**, *58*, 158–180. [CrossRef]
43. Dyldin, Y.V.; Orlov, A.M. Annotated List of Ichthyofauna of Inland and Coastal Waters of Sakhalin Island. Families Petromyzontidae—Salmonidae. *J. Ichthyol.* **2021**, *61*, 48–79. [CrossRef]

44. Misawa, R.; Orlov, A.M.; Orlova, S.Y.; Gordeev, I.I.; Ishihara, H.; Hamatsu, T.; Ueda, Y.; Fujiwara, K.; Endo, H.; Kai, Y. *Bathyraja* (*Arctoraja*) *sexoculata* sp. nov., a new softnose skate (Rajiformes: Arhynchobatidae) from Simushir Island, Kuril Islands (western North Pacific), with special reference to geographic variations in *Bathyraja* (*Arctoraja*) *smirnovi*. *Zootaxa* **2020**, *4861*, 515–543. [CrossRef]

45. Lindberg, G.U.; Legeza, M.I. *Fishes of the Japan Sea and Neighboring Waters of Okhotsk and Yellow Seas*; Izdatel'stvo Akademii Nauk SSSR: Moscow-Leningrad, Russia, 1959; pp. 1–208.

46. Dolganov, V.N. *Guide to the Diagnostic Characters of Cartilaginous Fishes from the Far East Sea of the U.S.S.R. and Neighboring Waters*; TINRO: Vladivostok, Russia, 1983; pp. 1–92.

47. Dolganov, V.N.; Tuponogov, V.N. Keys to Skates of Genus *Bathyraja* and *Rhinoraja* (Fam. Rajidae) of the Far Eastern Seas of Russia. *Izv. TINRO* **1999**, *126*, 657–664.

48. Stevenson, D.E.; Orr, J.W.; Hoff, G.R.; McEachran, J.D. *Sharks, Skates and Ratfish of Alaska*; Alaska Sea Grant, University of Alaska: Fairbanks, AK, USA, 2007; pp. 1–77.

49. Tuponogov, V.N.; Kodolov, L.S. *Field Guide for Identification of Commercial and Mass Species of Fishes from Far Eastern Seas of Russia*; Russkii Ostrov: Vladivostok, Russia, 2014; pp. 1–336.

50. Froese, R.; Pauly, D. (Eds.) *FishBase*; World Wide Web Electronic Publication. 2022. Available online: http://www.fishbase (accessed on 5 December 2022).

51. Fricke, R.; Eschmeyer, W.N.; van der Laan, R. (Eds.) *Eschmeyer's Catalog of Fishes: Genera, Species, References*; World Wide Web Publication. 2022. Available online: http://researcharchive.calacademy.org/research/ichthyology/catalog/fishcatmain.asp (accessed on 5 December 2022).

52. Ishiyama, R.; Ishihara, H. Five new species of skates in the genus *Bathyraja* from the western North Pacific, with reference to their interspecific relationships. *Jap. J. Ichthyol.* **1977**, *24*, 71–90.

53. Dolganov, V.N. New species of skates of the family Rajidae from the northwestern Pacific Ocean. *J. Ichthyol.* **1985**, *25*, 121–132.

54. Günther, A. Preliminary notes on new fishes collected in Japan during the expedition of H.M.S. "Challenger". *Ann. Mag. Nat. Hist. Ser.* **1877**, *20*, 433–446. [CrossRef]

55. Ishiyama, R. Studies on the rays and skates belonging to the family *Rajidae*, found in Japan and adjacent regions. A revision of three genera of Japanese rajids with descriptions of one new genus and four new species mostly occurred in northern Japan. *J. Shimonoseki Coll. Fish.* **1952**, *2*, 1–34.

56. Shuntov, V.P. *Zigzags of Fishery Science*; TINRO: Vladivostok, Russia, 1994; p. 368.

57. Shuntov, V.P. *Biology of the Far Eastern Seas of Russia*; TINRO-Center: Vladivostok, Russia, 2016; p. 604.

58. Orlov, A.M. Contemporary Ichthyological and Fisheries Research of Deepwater Fish: New Advances, Current Challenges, and Future Developments. *J. Mar. Sci. Eng.* **2022**, *10*, 166. [CrossRef]

59. Shuntov, V.P.; Dulepova, E.P.; Radchenko, V.I.; Lapko, V.V. New data about communities of plankton and nekton of the far-eastern seas in connection with climate-oceanological reorganization. *Fish. Oceanogr.* **1996**, *5*, 38–44. [CrossRef]

60. Shuntov, V.P.; Radchenko, V.I.; Dulepova, E.P.; Temnykh, O.S. Biological resources of the Russian Far East economic zone: The structure of pelagic and benthic communities, current status, long-term trends. *Izv. TINRO* **1997**, *122*, 3–15.

61. Shuntov, V.P.; Dulepova, E.P.; Temnykh, O.S.; Volkov, A.F.; Naidenko, S.V.; Chuchukalo, V.I.; Volvenko, I.V. The state of biological resources in connection with the dynamics of macroecosystems in the economic zone of the Far Eastern seas of Russia. In *Dynamics of Marine Ecosystems and Modern Problems of Conservation of Biological Resources of the Russian Seas*; Dalnauka: Vladivostok, Russia, 2007; pp. 75–176.

62. Shuntov, V.P. The state of biota and biological resources of marine macroecosystems of the Far Eastern economic zone of Russia. In *Bulletin No. 4 of the Implementation of the "Concepts of the Basin Program for the Study of Pacific Salmon"*; TINRO-center: Vladivostok, Russia, 2009; pp. 242–251.

63. Shuntov, V.P.; Volvenko, I.V. On rearrangements in bottom and near-bottom fish communities in the Russian far eastern seas under the fishing pressure. *Probl. Fish.* **2020**, *21*, 359–378.

Article

The Aleutians and Beyond: Distribution, Size Composition, and Catch Dynamics of the Aleutian Skate *Bathyraja aleutica* across the North Pacific

Igor V. Grigorov [1,*], Kirill K. Kivva [2] and Alexei M. Orlov [3,4,5,6,7,*]

[1] Fish Breeding Division, Central Branch of the Federal State-Funded Institution "Glav Basin Department of Fisheries and Conservation of Water Biological Resources", 117105 Moscow, Russia
[2] Department of Climate and Aquatic Ecosystems Dynamics, Russian Federal Research Institute of Fisheries and Oceanography, 105187 Moscow, Russia
[3] Laboratory of Oceanic Ichthyofauna, Shirshov Institute of Oceanology, Russian Academy of Sciences, 117997 Moscow, Russia
[4] Laboratory of Behavior of Lower Vertebrates, A.N. Severtsov Institute of Ecology and Evolution, Russian Academy of Sciences, 119071 Moscow, Russia
[5] Department of Ichthyology and Hydrobiology, Tomsk State University, 634050 Tomsk, Russia
[6] Department of Ichthyology, Dagestan State University, 367000 Makhachkala, Russia
[7] Laboratory of Marine Biology, Dagestan Federal Research Center, Caspian Institute of Biological Resources, Russian Academy of Sciences, 367000 Makhachkala, Russia
* Correspondence: grigoroviv@glavrybvod.ru (I.V.G.); orlov@vniro.ru (A.M.O.)

Simple Summary: Deep-water skates play an important role as top predators in the North Pacific, yet they are also considered promising targets of bottom-trawl and longline fisheries. Moreover, skates are highly vulnerable to over-fishing due to their very long lifespans, late maturation, small litter size, long incubation period, etc. Despite their ecological and commercial importance, the distribution, basic biological traits, and dynamics of abundance of many deep-water skate species remain poorly understood, information that is critical for their conservation and management. The Aleutian skate *Bathyraja aleutica* is one of the more common deep-water species and is widely distributed across the North Pacific, but it remains largely understudied. In this paper, we compiled and analyzed long-term data records of the Aleutian skate in the North Pacific from various databases, which revealed new information on its spatial and vertical distribution, size composition, reproductive biology, and interannual catch dynamics.

Citation: Grigorov, I.V.; Kivva, K.K.; Orlov, A.M. The Aleutians and Beyond: Distribution, Size Composition, and Catch Dynamics of the Aleutian Skate *Bathyraja aleutica* across the North Pacific. *Animals* **2022**, *12*, 3507. https://doi.org/10.3390/ani12243507

Academic Editor: Martina Francesca Marongiu

Received: 29 October 2022
Accepted: 6 December 2022
Published: 12 December 2022

Publisher's Note: MDPI stays neutral with regard to jurisdictional claims in published maps and institutional affiliations.

Abstract: The results of long-term (1948–2021) studies on the spatial and vertical distribution, dynamics of abundance, and size composition of the Aleutian skate *Bathyraja aleutica* in the North Pacific Ocean are presented. Maximum densities of this species were characteristic of the eastern Bering Sea slope, off the central Aleutian Islands, consisting of the Pacific waters off southeastern Kamchatka and the northern Kurils, and northeastern Sakhalin. This species was most abundant at depths of 100–600 m; in the cold months, *B. aleutica* migrates to greater depths for over-wintering, and in warm months it feeds at shallower depths. *Bathyraja aleutica* was most common at the bottom, at temperatures around 3 °C. The total length of individuals ranged from 9.6–170 cm, with a predominance of skates with a length of 50–100 cm. Males did not differ significantly from females in body weight and length. The maximum values of the condition factor were typical for the autumn–winter period. Across years, there was an increase in Aleutian skate catch rates from the western Bering Sea and the Sea of Okhotsk, and a decrease in the Pacific waters off the Kuril Islands and Kamchatka, as well as in Alaskan waters.

Keywords: spatial distribution; vertical distribution; length; body weight; sex ratio; condition factor; Bering Sea; Sea of Okhotsk; Kuril Islands; Aleutian Islands; Gulf of Alaska; British Columbia

---

1. Introduction

Skates and rays (Rajiformes) are an important component of the bottom-fish communities of the world's oceans. In the North Pacific, they are consumers of such ecologically and commercially important species as Pacific herring *Clupea pallasii*, walleye pollock *Gadus chalcogrammus*, flathead flounder *Hippoglossoides elassodon*, yellowfin flounder *Limanda aspera*, rock sole *Lepidopsetta bilineata*, Atka mackerel *Pleurogrammus monopterygius*, Pacific cod *Gadus macrocephalus*, shortraker rockfish *Sebastes borealis*, popeye grenadier *Coryphaenoides cinereus*, commander squid *Berryteuthis magister*, octopus (Octopoda), golden king crab *Lithodes aequispinis*, Tanner crabs *Chionoecetes* spp., and shrimps [1–8].

Skates are important targets of coastal fishing in many countries, especially in Southeast Asia, where their wings are dried for food, and specifically their meat is used for the production of crab sticks [9]. Fishing for skates in the North Pacific is still poorly developed, but quite promising [10,11]. Their biomass within Russian waters only amounts to about 677 thousand tons [12], but in some areas skates make up as much as 10% of the total groundfish biomass, and their abundance has increased markedly in recent years [13]. Technological studies have shown that the meat of skates contains almost a complete set of essential amino acids and is highly suitable for food production [14]. Moreover, the liver of skates is rich in vitamin A, and can serve as raw materials for the production of veterinary and medical oils [15]. Skates have a fairly high cost in Japan, where they are sold frozen at a price of JPY 650 per kg, and chilled at a price of JPY 800–1800 per kg. However, in the markets of the west coast of the USA, frozen skates only cost from USD 0.1 to 0.3 per pound [16]. Fishing for skates and the export of raw materials or other fishery products from Russia to Europe, Asia and America, where they are in high demand, was recognized in the early 1980s [17]. In recent years, several Russian companies that have focused on the supply of fish products to Japanese, Korean, and Chinese markets have also shown an interest in the skate fishery. It is therefore likely that in the following years, the volume of their catch of skates may increase significantly.

Until recently, skate fishing in the North Pacific was unregulated, with no established total allowable catches (TAC) and no catch statistics, since the skates were not retained and were discarded. The situation has changed in recent decades, when TAC of skates differentiated by area and species was introduced for both Russian and American waters [18,19].

The Aleutian skate *Bathyraja aleutica* (Gilbert, 1896) is widespread across the North Pacific Ocean, and it plays an important role in ecosystems [20]. It is a boreal species inhabiting the Okhotsk and Bering Seas, waters off the eastern Kamchatka, Japan, Kuril, and Aleutian Islands, as well as the Gulf of Alaska and along the American and Canadian Pacific coast down to California [12,20–31]. The information available from the literature on the distribution and biology of this species is not sufficient to understand its life history cycle, and it has so far been scattered and fragmentary [4–6,8,18,20,32–63].

The purpose of this paper is to summarize and analyze long-term data on the spatial and vertical distributions, size composition, biology, and dynamics of abundance of the Aleutian skate in the North Pacific.

2. Materials and Methods

Data for analysis of *Bathyraja aleutica* population and biological characteristics and trends were obtained from trawl, longline, and trap catches in the North Pacific from 1976 to 2021, obtained mostly by scientific surveys of the Pacific Branch of the Russian Federal Research Institute of Fisheries and Oceanography (TINRO, Vladivostok, Russia), the Alaska Fisheries Science Center (AFSC, Seattle, WA, USA), and the Northwestern Fisheries Science Center (NWFSC, Seattle, WA, USA), as well as by scientific observers on American commercial fishing vessels. In addition, we used the data on Aleutian skate records from ichthyological collections from a number of museums (Table 1). The materials used in this study consist entirely of information and data on those catches in which the Aleutian skate was recorded. Data from the USA bottom-trawl surveys were limited to the spring–summer sampling months of May–October. From the USA commercial fisheries observer data, other than year there was no information on the specific month or season collected, which precluded the data's full use in analyzing seasonal patterns in Aleutian skate abundance and distribution. In total, the data analyzed consisted of 15,377 catches of the Aleutian skate with the following fishing gear: mid-water, bottom, and pelagic trawls, as well as longlines and traps, including 6746 with records of fishing depth (Figure 1).

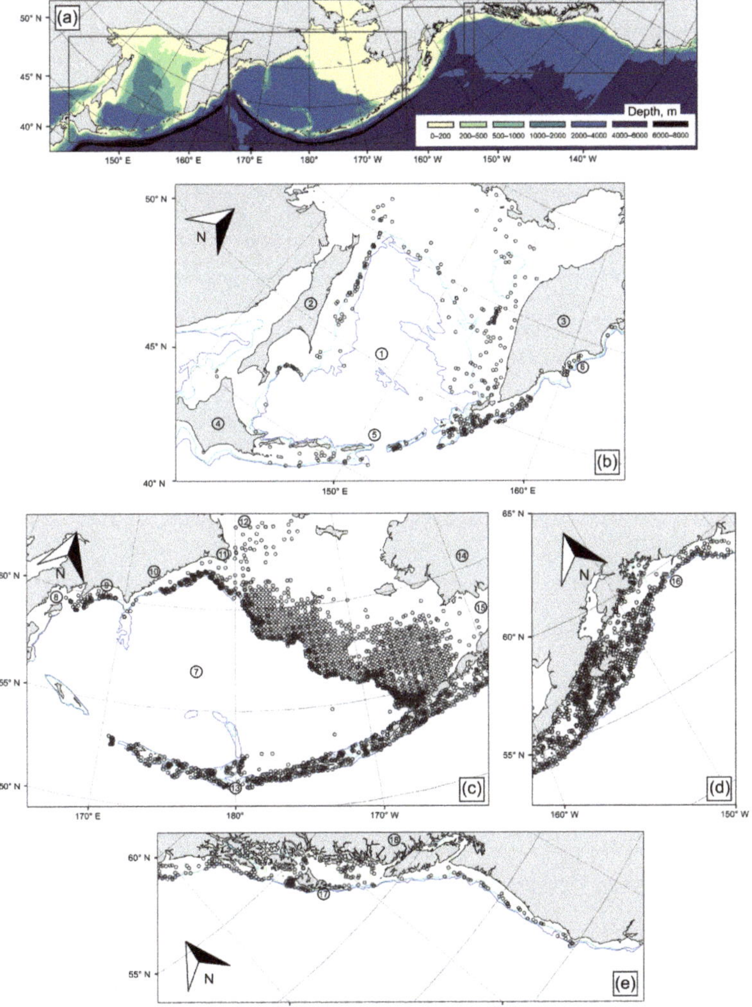

Figure 1. Study area (**a**) and location of trawl stations (small circles) in the survey areas: Sea of Okhotsk and North Pacific waters off Hokkaido, Kuril Islands, and eastern Kamchatka (**b**), Bering Sea (**c**), Gulf of Alaska (**d**), and US and Canada West Coast (**e**). Grey rectangles in (**a**) correspond to panels (**b**–**e**). The numbers in circles indicate the geographical names mentioned in the text: ①—Sea of Okhotsk, ②—Sakhalin Island, ③—Kamchatka Peninsula, ④—Hokkaido Island, ⑤—Kuril Islands, ⑥—Kronotsk Bay, ⑦—Bering Sea, ⑧—Karagin Bay, ⑨—Olyutor Bay, ⑩—Koryak Coast, ⑪—Navarin Cape, ⑫—Gulf of Anadyr, ⑬—Aleutian Islands, ⑭—Alaska, ⑮—Bristol Bay, ⑯—Gulf of Alaska, ⑰—Haida Gwaii, ⑱—British Columbia.

Table 1. Description of the data used for the analysis of spatial and vertical distribution of the Aleutian skate *Bathyraja aleutica* in the North Pacific, 1893–2021.

Data Type	No. Records	Data Source/ Organization	Link to Data Source
Scientific bottom-trawl surveys	3356	Pacific branch of Russian Federal Research Institute of Fisheries and Oceanography (TINRO), Vladivostok, Russia	The data received upon request on 1 October 2022
Scientific bottom-trawl surveys	3010	Alaska Fisheries Science Center (AFSC), National Oceanic and Atmospheric Administration (NOAA), Seattle, WA, USA	https://www.fisheries.noaa.gov/alaska/commercial-fishing/alaska-groundfish-bottom-trawl-survey-data accessed on 3 October 2022
Scientific bottom-trawl surveys	47	Northwestern Fisheries Science Center National Oceanic and Atmospheric Administration—NWFSC NOAA, Seattle, WA, USA.	https://www.webapps.nwfsc.noaa.gov/data/map accessed on 3 October 2022
Observations on commercial fishing vessels	8567	Alaska Fisheries Science Center (AFSC), National Oceanic and Atmospheric Administration (NOAA), Seattle, WA, USA	https://www.fisheries.noaa.gov/resource/map/spatial-data-collected-groundfish-observers-alaska accessed on 3 October 2022
Museum collection	40	Fish Collection of Hokkaido University—National Museum of Nature and Science, Hokkaido University, Sapporo, Hokkaido, Japan	https://www.gbif.org/ru/occurrence/search?dataset_key=848ae956-f762-11e1-a439-00145eb45e9a&taxon_key=2420280 accessed via GBIF.org on 3 October 2022
Institution collection	18	Ichthyology collection of the California Academy of Sciences (CAS), San Francisco, CA, USA	https://www.gbif.org/ru/occurrence/search?publishing_org=66522820-055c-11d8-b84e-b8a03c50a862&taxon_key=2420280 accessed via GBIF.org on 3 October 2022
Scientific bottom-trawl surveys	1	Pacific Biological Station (PBS), Fisheries and Oceans Canada, Nanaimo, BC, Canada	https://www.gbif.org/ru/occurrence/search?dataset_key=310dbb60-e369-4ec2-a6d8-8e4de843b208&publishing_org=c98a8d0d-cab2-4506-8c0f-c91a5e4996b9&taxon_key=2420280 accessed via GBIF.org on 3 October 2022
Scientific bottom-trawl surveys	128	Pacific Biological Station (PBS), Fisheries and Oceans Canada, Nanaimo, BC, Canada	https://www.gbif.org/ru/occurrence/search?dataset_key=9fcd8de5-a2b6-4c72-b863-debe63ee3a72&publishing_org=c98a8d0d-cab2-4506-8c0f-c91a5e4996b9&taxon_key=2420280 accessed via GBIF.org on 3 October 2022
Scientific bottom-trawl surveys	44	Pacific Biological Station (PBS), Fisheries and Oceans Canada, Nanaimo, BC, Canada	https://www.gbif.org/ru/occurrence/search?dataset_key=48060741-47d3-4093-86e6-2ec484323cc5&publishing_org=c98a8d0d-cab2-4506-8c0f-c91a5e4996b9&taxon_key=2420280 accessed via GBIF.org on 3 October 2022
Scientific bottom-trawl surveys	46	Pacific Biological Station (PBS), Fisheries and Oceans Canada, Nanaimo, BC, Canada	https://www.gbif.org/ru/occurrence/search?dataset_key=3d4cac0a-7441-4099-8eed-8a270d522f8f&publishing_org=c98a8d0d-cab2-4506-8c0f-c91a5e4996b9&taxon_key=2420280 accessed via GBIF.org on 3 October 2022
Scientific bottom-trawl surveys	4	Pacific Biological Station (PBS), Fisheries and Oceans Canada, Nanaimo, BC, Canada	https://www.gbif.org/ru/occurrence/search?dataset_key=f2189339-9098-4e0e-98d4-9c72e5e5f96d&publishing_org=c98a8d0d-cab2-4506-8c0f-c91a5e4996b9&taxon_key=2420280 accessed via GBIF.org on 3 October 2022
Scientific bottom-longline surveys	17	Pacific Biological Station (PBS), Fisheries and Oceans Canada, Nanaimo, BC, Canada	https://www.gbif.org/ru/occurrence/search?dataset_key=1677f287-3666-49b8-90f2-12a63ca58438&publishing_org=c98a8d0d-cab2-4506-8c0f-c91a5e4996b9&taxon_key=2420280 accessed via GBIF.org on 3 October 2022
Scientific bottom-longline surveys	7	Pacific Biological Station (PBS), Fisheries and Oceans Canada, Nanaimo, BC, Canada	https://www.gbif.org/ru/occurrence/search?dataset_key=e908605e-fb40-4db1-bff7-de1c241284a9&publishing_org=c98a8d0d-cab2-4506-8c0f-c91a5e4996b9&taxon_key=2420280 accessed via GBIF.org on 3 October 2022
Museum collection	2	Vertebrate, zoology, fishes collections of Smithsonian National Museum of Natural History—NMNH, Washington, DC, USA.	https://collections.nmnh.si.edu/search/fishes/ accessed on 3 October 2022
Museum collection	6	Fish collection of the Canadian Museum of Nature, Ottawa, ON, Canada	http://ipt.nature.ca/resource?r=cmn_fish&v=1.124 accessed on 3 October 2022

Table 1. *Cont.*

Data Type	No. Records	Data Source/ Organization	Link to Data Source
Museum collection	66	Ichthyology collection of Burke Museum of Natural History and Culture (BMNHC), Seattle, WA, USA	https://www.burkemuseum.org/collections-and-research/biology/ichthyology/collections-database/results.php?l=100&o=0&f=&g=&s=17h~IymyNd1yIp&d=gpA&w=PhIyNl1XhPYMY0j1yI1Kp0YndhYPhSh\T1\textbar{}b3NhsPYMY0jld1B0&wo=PhIyNl1XhPYjXhlE1Yj1yI1KpYndhYPhSh\T1\textbar{}b3NhsPYjXhlE1Yjld1B accessed on 3 October 2022
Museum collection	2	Ichthyology collection of Muséum national d'Histoire naturelle (MNHN), Paris, France	https://obis.org/taxon/271506 accessed via OBIS.org on 3 October 2022
Institution collection	1	Ichthyology collection University of Kansas Biodiversity Institute & Natural History Museum (KUBI), Lawrence, KS, USA	https://obis.org/taxon/271506 accessed via OBIS.org on 3 October 2022
Museum collection	1	Fish collection of the National Museum of Nature and Science (NMNS), Tokyo, Japan	https://obis.org/taxon/271506 accessed via OBIS.org on 3 October 2022
Institution collection	12	Texas Cooperative Wildlife Collection (TCWC), Department of Wildlife and Fisheries Sciences, Texas A&M University, College Station, TX, USA	https://www.gbif.org/ru/occurrence/search?publishing_org=2b8e48b0-812d-11de-86fe-b8a03c50a862&taxon_key=2420280 accessed via GBIF.org on 3 October 2022
Institution collection	1	Marine Vertebrate Collection of the Scripps Institution of Oceanography (SIO), San Diego, CA, USA	https://sioapps.ucsd.edu/collections/mv/collection/94-205/?q=Bathyraja+aleutica accessed on 3 October 2022
Museum collection	1	Fish collection of the Museum of the North, University of Alaska (UAM), Fairbanks, AK, USA	http://arctos.database.museum/guid/UAM:Fish:9455?seid=2666501 accessed on 3 October 2022
Total	15,377	22 sources/15 organizations	

To characterize the size composition, measurements of total length (TL) of 3541 individuals were used, including 770 females and 601 males, of which 455 and 376 specimens were weighed, respectively. Analysis of the size and weight composition is presented for the Bering Sea (422 ind.), the Pacific waters off the Kuril Islands and eastern Kamchatka (hereinafter NWPO) (228 ind.), the Sea of Okhotsk (111 ind.), and the Pacific coast of Canada (70 ind.). The analysis of the sex ratio within different size classes was also based on these data. The relationship between total length (TL, cm) and body weight (W, kg) for individuals of both sexes (831 ind.) was estimated from the data on the number of skates caught and their weight in the catch in the case of catching single individuals. The same data are the basis for analysis of Fulton's condition factor. Maps of spatial distributions were constructed using the GIS and R software.

Visualization of Aleutian skate spatial distribution was performed in R (https://www.R-project.org/, accessed on 24 October 2022). A discrete global hexagonal grid was created using the dggridR package (https://github.com/r-barnes/dggridR/, accessed on 24 October 2022). The distance between the centers of the hexagons was set to about 100 km. The main advantage of this approach is to maintain spatial bins of equal area across a wide range of latitudes and longitudes. Data were averaged over the cells containing >2 data points. For illustrative purposes, data points outside such cells are also shown on the maps. All maps were presented in the North Pole Lambert azimuthal equal-area projection for the Bering Sea (EPSG:3571).

Statistical analysis of the data was performed in R. Welch's *t*-test was used to compare two independent samples in cases when the distributional assumptions held (e.g., in the case of length-related samples) and when the sample sizes were greater than 40. Prior to performing *t*-tests, outliers were removed according to the interquartile range (IQR) method (threshold values $Q_1 - 1.5 \times$ IQR and $Q_3 + 1.5 \times$ IQR, where Q_1 and Q_3 are the first and third quartiles, respectively). In cases when violation of the distributional assumptions was suspected or when the sample size of at least one sample was less than 40, a Wilcoxon rank-sum test (also known as U-test or Mann–Whitney test) was performed. Holm adjustment of *p*-value was used in cases when pairwise comparison was required for both Welch's *t*-test and Wilcoxon rank-sum test. For the sex ratios, proportion tests (z-tests) were carried out (comparison to 1:1). In order to explore length–weight relationships as well as to compare them for males and females and across different regions, linear models were fitted to the data with and without interaction terms. An alpha level (α) of 0.05 was used in all statistical tests.

3. Results

3.1. Spatial Distribution of Catches and Their Seasonal and Long-Term Changes

The Aleutian skate is widely distributed in the North Pacific, with a known range extension from the northern part of the Gulf of Anadyr in the Bering Sea, to the Pacific coasts of Hokkaido and central California in the south, including the waters of the Bering and Okhotsk Seas (Figure 2). In its distribution, it tends to reside in the lower part of the shelf and the continental slope, and is most often found in the waters off western Kamchatka in the Sea of Okhotsk, the Pacific waters off the northern Kurils, the eastern Bering Sea slope, the Aleutian Islands, and the western part of the Gulf of Alaska.

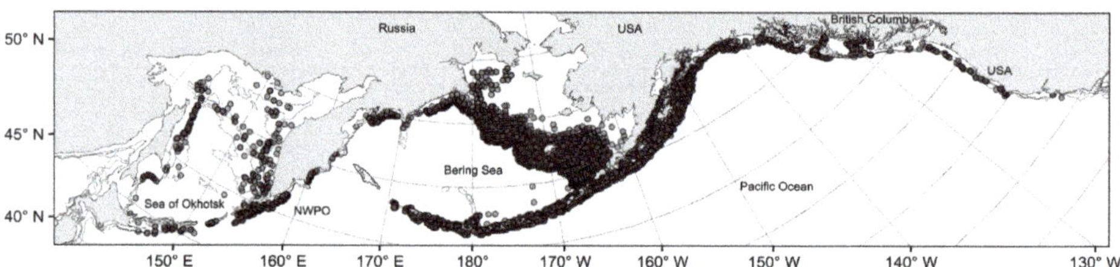

Figure 2. Sites of Aleutian skate *Bathyraja aleutica* captures (•) in the North Pacific Ocean.

According to the USA commercial fishing observer catch data from 1995 to 2021 (Figure 3), the Aleutian skate was the most numerous along the eastern Bering Sea slope, the Aleutian Islands, and the western part of the Gulf of Alaska, where its by-catch exceeded 0.6% of the total catch weight. From other areas in Alaskan waters, the by-catch of the Aleutian skate was considerably lower.

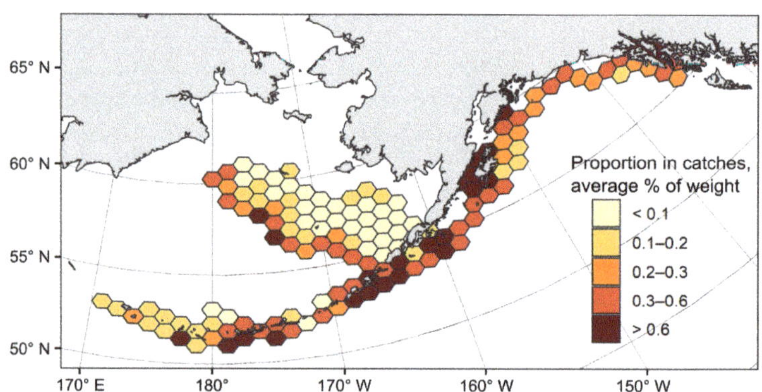

Figure 3. Distribution of catches (proportion in catches, % of weight) of the Aleutian skate *Bathyraja aleutica* in Alaskan waters in 1995–2021 according to data from American observers on board commercial fishing vessels.

Within-year seasonal variability in the spatial distribution of the Aleutian skate catch varied significantly. In winter months (December–February), as a whole, their values were minimal across the entire survey area (Figure 4a). Maximum catches during this period were recorded from only a few local areas, i.e., in the Pacific waters off the northern Kurils, off the central Aleutian Islands, the central part of the Koryak coast in the western Bering Sea, and in the southern part of the Bristol Bay of the eastern Bering Sea. During the spring, catches of Aleutian skates increased significantly across almost the entire surveyed area (Figure 4b). The highest densities were observed near northeastern Sakhalin, southwestern Kamchatka, the northern part of Karagin Bay, Olyutor Bay, the eastern part of the Koryak coast, along the entire eastern Bering Sea slope, along almost the entire Aleutian Islands, and in the western part of the Gulf of Alaska. In summer, the spatial distribution pattern of Aleutian skate catches changed little (Figure 4c), with maximum catches observed off northeastern Sakhalin, in the western part of the Gulf of Alaska, along the eastern Bering Sea slope, and in the northern part of Karagin Bay. At the same time, the density of aggregations along the Aleutian Islands decreased, and dense schoolings remained off the most western part. Maximum catches from the waters off western Kamchatka shifted to the Pacific waters off the northern Kurils and southwestern Kamchatka. Also, the maximum catches were recorded locally to the east of the southern Kurils. In the autumn, the density of aggregations and the areas with maximum catches significantly decreased (Figure 4d). Dense accumulations were observed in the waters off northeastern Sakhalin, in Olyutor Bay, in the northern part of Karagin Bay and along the eastern Bering Sea slope. In the waters off the northern Kuril and Aleutian Islands, maximum catches were occasional and recorded locally.

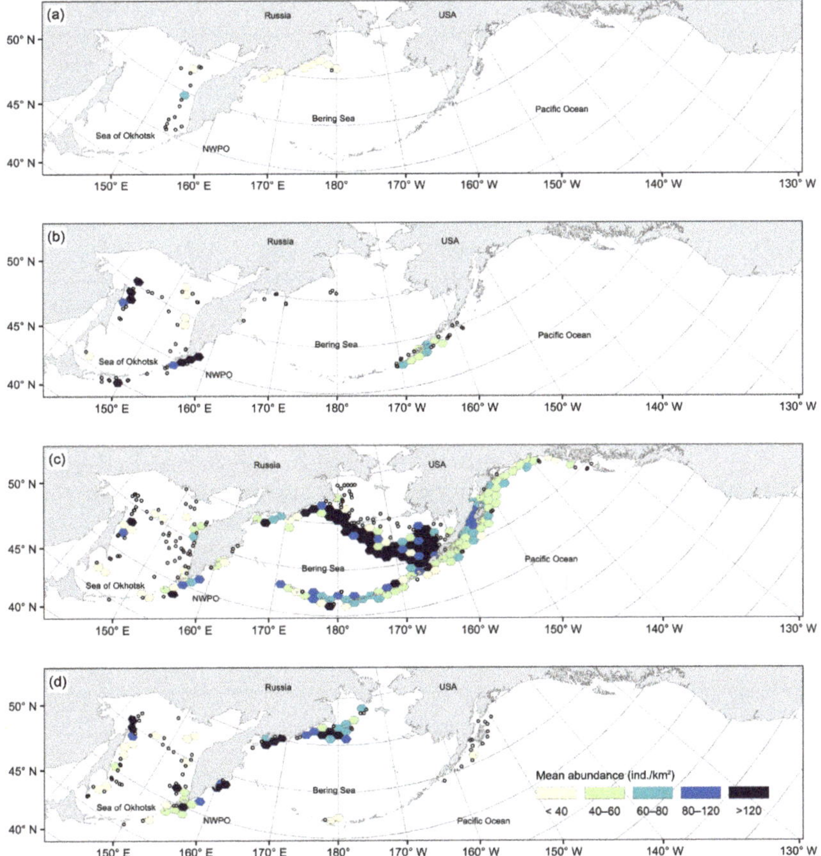

Figure 4. Mean abundance (ind./km^2) of the Aleutian skate *Bathyraja aleutica* in the North Pacific Ocean in (**a**) December–February, (**b**) March–May, (**c**) June–August, and (**d**) September–November. Small circles mark station locations outside of the hexagonal grid cells with at least three stations (see Methods).

Long-term interannual variation in the distribution of Aleutian skate catches was observed. From the 1990s, data on bottom-trawl surveys were available only from the Russian Exclusive Economic Zone (EEZ) (Figure 5a). During this period, catches of the species considered were recorded only in the western Bering Sea and in the Sea of Okhotsk along the coast of western Kamchatka and the northern Kurils. The maximum mean density of schoolings did not exceed 60–80 ind./km^2 and was recorded only in a local area of the central part of western Kamchatka. In 1991–2000s, catches of the Aleutian skate were recorded across the entire research area (Figure 5b). In Russian waters, it was observed from the southern Kurils in the south to Navarin Cape in the north. In the Sea of Okhotsk, catches were recorded along the entire eastern coast of Sakhalin, in the northern part of the sea and along western Kamchatka. In American waters, catches were recorded from the central Aleutian Islands to the western part of the Gulf of Alaska. Maximum mean catches exceeding 120 ind./km^2 were recorded off northeastern Sakhalin, the southern and northern Kurils, and southeastern Kamchatka. In 2001–2010, catches of the Aleutian skate in the study area were distributed most widely (Figure 5c). In Russian waters, it was found from the southern Kuril Islands in the south to the northern part of the Gulf of Anadyr in the north. In the Sea of Okhotsk, it was observed from nearly every site, with the exception of deep-water basins in the central and southwestern parts of the sea. In North American waters, it was registered in catches along the entire Aleutian Islands, in the Gulf of Alaska, and in waters off British Columbia south to Haida Gwaii. However, the Aleutian skate was most numerous along the eastern Bering Sea slope, where the maximum mean catches exceeded 120 ind./km^2. Single highest-density catches were also recorded off the eastern and central Aleutian Islands, in the northern part of Karagin Bay, in the Pacific waters off the northern Kurils, and off northeastern Sakhalin. After 2011, the overall distribution of the Aleutian skate in the North Pacific significantly decreased, especially in the waters off Alaska (Figure 5d). In Russian waters, the pattern of catch distribution practically did not change. Schoolings with maximum densities were still observed near northeastern Sakhalin and the northern Kurils, in the northern part of Karagin Bay, and along the Koryak coast. In addition, local areas with the maximum density aggregations were located in the waters off southwestern Kamchatka in the Kronotsk and Olyutor bays. In comparison with the previous period, there were practically no catches of this species along the Aleutian Islands, in most of the eastern Bering Sea slope waters, in the eastern part of the Gulf of Alaska, and in the waters off British Columbia.

3.2. Distribution Depending on Bottom Temperatures

Our data on bottom temperatures characterizing the thermal regime of the Aleutian skate habitat are limited to Alaskan waters (eastern part of the Bering Sea, Gulf of Alaska, and Aleutian Islands). They indicate that this species occurs at bottom temperatures of −1.8 to 10 °C (Figure 6). Maximum catches were observed in the range of 2–4 °C. At the same time, with an increase in temperature from negative values to 3 °C, catches grew, and then with its increase, they began to decrease. Thus, bottom temperatures of 2–4 °C should be considered as an optimal temperature for the Aleutian skate in the northeastern Pacific.

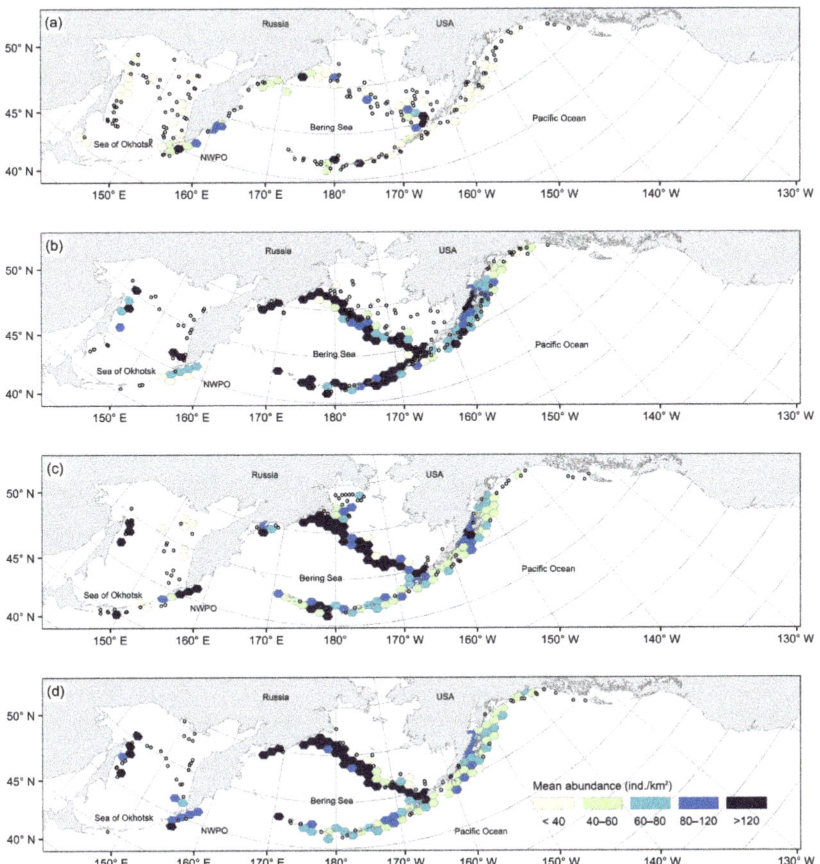

Figure 5. Mean abundance (ind./km^2) of the Aleutian skate *Bathyraja aleutica* in the North Pacific Ocean in 1976–1990 (**a**), 1991–2000 (**b**), 2001–2010 (**c**), and 2011–2019 (**d**). Small circles mark station locations outside of the hexagonal grid cells with at least three stations (see Methods).

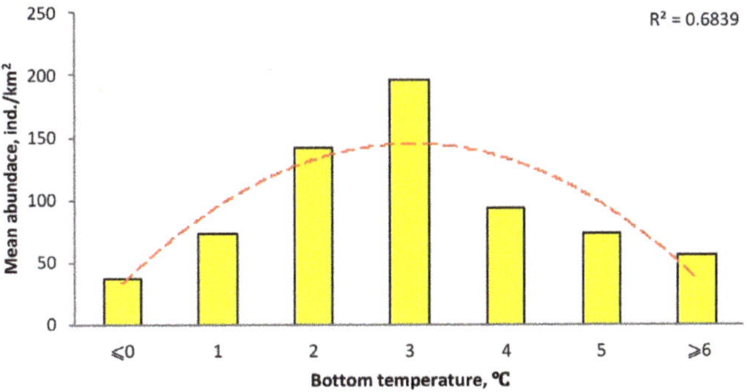

Figure 6. Catch distribution of the Aleutian skate *Bathyraja aleutica* according to bottom temperature in Alaskan waters: (yellow bar) mean value, (dash line) smoothed trend (R^2 = 0.6839).

3.3. Depth Distribution

The depth distribution of the Aleutian skate in different parts of its range (NWPO, the Sea of Okhotsk, the Bering Seas, the Aleutian Islands region, the Gulf of Alaska) varied significantly (pairwise two-sided Wilcoxon test with Holm *p*-value adjustment, $\alpha = 0.05$). In the Bering Sea (Figure 7a), the average catch and occurrence of this species at depths of 100–500 m varied between 66–188 ind./km^2 and 15–20%, respectively. This area was characterized by an increase in average catch with depth. In the NWPO (Figure 7b), the maximum occurrence of the Aleutian skate was observed at depths of 100–400 m (21–25%), and the highest rates of average catch were recorded within the depth range of 500–600 m (706 ind./km^2). In the Sea of Okhotsk (Figure 7c), the maximum occurrence was observed at depths from 400 to 600 m (26–30%), while the maximum value of the average catch (126 ind./km^2) was typical for depths of 100–200 m. For the waters off the Aleutian Islands (Figure 7d), the maximum density of accumulations and catch values were recorded at depths of 100–300 m (78% and 74–84 ind./km^2, respectively). Maximum occurrence and catches of the Aleutian skate in the Gulf of Alaska (Figure 7e) were recorded in the depth range of 100–300 m (27–49% and 60–80 ind./km^2, respectively).

The depth distribution of Aleutian skates across the entire area of research in summer statistically differed from that in other seasons; additionally, the depth distribution in winter was significantly different from that observed in the fall (pairwise two-sided Wilcoxon test with Holm *p*-value adjustment, $\alpha = 0.05$). Depth distributions during fall and spring, and during spring and winter were not significantly different based on the same test. In winter, the bulk of skates was observed in the range of 200–400 m (Figure 8a), where average catches varied between 15–24 ind./km^2, and the occurrence was between 28–33%. The maximum catch value (36 ind./km^2) was recorded at a depth of 600–700 m. In spring (Figure 8b), occurrence slightly increased for depths of 100–200 m and 400–500 m (23–24%), and the concentrations with maximum density (550 ind./km^2) were recorded at depths of 500–600 m. Summer (Figure 8c) was characterized by a higher occurrence of the Aleutian skate at lower depths of 100–300 m (21–33%) and a maximum density at depths over 800 m, where the average catches amounted to 341 ind./km^2. This period was characterized by an increase in the value of catches with depth. The autumn period (Figure 8d) was characterized by a maximum occurrence of the Aleutian skate at depths of 300–600 m (17–22%) and maximum catches at depths of 500–600 m (157 ind./km^2).

The data presented in Figure 9 provide a more detailed picture of the changes in the depth distribution of the Aleutian skate by year, characterized by a narrow bathymetric range of occurrence in the winter (December–February) and early spring (March–April). In general, across its entire range, the maximum average depths (523–550 m) were typical for January and December, and the minimum average depths fell in February and November (264–265 m). The minimum capture depths (24–50 m) were recorded from May to October, while the maximum depth of capture (1627 m) was recorded in May.

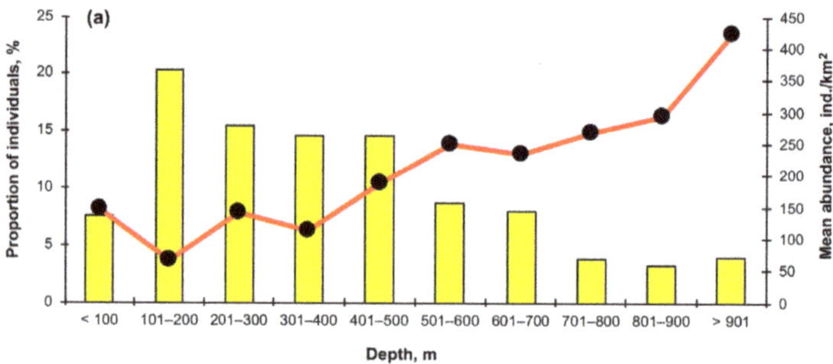

Figure 7. *Cont.*

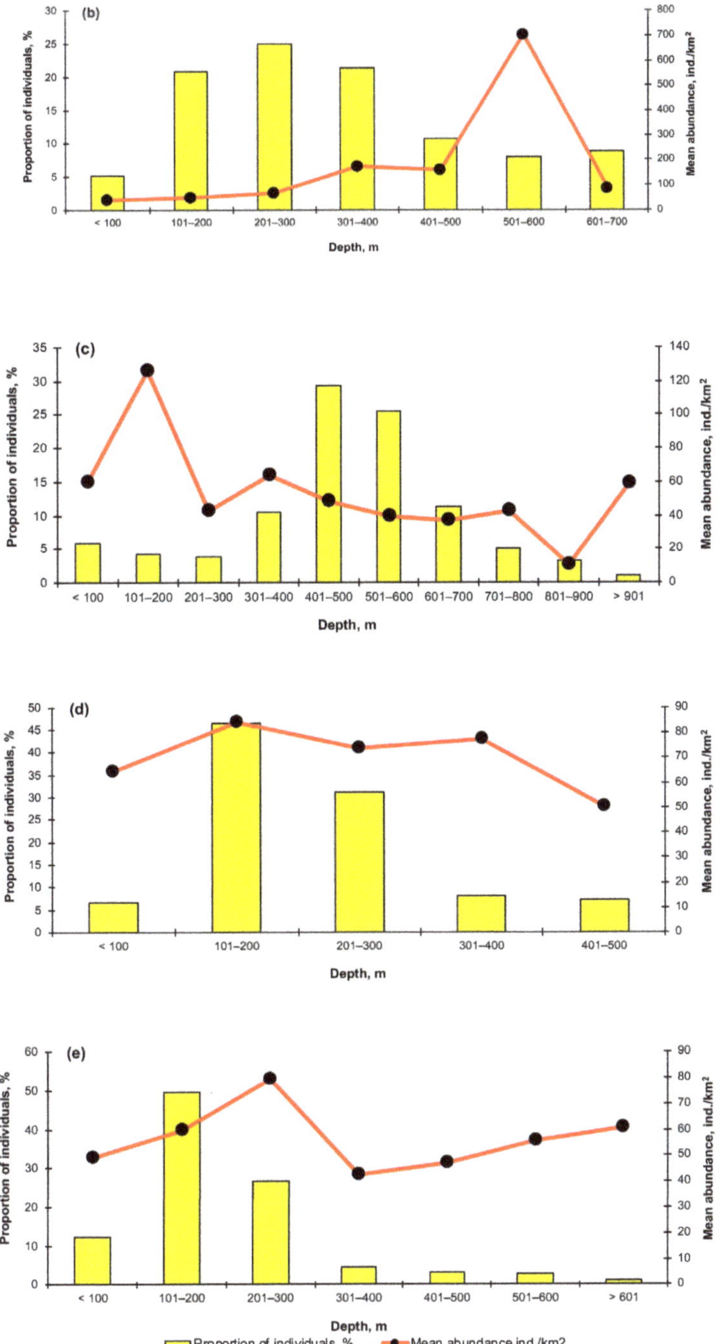

Figure 7. Depth distribution of the Aleutian skate *Bathyraja aleutica* from different regions of its range: (**a**) Bering Sea, (**b**) NWPO, (**c**) Sea of Okhotsk; (**d**) Aleutian islands, (**e**) Gulf of Alaska.

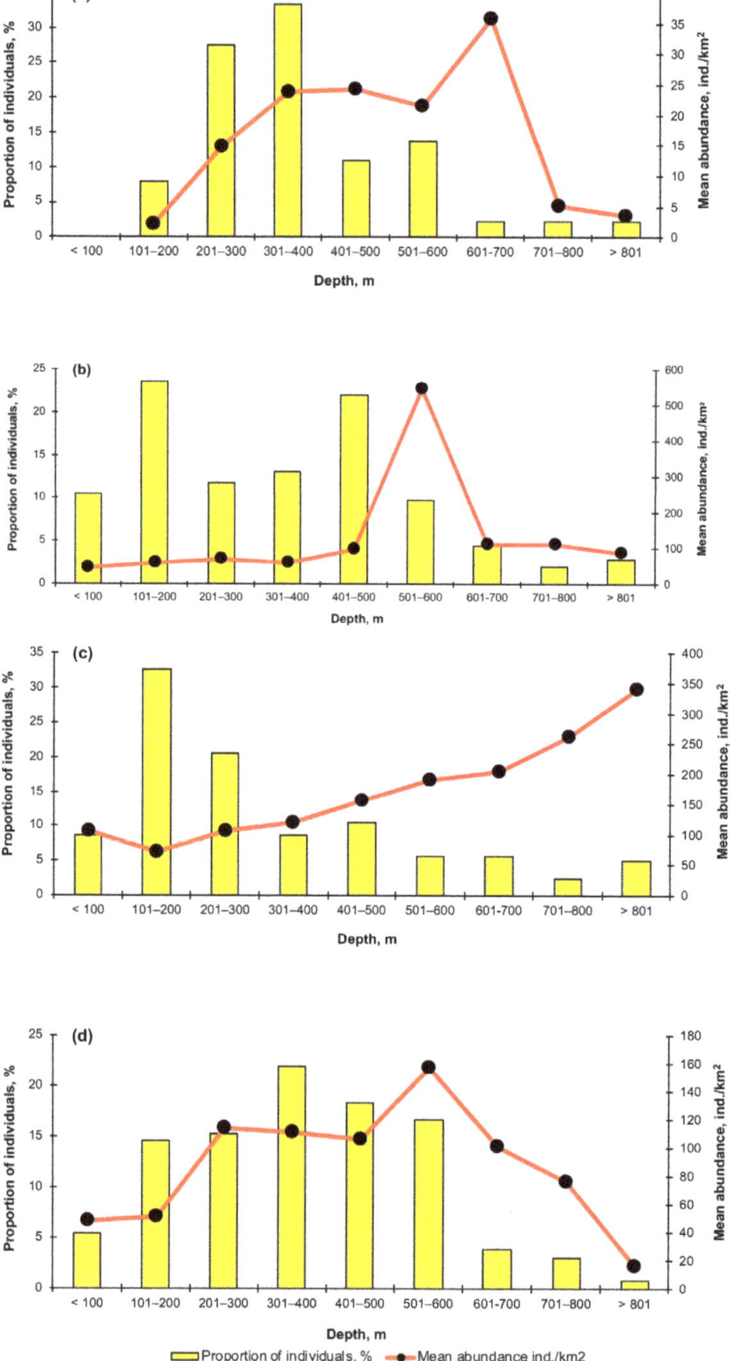

Figure 8. Depth distribution of the Aleutian skate *Bathyraja aleutica* across the North Pacific by season: (**a**) December–February, (**b**) March–May, (**c**) June–August, (**d**) September–November.

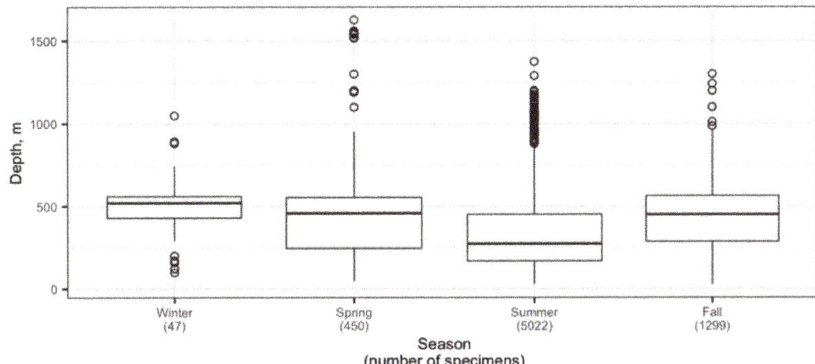

Figure 9. Boxplots of seasonal changes in the depth distribution of the Aleutian skate *Bathyraja aleutica* in the North Pacific Ocean. Circles denote outliers.

3.4. Length and Weight

The size distribution of Aleutian skate catches was represented by individuals' TL range of 10–170 (mean 74.97) cm, with a predominance (64.2%) of individuals having TL 50–100 cm (Figure 10a). Although visual analysis of the length histograms suggests a difference in length between males and females (Figure 10b), this difference was not statistically significant (two-sided Welch's *t*-test, $\alpha = 0.05$). Catches were characterized by a numerical predominance of individuals of the size group up to 70 cm (57% for males and 59% for females). The weight of females varied in the range of 0.04−21.55 (mean 3.65) kg, and males in catches had a body weight of 0.04−18.2 (mean 4.1) kg (significant difference, Welch's *t*-test, $\alpha = 0.05$).

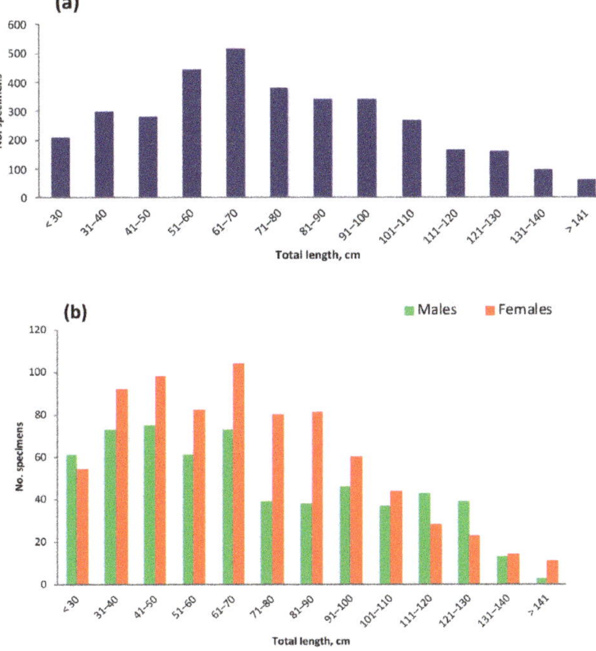

Figure 10. Total length of the Aleutian skate *Bathyraja aleutica* in catches in the North Pacific Ocean: (**a**) both sexes (M = 66.52 cm, 906 ind.), (**b**) males (M = 67.94 cm, 364 ind.) and females (M = 63.69 cm, 341 ind.).

The size of the Aleutian skate differed markedly between regions (Table 2). Note that the data on skate length is very limited in our database for the regions of Aleutian Islands, eastern Bering Sea, and Gulf of Alaska. Of all regions across its range, the Sea of Okhotsk and British Columbia waters showed the largest maximum mean length (significant differences with both NWPO and western Bering Sea in both cases, two-sided Welch's *t*-test, $\alpha = 0.05$). The smallest mean lengths were recorded from the NWPO and the western Bering Sea.

Table 2. Parameters of length–weight relationships (LWR) of the Aleutian skate *Bathyraja aleutica* by regions within its range (NWPO—the northwestern Pacific Ocean, WBS—the western Bering Sea, Okh—the Sea of Okhotsk, BC—waters off British Columbia, n—number of specimens examined, M—mean value, SE—standard error, TL—total length).

Sex	Area	Parameters of LWR			n	TL, cm		Weight, kg		Source
		a	b	R^2		Min–Max	M ± SE	Min–Max	M ± SE	
Males	NWPO	3.17×10^{-6}	3.12	0.99	83	20–145	64.34 ± 2.90	0.04–9.10	2.13 ± 0.27	Our data
	WBS	4.58×10^{-6}	3.06	0.97	213	11–135	67.30 ± 1.60	0.06–13.98	4.35 ± 0.26	Our data
		2.81×10^{-6}	3.15	0.99	74	22–130	-	-	-	[56]
	Okh	4.41×10^{-6}	3.04	0.97	50	20–146	90.51 ± 4.96	0.12–18.20	5.92 ± 0.79	Our data
	BC	3.51×10^{-6}	3.11	0.94	28	59–137	88.36 ± 4.06	1.02–15.52	4.75 ± 0.70	Our data
Females	NWPO	2.74×10^{-6}	3.17	0.99	145	24–115	64.83 ± 3.03	0.04–17.60	2.38 ± 0.26	Our data
	WBS	2.83×10^{-6}	3.16	0.96	207	13–143	66.28 ± 1.25	0.07–18.40	3.45 ± 0.26	Our data
		1.55×10^{-6}	3.28	0.99	69	28–134	-	-	-	[56]
	Okh	3.23×10^{-6}	3.12	0.98	61	27–153	85.91 ± 3.78	0.08–21.55	5.48 ± 0.75	Our data
	BC	1.31×10^{-5}	2.82	0.90	42	52–140	98.26 ± 3.39	0.82–17.15	6.35 ± 0.60	Our data
Both sexes	NWPO	2.61×10^{-6}	3.20	0.96	356	20–145	74.17 ± 1.54	0.04–18.20	3.89 ± 0.21	Our data
		3.03×10^{-6}	3.16	-	507	20–128	82.40 ± 0.93	0.06–14.10	4.42 ± 0.30	[45]
	WBS	2.79×10^{-6}	3.16	0.95	783	10–148	71.92 ± 0.55	0.04–31	4.16 ± 0.15	Our data
		2.13×10^{-6}	3.21	0.99	143	22–134	-	-	-	[56]
	Okh	5.51×10^{-6}	3.03	0.93	270	20–170	96.31 ± 1.73	0.08–35	8.21 ± 0.44	Our data
	BC	7.07×10^{-6}	2.95	0.92	71	10–140	96.83 ± 2.42	0.82–17.15	5.81 ± 0.47	Our data

The size composition of the Aleutian skate for the entire study area differed at various depths (Figure 11). Minimum mean values of length were recorded at depths of 100–300 m and 500–800 m, with maximum values being within depth ranges of 200–300 m and over 800 m. The widest size range was characteristic for depths of 200–600 m. Figure 11b shows results of pairwise Welch's *t*-tests between length samples from different depth intervals. Significant differences in length distribution ($p < 0.05$) were observed between most of the selected depth ranges.

The size composition of the Aleutian skate in the North Pacific was characterized by changes during the year (Figure 12). Total lengths of specimens caught in winter (December–February) were generally larger compared to other seasons, but note the relatively small sample size (44 specimens) which makes the result less robust. The proportion of smaller skates was larger in spring (March–May), summer (June–August), and fall (September–November) compared to winter. Overall, the total length of Aleutian skates caught in different seasons differed significantly, with the exception of spring and fall, for which the difference in total length was not significant (pairwise two-sided Welch's *t*-test, $\alpha = 0.05$).

According to the values of the linear coefficient *a* and the exponent *b*, the equations of the length–weight relationship (LWR) of Aleutian skates differed considerably between regions (Table 2), but not necessarily significantly (not shown in table). For males, no significant difference in both coefficients was observed between regions. However, for females, both coefficients showed significant differences across all considered regions ($p < 0.05$). When both sexes were considered together, no significant difference in either coefficient was observed between regions.

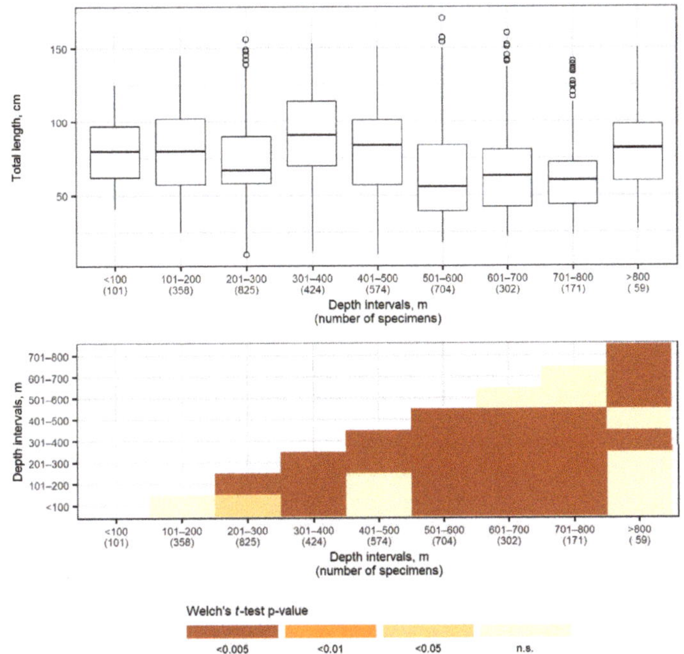

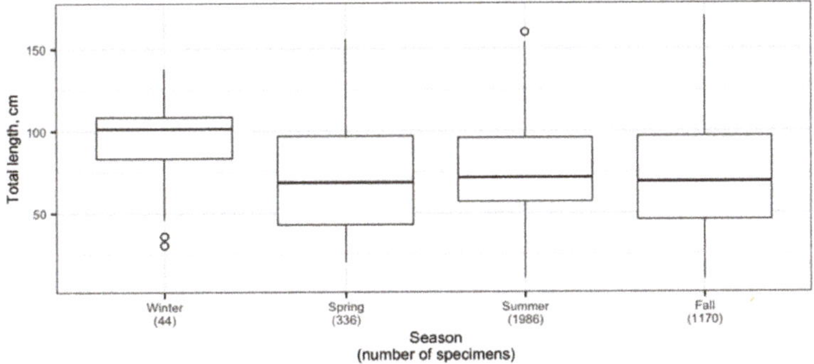

Figure 11. Distribution of the total length of Aleutian skate *Bathyraja aleutica* (cm) across selected depth intervals (m): box-plots of total length distribution (**upper** panel), circles denote outliers, and results of length distribution comparison at different depths by two-sided Welch's *t*-test with Holm *p*-value adjustment (**lower** panel).

Figure 12. Changes in total length of the Aleutian skate *Bathyraja aleutica* from the North Pacific by time of year. Circles denote outliers.

Analysis of the magnitude of the exponent of the LWR equation showed that males from the western Bering Sea and the Sea of Okhotsk are characterized by a growth pattern close to isometric (3.04–3.06). A similar growth pattern was also inherent in Aleutian skates in the Sea of Okhotsk when sex was not considered (3.03). For the female Aleutian skate and the bulk of skates from the waters off British Columbia, negative allometric growth was observed (2.82–2.95). From the remaining analyzed regions, individuals of the Aleutian skate showed positive allometric growth.

3.5. Sex Ratio and Condition Factor

The male-to-female ratio was close to 1:1 for small skates (TL < 30 cm) (53% of males), whereas females dominated in the 30–110 cm size group (54–68%), and then males predominated in the 110–130 cm size group (60–63%). An almost equal sex ratio was observed in the 130–140 cm size group, and among the largest individuals (TL > 140 cm), females dominated (79%). The male-to-female ratio did not significantly differ from 1:1 in all length classes with the exception of 61–70, 71–80, and 81–90 cm, where females significantly predominated (test of proportion 1:1, $\alpha = 0.05$, Figure 13).

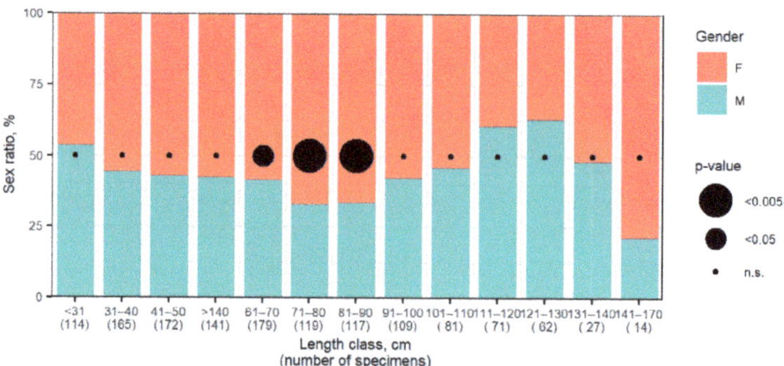

Figure 13. Sex ratios by size class of the Aleutian skate *Bathyraja aleutica* from the North Pacific. Black circles show the test of proportion (1:1) *p*-value.

Condition factor values, which characterize the ratio between length and body weight, in the Aleutian skate varied between 0.16 and 4.4 (mean 0.61) in the North Pacific as a whole. For the skates from the Bering Sea, it was in the range of 0.16 to 2.23 (0.58), the Sea of Okhotsk—0.34−1.42 (0.59), NWPO—0.26−4.40 (0.63), and in the Pacific waters of Canada—0.34–1.42 (0.59). In all areas considered, the relationship between total length and condition factor was not notably observed, although in the Bering Sea, the Sea of Okhotsk, and the NWPO the variability in condition factor for middle size classes (total length between 60 and 100 cm) was considerably higher than for smaller and larger individuals. Comparison of the condition factor within selected length classes across four regions for which sufficient data were available revealed an apparent increase of condition factor from east to west (the condition factor was lowest in the waters off British Columbia and the Bering Sea and highest in the northwest Pacific Ocean and the Sea of Okhotsk) for virtually all length classes. However, only in a few cases was this difference significant (pairwise two-sided Wilcoxon test with Holm *p*-value adjustment, $\alpha = 0.05$) (Figure 14).

The condition factor of the Aleutian skate is subject to some seasonal dynamics (Figure 15). Its maximum value typically occurred in January (0.99), from January to June it decreased, and from July to November it increased with a slight decrease in December.

3.6. Catch Dynamics

According to the bottom-trawl-survey catch data in the western Bering Sea, the catch rates of Aleutian skate during the entire period of research showed an upward trend, indicating a progressive increase in its abundance (Figure 16a). From the Sea of Okhotsk, there was a decrease in catches until the mid-1990s, after which their growth rates demonstrated positive trends. In the NWPO, until the early 2000s, catches showed a weak positive trend, after which their values varied little.

In the eastern Bering Sea, Aleutian skate catches steadily increased until the early 2000s, after which they also gradually began to decline (Figure 16b). Catches off the Aleutian Islands during the entire period of research showed a weak but positive trend. In the Gulf of Alaska, the catches gradually increased until the second half of the 2000s, after which they also gradually declined.

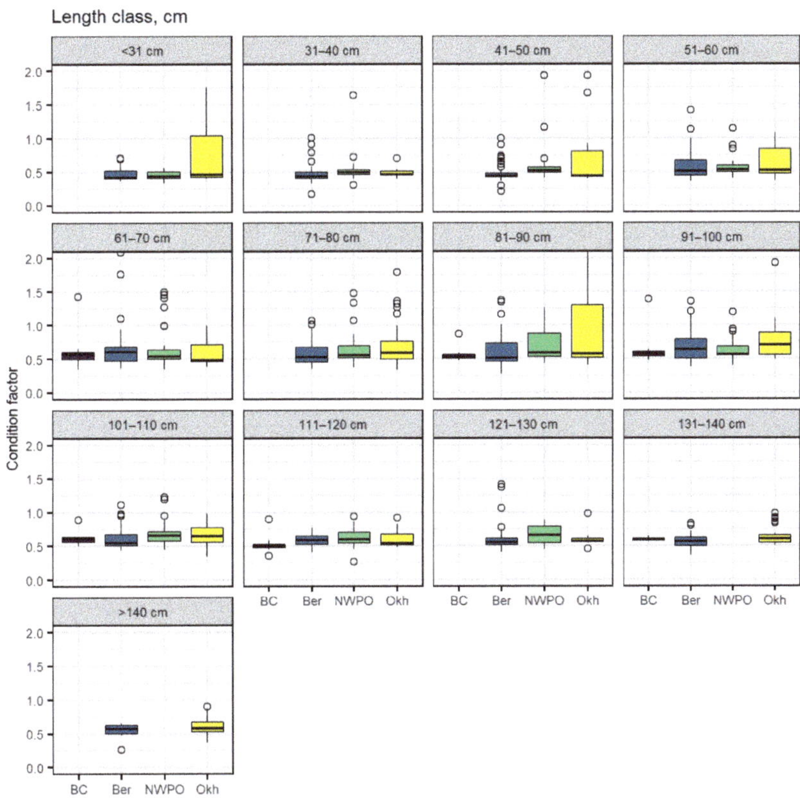

Figure 14. Boxplots of condition factor of the Aleutian skate *Bathyraja aleutica* of different size classes from different locations; BC—the waters off British Columbia, Ber—the Bering Sea, NWPO—the northwestern Pacific Ocean, Okh—the Sea of Okhotsk. Circles denote outliers.

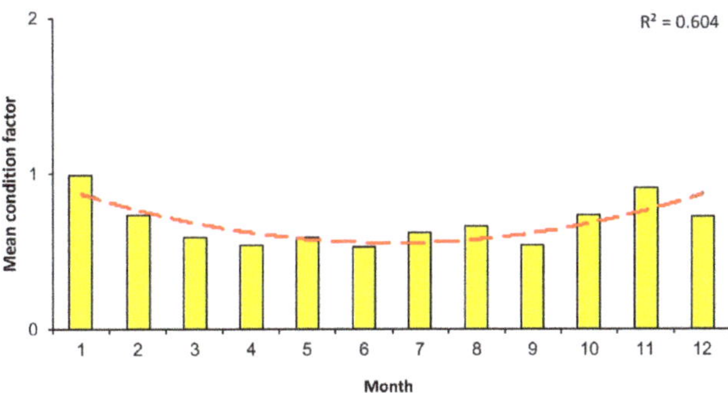

Figure 15. Changes in condition factor of the Aleutian skate *Bathyraja aleutica* from the North Pacific across the year; (yellow bar)—mean value, (dash line) smoothed trend ($R^2 = 0.604$).

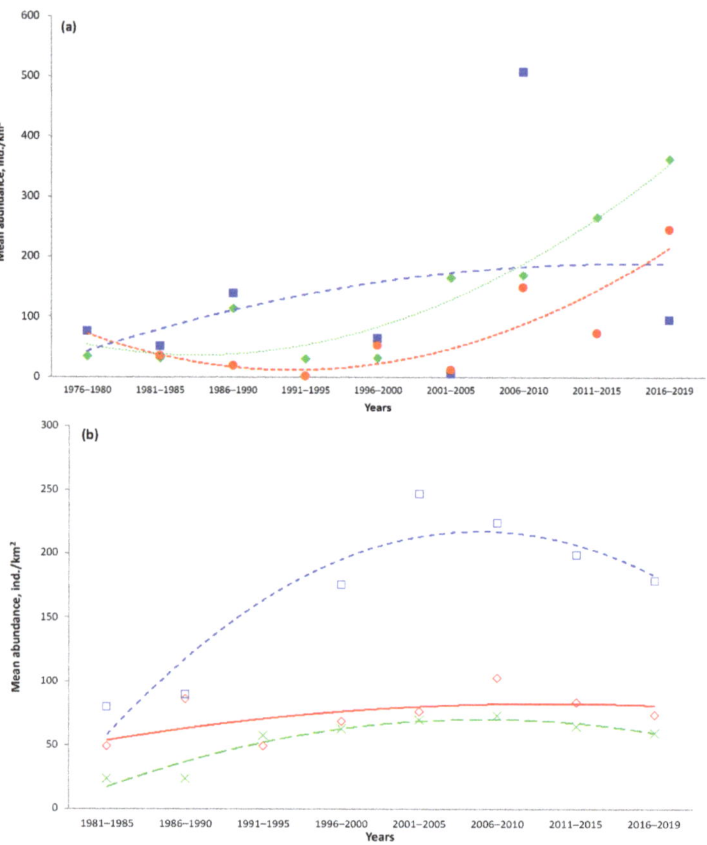

Figure 16. Region-specific dynamics of the abundance of the Aleutian skate *Bathyraja aleutica* from the North Pacific by year: (**a**) Russian waters and (**b**) Alaskan waters. Mean values and trends: (♦, ⋯) Western Bering Sea; (■, - -) NWPO; (●, —) Sea of Okhotsk; (◇) Aleutian Islands; (x, —) Gulf of Alaska; (□, ⋯) Eastern Bering Sea.

4. Discussion

4.1. Spatial Distribution

The spatial distribution of *B. aleutica* in particular areas of the North Pacific have been studied relatively well [12,39,45,51,64,65]. Our data generally do not contradict the previously obtained information on the spatial catch distribution of Aleutian skates in the north Pacific waters off the northern Kuril Islands, southeastern Kamchatka, Alaska, and British Columbia. In addition, the data presented fill a substantial the gap in our knowledge regarding the distribution of the Aleutian skate in other areas of the North Pacific and provide a more complete picture of its distribution throughout the species' range.

Thus far, seasonal patterns in the spatial distribution of the Aleutian skate have not been described. The catch distribution of this species in the north Pacific waters off the northern Kurils and southeastern Kamchatka during the summer and autumn of two adjacent years (1996–1997) were presented [39]. Our data allowed us to trace seasonal changes in the pattern of Aleutian skate distribution throughout its entire range.

Long-term changes in the spatial distribution of the Aleutian skate have not yet been analyzed. Our data allow us not only to assess the nature of changes that have occurred in the spatial distribution of this species over the past 40 years, but also to trace fluctuations in its abundance across various areas of the North Pacific during this period.

4.2. Depth Distribution

It is known that the Aleutian skate in the North Pacific occurs at depths from 15 to 1602 m [21,28,38,52,66,67]. At the same time, the optimal depth ranges for this species are 400–800 and 1400–1600 m [31]. According to our data, the species has been recorded in catches at depths of 14–1627 m, which somewhat expands the known bathymetric range of its habitat.

The Aleutian skate belongs to the lower bathyal (mesobenthic) species and it is constantly found throughout the year at depths of 500 (1000)–2000 (3000) m [36]. A characteristic feature of the lower bathyal skates is a very wide bathymetric habitat range caused by relatively homogeneous living conditions at great depths. As a result, in summer, the majority of individuals of the Aleutian skate adhere to depths of 250–500 m, and in winter they live at depths of 470–700 m [36]. The data obtained by us indicate that during the year the main concentrations of the Aleutian skate were within the depth range of 100–600 m. For all periods, except winter, the occurrence of individual Aleutian skates at depths shallower than 100 m was observed, which may be due to their dispersal during the feeding period in a wide bathymetric range in order to maximize their feeding ground and thereby reduce the level of intraspecific competition. Spring periods were characterized by an increase in the number of Aleutian skates at depths shallower than 100 m, which is probably explained by the beginning of the summer feeding migration. In contrast to [33], our materials indicate that during the summer period, the majority of skates feed at depths of 100–300 m, and not 250–550 m. The autumn was characterized by a shift of most individuals to depths of 300–600 m, which may mean the beginning of over-wintering.

According to literature data, in the western Bering Sea, the Aleutian skate occurs within the range of 157–1100 m [33]. The data obtained by us are very consistent with [33], since the main concentrations of this species were formed at depths of 100–500 m. However, our data indicate shallower depths of occurrence of this species in the Bering Sea in comparison with previously published data [33] (minimum depths of 24 m versus 157 m, respectively).

In the Pacific waters off the northern Kuril Islands and southeastern Kamchatka, in the period from May to October, the Aleutian skate occurs at depths from 75 to 800 m, while its main aggregations are observed at depths of 175–315 m [39]. According to [4], about 60% of the Aleutian skate in this area from spring to autumn are concentrated at depths of 250–600 m. In the Pacific waters off Japan, the range of occurrence of this species is within the range of 600–1210 m [33]. Our data show that the Aleutian skate is the most frequently considered species in the NWPO occurring in the depth range from 100 to 400 m, and their schoolings with maximum density are formed at depths of 500–600 m, which somewhat contradicts previously published data. The reason for this may be associated with the fact that the Pacific waters off Japan, the Kuril Islands, and eastern Kamchatka are characterized by a range of different biotopes, including both areas with a steep continental slopes and well-defined shelves, within which the nature of the depth distribution of *B. aleutica* can vary greatly. Our data represent the combined information from all waters of the northwestern Pacific, and therefore eliminate the differences inherent in particular areas.

4.3. Distribution Depending on Bottom Temperatures

Published data on bottom temperatures of the Aleutian skate habitat in the North Pacific indicate its eurythermicity. The bottom temperature range at which it occurs is quite wide: from −0.3 to 3.5 °C [34]. For example, in the Pacific waters off Kamchatka and the Kuril Islands, it was observed at bottom temperatures from 0 to 4 °C with maximum schoolings within the range of 3–4 °C [45], while off the west coast of the USA, captures were recorded within the range of bottom temperatures between 5.2 to 5.9 °C [22]. Our data indicate that in the waters off Alaska, the Aleutian skate occurs at bottom temperatures from −1.2 to 9.2 °C, and forms maximum concentrations at bottom temperatures around 3 °C. The data obtained on the thermal regime indicate a greater eurythermicity of the Aleutian skate than previously thought.

4.4. Length and Weight

Bottom trawl catches in the Pacific waters off the northern Kuril Islands and southeastern Kamchatka included Aleutian skates with TL 22–134 (mean 82.4) cm, with a predominance of individuals with a TL between 106 and 110 cm [45]. In the western Bering Sea this species was 22–134 cm long [56], while in the waters off Alaska its maximum length was 154 cm [43]. Our data showed that, in general, in the North Pacific, the Aleutian skate was represented in catches by individuals with TLs within the range of 9.6−170 cm, with the predominance of skates being 50–100 cm long.

The relationship between the length and body weight of the Aleutian skate is described for several regions (Table 2). A comparison of the exponents of this relationship obtained by us and previously published [45,56] values shows that they differed significantly, both in individuals of different sexes and from different areas. For males in the western Bering Sea, a positive allometric growth pattern was previously shown [56]. According to our data, from this area, as well as in the Sea of Okhotsk, males of the Aleutian skate show a growth pattern close to isometric. In the NWPO and British Columbia waters, they were also characterized by positive allometric growth. Positive allometric growth was previously shown for females of the Aleutian skate in the western Bering Sea [56]. We observed a similar growth pattern not only in this area, but also from the rest of the study areas, with the exception of the waters off British Columbia, where it was negative allometric. In general, positive allometric growth was previously observed for both sexes in the western Bering Sea and the Pacific waters off the northern Kurils and southeastern Kamchatka [45,56], which is confirmed by our data. In the Sea of Okhotsk their growth is close to isometric, and in the waters off British Columbia it has a negative allometric characteristic. The reasons for this difference may be associated with the fact that previously published data characterizing the LWR of the Aleutian skate was obtained during a short period of time (summer–autumn period), when its individuals were in a feeding condition. Our data were obtained in a more temporally protracted period (most of the year), when individuals of this species may be in varying physiological states. The obvious difference in the considered feature of the skates from British Columbia waters has not yet found a reasonable explanation. A smaller value of the LWR equation exponent indicates a lower condition factor [68,69], which may be due to suboptimal living conditions on the outskirts of its range (temperature, oxygen content, food availability, etc.).

4.5. Sex Ratio and Condition Factor

Data on the sex ratios of Aleutian skates across various size classes are few. Some authors [52,53] suggest that the sex of the Aleutian skate becomes distinguishable (claspers begin to be visible) only when TL > 70 mm is reached, below which all embryos resemble females, while in embryos TL > 70 mm, the sex ratio is approximately 1:1. An equal sex ratio of embryos is noted in egg capsules [36], while equal sex ratios are also characteristic of natural populations. However, females predominate in the older age groups of all Far Eastern skates due to earlier maturation of males, their shorter lifespan and lower growth rates [37]. Larger sizes of females are of adaptive importance for ensuring greater fecundity of the population [70]. It was previously shown [45] that in the Pacific waters off the northern Kuril Islands and southeastern Kamchatka, females predominate among the smallest individuals of the Aleutian skate (TL < 50 cm). They also dominate in the size classes 60–90 cm and >120 cm, while an equal sex ratio was observed in the size class of 110–120 cm. According to our data, males are more numerous among skates with TL < 30 cm, females dominate in the 30–110 cm size class, the sex ratio is almost equal in the 130–140 cm size class, and females predominate among individuals TL > 140 cm. Thus, our generalized data for the entire North Pacific on the sex ratio in different size classes of the Aleutian skate fully correspond to the previously published information for the Russian Far Eastern seas [37], and differ slightly from those for the Pacific waters off the northern Kurils and southeastern Kamchatka [45]. The reason for these differences probably include a number of limitations in the collection of data compared (bathymetric range, summer–autumn season, local geographical area, etc.).

Information regarding the condition factor of Aleutian skates is available only for the western Bering Sea [40]. Thus, according to these data, for male Aleutian skates, the maximum condition factor from different seasons was 8.8, and that for females was 9.6. According to our data, the maximum condition factor for the western Bering Sea was 4.4, while 95% of the individuals from all study areas had a condition factor < 1%. For the Aleutian skate, Alaska skate *B. parmifera*, Okhotsk skate *B. violacea*, and Pacific sleeper shark *Somniosus pacificus*, a general trend of increased condition factor in the autumn–winter period and a decrease in its value by the summer has been revealed [71–73], as well as in this study, which may be due to similar ecological and biological conditions of these species (ecological niche, biotopes, migrations, etc.). This trend indicates a feeding migration in the spring–summer from greater to shallower depths, and in the autumn–winter in the opposite direction.

4.6. Catch Dynamics

Among all the deep-water skates of the Russian Far Eastern seas, the Aleutian skate is the third in terms of biomass (87.7 thousand tons), after the Alaska skate *B. parmifera* and the Matsubara skate *B. matsubarae*, with biomasses of 273.6 and 117.6 thousand tons, respectively [12]. In the Russian part of our study area, the data on the catch dynamics of the Aleutian skate were presented only for the Pacific waters off the northern Kurils and southeastern Kamchatka for the period 1993–2000 [45], during which there was a fairly pronounced positive trend. The data presented by us for the NWPO differ from the results of the above-mentioned publications, demonstrating a possible downward trend in catch rates in the early 1990s and a rise only by the early 2000s. This difference may be due to the fact that the Pacific waters off southeastern Kamchatka and the northern Kurils are only part of our NWPO area, which, in addition, also includes the Pacific waters off Japan and northeastern Kamchatka.

Data on the Aleutian skate dynamics of the biomass in the Gulf of Alaska in the period from 2000 to 2019 [74] was quite comparable with our data on the dynamics of catch rates of this skate (an increase by 2005–2010, followed by a fall). When comparing the data obtained for the eastern Bering Sea and the Aleutians [75], differences were revealed, e.g., the USA data indicated an increase in biomass between 2005 and 2010, while our data showed a drop in catches, then since 2011 there was a positive trend in both the USA data and ours. The reasons for the differences between our and previously published data are still unclear and may be related to different methodological approaches to data analysis.

Currently, targeted fishing for skates in the Russian Far Eastern waters has not been developed, with catches occurring as by-catch of other target fisheries [10,11]. Historically, catch surveys did not separate skates by species, and they were not retained and utilized, but discarded after capture [76–78]. This situation has changed in recent years when several Russian companies started to export skate wings to China and the British Virgin Islands [10,11]. They also may be profitably marketed in Japan and Korea. In waters of the Russian Far East, 11.4–11.7 Kt were recommended for harvesting in the 1990s, 7.2–11.9 Kt in the 2000s, 11.2–14.0 Kt in the 2010s, and 11.2–11.3 Kt in the 2020s [13]. Since the fishing of deep-water skates, including the Aleutian skate, is currently developing in the North Pacific, an important role in their management should consider regular stock assessments and monitoring of the populations. The results of the present study have shown that in the western Bering Sea and the Sea of Okhotsk, the abundance of the Aleutian skate is increasing, to the extent that it would not require a reduction in the efforts of deep-water groundfish fisheries. From other areas of Aleutian skate habitat (the Pacific waters off the Kuril Islands and eastern Kamchatka, the eastern Bering Sea, the Aleutian Islands, and the Gulf of Alaska), in recent years there has been no increase in their abundance. Thus, the current state of its populations in these areas may require a reduction in fishing effort in deep-water groundfish fisheries, which will help to avoid overfishing and to contribute to the conservation of the species.

5. Conclusions

The results of the analysis of long-term data on the Aleutian skate *Bathyraja aleutica* records in the North Pacific Ocean allowed for obtaining the comprehensive information about species' spatial and vertical distributions, dynamics of abundance, and size composition in this area. This species is most abundant off the eastern Bering Sea slope, the central Aleutian Islands, in the Pacific waters off southeastern Kamchatka and the northern Kuril Island, and northeastern Sakhalin. The main habitat depths of the Aleutian skate are 100–600 m with shift to greater depths in cold period of the year and to shallower depths in warm seasons. *B. aleutica* prefers bottom temperatures around 3 °C. Bottom trawl catches are presented by individuals with total length ranged 9.6 to 170 cm with a predominance of skates 50–100 cm long. Sizes of male and female Aleutian skates were almost not different. Based on the analysis of *B. aleutica* catch rate dynamics it can be noted that the current state of its populations in the Pacific waters off the Kuril Islands and eastern Kamchatka, the eastern Bering Sea, the Aleutian Islands, and the Gulf of Alaska, requires a reduction in fishing effort in deep-water groundfish fisheries. At the same time, in the western Bering Sea and the Sea of Okhotsk, the abundance of this species is increasing, and therefore a reduction in the efforts of deep-water groundfish fisheries would not require. Such fishery management measures will help to avoid overfishing and to conserve the species considered.

Author Contributions: I.V.G., K.K.K. and A.M.O. equally contributed to the preparation of the paper. All authors have read and agreed to the published version of the manuscript.

Funding: Preparation of this paper was funded by the Ministry of Science and Higher Education of the Russian Federation (grant No. 13.1902.21.0012, contract No. 075-15-2020-796).

Institutional Review Board Statement: Not applicable.

Informed Consent Statement: Not applicable.

Data Availability Statement: Not applicable.

Acknowledgments: The authors are grateful to their colleagues from TINRO (Pacific Branch of VNIRO, Vladivostok, Russia) who collected material at sea. They also thank the Alaska Fisheries Science Center and Northwestern Fisheries Science Center (Seattle, WA, USA) for making accessible the primary data on bottom-trawl surveys in waters off Alaska and the American West Coast, as well as the data of observers on board American commercial fishing vessels. The authors are also grateful to OBIS, GBIF, and several museums and institutions (specified in Table 1) that provided access to their databases with records of *B. aleutica*. Todd Miller (Recruitment, Energetics, and Coastal Assessment Program, NOAA Alaska Fisheries Science Center, Juneau, AK, USA) is thanked for improvement of the manuscript's English. The authors also thank two anonymous reviewers for valuable comments and suggestions.

Conflicts of Interest: The authors declare no conflict of interest.

References

1. Mito, K. Food Relationships among Benthic Fish Populations in the Bering Sea on the *Theragra chalcogramma* Fishing Grounds in October and November of 1972. Master's Thesis, Hokkaido University Graduate School, Hakodate, Japan, 1974; pp. 1–135.
2. Brodeur, R.D.; Livingston, P.A. Food habits and diet overlap of various Eastern Bering Sea fishes. *U.S. Dep. Commer. NOAA Tech. Memo. NMFS F/NWC* **1988**, *127*, 1–76.
3. Livingston, P.A.; de Reynier, Y. Groundfish food habits and predation on commercially important prey species in the Eastern Bering Sea from 1990 to 1992. *U.S. Dep. Commer. NOAA/NMFS Alaska Fish. Sci. Cent. Proc. Rep.* **1996**, *96*, 1–214.
4. Orlov, A.M. The diets and feeding habits of some deep-water benthic skates (Rajidae) in the Pacific waters off the northern Kuril Islands and southeastern Kamchatka. *Alaska Fish. Res. Bull.* **1998**, *5*, 1–17.
5. Orlov, A.M. Feeding habits of some deep-benthic skates (Rajidae) in the western Bering Sea. In Proceedings of the MTS/IEEE Oceans 2001. An Ocean Odyssey, Honolulu, HI, USA, 5–8 November 2001; Volume 2, pp. 842–855. [CrossRef]

6. Orlov, A.M. Diets, feeding habits, and trophic relations of six deep-benthic skates (Rajidae) in the western Bering Sea. *Aqua. J. Ichthyol. Aquat. Biol.* **2003**, *7*, 45–60.

7. Chuchukalo, V.I.; Lapko, V.V.; Kuznetsova, N.A.; Slabinsky, A.M.; Napazakov, V.V.; Nadtochy, V.A.; Koblikov, V.N.; Pushchina, O.I. Feeding of groundfishes on the shelf and continental slope of the northern Sea of Okhotsk. *Izv. TINRO* **1999**, *126*, 24–57.

8. Chuchukalo, V.I.; Napazakov, V.V. Feeding and trophic state of the mass species of skates (Rajidae) from the western part of the Bering Sea. *Izv. TINRO* **2002**, *130*, 422–428.

9. Ishihara, H. The skates and rays of the western North Pacific: An overview of their fisheries, utilization, and classification. *U.S. Dep. Commer. NOAA Tech. Rep. NMFS* **1990**, *90*, 485–497.

10. Orlov, A.M. Bycatch reduction of multispecies bottom trawl fisheries: Some approaches to solve the problem in the Northwest Pacific. *Am. Fish. Soc. Symp.* **2007**, *49*, 241–258.

11. Orlov, A.M. Some approaches to reducing non-commercial bycatch of bottom trawl fisheries in the western Bering Sea. *Asian Fish. Sci.* **2011**, *24*, 397–412. [CrossRef]

12. Dolganov, V.N. Reserves of the skates in Far Eastern seas of Russia and their prospective fishery. *Izv. TINRO* **1999**, *126*, 650–652.

13. Orlov, A.M.; Volvenko, I.V. Long-term Changes of Distribution and Abundance of Nine Deep-water Skates (Arhynchobatidae: Rajiformes: Chondrichthyes) in the Northwestern Pacific. *Animals* **2022**, *12*, 3485. [CrossRef]

14. Nasedkina, E.A. Characteristics of flesh of skates. *Rybn. Khoz.* **1969**, *6*, 59–60.

15. Baidalova, G.F.; Dubnitskaya, G.M.; Nechaev, A.P.; Severinenko, S.M. Chemical composition and properties of thorny skate liver oil depending on the fishing season. *Rybn. Khoz.* **1984**, *10*, 60–62.

16. Orlov, A.M. Groundfish resources of the northern North Pacific continental slope: From science to sustainable fishery. In Proceedings of the Seventh North Pacific Rim Fisheries Conference, Busan, Republic of Korea, 18–20 May 2004; Alaska Pacific University: Anchorage, AK, USA, 2005; pp. 139–150.

17. Dvinin, Y.F.; Konstantinova, L.L.; Kuz'mina, V.I. Techno-chemical characteristics of some skates and chimaeras. *Tr. PINRO* **1981**, *47*, 29–51.

18. Orlov, A.M. Substantiation of commercial size of Far Eastern skates (*fam. Rajidae*) by example of widespread species from the Bering Sea. *Tr. VNIRO* **2006**, *146*, 252–264.

19. Ormseth, O.A. Assessment of the skate stock complex in the Bering Sea and Aleutian Islands. In *Stock Assessment and Fishery Evaluation Report for the Groundfish Resources of the Bering Sea/Aleutian Islands Regions*; North Pacific Fishery Management Council: Anchorage, AK, USA, 2012; pp. 1647–1734.

20. Dolganov, V.N. *Guide to the Diagnostic Characters of Cartilaginous Fishes from the Far East Seas of the U.S.S.R. and Neighboring Waters*; TINRO Vladivostok: Vladivostok, Russia, 1983; pp. 1–92.

21. Mecklenburg, C.; Mecklenburg, T.; Thorsteinson, L. *Fishes of Alaska*; American Fisheries Society: Bethesda, MA, USA, 2002; pp. 1–1037.

22. Hoff, G.R. New records of the Aleutian skate, *Bathyraja aleutica*, from northern California. *Calif. Fish Game* **2002**, *88*, 145–148.

23. Fedorov, V.V.; Chereshnev, I.A.; Nazarkin, M.V.; Shestakov, A.V.; Volobuev, V.V. *Catalogue of Marine and Freshwater Fishes in Northern Sea of Okhotsk*; Dal'nauka: Vladivostok, Russia, 2003; pp. 1–204.

24. Shinohara, G.; Narimatsu, Y.; Hattori, T.; Ito, M.; Takata, Y.; Matsuura, K. Annotated checklist of deep-sea fishes from the Pacific coast off Tohoku District, Japan. *Natl. Mus. Nat. Sci. Monogr.* **2009**, *39*, 683–735.

25. Lynghammar, A.; Christiansen, J.S.; Mecklenburg, C.W.; Karamushko, O.V.; Møller, P.R.; Gallucci, V.F. Species richness and distribution of chondrichthyan fishes in the Arctic Ocean and adjacent seas. *Biodiversity* **2013**, *14*, 57–66. [CrossRef]

26. Grigorov, I.V.; Orlov, A.M. Species diversity and conservation status of cartilaginous fishes (*Chondrichthyes*) of Russian waters. *J. Ichthyol.* **2013**, *53*, 923–936. [CrossRef]

27. Parin, N.V.; Evseenko, S.A.; Vasil'eva, E.D. *Fishes of Russian seas: Annotated Catalogue*; KMK Scientific Press: Moscow, Russia, 2014; pp. 1–733.

28. Tuponogov, V.N.; Kodolov, L.S. *Handbook for Identification of the Commercial and Mass Species of Fishes of Far Eastern Seas of Russia*; Russkiy Ostrov: Vladivostok, Russia, 2014; pp. 1–336.

29. Dyldin, Y.V. Annotated checklist of the sharks, batoids and chimaeras (Chondrichthyes: Elasmobranchii, Holocephali) from waters of Russia and adjacent areas. *Publ. Seto Mar. Biol. Lab.* **2015**, *43*, 40–91. [CrossRef]

30. Ivanov, O.A.; Sukhanov, V.V. Species structure of pelagic ichthyocenes in Russian waters of Far Eastern seas and the Pacific Ocean in 1980–2009. *J. Ichthyol.* **2015**, *55*, 497–526. [CrossRef]

31. Orlov, A.M.; Tokranov, A.M. Checklist of deep-sea fishes of the Russian northwestern Pacific Ocean found at depths below 1000 m. *Prog. Oceanogr.* **2019**, *176*, 102143. [CrossRef]

32. Dolganov, V.N. Feeding of skates of the Rajidae family and their role in ecosystems of the Far Eastern seas. *Izv. TINRO* **1998**, *124*, 417–424.

33. Dolganov, V.N. Distribution and migration of skates of the family Rajidae of the Far Eastern seas of Russia. *Izv. TINRO* **1998**, *124*, 433–437.

34. Dolganov, V.N. Abiotic conditions of the habitat of skates of the family Rajidae of the Far Eastern seas of Russia. *Izv. TINRO* **1998**, *124*, 429–432.

35. Dolganov, V.N. Reproduction of skates of the Rajidae family of the Far Eastern seas of Russia. *Izv. TINRO* **1998**, *124*, 425–428.

36. Dolganov, V.N. Geographical and bathymetric distribution of skates of the Rajidae family in the Far Eastern seas of Russia. *Vopr. Ikhtiol.* **1999**, *39*, 428–430.

37. Dolganov, V.N. The size, age and growth of the skates of family Rajidae of the Far Eastern seas of Russia. *Izv. TINRO* **2005**, *143*, 84–89.

38. Dudnik, Y.I.; Dolganov, V.N. Distribution and stocks of fish on the continental slope of the Sea of Okhotsk and the Kuril Islands in the summer of 1989. *Vopr. Ikhtiol.* **1992**, *32*, 83–98.

39. Fatykhov, R.N.; Poltev, Y.N.; Mukhametov, I.N.; Nemchinov, O.Y. Spatial distribution of mass species of skates of the genus *Bathyraja* in the area of the northern Kuril Islands and southeastern Kamchatka in various seasons of 1996–1997. In *Commercial and Biological Studies of Fish in the Pacific Waters of the Kuril Islands and Adjacent Areas of the Okhotsk and Bering Seas in 1992–1998*; VNIRO Publishing: Moscow, Russia, 2000; pp. 104–120.

40. Glubokov, A.I.; Orlov, A.M. Some morphophysiological indicators and feeding characteristics of the Aleutian skate *Bathyraja aleutica* from the western Bering Sea. *Probl. Fish.* **2000**, *1*, 126–149.

41. Orlov, A.M. Trophic interrelations in predatory fishes of Pacific waters circumambient the northern Kuril Islands and southeastern Kamchatka. *Hydrobiol. J.* **2004**, *40*, 1–19. [CrossRef]

42. Davis, C.D.; Ebert, D.A.; Ishihara, H.; Orlov, A.; Compagno, L.J.V.; Farrugia, T.J.; Tribuzio, C.A. Bathyraja aleutica. In *The IUCN Red List of Threatened Species 2015*; 2015; p. e.T161661A80674005. Available online: https://www.iucnredlist.org/species/161661/80674005 (accessed on 25 October 2022).

43. Ebert, D.A. Reproductive biology of skates, *Bathyraja* (Ishiyama), along the eastern Bering Sea continental slope. *J. Fish Biol.* **2005**, *66*, 618–649. [CrossRef]

44. Tokranov, A.M.; Orlov, A.M.; Sheiko, B.A. *Commercial Fishes of Continental Slope of Kamchatka Waters*; Kamchatpress: Petropavlovsk-Kamchatsky, Russia, 2005; pp. 1–52.

45. Orlov, A.; Tokranov, A.; Fatykhov, R. Common deep-benthic skates (Rajidae) of the northwestern Pacific: Basic ecological and biological features. *Cybium* **2006**, *30*, 49–65.

46. Spies, I.B.; Gaichas, S.; Stevenson, D.E.; Orr, J.W.; Canino, M.F. DNA-based identification of Alaska skates (*Amblyraja, Bathyraja* and *Raja*: Rajidae) using cytochrome c oxidase subunit I (*coI*) variation. *J. Fish Biol.* **2006**, *69*, 283–292. [CrossRef]

47. Spies, I.B.; Orr, J.W.; Stevenson, D.E.; Goddard, P.; Hoff, G.; Guthridge, J.; Hollowed, M.; Rooper, C. Skate egg nursery areas support genetic diversity of Alaska and Aleutian skates in the Bering Sea. *Mar. Ecol. Progr. Ser.* **2021**, *669*, 121–138. [CrossRef]

48. Ebert, D.A.; Bizzarro, J.J. Standardized diet compositions and trophic levels of skates (Chondrichthyes: Rajiformes: Rajoidei). *Environ. Biol. Fish.* **2007**, *80*, 221–237. [CrossRef]

49. Yang, M.-S. Food habits and diet overlap of seven skate species in the Aleutian Islands. *U.S. Dep. Commer. NOAA Tech. Memo. NMFS-AFSC* **2007**, *177*, 1–46.

50. Stevenson, D.E.; Orr, J.W.; Hoff, G.R.; McEachran, J.D. Emerging patterns of species richness, diversity, population density, and distribution in the skates (Rajidae) of Alaska. *Fish. Bull.* **2007**, *106*, 24–39.

51. Stevenson, D.E.; Orr, J.W.; Hoff, G.R.; McEachran, J.D. *Sharks, Skates and Ratfish of Alaska*; University of Alaska Sea Grant: Fairbanks, AK, USA, 2007; pp. 1–85.

52. Hoff, G.R. Embryo developmental events and the egg case of the Aleutian skate *Bathyraja aleutica* (Gilbert) and the Alaska skate *Bathyraja parmifera* (Bean). *J. Fish Biol.* **2009**, *74*, 483–501. [CrossRef]

53. Hoff, G.R. Skate *Bathyraja* spp. egg predation in the eastern Bering Sea. *J. Fish Biol.* **2009**, *74*, 250–269. [CrossRef]

54. Hoff, G.R. Identification of skate nursery habitat in the eastern Bering Sea. *Mar. Ecol. Prog. Ser.* **2010**, *403*, 243–254. [CrossRef]

55. Hoff, G.R. Identification of multiple nursery habitats of skates in the eastern Bering Sea. *J. Fish Biol.* **2016**, *88*, 1746–1757. [CrossRef] [PubMed]

56. Orlov, A.; Binohlan, C. Length-weight relationships of deep-sea fishes from the western Bering Sea. *J. Appl. Ichthyol.* **2009**, *25*, 223–227. [CrossRef]

57. Brown, S.C.; Bizzarro, J.J.; Cailliet, G.M.; Ebert, D.A. Breaking with tradition: Redefining measures for diet description with a case study of the Aleutian skate *Bathyraja aleutica* (Gilbert 1896). *Environ. Biol. Fish.* **2012**, *95*, 3–20. [CrossRef]

58. Teshima, K.; Tomonaga, S. Reproduction of Aleutian skate, Bathyraja aleutica, with comments on embryonic development. In *Indo-Pacific Fish Biology, Proceedings of the Second International Conference on Indo-Pacific Fishes, Conducted at the Tokyo National Museum, Ueno Park, Tokyo, Japan, 29 July–3 August 1985*; Tokyo National Museum: Tokyo, Japan, 1986; pp. 303–309.

59. Terentyev, D.A.; Zolotov, A.O. Fishing and long-term dynamics of biomass of skates (*Bathyraja*) off the western coast of Kamchatka. *Izv. TINRO* **2012**, *169*, 32–40.

60. Haas, D.L.; Ebert, D.A.; Cailliet, G.M. Comparative age and growth of the Aleutian skate, *Bathyraja aleutica*, from the eastern Bering Sea and Gulf of Alaska. *Environ. Biol. Fish.* **2016**, *99*, 813–828. [CrossRef]

61. Vinogradskaya, A.V.; Matveev, A.A.; Ryazanova, T.V.; Terentyev, D.A.; Kurbanov, Y.K. Methods of visualization of annual rings on the vertebrae of some species of rhombic rays (Rajidae Blainville, 1816). *Bull. Kamchat. Stat. Tech. Univ.* **2019**, *49*, 89–97. [CrossRef]

62. Vinogradskaya, A.V.; Matveev, A.A.; Terentyev, D.A. New data on fishing and the state of stocks of skates of the family Arhynchobatidae off the western coast of Kamchatka. *Bull. Kamchat. Stat. Tech. Univ.* **2022**, *59*, 49–61. [CrossRef]

63. Rooper, C.N.; Hoff, G.R.; Stevenson, D.E.; Orr, J.W.; Spies, I.B. Skate egg nursery habitat in the eastern Bering Sea: A predictive model. *Mar. Ecol. Prog. Ser.* **2019**, *609*, 163–178. [CrossRef]

64. Orlov, A.M. Trophic relationships of commercial fishes in the Pacific waters off southeastern Kamchatka and the northern Kuril Islands. In *Ecosystem Approaches for Fisheries Management*; Alaska Sea Grant College Program: Fairbanks, AK, USA, 1999; pp. 231–263.
65. Orlov, A.M. *Quantitative Distribution of Demersal Nekton in the Pacific Waters of the Northern Kuril Islands and Southeastern Kamchatka*; VNIRO Publishing: Moscow, Russia, 2010; pp. 1–335.
66. Stevenson, D.E. Identification of skates, sculpins, and smelts by observers in north Pacific groundfish fisheries (2002–2003). *U.S. Dep. Commer. Tech. Memo. NMFS-AFSC* **2004**, *142*, 1–67.
67. Dolganov, V.N.; Tuponogov, V.N. Keys to Skates of Genus *Bathyraja* and *Rhinoraja* (*Fam. Rajidae*) of the Far Eastern Seas of Russia. *Izv. TINRO* **1999**, *126*, 657–664.
68. Zotina, R.S.; Zotin, A.I. Quantitative ratio between weight, length, egg sizes, and fecundity of the animals. *Zh. Obshch. Biol.* **1967**, *28*, 82–92. [PubMed]
69. Krivobok, M.N.; Shatunovsky, M.I. Analysis methods of fishes using morphophysiological parameters. In *Methods of Morphophysiological and Biochemical Studies of Fishes*; VNIRO: Moscow, Russia, 1972; pp. 29–44.
70. Nikol'sky, G.V. *Ecology of Fishes*; Vysshaya Shkola: Moscow, Russia, 1974; pp. 1–367.
71. Orlov, A.M.; Baitalyuk, A.A. Spatial distribution and features of biology of Pacific sleeper shark *Somniosus pacificus* in the North Pacific. *J. Ichthyol.* **2014**, *54*, 526–546. [CrossRef]
72. Grigorov, I.V.; Orlov, A.M.; Baitalyuk, A.A. Spatial distribution, size composition, feeding habits, and dynamics of abundance of Alaska skate *Bathyraja parmifera* in the North Pacific. *J. Ichthyol.* **2015**, *55*, 644–663. [CrossRef]
73. Grigorov, I.V.; Baitalyuk, A.A.; Orlov, A.M. Spatial distribution, size composition, and dynamics of catches of the Okhotsk skate *Bathyraja violacea* in the North Pacific Ocean. *J. Ichthyol.* **2017**, *57*, 706–720. [CrossRef]
74. Ormseth, O.A. Partial Assessment of the Skate Stock Complex in the Bering Sea and Aleutian Islands. In *NPFMC Bering Sea and Aleutian Islands SAFE*; North Pacific Fisheries Management Council: Anchorage, AK, USA, 2021; pp. 1–9. Available online: https://www.fisheries.noaa.gov/resource/data/2021-assessment-skate-stock-complex-bering-sea-and-aleutian-islands (accessed on 3 October 2022).
75. Ormseth, O.A. Assessment of the skate stock complex in the Gulf of Alaska. In *NPFMC Gulf of Alaska SAFE*; North Pacific Fisheries Management Council: Anchorage, AK, USA, 2021; pp. 1–60. Available online: https://www.fisheries.noaa.gov/resource/data/2021-assessment-skate-stock-complex-gulf-alaska (accessed on 3 October 2022).
76. Antonov, N.P.; Klovatch, N.V.; Orlov, A.M.; Datsky, A.V.; Lepskaya, V.A.; Kuznetsov, V.V.; Yarzhombek, A.A.; Abramov, A.A.; Alekseev, D.O.; Moiseev, S.I.; et al. Fishing in the Russian Far East fishery basin in 2013. *Tr. VNIRO* **2013**, *160*, 133–211.
77. Badaev, O.Z. Estimation of bycatch and discards of longline fish in Far East seas. *Probl. Fish.* **2018**, *19*, 58–72.
78. Zolotov, A.O. Modern specialized fishery of sea fish in the western Bering Sea. *Izv. TINRO* **2021**, *201*, 76–101. [CrossRef]

Article

From Extermination to Conservation: Historical Records of Shark Presence during the Early and Development Phase of the Greek Fishery

Dimitrios K. Moutopoulos [1,*], Evridiki Lazari [1], George Katselis [1] and Ioannis Giovos [1,2,3,*]

[1] Department of Fisheries & Aquaculture, University of Patras, 30200 Mesolongi, Greece
[2] iSea, Environmental Organization for the Preservation of the Aquatic Ecosystems, 54645 Thessaloniki, Greece
[3] Department of Biology, University of Padova, 35122 Padova, Italy
* Correspondence: dmoutopo@upatras.gr (D.K.M.); ioannis.giovos@isea.com.gr (I.G.)

Simple Summary: A thorough search of traditional and digital libraries was conducted for retrieving issues on the interaction of the Greek fisheries with shark species. A significant contribution was made through newspaper collections covering both the Athenian and provincial press including Crete Island for a century (1883–1983). Our historical search showed that large species were common in the Greek Seas, a phenomenon not suspected from datasets of modern fishery surveys. The detection of sharks due to local media gradually increased up to 1969, with most records being more frequent for the Aegean Sea, whereas the number of sharks being sighted declined leading up to the middle of 1980s. The crucial point is that a large number of these observations were related to shark attacks on people, whereas this is not currently evident. The historical records presented here show us that the knowledge of the past can motivate us to identify strategies and policies that can be more acceptable for communities and thus succeed for conservation.

Citation: Moutopoulos, D.K.; Lazari, E.; Katselis, G.; Giovos, I. From Extermination to Conservation: Historical Records of Shark Presence during the Early and Development Phase of the Greek Fishery. *Animals* **2022**, *12*, 3575. https://doi.org/10.3390/ani12243575

Academic Editor: Martina Francesca Marongiu

Received: 16 November 2022
Accepted: 15 December 2022
Published: 17 December 2022

Publisher's Note: MDPI stays neutral with regard to jurisdictional claims in published maps and institutional affiliations.

Abstract: The lack of historical data on shark presence, distribution, and status in the Eastern Mediterranean undermines efforts to manage and protect their populations. An exhaustive review of anecdotal references related to shark presence during the early and development phase of Greek fisheries (1883–1983) was conducted. In the early-20th century (1912), the first sighting of the presence of a dead shark was reported in the Ionian Sea. Later on, the presence of sharks gradually increased up to 1969, with most records being more frequent for the Aegean Sea, whereas the number of sharks being sighted declined leading up to the middle of 1980s. The increase in shark attacks during the mid-20th century led to a calling for culling of sharks in co-operation with the competent authorities promoting the permission to hunt sharks with firearms and offering rewards for killed individuals. A high number of these observations potentially resulted from shark attacks on people, whereas this is not currently evident. This is an indicator of the lower abundance of sharks in modern times and subsequently an alteration in the way that our current modern society is approaching the protection of such vulnerable species.

Keywords: cartilaginous fish; historical accounts; shark occurrence; marine environmental history

1. Introduction

Historical anecdotal data can improve our understanding of past system dynamics, enabling us to determine whether contemporary systems are acting within the historical range of variability exhibited before large-scale human impacts [1]. Historical information can also help us with the phenomenon of shifting environmental baselines that each generation subconsciously views as "natural" given the way the environment appeared in their youth [2]. The transition of knowledge from earlier periods, when human impacts were limited [3], could be a useful "repository" of knowledge aiming to re-evaluate management thresholds [4]. In this direction, the compilation of historical and "forgotten" science with modern natural observations has significantly increased nowadays, under the framework of Marine Historical Ecology and Marine Environmental History [5,6], incorporating a wide range of multi-disciplinary fields of science (e.g., [7]).

The Mediterranean Sea's diverse biota provide a wide range of interactions with people and the local populations [8]. In this environment, sharks are among the oldest animals that are still alive today [9], and because of their biological traits (slow development, late maturity, and low fecundity), they are particularly vulnerable to overfishing [10]. In the present study, a thorough review of anecdotal references related to shark presence aims to evaluate shark–human interactions in the Greek waters during the early and developmental phase of the fisheries (1900-early 1980s). This effort will close the historical gap in knowledge on shark populations in the Eastern Mediterranean [11,12], and the data gathered could be used in the MEDLEM database, a repository of information on large elasmobranchs in the Mediterranean and Black seas [13]. It can also be used in assessments for the Red List by the International Union for Conservation of Nature (IUCN), which helps to understand the historical presence and distribution of species. Our results will also verify whether or not the reported shark attacks on people are well recorded in the International Shark Attack File (ISAF). This database was initiated in 1958, and more than 6500 individual investigations cover the period from the early 1500s to the present [14]. Evidence derived from the study of historical and archival documents, such as the one here, may be used for assessing the vulnerability of exploited species [15], including sharks [16]. Historical science may play an important role to comprehend present-day effects and conditions [17]. In this context, historical accounts on sharks may be extremely useful to add new data to their occurrence and distribution in poorly studied regions such as the Greek waters [18].

2. Materials and Methods

Greek fisheries officially started to getting organized in 1911 and, up to the early 1980s, these fisheries were shifting from an essentially pre-industrial stage toward the industrialization of fishing activities [19]. The present study dealt with research through Greek traditional and digital archives (i.e., newspapers, technical reports, and books) using the keyword "shark" during a century (1883–1983). Digital archives included the newspaper collection of the National Library of Greece (http://efimeris.nlg.gr/ns/main.html?fbclid=IwAR0 n__4AKJQ-ci7BFEwxCxZmu-90qQRZhlGhyMSmmcpkvB9gThXnwQmwi8E (accessed on 10 September 2022)) consisting of the following journals: *Eleftheria, Empros, Macedonia, Rizospastis, Scrip,* and *Acropolis*. These archives cover both the Athenian and provincial press from 1883 and up to 1983, depending on the journal. Duplicate records of the same report published in different newspapers were excluded. For the newspapers in Crete Island, old archives were found in the online database (http://vikelaia-epapers.heraklion.gr/%CE%B5%CF% 86%CE%B7%CE%BC%CE%B5%CF%81%CE%AF%CE%B4%CE%B5%CF%82/ (accessed on 10 September 2022)).

Elasmobranch records were also searched through the Mediterranean Elasmobranchs Citizen Observations (MECO) Project (www.facebook.com/pg/theMECOproject, accessed on 6 August 2022) [18], which launched in 2014 and aimed to collate knowledge on elasmobranch occurrence, seasonality, and distribution using citizen science and social media. The project involves the collaboration of local scientists who are searching the media, contacting the public, and creating a large, verified database of elasmobranch observations.

For the visualization of the reports, the open-access software QGIS ver 3.16.7-Hannover [20] was used. The percentage contribution of the reported sharks in Greek waters were estimated for overall and area disaggregated values across years.

3. Results

Overall, 197 historical records on shark presence in the Greek Seas were retrieved from Greek newspapers during a century (1883–1983). Based on the available archives, the first record mentioned the presence of a large 2 m shark in Ionian Sea in 1912 (Table A1, Figure 1a). Earlier from that record, two reports from the literature reported that a shark attacked a sailor in Corfu Island in 1847 [21], and that the bycatch of one specimen of *Carcharhinus* spp. of 4.5 m length occurred in Nisiros Island in 1905 (Dodecanese) (MECO database). Later on, the detection of sharks due to local media, journalists, photographs, better information travel, etc., gradually increased up to 1969, with most records being more frequent for the Aegean Sea (83.0%), whereas the number of records declined over the years leading up to the early 1980s (Table A1, Figure 1a). The spatial extent of the sharks' presence showed that a quarter of the shark sightings (25.7%) were mostly gathered near of the capital of Athens (Saronikos and Evoikos Gulfs). A significant part was also reported in enclosed gulfs adjacent to large ports, and mainly in cities on the mainland (Thessaloniki and Kavala Bays: 18.1%, Patras Bay: 4.1%) and on the big islands of the North and South Aegean with high touristic activities (Lesvos and Rhodes Islands, respectively) (Figure 1a).

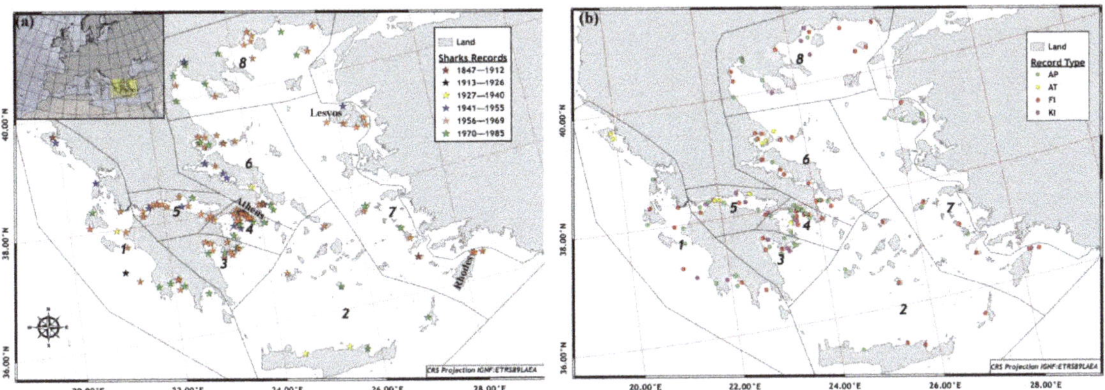

Figure 1. References on presence of sharks during the early fisheries period of the Greek fisheries in terms of: (**a**) year period and (**b**) type of reporting. 1: Ionian, 2: South Aegean–Cretan Seas, 3: Argolikos Gulf, 4: Evvoikos–Saronikos gulfs, 5: Gulf of Corinth, 6: Central Aegean Sea, 7: Eastern Aegean Sea, 8: North Aegean Sea. Record type: AP, appearance; AT, attack; FI, fishery; and KI, killing.

According to the taxonomic resolution of the shark reports, limited reports on this are available, either because during that time the knowledge on these species was in "early stages" or due to the fact that most of the incidents/catches were reported by non-scientific personnel. Thus, apart from many reports mentioning the presence of dog sharks, in very few the presence of *Hexanchus griseus* (three reports) and *Squalus acanthias* (one report; reported as *Acanthias* spp.) was mentioned (Table A1). With respect to the size of the sharks caught, the length was reported in a third of the total reports (33.0%). More than 70% of the sharks caught exhibited, according to the reports, a total length between 2 and 5 m, whereas in exceptional cases (7.0%) sharks of sizes greater than 8 m were also caught.

Nearly 80% of the reports of shark sightings (80.7%) mentioned either the appearance of a shark observed from the beach by a fishing vessel (43.3%) or as the bycatch by the fishery (37.4%), whereas only very small proportion of these reports (5.3%) spoke of the deliberate killing of sharks outside the context of the fishery (14.0%) (Figure 1b). Compared to fishing or shark's sightings, attacks on people by sharks were spatially restricted to the vicinity of large cities and the coastline (Figure 1b). Only three confirmed lethal attacks of sharks on people were reported between 1900 and 1983, whereas other minor shark attacks occurred mostly due to fishing (Table A1). Attacks on people were mostly reported in the Ionian and Central Aegean Seas, whereas the deliberate killing of sharks were largely reported in the North Aegean Sea (Figure 2a). More than half of the reported attacks on people (58.3%) occurred between 1951 and 1970, when 80% of the deliberate killings on sharks were reported (Figure 2c). Since there have been more shark attacks since 1959, there has been a call for shark culling in collaboration with the appropriate authorities, as well as requests for permission to use weapons to hunt sharks, rewards for those that are killed, and safety swimming recommendations (Figure 3). This was also marked by a gradually increase in the number of shark culling events that were reported during 1959–1967 (Table A1). In certain reports, it was evident that, despite being culled by the coast guard, many sharks belonged to species considered as harmless. Reports also included guidance about safety measures to be taken by sea-users. Moreover, in order to stop shark attacks, various protective measures have been suggested, such as putting special barbed wires in the sea and far enough from the beach (Table A1).

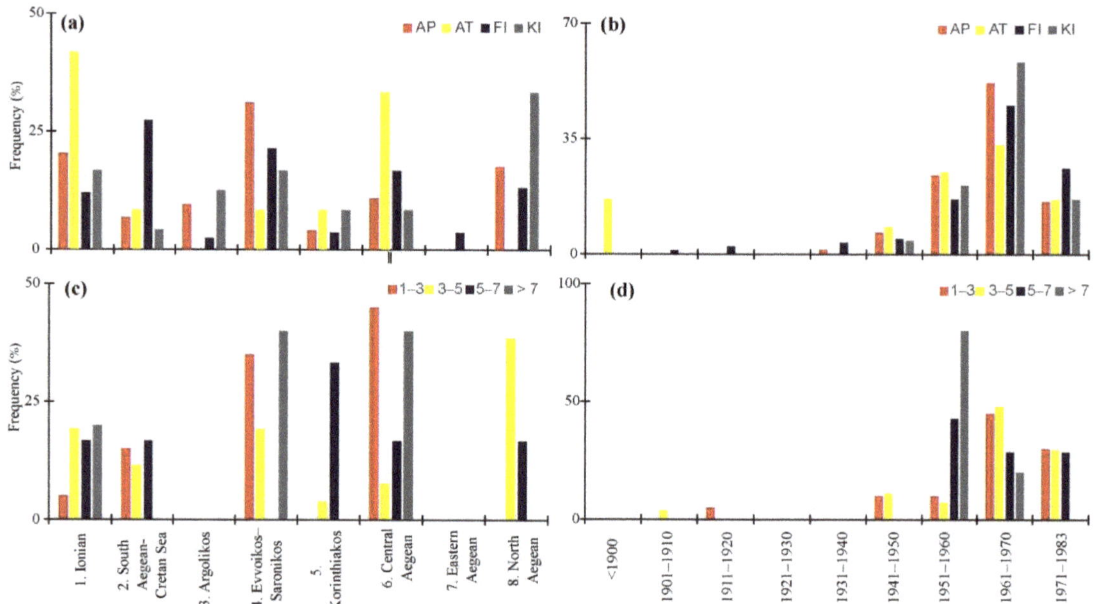

Figure 2. Percentage contribution (%) of shark reports per area and year period in terms of: (**a**,**b**) type of reporting and (**c**,**d**) reported length class (in m). Record type: AP, appearance; AT, attack; FI, fishery; and KI, deliberate killing.

In a third of the reports, the length of shark species was also recorded, indicating that the smallest (<2 m long) and largest (>7 m long) animals were mostly (contribution of more than 40% in each area) reported both in the Evvoikos–Saronikos gulfs and in the Central Aegean Sea (areas with Nos 4 and 6 in Figure 1) (Figure 2c). The largest animals were reduced by 75% during 1951–1960 and 1961–1971, and they completely vanished after 1981, according to the recorded length (Figure 2d). Cases of pregnant female sharks and of premature birth were also reported in certain cases of the sharks caught. However, during that time, there was neither any knowledge on premature care nor on post-release techniques of the embryos, and thus we assume that many embryos were killed. A significant report of a high number of sharks was also recorded after the great earthquake in Santorini Island (9 July 1956) (Table A1).

Figure 3. References on presence of sharks during the early fisheries period of the Greek fisheries: (**a**) *Hexanchus griseus* caught in Evvoikos gulf, (**b**) *Carcharodon carcharias* caught in North Aegean Sea, (**c**) Odontaspis ferox caught in Eastern Aegean Sea, (**d**) *Alopias superciliosus* caught in Argolikos gulf, (**e**) *Cetorhinus maximus* caught in Central Aegean Sea, (**f**) *Cetorhinus maximus* caught in North Aegean Sea.

4. Discussion

Historical accounts may be extremely useful to add new data to past shark occurrence and distribution, especially in poorly studied regions such as the waters of the Eastern Mediterranean [7,22]. This allow us to better understand temporal changes in species diversity and population trends, in a location where vast declines and local extinctions are taking place [16,23,24]. In our work, shark reports were spatially extended in most Greek waters, as is the case nowadays [18]. Our historical search showed that about 60 years ago, human beings and sharks were still competing for the same space and large shark species were common in the Greek Seas, a phenomenon not suspected from datasets of modern fishery surveys. However, in several locations, such as the South Aegean waters (i.e., Cyclades, Cretan Sea), there are limited references to shark sightings, especially after 1940. Although there are still significant gaps in knowledge in several Greek areas, such as the Cretan Sea [18], we consider that this issue is mostly attributed to a lack of information or due to the absence of local archives from these areas. Therefore, our work cannot be used for estimating species distribution and abundance during the studied period. The same is also true regarding the trend of the length classes reported across years. Even though there is evidence of a decrease in the length of sharks caught over time, particularly after 1970, this could not be confirmed by our findings because either the reported animals were not disaggregated by species and belonged to a variety of species with different sizes, or the length measurements were frequently based solely on personal observation rather than any scientific methodology.

In the past, some taxa, such as elasmobranchs, were richer in species and more abundant [25]. Currently, shark attacks incidents do not exist in Greece, but this was not true during the studied period. It is well known that the Mediterranean Sea is a hotspot for shark extinction globally [26,27], primarily as a result of overfishing [15,24], but also due to habitat degradation [15]. Although [28] underlines a direct link between shark attacks and the increasing density of sea-users, we must further consider the possibility that some shark species have shifted their distribution as a result of the increasing human pressure on the coastal ecosystems and urbanization (increase in population in the two biggest cities in Greece by more than 130% between 1951–2021: [29]). Such changes have been described elsewhere [23,30], while the recent continuous observations of blue shark sightings in ports and beaches in the Mediterranean (including Greece) during the COVID lockdowns (pers. Comm. with the MECO projects Spain, France, Italy, Croatia, Albania, Greece and Turkey) further support this theory.

Only three lethal attacks of sharks on people in Greek waters during the 20th century were confirmed by our study (i.e., in 1948 at Keratsini area, in 1951 at Corfu Island, and in 1963 at Pagasitikos Gulf), which were also included in the ISAF database. In this database, a couple of other unidentified cases of lethal attacks of sharks on people were also reported, but they were not supported by any publishing reports on the newspapers and thus we consider them as false cases.

Interestingly, the historical records showed that the presence of sharks was a significant issue for coastal communities, and the solutions suggested in the past were related to intentional killings, which are now largely regarded as socially unacceptable and, for some shark species, even illegal. These events were calling for the culling of sharks in co-operation with the competent authorities, requesting the permission to hunt sharks with firearms and offering rewards for killed animals. This was mostly evident in the North Aegean Sea, where local fishers' associations were leading the efforts of the deliberate killings of sharks and of the free use of guns (see also Table A1 and Figure 2a), whereas in other areas, the co-operation was absent, and all the caught specimens were the work of single fishers. In Greece, the first national legislation protecting two shark species (PD 67/1981) arrived 40 years ago. In contrast, shark-related policies did not arrive in the region until the 1990s, and even now, sharks are still treated and managed as fish stocks rather than as biodiversity. Given that their attitudes and perceptions were formed during the time when the intentional killing of sharks was a reality, citizens and fishers still fear sharks and perceive them as enemies and competitors. Although perceptions against sharks are shifting [5,31], their protection, which is supported by a variety of legal provisions, both national and European, is far from being taken for granted, both because of the delay in time by which these laws are enacted and/or ratified, as well as the lack of enforcement and compliance by professionals. This "back to the history" journey demonstrated that conservation requires patience and persistence, which will enable us to create eco-friendly conservation tactics and policies.

Last but not least, misidentifications, the dissemination of false information, depictions of sharks as prizes, and fear-inducing stories are still widely utilized in Greek media, illustrating the lack of understanding and misinformation about sharks [32]. For conservation priority of species such as sharks, where press coverage has the ability to influence public opinion, correct information is especially crucial; otherwise, the public will continue to associate sharks with negative traits that will hinder any conservation effort [33]. After all these decades of misinformation and fear around sharks in Greece, researchers and conservationists have to invest in training journalists in national and local media, as this can have an immediate impact on the species conservation in the country.

5. Conclusions

Our historical search revealed 197 historical records on shark sightings in the Greek Seas retrieved from Greek newspapers between 1883–1983. Such efforts are of particular importance, especially for understudied areas such as the Eastern Mediterranean where information about elasmobranch is lacking. Based on our results, we suspect that shark population in the past were either more abundant or distributed closer to the coast and further research is required to confirm or to reject these theories.

Shark conservation and policy-related instruments are still "young" and the endeavor needs to continue and be intensified in the future, given the gravel situation of their population both globally and in the Mediterranean. In this context, professionals involved in shark conservation need to focus more on the low compliance with current laws, which may be caused by the attitude that people developed when they were young that sharks are competitors and enemies. Based on our search and other available bibliography for Greece, misinformation about sharks has been a persistent phenomenon for over a century, shaping negative attitudes about sharks in the country and jeopardizing any conservation effort. Environmental journalists can serve as a link between people and governments, as well as between governmental and non-governmental organizations while tackling misinformation.

Author Contributions: Conceptualization, D.K.M.; methodology, D.K.M.; formal analysis, D.K.M., E.L. and I.G.; investigation, E.L.; data curation, D.K.M. and E.L.; writing—review and editing, D.K.M., I.G. and G.K.; visualization, G.K.; supervision, D.K.M. All authors have read and agreed to the published version of the manuscript.

Funding: This research received no external funding.

Institutional Review Board Statement: Not applicable.

Informed Consent Statement: Not applicable.

Data Availability Statement: Not applicable.

Acknowledgments: We would like to warmly thank all citizen scientists reporting to the M.E.C.O. project and additionally, the Shark Conservation Fund, the Save Our Seas Foundation, and the OceanCare and Shark Foundation for supporting data collection through funding other elasmobranch-related projects.

Conflicts of Interest: The authors declare no conflict of interest.

Appendix A

Table A1. References to shark sightings in the Greek Seas during 1900–1983.

Date	Details	Attacks (AT), Fishery (FI), Appearance (AP)
1905	One specimen of *Carcharhinus* spp. of 4.5 m length was caught in Nisiros Island (Dodecanese)	FI
1912	Stranding of a dead shark in Corfu (Ionian Sea)	FI
1913	Shark of 2 m length was bycaught (Ionian Sea)	FI
1932	Two sharks were seen in the port of Rethymnon (Cretan Sea)	AP
1933	Shark was caught (Ionian Sea)	FI
1934	Large shark was bycaught	FI
1940	Shark was bycaught (Ionian Sea)	FI
1948	Shark was seen (Saronikos Gulf)	AP
1948	Shark (*Hexanchus griseus*) of 4 m length and 546 kg weight was bycaught (Saronikos Gulf)	FI
1948	Shark attacked and killed a young man in Keratsini area (Saronikos Gulf)	AT
1948	Dog fish was seen (Saronikos Gulf)	AP

Table A1. *Cont.*

Date	Details	Attacks (AT), Fishery (FI), Appearance (AP)
1948	Shark of 4.5 length, 500 kg weight, and with a 65 cm mouth opening was by caught in Aigina Island (Saronikos Gulf). Stomach contents includes *Sarda* spp., *Pagrus pagrus* and *Raja* spp. individuals.	FI
1949	Dog fish of 4 m length was seen (Saronikos gulf)	AP
1949	Dog shark of 2.5 m length and 192 kg weight was bycaught (Saronikos gulf)	FI
1949	Dog fish was seen in Thermaikos gulf (North Aegean Sea)	AP
1949	Dog fish of 2 m length and 200 kg weight was bycaught (Saronikos gulf)	FI
1949	Shark was killed by commercial boat in Patraikos Gulf (Ionian Sea)	KI
1949	According to fishers' opinion, the number of dog sharks in Saronikos Gulf was estimated to 200	AP
1951	Large shark was seen in Lefkas Island	AP
1951	Shark attacked and killed a woman in Corfy Island (Ionian Sea)	AT
1951	Large shark was seen in Lefkas Island (Ionian Sea)	AP
1951	Shark was seen in Aigina Island (Saronikos Gulf)	AP
1951	Shark of 7 m length was seen in Kastela (Saronikos Gulf)	AP
1951	Shark was seen in Aigina Island (Saronikos Gulf)	AP
1951	Dog fish of 350 kg was by caught in North Evvoikos gulf	FI
1952	Shark of 10 m length was seen in Pagasitikos Gulf (Ionian Sea)	FI
1952	One specimen of Carcharodon carcharias was caught in Lesvos Island	FI
1953	Two sharks of 1500 kg each were bycaught in longlines in Evvoikos gulf	FI
1953	Shark was attacked by a fishing vessel in Pagasitikos Gulf (Central Aegean)	AT
1953	Shark of 10 m length was seen in Pagasitikos Gulf (Central Aegean)	AP
1954	Dog fish of 5 m length and 448 kg weight was bycaught in Gulf of Corinth	FI
1954	Two specimens of *Hexanchus griseus* were caught in Evvoia Island	FI
1955	Shark of 30 kg weight was bycaught in Evvoikos Gulf	FI
1956	Shark was seen in Freatida Island (Saronikos Gulf)	AP
1956	Many sharks were seen after a great earthquake in Santorini Island (9 July 1956)	AP
1956	Shark was seen in Gulf of Corinth	AP
1957	Shark of 897 kg weight was bycaught in trawl in Samothraki Island (North Aegean)	FI
1957	Shark of 5 m length was seen in Kavala Bay (North Aegean)	AP
1957	Dog fish was seen in Pagasitikos Gulf (Central Aegean)	AP
1957	Shark of 300 kg weight was seen in Pagasitikos gulf (Central Aegean)	AP
1957	Dog fish of 100 kg weight was bycaught in Pagasitikos gulf (Central Aegean)	FI
1957	Shark gave birth to 60 new-born sharks of 40–45 cm in Gulf of Corinth	KI
1957	Shark was seen in Patraikos Gulf	AP
1958	Shark of 450 kg weight was bycaught in Nafplion Bay (Aegean Sea)	FI
1958	Two sharks attached on purse seine catches in Zakynthos Island (Ionian Sea)	AP
1959	Shark of 1 m length was killed by harpoon in Pagasitikos Gulf (Central Aegean)	KI
1959	Dog fish of 4.6 m length and 200 kg weight was killed in Rhodes Island (Levantine Sea)	FI
1959	Shark was seen in Patras beach (Ionian Sea)	AT
1959	Dog fish of 8 m length and 76 kg weight was killed by recreational fishers in Patraikos Gulf	KI
1959	Dog fish of 1.4 m length and 90 kg weight was killed in Volos port (Central Aegean)	KI
1959	Shark of 6 m length and 200 kg weight was bycaught in Katerini Bay (North Aegean)	FI
1959	Shark of 4.2 m length and 450 kg weight was bycaught in Oxia Island (Ionian Sea)	FI
1959	School of sharks were killed by Navy in Lesvos Island (Eastern Aegean)	KI
1960	Shark was seen in Chalkis Bay (Evvoikos gulf)	AP
1960	Shark was seen in Lesvos Island (Eastern Aegean)	AP
1960	Shark was seen in Corfy Island (Ionian Sea)	AP
1960	Shark was seen in Salamina Island (Saronikos gulf)	AP
1960	Shark was seen in Rhodes Island (Levantine Sea)	AP
1961	Shark of 10 m length was attacked by fishers in Skiathos Island (Central Aegean)	AT
1961	Dog fish of 60 kg weight was bycaught in Pagasitikos Gulf (Central Aegean)	FI

Table A1. *Cont.*

Date	Details	Attacks (AT), Fishery (FI), Appearance (AP)
1961	Shark was seen in Saronikos Gulf	AP
1962	Shark attacked two kids in Mykonos Island (South Aegean)	AT
1962	Shark of 4 m length was seen in Lesvos Island (Eastern Aegean)	AP
1962	Dog fish was seen in Kalamata Bay (Ionian Sea)	AP
1962	Shark was seen in Kavala gulf (North Aegean)	AP
1962	School of shark was seen in Messiniakos Gulf (Ionian Sea)	AP
1962	Shark was seen in Peireaus (Saronikos Gulf)	AP
1962	Shark of 4 m length in Kalamata (Ionian Sea)	AP
1962	Dog fish was seen in Saronikos Gulf	AP
1962	Fishers caught 18 sharks and 21 dog fishes in Rhodes Island during 1963	FI
1962	Dog fish of 5 m length was seen in Pagasitikos Gulf (Central Aegean)	AP
1962	Shark of 300 kg weight was attacked to fish catches in Pagasitikos Gulf (Central Aegean)	AP
1962	Dog fish was killed by boat vessel in Saronikos gulf	KI
1962	Shark was seen in Kalamata Bay (Ionian Sea)	AP
1962	Shark was seen in Spetses Island (Aegean Sea)	AP
1962	Shark was seen in Thasos Island (North Aegean)	AP
1962	Shark of 3 m length and 360 kg weight was killed in Kassandra Bay (North Aegean)	KI
1962	Shark was seen in Gulf of Corinth	AP
1962	Shark was caught in Mykonos Island (South Aegean)	FI
1962	Shark of 4.8 m length and 2000 kg weight was bycaught in Patraikos Gulf	FI
1962	Shark of 4.8 m length and 2000 kg weight was caught in Patraikos Gulf	FI
1963	Shark of 4 m length and 400 kg weight was killed in Gulf of Corinth	KI
1963	Shark was seen in Milos Island (South Aegean)	AP
1963	Dog fish of 1.5 m length and 60 kg weight was bycaught in Trikeri Island (Pagasitikos Gulf)	FI
1963	Shark of 8 m length in Peireaus (Saronikos Gulf)	AP
1963	Shark of 4 m length in beaches of Kavouri (Saronikos Gulf)	AP
1963	Survey for the kill of sharks was organized in Athens	KI
1963	Shark was killed in Kavala Bay (North Aegean)	KI
1963	Shark was seen in Molivos Bay, Lesvos Island (Eastern Aegean)	AP
1963	Shark weighted of 350 kg was bycaught using longline in Rhodes Island (Levantine Sea)	FI
1963	Shark of 640 kg weight was seen in Lesvos Island (Eastern Aegean)	AP
1963	Shark was killed by coast guard in Kavala gulf (North Aegean)	KI
1963	Dog fish of 2 m length was caught in Rafina Bay (Saronikos Gulf)	FI
1963	Shark of 2 m length and 150 kg weight was caught in Pagasitikos Gulf (Central Aegean)	FI
1963	Shark was killed in Hydra Island (Aegean Sea)	KI
1963	Sharks (3) was seen in Pagasitikos Gulf (Central Aegean)	AP
1963	*Exarchus grizus* of 2.90 length and 180 kg weight was killed by researchers of Hydrological Institution in Rhodes Island (Levantine Sea)	KI
1963	Shark was caught by fishermen in Rhodes Island (Levantine Sea)	FI
1963	Shark of 2 m length and 200 kg weight was caught by fishermen in Pagasitikos Gulf (Central Aegean)	FI
1963	Shark breeding four embryos was bycaught in Alexandroupoli (North Aegean)	FI
1963	Shark was seen in Saronikos Gulf	AP
1963	Shark was seen in Patraikos Gulf (Ionian Sea)	AP
1963	Shark was seen in Kefalonia Island (Ionian Sea)	AP
1963	Sharks (2) were seen in Hydra Island (Aegean Sea)	AP
1963	Shark was fished in Rafina port (Saronikos Gulf)	AP
1963	Shark was seen in Gulf of Corinth	AP
1963	Shark was seen in Hydra Island (Aegean Sea)	AP
1963	Shark of 2.7 m length and of 200 kg weight was bycaught in Rhodes Island (Levantine Sea)	FI
1963	*Acanthias* spp. were caught in Rhodes Island (Levantine Sea)	FI
1963	Dog fish of 2 m length and of 150 kg weight was bycaught in Pagasitikos Gulf (Central Aegean)	FI
1963	Dog fish was killed by boat vessel in Agrolikos gulf (Aegean Sea)	KI
1963	Shark attacked and killed a woman in Trikeri Island in Pagasitikos gulf (Central Aegean)	AT
1963	Dog fish of 2 m length was seen in Pagasitikos Gulf (Central Aegean)	AP
1963	Shark was seen in Saronikos Gulf	AP
1963	Shark was seen in Hydra Island (Aegean Sea)	AP
1963	Dog fish was seen in Kyllini Bay (Ionian Sea)	AP

Table A1. *Cont.*

Date	Details	Attacks (AT), Fishery (FI), Appearance (AP)
1963	Shark sold as Galeus spp. in fish market in Athens	
1963	Shark was seen in Saronikos Gulf	AP
1963	Shark hunting was organized in Saronikos Gulf	KI
1963	Establishment of a committee of the Ministry of Merchant Shipping to protect against shark attacks	
1963	Shark was seen in Kefalonia Island (Ionian Sea)	AP
1963	Shark was seen in Argolikos Gulf (Aegean Sea)	AP
1963	Advice against sharks	
1964	Shark of 750 kg weight was bycaught in Lesvos Island (Eastern Aegean)	FI
1964	Shark of 230 kg weight was bycaught in Kavala gulf (North Aegean)	FI
1964	Shark was bycaught in Gulf of Corinth (in stomach, human bones were found)	FI
1965	Shark was seen in Corfu Island (Ionian Sea)	AP
1965	Shark of 4 m length was seen in Lesvos Island (Eastern Aegean)	AP
1965	Dog fish of 4 m length was caught in Pagasitikos Gulf (Central Aegean)	FI
1965	Three sharks were released from small-scale nets in Varkiza area (Saronikos Gulf)	FI
1965	Three sharks (sapounas) (3) were killed in 300 m distance from the coast of Rhodes Island (Levantine Sea)	FI
1965	Dog fish attacked a canoe in Patraikos Gulf (Ionian Sea)	AT
1965	Dog fish was killed in Argolikos Gulf (Aegean Sea)	KI
1965	Dog fish attacked to a fishing vessel in Pagasitikos Gulf (Central Aegean)	FI
1965	Two sharks were seen in 150 m distance from the beach in Saronikos Gulf	AP
1965	School of sharks was seen in Kalymnos Island (South Aegean)	AP
1965	Shark of 3.5 m length and 350 kg weight was killed with firegun in North Aegean Sea	KI
1965	Sharks were bycaught in Rhodes Island (Levantine Sea)	FI
1965	Shark of 130 kg weight was caught in Pagasitikos Gulf (Central Aegean)	FI
1965	Two sharks were caught in Saronikos Gulf	FI
1966	Two Sharks were caught in Rhodes Island (Levantine Sea)	FI
1966	Shark of 252 kg weight was caught by small-scale fisher in Santorini Island	FI
1966–1970	Sharks (29) weight from 100 to 500 kg weight were caught in 200–250 m of depth by recreational fisher in Santorini Island	FI
1967	Shark was seen in Kavala gulf (North Aegean)	AP
1967	Shark of 4.5 m length was killed by harpoon in Kavala gulf (North Aegean)	KI
1967	Shark of 4.5 m length was bycaught in Kavala gulf (North Aegean)	FI
1967	Shark of 4 m length was killed in Kavala (North Aegean)	KI
1967	Shark of 3.5 m length and 380 kg weight was bycaugth in Skiathos Island (Central Aegean)	FI
1967	Shark of 2.2 m length and 90 kg weight was caught in Attiki beach (Saronikos Gulf)	FI
1967	Shark of 310 kg weight was caught by small-scale fisher in Santorini Island	FI
1967	Shark of 400 kg weight was caught by small-scale fisher in Santorini Island	FI
1968	Dog fish of 250 kg weight was caught in Skopelos Island (Central Aegean)	FI
1968	Shark was caught in Kavala Gulf (North Aegean)	FI
1968	Shark of 6 m length and 600 kg weight was caught in Gulf of Corinth	FI
1969	Shark of 3 m length was caught. Live small sharks were found inside the shark in Saronikos Gulf	FI
1971	Dog fish of 300 kg weight was bycaught in Kardamili Bay (Ionian Sea)	FI
1971	Killing of sharks was funded by the Greek government	
1972	Dog fish of 3 m length was seen in Pagasitikos gulf (Central Aegean)	AP
1972	Dog fish were seen in Skiathos Island in 10 m from the coast (not confirmed-cetaceans)	FI
1972	Shark of 4.6 m length and 1 t weight was caught in North Aegean	FI
1972	Shark of 3 m length and 250 kg weight was caught in Samothraki Island (North Aegean)	FI
1972	Shark was caught in South Evvoikos Gulf	FI
1972	Shark of 600 kg weight was caught by small-scale fisher in Santorini Island (the shark was escaped)	FI

Table A1. *Cont.*

Date	Details	Attacks (AT), Fishery (FI), Appearance (AP)
1973	Two sharks were killed by coast guard authorities (even though authorities were considered as unharmed) in Thessaloniki Bay North Aegean	KI
1973	Shark of 6 m length and 1 t weight was caught in Lakonikos Gulf (Ionian Sea)	FI
1974	Shark of 600 kg weight was caught by small-scale fisher in Santorini Island (the shark was escaped)	FI
1974	One specimen of *Alopias* spp. superciliosus was caught in Argolikos gulf	FI
1975	Dog fish of 1.5 m length was seen in Pagasitikos Gulf (Central Aegean)	AP
1976	Shark of 2 m legnth and 200 kg weight was caught in South Evvoikos Gulf (Lavrio)	FI
1976	Shark was seen in Athens beach (Faliro)	AP
1976	Sharks were seen in Argolikos Gulf (Aegean Sea)	AP
1976	Sharks were seen in one nm distance from the coast in South Saronikos Gulf	AP
1977	Two sharks of 3.5 length and 500 kg weight were caught in Strymonikos Gulf (North Aegean)	FI
1977	Dog fish was found dead in Kalamata Bay (Ionian Sea)	KI
1978	Shark of 2.5 m length and 200 kg weight was bycaught in Karpathos Island (Levantine Sea)	FI
1978	Shark of 2.5 m length and 300 kg weight was bycaught in Saronikos Gulf	FI
1979	Sharks were seen in South Evvoikos Gulf	AP
1979	One specimen of Squatina aculeata was caught in Crete	FI
1980	Dog fish of 1.2 m length and 50 kg weight was bycaught in Pagasitikos Gulf (Central Aegean)	FI
1980	Dog fish attacked people in Pagasitikos Gulf (Central Aegean)	AT
1980	Shark of 3 m length and 300 kg weight was killed by knife in Messiniakos Gulf (Ionian Sea)	KI
1980	Shark of 4.5 m length was bycaught in Kefalonia Island (Ionian Sea)	FI
1980	Shark of 2.2 m length was killed with firearms in Aigina Island (Saronikos Gulf)	KI
1980	One specimen of *Odontaspis ferox* was caught in Ikaria Island	FI
1980	One specimen of *Cetorhinus maximus* was caught in Ikaria Island	FI
1981	Shark of 5.4 m length was caught in Leros Island (South Aegean)	FI
1981	Dog fish of 3 m length was caught in Navarino Bay (Aegean Sea)	AP
1981	Shark of 250 kg weight was caught by small-scale fisher in Santorini Island (the shark was escaped)	AP
1983	Shark of 3 m length and 250 kg weight fished with longline in 500 m of depth in Lakonikos Bay (Ionian Sea)	FI
1983	Shark seen in Thermaikos Gulf (Platamonas)	AP
1983	Shark seen in South Evvoikos Gulf (Petalioi Bay)	AP
1983	Shark seen in Argolikos Gulf (Methana)	AP
1983	Shark seen in Chalkidiki (Potidaia) (North Aegean)	AP
1983	Shark seen in Thermaikos Gulf (North Aegean)	AP
1983	Shark attacked divers in Antikyra Bay (Gulf of Corinth)	AT
1983	One specimen of *Cetorhinus maximus* was caught in Thracian Sea	FI

References

1. Thurstan, R.H. The potential of historical ecology to aid understanding of human–ocean interactions throughout the Anthropocene. *J. Fish. Biol.* **2022**, *101*, 351–364. [CrossRef] [PubMed]
2. Pauly, D. Anecdotes and the shifting baseline syndrome of fisheries. *TREE* **1995**, *10*, 430. [CrossRef] [PubMed]
3. Moutopoulos, D.K.; Stergiou, K.I. The evolution of the Greek fisheries during the 1928-1939 period. *Acta Adriat.* **2011**, *52*, 183–200.
4. Zeller, D.; Pauly, D. The 'presentist bias' in time-series data: Implications for fisheries science and policy. *Mar. Pol.* **2018**, *90*, 14. [CrossRef]
5. Fortibuoni, T.; Libralato, S.; Arneri, E. Fish and fishery historical data since the 19th century in the Adriatic Sea, Mediterranean. *Sci. Data* **2017**, *4*, 170104. [CrossRef]
6. Mazzoldi, C.; Bearzi, G.; Brito, C.; Carvalho, I.; Desiderà, E.; Endrizzi, L.; Freitas, L.; Giacomello, E.; Giovos, I.; Guidetti, P.; et al. From sea monsters to charismatic megafauna: Changes in perception and use of large marine animals. *PLoS ONE* **2019**, *14*, e0226810. [CrossRef]
7. Engelhard, G.H.; Thurstan, R.H.; MacKenzie, B.R.; Alleway, H.K.; Bannister, R.C.A.; Cardinale, M.; Clarke, M.W.; Lescrauwaet, A.-K. ICES meets marine historical ecology: Placing the history of fish and fisheries in current policy context. *ICES J. Mar. Sci.* **2016**, *73*, 1386–1403. [CrossRef]

8.   Coll, M.; Piroddi, C.; Steenbeek, J.; Kaschner, K.; Ben Rais Lasram, F.; Aguzzi, J.; Ballesteros, E.; Bianchi, C.N.; Corbera, J.; Dailianis, T.; et al. The Biodiversity of the Mediterranean Sea: Estimates, Patterns, and Threats. *PLoS ONE* **2010**, *5*, e11842. [CrossRef]
9.   Jacoby, D.M.; Croft, D.P.; Sims, D.W. Social behaviour in sharks and rays: Analysis, patterns and implications for conservation. *Fish Fish.* **2012**, *13*, 399–417. [CrossRef]
10. Myers, R.A.; Worm, B. Extinction, survival or recovery of large predatory fishes. *Philos. Trans. R. Soc. B Biol. Sci.* **2005**, *360*, 13–20. [CrossRef]
11. Bargnesi, F.; Gridelli, S.; Cerrano, C.; Ferretti, F. Reconstructing the history of the sand tiger shark (*Carcharias taurus*) in the Mediterranean Sea. *Aquat. Conserv.* **2020**, *30*, 915–927. [CrossRef]
12. Kabasakal, H.; Bayri, E. Great White Sharks, *Carcharodon carcharias*, hidden in the past: Three unpublished records of the species from Turkish waters. *Ann. Ser. Hist. Nat.* **2021**, *31*, 195–202.
13. Mancusi, C.; Baino, R.; Fortuna, C.; De Sola, L.G.; Morey, G.; Bradai, M.N.; Kallianotis, A.; Serena, F. MEDLEM database, a data collection on large Elasmobranchs in the Mediterranean and Black seas. *Medit. Mar. Sci.* **2020**, *21*, 276–288. [CrossRef]
14. International Shark Attack Files. Available online: https://www.floridamuseum.ufl.edu/shark-attacks/ (accessed on 22 November 2022).
15. Fortibuoni, T.; Giovarnani, O.; Pranovi, F.; Raicevich, S.; Solidoro, C.; Libralato, S. Analysis of Long-Term Changes in a Mediterranean Marine Ecosystem Based on Fishery Landings. *Front. Mar. Sci.* **2017**, *4*, 33. [CrossRef]
16. Dulvy, N.K.; Allen, D.J.; Ralph, G.M.; Walls, R.H.L. *The Conservation Status of Sharks, Rays and Chimaeras in the Mediterranean Sea*; International Union for Conservation of Nature (IUCN): Malaga, Spain, 2016; 14p.
17. Sokou, V.K.; Gonzalvo, J.; Giovos, I.; Brito, C.; Moutopoulos, D.K. Tracing dolphin-fisheries interactions in the early Greek fisheries. *Ann. Ser. Hist. Nat.* **2022**, *31*, 2.
18. Giovos, I.; Aga-Spyridopoulou, R.N.; Serena, F.; Soldo, A.; Barash, A.; Doumpas, N.; Gkafas, G.A.; Katsada, D.; Katselis, G.; Kleitou, P.; et al. An Updated Greek National Checklist of Chondrichthyans. *Fishes* **2022**, *7*, 199. [CrossRef]
19. Moutopoulos, D.K.; Stergiou, K.I. Spatial disentangling of Greek commercial fisheries landings by gear between 1928–2007. *J. Biol. Res.-Thessaloniki* **2012**, *18*, 265–279.
20. QGIS Development Team. QGIS Geographic Information System. *Open Source Geospatial Foundation Project.* 2020. Available online: http://qgis.osgeo.org (accessed on 5 October 2022).
21. Palamas, K. *Dionisiou Solomou Apanta, ta Evriskomena*; Sakelariou, P.D., Ed.; University of Crete: Athens, Greece, 1901.
22. Moro, S.; Jona-Lasinio, G.; Block, B.; Micheli, F.; De Leo, G.; Serena, F.; Bottaro, M.; Scacco, U.; Ferretti, F. Abundance and distribution of the white shark in the Mediterranean Sea. *Fish Fish.* **2020**, *21*, 338–349. [CrossRef]
23. Ferretti, F.; Myers, R.A.; Serena, F.; Lotze, H.K. Loss of large predatory sharks from the Mediterranean Sea. *Cons. Biol.* **2008**, *22*, 952–964. [CrossRef]
24. Serena, F.; Abella, A.J.; Bargnesi, F.; Barone, M.; Colloca, F.; Ferretti, F.; Fiorentino, F.; Jenrette, J.; Moro, S. Species diversity, taxonomy and distribution of Chondrichthyes in the Mediterranean and Black Sea. *Eur. Zool. J.* **2020**, *87*, 497–536. [CrossRef]
25. Damalas, D.; Maravelias, C.D.; Osio, G.C.; Maynou, F.; Sbrana, M.; Sartor, P. "Once upon a Time in the Mediterranean" Long Term Trends of Mediterranean Fisheries Resources Based on Fishers' Traditional Ecological Knowledge. *PLoS ONE* **2015**, *10*, e0119330. [CrossRef] [PubMed]
26. Dulvy, N.K.; Fowler, S.L.; Musick, J.A.; Cavanagh, R.D.; Kyne, P.M.; Harrison, L.R.; Pollock, C.M. Extinction risk and conservation of the world's sharks and rays. *ELife* **2014**, *3*, e00590. [CrossRef] [PubMed]
27. Geraci, M.L.; Ragonese, S.; Scannella, D.; Falsone, F.; Gancitano, V.; Mifsud, J.; Gambin, M.; Said, A.; Vitale, S. Batoid Abundances, Spatial Distribution, and Life History Traits in the Strait of Sicily (Central Mediterranean Sea): Bridging a Knowledge Gap through Three Decades of Survey. *Animals* **2021**, *11*, 2189. [CrossRef]
28. West, J.G. Changing patterns of shark attacks in Australian waters. *Mar. Freshw. Res.* **2011**, *62*, 744–754. [CrossRef]
29. HELSTAT (Hellenic Statistical Authority), 1955–2022. Resident population of Greece. Seven issues (for the period 1951–2021). Hellenic Statistical Authority, Athens. Available online: www.statistics.gr (accessed on 10 September 2022).
30. Hammerschlag, N.; Gutowsky, L.F.; Rider, M.J.; Roemer, R.; Gallagher, A.J. Urban sharks: Residency patterns of marine top predators in relation to a coastal metropolis. *Mar. Ecol. Prog. Ser.* **2022**, *691*, 1–17. [CrossRef]
31. Giovos, I.; Barash, A.; Barone, M.; Barría, C.; Borme, D.; Brigaudeau, C.; Mazzoldi, C. Understanding the public attitude towards sharks for improving their conservation. *Mar. Pol.* **2021**, *134*, 104811. [CrossRef]
32. Giovos, I.; Arculeo, M.; Doumpas, N.; Katsada, D.; Maximiadi, M.; Mitsou, E.; Moutopoulos, D.K. Assessing multiple sources of data to detect illegal fishing, trade and mislabelling of elasmobranchs in Greek markets. *Mar. Pol.* **2020**, *112*, 103730. [CrossRef]
33. Shiffman, D.S.; Bittick, S.J.; Cashion, M.S.; Colla, S.R.; Coristine, L.E.; Derrick, D.H.; Dulvy, N.K. Inaccurate and biased global media coverage underlies public misunderstanding of shark conservation threats and solutions. *iScience* **2020**, *23*, 101205. [CrossRef] [PubMed]

Article

Distribution and New Records of the Bluntnose Sixgill Shark, *Hexanchus griseus* (Hexanchiformes: Hexanchidae), from the Tropical Southwestern Atlantic

Jones Santander-Neto [1,*], Getulio Rincon [2], Bruno Jucá-Queiroz [3], Vanessa Paes da Cruz [4] and Rosângela Lessa [5]

[1] Instituto Federal de Educação, Ciência e Tecnologia do Espírito Santo, Campus Piúma, Rua Augusto Costa de Oliveira, 660, Praia Doce, Piúma 29285-000, Brazil
[2] UFMA Campus Pinheiro, Universidade Federal do Maranhão, Pinheiro 65200-000, Brazil
[3] SOD Geotecnia Construções, Rua do Sol, 300, Aleixo, Manaus 69060-084, Brazil
[4] Departamento de Biologia Estrutural e Funcional, Instituto de Biociências, Universidade Estadual Paulista Júlio de Mesquita Filho-UNESP, Botucatu 01049-010, Brazil
[5] Departamento de Pesca e Aquicultura, Rua Dom Manoel de Medeiros, s/n, Dois Irmãos, Universidade Federal Rural de Pernambuco-UFRPE, Recife 52171-900, Brazil
* Correspondence: jones.santander@ifes.edu.br

Simple Summary: The bluntnose sixgill shark, *Hexanchus griseus*, is a widely distributed species found in all oceans, even if irregularly, inhabiting continental shelves and slopes, islands, and mid-ocean ridges in deep seas. The species is currently classified by the International Union for Conservation of Nature (IUCN) as globally near-threatened, with a decreasing population. Despite some records in Brazil, the known distribution of this species in the Southwestern Atlantic is very patchy and, in some cases, not yet recorded in the world reference literature, such as in northeastern Brazil. This study, therefore, highlights new records for the species in the Tropical Southwestern Atlantic, particularly off the northeastern Brazilian coast. This information is paramount to establishing distribution maps for the species employed in a population status assessment by the IUCN.

Citation: Santander-Neto, J.; Rincon, G.; Jucá-Queiroz, B.; Paes da Cruz, V.; Lessa, R. Distribution and New Records of the Bluntnose Sixgill Shark, *Hexanchus griseus* (Hexanchiformes: Hexanchidae), from the Tropical Southwestern Atlantic. *Animals* **2023**, *13*, 91. https://doi.org/10.3390/ani13010091

Academic Editor: Martina Francesca Marongiu

Received: 27 October 2022
Revised: 7 December 2022
Accepted: 23 December 2022
Published: 26 December 2022

Abstract: The bluntnose sixgill shark, *Hexanchus griseus*, is a widely distributed demersal species found in tropical and temperate waters of the Pacific, Atlantic, and Indian Oceans, inhabiting continental shelves and slopes, islands, and mid-ocean ridges at depths ranging from 200 to 1100 m. In the Southwestern Atlantic, this species has been recorded from northeastern to southern Brazil, Argentina, and Uruguay. Despite this, the known distribution of this species in the Southwestern Atlantic is very patchy and, in some cases, still mostly ignored in the literature, such as in northeastern Brazil. This study, therefore, aimed to report 23 new records of *Hexanchus griseus* in the Tropical Southwestern Atlantic and highlight the presence of this species off the northeastern Brazilian coast. So far, *Hexanchus griseus* was officially reported from the Fernando de Noronha Archipelago, Saint Peter and Saint Paul Archipelago, and the state of Ceará along the northeast coast of Brazil. Herein, the known distribution is extended to the continental shelf breaks and upper slopes of other Brazilian states, reinforcing the previously reported occurrence of the species near the Saint Peter and Saint Paul Archipelago.

Keywords: fisheries; deep sea; DNA barcoding; elasmobranchs; Chondrichthyes

1. Introduction

Sixgill sharks belonging to the Hexanchidae Gray family, 1851, present unicuspidate front teeth in the upper jaw and comb-shaped and blade-like in the lower jaw, six or seven pairs of gill slits, and a fairly stocky body. The Hexanchidae family comprises three genera and five species. The bluntnose sixgill shark, *Hexanchus griseus* (Bonaterre, 1788), is a widely distributed demersal species found in tropical and temperate waters of the Pacific, Atlantic, and Indian Oceans [1], although with a patchy distribution, and can inhabit continental shelves and slopes, islands, and mid-ocean ridges at depths ranging from 200 to 1100 m, and able to extend down to at least 2500 m [2–4].

The bluntnose sixgill shark is a large-bodied shark ranging up to 550 cm in total length (TL), reaching maturity at about 310 cm and over 400 cm TL, for males and females, respectively [4,5]. Its reproductive cycle is still not well defined, but according to different studies [5–8], it is believed to be biannual, with a resting phase for females of 12 months followed by a gestation period of 12 months. This species is known to have very large litters of 47–108 pups born at around 55–74 cm TL [4–6,8]. A polyandrous mating system was detected in an analysis concerning the genetic relationship between a large female found on the beach and 71 individuals from a litter, with up to nine males contributing to her offspring, which confers a significant evolutionary success for the group [9]. This mating system can influence populations, leading to increased levels of genetic variability and decreasing inbreeding [10,11].

In the Southwestern Atlantic, bluntnose sixgill sharks have been recorded from northeastern to southern Brazil [12–14], Argentina, and Uruguay [15]. Despite this, the known distribution of this species in the Southwestern Atlantic is very patchy and, in some cases, still mostly ignored in the literature's main references, such as in northeastern Brazil [3,4]. Due to their biology and increasing anthropogenic actions, the bluntnose sixgill shark is currently classified by the International Union for Conservation of Nature (IUCN) as globally near-threatened, with a decreasing population trend [3].

In this context, this study aims to compile all available data on the species occurrence in this region including new records from the Tropical Southwestern Atlantic, highlighting the presence of this species off the northeastern Brazilian coast.

2. Materials and Methods

Data on the occurrence of the *Hexanchus* genus in the Tropical Southwestern Atlantic were obtained from three different sources, namely artisanal fishing landing sites, bottom longline fisheries (commercial fisheries), and handline fishing using line reels.

All analyzed specimens were identified based on specific literature [1,4]. Morphometric measurements, when possible, were determined as the length in centimeters and weight in kilograms. When the specimens were not available, total length (TL) and total weight (TW) was estimated by those who had access to the animals and corroborated by photographs or videos.

Care and use laws for experimental animal welfare were not applied in this study due to the nature of data collection from commercial fishing landings. The specimens were commercialized by fishers and we did not retain them for any other purposes. Even so, authorization from the Chico Mendes Institute for Biodiversity was obtained (ICMBio—SISBIO License n°49663-2). Concerning the sampling efforts conducted under the Program for Evaluation of Living Resources in the Exclusive Economic Zone (REVIZEE), a government program created to assess the potential for sustainable exploitation of all marine natural resources of Brazil's exclusive economic zone, the program was finalized before the Biodiversity Authorization and Information System (SISBIO) development, in 2007.

2.1. Bottom Longlines

Data from six scientific expeditions conducted in northeastern Brazil between 1997 and 1999 (two per year) were obtained under the REVIZEE program. These cruises operated with bottom longlines containing 500 hooks, each 13/0 in size (Mustad hook). Fishing activities took place at different depths, from 50 to 500 m every 50 m.

Further data from 53 commercial fishing trips along the northeastern Brazilian coast were obtained from November 2004 to August 2011. These fisheries operated with bottom longlines, with hooks numbering between 250 and 900 and hook sizes varying from 13/0 to 16/0 (Mustad hooks), with the predominant use of the 15/0 and 16/0 sizes. The fishing activities took place near the 100–200 m isobaths from latitude 1°20′ S to 11°30′ S.

2.2. Handline Using Line Reels

A video was recorded during a fishing trip around the Saint Peter and Saint Paul Archipelago (ASPSP) (coordinates: 00°55′03″ N/29°20′43″ W). After setting the longline, a handline using line reels was also prepared and set, from which one bluntnose sixgill shark specimen was caught. The line was approximately 200 m long employing 13/0 hooks.

2.3. Artisanal Fishing Landing Site

One bluntnose sixgill shark specimen was landed in an artisanal fishing landing site located in the Mucuripe Embayment, Fortaleza, in the state of Ceará, Brazil. The fishing area for this small-scale fishing fleet comprise 3217 km^2 bordered at 03°43′S/038°05′W; 03°23′ S/038°05′ W and 03°25′S/038°48′ W; 03°01′S/038°49′ W (Figure 1). According to the fishmonger that sent photos and a video, the specimen was caught with a hook size number 12/0 at depths between 70 and 120 m. The carcass (without head, fins, guts, and skin) was sold to a fish butcher. A tissue sample was obtained from this specimen to proceed with the DNA barcoding analysis.

2.4. DNA Barcode Analysis

For more accurate identification, identification was performed based on barcode DNA, using a small piece of the muscle tissue (40 mg) preserved in ethanol 95%. DNA was extracted using the one silica-based extraction method, the standard Canadian Centre for DNA Barcoding (CCDB) DNA extraction protocol using a glass fiber plate [16] following the corresponding instructions. One thousand and nine hundred microlitres of a lysis buffer (Lysis Buffer or T1) containing Proteinase K (Life Technologies; Invitrogen, Carlsbad, NM, USA) was added to the sample that was then ground with a pestle inside the tube and incubated overnight at 56 °C. Two-hundred-microlitre aliquots of the lysate were then transferred into Ultident tube racks, mixed with 400 µL of corresponding binding buffer, and then transferred to a binding plate [Bioinert membrane Acroprep plate (Pall Life Sciences, Ann Arbor, MI, USA)], which was then centrifuged at 6000× *g*. The first wash volume consisted of 400 µL, followed by a second wash of 750 µL using the corresponding buffer systems. The plate was then incubated at 56 °C for 30 min, and DNA was eluted with 50 µL of H$_2$O preheated to 56 °C.

A polymerase chain reaction (PCR) was performed using a previously published primer [17]. A Veriti 96 thermocycler (Applied Biosystems) was used following the protocol suggested by the Platinum® Taq DNA Polymerase manual (Invitrogen). PCR amplicons were visualized on a 1% agarose gel E-Gels (Invitrogen). Fragments of approximately ~650 base pairs (bp) barcode region of the cytochrome C oxidase subunit I COI gene were bi-directionally sequenced using a BigDye Terminator v.3.1 Cycle Sequencing Kit (Applied Biosystems, Inc.) employing an ABI 3730 capillary sequencer following the manufacturer's instructions.

Sequence files were assembled, and consensus sequences were constructed, aligned, and trimmed using the Geneious 4.8.5 software [18], and submitted to GenBank (Accession no. OP897150). The extended reference databases BOLD and GenBank were used, obtained through downloading sequences for *H. griseus*, *H. nakamurai*, and *H. vitulus*. The sequences were then aligned with the muscle algorithm [19] implemented in the Geneious 4.8 software. Interspecific and intraspecific distances were analyzed using the Kimura-2-parameter model (K2P) [20] available in the MEGA X software [21]. A maximum likelihood (ML) analysis was performed through the RAxML v8.2 software [22] with the GTR GAMMA model, and a posteriori bootstrap analysis was conducted employing the autoMRE function with Randomized Axelerated Maximum Likelihood (RAxML) criterion and the best-fitting ML tree was reconciled with the bootstrap replicates. The tree was viewed and edited in FigTree v1.4.3. software (http://tree.bio.ed.ac.uk/software/figtree/, accessed on 1 June 2022).

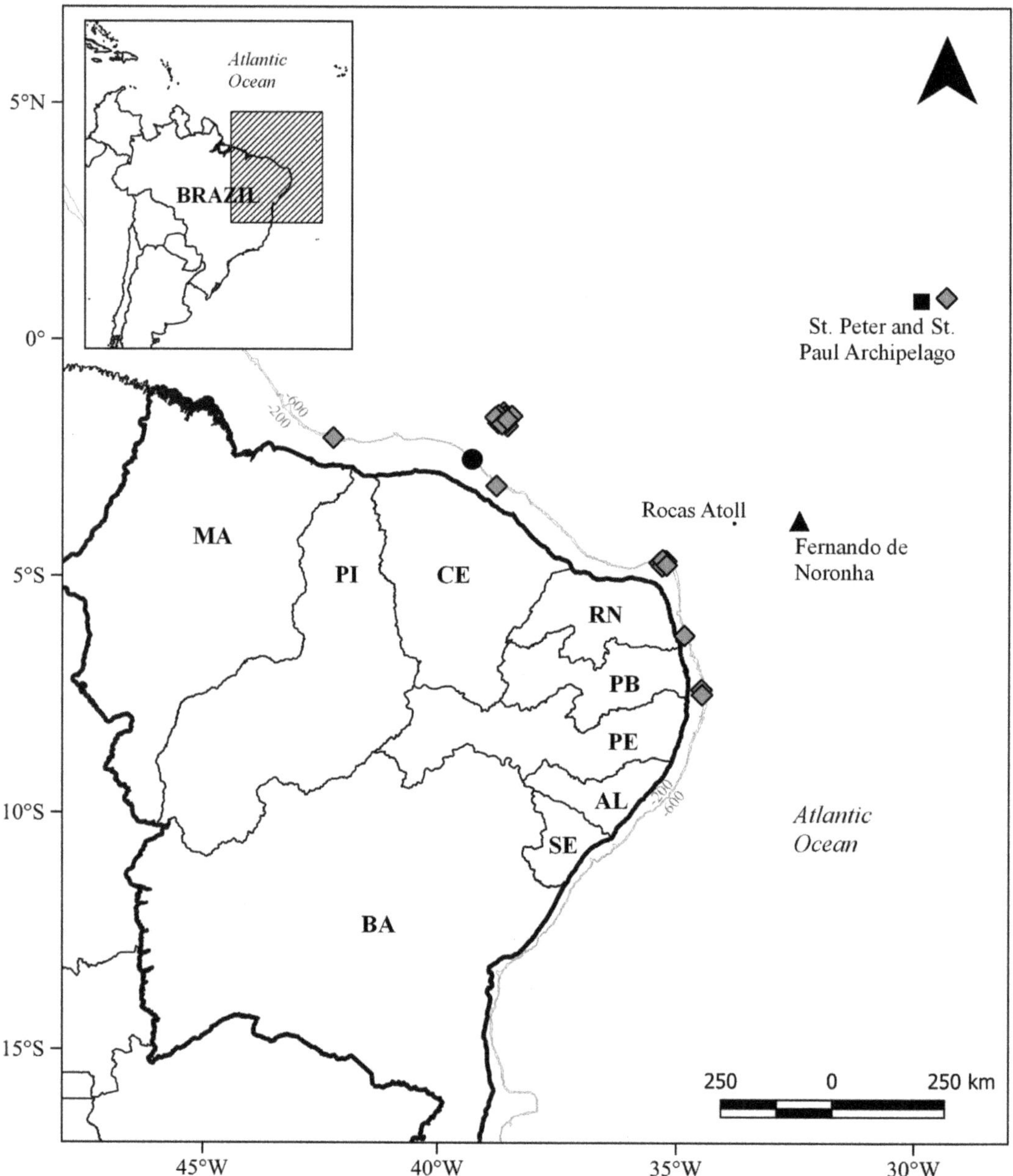

Figure 1. Map depicting previous and additional records for the bluntnose sixgill shark, *Hexanchus griseus*, from the Tropical Southwestern Atlantic. Gray diamonds indicate new records and black forms indicate previous records. Black triangle: Moreira-Junior (1993); black circle: Santander-Neto et al. (2007); black square: Pinheiro et al. (2020); MA: Maranhão; PI: Piauí; CE: Ceará; RN: Rio Grande do Norte; PB: Paraíba; PE: Pernambuco; AL: Alagoas; SE: Sergipe; BA: Bahia.

3. Results

From November 1998 to December 2020, 23 *Hexanchus griseus* specimens were recorded in the Tropical Southwestern Atlantic using bottom longlines, handlines using reels, and at an artisanal fishing landing site (Table 1).

Table 1. Information on the *Hexanchus griseus* specimens recorded in the Tropical Southwestern Atlantic from 1998 to 2020. Bottom longline, BL; artisanal fishing landing site, AFLS; handline using line reels, HLR; Maranhão, MA; Ceará, CE; Rio Grande do Norte, RN; Paraíba, PB; Saint Peter and Saint Paul Archipelago, ASPSP. N, number of specimens; TW, total weight in kg; TL, Total length in cm.

Date	Source	Coordenadas	Location and State	N	Depth	TW	TL	Sex
17 November 1998	BL	04°43′01″ S/035°18′39″ W	Cabo Calcanhar, RN	1	240			
17 November 1998	BL	04°43′01″ S/035°18′39″ W	Cabo Calcanhar, RN	1	240			F
17 November 1998	BL	04°43′01″ S/035°18′39″ W	Cabo Calcanhar, RN	1	240			F
17 November 1998	BL	04°43′22″ S/035°20′18″ W	Cabo Calcanhar, RN	1	82			
17 November 1998	BL	04°43′31″ S/035°19′58″ W	Cabo Calcanhar, RN	1	81			F
29 November 1998	BL	06°14′34″ S/034°51′06″ W	Baía Formosa, RN	1	250	116	294	F
2 December 1998	BL	07°24′10″ S/034°27′39″ W	Pitimbu, PB	1	260	82	248	F
2 December 1998	BL	07°24′33″ S/034°27′08″ W	Pitimbu, PB	1	450	85.5	267	F
August 2005	BL	02°04′39″ S/042°13′32″ W	Tutóia, MA	1	100–200	35		
29 March 2010	HLR	00°54′24″ N/029°21′54″ W	ASPSP	1	205		~280–300	F
June 2010	BL	~01°40′ S/038°40′ W	Oceanic banks, CE	2	100–200	300 *		
August 2010	BL	~01°40′ S/038°40′ W	Oceanic banks, CE	2	100–200	100 *		
September 2010	BL	~01°40′ S/038°40′ W	Oceanic banks, CE	3	100–200	200 *		
September 2010	BL	~01°40′ S/038°40′ W	Oceanic banks, CE	1	100–200	100		
December 2010	BL	~01°40′ S/038°40′ W	Oceanic banks, CE	1	100–200	110		
July 2011	BL	~01°40′ S/038°40′ W	Oceanic banks, CE	3	100–200	200*		
13 December 2020	AFLS	~03°05′ S/038°48′ W	CE	1	70–120		~180	F

* combined total weight of the specimens in the fishery.

3.1. Bottom Longlines

Two of six scientific expeditions during the REVIZEE program in 1998 provided the following eight records for *Hexanchus griseus* in northeastern Brazil (Table 1; Figure 1): Five specimens from Cabo Calcanhar, in the state of Rio Grande do Norte (RN) were caught in November 1998, ranging from 80 to 240 m in depths, one specimen captured at Baía Formosa, RN, in November 1998 at a depth of 250 m and two specimens captured at Pitimbu in the state of Paraíba, December 1998, ranging from 260 and 450 m in depth (Figure 2a,b).

Twelve *Hexanchus griseus* specimens were caught in seven out of the fifty-three commercial bottom longline fishing trips on the seamounts of the Brazilian North Chain off the state of Ceará from 2010 to 2011. One additional specimen was captured from the external continental shelves of the state of Maranhão in 2005 (Table 1).

3.2. Handline Using Line Reels

On 29 March 2010, a female specimen with an estimated total length between 280 and 300 cm was caught at Lat 0°54.410′ N, Long 29°21.892′ W (off the Saint Peter and Saint Paul Archipelago) at a depth of 205 m and only recorded in a video (Supplementary Material S1) before the animal was released by the fishermen due to difficulties in bringing it aboard.

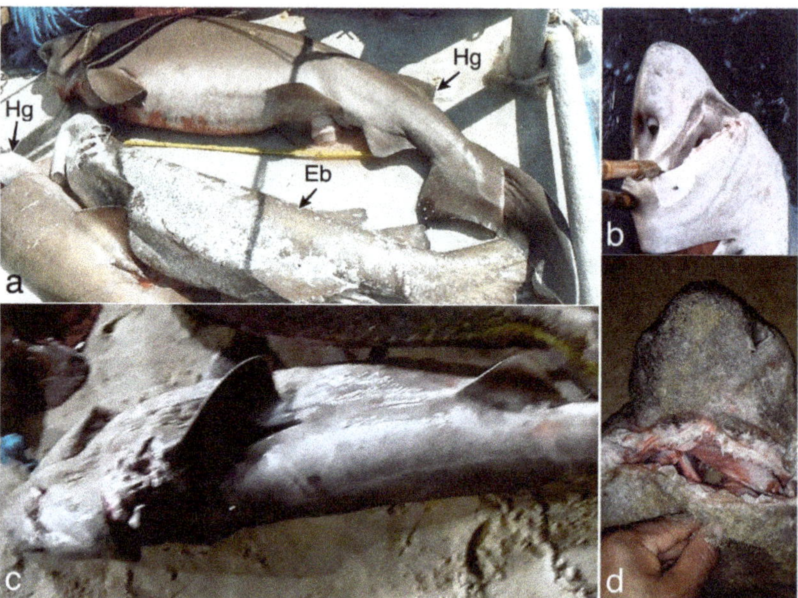

Figure 2. Female *Hexanchus griseus* specimen (**a**) caught off the state of Paraíba, Brazil, and (**b**) the specimen's head during the landing and; and another female specimen (**c**) landed at the Mucuripe Embayment, in the state of Ceará, Brazil, photographed by the fishmonger with a (**d**) ventral view of the head. Hg, *Hexanchus griseus*; Eb, *Echinorhinus brucus*.

3.3. Artisanal Fishing Landing Site

One specimen was captured 48 nautical miles (NM) off the coast of the state of Ceará, Brazil, and landed on 13 December 2020. A fishmonger sent photos and a video (Figure 2c,d; Supplementary Material S2). According to him, the female specimen was approximately 180 cm TL. A barcode sequence with more than 600 bp was obtained from this specimen, in addition to 36 sequences obtained from BOLD and GenBank, resulting in a final matrix of 37 sequences (23 *H. griseus*, 12 *H. vitulus,* and two *H. nakamurai*) (Figure 3). Insertions, deletions, and stop codons were not observed. Following alignment and editing, the final matrix contained 605 characters, 212 of which were variable, comprising 29.1% adenine, 21.8% cytosine, 31.8% thymine, and 17.2% guanine. The genetic distance analysis clearly splits the three Hexanchus species, with 0.605 ± 0.021 distance between *H. griseus* and *H. vitulus*, 0.640 ± 0.011 between *H. vitulus* and *H. nakamurai*, and 0.077 ± 0.012 between *H. griseus* and *H. nakamurai*. These findings reveal a low intraspecific genetic variation of <0.001 in all species. The RAxML tree displayed long branches structuring each of the three *Hexanchus* species, with the clear separation of *H. griseus*, *H. nakamurai*, and *H. vitulus*, with more than 85% bootstrap support, confirming the captured specimen as *H. griseus*.

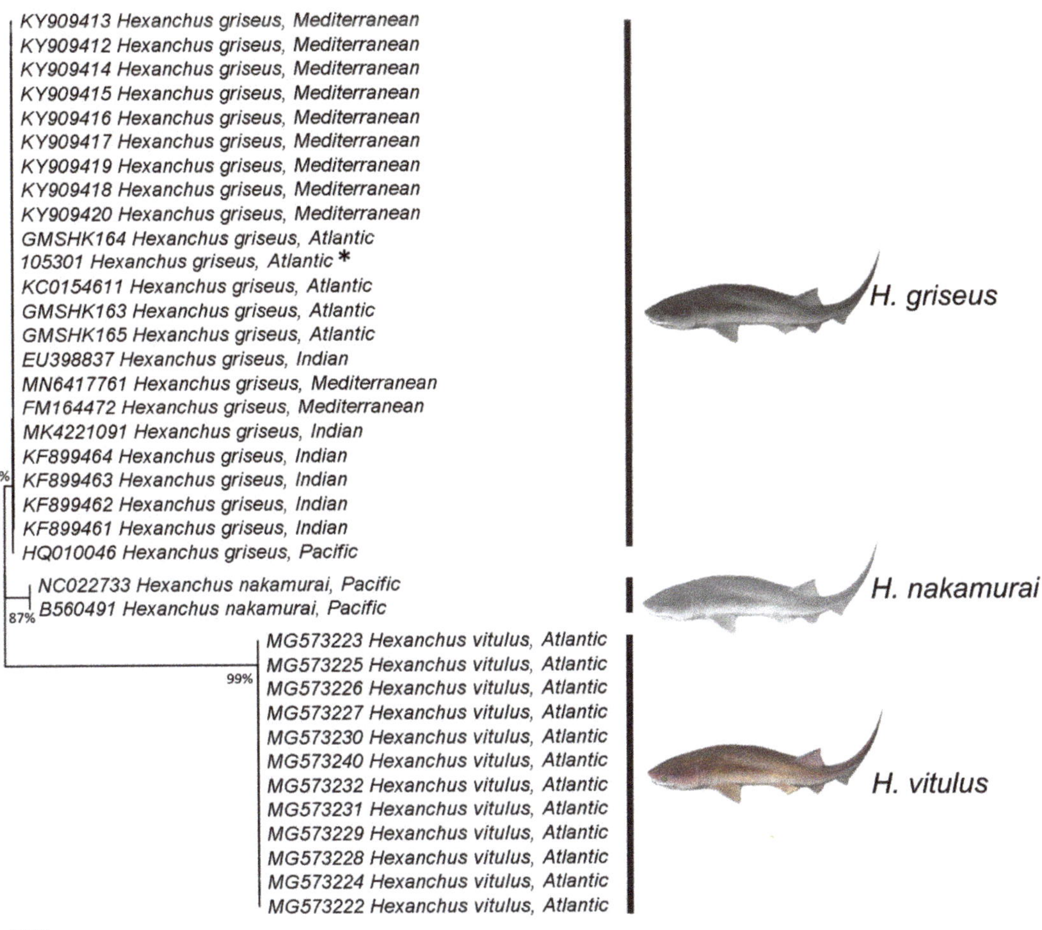

Figure 3. Relationships between *Hexanchus* spp. based on a maximum likelihood analysis and inferred from mitochondrial cytochrome c oxidase subunit I (COI) gene sequences. All nodes displayed bootstrap values above 85%. * indicates the sample collected in the present study.

4. Discussion

H. griseus's distribution in the Tropical Southwestern Atlantic is still not clear and presents some discrepancies in its literature-reported distribution where the species is widely reported for the region [12] or not reported at all [3,4]. Considering the records reported herein, the bluntnose sixgill shark seems to present a wide distribution range along the northeastern Brazilian continental slope, islands, and seamounts, rising to shallower waters in specific areas where specimens are caught from 70 to at least 450 m. The shelf-break demersal habitat of the species and the lack of specific deep-water fisheries, however, limit the understanding of the species distribution pattern [23] due to inherent difficulties in identifying and recording catches and landings [24]. The species is, thus, seldomly recorded.

The loose and oily muscles present in bluntnose sixgill sharks, typical of deep-water fishes, allied to the large size of these animals are the main reasons why fishermen avoid boarding them. Furthermore, fishermen indicate that the meat decomposes very fast and contaminates other fish, while also having no market value. It may, however, be used as bait when brought aboard, which makes it difficult to record captures.

The bluntnose sixgill shark has been reported at the Fernando de Noronha islands [25] and along the entire Brazilian Northeast coast [10] based on conference abstracts [26,27]. Considering the long extension of the Brazilian coast, we recovered the original data of these unreported captures and added new captures in order to provide a better interpretation of the species distribution. So far, *Hexanchus griseus* has been officially reported along the northeast coast of Brazil at the Fernando de Noronha Archipelago, the Saint Peter and Saint Paul Archipelago, and the state of Ceará [13,14,25].

With the data reported herein, the known distribution of the species is extended to the continental shelf breaks and upper slopes of the states of Maranhão, Rio Grande do Norte, and Paraíba. New reports are also provided for the Brazilian North Chain seamounts where twelve specimens were captured on different occasions. In addition, the present record of a female specimen captured near the Saint Peter and Saint Paul Archipelago (with the video recording provided herein as Supplementary Material) reinforces a previously reported occurrence of the species [14] in a biodiversity estimation study using baited remote underwater stereo-videos (stereo-BRUV's). These recurrent catches and video recordings suggest resident populations along the Brazilian North Chain and the oceanic islands of Fernando de Noronha and the SPSP Archipelago. However, only a larger study on the population genetics of this species would define if these animals belong to multiple or a single widely distributed population. Moreover, relevant literature still does not consider the presence of the species in the region [3,4], which is concerning with regard to local and regional biodiversity evaluation statuses as well as the fisheries interaction and potential population impacts.

The bluntnose sixgill shark presents an almost global continuous distribution alongside continental and insular shelf breaks and slopes, seamounts, and even bays connected to submarine canyons and trenches associated with colder waters from upwelling systems [28–30], also recorded in shallow waters of Tasmanian rivers [31]. It is possible that *H. griseus*, like other deep-sea species, may present wider distributions as they live in waters with restricted temperature variations that extend to different latitudes and depths. Therefore, this species may be expected to also occur in other states in northeastern Brazil where its occurrence has not yet been reported. In Brazil, a more common occurrence of the species has been reported for the southern region of the country, in the states of Santa Catarina and Rio Grande do Sul [12,23,32], corroborated by large captures later in southern Brazil [33]. However, the present data has raised suspicions about the possibility that the species may occur more frequently in northeastern Brazil than previously expected, due to the lack of occurrence records. Furthermore, it is important to maintain sampling efforts and use as many available tools as possible to elucidate the diversity and distribution of this genus in the Atlantic Ocean [34].

The *H. griseus* specimens reported herein were found at depths ranging between 70 and 450 m, corroborating the known depth range for the species [3], with most catches taking place at depths up to 250 m, probably due to the effective fishing gear performance depth in the monitored fisheries. It is possible that fisheries prospecting in waters deeper than 200 m may also capture the bluntnose sixgill shark.

All sex-identified specimens were females with total lengths ranging from 180 to 300 cm, indicating that they had not yet reached sexual maturity [4,5], similar to the male previously recorded in the state of Ceará [13]. The absence of adult specimens in the catches is probably due to fishing gear damage (cable breakage, broken hook, and others) due to the large size of adults, or fishermen's refusals to ship large specimens commercially devalued. This rarity concerning adult specimen records seems to be common worldwide [6,15,35]. However, it would be interesting to identify the species' size and sexual structure (life stages) along its bathymetric distribution, to better understand how the species uses the region. A recent study indicated that the sexual distinction observed by Sr/Ca *H. griseus* ratios could indicate spatial segregation, with the possibility that these ratios influence physiological differences between sexes [36]. Research on the spatial ecology of the species has already been suggested [31,36] but is still lacking.

Considering that *H. griseus* occupies a prominent position in the regulation of multiple prey in deep environments [31], further knowledge regarding the distribution of this species would aid in understanding how it can contribute to environmental equilibrium. In addition, these data will also aid in the construction of maps on the species distribution, which is paramount in regional population assessments aiming at Chondrichthyes conservation.

5. Conclusions

The new records provided herein for *Hexanchus griseus* in the Tropical Southwestern Atlantic extend the known distribution of the species to the states of Maranhão, Rio Grande do Norte and Paraíba in Brazil.

Supplementary Materials: The following supporting information can be downloaded at: https://www.mdpi.com/article/10.3390/ani13010091/s1, Video S1: Video of *Hexanchus griseus* caught with handline using line reels near to Saint Peter and Saint Paul's Archipelago; Video S2: Video of *Hexanchus griseus* landed from artisanal fisheries in Mucuripe Embayment, Ceará state, Brazil.

Author Contributions: Conceptualization, J.S.-N.; methodology, J.S.-N., G.R., B.J.-Q. and V.P.d.C.; formal analysis, J.S.-N., G.R. and V.P.d.C.; data curation, J.S.-N., G.R., B.J.-Q. and R.L.; writing—original draft preparation, J.S.-N.; writing—review and editing, J.S.-N., G.R., B.J.-Q., V.P.d.C. and R.L. All authors have read and agreed to the published version of the manuscript.

Funding: This research received no external funding.

Data Availability Statement: Not applicable.

Acknowledgments: The authors thank Fabio Freitas for providing photos, a video, and information about the specimen landed by the artisanal fleet. We thank the Program for Evaluation of Living Resources in the Exclusive Economic Zone—REVIZEE that supplied the analyzed data. R.L. thanks the Productivity Grant (PQ 310200/19) from the Conselho Nacional de Desenvolvimento Científico e Tecnológico (CNPq).

Conflicts of Interest: The authors declare no conflict of interest.

References

1. Compagno, L.J.V. FAO species catalogue. In *Sharks of the World. An Annotated and Illustrated Catalogue of Shark Species Known to Date. Part I Hexanchiformes to Lamniformes*; FAO Fisheries Synopsis; FAO: Rome, Italy, 1984; Volume 4, 249p.
2. Rodríguez-Cabello, C.; González-Pola, C.; Rodríguez, A.; Sánchez, F. Insights about depth distribution, occurrence and swimming behavior of *Hexanchus griseus* in the Cantabrian Sea (NE Atlantic). *Reg. Stud. Mar. Sci.* **2018**, *23*, 60–72. [CrossRef]
3. Finucci, B.; Barnett, A.; Bineesh, K.K.; Cheok, J.; Cotton, C.F.; Graham, K.J.; Kulka, D.W.; Neat, F.C.; Pacoureau, N.; Rigby, C.L.; et al. *Hexanchus griseus*. The IUCN Red List of Threatened Species 2020: E.T10030A495630. [CrossRef]
4. Ebert, D.A.; Dando, M.; Fowler, S. *Sharks of the World: A Complete Guide*; Princeton University Press: Princeton, UK, 2021; 607p.
5. Ebert, D.A. Some observations on the reproductive biology of the sixgill shark *Hexanchus griseus* (Bonnaterre, 1788) from South African waters. *Afr. J. Mar. Sci.* **2002**, *24*, 359–363. [CrossRef]
6. Ebert, D.A. Biological aspects of the sixgill shark, *Hexanchus griseus*. *Copeia* **1986**, *1*, 131–135. [CrossRef]
7. Ebert, D.A. Aspects on the biology of hexanchoid sharks along the California coast. In *Indo–Pacific Fish Biology: Proceedings of the Second International Conference on Indo–Pacific Fishes, Ichthyological Society of Japan, Tokyo, Japan, 29 July–3 August 1985*; Uyeno, T., Arai, R., Taniuchi, T., Matsuura, K., Eds.; Ichthyology Society of Japan: Tokyo, Japan, 1986; pp. 437–449.
8. Capapé, C.; Guélorget, O.; Barrull, J.; Mate, I.; Hemida, F.; Seridji, R.; Bensaci, J.; Bradai, M.N. Records of the bluntnose six-gill shark, *Hexanchus griseus* (Bonnaterre, 1788) (Chondrichthyes: Hexanchidae) in the Mediterranean Sea: A historical survey. *Annal. Ser. Hist. Nat.* **2003**, *13*, 157–166.
9. Larson, S.; Christiansen, J.; Griffing, D.; Ashe, J.; Lowry, D.; Andrews, K. Relatedness and polyandry of sixgill sharks, *Hexanchus griseus*, in an urban estuary. *Conserv. Genet.* **2011**, *12*, 679–690. [CrossRef]
10. Frankham, R. Genetics and extinction. *Biol. Conserv.* **2005**, *126*, 131–140. [CrossRef]
11. Lamarca, F.; Vianna, M.; Vilasboa, A. The first reproductive parameters and evidence of multiple paternity in one new spiny dogfish species, *Squalus albicaudus* (Squaliformes, Squalidae). *J. Fish Biol.* **2020**, *97*, 1268–1272. [CrossRef] [PubMed]
12. Soto, J.M.R. Annotated systematic checklist and bibliography of the coastal and oceanic fauna of Brazil. I. Shark. *Mare Magnum* **2001**, *1*, 51–120.

13. Santander-Neto, J.; Jucá-Queiroz, B.; Nascimento, F.C.P.; Basílio, T.H.; Medeiros, R.S.D.; Furtado-Neto, M.A.A.; Faria, V.V. On the occurrence of sevengill and sixgill sharks (Hexanchiformes: Hexanchidae) off Ceará state, Brazil, Western Equatorial Atlantic. *Arq. Ciênc. Mar.* **2007**, *40*, 59–63.
14. Pinheiro, H.T.; Macena, B.C.; Francini-Filho, R.B.; Ferreira, C.E.; Albuquerque, F.V.; Bezerra, N.P.; Carvalho-Filho, A.; Ferreira, R.C.P.; Luiz, O.J.; Mello, T.J.; et al. Fish biodiversity of Saint Peter and Saint Paul's Archipelago, Mid-Atlantic Ridge, Brazil: New records and a species database. *J. Fish Biol.* **2020**, *97*, 1143–1153. [CrossRef]
15. Soto, J.M.R. Sobre a presença de tubarões hexanquídeos (Chondrichthyes, Hexanchiformes) no sudoeste do Atlântico. *Acta Biol. Leopold.* **1999**, *21*, 241–251.
16. Ivanova, N.V.; Dewaard, J.R.; Hebert, P.D.N. An inexpensive, automation-friendly protocol for recovering high-quality DNA. *Mol. Ecol. Notes* **2006**, *6*, 998–1002. [CrossRef]
17. Hebert, P.D.; Ratnasingham, S.; Dewaard, J.R. Barcoding animal life: Cytochrome c oxidase subunit 1 divergences among closely related species. *Proc. R. Soc. Lond. B Biol.* **2003**, *270*, S96–S99. [CrossRef] [PubMed]
18. Kearse, M.; Moir, R.; Wilson, A.; Stones-Havas, S.; Cheung, M.; Sturrock, S.; Buxton, S.; Cooper, A.; Markowitz, S.; Duran, C.; et al. Geneious Basic: An integrated and extendable desktop software platform for the organization and analysis of sequence data. *Bioinformatics* **2012**, *28*, 1647–1649. [CrossRef] [PubMed]
19. Edgar, R.C. MUSCLE: Multiple sequence alignment with high accuracy and high throughput. *Nucleic Acids Res.* **2004**, *32*, 1792–1797. [CrossRef]
20. Kimura, M. A simple method for estimating evolutionary rates of base substitutions through comparative studies of nucleotide sequences. *J. Mol. Evol.* **1980**, *16*, 111–120. [CrossRef]
21. Kumar, S.; Stecher, G.; Li, M.; Knyaz, C.; Tamura, K. MEGA X: Molecular evolutionary genetics analysis across computing platforms. *Mol. Biol. Evol.* **2018**, *35*, 1547–1549. [CrossRef]
22. Stamatakis, A. RAxML version 8: A tool for phylogenetic analysis and post-analysis of large phylogenies. *Bioinformatics* **2014**, *30*, 1312–1313. [CrossRef]
23. Gadig, O.B.F. Tubarões da Costa Brasileira. Ph.D. Thesis, Instituto de Biociências de Rio Claro da Universidade Estadual Paulista, Rio Claro, Brazil, 2001.
24. Rincon, G.; Mazzoleni, R.C.; Palmeira, A.R.O.; Lessa, R. Deep-water sharks, rays, and chimaeras of Brazil. In *Chondrichthyes: Multidisciplinary Approach*; Rodrigues-Filho, L.F., Sales, J.B.L., Eds.; IntechOpen: London, UK, 2017; pp. 83–112. [CrossRef]
25. Moreira-Júnior, W. Presença de *Hexanchus griseus* (Bonnaterre, 1788), (Chondrichthyes, Hexanchidae) no nordeste do Brasil. In *Resumos do 10º Encontro Brasileiro de Ictiologia*; Sociedade Brasileira de Ictiologia: São Paulo, Brazil, 1993; p. 114.
26. Rincon, G.; Lessa, R. Tubarões do talude nordestino REVIZEE-NE. *Bol. Da Soc. Bras. Para O Estud. Elasmobrânquios* **1998**, *3*, 5–7.
27. Rincon, G.; Lessa, R. Elasmobrânquios demersais do talude do nordeste capturados nos cruzeiros do NPq. Prof. Martins Filho, Programa REVIZEE Score-NE. In *Resumos da 2ª Reunião da Sociedade Brasileira para o Estudo de Elasmobrânquios*; Sociedade Brasileira para o Estudo de Elasmobrânquios: Senandes, Brazil, 2000; p. 77.
28. Ebert, D.A. *Sharks, Rays and Chimaeras of California*; University of California: Oakland, CA, USA, 2003; 284p.
29. Compagno, L.; Dando, M.; Fowler, S. Sharks of the World. In *Princeton Field Guides*; Nathist Press: Princeton, NJ, USA, 2005; 368p.
30. Last, P.R.; Stevens, J.D. *Sharks and Rays of Australia*; Csiro, Division of Fisheries: Hobart, Australia, 2009.
31. Barnett, A.; Braccini, J.M.; Awruch, C.A.; Ebert, D.A. An overview on the role of Hexanchiformes in marine ecosystems: Biology, ecology and conservation status of a primitive order of modern sharks. *J. Fish Biol.* **2012**, *80*, 966–990. [CrossRef]
32. Vooren, C.M. Demersal elasmobranchs. In *Subtropical Convergence Environment: The Coast and the Sea in the Southwestern Atlantic*; Seeliger, U., Odebrecht, C., Castello, J.P., Eds.; Springer: Berlin/Heidelberg, Germany, 1997; pp. 141–146.
33. Perez, J.A.A.; Wahrlich, R. A bycatch assessment of the gillnet monkfish *Lophius gastrophysus* fishery off southern Brazil. *Fish. Res.* **2005**, *72*, 81–95. [CrossRef]
34. Daly-Engel, T.S.; Baremore, I.E.; Grubbs, R.D.; Gulak, S.J.; Graham, R.T.; Enzenauer, M.P. Resurrection of the sixgill shark *Hexanchus vitulus* Springer & Waller, 1969 (Hexanchiformes, Hexanchidae), with comments on its distribution in the northwest Atlantic Ocean. *Mar. Biodivers.* **2019**, *49*, 759–768. [CrossRef]
35. Springer, S.; Waller, R.A. *Hexanchus vitulus*, a new sixgill shark from the Bahamas. *Bull. Mar. Sci.* **1969**, *19*, 159–174.
36. Assemat, A.; Adnet, S.; Bayez, K.; Hassler, A.; Arnaud-Godet, F.; Mollen, F.H.; Girard, C.; Martin, J.E. Exploring diet shifts and ecology in modern sharks using calcium isotopes and trace metal records of their teeth. *J. Fish Biol.* **2022**, 1–13. [CrossRef] [PubMed]

 *animals* 

Article

Advancing DNA Barcoding to Elucidate Elasmobranch Biodiversity in Malaysian Waters

Kar-Hoe Loh [1,*], Kean-Chong Lim [1], Amy Yee-Hui Then [2], Serena Adam [3], Amanda Jhu-Xhin Leung [1], Wenjia Hu [4], Chui Wei Bong [1,2], Aijun Wang [5,6], Ahemad Sade [7], Jamil Musel [8] and Jianguo Du [4,9,*]

[1] Institute of Ocean and Earth Sciences, Universiti Malaya, Kuala Lumpur 50603, Malaysia
[2] Institute of Biological Sciences, Faculty of Science, Universiti Malaya, Kuala Lumpur 50603, Malaysia
[3] World Wide Fund for Nature Malaysia, Petaling Jaya 46150, Malaysia
[4] Key Laboratory of Marine Ecological Conservation and Restoration, Third Institute of Oceanography, Ministry of Natural Resources, Xiamen 361005, China
[5] Fujian Provincial Key Laboratory of Marine Physical and Geological Processes, Xiamen 361005, China
[6] Laboratory of Coastal and Marine Geology, Third Institute of Oceanography, Ministry of Natural Resources, Xiamen 361005, China
[7] Department of Fisheries Sabah, Kota Kinabalu 88624, Malaysia
[8] Fisheries Research Institute Sarawak, Department of Fisheries Malaysia, Kuching 93744, Malaysia
[9] Xiamen Ocean Vocational College, Xiamen 361100, China
* Correspondence: khloh@um.edu.my (K.-H.L.); dujianguo@tio.org.cn (J.D.)

Simple Summary: One-third of shark and ray species are threatened due to overfishing, but a lack of information on each species makes conservation decisions difficult. To address this issue, we conducted a study to identify the different species of sharks and rays in Malaysian waters using DNA barcoding of the CO1 gene, which is akin to DNA fingerprinting for species. We collected 175 individuals between June 2015 and June 2022, randomly selecting up to six specimens from each species. We successfully generated DNA barcodes for 67 species, belonging to 44 genera, 20 families, and 11 orders. Accurate species identification will improve species-specific catch landing data and accelerate the identification of use and illegal trade in Malaysia.

Citation: Loh, K.-H.; Lim, K.-C.; Then, A.Y.-H.; Adam, S.; Leung, A.J.-X.; Hu, W.; Bong, C.W.; Wang, A.; Sade, A.; Musel, J.; et al. Advancing DNA Barcoding to Elucidate Elasmobranch Biodiversity in Malaysian Waters. *Animals* **2023**, *13*, 1002. https://doi.org/10.3390/ani13061002

Academic Editor: Martina Francesca Marongiu

Received: 9 February 2023
Revised: 5 March 2023
Accepted: 7 March 2023
Published: 9 March 2023

Abstract: The data provided in this article are partial fragments of the Cytochrome c oxidase subunit 1 mitochondrial gene (CO1) sequences of 175 tissues sampled from sharks and batoids collected from Malaysian waters, from June 2015 to June 2022. The barcoding was done randomly for six specimens from each species, so as to authenticate the code. We generated barcodes for 67 different species in 20 families and 11 orders. DNA was extracted from the tissue samples following the Chelex protocols and amplified by polymerase chain reaction (PCR) using the barcoding universal primers FishF2 and FishR2. A total of 654 base pairs (bp) of barcode CO1 gene from 175 samples were sequenced and analysed. The genetic sequences were blasted into the NCBI GenBank and Barcode of Life Data System (BOLD). A review of the blast search confirmed that there were 68 valid species of sharks and batoids that occurred in Malaysian waters. We provided the data of the COI gene mid-point rooting phylogenetic relation trees and analysed the genetic distances among infra-class and order, intra-species, inter-specific, inter-genus, inter-familiar, and inter-order. We confirmed the addition of *Squalus edmundsi*, *Carcharhinus amboinensis*, *Alopias superciliosus*, and *Myliobatis hamlyni* as new records for Malaysia. The establishment of a comprehensive CO1 database for sharks and batoids will help facilitate the rapid monitoring and assessment of elasmobranch fisheries using environmental DNA methods.

Keywords: cytochrome c oxidase 1; species identification; DNA barcode; reference library; phylogenetic tree; shark and ray

1. Introduction

Elasmobranchs, a subclass of Chondrichthyans that include sharks and batoids (rays, skates, guitarfish, and sawfish), are widely distributed in the Southeast Asian (SEA) region and encompass more than 300 species from freshwater environments to deep seas [1]. Within the SEA region, Malaysia is one of the top countries with rich biodiversity of elasmobranch species. To date, there are at least 70 species (19 families) of sharks and 91 species (11 families) of batoids that have been reported to occur in Malaysian waters (Supplementary Materials Table S1) [2–17]. Although new species are continuously discovered, the populations of sharks and batoids have gradually decreased over the past decades [18]. The reduction in the population sizes of sharks and batoids is mainly due to high bycatch rates within commercial fisheries and recreational fishing activities [19]. Additionally, due to their K-selected life histories, some elasmobranch species are threatened with extinction [20,21]. Malaysia is the world's ninth largest producer of shark products, especially shark fins, and the second largest importer in terms of volume, as reported by the Food and Agriculture Organisation (FAO) of the United Nations [22]. It is also one of the major countries globally with high annual elasmobranch landings [8]. These sharks and batoids are generally landed whole and are mostly fully utilized. According to the reported annual fish landings in Malaysia (http://www.dof.gov.my, accessed on 5 January 2023), the biomass of sharks and batoids reduced around 15-fold (6487 tonnes to 438 tonnes) and 9-fold (11,993 tonnes to 1372 tonnes), respectively, from 2018 to 2021 [18,23,24]. These data suggested the depletion of shark and batoid populations in Malaysian waters, parallel to the findings on global shark and batoid landings [20,25]. However, as explained by the Department of Fisheries Malaysia (DOFM) [26], sharks are not specifically targeted by fishers in Malaysia, but they are caught as bycatch along with other commercially important species by various fishing gears.

Globally, the estimated species numbers of elasmobranchs threatened with extinction have increased from approximately one-quarter [20] to more than one-third [27] due to targeted and incidental overfishing. This can also be explained by their conservative life history traits, including slow growth, long lifespan, late maturity, and low fecundity, that render elasmobranch populations difficult to recover from anthropogenic pressures [20,21,28,29]. The decrease in elasmobranch populations demonstrated the urgent need for conservation and management plans [20,21,25,30,31]. Due to the lack of species-specific information and variability of population sizes among species, the current landing records fail to provide details as to which species require protection the most. Accurate species identification is one of the most important elements to take into consideration in order to carry out suitable conservation and management programmes. The field identification of several closely related sharks (including carcharhinid, sphyrnid, and triakid sharks) and batoids (Myliobatiformes and skates) is often challenging, which might result in inaccurate species compositions and diversity in catch reports [32,33]. Moreover, only a few detailed studies have been conducted on the taxonomy and diversity of elasmobranchs in Malaysia [9,34,35]. The lack of data in this field is mainly due to large specimen sizes, ethical reasons, lack of experienced taxonomists, and high field survey costs, which, in turn, render accurate identification more challenging [36]. Thus, there is the potential that many undiscovered species remain unknown. A simple and accurate species identification tool is therefore crucial to allow the species-specific development of fishery conservation and management plans, especially those pertaining to the mislabelling and misidentification of shark and batoid products [12].

Over the past two decades, DNA barcoding has been introduced for its efficiency and accuracy in the identification of challenging species from different taxa [37,38]. The idea of this approach is conceptually straightforward—it uses a short DNA fragment of a specific gene of an approximately 650 bp region to be compared against reference sequences for accurate animal species identification [38]. The Cytochrome c oxidase subunit 1 mitochondrial gene (CO1) is suggested as a highly suitable marker for DNA barcoding, as it can discriminate between closely related species across diverse animal phyla [36], including sharks and rays [39,40]. To date, DNA barcoding has been successful in identifying elasmobranch species from dried fins, tissues, and carcasses [41–47]. This is highly beneficial for non-taxonomists to identify species with reasonable confidence even in the absence of whole specimens. Species can be identified by comparing their DNA barcode sequences against an online repository of barcode references, such as the NCBI GenBank (https://www.ncbi.nlm.nih.gov/genbank, accessed on 5 January 2023) and Barcode of Life Data System (BOLD) (http://www.boldsystems.org, accessed on 5 January 2023) [38,48,49]. However, DNA barcoding has limitations, such as the possibility of errors due to poor DNA quality or contamination, the potential for hybridisation or incomplete lineage sorting, and the lack of universality of CO1 primers across different taxonomic groups.

DNA barcoding—more specifically, CO1 analysis—has been used widely to aid in the species identification of elasmobranchs from various regions, such as Australia [36,42,43], China [44], the Philippines [50], Indonesia [51], Singapore [45,47], India [19,52,53], Bangladesh [54], Southern Africa [46,55], the United Kingdom [56], the Mediterranean Sea [57], the North Atlantic Ocean [58], the United States [49], and Brazil [59–61]. To date, limited published DNA barcoding studies in Malaysia have focused on rays but not on sharks [9,34]. From the available records in BOLD [62], a total of 783 sequences representing 120 species (primarily from the orders Myliobatiformes and Carcharhiniformes) from Malaysian waters were found. These suggested that elasmobranchs in Malaysian waters are not completely barcoded with potential for undiscovered species. Therefore, the present study used DNA barcoding to elucidate the species diversity of sharks and batoids, as well as to update the online database of DNA sequences of elasmobranchs in Malaysian waters. The results could reveal a greater elasmobranch diversity and help support the implementation of elasmobranch conservation and management programs in Malaysian waters.

2. Materials and Methods

2.1. Tissue Sampling and Processing

Shark and batoid tissue samples were collected at various landing sites and markets from four major coastal areas: west coast of Peninsular Malaysia (WP), east coast of Peninsular Malaysia (EP), Sarawak (SR), and Sabah (SB) (Figure 1), from June 2015 to June 2022. These landing sites were chosen based on previous surveys conducted by the Department of Fisheries Malaysia (DOFM), which highlighted prime locations for local elasmobranch fisheries. Moreover, additional samples along EP, SR, and SB were obtained through opportunistic fisheries—independent demersal trawl surveys within the Exclusive Economic Zone (EEZ) organised by the Department of Fisheries Malaysia (July 2015 to July 2016). They were identified in the field and the laboratory by KCL and AJXL using authoritative species identification guides and taxonomic references of the region [4,6,13,14,63–66]. No live animals were collected or killed during this study. No permission to collect data from EP was needed at the time of the research, and permission to collect data from Sarawak waters was granted by the Fisheries Research Institute Sarawak, Department of Fisheries Malaysia. A Sabah sampling permit was provided by the Sabah Biodiversity Council (SaBC), Ref. No. JKM/MBS.1000-2/2 JLD.9 (22) and JKM/MBS.1000-2/3JLD.4 (18). At least one fin clip of each recorded species was collected and preserved either in absolute ethanol or gently squashed onto a Whatman FTA® Elute card before the subsequent molecular analysis.

2.2. Extraction, PCR Amplification, and DNA Sequencing

One to six tissue samples for each species were selected for barcoding and molecular analysis (Table 1). Particular attention was given to specimens showing morphological ambiguities during field identification and potential cryptic species, as well as species that have not been barcoded to date; molecular identification was used to clarify species identities in these cases. These includes samples of *Cephaloscyllium sarawakense* (2-013 and 3-640); *Carcharhinus leucas* (3-129); *Squalus altipinnis* (2-041 and 2-047); *Narcine maculata* (2-874 and 4-439); *Rhinobatos borneensis* (KK7, 2-070, 2-678, S6, S16, and S17); *Okamejei boesemani* (2-057, 2-093, and 4-065); *Hemitrygon parvonigra* (2-059, 2-127, 2-448, and T8);

Maculabatis gerrardi (TW5, 2-160, 2-162, 2-166, JHR5TC, and S27E); *Urogymnus lobistoma* (3-051, 3-821, and 3-876); *Myliobatis hamlyni* (T7, 2-027, and 2-029); and *Chimaera phantasma* (2-023 and 2-025). For samples stored in absolute ethanol, the total DNA was extracted using 10% Chelex resin incubated for two minutes at 60 °C, followed by 25 min at 103 °C, following a modified protocol of Hyde et al. [67]. For samples stored using Whatman FTA® Elute cards, the total DNA was extracted using the modified protocol in Rigby et al. [68].

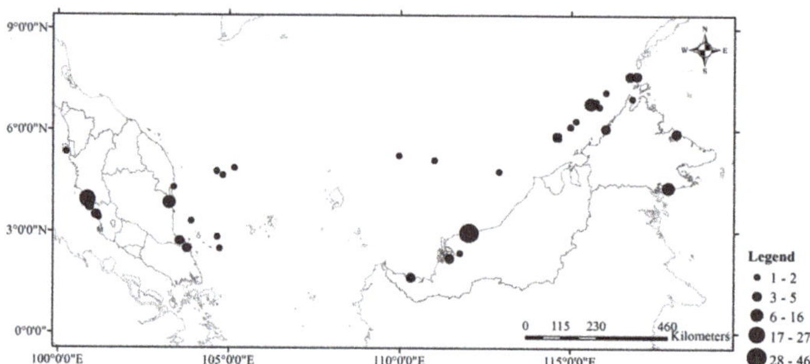

Figure 1. Map of the selected elasmobranch sampling sites in Malaysia.

Amplification of the partial cytochrome c oxidase subunit I (CO1) gene (650 base pairs, nucleotide position 51-701) was done using universal primers FishF2– 5′ TCG ACT AAT CAT AAA GAT ATC GGC AC 3′ and FishR2—5′ ACT TCA GGG TGA CCG AAG AAT CAG AA 3′ [38]. PCR reactions were performed in 25 µL volumes containing 12.5 µL of 2X PCR Master Mix (1st Base), 1 µL of 10 mM of both primers, 1 µL DNA templates, and 9.5 µL sterilized distilled water. The thermal conditions consisted of the initial preheating at 94 °C for 5 min, denaturation at 94 °C for 30 s, annealing at 44–54 °C for 30 s, and extension at 72 °C for 1 min, then repeated for 36–40 cycles, followed by a final extension at 72 °C of 5 min. The PCR products were checked using 1% agarose in TAE buffer. Only PCR products that showed good PCR amplification were sent for sequencing service at Apical Scientific Sdn Bhd (Selangor, Malaysia). The sequencing results were checked for confirmation of the species using the BLAST tool of the NCBI and compared with field identification. The species identification matching was based on the query cover (%) in the BLAST tool as the query cover that describes how similar the sample sequence is to the reference sequence in GenBank and the similarity in the BOLD system.

2.3. Data Analysis

For the data analysis, the obtained sequences were edited and aligned using BioEdit version 7.2.5 [69]. Molecular species identification was achieved using two approaches: Basic Local Alignment Search Tool (BLAST®) and phylogenetic tree reconstruction. For the first approach, reviewed sequences were uploaded to BLAST® (https://blast.ncbi.nlm.nih.gov/Blast.cgi, accessed on 5 January 2023) and BOLDSYSTEMS (http://www.boldsystems.org/index.php/IDS_OpenIdEngine, accessed on 5 January 2023) in search for local similarities with the available sequences in the National Center for Biotechnology Information (NCBI) GenBank and Barcode of Life Database (BOLD), respectively. The results from the searches were in the form of scores, query cover, and percentage identity of the 200 most similar sequences available in the databases. The list of sequences that produced significant alignments or best matches with each of our sequence was reviewed manually for validity of the suggested identity. This included their percentage similarity, query coverage, reliability of the sequence sources (from respectable authors in the field), congruency among sequences from different sources, etc. Suspicious sequences in the database were excluded in the process.

Table 1. Verification of the identity of barcoded shark and batoids species. Given are the sample code, sample location, NCBI accession number, field identification, best matched species from NCBI GenBank, similarity species and % from BOLD, tree-based identification, and the consensus identification. West coast of Peninsular Malaysia (WP): HM = Hutan Melintang, Pen = Penang, PP = Pasir Penambang, Sb = Sungai Besar, and Tk = Tanjung Karang; East coast of Peninsular Malaysia (EP): En = Endau, EP(D) = East coast of Peninsular Malaysia (DOF survey), Kt = Kuantan, Mer = Mersing, and Trg = Terengganu; Sarawak (SR): Btw = Bintawa, Kc = Kuching, Mk = Mukah, Sib = Sibu, SR(D) = Sarawak (DOF survey), and Srk = Sarikei; Sabah (SB): Kd = Kudat, KK = Kota Kinabalu, San = Sandakan, SB(D) = Sabah (DOF survey), and Tw = Tawau.

Samples	Location	Accession Number	Field Identification	NCBI Best Matched	%	BOLD Similarity	%	Tree Match	Consensus Identification
2-073	SB(D)	OQ384975	*Heterodontus zebra*	*H. zebra*	99.54	*H. zebra*	100	*H. zebra*	*H. zebra*
2-075	SB(D)	OQ384976	*Heterodontus zebra*	*H. zebra*	99.54	*H. zebra*	100	*H. zebra*	*H. zebra*
2-170	HM	OQ384977	*Chiloscyllium hasseltii*	*C. hasseltii*	100	*C. hasseltii/griseum*	100	*C. hasseltii*	*C. hasseltii*
2-502	Kt	OQ384978	*Chiloscyllium hasseltii*	*C. hasseltii*	100	*C. hasseltii griseum*	99.85	*C. hasseltii*	*C. hasseltii*
3-487	Trg	OQ384979	*Chiloscyllium hasseltii*	*C. hasseltii*	100	*C. hasseltii griseum*	100	*C. hasseltii*	*C. hasseltii*
S12	HM	OQ384980	*Chiloscyllium hasseltii*	*C. hasseltii*	100	*C. hasseltii griseum*	100	*C. hasseltii*	*C. hasseltii*
S25	HM	OQ384981	*Chiloscyllium hasseltii*	*C. hasseltii*	100	*C. hasseltii griseum*	100	*C. hasseltii*	*C. hasseltii*
3-215	Mk	OQ384982	*Chiloscyllium indicum*	*C. indicum*	100	*C. indicum*	100	*C. indicum*	*C. indicum*
SLG15	HM	OQ384983	*Chiloscyllium indicum*	*C. indicum*	100	*C. indicum*	100	*C. indicum*	*C. indicum*
2-452	SB(D)	OQ384984	*Chiloscyllium plagiosum*	*C. plagiosum*	99.85	*C. plagiosum*	100	*C. plagiosum*	*C. plagiosum*
4-357	EP(D)	OQ384985	*Chiloscyllium plagiosum*	*C. plagiosum*	99.85	*C. plagiosum*	100	*C. plagiosum*	*C. plagiosum*
J8	En	OQ384986	*Chiloscyllium punctatum*	*C. punctatum*	99.69	*C. punctatum*	99.85	*C. punctatum*	*C. punctatum*
J9	En	OQ384987	*Chiloscyllium punctatum*	*C. punctatum*	99.69	*C. punctatum*	100	*C. punctatum*	*C. punctatum*
Q11	Mk	OQ384988	*Chiloscyllium punctatum*	*C. punctatum*	99.69	*C. punctatum*	100	*C. punctatum*	*C. punctatum*
5-257	KK	OQ384989	*Alopias pelagicus*	*A. pelagicus*	100	*A. pelagicus*	100	*A. pelagicus*	*A. pelagicus*
Asup1	HM	OQ384990	*Alopias superciliosus*	*A. superciliosus*	100	*A. superciliosus*	100	*A. superciliosus*	*A. superciliosus*
2-842	HM	OQ384991	*Atelomycterus marmoratus*	*A. marmoratus*	100	*A. marmoratus*	100	*A. marmoratus*	*A. marmoratus*
4-883	San	OQ384992	*Atelomycterus marmoratus*	*A. marmoratus*	99.53	*A. marmoratus*	99.84	*A. marmoratus*	*A. marmoratus*
Q50	Sib	OQ384993	*Atelomycterus marmoratus*	*A. marmoratus*	99.37	*A. marmoratus*	99.85	*A. marmoratus*	*A. marmoratus*
2-013	SB(D)	OQ384994	*Cephaloscyllium sarawakense*	*C. umbratile* [LS]	98.93	*C. umbratile* [LS]	98.92	-	*C. sarawakense* [1]
3-640	SB(D)	OQ384995	*Cephaloscyllium sarawakense*	*C. umbratile* [DS]	99.85	*C. umbratile* [DS]	99.85	-	*C. sarawakense* [1]
2-011	SB(D)	OQ384996	*Halaelurus buergeri*	*H. buergeri*	99.69	*H. buergeri*	99.85	*H. buergeri*	*H. buergeri*
2-109	SB(D)	OQ384997	*Halaelurus buergeri*	*H. buergeri*	99.69	*H. buergeri*	99.85	*H. buergeri*	*H. buergeri*
2-536	Kt	OQ384998	*Hemigaleus microstoma*	*H. microstoma*	100	*H. microstoma*	100	*H. microstoma*	*H. microstoma*
3-367	Mk	OQ384999	*Hemigaleus microstoma*	*H. microstoma*	100	*H. microstoma*	100	*H. microstoma*	*H. microstoma*
4-419	EP(D)	OQ385000	*Hemigaleus microstoma*	*H. microstoma*	100	*H. microstoma*	100	*H. microstoma*	*H. microstoma*
TW12	Tw	OQ385001	*Hemigaleus microstoma*	*H. microstoma*	99.69	*H. microstoma*	100	*H. microstoma*	*H. microstoma*
2-746	Kt	OQ385002	*Hemipristis elongata*	*H. elongata*	100	*H. elongata*	100	*H. elongata*	*H. elongata*
3-423	Mk	OQ385003	*Hemipristis elongata*	*H. elongata*	100	*H. elongata*	100	*H. elongata*	*H. elongata*
5-082	San	OQ385004	*Hemipristis elongata*	*H. elongata*	99.69	*H. elongata*	100	*H. elongata*	*H. elongata*
3-754	Mk	OQ385005	*Carcharhinus amblyrhynchoides*	*C. amblyrhynchoides* [MM]	100	*C. amblyrhynchoides* [MM]	100	*C. amblyrhynchoides*	*C. amblyrhynchoides*
3-850	Mk	OQ385006	*Carcharhinus amblyrhynchoides*	*C. amblyrhynchoides* [MM]	100	*C. amblyrhynchoides* [MM]	100	*C. amblyrhynchoides*	*C. amblyrhynchoides*
3-960	Mk	OQ385007	*Carcharhinus amblyrhynchoides*	*C. amblyrhynchoides* [MM]	100	*C. amblyrhynchoides* [MM]	100	*C. amblyrhynchoides*	*C. amblyrhynchoides*
3-131	Mk	OQ385008	*Carcharhinus amblyrhynchoides*	*C. amblyrhynchoides* [MM]	100	*C. amblyrhynchoides* [MM]	100	*C. amblyrhynchoides*	*C. amblyrhynchoides*
3-133	Mk	OQ385009	*Carcharhinus amblyrhynchoides*	*C. amblyrhynchoides* [MM]	100	*C. amblyrhynchoides* [MM]	100	*C. amblyrhynchoides*	*C. amblyrhynchoides*

Table 1. *Cont.*

Samples	Location	Accession Number	Field Identification	NCBI Best Matched	%	BOLD Similarity	%	Tree Match	Consensus Identification
Q3E	Srk	OQ385010	*Carcharhinus amblyrhynchoides*	*C. amblyrhynchoides* [MM]	100	*C. amblyrhynchoides* [MM]	100	*C. amblyrhynchoides*	*C. amblyrhynchoides*
3-353	Mk	OQ385011	*Carcharhinus brevipinna*	*C. brevipinna*	100	*C. brevipinna*	99.85	*C. brevipinna*	*C. brevipinna*
3-535	HM	OQ385012	*Carcharhinus brevipinna*	*C. brevipinna*	100	*C. brevipinna*	100	*C. brevipinna*	*C. brevipinna*
Q19	Srk	OQ385013	*Carcharhinus brevipinna*	*C. brevipinna*	100	*C. brevipinna*	100	*C. brevipinna*	*C. brevipinna*
2-246	Pen	OQ385014	*Carcharhinus leucas*	*C. leucas*	100	*C. leucas*	100	*C. leucas*	*C. leucas*
2-530	Kt	OQ385015	*Carcharhinus leucas*	*C. leucas*	100	*C. leucas*	100	*C. leucas*	*C. leucas*
S1	Tk	OQ385016	*Carcharhinus leucas*	*C. leucas*	100	*C. leucas*	100	*C. leucas*	*C. leucas*
3-129	Mk	OQ385017	*Carcharhinus leucas*	*C. amboinensis* [DS]	99.69	*C. amboinensis* [DS]	100	*C. amboinensis*	*C. amboinensis*
3-038	Kc	OQ385018	*Carcharhinus limbatus*	*C. limbatus* [MM]	100	*C. limbatus* [MM]	100	*C. limbatus*	*C. limbatus*
2-588	Kt	OQ385019	*Carcharhinus limbatus*	*C. limbatus* [MM]	99.54	*C. limbatus* [MM]	99.84	*C. limbatus*	*C. limbatus*
3-331	Mk	OQ385020	*Carcharhinus melanopterus*	*C. melanopterus*	100	*C. melanopterus*	100	*C. melanopterus*	*C. melanopterus*
2-538	Kt	OQ385021	*Carcharhinus sealei*	*C. sealei*	100	*C. sealei*	100	*C. sealei*	*C. sealei*
3-103	Mk	OQ385022	*Carcharhinus sealei*	*C. sealei*	99.85	*C. sealei*	100	*C. sealei*	*C. sealei*
5-083	San	OQ385023	*Carcharhinus sealei*	*C. sealei*	99.85	*C. sealei*	100	*C. sealei*	*C. sealei*
2-380	HM	OQ385024	*Carcharhinus sorrah*	*C. sorrah*	99.85	*C. sorrah*	99.85	*C. sorrah*	*C. sorrah*
2-578	Kt	OQ385025	*Carcharhinus sorrah*	*C. sorrah*	99.85	*C. sorrah*	99.85	*C. sorrah*	*C. sorrah*
3-077	Mk	OQ385026	*Carcharhinus sorrah*	*C. sorrah*	100	*C. sorrah*	100	*C. sorrah*	*C. sorrah*
S23TC	Tk	OQ385027	*Carcharhinus sorrah*	*C. sorrah*	99.85	*C. sorrah*	99.85	*C. sorrah*	*C. sorrah*
3-740	Mk	OQ385028	*Lamiopsis tephrodes*	*L. tephrodes*	100	*L. tephrodes*	100	*L. tephrodes*	*L. tephrodes*
3-944	Mk	OQ385029	*Lamiopsis tephrodes*	*L. tephrodes*	100	*L. tephrodes*	100	*L. tephrodes*	*L. tephrodes*
3-988	Mk	OQ385030	*Lamiopsis tephrodes*	*L. tephrodes*	100	*L. tephrodes*	100	*L. tephrodes*	*L. tephrodes*
Q2	Srk	OQ385031	*Lamiopsis tephrodes*	*L. tephrodes*	100	*L. tephrodes*	100	*L. tephrodes*	*L. tephrodes*
Q21	Mk	OQ385032	*Lamiopsis tephrodes*	*L. tephrodes*	100	*L. tephrodes*	100	*L. tephrodes*	*L. tephrodes*
3-739	Mk	OQ385033	*Loxodon macrorhinus*	*L. macrorhinus*	100	*L. macrorhinus*	100	*L. macrorhinus*	*L. macrorhinus*
3-909	Mk	OQ385034	*Loxodon macrorhinus*	*L. macrorhinus*	100	*L. macrorhinus*	100	*L. macrorhinus*	*L. macrorhinus*
3-928	Mk	OQ385035	*Loxodon macrorhinus*	*L. macrorhinus*	99.85	*L. macrorhinus*	99.85	*L. macrorhinus*	*L. macrorhinus*
2-438	Kt	OQ385036	*Rhizoprionodon acutus*	*R. acutus*	100	*R. acutus*	100	*R. acutus*	*R. acutus*
3-321	Mk	OQ385037	*Rhizoprionodon acutus*	*R. acutus*	99.85	*R. acutus*	99.85	*R. acutus*	*R. acutus*
Q42	Mk	OQ385038	*Rhizoprionodon acutus*	*R. acutus*	99.85	*R. acutus*	99.85	*R. acutus*	*R. acutus*
Q43	Mk	OQ385039	*Rhizoprionodon acutus*	*R. acutus*	99.85	*R. acutus*	99.85	*R. acutus*	*R. acutus*
Q23	Mk	OQ385040	*Rhizoprionodon acutus*	*R. acutus*	100	*R. acutus*	100	*R. acutus*	*R. acutus*
3-211	Mk	OQ385041	*Rhizoprionodon oligolinx*	*R. oligolinx*	99.69	*R. oligolinx*	99.85	*R. oligolinx*	*R. oligolinx*
4-706	HM	OQ385042	*Rhizoprionodon oligolinx*	*R. oligolinx*	99.54	*R. oligolinx*	100	*R. oligolinx*	*R. oligolinx*
J10	Mer	OQ385043	*Rhizoprionodon oligolinx*	*R. oligolinx*	99.54	*R. oligolinx*	100	*R. oligolinx*	*R. oligolinx*
J11	Mer	OQ385044	*Rhizoprionodon oligolinx*	*R. oligolinx*	99.54	*R. oligolinx*	100	*R. oligolinx*	*R. oligolinx*
S13	HM	OQ385045	*Rhizoprionodon oligolinx*	*R. oligolinx*	99.54	*R. oligolinx*	100	*R. oligolinx*	*R. oligolinx*
Q6	Srk	OQ385046	*Scoliodon macrorhynchos*	*S. macrorhynchos*	100	*S. macrorhynchos*	100	*S. macrorhynchos*	*S. macrorhynchos*
Q8	Srk	OQ385047	*Scoliodon macrorhynchos*	*S. macrorhynchos*	100	*S. macrorhynchos*	100	*S. macrorhynchos*	*S. macrorhynchos*
Q41T	Mk	OQ385048	*Scoliodon macrorhynchos*	*S. macrorhynchos*	99.85	*S. macrorhynchos*	99.85	*S. macrorhynchos*	*S. macrorhynchos*
K01	Kd	OQ385049	*Triaenodon obesus*	*T. obesus*	99.69	*T. obesus*	100	*T. obesus*	*T. obesus*
2-664	Kt	OQ385050	*Galeocerdo cuvier*	*G. cuvier*	100	*G. cuvier*	100	*G. cuvier*	*G. cuvier*
3-115	Mk	OQ385051	*Galeocerdo cuvier*	*G. cuvier*	100	*G. cuvier*	100	*G. cuvier*	*G. cuvier*
4-996	San	OQ385052	*Galeocerdo cuvier*	*G. cuvier*	100	*G. cuvier*	100	*G. cuvier*	*G. cuvier*
2-226	HM	OQ385053	*Sphyrna lewini*	*S. lewini*	100	*S. lewini*	100	*S. lewini*	*S. lewini*
2-604	Kt	OQ385054	*Sphyrna lewini*	*S. lewini*	99.69	*S. lewini*	99.80	*S. lewini*	*S. lewini*
3-225	Mk	OQ385055	*Sphyrna lewini*	*S. lewini*	100	*S. lewini*	100	*S. lewini*	*S. lewini*
T3	Tw	OQ385056	*Sphyrna lewini*	*S. lewini*	100	*S. lewini*	100	*S. lewini*	*S. lewini*
2-041	SB(D)	OQ385057	*Squalus altipinnis*	*S. brevirostris* [LS]	98.78	*S. brevirostris* [LS]	98.92	-	*S. altipinnis* [1]
2-047	SB(D)	OQ385058	*Squalus altipinnis*	*S. edmundsi* [DS]	99.54	*S. edmundsi* [DS]	100	*S. edmundsi*	*S. edmundsi*
2-103	SB(D)	OQ385059	*Squatina tergocellatoides*	*S. tergocellatoides*	99.85	*S. tergocellatoides*	99.85	*S. tergocellatoides*	*S. tergocellatoides*
2-114	SB(D)	OQ385060	*Squatina tergocellatoides*	*S. tergocellatoides*	100	*S. tergocellatoides*	100	*S. tergocellatoides*	*S. tergocellatoides*
4-322	EP(D)	OQ385061	*Narcine brevilabiata*	*N. brevilabiata*	99.69	*N. brevilabiata*	99.85	*N. brevilabiata*	*N. brevilabiata*

Table 1. *Cont.*

Samples	Location	Accession Number	Field Identification	NCBI Best Matched	%	BOLD Similarity	%	Tree Match	Consensus Identification
4-368	EP(D)	OQ385062	*Narcine brevilabiata*	*N. brevilabiata*	99.85	*N. brevilabiata*	100	*N. brevilabiata*	*N. brevilabiata*
4-439	EP(D)	OQ385063	*Narcine maculata*	*N. maculata* [LS]	95.25	*N. maculata* [LS]	98.52	-	*N.* cf. *maculata* sp. 1 [1]
2-874	HM	OQ385064	*Narcine maculata*	*Narcine* sp.[MU]	99.35	*Narcine* cf. *oculifera* [MU]	99.69	-	*N.* cf. *maculata* sp. 2 [1]
2-762	Kt	OQ385065	*Rhina ancylostomus*	*R. ancylostomus*	99.85	*R. ancylostomus*	99.83	*R. ancylostomus*	*R. ancylostomus*
3-584	HM	OQ385066	*Rhina ancylostomus*	*R. ancylostomus*	100	*R. ancylostomus*	99.85	*R. ancylostomus*	*R. ancylostomus*
3-890	Mk	OQ385067	*Rhina ancylostomus*	*R. ancylostomus*	99.69	*R. ancylostomus*	99.67	*R. ancylostomus*	*R. ancylostomus*
T2	Tw	OQ385068	*Rhina ancylostomus*	*R. ancylostomus*	100	*R. ancylostomus*	100	*R. ancylostomus*	*R. ancylostomus*
2-347	Btw	OQ385069	*Rhynchobatus australiae*	*R. australiae*	100	*R. australiae*	100	*R. australiae*	*R. australiae*
2-514	Kt	OQ385070	*Rhynchobatus australiae*	*R. australiae*	100	*R. australiae*	100	*R. australiae*	*R. australiae*
3-570	HM	OQ385071	*Rhynchobatus australiae*	*R. australiae*	99.39	*R. australiae*	99.62	*R. australiae*	*R. australiae*
TW9	Tw	OQ385072	*Rhynchobatus australiae*	*R. australiae*	100	*R. australiae*	100	*R. australiae*	*R. australiae*
2-614	Kt	OQ385073	*Rhynchobatus australiae*	*R. australiae*	100	*R. australiae*	100	*R. australiae*	*R. australiae*
KK7	KK	OQ385074	*Rhinobatos borneensis*	*R. schlegelii* [DS]	100	*R. schlegelii* [DS]	100	-	*R. borneensis* [1]
2-070	SB(D)	OQ385075	*Rhinobatos borneensis*	*R. schlegelii* [DS]	99.85	*R. schlegelii* [DS]	99.85	-	*R. borneensis* [1]
2-678	SB(D)	OQ385076	*Rhinobatos borneensis*	*R. formosensis* [DS]	100	*R. formosensis* [DS]	99.85	-	*R. borneensis* [1]
S6	HM	OQ385077	*Rhinobatos* cf *borneensis*	*R. jimbaranensis* [LS]	97.54	*R. jimbaranensis* [LS]	97.64	-	*R.* cf. *borneensis* [1]
S16	HM	OQ385078	*Rhinobatos* cf *borneensis*	*R. jimbaranensis* [LS]	97.54	*R. jimbaranensis* [LS]	97.64	-	*R.* cf. *borneensis* [1]
S17	HM	OQ385079	*Rhinobatos* cf *borneensis*	*R. jimbaranensis* [LS]	97.54	*R. jimbaranensis* [LS]	97.64	-	*R.* cf. *borneensis* [1]
2-093	SB(D)	OQ385080	*Okamejei boesemani*	*O. boesemani* [MM]	99.38	*O. boesemani* [MM]	99.38	-	*O. boesemani* [1]
2-057	SB(D)	OQ385081	*Okamejei boesemani*	*O. boesemani* [MM]	99.38	*O. boesemani* [MM]	99.38	-	*O. boesemani* [1]
4-065	SR(D)	OQ385082	*Okamejei boesemani*	*O. boesemani* [MM]	99.54	*O. boesemani* [MM]	99.53	-	*O. boesemani* [1]
SK1	Trg	OQ385083	*Okamejei hollandi*	*O. hollandi*	99.52	*O. hollandi*	99.51	*O. hollandi*	*O. hollandi*
SK2	Trg	OQ385084	*Okamejei hollandi*	*O. hollandi*	99.84	*O. hollandi*	99.84	*O. hollandi*	*O. hollandi*
2-784	SR(D)	OQ385085	*Okamejei hollandi*	*O. hollandi*	99.52	*O. hollandi*	99.51	*O. hollandi*	*O. hollandi*
2-327	SR(D)	OQ385086	*Okamejei hollandi*	*O. hollandi*	99.68	*O. hollandi*	99.68	*O. hollandi*	*O. hollandi*
2-017	SB(D)	OQ385087	*Okamejei hollandi*	*O. hollandi*	99.84	*O. hollandi*	99.84	*O. hollandi*	*O. hollandi*
2-019	SB(D)	OQ385088	*Bathytoshia lata*	*B. lata*	100	*B. lata*	100	*B. lata*	*B. lata*
Q25TC	Mk	OQ385089	*Brevitrygon heterura*	*B. heterura*	99.69	*B. heterura*	99.69	*B. heterura*	*B. heterura*
S20	HM	OQ385090	*Brevitrygon heterura*	*B. heterura*	100	*B. heterura*	100	*B. heterura*	*B. heterura*
S21	HM	OQ385091	*Brevitrygon heterura*	*B. heterura*	99.85	*B. heterura*	99.85	*B. heterura*	*B. heterura*
Q20	Mk	OQ385092	*Hemitrygon bennettii*	*H. bennettii*	100	*H. bennettii*	100	*H. bennettii*	*H. bennettii*
3-862	Mk	OQ385093	*Hemitrygon bennettii*	*H. bennettii*	100	*H. bennettii*	100	*H. bennettii*	*H. bennettii*
2-138	Sb	OQ385094	*Hemitrygon bennettii*	*H. bennettii*	100	*H. bennettii*	100	*H. bennettii*	*H. bennettii*
2-059	SB(D)	OQ385095	*Hemitrygon parvonigra*	*H. parvonigra* [LS]	94.78	*H. fai* [LS]	100	-	*H.* cf. *parvonigra* [1]
2-448	SB(D)	OQ385096	*Hemitrygon parvonigra*	*H. parvonigra* [LS]	94.78	*H. fai* [LS]	100	-	*H.* cf. *parvonigra* [1]
2-127	SB(D)	OQ385097	*Hemitrygon parvonigra*	*H. parvonigra* [LS]	94.78	*H. fai* [LS]	100	-	*H.* cf. *parvonigra* [1]
T8	Tw	OQ385098	*Hemitrygon parvonigra*	*H. parvonigra* [LS]	94.78	*H. fai* [LS]	100	-	*H.* cf. *parvonigra* [1]
TW4	Tw	OQ385099	*Himantura leoparda*	*H. leoparda*	99.08	*H. leoparda*	99.19	*H. leoparda*	*H. leoparda*
K1	KK	OQ385100	*Himantura uarnak*	*H. uarnak*	100	*H. uarnak*	100	*H. uarnak*	*H. uarnak*
TW6	Tw	OQ385101	*Himantura undulata*	*H. undulata*	100	*H. undulata*	100	*H. undulata*	*H. undulata*
TW5	Tw	OQ385102	*Maculabatis gerrardi*	*M. macrura* [MM]	99.85	*M. macrura/H. gerrardi* [MM]	100	*M. gerrardi*	*M. gerrardi* [1]
2-160	Sb	OQ385103	*Maculabatisgerrardi*	*M. macrura/H. gerrardi* [MM]	100	*M. macrura/H. gerrardi* [MM]	100	*M. gerrardi*	*M. gerrardi* [1]
2-162	Sb	OQ385104	*Maculabatis gerrardi*	*M. macrura/H. gerrardi* [MM]	100	*M. macrura/H. gerrardi* [MM]	100	*M. gerrardi*	*M. gerrardi* [1]
2-166	Sb	OQ385105	*Maculabatis gerrardi*	*M. macrura/H. gerrardi* [MM]	100	*M. macrura/H. gerrardi* [MM]	100	*M. gerrardi*	*M. gerrardi* [1]
JHR5TC	Mer	OQ385106	*Maculabatis gerrardi*	*M. macrura/H. gerrardi* [MM]	99.69	*M. macrura/H. gerrardi* [MM]	100	*M. gerrardi*	*M. gerrardi* [1]
S27E	HM	OQ385107	*Maculabatis gerrardi*	*M. macrura/H. gerrardi* [MM]	100	*M. macrura/H. gerrardi* [MM]	100	*M. gerrardi*	*M. gerrardi* [1]
S10	HM	OQ385108	*Neotrygon malaccensis*	*N. kuhlii* [DS]	99.54	*N. kuhlii* [DS]	99.84	*N. malaccensis*	*N. malaccensis*

Table 1. *Cont.*

Samples	Location	Accession Number	Field Identification	NCBI Best Matched	%	BOLD Similarity	%	Tree Match	Consensus Identification
S11	HM	OQ385109	*Neotrygon malaccensis*	*N. kuhlii* [DS]	100	*N. kuhlii* [DS]	100	*N. malaccensis*	*N. malaccensis*
3-619	HM	OQ385110	*Neotrygon malaccensis*	*N. kuhlii* [DS]	99.85	*N. kuhlii* [DS]	100	*N. malaccensis*	*N. malaccensis*
4-579	HM	OQ385111	*Neotrygon malaccensis*	*N. kuhlii* [DS]	100	*N. kuhlii* [DS]	100	*N. malaccensis*	*N. malaccensis*
Q12	Mk	OQ385112	*Neotrygon orientalis*	*N. kuhlii* [DS]	100	*N. kuhlii* [DS]	100	*N. orientalis*	*N. orientalis*
3-459	Trg	OQ385113	*Neotrygon varidens*	*N. kuhlii* [DS]	100	*N. kuhlii* [DS]	100	*N. varidens*	*N. varidens*
4-421	EP(D)	OQ385114	*Neotrygon varidens*	*N. kuhlii* [DS]	99.85	*N. kuhlii* [DS]	100	*N. varidens*	*N. varidens*
Q17	Mk	OQ385115	*Pastinachus gracilicaudus*	*P. gracilicaudus*	100	*P. gracilicaudus/solocirostris*	100	*P. gracilicaudus*	*P. gracilicaudus*
3-263	Mk	OQ385116	*Pastinachus solocirostris*	*P. solocirostris*	100	*P. solocirostris*	100	*P. solocirostris*	*P. solocirostris*
3-746	Mk	OQ385117	*Pastinachus solocirostris*	*P. solocirostris*	100	*P. solocirostris*	100	*P. solocirostris*	*P. solocirostris*
3-841	Mk	OQ385118	*Pastinachus solocirostris*	*P. solocirostris*	100	*P. solocirostris*	100	*P. solocirostris*	*P. solocirostris*
Q16TC	Mk	OQ385119	*Pateobatis uarnacoides*	*P. uarnacoides*	99.85	*P. uarnacoides*	100	*P. uarnacoides*	*P. uarnacoides*
J12	En	OQ385120	*Taeniura lymma*	*T. lymma*	100	*T. lymma*	100	*T. lymma*	*T. lymma*
2-069	SB(D)	OQ385121	*Taeniurops meyeni*	*T. meyeni*	99.85	*T. meyeni*	100	*T. meyeni*	*T. meyeni*
S3	Tk	OQ385122	*Telatrygon biasa*	*T. biasa*	99.69	*T. zugei*	100	*T. biasa*	*T. biasa*
S18	HM	OQ385123	*Telatrygon biasa*	*T. biasa*	99.54	*T. zugei*	99.83	*T. biasa*	*T. biasa*
S19	HM	OQ385124	*Telatrygon biasa*	*T. biasa*	99.54	*T. zugei*	99.83	*T. biasa*	*T. biasa*
JHR2E	Mer	OQ385125	*Urogymnus asperrimus*	*U. asperrimus*	99.69	*U. asperrimus*	99.85	*U. asperrimus*	*U. asperrimus*
3-051	Kc	OQ385126	*Urogymnus lobistoma*	*H. chaophraya*	88.01	*H. uarnacoides*	100	-	*U. lobistoma* [I]
3-821	Mk	OQ385127	*Urogymnus lobistoma*	*H. chaophraya*	88.01	*H. uarmacoides*	100	-	*U. lobistoma* [I]
3-876	Mk	OQ385128	*Urogymnus lobistoma*	*H. chaophraya*	88.01	*H. uarmacoides*	100	-	*U. lobistoma* [I]
3-936	Mk	OQ385129	*Gymnura poecilura*	*G. poecilura*	100	*G. poecilura*	100	*G. poecilura*	*G. poecilura*
3-937	Mk	OQ385130	*Gymnura poecilura*	*G. poecilura*	100	*G. poecilura*	100	*G. poecilura*	*G. poecilura*
3-987	Mk	OQ385131	*Gymnura poecilura*	*G. poecilura*	99.85	*G. poecilura*	99.84	*G. poecilura*	*G. poecilura*
4-705	HM	OQ385132	*Gymnura poecilura*	*G. poecilura*	99.69	*G. poecilura*	99.69	*G. poecilura*	*G. poecilura*
G08c	En	OQ385133	*Gymnura zonura*	*G. zonura*	99.54	*G. zonura*	99.69	*G. zonura*	*G. zonura*
G09c	En	OQ385134	*Gymnura zonura*	*G. zonura*	99.85	*G. zonura*	100	*G. zonura*	*G. zonura*
2-021	SB(D)	OQ385135	*Plesiobatis daviesi*	*P. daviesi*	100	*P. daviesi*	100	*P. daviesi*	*P. daviesi*
TW14	Tw	OQ385136	*Urolophus* cf. *aurantiacus*	*U. expansus* [DS]	99.07	*U. expansus* [DS]	99.84	-	*U.* cf. *aurantiacus* [I]
2-656	Kt	OQ385137	*Aetobatus ocellatus*	*A. ocellatus*	100	*A. ocellatus*	100	*A. ocellatus*	*A. ocellatus*
2-946	Kc	OQ385138	*Aetobatus ocellatus*	*A. ocellatus*	100	*A. ocellatus*	100	*A. ocellatus*	*A. ocellatus*
S9	PP	OQ385139	*Aetobatus ocellatus*	*A. ocellatus*	100	*A. ocellatus*	100	*A. ocellatus*	*A. ocellatus*
K2	KK	OQ385140	*Aetobatus ocellatus*	*A. ocellatus*	99.85	*A. ocellatus*	99.85	*A. ocellatus*	*A. ocellatus*
Aves1	Kt	OQ385141	*Aetomylaeus vespertilio*	*A vespertilio*	99.07	*A vespertilio*	98.88	*A vespertilio*	*A vespertilio*
T7	Tw	OQ385142	*Myliobatis hamlyni*	*M. hamlyni* [R]	100	*M. tobijei*	100	*M. tobijei*	*M. hamlyni*
2-027	SB(D)	OQ385143	*Myliobatis hamlyni*	*M. hamlyni* [R]	100	*M. tobijei*	100	*M. tobijei*	*M. hamlyni*
2-029	SB(D)	OQ385144	*Myliobatis hamlyni*	*M. hamlyni* [R]	100	*M. tobijei*	100	*M. tobijei*	*M. hamlyni*
3-289	Mk	OQ385145	*Rhinoptera jayakari*	*R. jayakari*	100	*R. jayakari*	100	*R. jayakari*	*R. jayakari*
2-516	Kt	OQ385146	*Mobula thurstoni*	*M. thurstoni*	100	*M. thurstoni*	100	*M. thurstoni*	*M. thurstoni*
3-781	Mk	OQ385147	*Mobula kuhlii*	*M. kuhlii*	100	*M. kuhlii*	100	*M. kuhlii*	*M. kuhlii*
2-023	SB(D)	OQ385148	*Chimaera phantasma*	*C. phantasma* [LS]	97.39	*C. phantasma* [LS]	97.67	-	*C.* cf. *phantasma* [I]
2-025	SB(D)	OQ385149	*Chimaera phantasma*	*C. phantasma* [LS]	97.39	*C. phantasma* [LS]	97.67	-	*C.* cf. *phantasma* [I]

[R] = Revised identity based on White et al. [66]—reference sequence EU398924 *M. tobijei* was revised as *M. hamlyni*; [DS] = best match differed from field-identified species; [LS] = best match showed low similarity; [MM] = match with multiple species; [MU] = match with uncertain species; and [I] = inconclusive identity.

For the second approach, a phylogenetic tree was constructed using sequences from the current study and reference sequences available in the NCBI GenBank (Supplementary Materials Table S2). The selection of the reference sequences included the best matched sequences in the BLAST® search, as well as available sequences of elasmobranchs that are recorded in Malaysian waters. Reference sequences from Malaysian waters were prioritized over sequences from neighbouring waters. The list of elasmobranchs (70 sharks and 91 batoid species) and chimaera species that had been recorded in Malaysia (include some unverified records) is shown in Supplementary Materials Table S1. The number of reference sequences used for each species was limited to one or two sequences, except for the species complex of *Maculabatis gerardi* and *M. macrura*.

A midpoint rooted tree maximum likelihood (ML) based on Kimura-2-Parameter (K2P) distances was created using MEGA X [70] with 1000 bootstrap replicates. The general time-reversible model plus gamma distribution rate plus evolutionarily invariable model (GTR + G + I) was selected by MEGA X as the best-fitting substitution model based on the Akaike Information Criterion (AIC). Sequences of closely related cartilaginous fish (*Chimaera phantasma*) were used as the outgroup. In this approach, morphological identification was verified based on the clustering of the samples in the phylogenetic tree in comparison to the reference sequences. Furthermore, the genetic distance was calculated to evaluate the usefulness of COI across taxonomic rank. We also examined genetic distances (*p*-distance) within and between species and subspecies. The interindividual distances calculated in MEGA X were sorted into eight inter-rank categories (intraspecific, interspecific, inter-genus, inter-subfamily, inter-family, inter-order, inter-infraclass, and inter-class) based on the species classification in the Eschmeyer's Catalog of Fishes (CAS) [71] and the International Union for Conservation of Nature's (IUCN) [72] Red List of Threatened Species. The mean and ranges of p-distance for these categories were calculated and boxplot was generated using Statistica 7 [73]. In addition, the analysis of variance (ANOVA) supplemented with Tukey's HSD pairwise comparison was performed to determine the variations among inter-rank categories, between infraclass, and among orders of elasmobranchs.

3. Results

A total of 175 individuals belonging to 29 shark and 38 batoid species based on field identification were barcoded in this study. In addition, two individuals of chimaera *Chimaera phantasma* (closely related cartilaginous fishes) were also barcoded. All sequences were uploaded to GenBank, accession numbers: OQ384975–OQ385149 (n = 175). Another 231 sequences from NCBI GenBank were retrieved for phylogenetic tree reconstruction (Supplementary Materials Table S2). These sequences covered 82% of the total elasmobranch species recorded in Malaysia, i.e., there are 29 remaining elasmobranch species in Malaysian waters that have not been barcoded to date.

3.1. Matches with BLAST

Table 1 provides, for each barcoded individual, the percent matches to the best matched species in the NCBI GenBank and the similarity matched in the BOLD database. The results showed that 137 of the 175 samples strongly matched (>99%) with sequences of the same identity only (average 99.86%). The remaining individuals with uncertainty were noted as being matched with different species (13 individuals), low similarity (88.01–98.93%) with the best matched species (15 individuals), matched with multiple species at average genetic distance of 99.7% (17 individuals), or matched with unknown species (1 individual) (see Table 1).

3.2. Tree Matches

The ML tree (Figure 2a–e) matched most of the present sequences to reference sequences forming distinct clades by species (Supplementary Materials Table S3). Some of the exceptions in shark species were listed as follows. Field-identified *S. altipinnis* (samples 2-041 and 2-047) was found to be paraphyletic based on the ML tree; sample 2-041 formed a potential unique clade on its own (1.3–5.2% genetic difference with other highly similar *Squalus* species), while sample 2-047 clustered with the reference sequence of *S. edmundsi* (0.2% genetic distance). *Sphyrna lewini* (samples 2-226, 2-604, 3-225, and T3) formed two distinct clades, with sample 2-604 and T3 branching out from the other sequences, forming a unique clade (3.7–3.8% genetic distance). *Cephaloscyllium sarawakense* (samples 2-013 and 3-640) formed a clade with a reference sequence named *C. umbratile,* with genetic distance ranges from 0.6 to 1.3%. *Carcharhinus leucas* showed two separate clades: samples S1, 2-246, and 2-530 formed a clade with a reference sequence of *C. leucas*, while sample 3-129 formed a clade with a reference sequence of *C. amboinensis*. Sequences of *Carcharhinus amblyrhynchoides* and *C. limbatus* from the present study and the reference sequences used formed two separate clades; however, the genetic distance between the clades was low (0.6–0.8%).

(a)

Figure 2. *Cont.*

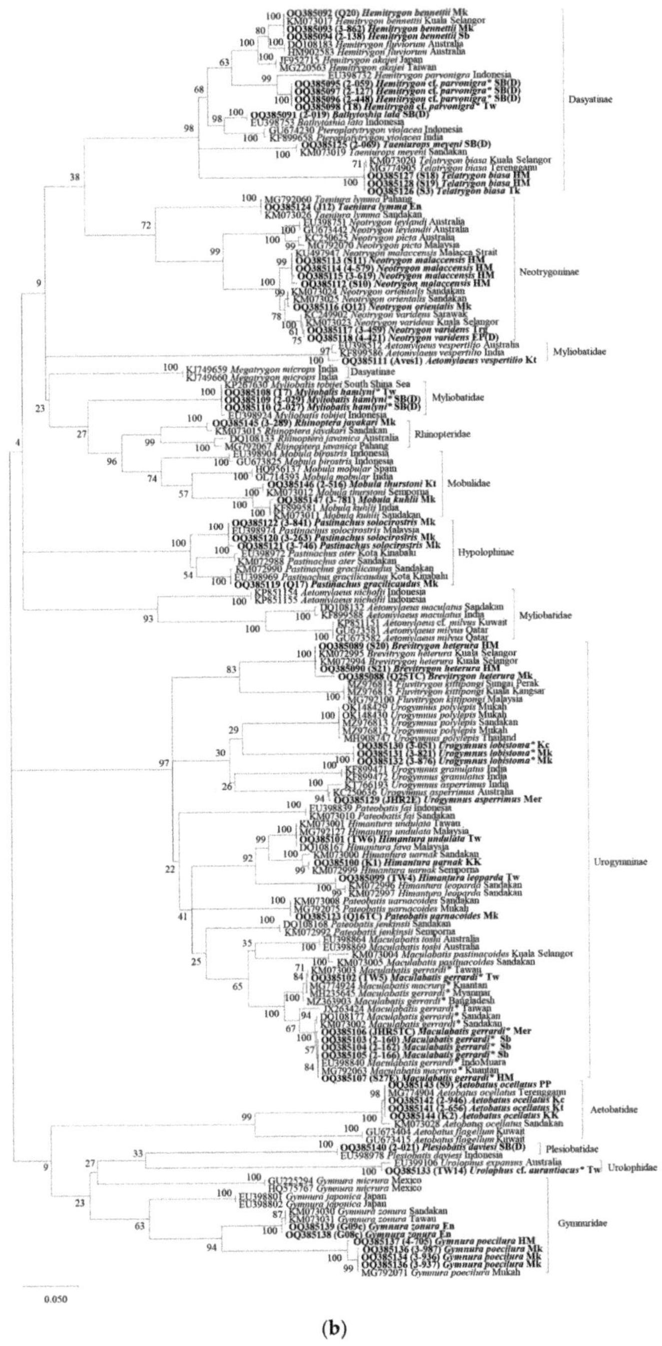

(b)

Figure 2. *Cont.*

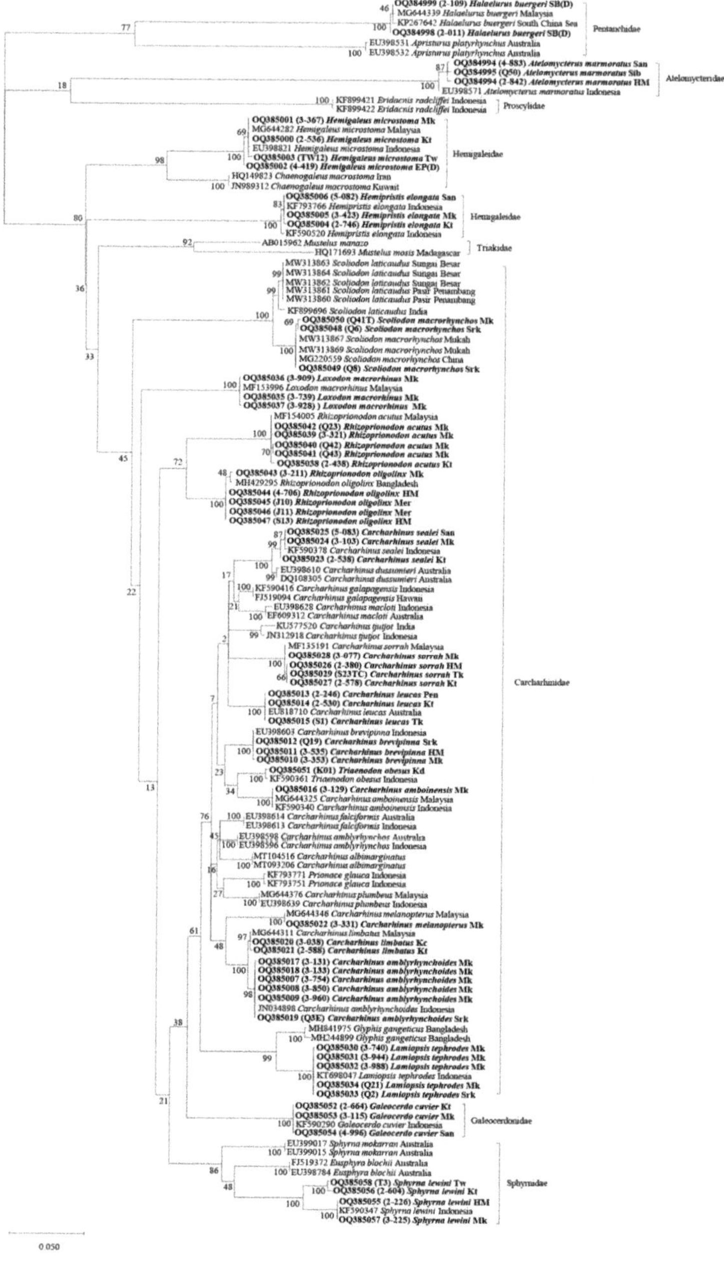

(**c**)

Figure 2. *Cont.*

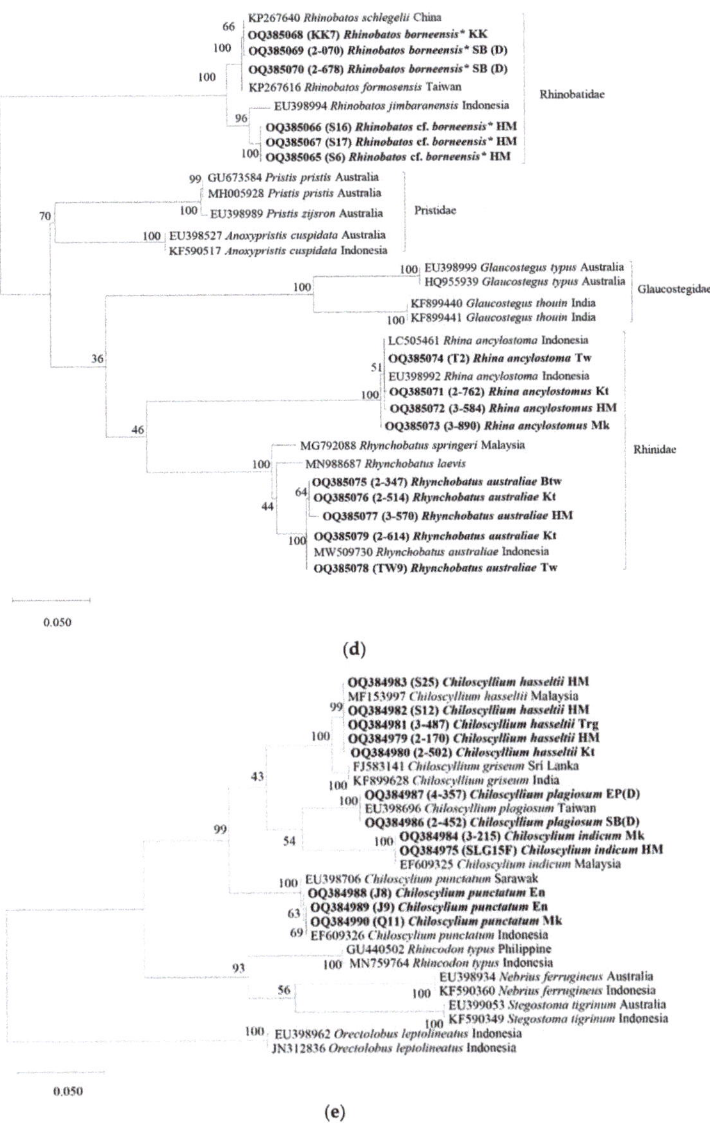

(d)

(e)

Figure 2. Maximum likelihood (ML) mid-point rooting tree based on Kimura-2-Parameter (K2P) distances of COI gene of (**a**) sharks and batoids. (**b**) Myliobatiformes species. (**c**) Carcharhiniformes species. (**d**) Rhinopristiformes species. (**e**) Orectolobiformes species. The bootstrap values (ML) are shown at branches. Sequence names in bold are from the present study. * Indicate specimens of concern. See Table 1 for the abbreviation of locations for sequences from the present study

The exceptions among batoids were listed as follows. *Hemitrygon parvonigra* (samples 2-059, 2-127, 2-448, and T8) was separated from the reference sequence of the species at 5.2% inter-clade genetic distance. The species identity of the *Neotrygon kuhlii* reference sequences found in Malaysian waters were updated to *N. malaccensis*, *N. orientalis,* and *N. varidens*, according to the assigned species names in Borsa et al. [74]. After renaming the references sequences, the *Neotrygon* sequences from the present study (samples S10, S11, Q12, 3-459, 3-619, 4-421, and 4-457) were found to cluster with the reference sequence of their respective species. *Urogymnus lobistoma* recorded in the present study showed a unique clade on its own as separate from other *Urogymnus* reference sequences (genetic distances ranged from 12.1 to 13.7%).

Sequences of *M. gerrardi* and *M. macrura* were paraphyletic showing multiple overlapping clades. The inter-individual genetic distances ranged from 0 to 5.1%. The sequence of *O. boesemani* formed a clade with reference sequences of *O. boesemani* from China (0.9–1.1% genetic distance with present sequences) and *O. cairae* from Vietnam (0.5–0.6% genetic distance with present sequences). Two *N. maculata* barcoded in this study were found to be different genetically and did not form clades with any of the best matched species; the genetic distance between samples 4-439 and *N. maculata* from India was 4.3% while that between samples 2-874 and *N. cf. oculifera* was 1.6%. Two or possibly three separate clades of field identified *Rhinobatos borneensis* were found with samples from Sabah (KK7, 2-070, and 2-678) forming clades with reference sequences named *R. formosensis* and *R. schlegelii* (0 to 0.3% within clade genetic distance), while samples from WP (S6, S16, and S17) were likely a unique clade separated from the aforementioned clade at 3.2 to 4.0% genetic distance and from *R. jimbaranensis* at 2.2% genetic distance. *Myliobatis hamlyni* (samples 2-027, 2-029, and T7) formed a clade with the reference sequence of the species (species name of sequence EU398924 was revised to *M. hamlyni* by White et al. [62]), as well as with *M. tobijei* sequence from South China Sea (0–0.3% genetic distance).

Although the sequences of *C. phantasma* were not the focus of this study, it was worth noting that the sequences obtained were separated from the reference sequences of *C. phantasma* at a genetic distance of 2.7 to 3.8%.

3.3. Final Taxonomic Identification

Considering the results of the BLAST matches and the ML tree, taxonomic identities of 121 individuals that showed consensus using both approaches were clearly verified. Seven samples of *Neotrygon* species were also confirmed after updated species identity of the reference sequences according to Borsa et al. [74]. Six samples of *C. amblyrhynchoides* (3-754, 3-850, 3-960, 3-131, 3-133, and Q3E) and two samples of *C. limbatus* (3-038 and 2-588) were cautiously treated as correctly identified due to the presence of misidentified sequences in NCBI GenBank. This suggests the need to review the identity of all submitted sequences in NCBI GenBank to prevent future confusion. Four samples of *S. lewini* (2-226, 2-604, 3-225, and T3) were also cautiously treated as correctly identified. However, the presence of two distinct clades suggested cryptic species of *S. lewini* that need further taxonomic evaluation. Three samples of *M. hamlyni* (T7, 2-027, and 2-029) were treated as correctly identified, backed by the reference sequence (EU398924) and specimens that were used in *M. hamlyni* redescription [66]. Two individuals, i.e., samples 2-047 and 3-129, were determined to be field misidentifications and were assigned as *S. edmundsi* and *C. amboinensis*, respectively, based on the phylogenetic tree information.

The identities of the remaining 30 individuals remained inconclusive based on the two molecular approaches. Field-identified *S. altipinnis* (2-041) and *U. lobistoma* (3-051, 3-821, and 3-876) were treated as the final species assignment as sequences of these species were not available in NCBI GenBank. The species identity of the two *C. sarawakense* samples was retained, as the morphological characteristics of the specimens did not match those of *C. umbratile*, suggesting a possible error in the submitted *C. umbratile* sequences in NCBI GenBank. Similarly, the identity of the *U. cf. aurantiacus* sample was retained, as the morphological characteristics of the sample were found to be different from the best matched species of *U. expansus*. Species identities of the sequences of three *R. borneensis* (from Sabah), three *O. boesemani*, and six *M. gerrardi* were also retained, as they belong to species complexes that need taxonomic clarification. The three other sequences of *R. borneensis* from WP were treated as *R. cf. borneensis* considering the low similarity with the available reference sequences of the genus. Eight sequences that matched at low similarity levels were each treated as *N. cf. maculata* sp1 (sample 4-439); *N. cf. maculata* sp2 (sample

2-874); *H.* cf. *parvonigra* (sample 2-059, 2-127, 2-448, and T8); and *C.* cf. *phantasma* (2-023 and 2-025).

3.4. Genetic Distance

Pairwise genetic distances were progressively increased as the inter-rank categories increased (Figure 3). These increases were significant ($p < 0.001$)-based overall and pairwise comparisons in the ANOVA and Tukey's HSD tests. However, the range for among inter-rank genetic distances were considered large and overlapped especially between the neighbouring categories.

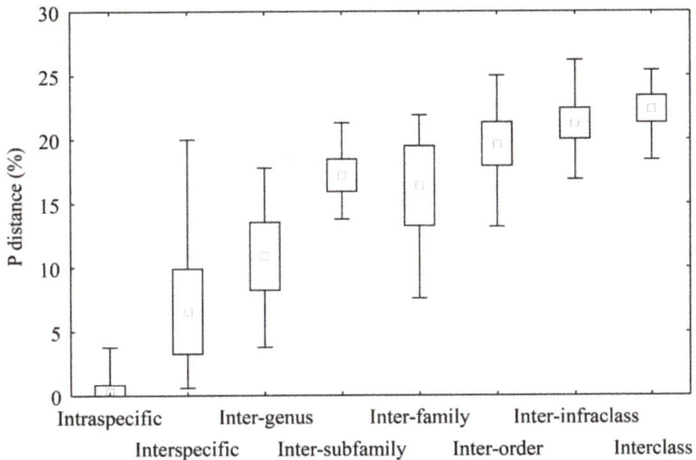

Figure 3. Genetic distance ranges among inter-rank categories. The middle point, box, and whiskers represent the mean, mean ± standard deviation, and total range, respectively.

Comparison across class/infraclass for each of the inter-rank categories is shown in Figure 4. Within intraspecific categories, the genetic distance of Holocephali was significantly higher than the elasmobranchs ($p < 0.001$). The genetic distance within batoids, on the other hand, was significantly higher than sharks for the interspecific, inter-genus, inter-family, and inter-order categories ($p < 0.001$). Comparisons across orders provide further details for the inter-individual distances (Figure 4). Within the intraspecific categories, it showed that most of the genetic distances were in the range from 0 to 1%, except in Hexanchiformes and Chimaeriformes. For interspecific distance, the range for most batoids was found above 8%, except Rhinopristiformes, while below 8% in most of the sharks, except Lamniformes and part of the Orectolobiformes. For the inter-genus category, the genetic distance within Myliobatiformes was significantly higher than the other comparable orders ($p < 0.001$). For inter-family, the genetic distances within Myliobatiformes and Squaliformes were each highest and lowest among the comparable orders ($p < 0.001$). Other significant comparisons can be found between Orectolobiformes and Lamniformes, Orectolobiformes and Rhinopristiformes, and Carcharhiniformes with the others, except Orectolobiformes ($p < 0.05$ or $p < 0.001$).

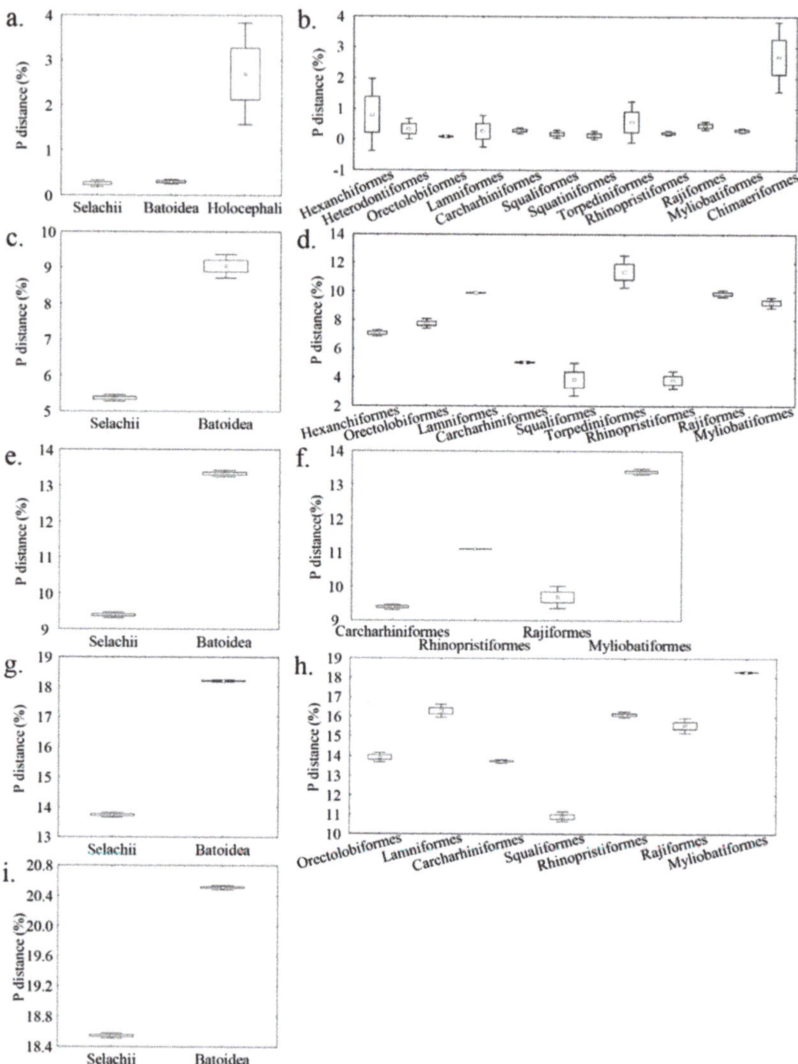

Figure 4. Genetic distances among infraclass (or class) (**a,c,e,g,i**) and order (**b,d,f,h**). (**a,b**) Intraspecific; (**c,d**) interspecific; (**e,f**) inter-genus; (**g,h**) inter-family; and (**i**) inter-order. The middle point, box, and whiskers represent the mean, mean ± standard error, and mean ± 1.96 standard error, respectively.

4. Discussion

A comprehensive DNA barcode reference library is one of the fundamental building blocks allowing application of the increasingly popular environmental DNA (eDNA) metabarcoding approach for the assessment and environmental monitoring of biodiversity [75]. While gaps remain in the current elasmobranch biodiversity reference library, collective efforts, including ours, have advanced barcoding data for both common and rarely occurring freshwater and marine species of sharks, batoids, and skate in Malaysia. Additionally, the field identification and molecular verification efforts added four new records for Malaysia, namely *Squalus edmundsi*, *Carcharhinus amboinensis*, *Alopias superciliosus*, and *Myliobatis hamlyni*. With the exception of *A. superciliosus* that was sampled from WP, the other three species were sampled from Borneo (Sabah and Sarawak). These findings further affirm that the waters of Malaysian Borneo remain understudied in terms of the full biodiversity characterization of local elasmobranchs.

From the collation of available studies, there are 70 species of sharks (19 families) and 91 species of batoids (11 families) recorded in Malaysian waters (Supplementary Materials Table S1). The combined records from various studies had excluded species that were unaccepted in World Register of Marine Species (WORMS) (*Cephaloscyllium circulopullum* and *Narcine indica*) and species that had been verified as absent from Malaysian waters (*Brevitrygon walga* and *B. imbricata* in Last et al. [13]). These records have been improved in terms of classification within the family Dasyatidae [13]; incorporation of the revision of *Telatrygon* species in Malaysian waters [14]; delimitation of cryptic species in *Neotrygon kuhlii* complexes [15,74,76]; new recorded species in Malaysian water, including *Carcharhinus tjutjot* [10], *Fluvitrygon kittipongi*, and *F. oxyrhyncha* [16,17]); and *Scoliodon laticaudus* [35] (previously revised as *S. macrorhynchos* in Last et al. [6]).

Among the new species records, there are two species that are field-misidentified based on their morphological characteristics, namely *S. edmundsi* and *C. amboinensis*. These misidentifications are unsurprising, given how uncommon these two species are worldwide; the former was only described in 2007 in specimens from Australia [77], while the latter is likely often misidentified as highly similar-looking bull shark *C. leucas*, resulting in poor distribution records [78]. The usefulness of the molecular approach is particularly evident in separating these two morphologically similar carcharhinid species that are genetically distinctive. The discovery of spurdog *S. edmundsi* was possible due to the Department of Fisheries Malaysia trawl survey within the deeper EEZ waters off the coast of Borneo. This relatively newly described species is currently known to occur only off the continental waters of Australia and Indonesia [79].

The new finding of the bigeye thresher *A. superciliosus* in a major fisheries landing site in Hutan Melintang, along the central-west coast of Peninsula Malaysia, was rather surprising. This thresher species is thought to occur in deeper waters and generally less catchable than the tropical congener *A. pelagicus* [80], which had only been reported off the open waters of Sabah. The trawl fisheries in the landing site where A. supercilious was found typically operates in the relatively shallow Strait of Malacca. Further investigations into this finding will help shed light on the ecology of the thresher.

For the eagle ray *M. hamlyni*, existing molecular evidence suggested a best match with *M. tobijei* from South China Sea (0% genetic difference) and *M. hamlyni* from Indonesia (0.3% genetic distance). The latest distributional assessment on both species by White et al. [66] showed that *M. tobijei* is restricted to the western North Pacific, while *M. hamlyni* occurs in neighbouring Indonesia and the Philippines. Moreover, the morphological characteristic of the present specimen best matches the species description of *M. hamlyni* in White et al. [66], especially the purplish brown dorsal coloration (compared to yellowish brown in *M. tobijei*), as well as the slightly convex cranial fontanelle that is visible in the dorsal view (nearly straight and broader in *M. tobijei*). More importantly, the *M. hamlyni* specimen corresponded to the reference sequence used in the present analysis (EU398924), i.e., one of the type of specimens in White et al. [66]. This suggests that the reference sequence for *M. tobijei* KP267630 (published in December 2015) was likely an outdated species identity. Prior to the White et al. [66] redescription study, *M. tobijei* was thought to be distributed from Japan down south to Indonesia and the Philippines [66]. The field-identified *U.* cf. *aurantiacus* showed the highest genetic similarity to *U. expansus* found only in Southwestern Australia, but the known distribution of *U. aurantiacus* in the Philippines makes the latter species identity more feasible. The *U.* cf. *aurantiacus* specimen may therefore be an undescribed species, and the significance of this finding is that no urolophids are known to occur within Malaysia to date.

One of the most fundamental requirements in conservation and fishery management of sharks and batoids is correct species identification. While some advances have been made from our study, improvement in the records is still needed, including verification of the 40 recorded species of concern (recorded by single/multiple reference/s by the same research team or absent in Malaysia according to the IUCN assessment). The use of available genetic sequences without careful morphological examination and corroboration from existing distributional records would result in erroneous species assignment. The BOLD species assignment of *Narcine* cf. *oculifera* and *Cephaloscyllium umbratile* are inconclusive, as the percentage match with the present sequences (except sample 3-640) was less than 99%, and they had not been previously recorded in Malaysian waters. Despite the use of multiple approaches, our work highlights the need to focus on select taxonomic groups that may be species complexes or cryptic species. *Sphyrna lewini* (which was also pointed out by Naylor et al. [81]), the *R. borneensis-schlegelli* group, and the *M. gerrardi-macrura* group are among those requiring further investigations that are feasible due to the relatively common occurrence locally from our past surveys. On the other hand, taxonomic work to resolve unclear identities of uncommon species of *S. altipinnis* (2-041); *C. sarawakense* (2-013 and 3-064); *N. maculata* (4-439 and 2-874); and *H. parvonigra* (2-059, 2-448, 2-127, and T8) may be hampered without extensive field samplings.

The use of barcoding approaches is also fundamental in describing species that are still unknown to the scientific communities. This is important to prevent the loss of species before even being discovered, as seen in *Carcharhinus obsolerus* [82]. However, to advance the use of barcoding for accurate biodiversity monitoring, there remains the urgent need for the scientific names in the NCBI database to be updated regularly. One example is the revision of the family Dasyatidae that has resulted in the renaming of *Himantura walga* to *Brevitrygon heterura* (the revised *B. walga* is found to be confined to the Indian Ocean) [13]. Another example is the identity of reference sequence EU398924 as *M. tobijei* in NCBI GenBank, which likely should be *M. hamlyni* after the updated study of White et al. [66]. It is imperative to review the identity of all submitted sequences in NCBI GenBank to prevent future confusion, preferably through collaboration with reputable taxonomist with molecular knowledge to further enhance the validity of the submitted sequences. Moving forward, it is also essential to increase the amount of localized data on individual species, and the correct identification of species through traditional morphology and the uploading of correct sequences for the right species are highly essential to allow full functionality of barcoding technology for species identification.

5. Conclusions

The barcoding of animal tissues has proven to be an important technique to identify species from animal remains that cannot be identified by conventional morphological means and therefore cannot be adequately traced. Accurate species identification will improve species-specific catch landing data and ease and accelerate the identification of their illegal trade and use in Malaysia. Overall, this will provide baseline and monitoring data for conservation and fisheries management. The sharks and batoids CO1 database establishment and improvement will also help facilitate the rapid monitoring and assessment of elasmobranch fish through environmental DNA methods. In the future, rapid and large-scale surveys can be conducted to obtain the distribution points of cartilaginous fishes and predict their potential habitat areas, thereby providing a scientific basis and basis for the identification of elasmobranch conservation priority areas.

Supplementary Materials: The following supporting information can be downloaded at: https://www.mdpi.com/article/10.3390/ani13061002/s1: Table S1: Compilation of elasmobranch records in Malaysia waters, latest IUCN status, and COI sequence availability in NCBI GenBank. Summary represent inferences based on available records. Table S2: The 231 references sequences from NCBI GenBank were retrieved for phylogenetic tree reconstruction. Table S3: Data for the estimates of evolutionary divergence between sequences.

Author Contributions: Designed the study: K.-H.L., K.-C.L., A.Y.-H.T., S.A. and J.D. Performed the field work: K.-H.L., K.-C.L., A.J.-X.L., J.M. and S.A. Conducted the statistical analysis of the data: K.-H.L., K.-C.L. and A.J.-X.L. Conducted the molecular analyses: K.-H.L., K.-C.L. and A.J.-X.L. Captured morphological and meristic data: A.J.-X.L., K.-C.L. and S.A. Wrote the manuscript: K.-H.L.,

A.Y.-H.T. and K.-C.L. Have checked and revised the manuscript: W.H. and A.S. Funding acquisition: K.-H.L., J.D., C.W.B., A.W. and S.A. All authors have read and agreed to the published version of the manuscript.

Funding: This research was funded by the Third Institute of Oceanography, Ministry of Natural Resources, China, grant number IF004-2022. Financial support for the lead author K.-H.L. was provided by the WWF-Malaysia, grant numbers PV049-2019 and PV023-2022; funding for renting the facility instruments for the molecular analyses was provided. The APC was funded by the China-ASEAN Maritime Cooperation fund projects "Marine Protected Areas Network in China-ASEAN Countries" and "Monitoring and Conservation of the coastal ecosystem in the South China Sea"; National Natural Science foundation of China (No: 41961144022). Field sampling was supported by Universiti Malaya grants BK018-2015, RP018C-16SUSC, and IF030B-2019.

Institutional Review Board Statement: Not applicable.

Informed Consent Statement: Not applicable.

Data Availability Statement: All sequence (n = 175) results are uploaded to GenBank, accession numbers: OQ384975–OQ385149.

Acknowledgments: The authors acknowledged Universiti Malaya (UM) and the Institute of Ocean and Earth Sciences (IOES), UM, for providing the necessary facilities. We are grateful to the Department of Fisheries Malaysia and Fisheries Research Institute (FRI) Bintawa, Sarawak, for the opportunity to participate in the fisheries-independent demersal trawl survey within the Malaysian Exclusive Economic Zone (EEZ). A Sabah sampling permit was provided by the Sabah Biodiversity Council (SaBC), Ref. No. JKM/MBS.1000-2/2 JLD.9 (22) and JKM/MBS. 1000-2/3 JLD. 4(18). We thank the anonymous reviewers for their comments that have significantly improved the manuscript.

Conflicts of Interest: The authors declare no conflict of interest.

References

1. Ahmad, A.; Lim, P.K.; Fahmi; Dharmadi; Krajangdara, T. *Identification Guide to Sharks, Rays and Skates of the Southeast Asian Region*; SEAFDEC/MFRDMD/SP/31; SEAFDEC/MFRDMD Institutional Repository: Kuala Terengganu, Malaysia, 2017; 33p.
2. Canton, T. Catalogue of Malaysian fishes. *J. R. Asiat. Soc. Bengal* **1894**, *18*, 983–991.
3. Mohsin, A.K.M.; Ambak, M.A. *Marine Fishes and Fisheries of Malaysia and Neighbouring Countries*; University Pertanian Malaysia Press: Serdang, Selangor, Malaysia, 1996; 744p.
4. Yano, K.; Ahmad, A.; Gambang, A.C.; Idris, A.H.; Solahuddin, A.R.; Aznan, Z. *Sharks and Rays of Malaysia and Brunei Darussalam*; SEAFDEC-MFRDMD/SP/12; SEAFDEC/MFRDMD Institutional Repository: Kuala Terengganu, Malaysia, 2005.
5. Ahmad, A.; Lim, A.P.K.; Abdul Rahman, M.; Raja Bindin, R.H.; Gambang, A.C.; Nor Azman, Z. *Panduan mengenali spesies ikan Yu di Malaysia, Brunei Darussalam, Indonesia dan Thailand. Kuala Terengganu: DPPSPM/DOF*; SEAFDEC/MFRDMD/SP/25; SEAFDEC/MFRDMD Institutional Repository: Kuala Terengganu, Malaysia, 2008; 294p.
6. Last, P.R.; White, W.T.; Caira, J.N.; Jensen, K.; Lim, A.P.; Manjaji-Matsumoto, B.M.; Naylor, G.J.; Pogonoski, J.J.; Stevens, J.D.; Yearsley, G.K. *Sharks and Rays of Borneo*; CSIRO Publishing: Clayton, VIC, Australia, 2010; 298p.
7. Ahmad, A.; Lim, A.P.K. *Field Guide to Sharks of the Southeast Asian Region*; SEAFDEC/MFRDMD/SP/18; SEAFDEC/MFRDMD Institutional Repository: Kuala Terengganu, Malaysia, 2012; 210p.
8. Ahmad, A.; Lim, A.P.; Krajangdara, T. *Field Guide to Rays, Sketes and Chimaeras of the Southeast Asian Region*; SEAFDEC/MFRDMD/SP/25 (84 sp); SEAFDEC/MFRDMD Institutional Repository: Kuala Terengganu, Malaysia, 2014; 289p.
9. Lim, K.C.; Lim, P.E.; Chong, V.C.; Loh, K.H. Molecular and morphological analyses reveal phylogenetic relationships of stingrays focusing on the family Dasyatidae (Myliobatiformes). *PLoS ONE* **2015**, *10*, e0120518.
10. Arai, T.; Azri, A. Diversity, occurrence and conservation of sharks in the southern South China Sea. *PLoS ONE* **2019**, *14*, e0213864. [CrossRef] [PubMed]
11. Booth, H.; Chaya, F.; Ng, S.; Tan, V.; Rao, M.; Teepol, B.; Matthews, E.; Lim, A.; Gumal, M. Elasmobranch fishing and trade in Sarawak, Malaysia, with implications for management. *Aquat. Conserv. Mar. Freshw. Ecosyst.* **2021**, *31*, 3056–3071. [CrossRef]
12. White, W.; Last, P. A redescription of *Carcharhinus dussumieri* and *C sealei*, with resurrection of *C. coatesi* and *C. tjutjot* as valid species (Chondrichthyes: Carcharhinidae). *Zootaxa* **2012**, *3241*, 1–34. [CrossRef]
13. Last, P.R.; Naylor, G.J.; Manjaji-Matsumoto, B.M. A revised classification of the family Dasyatidae (Chondrichthyes: Myliobatiformes) based on new morphological and molecular insights. *Zootaxa* **2016**, *4139*, 345–368. [CrossRef]
14. Last, P.R.; White, W.T.; Naylor, G. Three new stingrays (Myliobatiformes: Dasyatidae) from the Indo-West Pacific. *Zootaxa* **2016**, *4147*, 377–402. [CrossRef]

15. Borsa, P.; Arlyza, I.S.; Hoareau, T.B.; Shen, K.N. Diagnostic description and geographic distribution of four new cryptic species of the blue-spotted maskray species complex (Myliobatoidei: Dasyatidae; *Neotrygon* spp.) based on DNA sequences. *J. Ocean. Limnol* **2018**, *36*, 827–841. [CrossRef]

16. Hasan, V.; Gausmann, P.; Nafisyah, A.L.; Isroni, W.; Widodo, M.S.; Islam, I.; Chaidir, R.R.A. First record of longnose marbled whipray *Fluvitrygon oxyrhyncha* (Sauvage, 1878) (Myliobatiformes: Dasyatidae) in Malaysian waters. *Ecol. Montenegrina* **2021**, *40*, 75–79. [CrossRef]

17. Lim, K.C.; Then, A.Y. Environmental DNA approach complements social media reports to detect an endangered freshwater stingray species in the wild. *Endang. Species Res.* **2022**, *48*, 43–50. [CrossRef]

18. Du, J.; Ding, L.; Su, S.; Hu, W.; Wang, Y.; Loh, K.H.; Yang, S.; Chen, M.; Roeroe, K.A.; Songploy, S.; et al. Setting conservation priorities for marine sharks in China and the Association of Southeast Asian Nations (ASEAN) seas: What are the benefits of 30% conservation target? *Front. Mar. Sci. Mar. Ecosyst. Ecol.* **2022**, *9*, 933291. [CrossRef]

19. Bineesh, K.K.; Gopalakrishnan, A.; Akhilesh, K.V.; Sajeela, K.A.; Abdussamad, E.M.; Pillai, N.G.K.; Basheer, V.S.; Jena, J.K.; Ward, R.D. DNA barcoding reveals species composition of sharks and rays in the Indian commercial fishery. *Mitochondrial DNA Part A* **2017**, *28*, 458–472. [CrossRef]

20. Dulvy, N.K.; Fowler, S.L.; Musick, J.A.; Cavanagh, R.D.; Kyne, P.M.; Harrison, L.R.; Carlson, J.K.; Davidson, L.N.; Fordham, S.V.; Francis, M.P.; et al. Extinction risk and conservation of the world sharks and rays. *eLife* **2014**, *3*, e00590. [CrossRef] [PubMed]

21. Stevens, J.; Bonfi, R.; Dulvy, N.; Walker, P. The effects of fishing on sharks, rays, and chimaeras (chondrichthyans), and the implications for marine ecosystems. *ICES J. Mar. Sci.* **2000**, *57*, 476–494. [CrossRef]

22. Dent, F.; Clarke, S. *State of the Global Market for Shark Products*; FAO Fisheries and Aquaculture Technical Paper, No. 590; FAO: Rome, Italy, 2015; 187p.

23. Department of Fisheries Malaysia (DOFM). 2018. Available online: https://www.dof.gov.my/en/resources/i-extension-en/annual-statistics/ (accessed on 1 January 2023).

24. Department of Fisheries Malaysia (DOFM). 2021. Available online: https://www.dof.gov.my/en/resources/i-extension-en/annual-statistics/ (accessed on 5 January 2023).

25. Pacoureau, N.; Rigby, C.L.; Kyne, P.M.; Sherley, R.B.; Winker, H.; Carlson, J.K.; Fordham, S.V.; Barreto, R.; Fernando, D.; Francis, M.P.; et al. Half a century of global decline in oceanic sharks and rays. *Nature* **2021**, *589*, 567–571. [CrossRef] [PubMed]

26. Department of Fisheries Malaysia (DOFM). *Malaysia National Plan of Action for the Conservation and Management of Shark (Plan 2)*; Ministry of Agriculture and Agrobased Industry Malaysia: Putrajaya, Malaysia, 2014; 50p.

27. Dulvy, N.K.; Pacoureau, N.; Rigby, C.L.; Pollom, R.A.; Jabado, R.W.; Ebert, D.A.; Finucci, B.; Pollock, C.M.; Cheok, J.; Derrick, D.H.; et al. Overfishing drives over one-third of all sharks and rays toward a global extinction crisis. *Curr. Biol.* **2021**, *31*, 4773–4787. [CrossRef]

28. Stobutzki, I.C.; Miller, M.J.; Heales, D.S.; Brewer, D.T. Sustainability of elasmobranchs caught as bycatch in a tropical prawn (shrimp) trawl fishery. *Fishery Bulletin.* **2002**, *100*, 800–821.

29. Mardhiah, U.; Booth, H.; Simeon, B.M.; Muttaqin, E.; Ichsan, M.; Prasetyo, A.P.; Yulianto, I. Quantifying vulnerability of sharks and rays species in Indonesia: Is biological knowledge sufficient enough for the assessment? *IOP Conf. Ser. Earth Environ. Sci.* **2019**, *278*, 012043. [CrossRef]

30. Dulvy, N.K.; Baum, J.K.; Clarke, S.; Compagno, L.J.; Cortés, E.; Domingo, A.; Fordham, S.; Fowler, S.; Francis, M.P.; Gibson, C.; et al. You can swim but you can't hide: The global status and conservation of oceanic pelagic sharks. *Aquat. Conserv. Mar. Freshw. Ecosyst.* **2008**, *18*, 459–482. [CrossRef]

31. Jabado, R.W.; Kyne, P.M.; Pollom, R.A.; Ebert, D.A.; Simpfendorfer, C.A.; Ralph, G.M.; Dhaheri, S.S.; Akholesh, K.V.; Ali, K.; Ali, M.H.; et al. Troubled waters: Threats and extinction risk of the sharks, rays and chimaeras of the Arabian Sea and adjacent waters. *Fish Fish.* **2018**, *19*, 1043–1062. [CrossRef]

32. Tillett, B.J.; Field, I.C.; Bradshaw, C.J.A.; Johnson, G.; Buckworth, R.C.; Meekan, M.G.; Ovenden, J.R. Accuracy of species identification by fisheries observers in a north Australian shark fishery. *Fish. Res.* **2012**, *127–128*, 109–115. [CrossRef]

33. Verissimo, A.; Cotton, C.F.; Buch, R.H.; Guallart, J.; Burgess, G.H. Species diversity of the deep-water gulper sharks (Squaliformes: Centrophoridae: Centrophorus) in North Atlantic waters–current status and taxonomic issues. *Zool. J. Linn. Soc.* **2014**, *172*, 803–830. [CrossRef]

34. Akbar John, B.; Muhamad Asrul, M.A.; Mohd Arshaad, W.; Jalal, K.C.A.; Sheikh, H.I. DNA barcoding of rays from the South China Sea. In *DNA Barcoding and Molecular Phylogeny*; Springer: Cham, Switzerland, 2020; pp. 121–136. ISBN 978-3-030-50074-0.

35. Lim, K.C.; White, W.T.; Then, A.Y.H.; Naylor, G.J.P.; Arunrugstichai, S.; Loh, K.H. Integrated Taxonomy Revealed Genetic Differences in Morphologically Similar and Non-Sympatric *Scoliodon macrorhynchos* and *S. laticaudus*. *Animals* **2022**, *12*, 681. [CrossRef] [PubMed]

36. Cerutti-Pereyra, F.; Meekan, M.G.; Wei, N.-W.V.; O'Shea, O.R.; Bradshaw, C.J.A.; Austin, C.M. Identification of rays through DNA barcoding: An application for ecologists. *PLoS ONE* **2012**, *7*, e36479. [CrossRef]

37. Hebert, P.D.; Cywinska, A.; Ball, S.L.; DeWaard, J.R. Biological identifications through DNA barcodes. *Proc. R. Soc. London. Ser. B Biol. Sci.* **2003**, *270*, 313–321. [CrossRef] [PubMed]

38. Hebert, P.D.N.; Ratnasingham, S.; deWaard, J.R. Barcoding animal life: Cytochrome c oxidase subunit 1 divergences among closely related species. *Proc. R. Soc. Lond. Ser. B Biol. Sci.* **2003**, *270*, S96–S99. [CrossRef] [PubMed]

39. Hubert, N.; Hanner, R.; Holm, E.; Mandrak, N.; Taylor, E.; Burridge, M.; Watkinson, D.; Dumont, P.; Curry, A.; Bentzen, P.; et al. Identifying Canadian Freshwater Fishes through DNA Barcodes. *PLoS ONE* **2008**, *3*, e2490. [CrossRef] [PubMed]

40. Ward, R.D.; Zemlak, T.S.; Innes, B.H.; Last, P.R.; Hebert, P.D.N. DNA barcoding Australia's fish species. *Philos. Trans. R. Soc. London. Ser. B Biol. Sci.* **2005**, *360*, 1847–1857. [CrossRef]

41. Shivji, M.S.; Clarke, S.; Pank, M.; Natanson, L.; Kohler, N.; Stanhope, M. Genetic identification of pelagic shark body parts for conservation and trade monitoring. *Conserv. Biol.* **2002**, *16*, 1036–1047. [CrossRef]

42. Ward, R.D.; Holmes, B.H.; White, W.T.; Last, P.R. DNA barcoding Australasian chondrichthyans: Results and potential uses in conservation. *Mar. Freshw. Res.* **2008**, *59*, 57–71. [CrossRef]

43. Holmes, B.H.; Steinke, D.; Ward, R.D. Identification of shark and ray fins using DNA barcoding. *Fish. Res.* **2009**, *95*, 280–288. [CrossRef]

44. Cardeñosa, D.; Fields, A.; Abercrombie, D.; Feldheim, K.; Shea, S.K.H.; Chapman, D.D. A multiplex PCR mini-barcode assay to identify processed shark products in the global trade. *PLoS ONE* **2017**, *12*, e0185368. [CrossRef] [PubMed]

45. Wainwright, B.J.; Ip, Y.C.; Neo, M.L.; Chang, J.J.; Gan, C.Z.; Clark-Shen, N.; Huang, D.; Rao, M. DNA barcoding of traded shark fins, meat and mobulid gill plates in Singapore uncovers numerous threatened species. *Conserv. Genet.* **2018**, *19*, 1393–1399. [CrossRef]

46. Asbury, T.A.; Bennett, R.; Price, A.S.; da Silva, C.; Burgener, M.; Klein, J.D.; Maduna, S.N.; Sidat, N.; Fernando, S.; Bester-Van der Merwe, A.E. Application of DNA mini-barcoding reveals illegal trade in endangered shark products in southern Africa. *Afr. J. Mar. Sci.* **2021**, *43*, 511–520. [CrossRef]

47. Choy, C.P.P.; Wainwright, B.J. What Is in Your Shark Fin Soup? Probably an endangered shark species and a bit of mercury. *Animals* **2022**, *12*, 802. [CrossRef]

48. Zeng, Y.; Wu, Z.; Zhang, C.; Meng, Z.; Jiang, Z.; Zhang, J. DNA barcoding of mobulid ray gill rakers for implementing CITES on elasmobranch in China. *Sci. Rep.* **2016**, *6*, 37567. [CrossRef]

49. Steinke, D.; Bernard, A.M.; Horn, R.L.; Hilton, P.; Hanner, R.; Shivji, M.S. DNA analysis of traded shark fins and mobulid gill plates reveals a high proportion of species of conservation concern. *Sci. Rep.* **2017**, *7*, 9505. [CrossRef] [PubMed]

50. Asis, A.M.J.M.; Lacsamana, J.K.M.; Santos, M.D. Illegal trade of regulated and protected aquatic species in the Philippines detected by DNA barcoding. *Mitochondrial DNA Part A* **2016**, *27*, 659–666. [CrossRef]

51. Sembiring, A.; Pertiwi, N.P.D.; Mahardini, A.; Wulandari, R.; Kurniasih, E.M.; Kuncoro, A.W.; Cahyani, N.K.D.; Anggoro, A.W.; Ulfa, M.; Madduppa, H.; et al. DNA barcoding reveals targeted fisheries for endangered sharks in Indonesia. *Fish. Res.* **2015**, *164*, 130–134. [CrossRef]

52. Pavan-Kumar, A.; Gireesh-Babu, P.; Suresh Babu, P.P.; Jaiswar, A.K.; Prasad, K.P.; Chaudhari, A.; Raje, S.G.; Chakraborty, S.K.; Krishna, G.; Lakra, W.S. DNA barcoding of elasmobranchs from Indian Coast and its reliability in delineating geographically widespread specimens. *Mitochondrial DNA* **2015**, *26*, 92–100. [CrossRef]

53. Bineesh, K.K.; Akhilesh, K.V.; Sajeela, K.A.; Abdussamad, E.M.; Gopalakrishnan, A.; Basheer, V.S.; Jena, J.K. DNA Barcoding Confirms the Occurrence Rare Elasmobranchs in the Arabian Sea of Indian EEZ. *Middle East. J. Sci. Res.* **2014**, *19*, 1266–1271. [CrossRef]

54. Haque, A.B.; Das, S.A.; Biswas, A.R. DNA analysis of elasmobranch products originating from Bangladesh reveals unregulated elasmobranch fishery and trade on species of global conservation concern. *PLoS ONE* **2019**, *14*, e0222273. [CrossRef] [PubMed]

55. Kuguru, G.; Maduna, S.; da Silva, C.; Gennari, E.; Rhode, C.; Bester-van der Merwe, A. DNA barcoding of chondrichthyans in South African fisheries. *Fish. Res.* **2018**, *206*, 292–295. [CrossRef]

56. Hobbs, C.A.D.; Potts, R.W.A.; Walsh, M.B.; Usher, J.; Griffiths, A.M. Using DNA Barcoding to investigate patterns of species utilisation in UK shark products reveals threatened species on sale. *Sci. Rep.* **2019**, *9*, 1028. [CrossRef]

57. Cariani, A.; Messinetti, S.; Ferrari, A.; Arculeo, M.; Bonello, J.J.; Bonnici, L.; Cannas, R.; Carbonara, P.; Cau, A.; Charilaou, C.; et al. Improving the conservation of Mediterranean chondrichthyans: The Elasmomed DNA barcode reference library. *PLoS ONE* **2017**, *12*, e0170244. [CrossRef] [PubMed]

58. Serra-Pereira, B.; Moura, T.; Griffiths, A.M.; Gordo, L.S.; Figueiredo, I. Molecular barcoding of skates (Chondrichthyes: Rajidae) from the southern northeast Atlantic. *Zool. Scr.* **2011**, *40*, 76–84. [CrossRef]

59. Almerón-Souza, F.; Sperb, C.; Castilho, C.L.; Figueiredo, P.I.; Gonçalves, L.T.; Machado, R.; Oliveira, L.R.; Valiati, V.H.; Fagundes, N.J. Molecular identification of shark meat from local markets in Southern Brazil based on DNA barcoding: Evidence for mislabeling and trade of endangered species. *Front. Genet.* **2018**, *9*, 138. [CrossRef]

60. Feitosa, L.M.; Martins, A.P.; Giarrizzo, T.; Macedo, W.; Monteiro, I.L.; Gemaque, R.; Nunes, J.L.; Gomes, F.; Schneider, H.; Sampaio, I.; et al. DNA-based identification reveals illegal trade of threatened shark species in a global elasmobranch conservation hotspot. *Sci. Rep.* **2018**, *8*, 3347. [CrossRef]

61. Bernardo, C.; de Lima Adachi, A.M.C.; da Cruz, V.P.; Foresti, F.; Loose, R.H.; Bornatowski, H. The label "Cação" is a shark or a ray and can be a threatened species! Elasmobranch trade in southern Brazil unveiled by DNA barcoding. *Mar. Policy* **2020**, *116*, 103920. [CrossRef]

62. Ratnasingham, S.; Hebert, P.D.N. BOLD: The Barcode of Life Data System (www.barcodinglife.org). *Mol. Ecol. Notes* **2007**, *7*, 355–364. [CrossRef]

63. Last, P.R.; White, T.W.; Carvalho, M.R.; Seret, B.; Stehmann, N.F.W.; Naylor, G.J.P. *Rays of the World*; CSIRO Publishing: Melbourne, Australia, 2016.

64. Carpenter, K.E.; Niem, V.H. FAO species identification guide for fishery purposes. In *The Living Marine Resources of the Western Central Pacific*; Cephalopods, crustaceans, holothurians and sharks; FAO: Rome, Italy, 1998; Volume 2, pp. 687–1396.

65. Carpenter, K.E.; Niem, V.H. FAO species identification guide for fishery purposes. In *The living marine resources of the Western Central Pacific*; Batoid fishes, chimaeras and bony fishes part 1 (Elopidae to Linophrynidae); FAO: Rome, Italy, 1999; Volume 3, pp. 1397–2068.

66. White, W.T.; Kawauchi, J.; Corrigan, S.; Rochel, E.; Naylor, G.J. Redescription of the eagle rays *Myliobatis hamlyni* Ogilby, 1911 and *M. tobijei* Bleeker, 1854 (Myliobatiformes: Myliobatidae) from the East Indo-West Pacific. *Zootaxa* **2015**, *3948*, 521–548. [CrossRef]

67. Hyde, J.; Lynn, E.; Humphreys, R., Jr.; Musyl, M.; West, A.; Vetter, R. Shipboard identification of fish eggs and larvae by multiplex PCR, and description of fertilized eggs of blue marlin, shortbill spearfish, and wahoo. *Mar. Ecol. Prog. Ser.* **2005**, *286*, 269–277. [CrossRef]

68. Rigby, C.; Appleyard, S.; Chin, A.; Heupel, M.; Humber, F.; Simpfendorfer, C.; White, W.; Campbell, I. *Rapid Assessment Toolkit for Sharks and Rays*; WWF International and CSTFA, James Cook University: Townsville, Australia, 2019.

69. Hall, T.A. BioEdit: A user-friendly biological sequence alignment editor and analysis program for Windows 95/98/NT. *Nucl. Acids. Symp. Ser.* **1999**, *41*, 95–98.

70. Kumar, S.; Stecher, G.; Li, M.; Knyaz, C.; Tamura, K. MEGA X: Molecular Evolutionary Genetics Analysis across computing platforms. *Mol. Biol. Evol.* **2018**, *35*, 1547–1549. [CrossRef] [PubMed]

71. Fricke, R.; Eschmeyer, W.N.; Van der Laan, R. (Eds.) Eschmeyer's Catalog of Fishes: Genera, Species, References. 2022. Available online: http://researcharchive.calacademy.org/research/ichthyology/catalog/fishcatmain.asp (accessed on 2 February 2022).

72. IUCN. The IUCN Red List of Threatened Species. 2022. Version 2022-2. Available online: https://www.iucnredlist.org (accessed on 5 February 2023).

73. StatSoft, Inc. STATISTICA (Data Analysis Software System). 2007. Version 7.0. Available online: https://www.statsoft.com (accessed on 5 February 2023).

74. Borsa, P.; Shen, K.N.; Arlyza, I.S.; Hoareau, T.B. Multiple cryptic species in the blue-spotted maskray (Myliobatoidei: Dasyatidae: *Neotrygon* spp.): An update. *Comptes Rendus Biol.* **2016**, *339*, 9–10, 417–426. [CrossRef]

75. Valentini, A.; Taberlet, P.; Miaud, C.; Civade, R.; Herder, J.; Thomsen, P.F.; Bellemain, E.; Besnard, A.; Coissac, E.; Boyer, F.; et al. Next-generation monitoring of aquatic biodiversity using environmental DNA metabarcoding. *Mol. Ecol.* **2016**, *25*, 929–942. [CrossRef]

76. Last, P.R.; White, W.T.; Séret, B. Taxonomic status of maskrays of the *Neotrygon kuhlii* species complex (Myliobatidei: Dasyatidae) with the description of three new species from the Indo-West Pacific. *Zootaxa* **2016**, *4083*, 533–561. [CrossRef]

77. White, W.T.; Last, P.R.; Stevens, J.D. Part 7—Two new species of *Squalus* of the *mitsukurii* group from the Indo-Pacific. *CSIRO Mar. Atmos. Res. Pap.* **2007**, *14*, 71–81.

78. Spaet, J.L.; Cochran, J.E.; Berumen, M.L. First record of the pigeye shark, *Carcharhinus amboinensis* (Müller & Henle, 1839) (Carcharhiniformes: Carcharhinidae), in the Red Sea. *Zool. Middle East* **2011**, *52*, 118–121.

79. Finucci, B.; White, W.T. *Squalus edmundsi*. The IUCN Red List of Threatened Species 2019: E.T158617A68644689. Available online: https://www.iucnredlist.org/species/158617/68644689 (accessed on 5 February 2023).

80. Rigby, C.L.; Barreto, R.; Carlson, J.; Fernando, D.; Fordham, S.; Francis, M.P.; Herman, K.; Jabado, R.W.; Liu, K.M.; Marshall, A.; et al. Alopias superciliosus. The IUCN Red List of Threatened Species 2019: E.T161696A894216. Available online: https://www.iucnredlist.org/species/161696/894216 (accessed on 5 February 2023).

81. Naylor, G.J.P.; Caira, J.N.; Jensen, K.; Rosana, K.A.M.; White, W.T.; Last, P.R. A DNA sequence–based approach to the identification of shark and ray species and its implications for global elasmobranch diversity and parasitology. *Bull. Am. Mus. Nat. Hist.* **2012**, *367*, 1–262. [CrossRef]

82. White, W.T.; Kyne, P.M.; Harris, M. Lost before found: A new species of whaler shark *Carcharhinus obsolerus* from the Western Central Pacific known only from historic records. *PLoS ONE* **2019**, *14*, e0209387. [CrossRef] [PubMed]

Article

Aggregative Behaviour of Spiny Butterfly Rays (*Gymnura altavela*, Linnaeus, 1758) in the Shallow Coastal Zones of Gran Canaria in the Eastern Central Atlantic

Ana Espino-Ruano [1,*], Jose J. Castro [1], Airam Guerra-Marrero [1,*], Lorena Couce-Montero [1], Eva K. M. Meyers [2], Angelo Santana-del-Pino [3] and David Jimenez-Alvarado [1]

[1]   Biodiversidad y Conservación, IU-ECOAQUA, University of Las Palmas de Gran Canaria, Edf. Ciencias Básicas, Campus Universitario de Tafira, 35017 Las Palmas de Gran Canaria, Spain; lorena.couce@ulpgc.es (L.C.-M.)
[2]   Zoological Research Museum Alexander Koenig, 53113 Bonn, Germany
[3]   Department of Mathematics, University of Las Palmas of Gran Canaria, 35018 Las Palmas, Spain
*   Correspondence: anamaria.espino@ulpgc.es (A.E.-R.); airam.guerra@ulpgc.es (A.G.-M.)

Simple Summary: The presence of batoids in Gran Canaria, Canary Islands, Spain, in the Mid-Eastern Atlantic Ocean, is quite common depending on the species and time of year. For that reason, we examined the behaviour of spiny butterfly rays (*Gymnura altavela*) in the shallow waters of Gran Canaria, where the species' affinity to certain beaches was analysed according to the time of year and preference for the type of ocean environment. Such knowledge is important given the lack of information available on the species that is nevertheless vital for its sufficient management and conservation.

Citation: Espino-Ruano, A.; Castro, J.J.; Guerra-Marrero, A.; Couce-Montero, L.; Meyers, E.K.M.; Santana-del-Pino, A.; Jimenez-Alvarado, D. Aggregative Behaviour of Spiny Butterfly Rays (*Gymnura altavela*, Linnaeus, 1758) in the Shallow Coastal Zones of Gran Canaria in the Eastern Central Atlantic. *Animals* **2023**, *13*, 1455. https://doi.org/10.3390/ani13091455

Academic Editor: Martina Francesca Marongiu

Received: 1 March 2023
Revised: 16 April 2023
Accepted: 18 April 2023
Published: 25 April 2023

Abstract: The presence of spiny butterfly rays, *Gymnura altavela*, in waters less than 20 m deep off the Canary Islands shows marked seasonality, with relatively high abundances in the summer and autumn. Large aggregations of sometimes hundreds of individuals, primarily females, appear in specific shallow areas of the archipelago and seem to be associated with the seasonal variation in water temperature. This seasonal pattern of presence or absence in shallow areas suggests that spiny butterfly rays migrate into deeper waters or other unknown areas during the rest of the year. *G. altavela* shows sexual dimorphism; in our study, females were larger and more abundant than males, with a sex ratio of 1:18.9. The species' estimated asymptotic length, L_∞, was 183.75 cm and thus close to the common length reported for the species (200 cm). The von Bertalanffy growth constant (k) oscillated between 0.210 and 0.310 year^{-1}, as similarly described for the species in the Western North Atlantic off the U.S. coast. From June to November, the seawater temperature oscillated between 19 and 24 °C, and massive aggregations of females occurred at 22–24 °C and in a few specific sandy beaches on the islands. Spiny butterfly rays, mostly females, show a preference for aggregating in shallow waters during summertime, probably conditionate to mating or breeding behaviour.

Keywords: *Gymnura altavela*; butterfly ray; visual census; ecology; elasmobranchs; Canary Islands

1. Introduction

Worldwide, the current situation of chondrichthyans is critical. A quarter of the species have been confirmed to be threatened, and some have lost more than 70% of their population since the 1970s [1,2]. In this context of population decline, elasmobranchs play an important role in the regulation and structuring of the marine ecosystem as key predators [3–5]. Some studies have suggested that the decimation of a predator population changes the relative abundance of the prey population, which consequently suggests that elasmobranchs are an important element in the marine food web, one whose reduction can initiate trophic cascades via top-down effects [3,6–9].

In general, knowledge on the biology and ecology of elasmobranchs is sorely limited, primarily because many of the species are large, migrate rapidly, range widely in depth, have low commercial value, and are challenging to keep in captivity [10]. Even within such knowledge, to date, information about elasmobranchs is highly heterogeneous [11]. It also rarely concerns deep-sea sharks and most rays and skates, which has made the status of these species' populations rather uncertain [1,12–15].

Most such species of rays and skates inhabit coastal waters and are subject to significant anthropogenic pressure, hence their high representation among threatened species [1,16]. According to the Food and Agriculture Organization of the United Nations (2014) [17], global catches of the species decreased by nearly 20% between 2003 and 2012 [18]; however, that decrease may be far higher than indicated, for catches of the species are generally discarded, unreported, or misidentified [1]. According to Dulvy et al. (2000) [19], the decline in elasmobranchs is clearly reflected in the structure of bentho-demersal ecosystems. However, there is not enough information about the life cycles and parameters related to the distribution, longevity, reproductive periods, or age and size at maturity of many of the species to be able to realistically assess the status of their populations [20]. Regarding such data limitations and needs, the Scientific, Technical, and Economic Committee for Fisheries (2017) [3,21] has noted that the current data requirements and sampling efforts of the EU fisheries data collection programme on rays and skates cannot sufficiently provide robust estimates of various parameters (e.g., maturity, commercial catch composition, sex ratios, and indices of abundance) necessary for stock assessment and management. A case in point is the spiny butterfly ray, *Gymnura altavela* [22], a seasonally common ray in the shallow waters of the Canary Islands [23,24] that has been classified as critically endangered in the Mediterranean and Europe as well as endangered worldwide [25–27]. Nevertheless, the lack of data from other areas within the range of *G. atlavela* makes it difficult to define the conservation status of its population [27].

As a species of the Gymnuridae family, *G. altavela* exhibits demersal behaviour and inhabits the sandy, muddy bottom of shallow waters in tropical temperate regions [28] on both sides of the Atlantic Ocean, including in the Mediterranean and Black Seas [28–31]. The typical disc width of the largest species of the Gymnuridae family is 200 cm, and it can reach 60 kg in weight [32,33]. The coloration on its dorsal side allows the species to camouflage itself with the sandy bottom, where it spends a great part of the day resting [34].

Within the Canary Islands in the Eastern Central Atlantic, the spiny butterfly ray is as common as other benthic elasmobranchs [24,35,36] and, at the population level, seems to have a much better conservation status than in other European waters, including the Mediterranean Sea [37]. Two possible explanations for the species' seemingly favourable conservation status in the Canary Island archipelago are that it is not targeted by local artisanal fisheries and that trawling activity has been banned in the archipelago since 1986. Although no fisheries there target any ray species, some such species are nevertheless accidentally caught in the gillnets and handlines of artisanal and recreational fisheries [38–41].

Given the limited ecological and biological information that can be obtained using fishery-related surveys, citizen science programmes afford a valuable opportunity to provide insights into the ecological distribution and population structure of critically endangered species. A good example of the importance of citizen science programmes as a source of information is the conservation programme for the angel shark, *Squatina squatina*, in the Canary Islands [35,42]. Even so, the information provided by such programmes has many limitations for highly migratory species, including the periods that they can be observed in shallow coastal waters when sighted by divers. In this case, visual underwater censuses have emerged as an effective method of generating detailed biological, ecological, and behavioural data that can be used in management and conservation programmes [43]. Among the most highlighted behavioural patterns of several ray species, seasonal massive aggregations reported in many parts of the world make them highly vulnerable to human coastal activities (e.g., tourism and pollution). The environmental and/or social reasons behind this behaviour nevertheless remain somewhat unclear [44,45].

Against that background, in our study, we aimed to generate knowledge about the seasonal presence of spiny butterfly rays in shallow coastal areas. We especially sought data for an a priori assessment of the population structure and status of the species in the Canary Islands region that could be useful to create a baseline for the development of effective conservation measures for such a highly vulnerable fish species.

2. Material and Methods

2.1. Data Collection

Our study was conducted around the island of Gran Canaria, located in the centre of the Canary Island archipelago, Spain, in the mid-eastern Atlantic Ocean. Biological and behavioural data of spiny butterfly rays were obtained via visual census conducted from March 2018 to March 2020 near four different beaches (Sardina del Norte, Salinetas, El Cabrón, and Pasito Blanco) on the north-western, eastern, south-eastern, and southern sides of the island, respectively (Figure 1).

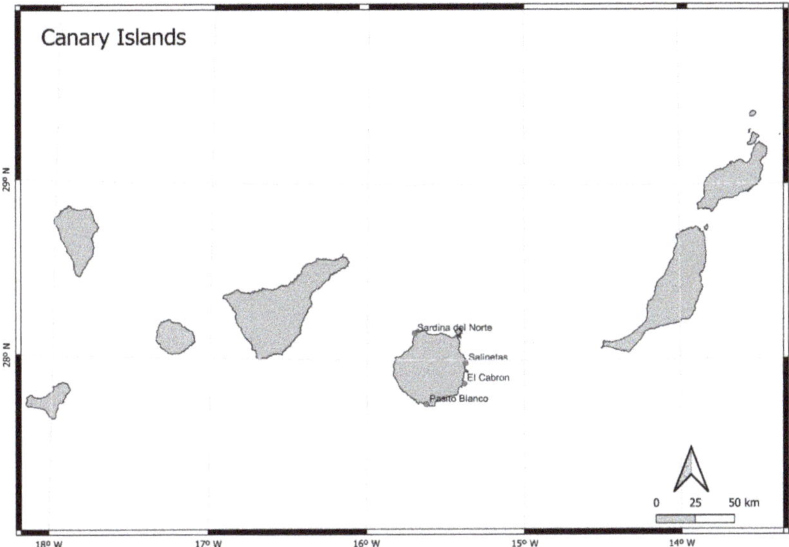

Figure 1. Geographical situation of the Canary Islands and the sampled areas (i.e., Sardina del Norte, Salinetas, El Cabrón, and Pasito Blanco).

In total, 108 visual censuses—27 surveys at each sampling site—were conducted at the different beaches over a 2 year period. Due to differences in the topographic and bathymetric characteristics of the sampling sites, visual censuses at Salinetas and Pasito Blanco, which have sandy bottoms with a low slope and high turbidity, were conducted at 0–5 m deep, whereas visual censuses at Sardina del Norte and El Cabrón beaches were conducted at 5–20 m deep due to the high slope and sandy bottoms in deeper waters.

Visual censuses were conducted by three divers using snorkelling equipment in areas 0–5 m deep and diving equipment in deeper areas, all following the method described by Labrosse et al. (2002) [46]. The researchers lined up at a maximum distance of 5 m from each other, a distance that could vary due to good or bad visibility in the different areas. The described areas were sampled in their entirety, which allowed obtaining precise data about the number of individuals of each species present in each area. During each survey, all individual spiny butterfly rays found in the area were recorded, sexed, and measured in terms of total length (i.e., from the tip of the nose to the end of the tail) and disc width (i.e., along the maximal axis), both in cm, estimated by approaching the rays with a tape measure as they rested on the bottom. We also recorded whether individuals were alone, aggregated, resting, feeding, swimming, and/or showed escape behaviour or any other characteristics that could be useful for the study, including body marks. Different environmental data (e.g., seawater temperature, turbidity, sea conditions, and, if present, companion species) were recorded as well (Table 1).

Table 1. Data provided by observers through the website.

Information
Diving centre or person who reported the sighting
Ability to distinguish the species
Date and location of sighting
Date or season of observation
Approximate length
Sex
Presence of offspring
Number of individuals observed
Aggregated or solitary status
Behaviour (e.g., resting, hunting or mating)
Body marks or injuries (e.g., by hook or spear gun)
Ability to reidentify the reported individuals due to those body marks

Added to the data recorded by the divers during the visual censuses, data concerning spiny butterfly ray sightings were obtained from the databases of the citizen science programmes Poseidon, at the University of Las Palmas de Gran Canaria (www.programaposeidon.eu, accessed on 17 September 2018), and RedPROMAR, within the government of the Canary Islands (https://redpromar.org/sightings?region_id=3, accessed on 15 October 2021), which collects marine biodiversity monitoring data for the Canary Islands. Whereas RedPROMAR provides data from sightings from 2015 to the present, Poseidon provides such data representing 2014–2018. Both databases include the number of individuals observed and the position of the sighting. In sum, we obtained data from 434 recorded sightings of *G. altavela*.

2.2. Data Analysis

Data obtained from the visual censuses were analysed using R version 4.2.2 (R Core Team, 2022). Generalised additive models (GAMs) for count data (i.e., Poisson and negative binomial models), including zero-inflated models, were considered in modelling the number of observed individuals. The Akaike information criteria (AIC) was used to select the model that best fit the data, which was found to be the zero-inflated Poisson GAM. The model assumes that the probability that no ray individuals are observed (or arrive) at a beach is p, while the number of rays arriving at the beach, b_i, in month m follows a truncated Poisson distribution of parameter μ (i.e., $P(N = k) = \frac{p\mu^k}{(\mu^k - 1)k!}$), a function of the given month and beach calculated as:

$$\mu = \mu(m, b_i) = exp$$

such that $I(b_i) = 1$ if the data comes from beach b_i and 0 otherwise.

Data obtained from both years sampled were standardised for joint analysis and compiled as the number of individuals recorded per square metre per minute of sampling. The data series presented a non-normal distribution (Kolmogorov–Smirnov test, $p < 0.05$); thus, nonparametric methods were used in data analyses. The growth parameters were also estimated using the von Bertalanffy growth function (VBGF; von Bertalanffy, 1938) in the R package TropFishR [47]. Growth parameters were analysed using different optimisation techniques as a function of length frequencies (i.e., 10 cm class interval), the data for which were collected monthly. The asymptotic length and the von Bertalanffy growth rate were calculated using the ELEFAN model with simulated annealing and a genetic algorithm, both as approaches to optimise the model. In the same way, the ELEFAN model with the best score fit (i.e., high R_n) was selected. Based on the Powell–Wetherall method described by Wetherall et al. (1987) [48], the asymptotic length, $L\infty$, was estimated.

3. Results

The information available on the geographical distribution of spiny butterfly rays in the Canary Islands, according to data extracted from sighting programmes provided by citizen science databases, is highly conditioned by the assiduity of divers at certain dive sites. For that reason, the entire visual census was conducted in four areas described in citizen science data. Figure 2 shows the coastal areas of the islands where sightings are most frequent, which possibly coincide with areas of the islands where the most frequent sightings occur and/or where massive aggregations of individuals of the species occur.

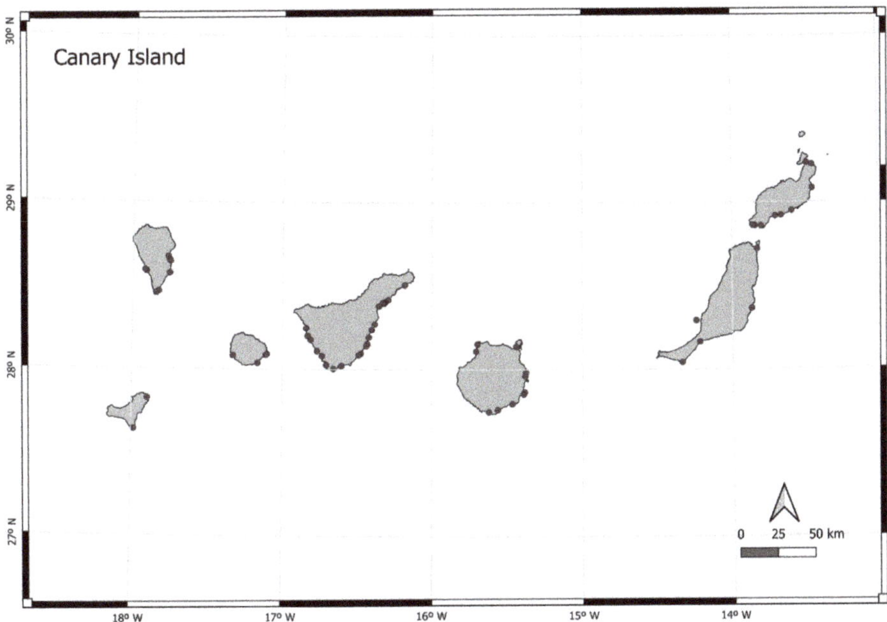

Figure 2. Map of the geographical situation of the Canary Islands, with areas where spiny butterfly rays were represented by grey dots.

The highest number of spiny butterfly rays per square metre was recorded at Pasito Blanco beach, at the leeward side of Gran Canaria (Kruskal–Wallis ANOVA, H = 60.794, $p = 0.047$; Figure 3). Several GAMs were fitted to the number of individuals observed per beach and per month, but the zero-inflated Poisson GAM fit the data best (AIC = 416.9). This model shows an annually cycled pattern of the number of individual spiny butterfly rays, with aggregations occurring from June to November and peaking in September (Figure 4). This pattern in the seasonal presence of individual spiny butterfly rays in the extremely shallow waters of some beaches seems to be influenced by the variation in seawater temperature (Spearman's correlation, $r = 0.81$; $p > 0.0001$, Table 2). Rays began arriving at the shallow beaches when the seawater reached 19 °C, i.e., at the beginning of summer (i.e., in June), and the number of individuals aggregating increased as the temperature reached 22–24 °C. Likewise, the aggregation decreased as the seawater decreased in temperature. Individuals began leaving the areas when the temperature fell below 22 °C in mid-autumn, i.e., at the end of October (Figure 5).

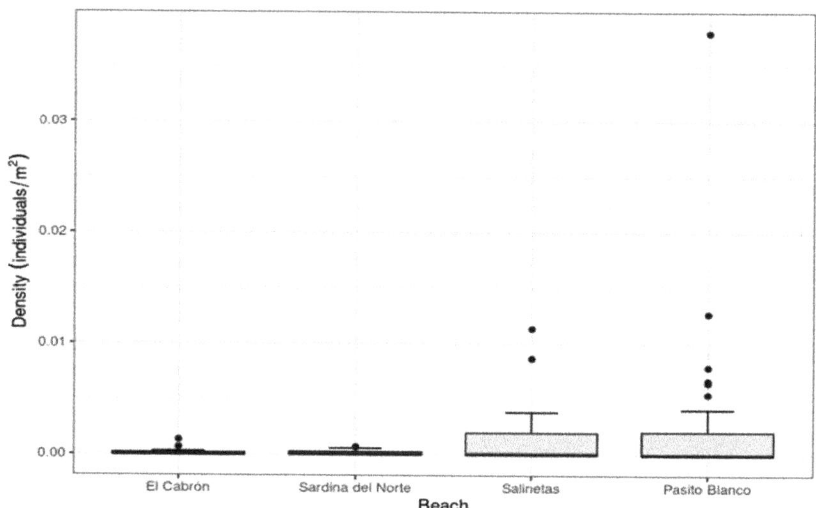

Figure 3. Density of spiny butterfly rays per unit of area at each beach surveyed. Black dots (out layers) represent massive aggregations out of the mean and standardized deviation.

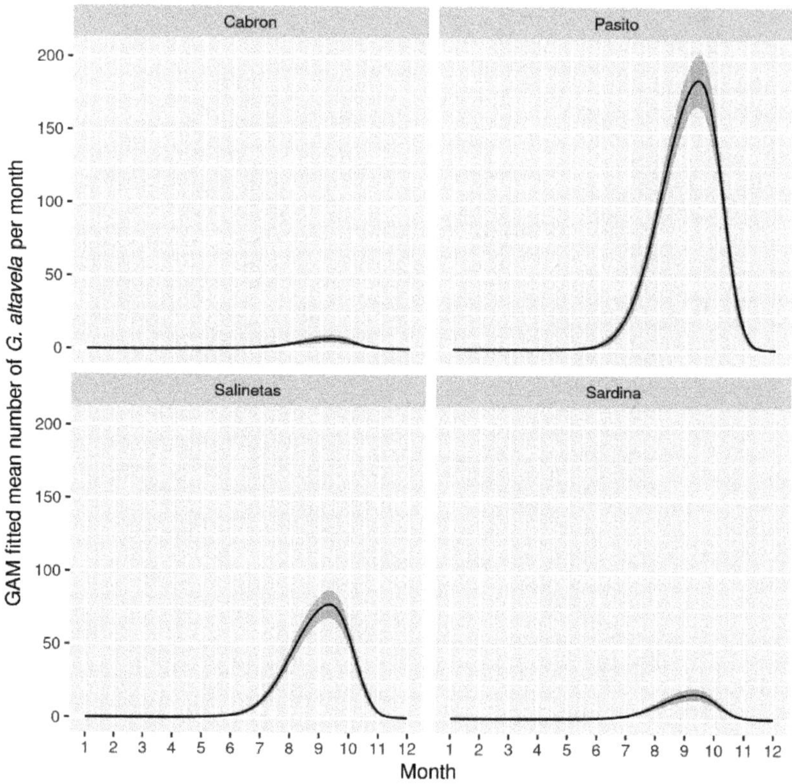

Figure 4. *Gymnura altavela*'s annual cycle pattern, drawn from a zero-inflated Poisson GAM (AIC = 416.9). Grey area (out layers) represent massive aggregations out of the fitted mean.

Table 2. Correlation between the number of individual spiny butterfly rays and mean temperature.

Correlation	Var1	Var2	Corr.	*p*	Conf. Low	Conf. High
Pearson	Temperature	N	0.86	0.0000000988	0.6906081	0.9359712
Spearman	Temperature	N	0.81	0.0000017800		

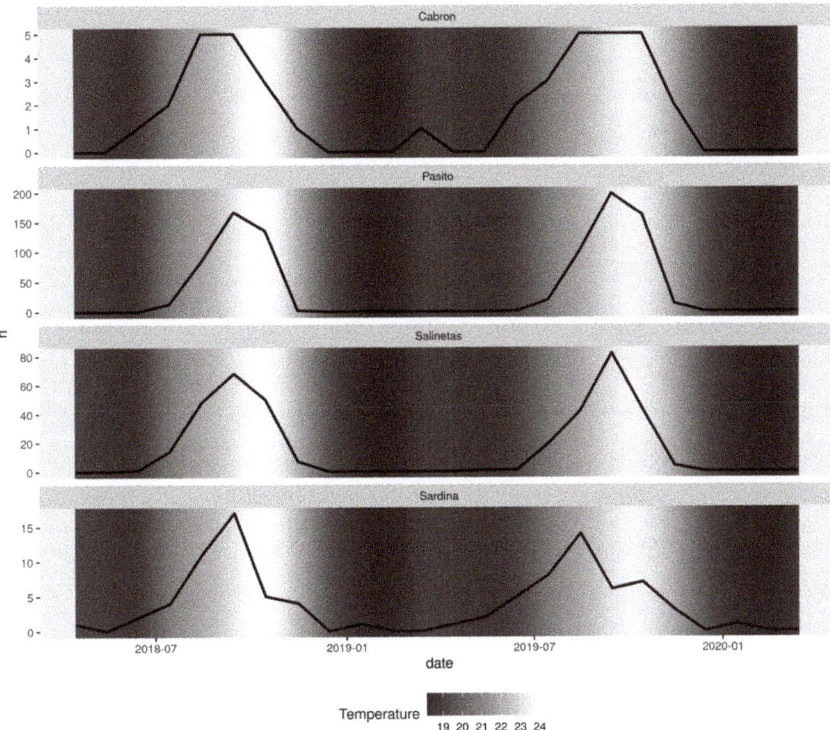

Figure 5. Number of spiny butterfly rays per month and year in relation to mean temperature (Spearman's correlation = 0.81, *p* > 0.0001).

The information collected shows that during these aggregations, spiny butterfly rays are distributed in shallow waters 0–20 m deep, but the highest abundances were reported between 0 and 5 m deep, with decreases in the number of individuals as the depth increased (Kruskal–Wallis ANOVA, $H = 12.995$, $p = 0.001$; Figure 6).

Most individuals (88%) were observed while buried, resting on the sand, and very close to each other, with some partly on top of other rays and piled up in a relatively short area (e.g., approx. 10% of the total sandy area of the beaches). Few rays (12%) were observed swimming and only one individual was observed feeding.

Of the 1424 individuals counted during the sampling period, females predominated significantly, with a female-to-male ratio of 1:18.9 (χ^2 test = 786.08, $p < 0.0001$). Males were significantly smaller than females (Mann–Whitney U test= 803.00, $Z = -12.1448$, $p < 0.0001$, Figure 7) and primarily observed at Salinetas to east of the island, although a few were found at Sardina del Norte. No males were reported at Pasito Blanco to the south, where the vast majority of females were observed. At the same time, significant differences in the frequency distribution of length were observed in different areas (Kruskal–Wallis ANOVA, $H = 215.955$, $p < 0.0001$, Figure 8a). The mean body size also varied between males (Kruskal–Wallis ANOVA, $H = 12.532$, $p = 0.0019$, Figure 8b) and females (Kruskal–Wallis

ANOVA, $H = 128.616$, $p < 0.0001$, Figure 8c) from beach to beach, with the largest at Pasito Blanco and the smallest at Salinetas on average.

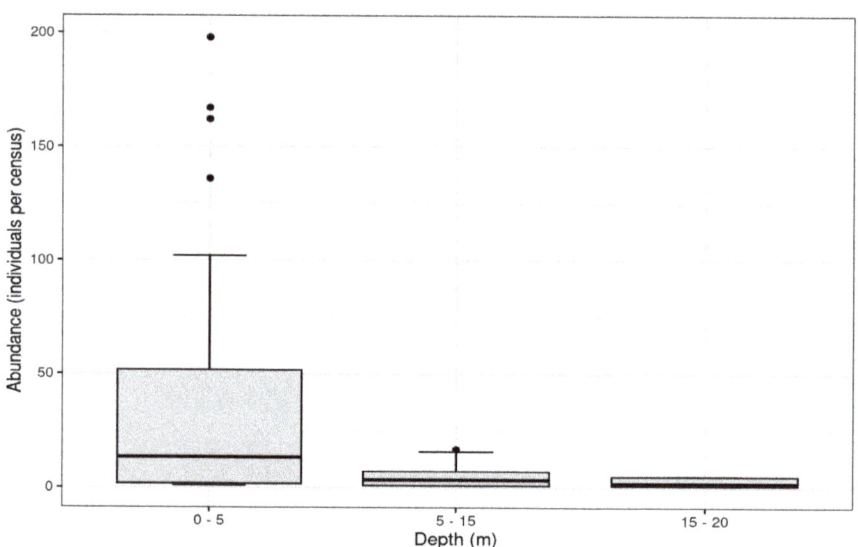

Figure 6. Number of individuals identified at different depths. Black dots (out layers) represent massive aggregations out of the mean and standardized deviation.

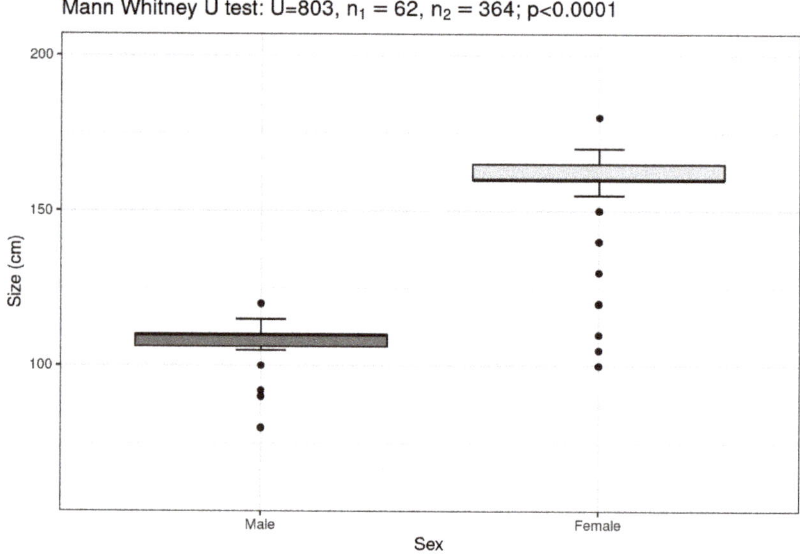

Figure 7. Mean total length of male and female spiny butterfly rays at Salinetas beach.

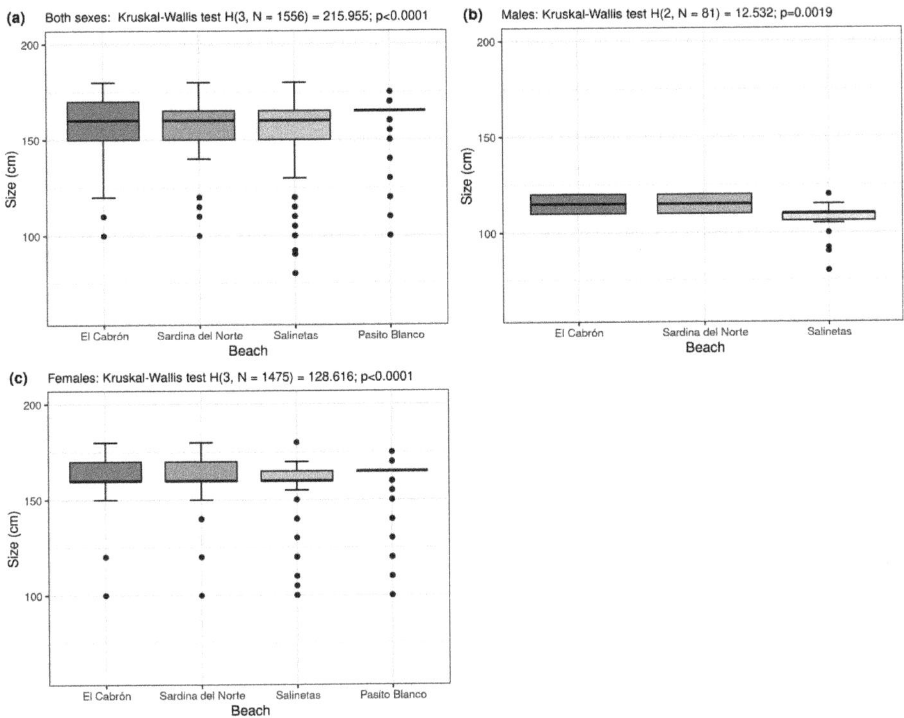

Figure 8. Mean total length of spiny butterfly rays in different sampled areas: (**a**) all individuals, (**b**) males, and (**c**) females.

We also measured parameters of the VBGF using two ELEFAN methods: a simulated annealing algorithm and a genetic annealing algorithm. With $L\infty = 200 \pm 7$ cm and k oscillating between 0.22 and 0.23 year^{-1}, the genetic algorithm showed the best goodness of fit (Table 3).

Table 3. Growth parameters and scores obtained from ELEFANs simulated annealing (SA) and genetic annealing (GA) algorithms for *Gymnura altavela* population. The selected model was based on the highest R_n_max values, marked in bold.

Parameter	ELEFAN S.A.	ELEFAN G.A.
Asymptotic length (cm)	206.07	199.28
Growth coefficient k (yr^{-1})	0.22	0.23
Summer point oscillation (ts)	0.22	0.39
Amplitude of growth oscillation (C)	0.35	0.79
Growth performance index (Φ')	3.96	3.95
t_anchor	0.31	0.60
Age_max	14	14
Goodness of fit (R_n_max) score	0.58	**0.60**

4. Discussion

The spiny butterfly ray, *Gymnura altavela*, seasonally aggregates in a few specific sandy beaches of the Canary Islands. The reasons why certain areas and not others are selected for these aggregations remains unknown, and beaches that seem to have similar environmental and oceanographic characteristics—beach orientation to waves or currents, sand granulometry, depth, water temperature, and food availability, among others—are not visited while others are [49]. Most likely, the presence of natural and/or artificial breakwater structures provides adequate shelter and resting areas for the rays [36]. Large aggregations of ray species have been observed at several beaches near breakwater structures on all of the islands of the archipelago. Likewise, refuging behaviour has also been observed in other rays and shark species (i.e., *Squatina squatina* or *Bathytoshia centroura*), often in physical proximity, if not contact, during regular periods of the year, not only in the Canary Islands but also in many other parts of the world [44,50,51].

In contrast, environmental factors such as seawater temperature seem to influence the presence of spiny butterfly rays in shallow waters and their aggregative behaviour near some sandy beaches. In this regard, our analysis revealed the existence of significant correlations between seawater temperature and patterns of the arrival and departure of rays from beaches. These relationships have also been considered to trigger movements and changes in the behaviour and habitat use of many elasmobranch species [35,52]. Along these lines, our GAM analyses revealed that spiny butterfly rays' arrival at some shallow beaches is a cyclical process—they begin arriving en masse in July and disappear in mid-November—which might suggest a certain philopatry, site fidelity and/or seasonal residency. Similar tendencies have been described by Jaine et al. (2014) [53] regarding reef manta rays in a Capricorn eddy region in Australia and by Gaspar et al. (2008) and Davy et al. (2015) [54,55] regarding *Urogymnus granulatus* and *Pateobatys fai*, who exhibited seasonal residency at Orpheus Island and the Mo'orea marine reserve, respectively. However, we did not detect any relationship between the numbers of rays aggregated per beach and seawater temperature. Nevertheless, water temperature can trigger a biological process that causes the species to approach the beach at the beginning of summer [56].

The largest numbers of spiny butterfly rays, primarily females, aggregated at specific beaches in the Canary Islands during the summer and early autumn, when the seawater temperature fluctuated between 19 and 24 °C. Such aggregation could be related to reproductive events. Mull et al. (2008) [57] have proposed that water temperature (18–20 °C) plays an important role in the regulation of testosterone in round stingray (*Urobatis halleri*) males and may ultimately cue the reproduction of the species. Along those lines, Taylan et al. (2019) [22] reported a pregnant *Gymnura altavela* caught in the Aegean Sea in April 2018 that had three well-developed embryos (i.e., 28.6–30.9 cm disc width), which indicated that fecundation had probably occurred 6 months prior (i.e., around October). However, the low number of males observed at sites of aggregation in our study did not indicate a mating aggregation, at least during the day, for only one male was observed chasing a female. The absence of males could be explained by their frequency in deeper waters, offshore, or in different areas far from the sampled areas. Mating events could have also happened at night. In fact, and as observed in other species such as *Carcharhinus melanopterus* and *Tetronarce californica* [58], data suggested that, at night, the spiny butterfly rays were more active and moved further away from the shallow areas where they were found resting during the day. Observations of mating among rays and skates are rarely reported in the literature, and copulation is normally preceded by chasing, in which males bite onto the anterior portion of the females' disc [59]. We observed some pregnant females in the areas studied, as well as the presence of newborns; thus, we cannot rule out that the aggregations of females are related with breeding seasons.

Similarly, large concentrations of individuals in very shallow waters have been described in other ray species, including *Pateobatys fai* in French Polynesia [60], *Aetobatus narinari* in the Gulf of México [61], and *Mobula alfredi* in West Papua [62]. However, significant differences arose in the abundance of spiny butterfly rays per unit area (individuals/m^2) across the four beaches surveyed. Those differences could relate to the topography of the beach strands and the abrupt changes in the depth of the subtidal zones. Rays were more abundant in waters 0–5 m deep at beaches with less sloped bottoms and more pleasant surf zones (e.g., Pasito Blanco and Salinetas). In contrast, they were less numerous in places that gain in depth quickly and where waves break more heavily against the shore (e.g., El Cabrón and Sardina del Norte). In general, rays aggregate in deeper areas from 10 to 20 metres deep.

G. altavela shows sexual dimorphism, with females being larger than males, a characteristic also reported in many other skates and ray species [63–66], including *G. altavela* in the Western Atlantic [67].

Elucidating the population dynamics of *G. altavela* first requires examining the parameters of the von Bertalanffy growth equation. In that sense, the estimated $L\infty$ for the spiny butterfly rays in Gran Canaria waters was 200 ± 7 cm and thus similar to the common length reported for the species but far from the maximal length reported [32]. In contrast, the growth constant (k) that we estimated oscillated between 0.22 and 0.23 year^{-1}, a growth coefficient similar to that described for other ray and skate species [68–71] and close to that reported by Parsons et al. (2018) [67] for individuals in the North Atlantic off the U.S. coast. Parsons et al. reported k values of 0.27 ± 0.04 for males and 0.6 ± 0.1 for females obtained from high-resolution X-ray computed tomography of the vertebral centre. These differences in the growth rhythms of spiny butterfly rays between the north-western Atlantic and the Canary Islands could be related to differences in the methodology of parameter estimation and/or with differences in environmental conditions, including the availability of food between highly productive areas and oligotrophic ones.

To be sure, extensive work is needed to obtain enough information to develop adequate strategies for conserving *G. altavela* and a management plan is required to guarantee the healthy status of its population around the Canary Islands. Implementing measures to protect shallow areas where the species seasonally aggregates to mate or breed are also crucial for their survival and future conservation.

5. Conclusions

Spiny butterfly rays aggregate in the shallow waters of the sandy beaches of Gran Canaria during the summer and autumn, with a peak of individuals in September. The beach of Pasito Blanco showed a significantly higher aggregation of individual rays per square metre than in the other beaches surveyed. The animals began to arrive at the beaches when the seawater temperature reached 19 °C at the beginning of summer (i.e., in June), and the aggregation of individuals increased until the temperature reached 23–24 °C and decreased as temperatures cooled, especially when the temperature reached 19 °C at the end of October. The spiny butterfly rays distributed in shallow waters 0–20 m deep but predominated at 0–5 m, thus decreasing in abundance with depth. The estimated female-to-male ratio was 1:18.9, with a significant predominance of females. Females were also significantly larger than males. Other significant differences emerged in the length of individuals between the areas observed, with rays from Pasito Blanco being the largest and those from Salinetas being the smallest on average. Among other results, the von Bertalanffy growth equation estimates were $L\infty = 183.75$ cm and $k = 0.210$–0.310 year^{-1}, while the total mortality rate was $z = 0.314$ year^{-1}.

Author Contributions: Conceptualization, A.E.-R., D.J.-A. and J.J.C.; methodology, A.E.-R., D.J.-A., J.J.C. and A.G.-M.; software, A.S.-d.-P., A.E.-R., A.G.-M. and L.C.-M.; validation, A.E.-R., D.J.-A. and J.J.C.; formal analysis, A.S.-d.-P., L.C.-M. and A.E.-R.; investigation, A.E.-R., A.G.-M., D.J.-A. and J.J.C.; resources, A.E.-R. and D.J.-A.; data curation, A.E.-R., A.G.-M. and J.J.C.; writing—original draft preparation, A.E.-R.; writing—review and editing, A.E.-R., J.J.C., D.J.-A. and E.K.M.M.; supervision, J.J.C. and D.J.-A.; project administration, D.J.-A. and J.J.C.; funding acquisition, D.J.-A. All authors have read and agreed to the published version of the manuscript.

Funding: This research was funded by The Save Our Seas Foundation, Keystone Grant number 391/2017 and CanBio Project (foundation registration number 269). Publication fees have been charged by the project ULPGC Excellence, funded by the Consejería de Economía, Conocimiento y Empleo del Gobierno de Canarias.

Institutional Review Board Statement: Not applicable.

Informed Consent Statement: Not applicable.

Data Availability Statement: The datasets generated and/or analysed during the current study are available from the corresponding author upon reasonable request.

Acknowledgments: This work was supported with funding from the Loro Parque Fundación and the Canary Islands Government, in the framework of the CanBIO Project. The authors wish to thank the Save Our Seas Foundation for its support of the Rays of Paradise project and for the opportunity to conduct the research, along with every collaborator who helped during the surveys over the years. Thanks also go to the citizen science programmes Poseidon at the University of Las Palmas de Gran Canaria and RedPROMAR of the government of the Canary Islands for their help and for the databases that they maintain.

Conflicts of Interest: The authors declare no conflict of interest.

References

1. Dulvy, K.N.; Fowler, S.L.; Musick, J.A.; Cavanagh, R.D.; Kyne, P.M.; Harrison, L.R.; Carlson, J.K.; Davidson, L.N.K.; Fordham, S.V.; Francis, M.P.; et al. Extinction risk and conservation of the world's sharks and rays. *eLife* **2014**, *3*, e00590. [CrossRef]
2. Pacoureau, N.; Rigby, C.L.; Kyne, P.M.; Sherley, R.B.; Winker, H.; Carlson, J.K.; Dulvy, N.K. Half a century of global decline in oceanic sharks and rays. *Nature* **2021**, *589*, 567–571. [CrossRef] [PubMed]
3. Stevens, J.D.; Bonfil, R.; Dulvy, N.K.; Walker, P.A. The effects of fishing on sharks, rays, and chimaeras (chondrichthyans), and the implications for marine ecosystems. *ICES J. Mar. Sci.* **2000**, *57*, 476–494. [CrossRef]
4. Bornatowski, H.; Navia, A.F.; Braga, R.R.; Abilhoa, V.; Corrêa, M.F.M. Ecological importance of sharks and rays in a structural foodweb analysis in southern Brazil. *ICES J. Mar. Sci.* **2014**, *71*, 1586–1592. [CrossRef]
5. Hammerschlag, N.; Williams, L.; Fallows, M.; Fallows, C. Disappearance of white sharks leads to the novel emergence of an allopatric apex predator, the seven gill shark. *Sci. Rep.* **2019**, *9*, 1908. [CrossRef]
6. Myers, R.A.; Baum, J.K.; Shepherd, T.D.; Powers, S.P.; Peterson, C.H. Cascading effects of the loss of apex predatory sharks from a coastal ocean. *Science* **2007**, *315*, 1846–1850. Available online: https://www.science.org/doi/10.1126/science.1138657 (accessed on 28 February 2023). [CrossRef] [PubMed]
7. Ferretti, F.; Worm, B.; Britten, G.L.; Heithaus, M.R.; Lotze, H.K. Patterns and ecosystem consequences of shark declines in the ocean. *Ecol. Lett.* **2010**, *13*, 1055–1071. [CrossRef]
8. Couce-Montero, L.; Christensen, V.; Castro, J.J. Simulating trophic impacts of fishing scenarios on two oceanic islands using Ecopath with Ecosim. *Mar. Environ. Res.* **2021**, *169*, 105341. [CrossRef]
9. Hammerschlag, N.; Fallows, C.; Meÿer, M.; Seakamela, S.M.; Orndorff, S.; Kirkman, S.; Kotze, D.; Creel, S. Loss of an apex predator in the wild induces physiological and behavioural changes in prey. *Biol. Lett.* **2022**, *18*, 20210476. [CrossRef]
10. Castro, J. The origins and rise of shark biology in the 20th century. *Mar. Fish. Rev.* **2016**, *78*, 14–33.
11. Dent, F.; Clarke, S.; State of the Global Market for Shark Products. FAO Fisheries and Aquaculture Technical Paper; I, III, IV, VII, VIII, 1–159, 161–167, 169–179, 181–185, 187. 2015. Available online: https://www.proquest.com/scholarly-journals/state-global-market-shark-products/docview/1708482071/se-2 (accessed on 25 February 2021).
12. McCully, S.R.; Scott, F.; Ellis, J.R. Lengths at maturity and conversion factors for skates (Rajidae) around the British Isles, with an analysis of data in the literature. *ICES J. Mar. Sci.* **2012**, *69*, 1812–1822. [CrossRef]
13. Bräutigam, A.; Callow, M.; Campbell, I.R.; Camhi, M.D.; Cornish, A.S.; Dulvy, N.K.; Fordham, S.V.; Fowler, S.L.; Hood, A.R.; McClennen, C.; et al. *Global Priorities for Conserving Sharks and Rays: A 2015–2025 Strategy*; Global Sharks and Rays Initiative, IUCN Species Survival Commission (SSC), Shark Specialist Group, The Shark Trust, UK, TRAFFIC International, Wildlife Conservation Society (WCS), US, WWF. 2015. Available online: https://portals.iucn.org/library/sites/library/files/documents/2016-007.pdf (accessed on 28 February 2023).
14. ICES. NEAFC and OSPAR joint request on the status and distribution of deep-water elasmobranchs. In *ICES Special Request Advice. Ecoregion in the Northeast Atlantic and Adjacent Seas*; ICES: Copenhagen, Denmark, 2020.
15. Chatzispyrou, A.; Koutsikopoulos, C. Tracing Patterns and Biodiversity Aspects of the Overlooked Skates and Rays (Subclass Elasmobranchii, Superorder Batoidea) in Greece. *Diversity* **2023**, *15*, 55. [CrossRef]
16. Liu, K.-M.; Huang, Y.-W.; Hsu, H.-H. Management Implications for Skates and Rays Based on Analysis of Life History Parameters. *Front. Mar. Sci.* **2021**, *8*, 664611. [CrossRef]
17. FAO. *The State of World Fisheries and Aquaculture*; Food and Agriculture Organization of the United Nations: Rome, Italy, 2014; 243p.
18. Davidson, L.N.K.; Krawchuk, M.A.; Duley, N.K. Why have global shark and ray landings declined: Improved management or overfishing. *Fish Fish.* **2016**, *17*, 438–458. [CrossRef]
19. Dulvy, N.K.; Reynolds, J.D.; Metcalfe, J.D.; Glanville, J. Fisheries stability, local extinctions and shifts in community structure in skates. *Conserv. Biol.* **2000**, *14*, 283–293. [CrossRef]
20. King, J.R.; McFarlane, G.A. Marine fish life history strategies: Applications to fishery management. *Fish. Manag. Ecol.* **2003**, *10*, 249–264. [CrossRef]
21. Scientific, Technical and Economic Committee for Fisheries (STECF). *Long-Term Management of Skates and Rays (STECF-17-21)*; JRC109366; Publications Office of the European Union: Luxembourg, 2017; ISBN 978-92-79-67493-8. [CrossRef]

22. Taylan, B.; Bayhan, B.; Saglam, C.; Kara, A. First observation of the embryos of spiny butterfly rayspiny butterfly rayspiny butterfly ray, *Gymnura altavela* (Linnaeus, 1758) (Chondrichthyes: Gymnuridae) from Eastern Mediterranean, a species critically endangered. *Fresenius Environ. Bull.* **2019**, *28*, 3147–3152.

23. Brito, A.; Pascual, P.J.; Falcón, J.M.; Sancho, A.; González, G. *Peces de las Islas Canarias. Catálogo Comentado e Ilustrado*; Francisco Lemus Editor: San Cristóbal de La Laguna, Spain, 2002.

24. Moreno, J.; Solleleit-Ferreira, S.E.; Riera Elena, R. Distribution and Abundance of Coastal Elasmobranchs in Tenerife (Canary Islands, NE Atlantic Ocean) with Emphasis on the Bull Ray, *Aetomylaeus bovinus*. *Thalass. Int. J. Mar. Sci.* **2021**, *38*, 723–731. [CrossRef]

25. Walls, R.; Vacchi, M.; Notarbartolo di Sciara, G.; Serena, F.; Dulvy, N. *Gymnura altavela* (Europe Assessment). The IUCN Red List of Threatened Species 2015: e.T63153A48931613. 2015. Available online: https://www.iucnredlist.org/ (accessed on 28 February 2023).

26. Walls, R.H.L.; Vacchi, M.; Notarbartolo di Sciara, G.; Serena, F.; Dulvy, N.K. *Gymnura altavela* (Mediterranean Assessment). The IUCN Red List of Threatened Species 2016: e.T63153A16527909. 2016. Available online: https://www.iucnredlist.org/ (accessed on 28 February 2023).

27. Dulvy, N.K.; Charvet, P.; Carlson, J.; Badji, L.; Blanco-Parra, M.P.; Chartrain, E.; De Bruyne, G.; Derrick, D.; Dia, M.; Doherty, P.; et al. *Gymnura altavela*. The IUCN Red List of Threatened Species 2021: e.T63153A3123409. 2021. Available online: https://dx.doi.org/10.2305/IUCN.UK.2021-1.RLTS.T63153A3123409.en (accessed on 28 February 2023).

28. McEachran, J.D.; Capape, C. Rhinobatidae. In *Fishes of the North-Eastern Atlantic and the Mediterranean*; Unesco: Paris, France, 1984; Volume 1, pp. 156–158.

29. McEachran, J.D.; De Carvalho, M.R.; Carpenter, K.E. Batoid fishes. In *The Living Marine Resources of the Western Central Atlantic*; FAO: Rome, Italy, 2002; Volume 1, pp. 507–589.

30. Menni, R.C.; Lucifora, L.O. *Chondrichthyes from Argentina and Uruguay*; ProBiota, Serie Técnica-Didáctica; UNLP: La Plata, Argentina, 2007; Volume 11, pp. 1–15. Available online: http://hdl.handle.net/1834/19532 (accessed on 28 February 2023).

31. Ray, C.; Robins, C.R. *A Field Guide to Atlantic Coast Fishes: North America*; Houghton Mifflin Harcourt: Boston, MA, USA, 2016.

32. Stehmann, M. Gymnuridae. In *FAO Species Identification Sheets for Fishery Purposes*; Fischer, W., Bianchi, G., Scott, W.B., Eds.; Eastern Central Atlantic (Fishing Areas 34, 47 (In Part)); FAO: Rome, Italy, 1981; Volume 5.

33. International Game Fish Association. *World Record Game Fishes*; International Game Fish Association: Dania Beach, FL, USA, 1991.

34. Dulčic, J.; Jardas, I.; Onofri, V.; Bolotin, J. The roughtail stingray *Dasyatis centroura* (Pisces: Dasyatidae) and spiny butterfly ray *Gymnura altavela* (Pisces: Gymnuridae) from the southern Adriatic. Marine Biological Association of the United Kingdom. *J. Mar. Biol. Assoc. UK* **2003**, *83*, 871–872. [CrossRef]

35. Meyers, E.K.M.; Tuya, F.; Barker, J.; Jiménez-Alvarado, D.; Castro-Hernández, J.J.; Haroun, R.; Rödder, D. Population structure, distribution and habitat use of the critically endangered angelshark, *Squatina squatina*, in the Canary Islands. *Aquat. Conserv. Mar. Freshw. Ecosyst.* **2017**, *27*, 1133–1144. [CrossRef]

36. Tuya, F.; Asensio, M.; Navarro, A. "Urbanite" rays and sharks: Presence, habitat use and population structure in an urban semi-enclosed lagoon. *Reg. Stud. Mar. Sci.* **2020**, *37*, 101342. [CrossRef]

37. IUCN. The IUCN Red List of Threatened Species. Version 2021-1. 2021. Available online: www.iucnredlist.org (accessed on 19 May 2022).

38. Bas, C.; Castro, J.J.; Hernandez-Garcia, V.; Lorenzo, J.M.; Moreno, T.; Pajuelo, J.M.; Gonzalez Ramos, A.J. *La Pesca en Canarias y Areas de Influencia*; Cabildo Insular de Gran Canaria: Las Palmas de Gran Canaria, Spain, 1995.

39. Falcón, J.M.; Bortone, S.A.; Brito, A.; Bundrick, C.M. Structure of and relationships within and between the littoral, rock-substrate fish communities off four islands in the Canarian Archipelago. *Mar. Biol.* **1996**, *125*, 215–231. [CrossRef]

40. Tuya, F.; Boyra, A.; Sanchez-Jerez, P.; Barbera, C.; Haroun, R.J. Relationships among fishes, the long-spined sea urchin *Diadema antillarum* and algae throughout the Canarian Archipelago. *Mar. Ecol. Prog. Ser.* **2004**, *278*, 157–169. [CrossRef]

41. Martínez Saavedra, J. Analysis of the State of Fishery Resources from Gran Canaria Study of the Historical Catch Series. Master's Thesis, University of Las Palmas of Gran Canaria, Las Palmas de Gran Canaria, Spain, 2011.

42. Noviello, N.; McGonigle, C.; Jacoby, D.M.P.; Meyers, E.K.M.; Jiménez-Alvarado, D.; Barker, J. Modelling Critically Endangered marine species: Bias-corrected citizen science data inform habitat suitability for the angelshark (*Squatina squatina*). *Aquat. Conserv. Mar. Freshw. Ecosyst.* **2021**, *31*, 3451–3465. [CrossRef]

43. Wetz, J.J.; Ajemian, M.J.; Shipley, B.; Stunz, G.W. An assessment of two visual survey methods for documenting fish community structure on artificial platform reefs in the Gulf of Mexico. *Fish. Res.* **2020**, *225*, 105492. [CrossRef]

44. Jacoby, D.M.P.; Croft, D.P.; Sis, D.W. Social behaviour in sharks and rays: Analysis, patterns and implications for conservation. *Fish Fish.* **2011**, *13*, 399–417. [CrossRef]

45. Mendonça, S.A.; Macena, B.C.L.; Afonso, A.S.; Hazin, H.V. Seasonal aggregation and diel activity by the sicklefin devil ray *Mobula tarapacana* off a small, equatorial outcrop of the Mid-Atlantic Ridge. *J. Fish Biol.* **2018**, *93*, 1121–1129. [CrossRef] [PubMed]

46. Labrosse, P.; Kulbicki, M.; Ferraris, J. *Underwater Visual Fish Census Surveys: Proper Use and Implementation*; Secretariat of Pacific Community: Noumea, New Caledonia, 2002; Volume 61.

47. Mildenberger, T.K.; Taylor, M.H.; Wolff, M. TropFishR: An R package for fisheries analysis with length-frequency data. *Methods Ecol. Evol.* **2017**, *8*, 1520–1527. [CrossRef]

48. Wetherall, J.A.; Polovina, J.J.; Ralston, S. Estimating growth and mortality in steady-state fish stocks from length-frequency data. *ICLARM Conf. Proc.* **1987**, *13*, 53–74.

49. Aguiar, A.A.; Valentin, J.L.; Rosa, R.S. Habitat use by *Dasyatis americana* in a south-western Atlantic oceanic island. *J. Mar. Biol. Assoc. UK* **2009**, *89*, 1147–1152. [CrossRef]

50. Sims, D.W. *Differences in Habitat Selection and Reproductive Strategies of Male and Female Sharks*; Cambridge University Press: Cambridge, UK, 2005; pp. 127–147.

51. Powter, D.M.; Gladstone, W. Habitat-mediated use of space by juvenile and mating adult port Jackson sharks, *Heterodontus portusjacksoni*, in Eastern Australia. *Pac. Sci.* **2009**, *63*, 1–14. [CrossRef]

52. Schlaff, A.; Heupel, M.R.; Simpfendorfer, C.A. Influence of environmental factors on the shark and ray movement, behaviour and habitat use: A review. *Rev. Fish Biol. Fish.* **2014**, *24*, 1089–1103. Available online: https://link.springer.com/content/pdf/10.1007/s11160-014-9364-8.pdf (accessed on 28 February 2023). [CrossRef]

53. Jaine, F.R.A.; Rohner, C.A.; Weeks, S.J.; Couturier, L.I.E.; Bennett, M.B.; Townsend, K.A.; Richardson, A.J. Movements and habitat use of reef manta rays off eastern Australia: Offshore excursions, deep diving and eddy affinity revealed by satellite telemetry. *Mar. Ecol. Prog. Ser.* **2014**, *510*, 73–86. [CrossRef]

54. Gaspar, C.; Chateau, O.; Galzin, R. Feeding sites frequentation by the pink whipray *Himantura fai* in Moorea (French Polynesia) as determined by acoustic telemetry. *Cybium* **2008**, *32*, 153–164.

55. Davy, L.E.; Simpfendorfer, C.A.; Heupel, M.R. Movement patterns and habitat use of juvenile mangrove whiprays (*Himantura granulata*). *Mar. Freshw. Res.* **2015**, *66*, 481–492. [CrossRef]

56. Martins, A.P.B.; Heupel, M.R.; Chin, A.; Simpfendorfer, C.A. Batoid nurseries: Definition, use and importance. *Mar. Ecol. Prog. Ser.* **2018**, *595*, 253–267. [CrossRef]

57. Mull, C.G.; Lowe, C.G.; Young, K.A. Photoperiod and water temperature regulation of seasonal reproduction in male round stingrays (*Urobatis halleri*). *Comp. Biochem. Physiol. Part A Mol. Integr. Physiol.* **2008**, *151*, 717–725. [CrossRef]

58. Hammerschlag, N.; Skubel, R.A.; Clich, H.; Nelson, E.R.; Shiffman, D.S.; Wester, J.; Macdonald, C.C.; Cain, S.; Jennings, L.; Enchelmaier, A.; et al. Nocturnal and crepuscular behavior in elasmobranchs: A review of movement, habitat use, foraging, and reproduction in the dark. *Bull. Mar. Sci.* **2017**, *93*, 355–374. [CrossRef]

59. Nordell, S.E. Observations of the mating behavior and dentition of the stingray, *Urolophus helleri*. *Environ. Biol. Fishes* **1994**, *39*, 219–229. [CrossRef]

60. Kiszka, J.J.; Mourier, J.; Gastrich, K.; Heithaus, M.R. Using unmanned aerial vehicles (UAVs) to investigate shark and ray densities in a shallow coral lagoon. *Mar. Ecol. Prog. Ser.* **2016**, *560*, 237–242. [CrossRef]

61. Bassos-Hull, K.; Wilkinson, K.A.; Hull, P.T.; Dougherty, D.A.; Omori, K.L.; Ailloud, L.E.; Morris, J.J.; Hueter, R.E. Life history and seasonal occurrence of the spotted eagle *Aetobatus narinari*, in the eastern Gulf of Mexico. *Environ. Biol. Fishes* **2014**, *97*, 1039–1056. [CrossRef]

62. Perryman RJYVenables, S.K.; Tapilatu, R.F.; Marshall, A.D.; Brown, C.; Frank, D.W. Social preferences and network structure in a population reef manta rays. *Behav. Ecol. Sociobiol.* **2019**, *73*, 114. [CrossRef]

63. Castillo-Geniz, J.L.; Sosa-Nishizaki, O.; Pérez-Jiménez, J.C. Morphological variation and sexual dimorphism in the California skate, *Raja inornate* Jordan and Gilbert, 1881, from the Gulf of California, Mexico. *Zootaxa* **2007**, *1545*, 1–16. [CrossRef]

64. Orlov, A.M.; Cotton, C.F. Sexually dimorphic morphological characters in five north Atlantic deepwater skates (Chondrichthyes: Rajiformes). *J. Mar. Biol.* **2011**, *2011*, 842821. [CrossRef]

65. Orlov, A.M.; Smirnov, A.A. New data on sexual dimorphism and reproductive biology of Alaska skate *Bathyraja parmifera* from the northwestern Pacific Ocean. *J. Ichthyol.* **2011**, *51*, 590–603. [CrossRef]

66. Orlov, A.M.; Cotton, C.F.; Shevernitsky, D.A. Sexual dimorphism of external morphological characters in some deepwater skates (Rajidae, Rajifoems, Chondrichthyes) of the North Atlantic. *Mosc. Univ. Biol. Sci. Bull.* **2010**, *65*, 40–44. [CrossRef]

67. Parsons, K.T.; Maisano, J.; Gregg, J.; Cotton, C.F.; Latour, R.J. Age and growth assessment of western North Atlantic spiny butterfly ray *Gymnura altavela* (L. 1758) using computed tomography of vertebral centra. *Environ. Biol. Fishes* **2018**, *101*, 137–151. [CrossRef]

68. Coelho, R.; Erzini, K. Age and growth of the undulate ray, *Raja undulata*, in the Algarve (southern Portugal). *J. Mar. Biol. Assoc. United Kingd.* **2002**, *82*, 987–990. [CrossRef]

69. Neer, J.A. *Aspects of the Life History, Ecophysiology, Bioenergetics, and Population Dynamics of the Cownose Ray, Rhinoptera bonasus, in the Northern Gulf of Mexico*; UMI Number: 3329085; Louisiana State University and Agricultural & Mechanical College: Baton Rouge, LA, USA, 2005.

70. Ainsley, S.A.; Ebert, D.A.; Cailliet, G.M. Age, growth, and maturity of the whitebrow skate, *Bathyraja minispinosa*, from the eastern Bearing Sea. *ICES J. Mar. Sci.* **2011**, *68*, 1426–1434. [CrossRef]

71. Fisher, R.A.; Call, G.C.; Grubbs, R.D. Age, growth, and reproductive biology of cownose rays in Chesapeake Bay. *Mar. Coast. Fish.* **2013**, *5*, 224–235. [CrossRef]

Article

To Be, or Not to Be: That Is the Hamletic Question of Cryptic Evolution in the Eastern Atlantic and Mediterranean *Raja miraletus* Species Complex

Alice Ferrari [1,†], Valentina Crobe [1,†], Rita Cannas [2], Rob W. Leslie [3], Fabrizio Serena [4], Marco Stagioni [5], Filipe O. Costa [6], Daniel Golani [7], Farid Hemida [8], Diana Zaera-Perez [9], Letizia Sion [10], Pierluigi Carbonara [11], Fabio Fiorentino [4,12], Fausto Tinti [1,*] and Alessia Cariani [1]

1   Department of Biological, Geological and Environmental Sciences, University of Bologna, 40126 Bologna, Italy; alice.ferrari6@unibo.it (A.F.); valentina.crobe2@unibo.it (V.C.); alessia.cariani@unibo.it (A.C.)
2   Department of Life and Environmental Sciences, University of Cagliari, 09126 Cagliari, Italy; rcannas@unica.it
3   Branch Fisheries Management, Department Agriculture, Forestry and Fisheries, Cape Town 8018, South Africa; roblesliesa@hotmail.com
4   Institute for Biological Resources and Marine Biotechnology, National Research Council, 91026 Trapani, Italy; fabrizio.serena@irbim.cnr.it (F.S.); fabio.fiorentino@irbim.cnr.it (F.F.)
5   Laboratory of Marine Biology and Fisheries, Department Biological, Geological and Environmental Sciences, University of Bologna, 61032 Fano, Italy; marco.stagioni3@unibo.it
6   Centre of Molecular and Environmental Biology (CBMA) and ARNET-Aquatic Research Network, Department of Biology, University of Minho, Campus de Gualtar, 4710-057 Braga, Portugal; fcosta@bio.uminho.pt
7   Department of Evolution, Systematics and Ecology, The Hebrew University of Jerusalem, Jerusalem 9190401, Israel; dani.golani@mail.huji.ac.il
8   Ecole Nationale Supérieure des Sciences de la Mer et de l'Aménagement du Littoral, Campus Universitaire de Dely Ibrahim, Algiers 16320, Algeria; hemidafarid@yahoo.fr
9   Institute of Marine Research, 5817 Bergen, Norway; diana.zaera-perez@hi.no
10   Department of Biosciences, Biotechnologies and Environment, University of Bari Aldo Moro, 70125 Bari, Italy; letizia.sion@uniba.it
11   COISPA Technology and Research, 70126 Bari, Italy; carbonara@coispa.it
12   Stazione Zoologica Anton Dohrn, 90149 Palermo, Italy
*   Correspondence: fausto.tinti@unibo.it
†   These authors contributed equally to this work.

Citation: Ferrari, A.; Crobe, V.; Cannas, R.; Leslie, R.W.; Serena, F.; Stagioni, M.; Costa, F.O.; Golani, D.; Hemida, F.; Zaera-Perez, D.; et al. To Be, or Not to Be: That Is the Hamletic Question of Cryptic Evolution in the Eastern Atlantic and Mediterranean *Raja miraletus* Species Complex. *Animals* **2023**, *13*, 2139. https://doi.org/10.3390/ani13132139

Academic Editor: James Albert

Received: 31 May 2023
Revised: 25 June 2023
Accepted: 25 June 2023
Published: 28 June 2023

Simple Summary: The *Raja miraletus* species complex exhibits high levels of morphological and ecological stasis along with the antipodean distribution in the Eastern Atlantic and Indian Oceans. We investigated genetic variability and differentiation between taxa and geographical populations by integrating mitochondrial and nuclear DNA markers. The extraordinary occurrence of at least five different sibling taxa in the Northeastern Atlantic Ocean and Mediterranean Sea is documented, supporting cryptic speciation and stabilising selection.

Abstract: Despite a high species diversity, skates (Rajiformes) exhibit remarkably conservative morphology and ecology. Limited trait variations occur within and between species, and cryptic species have been reported among sister and non-sister taxa, suggesting that species complexes may be subject to stabilising selection. Three sibling species are currently recognised in the *Raja miraletus* complex: (i) *R. miraletus* occurring along the Portuguese and Mediterranean coasts, (ii) *R. parva* in the Central-Eastern Atlantic off West Africa and (iii) *R. ocellifera* in the Western Indian Ocean off South Africa. In the present study, the genetic variation at mitochondrial and nuclear markers was estimated in the species complex by analysing 323 individuals sampled across most of its geographical distribution area to test the hypothesis that restricted gene flow and genetic divergence within species reflect known climate and bio-oceanographic discontinuities. Our results support previous morphological studies and confirm the known taxonomic boundaries of the three recognised species. In addition, we identified multiple weakly differentiated clades in the Northeastern Atlantic Ocean and Mediterranean, at least two additional cryptic taxa off Senegal and Angola, a pronounced

---

differentiation of ancient South African clades. The hidden genetic structure presented here may represent a valuable support to species' conservation action plans.

Keywords: cartilaginous fish; brown skate; conservation biology; population genetics; mtDNA; microsatellite loci

1. Introduction

The evolutionary debate on the nature of species boundaries [1,2] is based on paradigms such as Mayr's discontinuous variation and reproductive isolation of species and Darwin's continuity between varieties, geographical populations and species [2]. Natural hybrid zones and secondary contacts with gene introgression unequivocally show that species boundaries have a semipermeable nature [3] and that intrinsic barriers to gene flow (i.e., pre- and postzygotic barriers) are in some cases incomplete. Species life cycles, ecological features and adaptive phenotypes are particular key points influencing species distribution and dispersal within the marine environment [4,5], in which permanent and intermittent breaks (e.g., landmasses, unsuitable habitats, upwelling areas and oceanographic fronts) may isolate populations, enabling ecological differentiation [6,7], genetic divergence [8], reproductive isolation and speciation [9–11]. Molecular systematics and phylogenetics have greatly contributed to the assessment of relationships among taxa and effectively contributed to delineate the hierarchy of evolutionary frames in which recently diverged taxa exhibit, on average, lower divergence than taxa in the later stages of the speciation process [6]. Over the last two decades, increased evidence has emerged for speciation governed by entirely different mechanisms, leading to so-called sibling or cryptic species (sensu Bickford [12]). The idea that species can evolve into similar morphologies is well established [13], but the use of molecular delimitation methods has now brought cryptic species to the forefront in many research arenas [6,14,15]. Bickford et al. [12] identified at least two recurrent themes in animals wherein morphological distinctiveness and reproductive isolation are unpaired: in groups using non-visual mate-recognition signals (e.g., chemical, olfactory, acoustic, electric-field senses) and in groups living under environmental conditions that promote the stabilising selection of morphological traits (e.g., extreme habitats, specialised host–parasite relationships, deep-sea environments, fishery pressure). Among elasmobranchs, skates and rays exhibit highly effective modulation of electro-sensory signals depending on behaviour ([16] and references therein). At the same time, they display a marked conservation of ecological and morphological traits [17–23], especially between recently diverged species [12]; a strong evolutionary success in terms of resilience at the evolutionary scale [24] and a high degree of endemism [25,26] and species richness [27–29].

Investigations into the role of biogeographical barriers on the speciation of marine organisms have increasingly concentrated across several taxa [30–32]. Prior to 2016, the brown skate *Raja miraletus* Linnaeus, 1758 was thought to be distributed throughout the Mediterranean Sea and from northern Portugal, along the western and south-eastern coasts of Africa [33,34]. This distributional range is much wider than expected for a small-sized rajid, given the limited potential for dispersal in a species with a relatively sedentary behaviour of adults and juveniles [35–39] and the lack of egg dispersal [40]. Nominal *R. miraletus* exhibits a pronounced benthic ecology, with most records from 10 m to 150 m on sandy and hard bottoms [25,33] and a generalist feeding behaviour [41,42]. Due to its high and stable abundance over its distribution, small body size and early maturation (age at maturity estimated at 2.7 years; [43]), it was considered highly resilient to exploitation and was assessed globally as Least Concern in the International Union for Conservation of Nature Red List [44,45]. Since then, *Raja ocellifera* Regan 1906 has been resurrected for the southern African population [28,46] and was assessed as Endangered in 2020 [46]. The newly described *R. parva* Last and Séret 2016 from west Africa has not been assessed [28].

The *R. miraletus* complex exhibits a high level of stasis of the general external morphology over its range; all populations exhibit a distinctive bright tricolored (blue, black and yellow) eyespot on the upper ochre-brownish surface at the base of each pectoral fin [34,47]. After the first identification of three parapatric or allopatric groups (Mediterranean, West Africa and South Africa) based on the variation of morphometric and meristic characters [48], preliminary evidence of cryptic speciation in *R. miraletus* was observed by integrating results from mitochondrial DNA analysis, morphology and host–parasite relationships from specimens collected in Central-Southern Africa [49,50]. *Raja miraletus* was then recognised as a species complex of at least four valid taxa based on combined data for *COI* and *NADH2* [51]: (1) the northernmost *R. miraletus*, occurring in the Mediterranean Sea and adjacent North-Eastern Atlantic waters, (2) the southernmost and resurrected *R. ocellifera* (associated with mtDNA data from Naylor et al. [49]; *Raja* cf. *miraletus 1*, NCBI Accession Number JQ518895, [49]) occurring, in the Western Indian Ocean, off South Africa, and in the Indian Ocean, from False Bay to Durban, (3) the central African *R. parva*, distributed from Senegal to Angola (associated with mtDNA data from Naylor et al. [49]; *Raja* cf. *miraletus 2*, NCBI Accession Number JQ518890, [49]) and (4) a still undescribed taxon (*Raja miraletus*, NCBI Accession Number JQ518891, [49]), occurring from Mauritania to Senegal where it is therefore sympatric with *R. parva*.

The advent of high-throughput DNA sequencing technologies and the launch of global DNA-based biodiversity assessments (e.g., DNA barcoding; [52]) has provided raw data, enabling the determination of taxonomic, ecological and evolutionary aspects of cryptic and sibling species, where the term "sibling" denotes a cryptic species with a recent common ancestor, implying a sister species relationship [53] and even more challenging conservation issues [12]. Moreover, molecular methods coupling markers obtained from mitochondrial DNA (mtDNA) and nuclear DNA (nuDNA) have improved the resolution of species boundaries and revealed gene introgression/hybridisation phenomena in marine fish including elasmobranchs [10,54–56]. Understanding the liaison between species life-history traits, ecology and the adaptive phenotypes leading to hidden population divergence and reproductive isolation is of utmost importance for skates, whose conservation is often hampered by the lack of species-specific data [57].

This study uses the integrative support of the mtDNA *cytochrome oxidase subunit I* barcode region (*COI*) and eight nuDNA *EST-linked* polymorphic microsatellite loci (*EST-SSRs*; [58]) to estimate the genetic variation among 323 specimens collected across almost the full geographic range of the *R. miraletus* species complex [40,47] and exhibiting the typical tricoloured eyespots. We tested hypotheses of the relationship between restricted gene flow and genetic divergence within the species complex, specifically in relation to climatic and oceanographic discontinuities. Additionally, we sought to establish parallel patterns between our findings and variations in morphology and parasite prevalence, which were independently assessed [48,50]. As compared to previous knowledge, our findings contributed to describe a richer scenario concerning the taxonomic units and zoogeographic boundaries characterising the *R. miraletus* complex.

2. Materials and Methods

2.1. Sampling

A total of 323 specimens of from the *R. miraletus* species complex were collected between 2000 and 2014 (Tables 1 and S1) during scientific research programs (South Africa, Angola and Mediterranean Sea) by contracted commercial fishermen (Senegal, Levantine Sea and Israel) or at local fish markets (Algeria). Scientific trawl surveys were carried out in South Africa (2006 and 2011, Africana cruises), Angola (2006, Nansen cruises), the whole Mediterranean Sea (2000–2014 Mediterranean International Trawl scientific Surveys, MedITS; [59]) and national scientific trawl surveys (2000–2010 Italian Gruppo Nazionale Demersali surveys, GruND; [60]; the 2007 Portuguese scientific surveys of the Instituto Português de Investigação do Mar) allowed for a comprehensive sampling, covering most of the wide geographical distribution of the *R. miraletus* complex (Figure 1). All individuals were easily assigned to the *R. miraletus* complex based on their very distinctive morphotype and species-specific diagnostic characters [25,40]. Fin clips and muscle tissues were cut from each individual using sterile tweezers and clippers, transferred to a clean tube filled with 96% ethanol and stored at $-20\ ^{\circ}$C for subsequent DNA analyses.

Table 1. Sampling data and locations. "N (tot)", refers to the total number of individuals sampled in this study, according to the methods indicated in the Source column. Of these "N (*COI*)" and "N (*EST-SSRs*)" refer to individuals *COI*-sequenced and genotyped in this study. N (tot) = 0 when available *COI* sequences were retrieved from public databases and integrated into the mtDNA dataset, as specified in the "Source" column. The last row refers to geographical samples previously described in McEachran et al. 1989 [48]. 1—Mediterranean group. 2—Mauritania and Senegal group. 3—Gulf of Guinea–equatorial African group. 4—Angolan sample. 5—South African sample.

Sampling Area	Macro Area Code	N (tot)	N (*COI*)	N (*EST-SSRs*)	Years	Source (Trawl Survey Program)	McEachran et al., 1989 [48]
Central-Southern Africa (CSA)							
South Africa—South Coast		8	6	8	2006	ST (Africana)	5
South Africa—South Coast	SAF	0	5	0	2007, 2012	GB/BOLD	5
South Africa—South Coast		32	30	31	2011	ST (Africana)	5
Namibia	NAM	0	3	0	2009, 2010	GB/BOLD	5
Angola	ANG	27	27	26	2006	ST (Nansen)	4
Senegal	SEN	5	5	5	2007	CF	2
Northeastern Atlantic–Mediterranean Sea (NEAM)							
Portugal	POR	3	3	3	2007	ST (IPIMAR)	n.a.
Portugal		0	7	0	2005, 2007	GB/BOLD	n.a.
Balearic Islands	BAL	19	19	16	2006	ST (MedITS)	1
Balearic Islands		0	3	0	2013	ST (MedITS)	1
Sardinia	SAR	11	11	8	2002, 2005	ST (MedITS)	1
Tuscany	TUS	26	22	21	2005, 2006	ST (MedITS)	1
Tuscany		16	6	13	2008, 2010	ST (MedITS)	1
Algeria	ALG	8	8	5	2002, 2003	FM (Algiers)	1
Algeria		10	9	8	2009, 2010	FM (Algiers)	1
Strait of Sicily—Adventura Bank		22	22	22	2014	ST (MedITS)	1
Strait of Sicily—Maltese Bank	SIC	16	12	8	2000, 2002	ST (MedITS)	1
Strait of Sicily—Maltese Bank		0	11	0	2007	GB/BOLD	1
Adriatic Sea—Northern Italian coast		39	31	20	2006, 2007	ST (MedITS; GruND)	1
Adriatic Sea—Croatian coast		24	24	8	2002, 2004	ST (MedITS; GruND)	1
Adriatic Sea—Southern Italian coast	ADR	19	16	19	2004	ST (MedITS; GruND)	1
Adriatic Sea—Albanian coast		19	13	17	2004	ST (MedITS; GruND)	1
Ionian Sea		4	3	4	2004	ST (MedITS; GruND)	1
Greece—Aegean coast	GRE	0	3	0	2014	GB/BOLD	1
Israel		8	7	7	2009	CF	1
Israel	ISR	0	3	0	2012	BOLD	1
Israel		0	4	0	2014	GB/BOLD	1
Levantine Sea	LEV	0	2	0	2016	GB/BOLD	1
Levantine Sea		7	7	7	2009	CF	1

n.a.: not available. ST: Scientific Trawl survey. CF: Contracted Fishermen. FM: Fishery Market. GB: GenBank Database. BOLD: Barcoding of Life Database.

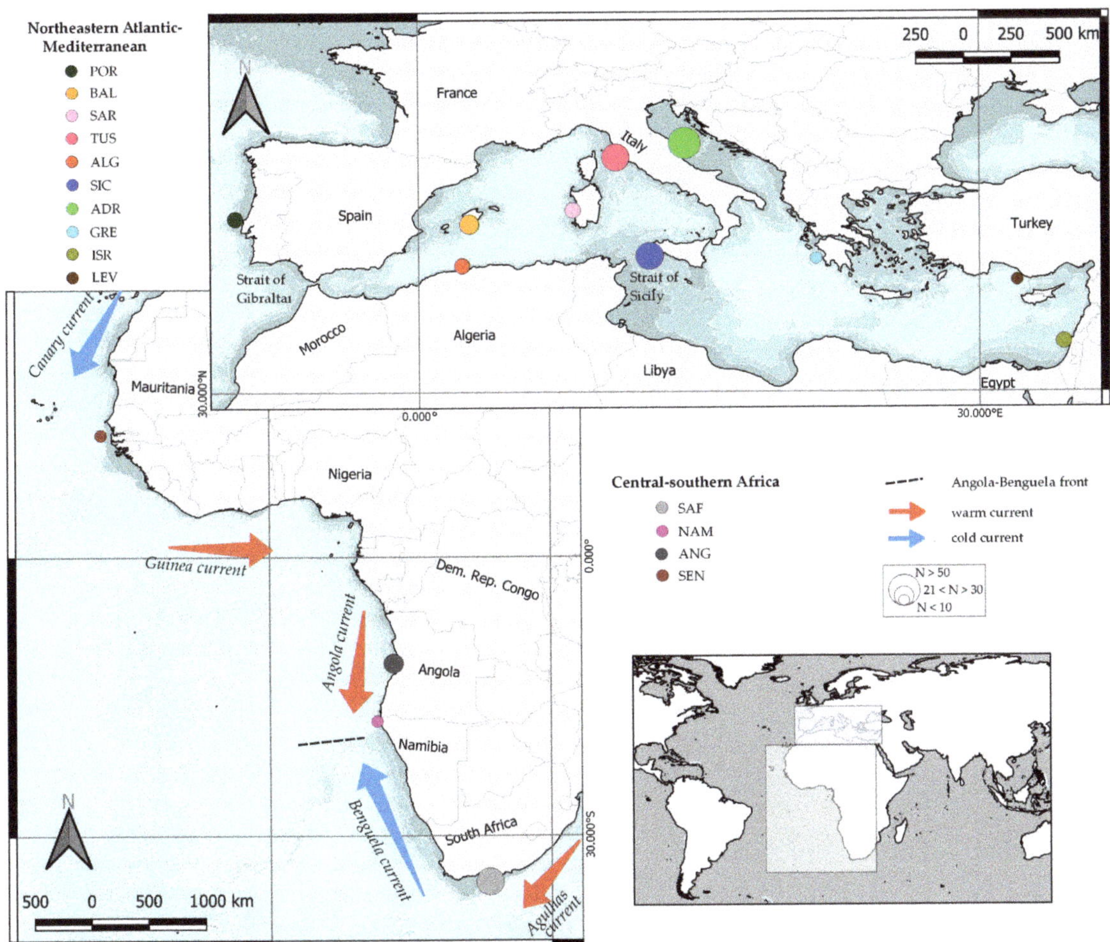

Figure 1. Sampling sites of *Raja miraletus* species complex collected in the Northeastern Atlantic–Mediterranean and the Central–Southern Africa regions. A simplified representation of the main oceanic currents of the Eastern Atlantic is indicated. Acronyms of the Macro Area Codes are given as in Table 1. Colours in agreement with the haplotype network legend of Figure 2. Different dimensions of circles are related to sample size.

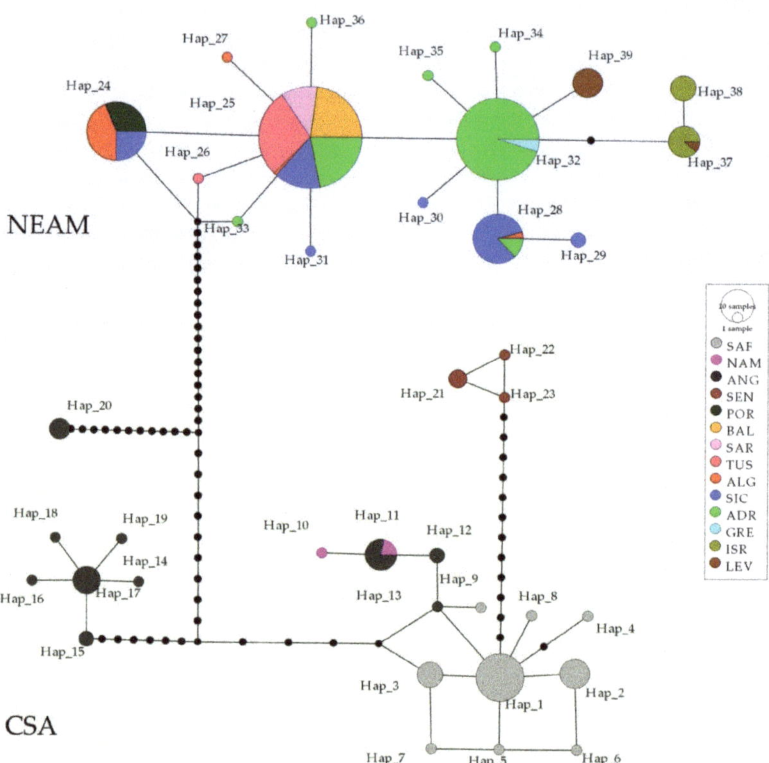

Figure 2. TCS network of *cytochrome oxidase subunit I (COI)* haplotypes shown by *Raja miraletus* across most of its distribution area. CSA, Central–Southern Africa; NEAM, Northeastern Atlantic–Mediterranean Sea. Circles are proportional to haplotype frequencies. Black dots between branch nodes indicate substitutions. Black circles at network nodes represent unsampled haplotypes. Acronyms of the Macro Area Codes are given in Table 1.

2.2. Genetic Data Analysis

Detailed protocols used for DNA extraction, PCR amplification, DNA sequencing and genotyping of mitochondrial and nuclear markers [56,58,61,62] are described in the Supplementary Material Text S1.

2.2.1. Genetic Diversity

A total of 281 *COI* newly generated sequence electropherograms were manually edited and aligned by CLUSTAL W software [63] implemented in MEGA v.11 [64]. The presence of stop codons and sequencing error was verified through amino acidic translation [65]. Individual *COI* sequences were first compared with sequences deposited in public repositories in order to confirm their phylogenetic identity and rule out any error due to mishandling of samples on board or during the laboratory activities, namely GenBank (http://www.ncbi.nlm.nih.gov/genbank/, accessed on 21 May 2023) through the BLAST algorithm (http://blast.ncbi.nlm.nih.gov/Blast.cgi, accessed on 21 May 2023) and the Barcode of Life Data System (BOLD), using the BOLD identification engine ([66]; http://www.boldsystems.org, accessed on 21 May 2023). A total of 41 additional homologous *COI* sequences of *R. miraletus* complex were retrieved from both databases selecting records from different geographical locations (South Africa, Namibia, Strait of Sicily,

Aegean Sea and Israel) when metadata giving the collection area were accessible ([26,67–79]; Table 1; see Table S1 for more detail). The retrieved sequences were aligned with those newly generated and a final dataset of 322 homologous *COI* sequences was obtained.

The number of polymorphic sites (S), the number of haplotypes (H), the haplotype diversity (Hd), the nucleotide diversity (π; [80]) and their standard deviations were calculated using DNASP v.6 [81]. The haplotype frequencies were estimated using ARLEQUIN v.3.5.2.2. [82].

The average genetic distances observed within and between the two identified Central–Southern African and the Northeastern Atlantic–Mediterranean clades of *R. miraletus* complex were calculated using the Tamura-Nei (1993) model implemented in MEGA as the best evolutionary substitution model following the corrected Akaike Information Criterion (AICc; [83]). Genetic distances were then compared with the range of those estimated among other congeneric species. For this, we retrieved homologous *COI* sequences of the following species public databases (NCBI and BOLD): *Raja straeleni* Poll 1951, *Raja microocellata* Montagu 1818, *Raja asterias* Delaroche 1809, *Raja brachyura* Lafont 1873, *Raja clavata* Linnaeus 1758, *Raja montagui* Fowler 1910, *Raja polystigma* Regan 1923, *Raja radula* Delaroche 1809 and *Raja undulata* Lacepède 1802 ([26,84]; Table S2).

A total of 256 chromatograms for each of the eight *EST-SSR* loci were obtained and manually inspected using GENEMAPPER v.5 (Applied Biosystems, Waltham, MA, USA). Allele calling and binning were performed with GENEMAPPER. The presence of null alleles, stuttering and allele drop-out was tested using MICRO-CHECKER v.2.2.3 [85] with 1000 randomisations on Bonferroni correction. The multilocus *EST-SSR* genotypes were analysed using GENETIX v.4.05 [86] to estimate observed (H_O) and expected heterozygosity (H_E) and the number of alleles (A). Jack-knifing over loci was performed to assess the single-locus effects on Weir and Cockerham's F-statistics estimators. The deviation from the Hardy–Weinberg equilibrium (HWE) and Linkage Disequilibrium (LD) was investigated using GENEPOP on the web v.4.2 [87]. Allelic richness (Ar) and inbreeding coefficient (F_{IS}) were estimated using FSTAT v.2.9.3.2 [88].

2.2.2. Population Connectivity and Phylogenetic Signals

The phylogenetic relationships among individual haplotypes were inferred by the TCS method implemented in the software POPART [89]. The graphical representation of the resulting network has been modified with Adobe Photoshop.

Population connectivity within the *R. miraletus* species complex was investigated by estimates of Φst and Fst values using ARLEQUIN with 10,000 permutations, $p < 0.05$. The Tamura-Nei (1993) substitution model was applied to the mtDNA dataset to estimate Φst values. Genetic heterogeneity among the geographical samples was also assessed by a hierarchical analysis of molecular variance (AMOVA, [90]). Significance was assessed using a null distribution of the test statistic generated by 10,000 random permutations of the individuals in the samples. The significance threshold of the pairwise comparisons ($p < 0.05$) was adjusted with the sequential Bonferroni correction for multiple simultaneous comparisons [91] implemented in the R package "sgof" [92].

To unravel the individual-based genetic clustering, the *COI* and the *EST-SSR* datasets were analysed using the Bayesian algorithm implemented in BAPS v.6.0 [93] and STRUC-TURE v.2.3.4 [94], respectively. The latter analysis on SSRs loci was carried out assuming an admixture ancestry model with the geographical origin of samples as prior information (LOCPRIOR models), associated with a correlated allele frequencies model. For each simulation of K (1–20), five independent replicates were run, setting a burn-in of 200,000 iterations and 500,000 iterations for the Markov Chain Monte Carlo (MCMC) simulation. Cluster matching and permutation were performed using CLUMPAK [95], while the most likely value for K was estimated from the mean log probability of the data using four alternative statistics (medmedk, medmeak, maxmedk and maxmeak) carried out using STRUCTURESELECTOR [96].

Discriminant analysis of principal components (DAPC) using the R package Adegenet v.2.0.1 [97] was implemented in R v.4.0.5 (R Core Team [98]) using sampling locations as a priori groups (K = 5). Then, the optimal number of PCs to use in the DAPC was determined using the optim.a.score() command.

Phylogenetic relationships between and within the Central–Southern African and Northeastern Atlantic–Mediterranean *COI* lineages were estimated using a Bayesian coalescent approach, implemented in BEAST v.1.10.4 [99]. Sequences of *R. undulata* (BOLD record ELAME177-09, NCBI accession numbers KT307412, KT307413, KT307414), the closest related species to the *R. miraletus* complex, were used as outgroups. The Bayesian reconstruction was obtained using the Hasegawa, Kishino and Yano (HKY + G) model of evolution [100], as the most appropriated model inferred by MEGA software, a strict molecular clock model, the Yule Process as species tree prior, the Piecewise linear and constant root as population size prior. To ensure convergence of the posterior distributions, an MCMC run of 60,000,000 generations sampled every 1000 generations with the first 25% of the sampled points removed as burn-in was performed. We analysed the log file using TRACER v.1.7.2 [101] to calculate the robustness of the posterior distributions for all parameters and recover average divergence time and 95% confidence intervals. The plausible trees obtained with BEAST were summarised using the program TREEANNOTATOR and the resulting phylogenetic relationships among population samples and the posterior probabilities at nodes were visualised with FigTree v.1.4.4 (available at http://tree.bio.ed.ac.uk/software/figtree/, accessed on 21 May 2023) and edited with the iTol v.6.7.5 online tool [102].

Cryptic species were also delimited by using two different methods: the distance-based method "Automatic Barcode Gap Discovery" (ABGD; [103]) computed on the online web application (http://wwwabi.snv.jussieu.fr/public/abgd/abgdweb.html, accessed on 23 May 2023), using default values, and the phylogenetic-based method Bayesian Poisson Tree Process (bPTP, [104]), conducted on the web server (available at http://species.h-its.org/ptp/, accessed on 23 May 2023) with 100,000 MCMC generations, a thinning interval of 100% and 10% of burn-in.

3. Results

3.1. Genetic Diversity

The *COI* dataset was a total of 322 sequences, while the *EST-SSR* dataset was made up of a total of 256 individuals overall distributed in 14 geographic samples (Tables 1 and S1). The final *COI* alignment consisted of 529 nucleotide positions and included 76 variable sites (14.3%) and 64 parsimony informative sites (12.1%). On average, *COI* polymorphism showed low estimates of nucleotide diversity (π) and very high haplotype diversity (Hd), with ANG being the most polymorphic sample (Hd = 0.858 ± 0.041 SD, π = 0.02543 ± 0.00380 SD, K = 13.453; Table S3). Thirty-nine haplotypes were found and none were shared between samples from the Northeastern Atlantic–Mediterranean and Central–Southern African Regions (Figure 2 and Table S4 in Supplementary Material). The average Tamura-Nei genetic distances (DTN) among geographical samples of the Northeastern Atlantic–Mediterranean were extremely low (DTN = 0.0025 ± 0.0011 SE; Table 2), while those observed among geographical samples of the Central–Southern Africa were an order of magnitude higher (mean DTN = 0.0183 ± 0.0029 SE). The DTN between Northeastern Atlantic–Mediterranean and Central–Southern Africa samples were much higher (mean DTN = 0.0733 ± 0.0117 SE) and is comparable between species distances recorded among species in the genus *Raja* (Table 2).

Summary statistics of the eight polymorphic microsatellite loci per geographical sample and over all the loci considered are shown in Table S5 in the Supplementary Material. Mean allelic richness (Ar_{mean}) ranged from 1.242 (POR) to 1.724 (SEN). After Bonferroni correction, significant LD was not detected between any pair of loci, and the average mean observed and expected heterozygosity (H_O/H_E) for the eight loci was 0.259/0.392. After applying the Bonferroni correction, significant HWE departures were found over all loci in several geographical samples, apart from SEN, POR, SIC and ISR. The Portuguese sample was monomorphic at five loci (LERI 26, LERI 34, LERI 63, LERI 40 and LERI 44). Overall, MICRO-CHECKER results detected the presence of scoring errors such as stuttering and null alleles in all loci (Table S5), regardless of the geographical sample. Nevertheless, we did not exclude any of them since Jack-knife analysis did not reveal outliers outside of the confidence interval (Table S6).

Table 2. Mean genetic distances within (in grey, in diagonal) and between congeneric *Raja* species. Standard error values for distances between species are reported above the diagonal. CSA, Central–Southern Africa; NEAM, Northeastern Atlantic–Mediterranean Sea. The mean genetic distance between NEAM and CSA is indicated in bold.

	Raja asterias	*Raja brachyura*	*Raja clavata*	*Raja microocellata*	*Raja montagui*	*Raja polystigma*	*Raja radula*	*Raja straeleni*	*Raja undulata*	*Raja miraletus* (CSA)	*Raja miraletus* (NEAM)
Raja asterias	0.0025 ± 0.0012	0.0139	0.0106	0.0132	0.0135	0.0121	0.0095	0.0111	0.0121	0.0155	0.0189
Raja brachyura	0.0863	0.0030 ± 0.0016	0.0107	0.0094	0.0119	0.0114	0.0121	0.0125	0.0136	0.0141	0.0172
Raja clavata	0.0591	0.0514	0.0038 ± 0.0000	0.0115	0.0103	0.0113	0.0070	0.0059	0.0141	0.0133	0.0169
Raja microocellata	0.0877	0.0465	0.0598	0.0000 ± 0.0000	0.0125	0.0107	0.0114	0.0115	0.0127	0.0140	0.0172
Raja montagui	0.0829	0.0606	0.0511	0.0638	0.0000 ± 0.0000	0.0066	0.0126	0.0116	0.0128	0.0132	0.0172
Raja polystigma	0.0748	0.0526	0.0514	0.0596	0.0229	0.0000 ± 0.0000	0.0108	0.0112	0.0123	0.0119	0.0164
Raja radula	0.0487	0.0685	0.0280	0.0648	0.0646	0.0564	0.0010 ± 0.0012	0.0071	0.0119	0.0137	0.0173
Raja straeleni	0.0593	0.0593	0.0150	0.0601	0.0590	0.0558	0.0282	0.0019 ± 0.0010	0.0122	0.0131	0.0174
Raja undulata	0.0767	0.0736	0.0733	0.0799	0.0790	0.0711	0.0742	0.0735	0.0009 ± 0.0010	0.0122	0.0170
Raja miraletus (CSA)	0.1053	0.0857	0.0861	0.0947	0.0830	0.0778	0.0908	0.0907	0.0770	0.0183 ± 0.0029	0.0117
Raja miraletus (NEAM)	0.1072	0.1020	0.0989	0.1007	0.0921	0.0897	0.0974	0.1033	0.0959	**0.0733**	0.0025 ± 0.0010

3.2. Population Connectivity and Phylogenetic Signals

Caution should be applied when interpreting the results obtained here due to the small sample size for some localities and the subsequent decrease in the discriminatory power of the analyses.

The TCS network of the *COI* haplotypes (Figure 2) identified two main haplogroups, differentiated by at least 30 mutations and corresponding to the Central–Southern African and the Northeastern Atlantic–Mediterranean samples. The former haplogroup included 23 haplotypes that grouped into four largely differentiated geographic clusters located off Senegal, Angola/Namibia/South Africa and two off Angola. The Senegalese cluster formed only by the SEN sample (N = 5) showed three slightly differentiated private haplotypes. In contrast, the Angolan sample (ANG, N = 27) showed strongly differentiated haplotypes grouped into two endemic Angolan subclusters together and a third cluster shared with the South African (SAF, N = 40) and Namibian (NAM = 3) samples. The Northeastern Atlantic–Mediterranean haplogroup included 16 weakly divergent haplotypes (Figure 2 and Table S4). Four of them were shared by several samples and areas: (i) the haplotype Hap_24 was shared by Portuguese, Algerian and Strait of Sicily samples; (ii) the most frequent Hap_25 was shared by samples from Algeria, Balearic Islands, Sardinia, Strait of Sicily, Tuscany and Adriatic Sea; (iii) the Hap_28 was shared by samples from Algeria, Strait of Sicily and Adriatic Sea; iv) the Hap_32 was shared by Adriatic and Greek samples. In contrast, the Eastern Mediterranean samples from the Israeli coast (Hap_37 and Hap_38) and Levantine Sea (Hap_39) yielded only three endemic haplotypes.

Most of the pairwise Φst values among 14 geographical samples based on *COI* data were significant even after the Bonferroni correction was applied (Table S7). High levels of differentiation were observed between the African and Northeastern Atlantic–Mediterranean samples, as well as between the Western and Eastern Mediterranean. The

EST-SSR data showed a similar pattern of genetic differentiation in terms of Fst calculated over 12 macro areas, even after the Bonferroni correction was applied (Table S8).

The hierarchical AMOVA found the highest percentage of molecular variation among groups with four sample groupings (AMOVA 3: Southern Africa vs. Angola vs. Senegal vs. Northeastern Atlantic–Mediterranean Sea; Table S9) when using the *COI* dataset, and with five sample groups (40.06%, AMOVA 4: Southern Africa vs. Angola vs. Senegal vs. Portugal–West Mediterranean vs. Eastern Mediterranean) when using the AMOVA 5 (six groups) explained the genetic variation among samples with the proportion of the genetic variation among populations within very low groups (2.58% with mitochondrial data and 7.72% with the *EST-SSR* data).

The Bayesian clustering analysis based on mtDNA data (Figure 3a) revealed six genetic clusters: the first cluster (green) included individuals from SAF, NAM and ANG; the second (purple) and third clusters (light blue) ANG; the fourth cluster (red) was unique to SEN; the fifth (yellow) and sixth (blue) clusters were unique to the Northeastern Atlantic and Mediterranean with individuals from POR, BAL, SAR and TUS in the fifth cluster and those from GRE, ISR and LEV in the sixth cluster. Individuals from ALG, SIC and ADR were randomly associated with the fifth and the sixth clusters, suggesting the southern Mediterranean, especially the Siculo-Tunisian area, as a potential admixture zone of the latter clusters.

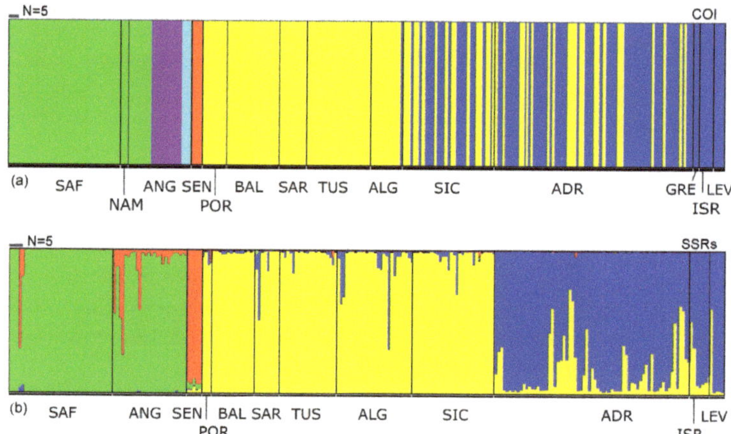

Figure 3. Bayesian admixture analysis among individuals belonging to the *Raja miraletus* species complex across most of its geographical distribution area. (**a**) Distribution of the *cytochrome oxidase subunit I* clades in the population samples inferred with BAPS; each colour represents one distinct haplogroup (cluster), and each bar represents a different individual. (**b**) Results of the Bayesian individual clustering using STRUCTURE results for K = 4; each vertical bar represents one individual, in which a different colour represents the estimated cluster membership. Acronyms of the Macro Area Codes are given in Table 1.

The outputs of the STRUCTURE analysis based on *EST-SSR* data analysed with STRUCTURESELECTOR did not provide clear-cut evidence of the most likely number of clusters using four alternative statistics (medmedk, medmeak, maxmedk and maxmeak), while the maximum value of ΔK was verified with K = 3 (Figures S1 and S2). Thus, the results from K = 2 to K = 7 were assessed with CLUMPAK (Figure S3). The barplot of the clustering K = 2 supported the separation of the Central–Southern Africa samples from Northeastern Atlantic–Mediterranean Sea and with an admixed genetic composition of the Senegalese individuals (Figure S3). The clustering K = 3 further discriminated between the Angolan sample from those of South Africa, as well as samples from the Western and Eastern Mediterranean Sea (Figure S3). At the same time, the Angolan sample displayed an intermediate genetic composition between the South African and Senegalese clusters. This trend was resolved by the clustering K = 4 (Figures 3b and S3), corresponding to the best grouping revealed by AMOVA (Table S9). The plot showed a deep differentiation of the Northeastern Atlantic–Mediterranean samples east to the Strait of Sicily.

The DAPC computed on *EST-SSR* data with sampling locations used as a priori group identified 10 optimal numbers of PCs and separated five main clusters: South African, Angolan, Senegalese, Northeastern Atlantic–Western Mediterranean and Eastern Mediterranean clusters, with the South African and Angolan clusters partially overlapping (Figure S4).

Bayesian approach using MCMC simulation was used to test for a speciation signal (Yule process) between lineages from Central–Southern Africa and Northeastern Atlantic–Mediterranean (Figure 4, see Table S4 for haplotype distribution among samples). All effective sample size (ESS) values exceeded 200, indicating a solid evaluation of all parameters. The model based on the substitution rate estimated for mtDNA showed a clear separation between haplotypes from Central–Southern Africa (Hap_1 to Hap_23) and the Northeastern Atlantic–Mediterranean (Hap_24 to Hap_39; Figure 4). The phylogenetic relationships among lineages and haplotypes were congruent with the relationships obtained with the parsimony network results (Figure 2). Furthermore, within the main Central–Southern African lineage, four clusters of haplotypes were reconstructed with high posterior probability (PP = 1): the most basal Angolan haplotype H_20; a second cluster formed by six Angolan haplotypes (Hap_14 to Hap_19); a Senegalese cluster (Hap_21 to Hap_23); and a South African/Angolan/Namibian cluster formed by all the South African haplotypes (Hap_1–9), the Namibian haplotype Hap_10, the Angolan haplotypes (Hap_12 and Hap_13) and the Angolan/Namibian haplotype (Hap_11).

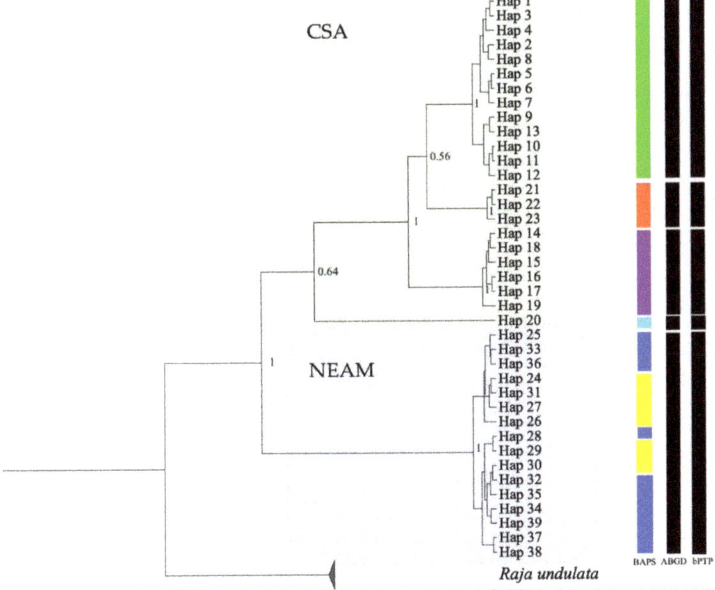

Figure 4. Bayesian coalescent tree summarising phylogenetic relationships between the Central–Southern African (CSA) and Northeastern Atlantic–Mediterranean (NEAM) *COI* lineages. *Raja undulata* was used as an outgroup (BOLD record ELAME177-09, NCBI accession numbers KT307412, KT307413, KT307414). Posterior probability (≥0.5) values are reported on nodes. Coloured bars near haplotype nodes refer to the genetic clusters identified by BAPS results (Figure 3a). The outcomes of species delimitation analyses using ABGD and bPTP methods are shown as vertical bars on the right.

The two species delimitation approaches (the ABGD and bPTP methods) yielded the same result (Figure 4). Five groups have been identified: four are formed by samples from the Central–Southern Atlantic, in agreement with the phylogenetic reconstruction, and one formed by samples from the Northeastern Atlantic and Mediterranean.

4. Discussion

The *R. miraletus* species complex is widely distributed, occurring from the Mediterranean Sea down the west coast of Africa to South Africa. The geographically isolated population off the south coast of South Africa was originally described as *R. ocellifera* Regan, 1906, but was synonymised with *R. miraletus* [105]. An extensive morphological study by McEachran et al. [48] found a marked differentiation between specimens from South Africa and those from Mediterranean elements, whilst the West African samples, in particular those from Angola, showed intermediate meristic features. However, they considered the differences between the Mediterranean and South African populations to be clinal and concluded that *R. miraletus* is a polymorphic species with three partially separated populations. Strikingly stable gross morphology has always been misleading for taxonomists. Subsequent integrated molecular and morphometric studies have shown that the three partially separated populations are valid species [51]. *Raja ocellifera* has been resurrected [44] and Last and Séret [51] described a new species, *R. parva* from Liberia and Angola. Last and Séret [51] stated that *R. parva* differed from the Senegalese *Raja* cf. *miraletus* 1 (sensu Naylor et al. [49]) in total body length, a shorter snout and a smaller tail, and they suggest that two further putative species (or taxa) occur off Senegal, Guinea, Liberia, down to Angola. Among the West African samples, the *R. parva* were the most distinct, even though some characters were intermediate between the Mediterranean and South African specimens [48].

The recent designation of the *R. miraletus* species complex [51] increased interest in an evolutionary and phylogenetic investigation of the complex based on more extensive sampling and analysis of combined nuclear and mitochondrial genetic data. Ferrari et al. [73] inferred population structure within *R. miraletus* (sensu stricto) across the Mediterranean Sea based on an analysis of nucleotide variation in three mtDNA markers, while Crobe et al. [67] preliminarily recognised four divergent *COI* lineages from the Eastern Atlantic populations of this morphologically conserved taxon. It should be noted that our study (i) was based on an unprecedently high number of specimens unequivocally assigned to the *R. miraletus* species complex; (ii) included specimens collected from the localities within the distribution ranges of the four putative taxa occur and (iii) was based on an integrated analytical approach that combines sequence variation of both mtDNA (the universal *COI* barcode region) and nuDNA(allelic variation in eight polymorphic *EST-linked* microsatellite loci) markers. This methodology enables our study to advance the molecular characterisation of this Hamletic taxon and to increase the knowledge on its status in the Mediterranean Sea.

The mitochondrial and microsatellite data consistently agreed in genetically defining the taxonomic and geographical boundaries of *R. miraletus* (sensu stricto), which are distributed in the entire Mediterranean Sea and the adjacent Northeastern Atlantic Ocean, at least in the Portuguese coastal waters. The estimated mean Tamura-Nei genetic distance between the Northeastern Atlantic–Mediterranean clade and the Central–Southern African clade (DTN = 0.0735) was in the upper range of corresponding pairwise interspecific estimates among several congeneric *Raja* species supporting a specific level of differentiation.

Surprisingly, the host–parasite specificity established between different African population of the *R. miraletus* species complex, with species from the genus *Echinobothrium* (Cestoda: Diphyillidea) highlighted by Caira et al. [50], supports the clades identified by Naylor et al. [49] based on molecular data. Specimens of *R. ocellifera* (*R.* cf. *miraletus* 1 from South Africa from [49]) hosted *E. yiae* Caira, Rodriguez and Pickering, 2013, those of *R. miraletus* from Senegal hosted *E. mercedesae* Caira, Rodriguez and Pickering, 2013 and specimens of a second clade off Senegal likely belonging to *R. parva* (*R.* cf. *miraletus*

2 sensu Naylor et al. [49]) hosted two additional new species [50]. These findings were partially supported by our study, where the Senegalese sample (SEN) is a distinct subunit (Figures 2 and 3; Hap_21, Hap_22 and Hap_23 in Figure 4; Φst in Table S7). Unfortunately, the very limited sampling along the Senegalese coast (N = 5) prevents any definitive conclusions.

Overall, our results emphasised the strong differentiation between South Atlantic-Indian *R. ocellifera* and the Northeastern Atlantic–Mediterranean *R. miraletus* and justifies the resurrection of the former taxon. The complex oceanographic conditions along the African coast with alternating cold and warm currents, from north to south cold Canary Current, warm Angola Current, cold Benguela Current and warm Agulhas Current, undoubtedly played a role in speciation along the African coast. Likewise, the complexity oceanographic and geological discontinuities characterising the Eastern Atlantic and the Mediterranean Basin may likely influence phylogeography, population structure and connectivity as well as evolution at multiple taxonomic levels [48,106]. Oceanographic heterogeneities, such as current systems, play a key role for ecologically divergent natural selection in elasmobranchs, such as the ecological radiation of the genus *Pseudobatos* Last, Séret and Naylor, 2016 in the Gulf of California, strongly influenced by habitat heterogeneity and the geological history of the region [107]. This condition seemed to be true not only for skates, but even more so for African coastal bony fish (i.e., genus *Argynosomus* De la Plyaie, 1835), whose evolutionary histories, including the dispersal phase, have been influenced by the Benguela Current [108,109]. The Benguela Current region ranges from Cape Agulhas to Cape Frio, where the north-flowing cold Benguela Current meets the south-flowing warm Angola current (see Hirschfeld et al. [106], Figure 2d,f for current model map details). The cold waters of the Benguela system are likely to have strongly reduced gene flow between *R. ocellifera* and *Raja* cf. *miraletus* (sensu Naylor et al. [49]). This diversification was evident from the level of mean sequence divergence observed between the two geographical populations estimated at 7.3%, a value comparable to, or higher than, any divergence measured between *R. undulata* and other congeneric taxa. On the other hand, the intertropical Canary current inflowing from the northeast could have influenced the diversification of the Senegalese taxa, whose migration southwards would be hampered by the intermittent Cape Blanc upwelling area. This upper boundary could have influenced and limited gene flow in Northern Mauritania. Like other skates [10,110,111] and teleost species [112], no genetic differentiation was observed between Northeastern Atlantic and Mediterranean populations of *R. miraletus*. This suggests that the Strait of Gibraltar did not represent an effective barrier to gene flow, rather than an accession gate to ancient *refugia*. In contrast, Melis et al. [113] found moderate significant population differentiation between the Mediterranean and the Atlantic Ocean in the congeneric thornback skate *R. clavata*, suggesting an effective role of the Strait in limiting the dispersal of individuals.

The slight genetic population structure observed within the Mediterranean Sea represents a true novelty for this species complex. The unforeseen regional East–West Mediterranean structure highlighted by nuclear markers (Figures 3b and S4) could be linked to bathymetry and hydrogeological fronts or discontinuities. In particular, the shallow bathymetry characterising the Southwest part of the Mediterranean, coupled with the species' preferences for continental shelf habitats, may likely enhance the dispersal of brown skate. The area ranging from the easternmost part of Sicily and the adjacent geomorphological depression of the Calabrian Arc (down to 3000 m) is dominated by cyclonic/anticyclonic inversions of water masses. The combination of these features could have limited gene flow between western and eastern populations and driven the differentiation of the Eastern Mediterranean samples (ADR, ISR and LEV; Figure 1). The specific habitat and depth preferences, the less pronounced migratory behaviour and the limited dispersal capability of taxa belonging to the *R. miraletus* species complex are rather common characteristics among skates [62,73,113,114], although other congeneric species did not show such evidence of deep differentiation at both nuDNA and mtDNA markers [10,56,72,110]. In detail, the Mediterranean starry ray *R. asterias* showed a strongly structured population with three geographical clades corresponding to the western, central–western and central–eastern Mediterranean areas [62]. *R. clavata* showed a weak but detectable phylogeographic structure in the Levantine Sea [73] and a finer structuring located off the Algerian coasts and Tyrrhenian basins, suggesting the occurrence of additional barriers to dispersal [113]. On the contrary, *R. polystigma* showed a slightly differentiated Adriatic haplotype but a near panmictic population in the central–western part of the basin [56]. These different patterns of population structure in such closely related species can be explained by their bathymetric range which drive different ecological features [72].

5. Conclusions

This study coupled a massive and extensive sampling effort covering the full distribution of the *R. miraletus* species complex with analysis of genetic variability in both mtDNA and nuDNA markers, individual clustering, phylogeography and variance at different population levels. The results were partially congruent with previous taxonomic and meristic analyses. The use of both nuclear and mitochondrial markers resulted in identifying signals of species differentiation and in supporting the existence of at least five cryptic taxa within the *R. miraletus* species complex, four of which have been previously suggested with scattered genetic data. In addition to the evolutionary meaning of this evidence, genetics is shown to aid conservation efforts by revealing hidden diversity that deserves special attention and in the monitoring of taxa that are important fishing bycatch species. The new insights highlighted in the present research paper suggest that the extraordinary intraspecific diversity observed across such a wide geographical scale should be carefully considered to update or set dedicated and effective measures to reduce the impact of skate bycatch during fishing activities and improve their conservation.

Supplementary Materials: The following supporting information can be downloaded at: https://www.mdpi.com/article/10.3390/ani13132139/s1, Text S1: Molecular methods; Table S1: List of samples with detail of individual code, location, geographical coordinates, biological data and indication of available sequences integrated to the original dataset (BOLD records and NCBI Accession Numbers); Table S2: List of public sequences used for the estimation of genetic distances among the genus *Raja*. Details of BOLD Record and NCBI Accession Numbers are also listed; Table S3: mtDNA polymorphism and its parameters. N, number of individuals; Nh, number of haplotypes; S, number of polymorphic sites; Hd, haplotype diversity; π, nucleotide diversity; k, average number of nucleotide differences and SD, standard deviation. The locations are indicated by codes as given in Table 1; Table S4: Distribution of the *COI* haplotypes in the *Raja miraletus* complex population samples; Table S5: Summary statistics of the *EST-SSRs* polymorphism per geographical sample and over all the loci considered. Table S6: Jack-knifing over all loci at 95% confidence, 1000 bootstraps; Table S7: Pairwise Φst values (below the diagonal) and associated significance (above the diagonal). Significant *p*-values are highlighted in bold ($p < 0.05$), * *p* significant after sequential Bonferroni; Table S8: Pairwise Fst values (below the diagonal) and associated significance (above the diagonal). Significant *p*-values are highlighted in bold ($p < 0.05$), * *p* significant after sequential Bonferroni; Table S9: Five-levels AMOVA performed from two and six sample groupings on both mtDNA and nuDNA; Figure S1: Processing of STRUCTURE results from the analysis conducted on *R. miraletus* individuals using medmedk, medmeak, maxmedk and maxmeak statistics obtained with STRUCTURESELECTOR; Figure S2: Absolute value of derived delta K (ΔK) for each cluster number (K) from one to 12; Figure S3: Results of the Bayesian clustering using STRUCTURE from K = 2 to K = 7 obtained with CLUMPAK; each vertical bar represents one individual, in which a different colour represents the estimated cluster membership. Acronyms of the samples are given in Table 1; Figure S4: Discriminant analysis of principal components (DAPCs) scatter plot conducted on 256 *Raja miraletus* complex individuals based on *EST-SSRs* loci. Clusters refer to SAF, South Africa; ANG, Angola; SEN, Senegal; NATL-WMED, Northeastern Atlantic Ocean and Western Mediterranean Sea; EMED, Eastern Mediterranean Sea. Colours refer to the genetic clusters observed in BAPS plot (Figure 2).

Author Contributions: Conceptualisation, F.T. and A.C.; data curation, A.F.; investigation, A.F. and V.C.; methodology, A.F.; resources, R.C., R.W.L., F.S., M.S., F.O.C., D.G., F.H., D.Z.-P., L.S., P.C. and F.F.; supervision, F.T. and A.C.; visualisation, A.F.; writing—original draft, A.F. and V.C.; writing—review and editing, V.C., R.C., R.W.L., F.S., M.S., F.O.C., D.G., F.H., D.Z.-P., L.S., P.C., F.F., F.T. and A.C. All authors have read and agreed to the published version of the manuscript.

Funding: This research received no dedicated funding and was supported by institutional funding from University of Bologna assigned to FT and AC (RFO and Canziani grants).

Institutional Review Board Statement: The authors declare that the sample of brown skate individuals analysed in the present work were obtained from commercial and scientific fisheries. The activity was conducted with the observation of the Regulation of the European Parliament and the Council for fishing in the General Fisheries Commission for the Mediterranean (GFCM) Agreement area and amending Council Regulation (EC) No. 1967/2006. This Regulation is de facto the unique authorisation needed to conduct this type of activities. Dead fish vertebrates are out of scope of "the University of Bologna—Animal Care and Use Committee" since it authorises/not authorises experimental procedures on living vertebrates.

Informed Consent Statement: Not applicable.

Data Availability Statement: The newly produced sequence data that support the findings of this study are openly available in GenBank at https://www.ncbi.nlm.nih.gov/genbank (accessed on 27 June 2023) reference number OR193802–OR194004. The *EST-SSR* data that support the findings of this study are available as a genotype matrix from the corresponding author upon request.

Acknowledgments: We are deeply grateful to all the Mediterranean Trawl Survey, the South African Trawl Survey and the EAF-Nansen Programme partners for their collaboration in the *R. miraletus* complex sample collection. Moreover, we thank Samira Vinjau and Alessia Crosara for their contribution in the analysis of specimens.

Conflicts of Interest: The authors declare no conflict of interest.

References

1. Harrison, R.G. Linking evolutionary pattern and process. In *Endless Forms*; Oxford University Press: Oxford, UK, 1998; pp. 19–31.
2. Mallet, J. Hybridization, ecological races and the nature of species: Empirical evidence for the ease of speciation. *Philos. Trans. R. Soc. B Biol. Sci.* **2008**, *363*, 2971–2986. [CrossRef] [PubMed]
3. Harrison, R.G.; Larson, E.L. Hybridization, introgression, and the nature of species boundaries. *J. Hered.* **2014**, *105*, 795–809. [CrossRef] [PubMed]
4. Norris, R.D. Pelagic species diversity, biogeography, and evolution. *Paleobiology* **2000**, *26*, 236–258. [CrossRef]
5. Pinheiro, H.T.; Bernardi, G.; Simon, T.; Joyeux, J.-C.; Macieira, R.M.; Gasparini, J.L.; Rocha, C.; Rocha, L.A. Island biogeography of marine organisms. *Nature* **2017**, *549*, 82–85. [CrossRef] [PubMed]
6. Palumbi, S.R.; Lessios, H.A. Evolutionary animation: How do molecular phylogenies compare to Mayr's reconstruction of speciation patterns in the sea? *Proc. Natl. Acad. Sci. USA* **2005**, *102*, 6566–6572. [CrossRef]
7. Lynghammar, A.; Christiansen, J.S.; Griffiths, A.M.; Fevolden, S.-E.; Hop, H.; Bakken, T. DNA barcoding of the Northern Northeast Atlantic skates (Chondrichthyes, Rajiformes), with remarks on the widely distributed starry ray. *Zool. Scr.* **2014**, *43*, 485–495. [CrossRef]
8. Palumbi, S.R. Genetic divergence, reproductive isolation, and marine speciation. *Annu. Rev. Ecol. Syst.* **1994**, *25*, 547–572. [CrossRef]
9. Avise, J.C.; Neigel, J.E.; Arnold, J. Demographic influences on mitochondrial DNA lineage survivorship in animal populations. *J. Mol. Evol.* **1984**, *20*, 99–105. [CrossRef]
10. Pasolini, P.; Ragazzini, C.; Zaccaro, Z.; Cariani, A.; Ferrara, G.; Gonzalez, E.G.; Landi, M.; Milano, I.; Stagioni, M.; Guarniero, I.; et al. Quaternary geographical sibling speciation and population structuring in the eastern Atlantic skates (Suborder Rajoidea) *Raja clavata* and *R. straeleni*. *Mar. Biol.* **2011**, *158*, 2173–2186. [CrossRef]
11. Milá, B.; Van Tassell, J.L.; Calderón, J.A.; Rüber, L.; Zardoya, R. Cryptic lineage divergence in marine environments: Genetic differentiation at multiple spatial and temporal scales in the widespread intertidal goby *Gobiosoma bosc*. *Ecol. Evol.* **2017**, *7*, 5514–5523. [CrossRef]
12. Bickford, D.; Lohman, D.J.; Sodhi, N.S.; Ng, P.K.; Meier, R.; Winker, K.; Ingram, K.K.; Das, I. Cryptic species as a window on diversity and conservation. *Trends Ecol. Evol.* **2007**, *22*, 148–155. [CrossRef] [PubMed]
13. Mayr, E. *Systematics and the Origin of Species from the Viewpoint of a Zoologist*; Columbia University Press: New York, NY, USA, 1942; p. 372, ISBN 9780674862500.
14. Knowlton, N. Sibling species in the sea. *Annu. Rev. Ecol. Syst.* **1993**, *24*, 189–216. [CrossRef]
15. Nygren, A. Cryptic polychaete diversity: A Review. *Zool. Scr.* **2014**, *43*, 172–183. [CrossRef]
16. Bellono, N.W.; Leitch, D.B.; Julius, D. Molecular tuning of electroreception in sharks and skates. *Nature* **2018**, *558*, 122–126. [CrossRef] [PubMed]

17. McEachran, J.D.; Dunn, K.A. Phylogenetic analysis of skates, a morphologically conservative clade of Elasmobranchs (Chondrichthyes: Rajidae). *Copeia* **1998**, *1998*, 271–290. [CrossRef]

18. Tinti, F.; Ungaro, N.; Pasolini, P.; De Panfilis, M.; Garoia, F.; Guarniero, I.; Sabelli, B.; Marano, G.; Piccinetti, C. Development of molecular and morphological markers to improve species-specific monitoring and systematics of Northeast Atlantic and Mediterranean skates (Rajiformes). *J. Exp. Mar. Biol. Ecol.* **2003**, *288*, 149–165. [CrossRef]

19. Iglésias, S.P.; Toulhoat, L.; Sellos, D.Y. Taxonomic confusion and market mislabelling of threatened skates: Important consequences for their conservation status. *Aquat. Conserv. Mar. Freshw. Ecosyst.* **2010**, *20*, 319–333. [CrossRef]

20. Carugati, L.; Melis, R.; Cariani, A.; Cau, A.; Crobe, V.; Ferrari, A.; Follesa, M.C.; Geraci, M.L.; Iglésias, S.P.; Pesci, P.; et al. Combined COI barcode-based methods to avoid mislabelling of threatened species of deep-sea skates. *Anim. Conserv.* **2022**, *25*, 38–52. [CrossRef]

21. Griffiths, A.M.; Sims, D.W.; Cotterell, S.P.; El Nagar, A.; Ellis, J.R.; Lynghammar, A.; McHugh, M.; Neat, F.C.; Pade, N.G.; Queiroz, N.; et al. Molecular markers reveal spatially segregated cryptic species in a critically endangered fish, the common skate (*Dipturus batis*). *Proc. R. Soc. B Biol. Sci.* **2010**, *277*, 1497–1503. [CrossRef]

22. Cannas, R.; Follesa, M.C.; Cabiddu, S.; Porcu, C.; Salvadori, S.; Iglésias, S.P.; Deiana, A.M.; Cau, A. Molecular and morphological evidence of the occurrence of the norwegian skate *Dipturus nidarosiensis* (Storm, 1881) in the Mediterranean Sea. *Mar. Biol. Res.* **2010**, *6*, 341–350. [CrossRef]

23. Carbonara, P.; Bellodi, A.; Zupa, W.; Donnaloia, M.; Gaudio, P.; Neglia, C.; Follesa, M.C. Morphological traits and capture depth of the Norwegian skate (*Dipturus nidarosiensis* (Storm, 1881)) from two Mediterranean populations. *J. Mar. Sci. Eng.* **2021**, *9*, 1462. [CrossRef]

24. Domingues, R.R.; Hilsdorf, A.W.S.; Gadig, O.B.F. The importance of considering genetic diversity in shark and ray conservation policies. *Conserv. Genet.* **2018**, *19*, 501–525. [CrossRef]

25. Serena, F.; Mancusi, C.; Barone, M. Field identification guide to the skates (Rajidae) of the Mediterranean Sea. Guidelines for data collection and analysis. *Biol. Mar. Mediterr.* **2010**, *17*, 204.

26. Cariani, A.; Messinetti, S.; Ferrari, A.; Arculeo, M.; Bonello, J.J.; Bonnici, L.; Cannas, R.; Carbonara, P.; Cau, A.; Charilaou, C.; et al. Improving the conservation of Mediterranean Chondrichthyans: The ELASMOMED DNA barcode reference library. *PLoS ONE* **2017**, *12*, e0170244. [CrossRef]

27. Ebert, D.A.; Compagno, L.J. Biodiversity and systematics of skates (Chondrichthyes: Rajiformes: Rajoidei). *Biol. Skates* **2007**, *27*, 5–18.

28. Last, P.R.; White, W.T.; de Carvalho, M.R.; Séret, B.; Stehmann, M.F.W.; Naylor, G.J.P. (Eds.) *Rays of the World*; CSIRO Publishing: Clayton, Australia; Comstock Publishing Associates: Sacramento, CA, USA; pp. i–ix + 1–790. ISBN 978-0-643-10914-8.

29. Serena, F.; Abella, A.J.; Bargnesi, F.; Barone, M.; Colloca, F.; Ferretti, F.; Fiorentino, F.; Jenrette, J.; Moro, S. Species diversity, taxonomy and distribution of Chondrichthyes in the Mediterranean and Black Sea. *Eur. Zool. J.* **2020**, *87*, 497–536. [CrossRef]

30. Grant, V. The systematic and geographical distribution of hawkmoth flowers in the temperate North American flora. *Bot. Gaz.* **1983**, *144*, 439–449. [CrossRef]

31. Quintero, I.; Keil, P.; Jetz, W.; Crawford, F.W. Historical biogeography using species geographical ranges. *Syst. Biol.* **2015**, *64*, 1059–1073. [CrossRef]

32. Henriques, S.; Guilhaumon, F.; Villéger, S.; Amoroso, S.; França, S.; Pasquaud, S.; Cabral, H.N.; Vasconcelos, R.P. Biogeographical region and environmental conditions drive functional traits of estuarine fish assemblages worldwide. *Fish Fish.* **2017**, *18*, 752–771. [CrossRef]

33. Neat, F.; Pinto, C.; Burrett, I.; Cowie, L.; Travis, J.; Thorburn, J.; Gibb, F.; Wright, P.J. Site fidelity, survival and conservation options for the threatened flapper skate (*Dipturus* cf. *intermedia*). *Aquat. Conserv. Mar. Freshw. Ecosyst.* **2015**, *25*, 6–20. [CrossRef]

34. Frisk, M.G.; Jordaan, A.; Miller, T.J. Moving beyond the current paradigm in marine population connectivity: Are adults the missing link? *Fish Fish.* **2014**, *15*, 242–254. [CrossRef]

35. Wearmouth, V.J.; Sims, D.W. Movement and behaviour patterns of the critically endangered common skate *Dipturus batis* revealed by electronic tagging. *J. Exp. Mar. Biol. Ecol.* **2009**, *380*, 77–87. [CrossRef]

36. Hunter, E.; Buckley, A.A.; Stewart, C.; Metcalfe, J.D. Repeated seasonal migration by a thornback ray in the southern North Sea. *J. Mar. Biol. Assoc. UK* **2005**, *85*, 1199–1200. [CrossRef]

37. Hunter, E.; Buckley, A.A.; Stewart, C.; Metcalfe, J.D. Migratory behaviour of the thornback ray, *Raja clavata*, in the southern North Sea. *J. Mar. Biol. Assoc. UK* **2005**, *85*, 1095–1105. [CrossRef]

38. Musick, J.A.; Ellis, J.K.; Hamlett, W. Reproductive evolution of chondrichthyans. In *HAMLETT, WC, Reproductive Biology and Phylogeny of Chondrichthyes, Sharks, Batoids and Chimaeras*; CRC Press: Boca Raton, FL, USA, 2005; pp. 45–71.

39. Compagno, L.J.V.; Ebert, D.A. Southern African Skate Biodiversity and Distribution. In *Biology of Skates*; Ebert, D.A., Sulikowski, J.A., Eds.; Developments in Environmental Biology of Fishes 27; Springer: Dordrecht, The Netherlands, 2007; pp. 19–39, ISBN 978-1-4020-9703-4.

40. Stehmann, M.F.W.; Bürkel, D.L. Rajidae. In *Fishes of the North-Eastern Atlantic and the Mediterranean*; Whitehead, P.J.P., Bauchot, M., Hureau, J., Nielsen, J., Eds.; Unesco: Paris, France, 1984; Volume 1, ISBN 9789230022150.

41. Kadri, H.; Marouani, S.; Bradai, M.N.; Bouaïn, A. Food habits of the brown ray *Raja miraletus* (Chondrichthyes: Rajidae) from the Gulf of Gabès (Tunisia). *Mar. Biol. Res.* **2014**, *10*, 426–434. [CrossRef]

42. Šantić, M.; Radja, B.; Pallaoro, A. Feeding habits of brown ray (*Raja miraletus* Linnaeus, 1758) from the eastern central Adriatic Sea. *Mar. Biol. Res.* **2013**, *9*, 301–308. [CrossRef]
43. Tsikliras, A.C.; Stergiou, K.I. Age at maturity of Mediterranean marine fishes. *Mediterr. Mar. Sci.* **2015**, *16*, 5–20. [CrossRef]
44. Dulvy, N.K. *Raja miraletus*. IUCN Red List of Threatened Species. Available online: https://doi.org/10.2305/IUCN.UK.2019-3 .RLTS.T124569516A124512700.en (accessed on 17 May 2023).
45. Dulvy, N.K.; Walls, R.H.L.; Abella, A.; Serena, F.; Bradai, M.N. *Raja miraletus* (Mediterranean Assessment). The IUCN Red List of Threatened Species. Available online: https://doi.org/10.2305/IUCN.UK.2020-3.RLTS.T124569516A176535719.en (accessed on 17 May 2023).
46. Ebert, D.A.; Wintner, S.P.; Kynes, P.M. An annotated checklist of the chondrichthyans of South Africa. *Zootaxa* **2021**, *4947*, 1–127. [CrossRef]
47. Compagno, L.J.V.; Ebert, D.A.; Smale, M.J. *Guide to the Sharks and Rays of Southern Africa*; Struik: Cape Town, South Africa, 1989.
48. McEachran, J.D.; Séret, B.; Miyake, T. Morphological variation within *Raja miraletus* and status of *R. ocellifera* (Chondrichthyes, Rajoidei). *Copeia* **1989**, *1989*, 629–641. [CrossRef]
49. Naylor, G.; Caira, J.; Jensen, K.; Rosana, K.; Straube, N.; Lakner, C. Elasmobranch phylogeny: A mitochondrial estimate based on 595 species. In *Biology of Sharks and Their Relatives*, 2nd ed.; CRC Press: Boca Raton, FL, USA, 2012; pp. 31–56, ISBN 978-1-4398-3924-9.
50. Caira, J.N.; Rodriguez, N.; Pickering, M. New African species of *Echinobothrium* (Cestoda: Diphyllidea) and implications for the identities of their skate hosts. *J. Parasitol.* **2013**, *99*, 781–788. [CrossRef]
51. Last, P.R.; Séret, B. A New eastern central Atlantic skate *Raja parva* sp. nov. (Rajoidei: Rajidae) belonging to the *Raja miraletus* species complex. *Zootaxa* **2016**, *4147*, 477–489. [CrossRef]
52. Hebert, P.D.N.; Cywinska, A.; Ball, S.L.; de Waard, J.R. Biological identifications through DNA barcodes. *Proc. R. Soc. Lond. B Biol. Sci.* **2003**, *270*, 313–321. [CrossRef]
53. Knowlton, N. Cryptic and sibling species among the decapod Crustacea. *J. Crustac. Biol.* **1986**, *6*, 356–363. [CrossRef]
54. Morgan, J.A.T.; Harry, A.V.; Welch, D.J.; Street, R.; White, J.; Geraghty, P.T.; Macbeth, W.G.; Tobin, A.; Simpfendorfer, C.A.; Ovenden, J.R. Detection of interspecies hybridisation in Chondrichthyes: Hybrids and hybrid offspring between Australian (*Carcharhinus tilstoni*) and common (*C. limbatus*) blacktip shark found in an Australian fishery. *Conserv. Genet.* **2012**, *13*, 455–463. [CrossRef]
55. Arlyza, I.S.; Shen, K.-N.; Solihin, D.D.; Soedharma, D.; Berrebi, P.; Borsa, P. Species boundaries in the *Himantura uarnak* species complex (Myliobatiformes: Dasyatidae). *Mol. Phylogenet. Evol.* **2013**, *66*, 429–435. [CrossRef]
56. Frodella, N.; Cannas, R.; Velonà, A.; Carbonara, P.; Farrell, E.; Fiorentino, F.; Follesa, M.; Garofalo, G.; Hemida, F.; Mancusi, C.; et al. Population connectivity and phylogeography of the mediterranean endemic skate *Raja polystigma* and evidence of its hybridization with the parapatric sibling *R. montagui*. *Mar. Ecol. Prog. Ser.* **2016**, *554*, 99–113. [CrossRef]
57. Siskey, M.R.; Shipley, O.N.; Frisk, M.G. Skating on thin ice: Identifying the need for species-specific data and defined migration ecology of *Rajidae* Spp. *Fish Fish.* **2019**, *20*, 286–302. [CrossRef]
58. El Nagar, A.; McHugh, M.; Rapp, T.; Sims, D.W.; Genner, M.J. Characterisation of polymorphic microsatellite markers for skates (Elasmobranchii: Rajidae) from expressed sequence tags. *Conserv. Genet.* **2010**, *11*, 1203–1206. [CrossRef]
59. Spedicato, M.T.; Massutí, E.; Mérigot, B.; Tserpes, G.; Jadaud, A.; Relini, G. The MEDITS trawl survey specifications in an ecosystem approach to fishery management. *Sci. Mar.* **2019**, *83*, 9–20. [CrossRef]
60. Relini, G. Demersal trawl surveys in Italian Seas: A short review. *Actes Colloq. IFREMER* **2000**, *26*, 76–93. Available online: https://archimer.ifremer.fr/doc/00716/82814/87636.pdf (accessed on 20 March 2023).
61. Ivanova, N.V.; Zemlak, T.S.; Hanner, R.H.; Hebert, P.D.N. Universal primer cocktails for fish DNA barcoding. *Mol. Ecol. Notes* **2007**, *7*, 544–548. [CrossRef]
62. Catalano, G.; Crobe, V.; Ferrari, A.; Baino, R.; Massi, D.; Titone, A.; Mancusi, C.; Serena, F.; Cannas, R.; Carugati, L.; et al. Strongly structured populations and reproductive habitat fragmentation increase the vulnerability of the Mediterranean starry ray *Raja asterias* (Elasmobranchii, Rajidae). *Aquat. Conserv. Mar. Freshw. Ecosyst.* **2022**, *32*, 66–84. [CrossRef]
63. Thompson, J.D.; Higgins, D.G.; Gibson, T.J. CLUSTAL W: Improving the sensitivity of progressive multiple sequence alignment through sequence weighting, position-specific gap penalties and weight matrix choice. *Nucleic Acids Res.* **1994**, *22*, 4673–4680. [CrossRef]
64. Tamura, K.; Stecher, G.; Kumar, S. MEGA11: Molecular Evolutionary Genetics Analysis version 11. *Mol. Biol. Evol.* **2021**, *38*, 3022–3027. [CrossRef]
65. Moulton, M.J.; Song, H.; Whiting, M.F. Assessing the effects of primer specificity on eliminating numt coamplification in DNA barcoding: A case study from Orthoptera (Arthropoda: Insecta). *Mol. Ecol. Resour.* **2010**, *10*, 615–627. [CrossRef]
66. Ratnasingham, S.; Hebert, P.D.N. Bold: The Barcode of Life Data System (http://www.barcodinglife.org). *Mol. Ecol. Notes* **2007**, *7*, 355–364. [CrossRef]
67. Crobe, V.; Ferrari, A.; Hanner, R.; Leslie, R.W.; Steinke, D.; Tinti, F.; Cariani, A. Molecular taxonomy and diversification of Atlantic skates (Chondrichthyes, Rajiformes): Adding more pieces to the puzzle of their evolutionary history. *Life* **2021**, *11*, 596. [CrossRef]
68. Steinke, D.; Connell, A.D.; Hebert, P.D.N. Linking adults and immatures of South African marine fishes. *Genome* **2016**, *59*, 959–967. [CrossRef]

69. Van Der Bank, H. DNA barcoding results for some Southern African elephantfish, guitarfish, rattails, rays, sharks and skates. *Int. J. Oceanogr. Aquac.* **2019**, *3*, 000163. [CrossRef]

70. Serra-Pereira, B.; Moura, T.; Griffiths, A.M.; Serrano Gordo, L.; Figueiredo, I. Molecular barcoding of skates (Chondrichthyes: Rajidae) from the Southern Northeast Atlantic. *Zool. Scr.* **2011**, *40*, 76–84. [CrossRef]

71. Costa, F.O.; Landi, M.; Martins, R.; Costa, M.H.; Costa, M.E.; Carneiro, M.; Alves, M.J.; Steinke, D.; Carvalho, G.R. A ranking system for reference libraries of DNA barcodes: Application to marine fish species from Portugal. *PLoS ONE* **2012**, *7*, e35858. [CrossRef]

72. Ramírez-Amaro, S.; Ordines, F.; Picornell, A.; Castro, J.A.; Ramon, C.; Massutí, E.; Terrasa, B. The evolutionary history of mediterranean batoidea (Chondrichthyes: Neoselachii). *Zool. Scr.* **2018**, *47*, 686–698. [CrossRef]

73. Ferrari, A.; Tinti, F.; Maresca, V.B.; Velonà, A.; Cannas, R.; Thasitis, I.; Costa, F.O.; Follesa, M.C.; Golani, D.; Hemida, F.; et al. Natural history and molecular evolution of demersal Mediterranean sharks and skates inferred by comparative phylogeographic and demographic analyses. *PeerJ* **2018**, *6*, e5560. [CrossRef]

74. Landi, M.; Dimech, M.; Arculeo, M.; Biondo, G.; Martins, R.; Carneiro, M.; Carvalho, G.R.; Brutto, S.L.; Costa, F.O. DNA barcoding for species assignment: The case of Mediterranean marine fishes. *PLoS ONE* **2014**, *9*, e106135. [CrossRef]

75. Vella, A.; Vella, N.; Schembri, S. A molecular approach towards taxonomic identification of elasmobranch species from Maltese fisheries landings. *Mar. Genomics* **2017**, *36*, 17–23. [CrossRef]

76. Gkafas, G.A.; Megalofonou, P.; Batzakas, G.; Apostolidis, A.P.; Exadactylos, A. Molecular phylogenetic convergence within Elasmobranchii revealed by Cytochrome Oxidase Subunits. *Biochem. Syst. Ecol.* **2015**, *61*, 510–515. [CrossRef]

77. Zambounis, A.G.; Ekonomou, G.; Megalofonou, P.; Batzakas, G.; Malandrakis, E.; Martsicalis, P.; Panagiotaki, P.; Neofitou, C.; Exadactylos, A. Molecular Phylogenetic Interrelations between Species of the *Elasmobranchii subclass*. 2010; unpublished; submitted to the EMBL/GenBank/DDBJ databases.

78. Shirak, A.; Dor, L.; Seroussi, E.; Ron, M.; Hulata, G.; Golani, D. DNA barcoding of fish species from the Mediterranean coast of Israel. *Mediterr. Mar. Sci.* **2016**, *17*, 459–466. [CrossRef]

79. Yokes, M.B. DNA Barcoding of Marine Fish Species from Turkish Coastline. 2016; unpublished; submitted to the EMBL/GenBank/DDBJ databases.

80. Nei, M. *Molecular Evolutionary Genetics*; Columbia University Press: New York, NY, USA, 1987. [CrossRef]

81. Rozas, J.; Ferrer-Mata, A.; Sánchez-DelBarrio, J.C.; Guirao-Rico, S.; Librado, P.; Ramos-Onsins, S.E.; Sánchez-Gracia, A. DnaSP 6: DNA sequence polymorphism analysis of large data sets. *Mol. Biol. Evol.* **2017**, *34*, 3299–3302. [CrossRef]

82. Excoffier, L.; Lischer, H.E.L. Arlequin Suite ver 3.5: A new series of programs to perform population genetics analyses under Linux and Windows. *Mol. Ecol. Resour.* **2010**, *10*, 564–567. [CrossRef]

83. Akaike, H. A new look at the statistical model identification. *Curr. Contents Eng. Technol. Appl. Sci.* **1981**, *12*, 42.

84. Knebelsberger, T.; Landi, M.; Neumann, H.; Kloppmann, M.; Sell, A.F.; Campbell, P.D.; Laakmann, S.; Raupach, M.J.; Carvalho, G.R.; Costa, F.O. A reliable DNA barcode reference library for the identification of the North European shelf fish fauna. *Mol. Ecol. Resour.* **2014**, *14*, 1060–1071. [CrossRef]

85. Van Oosterhout, C.; Hutchinson, W.F.; Wills, D.P.M.; Shipley, P. MICRO-CHECKER: Software for identifying and correcting genotyping errors in microsatellite data. *Mol. Ecol. Notes* **2004**, *4*, 535–538. [CrossRef]

86. Belkhir, K.; Borsa, P.; Chikhi, L.; Raufaste, N.; Bonhomme, F. GENETIX 4.05, Logiciel Sous Windows TM Pour la Génétique des Populations. 1996. Available online: https://www.scienceopen.com/document?_vid=7cfcd230-1958-4cfc-a571-ce0ba003e63f (accessed on 1 April 2022).

87. Rousset, F. GENEPOP'007: A complete re-implementation of the GENEPOP software for Windows and Linux. *Mol. Ecol. Resour.* **2008**, *8*, 103–106. [CrossRef]

88. Goudet, J. FSTAT 2.9.3, a Program to Estimate and Test Gene Diversities and Fixation Indices. 2001. Available online: http://www2.unil.ch/popgen/softwares/fstat.htm (accessed on 15 March 2023).

89. Leigh, J.W.; Bryant, D. POPART: Full-Feature software for haplotype network construction. *Methods Ecol. Evol.* **2015**, *6*, 1110–1116. [CrossRef]

90. Excoffier, L.; Smouse, P.E.; Quattro, J.M. Analysis of molecular variance inferred from metric distances among DNA haplotypes: Application to human mitochondrial DNA restriction Data. *Genetics* **1992**, *131*, 479–491. [CrossRef]

91. Rice, W.R. Analyzing tables of statistical tests. *Evolution* **1989**, *43*, 223. [CrossRef]

92. Castro-Conde, I.; de Uña-Álvarez, J. Sgof: An R package for multiple testing problems. *R J.* **2014**, *6*, 96–113. [CrossRef]

93. Cheng, L.; Connor, T.R.; Sirén, J.; Aanensen, D.M.; Corander, J. Hierarchical and spatially explicit clustering of DNA sequences with BAPS software. *Mol. Biol. Evol.* **2013**, *30*, 1224–1228. [CrossRef]

94. Hubisz, M.J.; Falush, D.; Stephens, M.; Pritchard, J.K. Inferring weak population structure with the assistance of sample group information. *Mol. Ecol. Resour.* **2009**, *9*, 1322–1332. [CrossRef]

95. Kopelman, N.M.; Mayzel, J.; Jakobsson, M.; Rosenberg, N.A.; Mayrose, I. Clumpak: A program for identifying clustering modes and packaging population structure inferences across K. *Mol. Ecol. Resour.* **2015**, *15*, 1179–1191. [CrossRef]

96. Li, Y.-L.; Liu, J.-X. StructureSelector: A web-based software to select and visualize the optimal number of clusters using multiple methods. *Mol. Ecol. Resour.* **2018**, *18*, 176–177. [CrossRef]

97. Jombart, T.; Devillard, S.; Balloux, F. Discriminant analysis of principal components: A new method for the analysis of genetically structured populations. *BMC Genet.* **2010**, *11*, 94. [CrossRef]

98. R Core Team R: A language and environment for statistical computing 2021.

99. Suchard, M.A.; Lemey, P.; Baele, G.; Ayres, D.L.; Drummond, A.J.; Rambaut, A. Bayesian phylogenetic and phylodynamic data integration using BEAST 1.10. *Virus Evol.* **2018**, *4*, vey016. [CrossRef]

100. Hasegawa, M.; Kishino, H.; aki Yano, T. Dating of the human-ape splitting by a molecular clock of mitochondrial DNA. *J. Mol. Evol.* **1985**, *22*, 160–174. [CrossRef]

101. Rambaut, A.; Drummond, A.J.; Xie, D.; Baele, G.; Suchard, M.A. Posterior summarization in Bayesian phylogenetics using tracer 1.7. *Syst. Biol.* **2018**, *67*, 901–904. [CrossRef]

102. Letunic, I.; Bork, P. Interactive Tree Of Life (ITOL): An online tool for phylogenetic tree display and annotation. *Bioinformatics* **2007**, *23*, 127–128. [CrossRef]

103. Puillandre, N.; Lambert, A.; Brouillet, S.; Achaz, G. ABGD, Automatic Barcode Gap Discovery for primary species delimitation. *Mol. Ecol.* **2012**, *21*, 1864–1877. [CrossRef]

104. Zhang, J.; Kapli, P.; Pavlidis, P.; Stamatakis, A. A general species delimitation method with applications to phylogenetic placements. *Bioinforma. Oxf. Engl.* **2013**, *29*, 2869–2876. [CrossRef]

105. Wallace, J.H. The batoid fishes of the east coast of southern Africa. Part III: Skates and electric rays. *S. Afr. Assoc. Mar. Biol. Res. Investig. Rep.* **1967**, *17*, 1–62.

106. Hirschfeld, M.; Dudgeon, C.; Sheaves, M.; Barnett, A. Barriers in a sea of elasmobranchs: From fishing for populations to testing hypotheses in population genetics. *Glob. Ecol. Biogeogr.* **2021**, *30*, 2147–2163. [CrossRef]

107. Sandoval-Castillo, J.; Beheregaray, L.B. Oceanographic heterogeneity influences an ecological radiation in elasmobranchs. *J. Biogeogr.* **2020**, *47*, 1599–1611. [CrossRef]

108. Henriques, R.; Potts, W.M.; Sauer, W.H.; Santos, C.V.; Kruger, J.; Thomas, J.A.; Shaw, P.W. Molecular genetic, life-history and morphological variation in a coastal warm-temperate sciaenid fish: Evidence for an upwelling-driven speciation event. *J. Biogeogr.* **2016**, *43*, 1820–1831. [CrossRef]

109. Henriques, R.; Potts, W.M.; Santos, C.V.; Sauer, W.H.; Shaw, P.W. Population connectivity and phylogeography of a coastal fish, *Atractoscion aequidens* (Sciaenidae), across the Benguela current region: Evidence of an ancient vicariant event. *PLoS ONE* **2014**, *9*, e87907. [CrossRef]

110. Chevolot, M.; Hoarau, G.; Rijnsdorp, A.D.; Stam, W.T.; Olsen, J.L. Phylogeography and population structure of thornback ray (*Raja clavata* L., Rajidae). *Mol. Ecol.* **2006**, *15*, 3693–3705. [CrossRef]

111. Valsecchi, E.; Pasolini, P.; Bertozzi, M.; Garoia, F.; Ungaro, N.; Vacchi, M.; Sabelli, B.; Tinti, F. Rapid Miocene-Pliocene dispersal and evolution of mediterranean rajid fauna as inferred by mitochondrial gene variation. *J. Evol. Biol.* **2005**, *18*, 436–446. [CrossRef]

112. Patarnello, T.; Volckaert, F.A.M.J.; Castilho, R. Pillars of Hercules: Is the Atlantic–Mediterranean transition a phylogeographical break? *Mol. Ecol.* **2007**, *16*, 4426–4444. [CrossRef]

113. Melis, R.; Vacca, L.; Cariani, A.; Carugati, L.; Charilaou, C.; Di Crescenzo, S.; Ferrari, A.; Follesa, M.C.; Mancusi, C.; Pinna, V.; et al. Baseline genetic distinctiveness supports structured populations of thornback ray in the Mediterranean Sea. *Aquat. Conserv. Mar. Freshw. Ecosyst.* **2023**, *33*, 458–471. [CrossRef]

114. Serena, F. *Field Identification Guide to the Sharks and Rays of the Mediterranean and Black Sea*; Food & Agriculture Organisation of The United Nations: Rome, Italy, 2005; 97 pp. + 11 colour plates + egg cases.

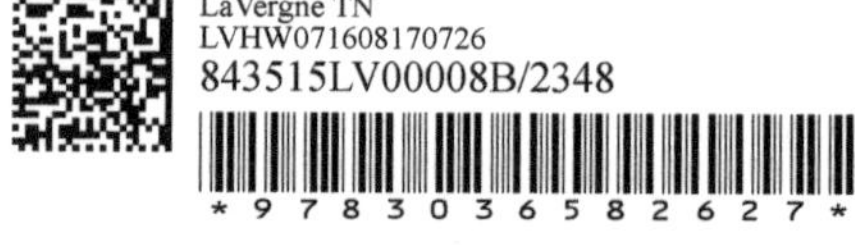

MDPI
St. Alban-Anlage 66
4052 Basel
Switzerland
Tel. +41 61 683 77 34
Fax +41 61 302 89 18
www.mdpi.com

Animals Editorial Office
E-mail: animals@mdpi.com
www.mdpi.com/journal/animals